# Digital Signal Processing

## *A Modern Introduction*

## International Student Edition



THOMSON



Australia Brazil Canac ited Kingdom United States

THOMSON

**Digital Signal Processing: *A Modern Introduction* (International Student Edition)**
**by Ashok Ambardar**

**Associate Vice-President and Editorial Director:**
Evelyn Veitch

**Publisher:** Chris Carson

**Developmental Editor:**
Kamilah Reid Burrell/Hilda Gowaus

**Permissions Coordinator:**
Vicki Gould

**Production Services:**
RPK Editorial Services

**Copy Editor:**
Shelly Gerger-Knechtl

**Proofreader:**
Erin Wagner

**Indexer:**
RPK Editorial Services

**Production Manager:**
Renate McCloy

**Creative Director:**
Angela Cluer

**Interior Design:**
Carmela Pereira

**Cover Design:**
Andrew Adams

**Compositor:**
International Typesetting and Composition

**Printer:**
Quebecor World



Printed and bound in United States of America
1 2 3 4 07 06

For more information contact Nelson, 1120 Birchmount Road, Toronto, Ontario, Canada, M1K 5G4.
Or you can visit our Internet site at http://www.nelson.com

Library of Congress Control Number: 2006900028

ISBN 0-495-08238-4



**North America**
Nelson
1120 Birchmount Road
Toronto, Ontario M1K 5G4
Canada

**Asia**
Thomson Learning
5 Shenton Way #01-01
UIC Building
Singapore 068808

**Australia/New Zealand**
Thomson Learning
102 Dodds Street
Southbank, Victoria
Australia 3006

**Europe/Middle East/Africa**
Thomson Learning
High Holborn House
50/51 Bedford Row
London WC1R 4LR
United Kingdom

**Latin America**
Thomson Learning
Seneca, 53
Colonia Polanco
11569 Mexico D.F.
Mexico

**Spain**
Paraninfo
Calle/Magallanes, 25
28015 Madrid, Spain

*For keeping the memories alive!*

To Nancy and Shyamaji

# Contents

## Appendix A Useful Concepts from Analog Theory

# Preface

This book provides a modern and self-contained introduction to digital signal processing (DSP) and is written with several audiences in mind. First and foremost, it is intended to serve as a textbook suitable for a one-semester junior or senior level undergraduate course. To this extent, it includes the relevant topics covered in a typical undergraduate curriculum and is supplemented by a vast number of worked examples, drill exercises and problems. It also attempts to provide a broader perspective by introducing useful applications and additional special topics in each chapter. These form the background for more advanced graduate courses in this area and also allow the book to be used as a source of basic reference for professionals across various disciplines interested in DSP.

## Scope

The text stresses the fundamental principles and applications of digital signal processing. The relevant concepts are explained and illustrated by worked examples. Applications are introduced in each chapter and two chapters concentrate solely on the applications of DSP. Since many applications of DSP relate to the processing of analog signals, some familiarity with basic analog theory, at the level taught in a typical undergraduate signals and systems course, is assumed and expected. In order to make the book self-contained, the key concepts and results from analog theory that are relevant to a study of DSP are outlined in an appendix. The topics covered in this book may be grouped into the following broad areas:

1. The first chapter starts with a brief overview. An introduction to **discrete signals**, their representation and their classification is provided in Chapter 2.
2. Chapter 3 details the analysis of digital filters in the time-domain using the solution of difference equations or the process of **convolution** that also serves to link the time domain and the frequency domain.
3. Chapter 4 covers the analysis in the transformed domain using the ***z*-transform** that forms a powerful tool for studying discrete-time signals and systems.
4. Chapter 5 describes the analysis of discrete signals and digital filters in the frequency domain using the discrete-time Fourier transform (**DTFT**) that arises as a special case of the ***z*-transform**.

5. Chapter 6 introduces the terminology of digital filters and studies and compares a variety of filters and the various methods of studying them.
6. Chapter 7 is an application-oriented chapter that discusses the digital processing of analog signals based on the concepts of **sampling** and **quantization** and the spectral representation of sampled signals.
7. Chapter 8 is also an application oriented chapter that provides an introduction to the spectral analysis of both analog and discrete signals based on numerical computation of the **DFT** and the **FFT** and its applications.
8. Chapter 9 and Chapter 10 describe the the design of IIR and FIR filters for various applications using well established techniques.
9. Chapter 11 provides examples of MATLAB® code for implementing and visualizing many of the DSP concepts discussed in the book. Putting the code in a separate chapter allows us to examine interesting problems whose solution requires concepts across chapters. This approach also preserves the continuity and logical flow of the textual material.

This book attempts to preserve a rational approach and includes all the necessary mathematical details. At the same time, the results are also described on the basis of simple heuristic explanations in order to to reinforce the understanding of major DSP concepts.

In each chapter, a short opening section outlines the objectives and topical coverage. Central concepts are illustrated by worked examples. Many figures have been included to help the student grasp and visualize critical concepts. Results are tabulated and summarized for easy reference and access. End-of-chapter problems include a variety of drills and exercises including application oriented problems that require the use of computational resources such as MATLAB. A suite of MATLAB-based routines that illustrate the principles and concepts presented in the book are available on the author's website. These routines are supplied in good faith and the author is not responsible for any consequences arising from their use! A solutions manual for instructors is available from the publisher.

# Acknowledgments

This book has gained immensely from the constructive criticism of many reviewers. Thanks go, in particular, to Proressor Vijayakumar Bhagavatula of Carnegie Mellon University; Professor Dimitrios Hatzinakos of the University of Toronto; and Professor Aryan Saadat Mehr of the University of Saskatchewan.

The individuals who have contributed in various ways behind the scenes include Chris Carson (the publisher), Kamilah Reid Burrell//Hilda Gowans (developmental editors), Rose Kernan (production services and interior design), Shelly Gerger-Knechtl (copy editor), Erin Wagner (proofreader), Renate McCloy (production manager), Angela Cluer (creative director), and Andrew Adams (cover design).

If you come across any errors in the text or discover any bugs in the software, we would appreciate hearing from you. Any errata will be posted on the author's website.

| | |
|---|---|
| Ashok Ambardar | Michigan Technological University |
| Internet: | **http://www.ee.mtu.edu/faculty/akambard.html** |
| e-mail: | **akambard@mtu.edu** |

CHAPTER 1

# Overview

## 1.0 Introduction

Few other technologies have revolutionized the world as profoundly as those based on digital signal processing. For example, the technology of recorded music was, until recently, completely analog from end to end and the most important commercial source of recorded music used to be the LP (long-playing) record. The advent of the digital compact disc changed all that in the span of just a few short years and made the long-playing record practically obsolete. With the proliferation of high speed, low cost computers and powerful, user-friendly software packages, digital signal processing (DSP) has truly come of age. This chapter provides an overview of the terminology of digital signal processing and of the connections between the various topics and concepts covered in the text.

## 1.1 Signals

Our world is full of signals, both natural and man-made. Examples are the variation in air pressure when we speak, the daily highs and lows in temperature, and the periodic electrical signals generated by the heart. Signals represent information. Often, signals may not convey the required information directly and may not be free from disturbances. It is in this context that signal processing forms the basis for enhancing, extracting, storing, or transmitting useful information. Electrical signals perhaps offer the widest scope for such manipulations. In fact, it is commonplace to convert signals to electrical form for processing.

The signals we encounter in practice are often very difficult to characterize. So, we choose simple mathematical models to approximate their behavior. Such models also give us the ability to make predictions about future signal behaviour. Of course, an added advantage of using models is that they are much easier to generate and manipulate. What is more, we can gradually increase the complexity of our model to obtain better approximations, if needed.

The simplest signal models are a constant variation, an exponential decay and a sinusoidal or periodic variation. Such signals form the building blocks from which we can develop representations for more complex forms.

This book starts with a quick overview of discrete signals, how they arise, and how they are modeled. We review some typical measures (such as power and energy) used to characterize discrete signals and the operations of interpolation and decimation which are often used to change the sampling rate of an already sampled signal.

## 1.2 Digital Filters

The processing of discrete signals is accomplished by discrete-time systems, also called digital filters. In the time domain, such systems may be modeled by difference equations in much the same way that analog systems are modeled by differential equations. We concentrate on models of linear time-invariant (LTI) systems whose difference equations have constant coefficients. The processing of discrete signals by such systems can be achieved by resorting to the well known mathematical techniques. For input signals that can be described as a sum of simpler forms, linearity allows us to find the response as the sum of the response to each of the simpler forms. This is **superposition**. Many systems are actually nonlinear. The study of nonlinear systems often involves making simplifying assumptions, such as linearity. The system response can also be obtained using convolution, a method based on superposition: If the response of a system is known to a unit sample (or impulse) input, then it is also known to any arbitrary input which can be expressed as a sum of such impulses.

Two important classes of digital filters are finite impulse response (FIR) filters whose impulse response (response to an impulse input) is a finite sequence (lasts only for finite time) and infinite impulse response (IIR) filters whose response to an impulse input lasts forever.

### 1.2.1 The *z*-Transform

The $z$-transform is a powerful method of analysis for discrete signals and systems. It is analogous to the Laplace transform used to study analog systems. The transfer function of an LTI system is a ratio of polynomials in the complex variable $z$. The roots of the numerator polynomial are called **zeros** and of the denominator polynomial are called **poles**. The pole-zero description of a transfer function is quite useful if we want a qualitative picture of the frequency response. For example, the frequency response goes to zero if $z$ equals one of the zero locations and becomes unbounded if $z$ equals one of the pole locations.

### 1.2.2 The Frequency Domain

It turns out that discrete sinusoids and harmonic signals differ from their analog cousins in some striking ways. A discrete sinusoid is not periodic for any choice of frequency. Yet it has a *periodic* spectrum. An important consequence of this result is that if the spectrum is periodic for a sampled sinusoid, it should also be periodic for a sampled combination of sinusoids. This concept forms the basis for the frequency domain description of discrete signals called the **Discrete-Time Fourier Transform** (DTFT). And since analog signals can be described as a combination of sinusoids (periodic ones by their Fourier series and others by their Fourier transform), their sampled combinations (and consequently any sampled signal) have a periodic spectrum in the frequency domain. The central period corresponds to the true spectrum of the analog signal if the sampling rate exceeds the Nyquist rate.

### 1.2.3 Filter Concepts

The term *filter* is often used to denote systems that process the input in a specified way. In this context, filtering describes a signal-processing operation that allows signal enhancement, noise reduction, or increased **signal-to-noise ratio**. Systems for the processing of discrete-time signals are also called **digital filters**. Depending on the requirements and application, the analysis of a digital filter may be carried out in the time domain, the $z$-domain, or the frequency domain. A common application of digital filters is to modify the frequency response in some specified way. An ideal lowpass filter passes frequencies up to a specified value and totally blocks all others. Its spectrum shows an abrupt transition from unity (perfect transmission) in the passband to zero (perfect suppression) in the stopband. An important consideration is that a symmetric impulse-response sequence possesses linear phase (in its frequency response), which results only in a *constant delay* and no distortion. An ideal lowpass filter possesses linear phase because its impulse response happens to be a symmetric sequence, but unfortunately, it cannot be realized in practice.

One way to approximate an ideal lowpass filter is by symmetric truncation of its impulse response (which ensures linear phase). Truncation is equivalent to multiplying (windowing) the impulse response by a finite duration sequence (window) of unit samples. The abrupt truncation imposed by such a window results in an overshoot and oscillation in the frequency response that persists no matter how large the length of the truncated sequence. To eliminate overshoot and reduce the oscillations, we use *tapered* windows. The impulse response and frequency response of highpass, bandpass, and bandstop filters may be related to those of a lowpass filter using frequency transformations based on the properties of the DTFT.

Filters that possess constant gain but whose phase varies with frequencies are called allpass filters and may be used to modify the phase characteristics of a system. A filter whose gain is zero at a selected frequency is called a notch filter and may be used to remove the unwanted frequency from a signal. A filter whose gain is zero at multiples of a selected frequency is called a comb filter and may be used to remove an unwanted frequency and its harmonics from a signal.

## 1.3 Signal Processing

Two conceptual schemes for the processing of signals are illustrated in Figure 1.1. The digital processing of analog signals requires that we use an analog-to-digital converter (ADC) for sampling the analog signal prior to processing and a digital-to-analog converter (DAC) to convert the processed digital signal back to analog form.

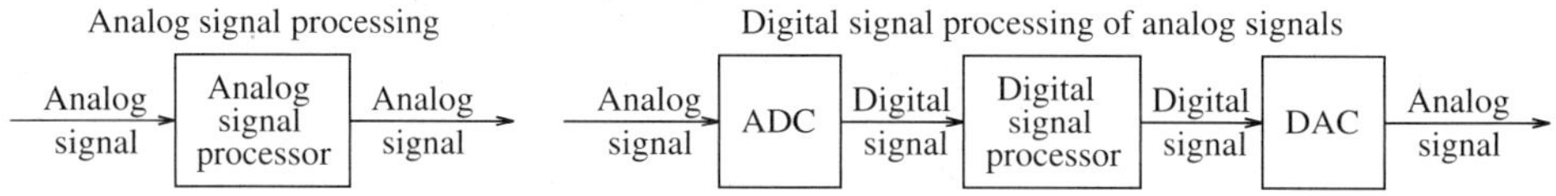


**FIGURE 1.1** Analog and digital signal processing

### 1.3.1 Digital Processing of Analog Signals

Many DSP applications involve the processing of digital signals obtained by sampling analog signals and the subsequent reconstruction of analog signals from their samples. For example, the music you hear from your compact disc (CD) player is due to changes in the air pressure caused by the vibration of the speaker diaphragm. It is an **analog** signal because the pressure variation is a *continuous* function of time. However, the information stored on the compact disk is in digital form. It must be processed and converted to analog form before you can hear the music. A record of the yearly increase in the world population describes time measured in increments of one (year) while the population increase is measured in increments of one (person). It is a **digital** signal with discrete values for both time and population.

For digital signal processing we need digital signals. To process an analog signal by digital means, we must convert it to a digital signal in two steps. First, we must **sample** it, typically at uniform intervals $t_s$ (every 2 ms, for example). The discrete quantity $nt_s$ is related to the integer index $n$. Next, we must **quantize** the sample values (amplitudes) (by rounding to the nearest millivolt, for example). The central concept in the digital processing of analog signals is that the sampled signal must be a unique representation of the underlying analog signal. Even though sampling leads to a potential loss of information, all is not lost! Often, it turns out that if we choose the sampling interval wisely, the processing of an analog signal is entirely equivalent to the processing of the corresponding digital signal; there is no loss of information! This is one of the wonders of the **sampling theorem** that makes digital signal processing such an attractive option. For a unique correspondence between an analog signal and the version reconstructed from its samples, the sampling rate $S$ must *exceed twice* the highest signal frequency $f_0$. The value $S = 2f_0$ is called the Nyquist sampling rate. If the sampling rate is less than the Nyquist rate, a phenomenon known as **aliasing** manifests itself. Components of the analog signal at high frequencies appear at (alias to) lower frequencies in the sampled signal. This results in a sampled signal with a smaller highest frequency. Aliasing effects are impossible to undo once the samples are acquired. It is thus commonplace to band-limit the signal before sampling (using lowpass filters).

Numerical processing using digital computers requires finite data with finite precision. We must limit signal amplitudes to a finite number of levels. This process, called **quantization**, produces nonlinear effects that can be described only in statistical terms. Quantization also leads to an irreversible loss of information and is typically considered only in the final stage in any design.

A typical system for the digital processing of analog signals consists of the following:

- An analog lowpass **pre-filter** or **anti-aliasing filter** which limits the highest signal frequency to ensure freedom from aliasing.

- A **sampler**, which operates above the Nyquist sampling rate.
- A **quantizer**, which quantizes the sampled signal values to a finite number of levels. Currently, 16-bit quantizers are quite commonplace.
- An **encoder**, which converts the quantized signal values to a string of *binary* **bits** or zeros and ones (**words**) whose length is determined by the number of quantization levels of the quantizer.
- The digital processing system itself (hardware or software), which processes the encoded digital signal (or **bit stream**) in a desired fashion.
- A **decoder**, which converts the processed bit stream to a discrete signal with quantized signal values.
- A **reconstruction filter**, which reconstructs a staircase approximation of the discrete time signal.
- A lowpass analog **anti-imaging filter**, which extracts the central period from the periodic spectrum, removes the unwanted replicas, and results in a smoothed reconstructed signal.

### 1.3.2 Filter Design

The design of filters is typically based on a set of specifications in the frequency domain corresponding to the magnitude spectrum or filter gain. The design of IIR filters typically starts with a lowpass prototype from which other forms may be developed readily using frequency transformations.

### 1.3.3 The Design of IIR Filters

The design of IIR filters starts with an analog *lowpass prototype* based on the given specifications. Classical analog filters include Butterworth (maximally flat passband), Chebyshev I (rippled passband), Chebyshev II (rippled stopband) and elliptic (rippled passband and stopband). The analog lowpass prototype is then converted to a lowpass digital filter using an appropriate mapping and finally to the required form using an appropriate *spectral transformation*. Practical mappings are based on response-invariance or equivalence of ideal operations, such as integration and their numerical counterparts. Not all of these avoid the effects of aliasing. The most commonly used mapping is based on the trapezoidal rule for numerical integration and is called the **bilinear transformation**. It compresses the entire infinite analog frequency range into a finite range and thus avoids aliasing at the expense of *warping* (distorting) the analog frequencies. We can compensate for this warping if we **prewarp** (stretch) the analog frequency specifications before designing the analog filter.

### 1.3.4 The Design of FIR Filters

FIR filters are inherently stable and can be designed with linear phase leading to no distortion, but their realization often involves a large filter length to meet given requirements. Their design is typically based on selecting a symmetric (linear phase) impulse response sequence of the smallest length that meets design specifications and involves iterative techniques. Even though the spectrum of the truncated ideal filter is, in fact, the best approximation (in the mean square sense) compared to the spectrum of any other filter of the same length, it shows the undesirable oscillations and overshoot which can be eliminated by modifying (windowing) the impulse response sequence using tapered windows. The smallest length that meets specifications depends on the choice of window and is often estimated by empirical means.

## 1.4 The DFT and FFT

The periodicity of the DTFT is a consequence of the fundamental result that sampling a signal in one domain leads to periodicity in the other. Just as a periodic signal has a discrete spectrum, a discrete-time signal has a periodic spectrum. This duality also characterizes several other transforms. If the time signal is both discrete and periodic, its spectrum is also discrete and periodic and describes the **discrete Fourier transform** (DFT). The DFT is essentially the DTFT evaluated at a finite number of frequencies and is also periodic. The DFT can be used to approximate the spectrum of analog signals from their samples, provided the relations are understood in their proper context using the notion of implied periodicity. The **Fast Fourier Transform** (FFT) is a set of fast practical algorithms for computing the DFT. The DFT and FFT find extensive applications in fast convolution, signal interpolation, spectrum estimation, and transfer function estimation.

## 1.5 Advantages of DSP

In situations where signals are encountered in digital form, their processing is performed digitally. In other situations that relate to the processing of analog signals, DSP offers many advantages.

### Processing

DSP offers a wide variety of processing techniques that can be implemented easily and efficiently. Some techniques (such as processing by linear phase filters) have no counterpart in the analog domain.

### Storage

Digital data can be stored and later retrieved with no degradation or loss of information. Data recorded by analog devices is subject to the noise inherent in the recording media (such as tape) and degradation due to aging and environmental effects.

### Transmission

Digital signals are more robust and offer much better noise immunity during transmission as compared to analog signals.

### Implementation

A circuit for processing analog signals is typically designed for a specific application. It is sensitive to component tolerances, aging, and environmental effects (such as changes in the temperature and humidity) and not easily reproducible. A digital filter, on the other hand, is extremely easy to implement and highly reproducible. It may be designed to perform a variety of tasks without replacing or modifying any hardware but simply by changing the filter coefficients on the fly.

### Cost

With the proliferation of low-cost, high-speed digital computers, DSP offers effective alternatives for a wide variety of applications. High-frequency analog applications may still require analog signal processing but their number continues to shrink. As long as the criteria of the sampling theorem are satisfied and quantization is carried out to the desired precision (using the devices available), the digital processing of analog signals has become the method of choice unless compelling reasons dictate otherwise.

In the early days of the digital revolution, DSP did suffer from disadvantages (such as speed, cost, and quantization effects), but these continue to pale into insignificance with advances in semiconductor technology and processing and computing power.

## 1.5.1 Applications of DSP

Digital signal processing finds applications in almost every conceivable field. Its impact on consumer electronics is evidenced by the proliferation of digital communication, digital audio, digital (high-definition) television, and digital imaging (cameras). Its applications to biomedical signal processing include the enhancement and interpretation of tomographic images and analysis of the electrical signals that describe the activity of the heart and brain. Space applications include satellite navigation and guidance systems and the analysis of satellite imagery obtained by various means. The list goes on and continues to grow.

CHAPTER 2

# Discrete Signals

## 2.0 Scope and Overview

This chapter begins with an overview of discrete signals. It starts with various ways of signal classification, shows how discrete signals can be manipulated by various operations, and quantifies the measures used to characterize such signals. It introduces the concept of sampling and describes the sampling theorem as the basis for sampling analog signals without loss of information. It concludes with an introduction to random signals.

### 2.0.1 Goals and Learning Objectives

The goals of this chapter are to provide a framework for the description and analysis of discrete signals in the context of digital signal processing. After going through this chapter, the reader should:

1. Know how to transform a discrete signals using time shift and time reversal.
2. Understand even symmetry and odd symmetry and know how to find the even and odd parts of an unsymmetric signal.
3. Understand how to classify a discrete signal based on energy and power.
4. Understand the basics of interpolation and decimation.
5. Know the definitions of standard signal forms.
6. Know how to find the digital frequency of a sinusoid or complex harmonic.
7. Know how to find the common period of a sum of sinusoids or complex harmonics.
8. Understand the sampling theorem and know how to determine the sampling rate.
9. Understand the concept of aliasing and know how to find aliased frequencies.
10. Understand the measures used to describe random signals.

# 2.1 Discrete Signals

Discrete signals (such as the annual population) may arise naturally or as a consequence of sampling continuous signals (typically at a uniform sampling interval $t_s$). A sampled or discrete signal $x[n]$ is just an *ordered* sequence of values corresponding to the integer index $n$ that embodies the time history of the signal. It contains no direct information about the **sampling interval** $t_s$, except through the index $n$ of the sample locations. The sample values of a discrete signal $x[n]$ are plotted as lines against the index $n$.

## Comparing Analog and Discrete Signals

When comparing analog signals with their sampled versions, we shall assume that the origin $t = 0$ also corresponds to the instant $n = 0$. We need information about the sampling interval $t_s$ only in a few situations, such as plotting the signal explicitly against the variable $t$ (at the instants $t = nt_s$) or approximating the area of the underlying analog signal from its samples.

### Notation for a Numeric Sequence *x*[*n*]

A marker (⇓) indicates the origin $n = 0$.

***Example:*** $x[n] = \{1,\ 2,\ \overset{\Downarrow}{4},\ 8\}$

Ellipses (...) denote infinite extent on either side.

***Example:*** $x[n] = \{2,\ 4,\ \overset{\Downarrow}{6},\ 8,\ \ldots\}$

An infinite-length discrete signal $x[n]$ is called **right-sided** if it is zero for $n < N$ (where $N$ is finite), **causal** if it is zero for $n < 0$, **left-sided** if it is zero for $n > N$, and **anti-causal** if it is zero for $n \geq 0$. This is illustrated in Figure 2.1.

A discrete periodic signal repeats every $N$ samples and is described by

$$x[n] = x[n \pm kN], \quad k = 0, 1, 2, 3, \ldots \tag{2.1}$$

The period $N$ is the *smallest* number of samples that repeats and is always an integer.

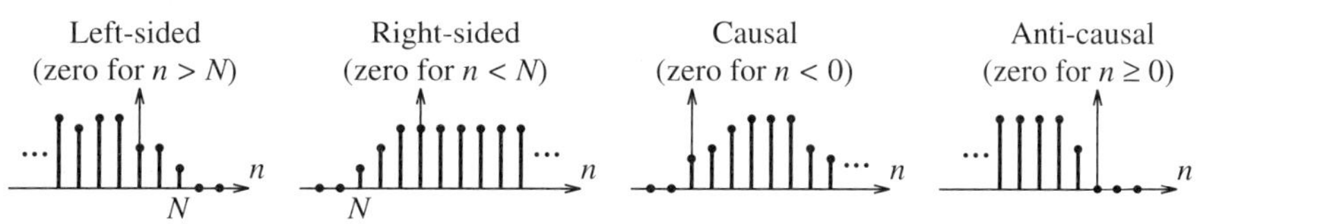


**FIGURE 2.1** An infinite-length discrete signal may be left-sided, right-sided, causal, or anti-causal

# Drill Problem 2.1

**(a)** Let $x[n] = \{\ldots,\ \overset{\Downarrow}{1},\ 2,\ 0,\ 0,\ 4,\ 1,\ 2,\ 0,\ 0,\ 4,\ 1,\ 2,\ 0,\ 0,\ 4,\ 1,\ 2,\ 0,\ 0,\ 4,\ \ldots\}$.
What is its period?

**(b)** Let $x[n] = \{\ldots,\ \overset{\Downarrow}{1},\ 2,\ 1,\ 2,\ 1,\ 2,\ 1,\ 2,\ 1,\ 2,\ 1,\ 2,\ 1,\ 2,\ 1,\ 2,\ 1,\ 2,\ 1,\ 2,\ \ldots\}$
and $y[n] = \{\ldots,\ \overset{\Downarrow}{1},\ 2,\ 3,\ 1,\ 2,\ 3,\ 1,\ 2,\ 3,\ 1,\ 2,\ 3,\ 1,\ 2,\ 3,\ 1,\ 2,\ 3,\ 1,\ 2,\ \ldots\}$.

Let $g[n] = x[n] + y[n]$. What is the period of $g[n]$? What are the sample values in one period of $g[n]$?

**Answers:** (a) 5 (b) 6, $\{\overset{\Downarrow}{2},\ 4,\ 4,\ 3,\ 3,\ 5\}$

## 2.1.1 Signal Measures

Signal measures for discrete signals are often based on summations. Summation is the discrete-time equivalent of integration. The **discrete sum** $S_D$, the *absolute sum* $S_A$, and the *cumulative sum* (running sum) $s_C[n]$ of a signal $x[n]$ are defined by

$$S_D = \sum_{n=-\infty}^{\infty} x[n] \qquad S_A = \sum_{n=-\infty}^{\infty} |x[n]| \qquad s_C[n] = \sum_{k=-\infty}^{n} x[k] \tag{2.2}$$

Signals for which the absolute sum $|x[n]|$ is finite are called **absolutely summable**. The **instantaneous power** of a signal $x[n]$ is defined as $p[n] = |x[n]|^2$. For nonperiodic signals, the **signal energy** $E$ is a useful measure. It is defined as the sum of the instantaneous power

$$E = \sum_{m=-\infty}^{\infty} |x[m]|^2 = \sum_{m=-\infty}^{\infty} |x[m]|^2 \tag{2.3}$$

The absolute value allows us to extend this relation to complex-valued signals. For periodic signals, the signal energy is infinite and provides no useful information. Measures for periodic signals are therefore based on averages. The **average value** $x_{\text{av}}$ of a periodic signal $x[n]$ with period $N$ is defined as the average sum per period:

$$x_{\text{av}} = \frac{1}{N}\sum_{m=0}^{N-1} x[m] \quad \text{(average value)} \tag{2.4}$$

The **average power** $P$ of a periodic signal $x[n]$ with period $N$ is defined as the average energy per period:

$$P = \frac{1}{N}\sum_{m=0}^{N-1} |x[m]|^2 \quad \text{(average power)} \tag{2.5}$$

Note that the index runs from $m = 0$ to $m = N - 1$ and includes all $N$ samples in one period. Only for nonperiodic signals is it useful to use the following limiting forms:

$$x_{\text{av}} = \lim_{L\to\infty} \frac{1}{2L+1}\sum_{m=-L}^{L} x[m] \qquad P = \lim_{L\to\infty} \frac{1}{2L+1}\sum_{m=-L}^{L} |x[m]|^2 \tag{2.6}$$

Signals with finite energy are called *energy signals* (or square summable). Signals with finite average power are called *power signals*. All periodic signals are power signals.

### A Note on the Units of Energy and Power

We have chosen not to associate the units of joules with energy or watts with power because, in digital signal processing, a signal may not always represent a physical entity (such as a voltage) for which such units are appropriate and meaningful.

**Example 2.1**

### Signal Energy and Power

**(a)** Find the energy in the signal $x[n] = 3(0.5)^n$, $n \geq 0$.

This describes a one-sided decaying exponential. Its signal energy is

$$E = \sum_{m=-\infty}^{\infty} x^2[m] = \sum_{m=0}^{\infty} |3(0.5)^m|^2$$

$$= \sum_{m=0}^{\infty} 9(0.25)^m = \frac{9}{1-0.25} = 12 \quad \left(Note\text{:}\ \sum_{m=0}^{\infty} \alpha^m = \frac{1}{1-\alpha}\right)$$

**(b)** Consider the periodic signal $x[n] = 6\cos(2\pi n/4)$ whose period is $N = 4$.

One period of this signal is $x_1[n] = \{\overset{\Downarrow}{6},\ 0,\ -6,\ 0\}$. The average value and average power of $x[n]$ is

$$x_{\text{av}} = \frac{1}{4}\sum_{m=0}^{3} x[m] = 0 \qquad P = \frac{1}{4}\sum_{m=0}^{3} x^2[m] = \frac{1}{4}(36+36) = 18$$

**(c)** Consider the periodic signal $x[n] = 6e^{j2\pi n/4}$ whose period is $N = 4$.

This signal is complex-valued, with $|x[n]| = 6$. One period of this signal is $x_1[n] = \{\overset{\Downarrow}{6}, j6, -6, -j6\}$. The average power of $x[n]$ is

$$P = \frac{1}{4}\sum_{m=0}^{3} |x[m]|^2 = \frac{1}{4}(36+36+36+36) = 36$$

# Drill Problem 2.2

**(a)** Let $x[n] = \{\overset{\Downarrow}{1},\ 2,\ 0,\ 0,\ 4\}$. What is the energy in $x[n]$?

**(b)** Let $x[n] = \{\ldots,\ \overset{\Downarrow}{1},\ 2,\ 0,\ 0,\ 4,\ 1,\ 2,\ 0,\ 0,\ 4,\ 1,\ 2,\ 0,\ 0,\ 4,\ 1,\ 2,\ \ldots\}$. Find its average power.

**(c)** Let $x[n] = \{\ldots,\ \overset{\Downarrow}{1},\ 2,\ 1,\ 2,\ 1,\ 2,\ 1,\ 2,\ 1,\ 2,\ 1,\ 2,\ 1,\ 2,\ 1,\ 2,\ 1,\ 2,\ 1,\ 2,\ \ldots\}$ and $y[n] = \{\ldots,\ \overset{\Downarrow}{1},\ 2,\ 2,\ 1,\ 2,\ 2,\ 1,\ 2,\ 2,\ 1,\ 2,\ 2,\ 1,\ 2,\ 2,\ 1,\ 2,\ 2,\ 1,\ 2,\ \ldots\}$.

Let $g[n] = x[n] + y[n]$. Find the average value of $g[n]$ and the average power for $x[n]$, $y[n]$ and $g[n]$.

**Answers:** **(a)** 21 **(b)** 4.2 **(c)** 3, 2.5, 2.5, 9.5

## 2.2 Operations on Discrete Signals

Common operations on discrete signals include element-wise addition and multiplication. Two other useful operations are **shifting** and **reversal**.

**Time Shift:** The signal $y[n] = x[n - \alpha]$ describes a delayed version of $x[n]$ for $\alpha > 0$. In other words, if $x[n]$ starts at $n = N$, then its shifted version $y[n] = x[n - \alpha]$ starts at $n = N + \alpha$. Thus, the signal $y[n] = x[n - 2]$ is a delayed (shifted right by 2) version of $x[n]$, and the signal $g[n] = x[n + 2]$ is an advanced (shifted left by 2) version of $x[n]$. A useful consistency check for sketching shifted signals is based on the fact that if $y[n] = x[n - \alpha]$, a sample of $x[n]$ at the original index $n$ gets relocated to the new index $n_N$ based on the operation $n = n_N - \alpha$.

**Time Reversal:** The signal $y[n] = x[-n]$ represents a time-reversed version of $x[n]$, a *mirror image* of the signal $x[n]$ about the origin $n = 0$. The signal $y[n] = x[-n - \alpha]$ may be obtained from $x[n]$ in one of two ways:

$$x[n] \longrightarrow \boxed{\text{delay (shift right) by } \alpha} \longrightarrow x[n-\alpha] \longrightarrow \boxed{\text{reversal}} \longrightarrow x[-n-\alpha]$$

$$x[n] \longrightarrow \boxed{\text{reversal}} \longrightarrow x[-n] \longrightarrow \boxed{\text{advance (shift left) by } \alpha} \longrightarrow x[-n-\alpha]$$

In either case, a sample of $x[n]$ at the original index $n$ will be plotted at a new index $n_N$ given by $n = -n_N - \alpha$, and this can serve as a consistency check in sketches.

**Example 2.2**

### Operations on Discrete Signals

Let $x[n] = \{2,\ 3,\ \overset{\Downarrow}{4},\ 5,\ 6,\ 7\}$. Find and sketch the following:

$$y[n] = x[n-3], \qquad f[n] = x[n+2], \qquad g[n] = x[-n], \qquad h[n] = x[-n+1],$$
$$s[n] = x[-n-2]$$

**Solution:** Here is how we obtain the various signals:

$$y[n] = x[n-3] = \{\overset{\Downarrow}{0},\ 2,\ 3,\ 4,\ 5,\ 6,\ 7\} \text{ (shift } x[n] \text{ right 3 units)}$$

$$f[n] = x[n+2] = \{2,\ 3,\ 4,\ 5,\ \overset{\Downarrow}{6},\ 7\} \text{ (shift } x[n] \text{ left 2 units)}$$

$$g[n] = x[-n] = \{7,\ 6,\ 5,\ \overset{\Downarrow}{4},\ 3,\ 2\} \text{ (reverse } x[n])$$

$$h[n] = x[-n+1] = \{7,\ 6,\ \overset{\Downarrow}{5},\ 4,\ 3,\ 2\} \text{ (reverse } x[n] \text{, then delay by 1)}$$

$$s[n] = x[-n-2] = \{7,\ 6,\ 5,\ 4,\ 3,\ \overset{\Downarrow}{2}\} \text{ (reverse } x[n] \text{, then advance by 2)}$$

Refer to Figure E2.2 for the sketches.

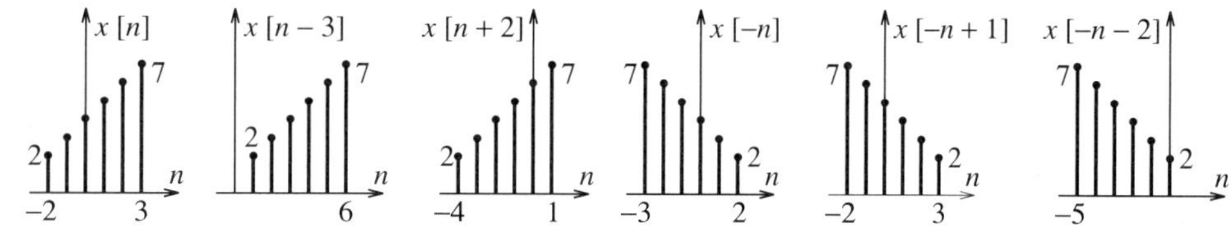


**FIGURE E.2.2** The signals for Example 2.2

# Drill Problem 2.3

**(a)** Let $x[n] = \{\overset{\Downarrow}{1}, 4, 2, 3\}$. Express $g[n] = x[-n+2]$ as a sequence and sketch.

**(b)** Let $x[n] = \{3, \overset{\Downarrow}{1}, 4\}$. Express $y[n] = x[n+2]$ and $f[n] = x[-n-1]$ as sequences and sketch?

**Answers: (a)** $g[n] = \{3, \overset{\Downarrow}{2}, 4, 1\}$ **(b)** $y[n] = \{3, 1, 4, \overset{\Downarrow}{0}\}$ $f[n] = \{4, 1, \overset{\Downarrow}{3}\}$

## 2.2.1 Symmetry

If a signal $x[n]$ is identical to its mirror image $x[-n]$, it is called an **even symmetric signal**. If $x[n]$ differs from its mirror image $x[-n]$ only in sign, it is called an **odd symmetric** or **antisymmetric signal**. Mathematically,

$$x_e[n] = x_e[-n] \quad \text{(even symmetry)} \qquad x_o[n] = -x_o[-n] \quad \text{(odd symmetry)} \tag{2.7}$$

In either case, the signal extends over symmetric limits $-N \le n \le N$. For an odd symmetric signal, note that $x_o[0] = 0$ and the sum of samples in $x_o[n]$ over symmetric limits $(-\alpha, \alpha)$ equals zero:

$$\sum_{m=-L}^{L} x_o[m] = 0 \tag{2.8}$$

Even symmetric and odd symmetric signals are illustrated in Figure 2.2.

## 2.2.2 Even and Odd Parts of Signals

Even symmetry and odd symmetry are mutually exclusive. Consequently, if a signal $x[n]$ is formed by summing an even symmetric signal $x_e[n]$ and an odd symmetric signal $x_o[n]$, it

Even symmetry: $x_e[n] = x_e[-n]$ Odd symmetry: $x_o[n] = -x_o[-n]$ and $x_o[0] = 0$

**FIGURE 2.2** An even symmetric signal shows mirror symmetry about $n = 0$. The value of an odd symmetric signal is zero at $n = 0$. The sum of an odd symmetric signal and its time-reversed version is zero everywhere

will be devoid of either symmetry. Turning things around, any signal $x[n]$ may be expressed as the sum of an even symmetric part $x_e[n]$ and an odd symmetric part $x_o[n]$:

$$x[n] = x_e[n] + x_o[n] \tag{2.9}$$

To find $x_e[n]$ and $x_o[n]$ from $x[n]$, we reverse $x[n]$ and invoke symmetry to get

$$x[-n] = x_e[-n] + x_o[-n] = x_e[n] - x_o[n] \tag{2.10}$$

Adding and subtracting the two preceding equations, we obtain

$$x_e[n] = 0.5x[n] + 0.5x[-n] \qquad x_o[n] = 0.5x[n] - 0.5x[-n] \tag{2.11}$$

This means that the even part $x_e[n]$ equals half the sum of the original and reversed version and the odd part $x_o[n]$ equals half the difference between the original and reversed version. Naturally, if a signal $x[n]$ has even symmetry, its odd part $x_o[n]$ will equal zero, and if $x[n]$ has odd symmetry, its even part $x_e[n]$ will equal zero.

**Example 2.3**

### Signal Symmetry

**(a)** Let $x[n] = \{4,\ -2,\ \overset{\Downarrow}{4},\ -6\}$. Find and sketch its odd and even parts.

We zero-pad the signal to $x[n] = \{4,\ -2,\ \overset{\Downarrow}{4},\ -6,\ 0\}$ so that it covers symmetric limits. Then

$$0.5x[n] = \{2,\ -1,\ \overset{\Downarrow}{2},\ -3,\ 0\} \qquad 0.5x[-n] = \{0,\ -3,\ \overset{\Downarrow}{2},\ -1,\ 2\}$$

Zero-padding, though not essential, allows us to perform element-wise addition or subtraction with ease to obtain

$$x_e[n] = 0.5x[n] + 0.5x[-n] = \{2,\ -4,\ \overset{\Downarrow}{4},\ -4,\ 2\}$$

$$x_o[n] = 0.5x[n] - 0.5x[-n] = \{2,\ 2,\ \overset{\Downarrow}{0},\ -2,\ -2\}$$

The various signals are sketched in Figure E2.3a. As a consistency check, you should confirm that $x_o[0] = 0$, $\sum x_o[n] = 0$, and that the sum $x_e[n] + x_o[n]$ recovers $x[n]$.

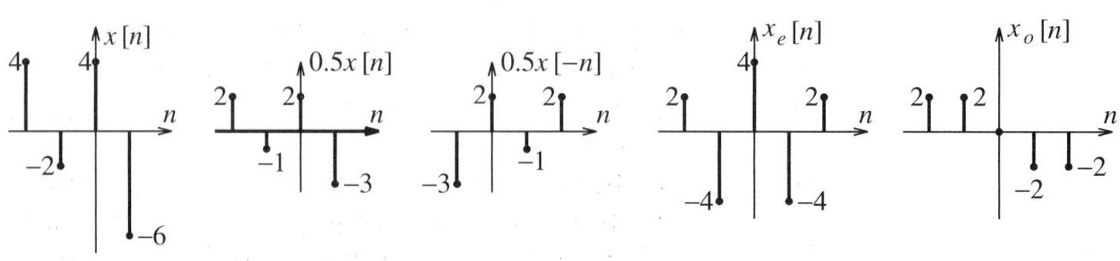


**FIGURE E 2.3.a** The signal $x[n]$ and its odd and even parts for Example 2.3(a)

**(b)** Let $x[n] = u[n] - u[n-5]$. Find and sketch its odd and even parts.

The signal $x[n]$ and the genesis of its odd and even parts are shown in Figure E2.3b. Note the value of $x_e[n]$ at $n = 0$ in the sketch.

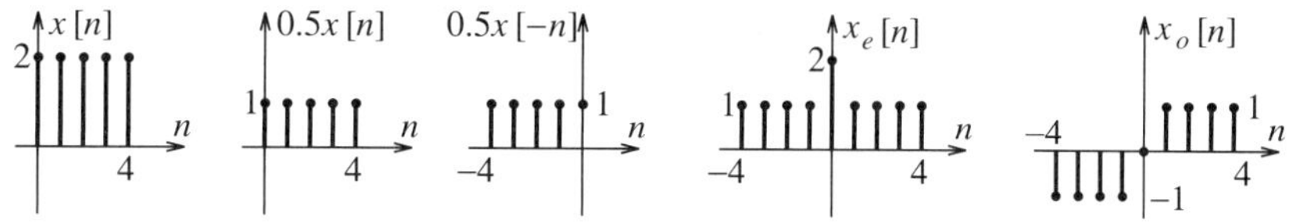


**FIGURE E 2.3.b** The signal $x[n]$ and its odd and even parts for Example 2.3(b)

# Drill Problem 2.4

**(a)** Find and sketch the even and odd parts of $x[n] = \{\overset{\Downarrow}{8},\ 4,\ 2\}$.

**(b)** Find and sketch the odd part of $y[n] = \{8,\ \overset{\Downarrow}{4},\ 2\}$.

**Answers: (a)** $x_e[n] = \{1,\ 2,\ \underset{\Uparrow}{8},\ 2,\ 1\}$ $\quad x_o[n] = \{-1,\ -2,\ \underset{\Uparrow}{0},\ 2,\ 1\}$
**(b)** $y_o[n] = \{3,\ \underset{\Uparrow}{0},\ -3\}$

## 2.3 Decimation and Interpolation

The time scaling of discrete-time signals is equivalent to decreasing or increasing the signal length. Decimation refers to a process of reducing the signal length by discarding signal samples. Suppose $x[n]$ corresponds to an analog signal $x(t)$ sampled at intervals of $t_s$. If the time-compressed signal $y(t) = x(2t)$ is sampled at $t_s$, the samples correspond to the samples of $x(t)$ taken at intervals of $2t_s$. The sampled signal $y[n]$ thus contains only alternate sample values of $x[n]$ (corresponding to $x[0], x[2], x[4], \ldots$). The length of $y[n]$ is thus reduced by a factor of two. This is decimation. We could also obtain $y[n]$ directly from $x(t)$ (not its compressed version) if we sample it at the larger intervals of $2t_s$. This corresponds to a twofold reduction in the sampling rate. Decimation by a factor of $N$ is equivalent to sampling $x(t)$ at intervals of $Nt_s$ and implies an *$N$-fold reduction in the sampling rate*. The decimated signal $y[n]$ is generated from the samples of $x[n]$ by retaining every $N$th sample corresponding to the indices $k = Nn$ and discarding all others. Mathematically,

$$y[n] = x(Nn)$$

Interpolation refers to a process of increasing the signal length by inserting signal samples. Let the signal $x[n]$ correspond to samples of $x(t)$ obtained at intervals of $t_s$. Sampling the time-stretched signal $y(t) = x(t/2)$ is equivalent to sampling the signal $x(t)$ at the smaller intervals of $t_s/2$. The sampled signal $y[n]$ thus contains samples of $x(t)$ sampled at $t_s/2$ and has twice the length of $x[n]$ with one new sample between adjacent samples of the original signal $x[n]$. This describes interpolation by a factor of two. For interpolation by $N$, we insert $N-1$ new values after each sample of $x[n]$. This results in an $N$-fold increase in the length. If an expression for $x[n]$ (or the underlying analog signal) were known, it would be no problem to determine these new sample values. However, if we are only given the sample values of $x[n]$ (without its underlying analytical form), the interpolated values can be chosen (or even estimated) in many different ways. For example, we may choose each new sample value as zero (*zero interpolation*), a constant equal to the previous sample value (*step interpolation*), or the average of adjacent sample values (*linear interpolation*). Zero interpolation is referred to as **up-sampling** and plays an important role in practical interpolation schemes. Mathematically, a zero interpolated signal $y[n]$ may be described in terms of the original signal $x[n]$ by

$$y[n] = \begin{cases} x(n/N), & n = 0, \pm N, \pm 2N, \ldots \\ 0, & \text{otherwise} \end{cases} \qquad \text{(zero interpolation by a factor of } N\text{)}$$

Interpolation by a factor of $N$ is equivalent to sampling $x(t)$ at intervals $t_s/N$ and implies an *N-fold increase in both the sampling rate and the signal length*. For convenience, we shall use the notation $x^{\uparrow}[n/N]$ to describe an interpolated version of the signal $x[n]$.

### Some Caveats

It may appear that decimation (discarding signal samples) and interpolation (inserting signal samples) are inverse operations, but this is not always the case. Consider the two sets of operations shown below:

$$x[n] \longrightarrow \boxed{\text{decimate by 2}} \longrightarrow x[2n] \longrightarrow \boxed{\text{interpolate by 2}} \longrightarrow x[n]$$

$$x[n] \longrightarrow \boxed{\text{interpolate by 2}} \longrightarrow x^{\uparrow}[n/2] \longrightarrow \boxed{\text{decimate by 2}} \longrightarrow x[n]$$

On the face of it, both sets of operations start with $x[n]$ and appear to recover $x[n]$, suggesting that interpolation and decimation are inverse operations. In fact, only the second sequence of operations (interpolation followed by decimation) recovers $x[n]$ exactly. To see why, let $x[n] = \{\overset{\Downarrow}{1}, 2, 6, 4, 8\}$. Using step interpolation, for example, the two sequences of operations result in

$$\{\overset{\Downarrow}{1}, 2, 6, 4, 8\} \xrightarrow[n \to 2n]{\text{decimate}} \{\overset{\Downarrow}{1}, 6, 8\} \xrightarrow[n \to n/2]{\text{step interpolation}} \{\overset{\Downarrow}{1}, 1, 6, 6, 8, 8\}$$

$$\{\overset{\Downarrow}{1}, 2, 6, 4, 8\} \xrightarrow[n \to n/2]{\text{step interpolation}} \{\overset{\Downarrow}{1}, 1, 2, 2, 6, 6, 4, 4, 8, 8\} \xrightarrow[n \to 2n]{\text{decimate}} \{\overset{\Downarrow}{1}, 2, 6, 4, 8\}$$

We see that decimation is indeed the inverse of interpolation, but the converse is not necessarily true. After all, it is highly unlikely for any interpolation scheme to recover or predict the exact value of the samples that were discarded during decimation. In situations where both interpolation and decimation are to be performed in succession, it is therefore best to interpolate first. In practice, of course, interpolation or decimation should preserve the information content of the original signal, and this imposes constraints on the rate at which the original samples were acquired.

## 2.3.1 Fractional Delays

Fractional (typically half-sample) delays are sometimes required in practice and can be implemented using interpolation and decimation. If we require that interpolation be followed by decimation and integer shifts, the correct result is obtained by using interpolation followed by an integer shift and decimation. To generate the signal $y[n] = x[n - \frac{M}{N}] = x[\frac{Nn-M}{N}]$ from $x[n]$, we use the following sequence of operations.

$$x[n] \longrightarrow \boxed{\text{interpolate by } N} \longrightarrow x^{\uparrow}[\tfrac{n}{N}] \longrightarrow \boxed{\text{delay by } M} \longrightarrow x[\tfrac{n-M}{N}]$$

$$X[\tfrac{n-M}{N}] \longrightarrow \boxed{\text{decimate by } N} \longrightarrow x[\tfrac{Nn-M}{N}] = y[n]$$

The idea is to ensure that each operation (interpolation, shift, and decimation) involves integers.

**Example 2.4**

### Decimation and Interpolation

**(a)** Let $x[n] = \{1, \overset{\Downarrow}{2}, 5, -1\}$. Generate $x[2n]$ and various interpolated versions of $x^{\uparrow}[n/3]$.

To generate $y[n] = x[2n]$, we remove samples at the odd indices to obtain $x[2n] = \{\overset{\Downarrow}{2}, -1\}$.

The zero-interpolated signal:

$$g[n] = x^{\uparrow}[\tfrac{n}{3}] = \{1, 0, 0, \overset{\Downarrow}{2}, 0, 0, 5, 0, 0, -1, 0, 0\}$$

The step-interpolated signal:

$$h[n] = x^{\uparrow}[\tfrac{n}{3}] = \{1, 1, 1, \overset{\Downarrow}{2}, 2, 2, 5, 5, 5, -1, -1, -1\}$$

The linearly interpolated signal:

$$s[n] = x^{\uparrow}[\tfrac{n}{3}] = \{1, \tfrac{4}{3}, \tfrac{5}{3}, \overset{\Downarrow}{2}, 3, 4, 5, 3, 1, -1, -\tfrac{2}{3}, -\tfrac{1}{3}\}$$

In linear interpolation, note that we interpolated the last two values toward zero.

**(b)** Let $x[n] = \{3, 4, \overset{\Downarrow}{5}, 6\}$. Find $g[n] = x[2n - 1]$ and the step-interpolated signal $h[n] = x[0.5n - 1]$. In either case, we first find $y[n] = x[n - 1] = \{3, \overset{\Downarrow}{4}, 5, 6\}$. Then,

$$g[n] = y[2n] = x[2n - 1] = \{\overset{\Downarrow}{4}, 6\}$$

$$h[n] = y^{\uparrow}[\tfrac{n}{2}] = x[0.5n - 1] = \{3, 3, \overset{\Downarrow}{4}, 4, 5, 5, 6, 6\}$$

**(c)** Let $x[n] = \{3, 4, \overset{\Downarrow}{5}, 6\}$. Find $y[n] = x[2n/3]$ assuming step interpolation where needed. Since we require both interpolation and decimation, we first interpolate and then decimate to get

$$\text{After interpolation: } g[n] = x^{\uparrow}[\tfrac{n}{3}] = \{3, 3, 3, 4, 4, 4, \overset{\Downarrow}{5}, 5, 5, 6, 6, 6\}$$

$$\text{After decimation: } y[n] = g[2n] = x[\tfrac{2}{3}n] = \{3, 3, 4, \overset{\Downarrow}{5}, 5, 6\}$$

**(d)** Let $x[n] = \{2, 4, \overset{\Downarrow}{6}, 8\}$. Find the signal $y[n] = x[n - 0.5]$, assuming linear interpolation where needed. We first interpolate by 2, then delay by 1, and then decimate by 2 to get

$$\text{After interpolation:}\, g[n] = x^{\uparrow}[\tfrac{n}{2}] = \{2,\ 3,\ 4,\ 5,\ \overset{\Downarrow}{6},\ 7,\ 8,\ 4\}$$
(last sample interpolated to zero)

$$\text{After delay: } h[n] = g[n-1] = x[\tfrac{n-1}{2}] = \{2,\ 3,\ 4,\ \overset{\Downarrow}{5},\ 6,\ 7,\ 8,\ 4\}$$

$$\text{After decimation: } y[n] = h[2n] = x[\tfrac{2n-1}{2}] = x[n-0.5] = \{3,\ \overset{\Downarrow}{5},\ 7,\ 4\}$$

# Drill Problem 2.5

Let $x[n] = \{\overset{\Downarrow}{8},\ 4,\ 2,\ 6\}$. Find $y[n] = x[2n]$, $g[n] = x[2n+1]$, $h[n] = x^{\uparrow}[n/2]$, and $f[n] = x[n+0.5]$. Assume linear interpolation where required.

**Answers:** $y[n] = \{\overset{\Downarrow}{8},\ 2\}$ $\quad g[n] = \{\overset{\Downarrow}{4},\ 6\}$ $\quad h[n] = \{\overset{\Downarrow}{8},\ 6,\ 4,\ 3,\ 2,\ 4,\ 6,\ 3\}$
$f[n] = \{\overset{\Downarrow}{6},\ 3,\ 4,\ 3\}$

## 2.4 Some Standard Discrete Signals

The **unit impulse** (or **unit sample**) $\delta[n]$, the **unit step** $u[n]$, and the **unit ramp** $r[n]$ are defined as

$$\delta[n] = \begin{cases} 0, & n \neq 0 \\ 1, & n = 0 \end{cases} \qquad u[n] = \begin{cases} 0, & n < 0 \\ 1, & n \geq 0 \end{cases} \qquad r[n] = nu[n] = \begin{cases} 0, & n < 0 \\ n, & n \geq 0 \end{cases} \tag{2.12}$$

These signals are illustrated in Figure 2.3. The discrete impulse is just a unit sample at $n = 0$. It is completely free of the kind of ambiguities associated with the analog impulse $\delta(t)$ at $t = 0$. The discrete unit step $u[n]$ also has a well-defined, unique value of $u[0] = 1$ (unlike its analog counterpart $u(t)$). The signal $x[n] = Anu[n] = Ar[n]$ describes a discrete ramp whose "slope" $A$ is given by $x[k] - x[k-1]$, which is the difference between adjacent sample values.

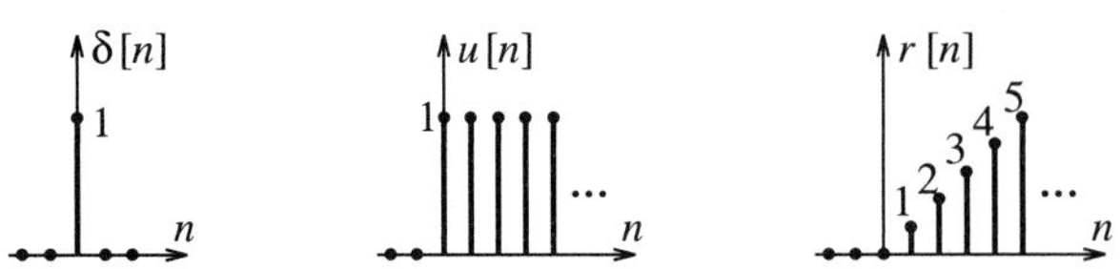


**FIGURE 2.3** The discrete unit impulse $\delta[n]$ is also called the unit sample. The standard unit step $u[n]$ and unit ramp $r[n] = nu[n]$ are both causal signals

### 2.4.1 Properties of the Discrete Impulse

The product of a signal $x[n]$ with the impulse $\delta[n-k]$ results in

$$x[n]\delta[n-k] = x[k]\delta[n-k] \tag{2.13}$$

This is because $\delta[n-k]$ is nonzero only at $n = k$, where the value of $x[n]$ corresponds to $x[k]$. The result is an impulse with strength $x[k]$. The product property leads directly to

$$\sum_{n=-\infty}^{\infty} x[n]\delta[n-k] = x[k] \tag{2.14}$$

This is the **sifting property**. The impulse extracts (sifts out) the value $x[k]$ from $x[n]$ at the impulse location $n = k$.

### 2.4.2 Signal Representation by Impulses

A discrete signal $x[n]$ may be expressed as a sum of shifted impulses $\delta[n-k]$ whose sample values correspond to $x[k]$, with the values of $x[n]$ at $n = k$. Thus,

$$x[n] = \sum_{k=-\infty}^{\infty} x[k]\delta[n-k] \tag{2.15}$$

For example, the signals $u[n]$ and $r[n]$ may be expressed as a train of shifted impulses:

$$u[n] = \sum_{k=0}^{\infty} \delta[n-k] \qquad r[n] = \sum_{k=0}^{\infty} k\delta[n-k] \tag{2.16}$$

The signal $u[n]$ may also be expressed as the cumulative sum of $\delta[n]$, and the signal $r[n]$ may be described as the cumulative sum of $u[n]$:

$$u[n] = \sum_{k=-\infty}^{n} \delta[k] \qquad r[n] = \sum_{k=-\infty}^{n} u[k] \tag{2.17}$$

### 2.4.3 Discrete Pulse Signals

The discrete rectangular pulse $\mathbf{rect}(n/2N)$ and the discrete triangular pulse $\mathbf{tri}(n/N)$ are defined by

$$\text{rect}\left(\frac{n}{2N}\right) = \begin{cases} 1, & |n| \le N \\ 0, & \text{elsewhere} \end{cases} \qquad \text{tri}\left(\frac{n}{N}\right) = \begin{cases} 1 - \frac{|n|}{N}, & |n| \le N \\ 0, & \text{elsewhere} \end{cases} \tag{2.18}$$

These signals are illustrated in Figure 2.4. The signal $\text{rect}(\frac{n}{2N})$ has $2N+1$ unit samples over $-N \le n \le N$. The factor $2N$ in $\text{rect}(\frac{n}{2N})$ gets around the problem of having to deal with

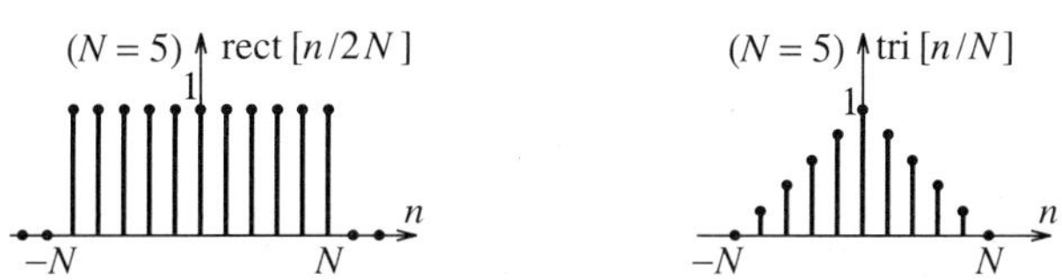


**FIGURE 2.4** The definition of $\text{rect}(\frac{n}{2N})$ implies even symmetry and $2N+1$ samples. The signal $\text{tri}(n/N)$ has even symmetry and also contains $2N+1$ samples (if we include the two zero-valued end samples)

half-integer values of $n$ when $N$ is odd. The signal $x[n] = \text{tri}(n/N)$ also has $2N+1$ samples over $-N \le n \le N$, with its end samples $x[N]$ and $x[-N]$ being zero. It is sometimes convenient to express pulse-like signals in terms of these standard forms.

**Example 2.5**

### Describing Sequences and Signals

**(a)** Let $x[n] = (2)^n$ and $y[n] = \delta[n-3]$. Find $z[n] = x[n]y[n]$ and evaluate the sum $A = \sum z[n]$.
The product, $z[n] = x[n]y[n] = (2)^3\delta[n-3] = 8\delta[n-3]$, is an impulse.
The sum, $A = \sum z[n]$, is given by $\sum (2)^n\delta[n-3] = (2)^3 = 8$.

**(b)** Mathematically describe the signals of Figure E2.5b in at least two different ways.

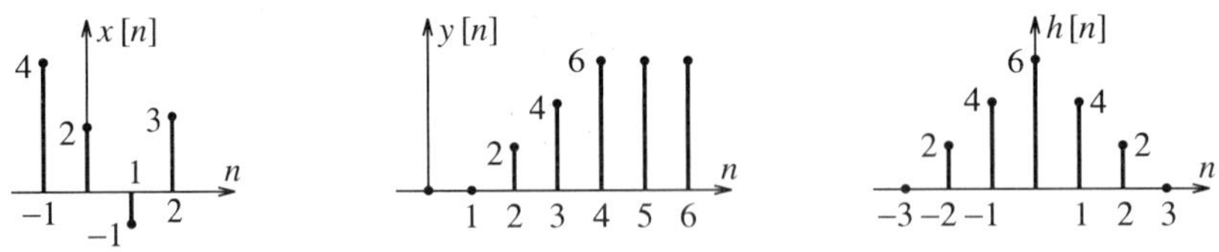


**FIGURE E 2.5.b** The signals for Example E2.5(b)

1. The signal $x[n]$ may be described as the sequence $x[n] = \{4, \overset{\Downarrow}{2}, -1, 3\}$
   It may also be written as $x[n] = 4\delta[n+1] + 2\delta[n] - \delta[n-1] + 3\delta[n-2]$
2. The signal $y[n]$ may be represented variously as
   A numeric sequence: $y[n] = \{0, 0, 2, 4, 6, 6, 6\}$
   A sum of shifted impulses:
   $y[n] = 2\delta[n-2] + 4\delta[n-3] + 6\delta[n-4] + 6\delta[n-5] + 6\delta[n-6]$
   A sum of steps and ramps: $y[n] = 2r[n-1] - 2r[n-4] - 6u[n-7]$
   Note carefully that the argument of the step function is $[n-7]$ (and not $[n-6]$)
3. The signal $h[n]$ may be described as $h[n] = 6\,\text{tri}(n/3)$ or variously as
   A numeric sequence: $h[n] = \{0, 2, 4, \overset{\Downarrow}{6}, 4, 2, 0\}$
   A sum of impulses: $h[n] = 2\delta[n+2] + 4\delta[n+1] + 6\delta[n] + 4\delta[n-1] + 2\delta[n-2]$
   A sum of steps and ramps: $h[n] = 2r[n+3] - 4r[n] + 2r[n-3]$

# Drill Problem 2.6

**(a)** Sketch the signals $x[n] = \delta[n+2] + 2\delta[n-1]$ and $y[n] = 2u[n+1] - u[n-3]$.

**(b)** Express the signal $h[n] = \{3, 3, \overset{\Downarrow}{3}, 5, 5, 5\}$ using step functions.

**(c)** Express the signal $g[n] = \{\overset{\Downarrow}{2}, 4, 6, 4, 2\}$ using tri functions.

**Answers: (b)** $h[n] = 3u[n+2] + 2u[n-1] - 5u[n-4]$ **(c)** $g[n] = 6\,\text{tri}(\frac{n-2}{3})$

### 2.4.4 The Discrete Sinc Function

The discrete sinc function is defined by

$$\mathrm{sinc}\left(\frac{n}{N}\right) = \frac{\sin(n\pi/N)}{(n\pi/N)}, \qquad \mathrm{sinc}(0) = 1 \tag{2.19}$$

The signal $\mathrm{sinc}(n/N)$ equals zero at $n = kN, k = \pm 1, \pm 2, \ldots$ . At $n = 0$, $\mathrm{sinc}(0) = 0/0$ and cannot be evaluated in the limit since $n$ can take on only integer values. *We therefore define* $\mathrm{sinc}(0) = 1$. The *envelope* of the sinc function shows a mainlobe and gradually decaying sidelobes. The definition of $\mathrm{sinc}(n/N)$ also implies that $\mathrm{sinc}(n) = \delta[n]$.

### 2.4.5 Discrete Exponentials

Discrete exponentials are often described using a rational base. For example, the signal $x[n] = 2^n u[n]$ shows exponential growth, while $y[n] = (0.5)^n u[n]$ is a decaying exponential. The signal $f[n] = (-0.5)^n u[n]$ shows values that alternate in sign. The exponential $x[n] = \alpha^n u[n]$—where $\alpha = re^{j\theta}$ is complex—may be described using the various formulations of a complex number as

$$x[n] = \alpha^n u[n] = (re^{j\theta})^n u[n] = r^n e^{jn\theta} u[n] = r^n[\cos(n\theta) + j\sin(n\theta)]u[n] \tag{2.20}$$

This complex-valued signal requires two separate plots (the real and imaginary parts, for example) for a graphical description. If $0 < r < 1$, $x[n]$ describes a signal whose real and imaginary parts are exponentially decaying cosines and sines. If $r = 1$, the real and imaginary parts are pure cosines and sines with a peak value of unity. If $r > 1$, we obtain exponentially growing sinusoids.

## 2.5 Discrete-Time Harmonics and Sinusoids

If we sample an analog sinusoid $x(t) = \cos(2\pi f_0 t)$ at intervals of $t_s$ corresponding to a sampling rate of $S = 1/t_s$ samples/s (or $S$ Hz), we obtain the sampled sinusoid:

$$x[n] = \cos(2\pi f n t_s + \theta) = \cos(2\pi n \tfrac{f}{S} + \theta) = \cos(2\pi n F + \theta) \tag{2.21}$$

More generally, we find it useful to deal with complex valued signals of the form:

$$\tilde{x}[n] = e^{j(2\pi n F + \theta)} = \cos(2\pi n F + \theta) + j\sin(2\pi n F + \theta) \tag{2.22}$$

This allows us to regard the real sinusoid $x[n] = \cos(2\pi n F + \theta)$ as the real part of the complex-valued signal $\tilde{x}[n]$.

The quantities $f$ and $\omega = 2\pi f$ describe analog frequencies. The normalized frequency $F = f/S$ is called the **digital frequency** and has units of cycles/sample. The frequency $\Omega = 2\pi F$ is the digital *radian* frequency with units of radians/sample. Note that the analog frequency $f = S$ (or $\omega = 2\pi S$) corresponds to the digital frequency $F = 1$ (or $\Omega = 2\pi$). The various analog and digital frequencies are compared in Figure 2.5.

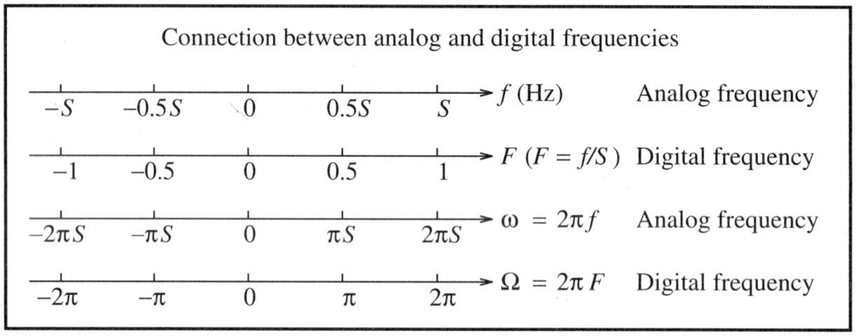


**FIGURE 2.5** Comparison of analog and digital frequencies. Note that the digital frequency $F = 1$ corresponds to the sampling frequency $f = S$ because, by definition, the digital frequency $F = f/S$ is the ratio of the analog frequency $f$ and the sampling rate $S$

**The Digital Frequency is the Analog Frequency Divided by Sampling Rate S**

$$F(\text{cycles/sample}) = \frac{f(\text{cycles/sec})}{S(\text{samples/sec})} \qquad \Omega(\text{radians/sample}) = \frac{\omega(\text{radians/sec})}{S(\text{samples/sec})} = 2\pi F$$

### 2.5.1 Discrete-Time Harmonics are not Always Periodic in Time

An analog sinusoid $x(t) = \cos(2\pi f_0 t + \theta)$ has two remarkable properties. It is unique for every frequency. And it is periodic in time for every choice of the frequency $f_0$. Its sampled version, however, is a beast of a different kind.

Are all discrete-time sinusoids and harmonics periodic in time? Not always! To understand this idea, suppose $x[n]$ is periodic with period $N$ such that $x[n] = x[n + N]$. This leads to

$$\cos(2\pi n F_0 + \theta) = \cos[2\pi(n + N)F_0 + \theta] = \cos(2\pi n F_0 + \theta + 2\pi N F_0) \qquad (2.23)$$

The two sides are equal provided $NF_0$ equals an integer $k$. In other words, $F_0$ must be a *rational fraction* (ratio of integers) of the form $k/N$. What we are really saying is that a discrete sinusoid is *not* always periodic but only if its digital frequency is a ratio of integers or a *rational fraction*. The period $N$ equals the denominator of $k/N$, provided common factors have been canceled from its numerator and denominator. The significance of $k$ is that it takes $k$ full periods of the analog sinusoid to yield one full period of the sampled sinusoid. The common period of a combination of periodic discrete sinusoids equals the least common multiple (LCM) of their individual periods. If $F_0$ is not a rational fraction, there is no periodicity, and the discrete sinusoid is classified as nonperiodic or *almost periodic*. Examples of periodic and nonperiodic discrete sinusoids appear in Figure 2.6. Even though a discrete sinusoid may not always be periodic, it will always have a periodic *envelope*. This discussion also applies to complex-valued harmonics of the type $e^{j(2\pi n F_0 + \theta)}$.

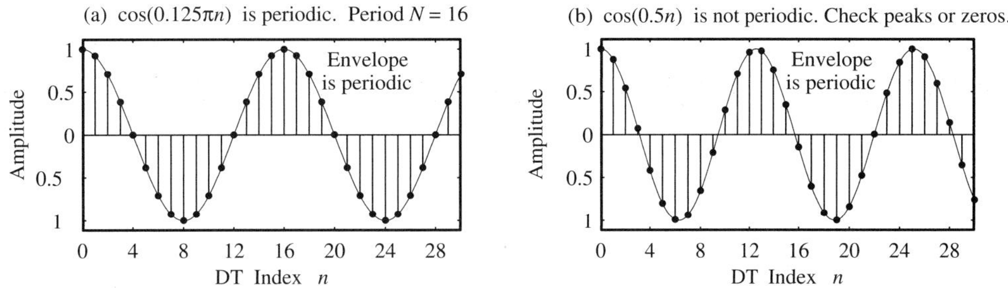


**FIGURE 2.6** Discrete-time sinusoids are not always periodic. The first panel shows the signal $\cos(0.125n\pi)$ (whose period is $N = 16$) and its envelope. The second panel shows the signal $\cos(0.5n)$ and its envelope. Even though their envelopes are periodic, look carefully at the peaks and troughs and confirm that the first signal is periodic while the second signal is not periodic

**The Discrete Harmonic $\cos(2\pi nF_0 + \theta)$ or $e^{j(2\pi nF_0+\theta)}$ is not Always Periodic in Time**

It is periodic only if its digital frequency $F_0 = k/N$ can be expressed as a ratio of integers. Its period equals $N$ if common factors have been canceled in $k/N$. One period of the sampled sinusoid is obtained from $k$ full periods of the analog sinusoid. For a combination of periodic sinusoids, the common period $N$ is the LCM of their individual periods.

**Example 2.6**

### Discrete-Time Harmonics and Periodicity

**(a)** Is $x[n] = \cos(2\pi Fn)$ periodic if $F = 0.32$? If $F = \sqrt{3}$? If periodic, what is its period $N$?
If $F = 0.32$, $x[n]$ is periodic because $F = 0.32 = \frac{32}{100} = \frac{8}{25} = \frac{k}{N}$. The period is $N = 25$.
If $F = \sqrt{3}$, $x[n]$ is *not* periodic because $F$ is irrational and cannot be expressed as a ratio of integers.

**(b)** What is the period of the harmonic signal $x[n] = e^{j0.2n\pi} + e^{-j0.3n\pi}$?
The digital frequencies in $x[n]$ are $F_1 = 0.1 = \frac{1}{10} = \frac{k_1}{N_1}$ and $F_2 = 0.15 = \frac{3}{20} = \frac{k_2}{N_2}$.
Their periods are $N_1 = 10$ and $N_2 = 20$.
The common period is thus $N = \text{LCM}(N_1, N_2) = \text{LCM}(10, 20) = 20$.

**(c)** The signal $x(t) = 2\cos(40\pi t) + \sin(60\pi t)$ is sampled at 75 Hz. What is the common period of the sampled signal $x[n]$, and how many full periods of $x(t)$ does it take to obtain one period of $x[n]$?
The frequencies in $x(t)$ are $f_1 = 20$ Hz and $f_2 = 30$ Hz.

The digital frequencies of the individual components are $F_1 = \frac{20}{75} = \frac{4}{15} = \frac{k_1}{N_1}$ and $F_2 = \frac{30}{75} = \frac{2}{5} = \frac{k_2}{N_2}$.
Their periods are $N_1 = 15$ and $N_2 = 5$.
The common period is thus $N = \text{LCM}(N_1, N_2) = \text{LCM}(15, 5) = 15$.
The fundamental frequency of $x(t)$ is $f_0 = \text{GCD}(20, 30) = 10$ Hz. One period of $x(t)$ is $T = \frac{1}{f_0} = 0.1$ s.
Since $N = 15$ corresponds to a duration of $Nt_s = \frac{N}{S} = 0.2$ s, it takes two full periods of $x(t)$ to obtain one period of $x[n]$. We also get the same result by computing $\text{GCD}(k_1, k_2) = \text{GCD}(4, 2) = 2$.

# Drill Problem 2.7

**(a)** What is the digital frequency of $x[n] = 2e^{j(0.25n\pi + 30^\circ)}$? Is $x[n]$ periodic?
**(b)** What is the digital frequency of $y[n] = \cos(0.5n + 30^\circ)$? Is $y[n]$ periodic?
**(c)** What is the common period $N$ of the signal $f[n] = \cos(0.4n\pi) + \sin(0.5n\pi + 30^\circ)$?

**Answers:** **(a)** $F_0 = 0.125$, yes **(b)** $F_0 = 0.25/\pi$, no **(c)** $N = 20$

## 2.5.2 Discrete-Time Harmonics are Always Periodic in Frequency

Unlike analog sinusoids, discrete-time sinusoids and harmonics are *always* periodic in frequency. If we start with the sinusoid $x[n] = \cos(2\pi nF_0 + \theta)$ and add an integer $m$ to $F_0$, we get

$$\cos[2\pi n(F_0 + m) + \theta] = \cos(2\pi nF_0 + \theta + 2\pi nm) = \cos(2\pi nF_0 + \theta) = x[n]$$

This result says that discrete sinusoids at the frequencies $F_0 \pm m$ are *identical*. Put another way, a discrete sinusoid is *periodic* in frequency (has a periodic spectrum) with unit period. The range $-0.5 \le F \le 0.5$ defines the **central period** or **principal range**. A discrete sinusoid can be uniquely identified only if its frequency falls in the principal range. A discrete sinusoid with a frequency $F_0$ outside this range can always be expressed as a discrete sinusoid with a frequency that falls in the central period by subtracting out an integer $M$ from $F_0$ such that the new frequency $F_a = F_0 - M$ satisfies $-0.5 \le F_u \le 0.5$). The frequency $F_a$ is called the aliased digital frequency and it is always smaller than original frequency $F_0$. This discussion also applies to complex-valued harmonics of the type $e^{j(2\pi nF_0+\theta)}$.

To summarize, *a discrete-time sinusoid or harmonic is periodic in time only if its digital frequency $F_0$ is a rational fraction, but it is always periodic in frequency*. The following panel summarizes these fundamental results.

**The Discrete Harmonic $\cos(2\pi n F_0 + \theta)$ or $e^{j(2\pi n F_0+\theta)}$ is Periodic in Frequency**

The period is $F = 1$. Discrete harmonics at $F_0$ and $F_0 \pm K$ are identical for *integer* $K$).
A discrete harmonic is **unique** only if its frequency $F_0$ lies in the *central period* $-0.5 < F_0 \le 0.5$.
If $F_0 > 0.5$, the unique frequency is $F_a = F_0 - M$, where the integer $M$ is chosen to ensure $-0.5 < F_a \le 0.5$.

# Drill Problem 2.8

**(a)** Let $x[n] = e^{j1.4n\pi} = e^{j2\pi F_u n}$ where $F_u$ is in the principal range. What is the value of $F_u$?
**(b)** Let $y[n] = \cos(2.4n\pi + 30^\circ)$. Rewrite $y[n]$ in terms of its frequency in the principal range.
**(c)** Rewrite $f[n] = \cos(1.4n\pi + 20^\circ) + \cos(2.4n\pi + 30^\circ)$ using frequencies in the principal range.

**Answers: (a)** $-0.3$ **(b)** $\cos(0.4n\pi + 30^\circ)$ **(c)** $\cos(0.6n\pi - 20^\circ) + \cos(0.4n\pi + 30^\circ)$

## 2.6 The Sampling Theorem

The central concept in the digital processing of analog signals is that the sampled signal must be a unique representation of the underlying analog signal. When the sinusoid $x(t) = \cos(2\pi f_0 t + \theta)$ is sampled at the sampling rate $S$, the digital frequency of the sampled signal is $F_0 = f_0/S$. In order for the sampled sinusoid to permit a unique correspondence with the underlying analog sinusoid, the digital frequency $F_0$ must lie in the principal range, i.e., $|F_0| < 0.5$. This implies $S > 2|f_0|$ and suggests that we must choose a sampling rate $S$ that exceeds $|2f_0|$. More generally, the **sampling theorem** says that for a unique correspondence between an analog signal and the version reconstructed from its samples (using the *same* sampling rate), the sampling rate must *exceed* **twice** the highest signal frequency $f_{\max}$. The value $S = 2f_{\max}$ is called the **critical sampling rate** or **Nyquist rate** or **Nyquist frequency**. The time interval $t_s = \frac{1}{2f_{\max}}$ is called the **Nyquist interval**. For the sinusoid $x(t) = \cos(2\pi f_0 t + \theta)$, the Nyquist rate is $S_N = 2f_0 = \frac{2}{T}$ and this rate is equivalent to taking exactly *two samples per period* (because the sampling interval is $t_s = \frac{T}{2}$). In order to exceed the Nyquist rate, we should obtain *more than* two signal samples per period.

**The Sampling Theorem: How to Sample an Analog Signal without Loss of Information**

For an analog signal band-limited to $f_{max}$ Hz, the sampling rate $S > 2f_{max}$ must *exceed* $2f_{max}$.
$S = 2f_{max}$ defines the **Nyquist rate.**
$t_s = \frac{1}{2f_{max}}$ defines the **Nyquist interval.**
**For an Analog Sinusoid:** The Nyquist rate corresponds to taking *two samples per period.*

# Drill Problem 2.9

**(a)** What is the critical sampling rate in Hz for the following signals:

$$x(t) = \cos(10\pi t) \quad y(t) = \cos(10\pi t) + \sin(15\pi t)$$

$$f(t) = \cos(10\pi t)\sin(15\pi t) \quad g(t) = \cos^2(10\pi t)$$

**(b)** A 50 Hz sinusoid is sampled at twice the Nyquist rate. How many samples are obtained in 3 s?

**Answers: (a)** 10, 15, 25, 20 **(b)** 600

## 2.6.1 Signal Reconstruction and Aliasing

Consider an analog signal $x(t) = \cos(2\pi f_0 t + \theta)$ and its sampled version $x[n] = \cos(2\pi n F_0 + \theta)$, where $F_0 = f_0/S$. If $x[n]$ is to be a unique representation of $x(t)$, we must be able to reconstruct $x(t)$ from $x[n]$. In practice, reconstruction uses only the central copy or **image** of the periodic spectrum of $x[n]$ in the **central period** $-0.5 \le F \le 0.5$, which corresponds to the analog frequency range $-0.5S \le f \le 0.5S$. We use a lowpass filter to remove all other replicas or images, and the output of the lowpass filter corresponds to the reconstructed analog signal. As a result, the highest frequency $f_H$ we can identify in the signal reconstructed from its samples is $f_H = 0.5S$.

Whether the reconstructed analog signal matches $x(t)$ or not depends on the sampling rate $S$. If we exceed the Nyquist rate (i.e., $S > 2f_0$), the digital frequency $F_0 = f_0/S$ is always in the principal range $-0.5 \le F \le 0.5$, and the reconstructed analog signal is *identical* to $x(t)$. If the sampling rate is below the Nyquist rate (i.e., $S < 2f_0$), the digital frequency exceeds 0.5. Its image in the principal range appears at the *lower* digital frequency $F_a = F_0 - M$ (corresponding to the *lower* analog frequency $f_a = f_0 - MS$), where $M$ is an integer that places the aliased digital frequency $F_a$ between $-0.5$ and 0.5 (or the aliased analog frequency $f_a$ between $-0.5S$ and $0.5S$). The reconstructed aliased signal $x_a(t) = \cos(2\pi f_a t + \theta)$ is at a *lower* frequency $f_a = SF_a$ than $f_0$ and is no longer a replica of $x(t)$. The phenomenon, where a reconstructed sinusoid appears at a lower frequency than the original, is what **aliasing** is all about. The real problem is that the original signal $x(t)$ and the aliased signal $x_a(t)$ yield *identical* sampled representations at the sampling frequency $S$ and prevent unique identification of the original signal $x(t)$ from its samples!

---

**Aliasing Occurs if an Analog Signal $\cos(2\pi f_0 t + \theta)$ is Sampled Below the Nyquist Rate**

If $S < 2f_0$, the *reconstructed* analog signal is aliased to a *lower* frequency $|f_a| < 0.5S$. We find $f_a$ as $f_a = f_0 - MS$, where $M$ is an integer that places $f_a$ in the *central period* $(-0.5S < f_a \leq 0.5S)$.

**The aliased or reconstructed signal is based on frequencies in the principal range.** The highest frequency in the reconstructed signal cannot exceed half the reconstruction rate.

---

**Example 2.7**

### Aliasing and Its Effects

**(a)** A 100-Hz sinusoid $x(t)$ is sampled at 240 Hz. Has aliasing occurred? How many full periods of $x(t)$ are required to obtain one period of the sampled signal?

The sampling rate exceeds 200 Hz, so there is no aliasing. The digital frequency is $F = \frac{100}{240} = \frac{5}{12}$. Thus, five periods of $x(t)$ yield 12 samples (one period) of the sampled signal.

**(b)** A 100-Hz sinusoid is sampled at rates of 240 Hz, 140 Hz, 90 Hz, and 35 Hz. In each case, has aliasing occurred, and if so, what is the aliased frequency?

To avoid aliasing, the sampling rate must exceed 200 Hz. If $S = 240$ Hz, there is no aliasing, and the reconstructed signal (from its samples) appears at the original frequency of 100 Hz. For all other choices of $S$, the sampling rate is too low and leads to aliasing. The aliased signal shows up at a lower frequency. The aliased frequencies corresponding to each sampling rate $S$ are found by subtracting out multiples of $S$ from 100 Hz to place the result in the range $-0.5S \leq f \leq 0.5S$. If the original signal has the form $x(t) = \cos(200\pi t + \theta)$, we obtain the following aliased frequencies and aliased signals:

$$S = 140\text{ Hz},\quad f_a = 100 - 140 = -40\text{ Hz},\quad x_a(t) = \cos(-80\pi t + \theta) = \cos(80\pi t - \theta)$$

$$S = 90\text{ Hz},\quad f_a = 100 - 90 = 10\text{ Hz},\quad x_a(t) = \cos(20\pi t + \theta)$$

$$S = 35\text{ Hz},\quad f_a = 100 - 3(35) = -5\text{ Hz},\quad x_a(t) = \cos(-10\pi t + \theta) = \cos(10\pi t - \theta)$$

We thus obtain a 40-Hz sinusoid (with reversed phase), a 10-Hz sinusoid, and a 5-Hz sinusoid (with reversed phase), respectively. Notice that negative aliased frequencies simply lead to a phase reversal and do not represent any new information. Finally, had we used a sampling rate exceeding the Nyquist rate of 200 Hz, we would have recovered the original 100-Hz signal every time. Yes, it pays to play by the rules of the sampling theorem!

**(c)** Two analog sinusoids $x_1(t)$ (shown light) and $x_2(t)$ (shown dark) lead to an identical sampled version as illustrated in Figure E2.7c. Has aliasing occurred? Identify the original and aliased signal. Identify the digital frequency of the sampled signal corresponding to each sinusoid. What is the analog frequency of each sinusoid if $S = 50$ Hz? Can you provide exact expressions for each sinusoid?

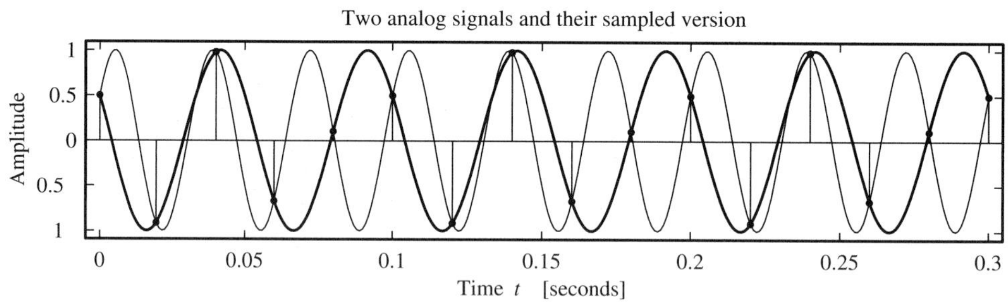


**FIGURE E 2.7.c** The sinusoids for Example 2.7(c). We see five samples per period. These five samples are obtained from three full cycles of $x_1(t)$ (shown light) or two full cycles of $x_2(t)$ (shown dark). The signal $x_2(t)$ has the lower frequency and thus corresponds to the aliased version

The samples are 0.02 s apart. We observe a period of $N = 5$ (corresponding to 0.1 s). The five samples correspond to three full periods of $x_1(t)$ and so $F_1 = \frac{3}{5}$. They also correspond to two full periods of $x_2(t)$, and so $F_2 = \frac{2}{5}$. Clearly, $x_1(t)$ (with $|F_1| > 0.5$) is the original signal that is aliased to $x_2(t)$. The sampling interval is 0.02 s. So, the sampling rate is $S = 50$ Hz. The original and aliased frequencies are $f_1 = SF_1 = 30$ Hz and $f_2 = SF_2 = 20$ Hz.

From Figure E 2.7C, we can identify exact expressions for $x_1(t)$ and $x_2(t)$ as follows. Since $x_1(t)$ is a delayed cosine with $x_1(0) = 0.5$, we have $x_1(t) = \cos(60\pi t - \frac{\pi}{3})$. With $S = 50$ Hz, the frequency $f_1 = 30$ Hz actually aliases to $f_2 = -20$ Hz, and thus $x_2(t) = \cos(-40\pi t - \frac{\pi}{3}) = \cos(40\pi t + \frac{\pi}{3})$. With $F = \frac{30}{50} = 0.6$ (or $F = -0.4$), the expression for the sampled signal is $x[n] = \cos(2\pi nF - \frac{\pi}{3})$.

**(d)** A 100-Hz sinusoid is sampled, and the reconstructed signal (from its samples) shows up at 10 Hz. What was the sampling rate $S$?

One choice is to set $100 - S = 10$ and obtain $S = 90$ Hz. Another possibility is to set $100 - S = -10$ to give $S = 110$ Hz. In fact, we can also subtract out integer multiples of $S$ from 100 Hz, set $100 - MS = 10$ and compute $S$ for various choices of $M$. For example, if $M = 2$, we get $S = 45$ Hz and if $M = 3$, we get $S = 30$ Hz. We can also set $100 - NS = -10$ and get $S = 55$ Hz for $N = 2$. Which of these sampling rates was actually used? We have no way of knowing!

# Drill Problem 2.10

**(a)** A 60-Hz sinusoid $x(t)$ is sampled at 200 Hz. What is the period $N$ of the sampled signal? How many full periods of $x(t)$ are required to obtain these $N$ samples? What is the frequency (in Hz) of the analog signal reconstructed from the samples?

**(b)** A 160-Hz sinusoid $x(t)$ is sampled at 200 Hz. What is the period $N$ of the sampled signal? How many full periods of $x(t)$ are required to obtain these $N$ samples? What is the frequency (in Hz) of the analog signal reconstructed from the samples?

**(c)** The signal $x(t) = \cos(60\pi t + 30^\circ)$ is sampled at 50 Hz. What is the expression for the analog signal $y(t)$ reconstructed from the samples?

**(d)** A 150-Hz sinusoid is to be sampled. Pick the range of sampling rates (in Hz) closest to 150 Hz that will cause aliasing but prevent phase reversal of the analog signal reconstructed from the samples.

**Answers: (a)** 10, 3, 60 **(b)** 5, 4, −40
**(c)** $y(t) = \cos(40\pi t - 30^\circ)$ **(d)** $100 < S < 150$

## 2.6.2 Reconstruction at Different Sampling Rates

There are situations when we sample a signal using one sampling rate $S_1$ but reconstruct the analog signal from samples using a different sampling rate $S_2$. In such situations, a frequency $f_0$ in the original signal will result in a digital frequency $F_0 = f_0/S_1$. If $S_1 > 2f_0$, there is no aliasing, $F_0$ is in the principal range and the recovered frequency is just $f_r = F_0 S_2 = f_0(S_2/S_1)$. If $S_1 < 2f_0$, there is aliasing and the recovered frequency is $f_r = F_a S_2 = f_a(S_2/S_1)$, where $F_a$ corresponds to the aliased digital frequency in the principal range. In other words, all frequencies, aliased or recovered, should be identified by their central period.

**Example 2.8**

### Signal Reconstruction at Different Sampling Rates

A 100-Hz sinusoid is sampled at $S$ Hz, and the sampled signal is then reconstructed at 420 Hz. What is the frequency of the reconstructed signal if $S = 210$ Hz? If $S = 140$ Hz?

1. If $S = 210$ Hz, the digital frequency of the sampled signal is $F = \frac{100}{210} = \frac{10}{21}$, which lies in the central period. The frequency $f_r$ of the reconstructed signal is then $f_r = SF = 420F = 200$ Hz.
2. If $S = 140$ Hz, the digital frequency of the sampled signal is $F = \frac{100}{140}$, which does not lie in the central period. The frequency in the principal range is $F = \frac{100}{140} - 1 = -\frac{40}{140}$, and the frequency $f_r$ of the reconstructed signal is then $f_r = SF = 420F = -120$ Hz. The negative sign simply translates to a phase reversal in the reconstructed signal.

## Drill Problem 2.11

**(a)** A 60-Hz signal is sampled at 200 Hz. What is the frequency (in Hz) of the signal reconstructed from the samples if the reconstruction rate is 300 Hz?
**(b)** A 160-Hz signal is sampled at 200 Hz. If the frequency of the signal reconstructed from the samples is also 160 Hz, what reconstruction rate (in Hz) was used?

**Answers:** **(a)** 90 **(b)** 800

# 2.7 An Introduction to Random Signals

The signals we have studied so far are called **deterministic** or predictable. They are governed by a unique mathematical representation that, once established, allows us to completely characterize the signal for all time, past, present, or future. In contrast to this is the class of signals known as **random** or **stochastic**, whose precise value cannot be predicted in advance. We stress that only the future values of a random signal pose a problem since past values of any signal, random or otherwise, are known exactly once they have occurred. Randomness or uncertainty about future signal values is inherent in many practical situations. In fact, a degree of uncertainty is essential for communicating information. The longer we observe a random signal, the more the additional information we gain and the less the uncertainty. To fully understand the nature of random signals requires the use of probability theory, random variables, and statistics. Even with such tools, the best we can do is to characterize random signals only *on the average* based on their past behavior.

## 2.7.1 Probability

Figure 2.7 shows the results of two experiments, each repeated under identical conditions. The first experiment always yields identical results—no matter how many times it is run—and yields a deterministic signal. We need to run the experiment only once to predict what the next or any other run will yield.

The second experiment gives a different result or **realization** $x(t)$ every time the experiment is repeated and describes a **stochastic** or random system. A **random signal** or **random process** $X(t)$ comprises the family or **ensemble** of all such realizations obtained by repeating the experiment many times. Each realization $x(t)$, once obtained, ceases to be random and can be subjected to the same operations as we use for deterministic signals (such as derivatives, integrals, and the like). The randomness of the signal stems from the fact that one realization provides no clue as to what the next, or any other, realization might yield. At a given instant $t$, each realization of a random signal can assume a different value, and the collection of all such values defines a **random variable**. Some values are more likely to occur, or more probable, than others. The concept of probability is tied to the idea of repeating an experiment a large number of times in order to estimate this probability. Thus, if the value 2 V occurs 600 times in 1000 runs, we say that the probability of occurrence of 2 V is 0.6.

The **probability of an event** $A$, denoted $\Pr(A)$, is the proportion of successful outcomes to the (very large) number of times the experiment is run and is a fraction between 0 and 1

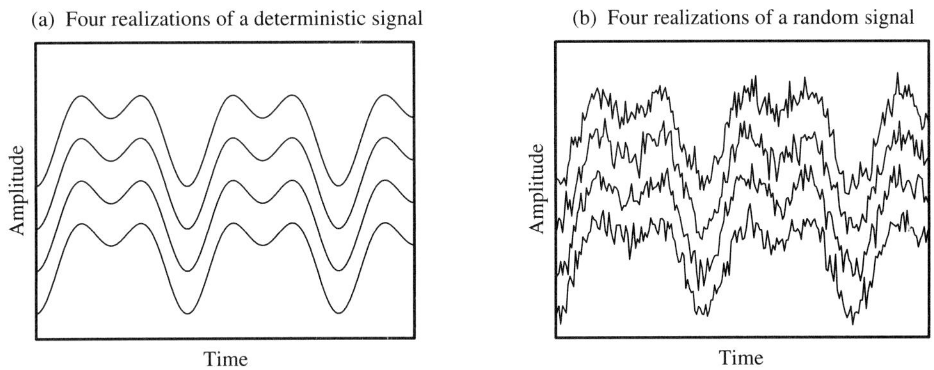


**FIGURE 2.7** A deterministic process results in identical realizations but every realization of a stochastic or random process is different

since the number of successful runs cannot exceed the total number of runs. The larger the probability $\Pr(A)$, the more the chance of event $A$ occurring. To fully characterize a random variable, we must answer two questions:

1. What is the range of all possible (nonrandom) values it can acquire? This defines an **ensemble space**, which may be finite or infinite.
2. What are the probabilities for all the possible values in this range? This defines the **probability distribution function** $F(x)$. Clearly, $F(x)$ must always lie between 0 and 1.

It is common to work with the derivative of the probability distribution function called the **probability density function** $f(x)$. The distribution function $F(x)$ is simply the running integral of the density $f(x)$:

$$f(x) = \frac{dF(x)}{dx} \quad \text{or} \quad F(x) = \int_{-\infty}^{x} f(\lambda)\,d\lambda \tag{2.24}$$

The probability $F(x_1) = \Pr[X \le x_1]$ that $X$ is less than $x_1$ is given by

$$\Pr[X \le x_1] = \int_{-\infty}^{x_1} f(x)\,dx \tag{2.25}$$

The probability that $X$ lies between $x_1$ and $x_2$ is $\Pr[x_1 < X \le x_2] = F(x_2) - F(x_1)$. The area of $f(x)$ is 1.

## 2.7.2 Measures for Random Variables

Measures or features of a random variable $X$ are based on its distribution. The mean, or **expectation**, is a measure of where the distribution is *centered* and is defined by

$$\hat{E}(x) = m_x = \int_{-\infty}^{\infty} x f(x)\,dx \quad \text{(mean or expectation)} \tag{2.26}$$

The **mean square value** is similarly defined by

$$\hat{E}(x^2) = \int_{-\infty}^{\infty} x^2 f(x)\,dx \quad \text{(mean square value)} \tag{2.27}$$

Many of the features of deterministic or random signals are based on **moments**. The $n$th moment $m_n$ is defined by

$$m_n = \int_{-\infty}^{\infty} x^n f(x)\,dx \quad (n\text{th moment}) \tag{2.28}$$

We see that the zeroth moment $m_0$ gives the signal area, the first moment $m_1$ corresponds to the mean, and the second moment $m_2$ defines the mean square value. Moments about the mean are called **central moments** and also find widespread use. The $n$th central moment $\mu_n$ is defined by

$$\mu_n = \int_{-\infty}^{\infty} (x - m_x)^n f(x)\,dx \quad (n\text{th central moment}) \tag{2.29}$$

A very commonly used feature is the second central moment $\mu_2$. It is also called the **variance**, denoted $\sigma_x^2$, and defined by

$$\sigma_x^2 = \hat{E}[(x - m_x)^2] = \mu_2 = \int_{-\infty}^{\infty} (x - m_x)^2 f(x)\,dx \quad (\text{variance}) \tag{2.30}$$

The variance may be expressed in terms of the mean and the mean square values as

$$\sigma_x^2 = \hat{E}[(x - m_x)^2] = \hat{E}(x^2) - m_x^2 = \int_{-\infty}^{\infty} x^2 f(x)\,dx - m_x^2 \tag{2.31}$$

The **variance** measures the spread (or dispersion) of the distribution about its mean. The less the spread, the smaller is the variance. The quantity $\sigma$ is known as the **standard deviation** and provides a measure of the uncertainty in a physical measurement. The variance is also a measure of the ac power in a signal. For a periodic deterministic signal $x(t)$ with period $T$, the variance can be found readily by evaluating the signal power (and subtracting the power due to the dc component if present)

$$\sigma_x^2 = \underbrace{\frac{1}{T}\int_0^T x^2(t)\,dt}_{\text{total signal power}} - \underbrace{\left[\frac{1}{T}\int_0^T x(t)\,dt\right]^2}_{\text{dc power}} \tag{2.32}$$

This equation can be used to obtain the results listed in the following review panel.

**The Variance of Some Useful Periodic Signals with Period $T$**

**Sinusoid:** If $x(t) = A\cos(2\pi\frac{t}{T} + \theta)$, then $\sigma^2 = \frac{A^2}{2}$

**Triangular Wave:** If $x(t) = A\frac{t}{T}$, $0 \le t < T$, or $x(t) = A(1 - \frac{|t|}{0.5T})$, $|t| \le 0.5T$, then $\sigma^2 = \frac{A^2}{12}$

**Square Wave:** If $x(t) = A$, $0 \le t < 0.5T$ and $x(t) = 0$, $0.5 \le t < T$, then $\sigma^2 = \frac{A^2}{4}$

# Drill Problem 2.12

**(a)** Find the variance of the periodic signal $x(t) = A$, $0 \le t < 0.25T$ and $x(t) = 0$, $0.25T \le t < T$.

**(b)** Find the variance of the raised cosine signal $x(t) = A[1 + \cos(2\pi\frac{t}{T} + \theta)]$.

**Answers: (a)** $\frac{3}{16}A^2$ **(b)** $\frac{5}{4}A^2$

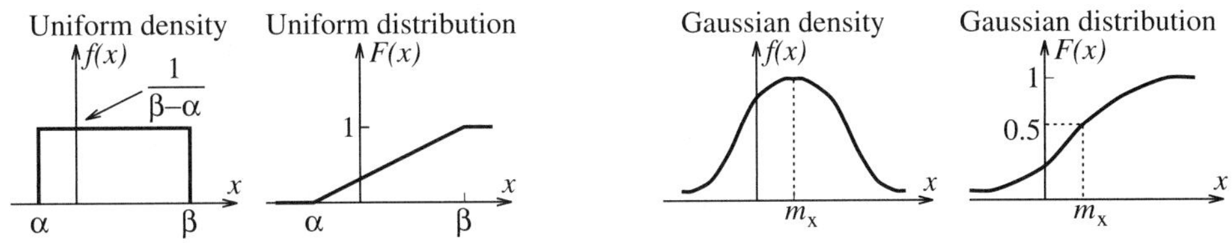


**FIGURE 2.8** The uniform and normal probability distributions are obtained by integrating their respective density functions

## 2.7.3 The Chebyshev Inequality

The measurement of the variance or standard deviation gives us some idea of how much the actual values will deviate from the mean but provides no indication of how often we might encounter large deviations from the mean. The **Chebyshev inequality** allows us to estimate the probability for a deviation to be within certain bounds (given by $\alpha$) as

$$\Pr(|x - m_x| > \alpha) \le (\sigma_x/\alpha)^2 \quad \text{or} \quad \Pr(|x - m_x| \le \alpha) > 1 - \frac{\sigma_x^2}{\alpha^2} \quad \text{(Chebyshev inequality)} \tag{2.33}$$

It assumes that the variance or standard deviation is known. To find the probability for the deviation to be within $k$ standard deviations, we set $\alpha = k\sigma_x$ to give

$$\Pr(|x - m_x| \le k\sigma_x) > 1 - \frac{1}{k^2}$$

### The Law of Large Numbers

Chebyshev's inequality, in turn, leads to the so called **law of large numbers** which, in essence, states that while an individual random variable may take on values quite far from its mean (show a large spread), the arithmetic mean of a large number of random values shows little spread, taking on values very close to its common mean with a very high probability.

## 2.7.4 Probability Distributions

Two of the most commonly encountered probability distributions are the uniform distribution and normal (or Gaussian) distribution. These are illustrated in Figure 2.8.

## 2.7.5 The Uniform Distribution

In a uniform distribution, every value is equally likely, since the random variable shows no preference for a particular value. The density function $f(x)$ of a typical uniform distribution is just a rectangular pulse with unit area defined by

$$f(x) = \begin{cases} \frac{1}{\beta-\alpha}, & \alpha \le x \le \beta \\ 0, & \text{otherwise} \end{cases} \quad \text{(uniform density function)} \tag{2.34}$$

Its mean and variance are given by

$$m_x = 0.5(\alpha + \beta) \qquad \sigma_x^2 = \frac{(b-a)^2}{12} \tag{2.35}$$

The distribution function $F(x)$ is given by

$$F(x) = \begin{cases} 0 & x < a \\ \frac{x-a}{b-a} & a \le x \le b \\ 1 & x > b \end{cases} \quad \text{(uniform distribution function)} \tag{2.36}$$

This is a finite ramp that rises from a value of zero at $x = \alpha$ to a value of unity at $x = \beta$ and equals unity for $x \geq \alpha$. For a uniform distribution in which values are equally likely to fall between $-0.5$ and $0.5$, the density function is $f(x) = 1, \; -0.5 \leq x < 0.5$ with a mean of zero and a variance of $\frac{1}{12}$.

Uniform distributions occur frequently in practice. When quantizing signals in uniform steps, the error in representing a signal value is assumed to be uniformly distributed between $-0.5\Delta$ and $0.5\Delta$, where $\Delta$ is the quantization step. The density function of the phase of a sinusoid with random phase is also uniformly distributed between $-\pi$ and $\pi$.

# Drill Problem 2.13

**(a)** The phase of a sinusoid with random phase is uniformly distributed between $-\pi$ and $\pi$. Compute the variance.

**(b)** If the quantization error $\epsilon$ is assumed to be uniformly distributed between $-0.5\Delta$ and $0.5\Delta$, sketch the density function $f(\epsilon)$ and compute the variance.

**Answers: (a)** $\frac{\pi^2}{3}$ **(b)** $\sigma_\epsilon^2 = \frac{\Delta^2}{12}$

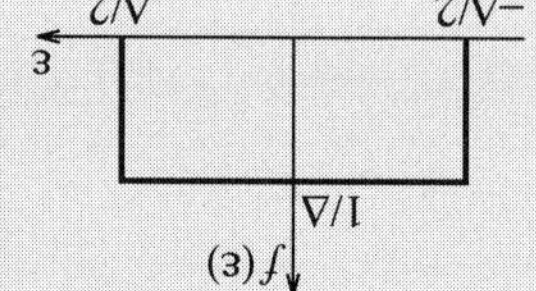


## 2.7.6 The Gaussian or Normal Distribution

The Gaussian or **normal** probability density is bell shaped and defined by

$$f(x) = \frac{1}{\sqrt{2\pi\sigma_x^2}} \exp\left[-\frac{(x - m_x)^2}{2\sigma_x^2}\right] \quad \text{(normal distribution)} \tag{2.37}$$

It has a mean of $m_x$ and a variance of $\sigma_x^2$. This density function has a single peak at $x = m_x$, even symmetry about the peak ($x = m_x$), and inflection points at $m_x \pm \sigma$.

The Gaussian process exhibits several useful properties

1. The variable $ax + b$ where $a > 0$ is also normally distributed.
2. The sum of normally distributed random signals is also **normally distributed**. The mean of the sum equals the sum of the individual means and, if the statistically independent, the variance of the sum equals the sum of the individual variances.
3. The ratio of the mean deviation $(x - m_x)$ and the standard deviation $\sigma_x$ for all normal distributions equals $\sqrt{2/\pi}$. Thus, for normal distributions, one may equally well work with the mean deviation rather than the standard deviation.
4. All of the higher order moments of a Gaussian random variable may be obtained from a knowledge of the first two moments alone. In particular, the $n$th central moments are zero for odd values of $n$ and the following relation then obtains for the

even order ($n = 2k$) central moments

$$E[(x - m_x)^{2k}] = (1)(3)(5)\ldots(2k-1)\sigma^{2k} = \frac{(2k)!}{k!2^k}, \qquad n = 2k$$

(central moments of Gaussian) (2.38)

### The Gaussian Distribution Function

The distribution function $F(x)$ of a Gaussian distribution cannot be written in closed form and is given by

$$F(x) = \frac{1}{\sqrt{2\pi\sigma^2}} \int_{-\infty}^{x} \exp\left[-\frac{(\lambda - \hat{m}_x)^2}{2\sigma^2}\right] d\lambda \quad \text{(Gaussian distribution function)} \tag{2.39}$$

### The Standard Gaussian Distribution

A Gaussian distribution with a mean of zero and a variance of unity ($m_x = 0, \sigma^2 = 1$) is called a **standard** Gaussian and denoted $P(x)$.

$$P(x) = \frac{1}{\sqrt{2\pi}} \int_{-\infty}^{x} e^{-\lambda^2/2}\, d\lambda \quad \text{(standard Gaussian distribution)} \tag{2.40}$$

### The Error Function

Another function that is used extensively is the **error function** defined by

$$\operatorname{erf}(x) = \frac{2}{\sqrt{\pi}} \int_{0}^{x} e^{-\lambda^2}\, d\lambda \quad \text{(error function)} \tag{2.41}$$

The Gaussian distribution $F(x)$ may be expressed in terms of the standard form or the error function as

$$F(x) = P\left(\frac{x - m_x}{\sigma}\right) = 0.5 + 0.5\operatorname{erf}\left(\frac{x - m_x}{\sigma\sqrt{2}}\right) \tag{2.42}$$

The probability that $x$ lies between $x_1$ and $x_2$ may be expressed in terms of the error function as

$$\Pr[x_1 \le x \le x_2] = F(x_2) - F(x_1) = 0.5\operatorname{erf}\left(\frac{x_2 - m_x}{\sigma\sqrt{2}}\right) - 0.5\operatorname{erf}\left(\frac{x_1 - m_x}{\sigma\sqrt{2}}\right) \tag{2.43}$$

This is a particularly useful form since tables of error functions are widely available. A note of caution to the unwary, however. Several different (though functionally equivalent) definitions of $P(x)$, $\operatorname{erf}(x)$ and related functions are also prevalent in the literature.

### The Q-Function

The **Q-function** describes the area of the tail of a Gaussian distribution. For the standard Gaussian, the area $Q(x)$ of the tail is given by

$$Q(x) = 1 - P(x) = \frac{1}{\sqrt{2\pi}} \int_{x}^{\infty} e^{-\lambda^2/2}\, d\lambda$$

The Q-function and may also be expressed in terms of the error function as

$$Q(x) = 0.5 - 0.5\operatorname{erf}(x/\sqrt{2}) \tag{2.44}$$

The probability that $x$ lies between $x_1$ and $x_2$ may also be expressed in terms of the Q-function as

$$\Pr[x_1 \le x \le x_2] = Q\left(\frac{x_1 - m_x}{\sigma}\right) - Q\left(\frac{x_2 - m_x}{\sigma}\right) \tag{2.45}$$

The results for the standard distribution may be carried over to a distribution with arbitrary mean $m_x$ and arbitrary standard deviation $\sigma_x$ via the simple change of variable $x \to (x - m_x)/\sigma$.

### The Central Limit Theorem

The central limit theorem asserts that the probability density function of the sum of many random signals approaches a Gaussian as long as their means are finite and their variance is small compared to the total variance (but nonzero). The individual processes need not even be Gaussian.

## 2.7.7 Discrete Probability Distributions

The central limit theorem is even useful for discrete variables. When the variables that make up a given process $s$ are discrete and the number $n$ of such variables is large, that is,

$$s = \sum_{i=1}^{n} x_i, \quad n \gg 1$$

we may approximate $s$ by a Gaussian whose mean equals $nm_x$ and whose variance equals $n\sigma_x^2$. Thus,

$$f_n(s) \approx \frac{1}{\sqrt{2\pi n\sigma_x^2}} \exp\left[-\frac{(s - nm_x)^2}{2n\sigma_x^2}\right], \quad n \gg 1 \tag{2.46}$$

Its distribution allows us to compute the probability $\Pr[s_1 \le s \le s_2]$ in terms of the error function or Q-function as

$$\begin{aligned} \Pr[s_1 \le s \le s_2] &\approx 0.5 \,\mathrm{erf}\left(\frac{s_2 - nm_x}{\sigma\sqrt{2n}}\right) - 0.5 \,\mathrm{erf}\left(\frac{s_1 - nm_x}{\sigma\sqrt{2n}}\right) \\ &= Q\left(\frac{s_1 - nm_x}{\sigma\sqrt{n}}\right) - Q\left(\frac{s_2 - nm_x}{\sigma\sqrt{n}}\right) \end{aligned} \tag{2.47}$$

This relation forms the basis for numerical approximations involving discrete probabilities.

### The Binomial Distribution

Consider an experiment with two outcomes which result in mutually independent and complementary events. If the probability of a success is $p$, and the probability of a failure is $q = 1 - p$, the probability of exactly $k$ successes in $n$ trials follows the binomial distribution and is given by

$$\Pr[s = k] = p_n(k) = C_k^n (p)^k (1 - p)^{n-k} \quad \text{(binomial probability)} \tag{2.48}$$

Here, $C_k^n$ represents the **binomial coefficient** and may be expressed in terms of factorials or gamma functions as

$$C_k^n = \frac{n!}{k!\,(n-k)!} = \frac{\Gamma(n+1)}{\Gamma(k+1)\,\Gamma(n-k+1)}$$

The probability of *at least* $k$ successes in $n$ trials is given by

$$\Pr[s \geq k] = \sum_{i=k}^{n} p_n(i) = I_p(k, n-k+1)$$

where $I_x(a,b)$ is the **incomplete beta function** defined by

$$I_x(a,b) = \frac{1}{B(a,b)} \int_0^x t^{a-1}(1-t)^{b-1}dt \qquad B(a,b) = \frac{\Gamma a \Gamma b}{\Gamma(a+b)}, \qquad a > 0, b > 0$$

The probability of getting between $k_1$ and $k_2$ successes in $n$ trials describes a **cumulative probability**, because we must sum the probabilities of all possible outcomes for the event (the probability of exactly $k_1$, then $k_1+1$, then $k_1+2$ successes, and so on to $k_2$ successes) and is given by

$$\Pr[k_1 \leq s \leq k_2] = \sum_{i=k_1}^{k_2} p_n(i)$$

For large $n$, its evaluation can become a computational nightmare.

## Some Useful Approximations

When $n$ is large and neither $p \to 0$ nor $q \to 0$, the probability $p_n(k)$ of exactly $k$ successes in $n$ trials may be approximated by the Gaussian

$$p_n(k) = C_k^n(p)^k(1-p)^{n-k} \simeq \frac{1}{\sqrt{2\pi\sigma^2}} \exp\left[-\frac{(k-m)^2}{2\sigma^2}\right], \qquad m = np, \; \sigma^2 = np(1-p)$$

This result is based on the central limit theorem and called the **de Moivre-Laplace approximation**. It assumes that $\sigma^2 \gg 1$ and $|k-m|$ is of the same order as $\sigma$ or less. Using this result, the probability of at least $k$ successes in $n$ trials may be written in terms of the error function or Q-function as

$$\Pr[s \geq k] \approx 0.5 - 0.5\,\text{erf}\left(\frac{k-m}{\sigma\sqrt{2}}\right) = Q\left(\frac{k-m}{\sigma}\right)$$

Similarly, the cumulative probability of between $k_1$ and $k_2$ successes in $n$ trials is

$$\Pr[k_1 \leq s \leq k_2] \approx 0.5\,\text{erf}\left(\frac{k_2-m}{\sigma\sqrt{2}}\right) - 0.5\,\text{erf}\left(\frac{k_1-m}{\sigma\sqrt{2}}\right) = Q\left(\frac{k_1-m}{\sigma}\right) - Q\left(\frac{k_2-m}{\sigma}\right)$$

## The Poisson Probability

If the number of trials $n$ approaches infinity and the probability $p$ of success in each trial approaches zero (but their product $m = np$ remains constant), the binomial probability is well approximated by

$$\Pr[s = k] = C_k^n(p)^k(1-p)^{n-k} \approx \frac{m^k e^{-m}}{k!} = p_m(k) \quad \text{(Poisson probability)} \qquad (2.49)$$

where $n \gg 1$, $p \ll 1$, and $m = np$. This is known as the **Poisson probability**. The mean and variance of this distribution are both equal to $m$. In practical situations, the Poisson approximation is often used whenever $n$ is large and $p$ is small and not just under the stringent limiting conditions imposed in its derivation. Unlike the binomial distribution which requires probabilities for both success and failure, the Poisson distribution requires only the probability of a success $p$ (through the parameter $m = np$ that describes the expected number of successes) and may thus be used even when the number of unsuccessful outcomes is unknown. The probability that the number of successes will lie between 0 and $k$ is **inclusive**, if the expected number is $m$, is given by the summation:

$$\Pr[s \le k] = \sum_{i=0}^{k} kp_m(i) = \sum_{i=0}^{k} \frac{m^i e^{-m}}{i!} = 1 - P(k+1, m), \quad k \ge 1$$

where $P(a, x)$ is called the **incomplete gamma function** and is defined by

$$P(a, x) = \frac{1}{\Gamma(a)} \int_0^x t^{a-1} e^{-t}\, dt \qquad a > 0$$

Note that $\Pr[s \le 0] = e^{-m}$. For large $m$, the Poisson probability of exactly $k$ successes also may be approximated using the central limit theorem to give

$$p_m(k) \simeq \frac{1}{\sqrt{2\pi m}} \exp\left[-\frac{(k-m)^2}{2m}\right]$$

Using this approximation, the probability of between $k_1$ and $k_2$ successes is

$$\Pr[k_1 \le s \le k_2] \approx 0.5\,\mathrm{erf}\left(\frac{k_2 - m}{\sqrt{2m}}\right) - 0.5\,\mathrm{erf}\left(\frac{k_1 - m}{\sqrt{2m}}\right) = Q\left(\frac{k_1 - m}{\sqrt{m}}\right) - Q\left(\frac{k_2 - m}{\sqrt{m}}\right)$$

### 2.7.8 Distributions for Deterministic Signals

The idea of distributions also applies to *deterministic* periodic signals for which they can be found as exact analytical expressions. Consider the periodic signal $x(t)$ of Figure 2.9. The probability $\Pr[X < 0]$ that $x(t) < 0$ is zero. The probability $\Pr[X < 3]$ that $x(t)$ is less than 3 is 1. Since $x(t)$ varies linearly over one period ($T = 3$), all values in this range are equally likely. This means that the density function is constant over the range $0 \le x \le 3$ with an area of unity, and the distribution $F(x)$ rises linearly from zero to unity over this range. The distribution $F(x)$ and density $f(x)$ also are shown in Figure 2.9. The variance can be computed either from the signal $x(t)$ itself or from its density function $f(x)$. The results are repeated for convenience.

$$\sigma_x^2 = \underbrace{\frac{1}{T}\int_0^T x^2(t)\, dt}_{\text{total signal power}} - \underbrace{\left[\frac{1}{T}\int_0^T x(t)\, dt\right]^2}_{\text{dc power}} \quad \text{or} \quad \sigma_x^2 = \int_{-\infty}^{\infty} x^2 f(x)\, dx - m_x^2 \qquad (2.50)$$

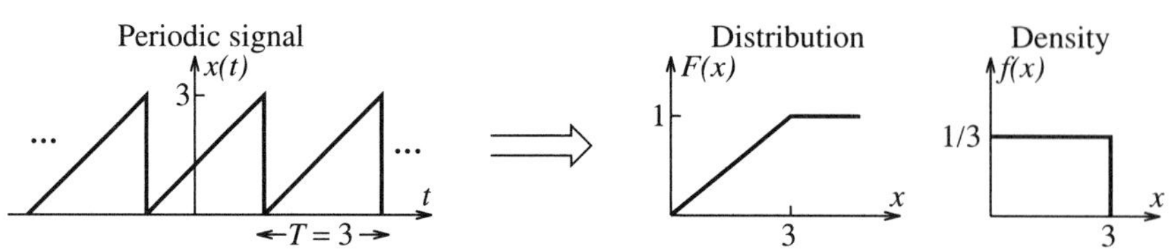


**FIGURE 2.9** A periodic signal and its distribution and density functions. The signal varies linearly from 0 to 3. Thus, all values in the range are equally likely and the density function is a constant over this range. The constant is chosen to give unit area. The distribution is found by integrating the density function

# Drill Problem 2.14

**(a)** Refer to Figure 2.5. Compute the variance from $x(t)$ itself.
**(b)** Refer to Figure 2.5. Compute the variance from its density function $f(x)$.

**Answers:** **(a)** 0.75 **(b)** 0.75

### 2.7.9 Stationary, Ergodic, and Pseudorandom Signals

A random signal is called **stationary** if its statistical features do not change over time. Thus, different (nonoverlapping) segments of a single realization are more or less identical in the statistical sense. Signals that are **non-stationary** do not possess this property and may indeed exhibit a trend (a linear trend, for example) with time. Stationarity suggests a state of statistical equilibrium akin to the steady-state for deterministic situations. A stationary process is typically characterized by a constant mean and constant variance. The statistical properties of a stationary random signal may be found as **ensemble averages** *across* the process by averaging over all realizations at one specific time instant or as **time averages** *along* the process by averaging a single realization over time. The two are not always equal. If they are, a stationary process is said to be **ergodic**. The biggest advantage of ergodicity is that we can use features from a single realization to describe the whole process. It is very difficult to establish whether a stationary process is ergodic but, because of the advantages it offers, ergodicity is often assumed in most practical situations! For an ergodic signal, the mean equals the time average, and the variance equals the ac power (the power in the signal with its dc component removed).

### 2.7.10 Statistical Estimates

Probability theory allows us to fully characterize a random signal from an *a priori* knowledge of its probability distribution. This yields features like the mean and variance of the random variable. In practice, we are faced with exactly the opposite problem of finding such features from a set of discrete data, often in the absence of a probability distribution. The best we can do is get an estimate of such features and perhaps even the distribution itself. This is what statistical estimation achieves. The mean and variance are typically estimated directly from the observations $x_k$, $k = 0, 1, 2, \ldots, N-1$ as

$$m_x = \frac{1}{N}\sum_{k=0}^{N-1} x_k \qquad \sigma_x^2 = \frac{1}{N-1}\sum_{k=0}^{N-1}(x_k - m_x)^2 \tag{2.51}$$

#### Histograms

The estimates $f_k$ of a probability distribution are obtained by constructing a **histogram** from a large number of observations. A histogram is a bar graph of the number of observations falling within specified amplitude levels, or *bins*, as illustrated in Figure 2.10.

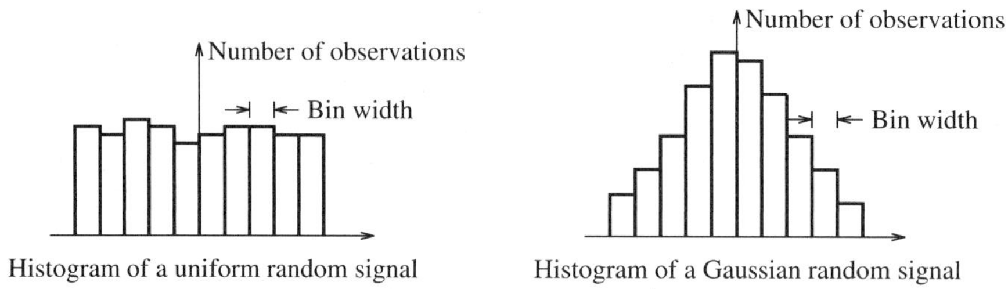


**FIGURE 2.10** Histograms of a uniformly distributed and a normally distributed random signal

### Pseudorandom Signals

In many situations, we use artificially generated signals (which can never be truly random) with prescribed statistical features called **pseudorandom signals**. Such signals are actually periodic (with a *very* long period), but over one period their statistical features approximate those of random signals.

## 2.7.11 Random Signal Analysis

If a random signal forms the input to a system, the best we can do is to develop features that describe the output *on the average* and **estimate** the response of a system under the influence of random signals. Such estimates may be developed either in the time domain or in the frequency domain.

### Signal-to-Noise Ratio

For a noisy signal $x(t) = s(t) + An(t)$ with a signal component $s(t)$ and a noise component $An(t)$ (with noise amplitude $A$), the **signal-to-noise ratio** (SNR) is the ratio of the signal power $\sigma_s^2$ and noise power $A^2\sigma_n^2$ and is usually defined in **decibels** (dB) as

$$\text{SNR} = 10\log\left(\frac{\sigma_s^2}{A^2\sigma_n^2}\right) \text{ dB} \tag{2.52}$$

The decibel value of $\alpha$ is defined as $20\log\alpha$. We can adjust the SNR by varying the noise amplitude $A$.

### Application: Coherent Signal Averaging

Coherent signal averaging is a method of extracting signals from noise and assumes that the experiment can be repeated and the noise corrupting the signal is random (and uncorrelated). Averaging the results of many runs tends to average out the noise to zero, and the signal quality (or signal-to-noise ratio) improves. The more the number of runs, the smoother and less noisy the averaged signal. We often remove the mean or any linear trend before averaging. Figure 2.11 shows one realization of a noisy sine wave and the much smoother results of averaging 8 and 48 such realizations. This method is called **coherent** because it requires *time coherence* (time alignment of the signal for each run). It relies, for its success, on perfect synchronization of each run and on the statistical independence of the contaminating noise.

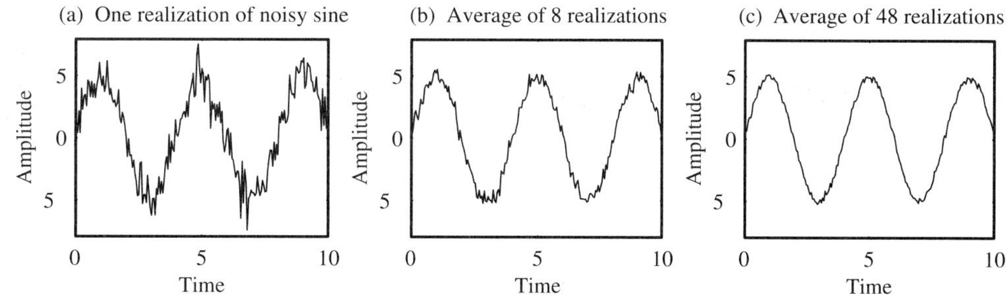


**FIGURE 2.11** Coherent averaging of a noisy sine wave. Notice how the signal quality improves as the number of realizations to be averaged increases

## 2.8 Problems

**2.1. (Discrete Signals)** Sketch each signal and find its energy or average power as appropriate.

**(a)** $x[n] = \{\overset{\Downarrow}{6}, 4, 2, 2\}$ **(b)** $y[n] = \{-3, -2, \overset{\Downarrow}{-1}, 0, 1\}$

**(c)** $f[n] = \{\overset{\Downarrow}{0}, 2, 4, 6\}$ **(d)** $g[n] = u[n] - u[n-4]$

**(e)** $p[n] = \cos(n\pi/2)$ **(f)** $q[n] = 8(0.5)^n u[n]$

[**Hints and Suggestions:** Only $p[n]$ is a power signal. The rest have finite energy.]

**2.2. (Signal Duration)** Use examples to argue that the product of a right-sided and a left-sided discrete-time signal is always time-limited or identically zero.

[**Hints and Suggestions:** Select simple signals that either overlap or do not overlap.]

**2.3. (Operations)** Let $x[n] = \{\overset{\Downarrow}{6}, 4, 2, 2\}$. Sketch the following signals and find their signal energy.

**(a)** $y[n] = x[n-2]$ **(b)** $f[n] = x[n+2]$

**(c)** $g[n] = x[-n+2]$ **(d)** $h[n] = x[-n-2]$

[**Hints and Suggestions:** Note that $g[n]$ is a time-reversed version of $f[n]$.]

**2.4. (Operations)** Let $x[n] = 8(0.5)^n(u[n+1] - u[n-3])$. Sketch the following signals.

**(a)** $y[n] = x[n-3]$ **(b)** $f[n] = x[n+1]$

**(c)** $g[n] = x[-n+4]$ **(d)** $h[n] = x[-n-2]$

[**Hints and Suggestions:** Note that $x[n]$ contains five samples (from $n = -1$ to $n = 3$). To display the marker for $y[n]$ (which starts at $n = 2$), we include two zeros at $n = 0$ (the marker) and $n = 1$.]

**2.5. (Energy and Power)** Classify the following as energy signals, power signals, or neither and find the energy or average power as appropriate.

**(a)** $x[n] = 2^n u[-n]$ **(b)** $y[n] = 2^n u[-n-1]$ **(c)** $f[n] = \cos(n\pi)$

**(d)** $g[n] = \cos(n\pi/2)$ **(e)** $p[n] = \dfrac{1}{n}u[n-1]$ **(f)** $q[n] = \dfrac{1}{\sqrt{n}}u[n-1]$

**(g)** $r[n] = \dfrac{1}{n^2}u[n-1]$ **(h)** $s[n] = e^{jn\pi}$ **(i)** $d[n] = e^{jn\pi/2}$

**(j)** $t[n] = e^{(j+1)n\pi/4}$ **(k)** $v[n] = j^{n/4}$ **(l)** $w[n] = (\sqrt{j})^n + (\sqrt{j})^{-n}$

[**Hints and Suggestions:** For $x[n]$ and $y[n]$, $2^{2n} = 4^n = (0.25)^{-n}$. Sum this from $n = -\infty$ to $n = 0$ (or $n = -1$) using a change of variable ($n \to -n$) in the summation. For $p[n]$, sum $1/n^2$ over $n = 1$ to $n = \infty$ using tables. For $q[n]$, the sum of $1/n$ from $n = 1$ to $n = \infty$ does not converge! For $t[n]$, separate the exponentials. To compute the average power for $s[n]$ and $d[n]$, note that $|s[n]| = |d[n]| = 1$. For $v[n]$, use $j = e^{j\pi/2}$. For $w[n]$, set $\sqrt{j} = e^{j\pi/4}$ and use Euler's relation to convert to a sinusoid.]

2.6. **(Energy and Power)** Sketch each of the following signals, classify as an energy signal or power signal, and find the energy or average power as appropriate.

**(a)** $x[n] = \sum_{k=-\infty}^{\infty} y[n - kN]$, where $y[n] = u[n] - u[n-3]$ and $N = 6$

**(b)** $f[n] = \sum_{k=-\infty}^{\infty} (2)^{n-5k}(u[n-5k] - u[n-5k-4])$

[**Hints and Suggestions:** The period of $x[n]$ is $N = 6$. With $y[n] = u[n] - u[n-3]$, one period of $x[n]$ (starting at $n = 0$) is $\{1, 1, 1, 0, 0, 0\}$. The period of $f[n]$ is $N = 5$. Its one period (starting at $n = 0$) contains four samples from $2^n(u[n] - u[n-4])$ and one trailing zero.]

2.7. **(Decimation and Interpolation)** Let $x[n] = \{4,\ 0,\ \overset{\Downarrow}{2},\ -1,\ 3\}$. Find and sketch the following signals and compare their signal energy with the energy in $x[n]$.

**(a)** The decimated signal $d[n] = x[2n]$
**(b)** The *zero-interpolated* signal $f[n] = x^{\uparrow}[\frac{n}{2}]$
**(c)** The *step-interpolated* signal $g[n] = x^{\uparrow}[\frac{n}{2}]$
**(d)** The *linearly interpolated* signal $h[n] = x^{\uparrow}[\frac{n}{2}]$

[**Hints and Suggestions:**To get $d[n]$, retain the samples of $x[n]$ at $n = 0, \pm 2, \pm 4, \ldots$. Assuming that the interpolated signals will be twice the length of $x[n]$, the last sample will be 0 for $f[n]$, 3 for $g[n]$ and 1.5 (the linearly interpolated value with $x[n] = 0$, $n > 2$) for $h[n]$.]

2.8. **(Interpolation and Decimation)** Let $x[n] = 4\,\mathrm{tri}(n/4)$. Sketch the following signals and describe how they differ.

**(a)** $x[\frac{2}{3}n]$ (using zero interpolation followed by decimation)
**(b)** $x[\frac{2}{3}n]$ (using step interpolation followed by decimation)
**(c)** $x[\frac{2}{3}n]$ (using decimation followed by zero interpolation)
**(d)** $x[\frac{2}{3}n]$ (using decimation followed by step interpolation)

2.9. **(Fractional Delay)** Starting with $x[n]$, we can generate the signal $x[n-2]$ (using a delay of 2) or $x[2n-3]$ (using a delay of 3 followed by decimation). However, to generate a *fractional delay* of the form $x[n - \frac{M}{N}]$ requires a delay, interpolation, and decimation!

**(a)** Describe the sequence of operations required to generate $x[n - \frac{2}{3}]$ from $x[n]$.
**(b)** Let $x[n] = \{\overset{\Downarrow}{1}, 4, 7, 10, 13\}$. Sketch $x[n]$ and $x[n - \frac{2}{3}]$. Use *linear* interpolation where required.
**(c)** Generalize the results of part (a) to generate $x[n - \frac{M}{N}]$ from $x[n]$. Any restrictions on $M$ and $N$?

[**Hints and Suggestions:** In part (a) the sequence of operations requires interpolation, delay (by 2) and decimation. The interpolation and decimation factors are identical.]

**2.10.** **(Symmetry)** Sketch each signal and its even and odd parts.

**(a)** $x[n] = 8(0.5)^n u[n]$ **(b)** $y[n] = u[n]$ **(c)** $f[n] = 1 + u[n]$

**(d)** $g[n] = u[n] - u[n-4]$ **(e)** $p[n] = \text{tri}(\frac{n-3}{3})$ **(f)** $q[n] = \{6, \overset{\Downarrow}{4}, 2, 2\}$

[**Hints and Suggestions:** Confirm the appropriate symmetry for each even part and each odd part. For each even part, the sample at $n = 0$ must equal the original sample value. For each odd part, the sample at $n = 0$ must equal zero.]

**2.11.** **(Sketching Discrete Signals)** Sketch each of the following signals:

**(a)** $x[n] = r[n+2] - r[n-2] - 4u[n-6]$ **(b)** $y[n] = \text{rect}(\frac{n}{6})$

**(c)** $f[n] = \text{rect}(\frac{n-2}{4})$ **(d)** $g[n] = 6\,\text{tri}(\frac{n-4}{3})$

[**Hints and Suggestions:** Note that $f[n]$ is a rectangular pulse centered at $n = 2$ with five samples. Also, $g[n]$ is a triangular pulse centered at $n = 4$ with seven samples (including the zero-valued end samples).]

**2.12.** **(Sketching Signals)** Sketch the following signals and describe how they are related.

**(a)** $x[n] = \delta[n]$ **(b)** $f[n] = \text{rect}(n)$ **(c)** $g[n] = \text{tri}(n)$ **(d)** $h[n] = \text{sinc}(n)$

**2.13.** **(Signal Description)** For each signal shown in Figure P2.13,

**(a)** Write out the numeric sequence, and mark the index $n = 0$ by an arrow.
**(b)** Write an expression for each signal using impulse functions.
**(c)** Write an expression for each signal using steps and/or ramps.
**(d)** Find the signal energy.
**(e)** Find the average power, assuming that the sequence shown repeats itself.

**FIGURE P.2.13**
Signals for Problem 2.13

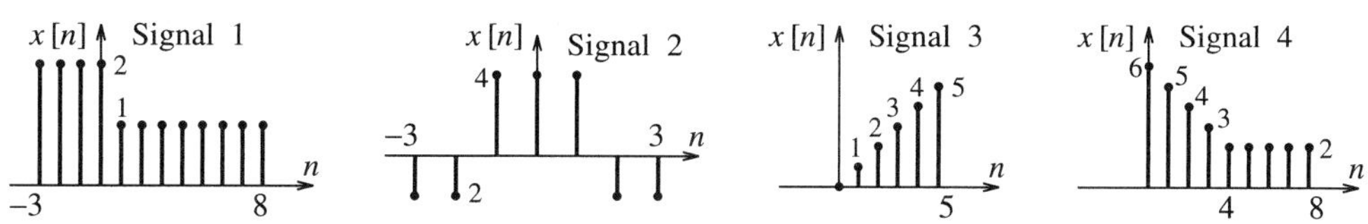


[**Hints and Suggestions:** In part (c), all signals must be turned off (by step functions) and any ramps must be first flattened out (by other ramps). For example, signal 3 = $r[n] - r[n-5] - 5u[n-6]$. The second term flattens out the first ramp and last term turns the signal off after $n = 5$.]

**2.14.** **(Discrete Exponentials)** A causal discrete exponential has the form $x[n] = \alpha^n u[n]$.

**(a)** Assume that $\alpha$ is real and positive. Pick convenient values for $\alpha > 1$, $\alpha = 1$, and $\alpha < 1$; sketch $x[n]$; and describe the nature of the sketch for each choice of $\alpha$.
**(b)** Assume that $\alpha$ is real and negative. Pick convenient values for $\alpha < -1$, $\alpha = -1$, and $\alpha > -1$; sketch $x[n]$; and describe the nature of the sketch for each choice of $\alpha$.
**(c)** Assume that $\alpha$ is complex and of the form $\alpha = Ae^{j\theta}$, where $A$ is a positive constant. Pick convenient values for $\theta$ and for $A < 1$, $A = 1$, and $A > 1$; sketch the real part and imaginary part of $x[n]$ for each choice of $A$; and describe the nature of each sketch.

**(d)** Assume that $\alpha$ is complex and of the form $\alpha = Ae^{j\theta}$, where $A$ is a positive constant. Pick convenient values for $\theta$ and for $A < 1$, $A = 1$, and $A > 1$; sketch the magnitude and imaginary phase of $x[n]$ for each choice of $A$; and describe the nature of each sketch.

**2.15. (Signal Representation)** The two signals shown in Figure P2.15 may be expressed as

**(a)** $x[n] = A\alpha^n(u[n] - u[n-N])$ **(b)** $y[n] = A\cos(2\pi Fn + \theta)$

Find the constants in each expression and then find the signal energy or average power as appropriate.

**FIGURE P.2.15**
Signals for Problem 2.15

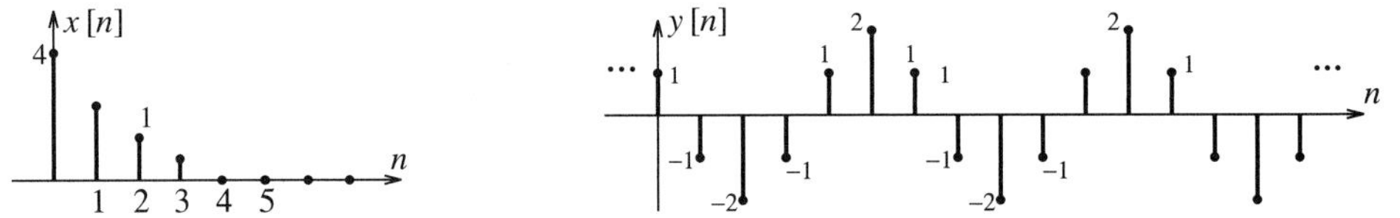


[**Hints and Suggestions:** For $y[n]$, first find the period to compute $F$. Then, evaluate $y[n]$ at two values of $n$ to get two equations for, say $y[0]$ and $y[1]$. These will yield $\theta$ (from their ratio) and $A$.]

**2.16. (Discrete-Time Harmonics)** Check for the periodicity of the following signals, and compute the common period $N$ if periodic.

**(a)** $x[n] = \cos(\frac{n\pi}{2})$ **(b)** $y[n] = \cos(\frac{n}{2})$
**(c)** $f[n] = \sin(\frac{n\pi}{4}) - 2\cos(\frac{n\pi}{6})$ **(d)** $g[n] = 2\cos(\frac{n\pi}{4}) + \cos^2(\frac{n\pi}{4})$
**(e)** $p[n] = 4 - 3\sin(\frac{7n\pi}{4})$ **(f)** $q[n] = \cos(\frac{5n\pi}{12}) + \cos(\frac{4n\pi}{9})$
**(g)** $r[n] = \cos(\frac{8}{3}n\pi) + \cos(\frac{8}{3}n)$ **(h)** $s[n] = \cos(\frac{8n\pi}{3})\cos(\frac{n\pi}{2})$
**(i)** $d[n] = e^{j0.3n\pi}$ **(j)** $e[n] = 2e^{j0.3n\pi} + 3e^{j0.4n\pi}$
**(k)** $v[n] = e^{j0.3n}$ **(l)** $w[n] = (j)^{n/2}$

[**Hints and Suggestions:** There is no periodicity if $F$ is not a rational fraction for any component. Otherwise, work with the periods and find their LCM. For $w[n]$, note that $j = e^{j\pi/2}$.]

**2.17. (The Roots of Unity)** The $N$ roots of the equation $z^N = 1$ can be found by writing it as $z^N = e^{j2k\pi}$ to give $z = e^{j2k\pi/N}$, $k = 0, 1, \ldots, N-1$. What is the magnitude and angle of each root? The roots can be displayed as vectors directed from the origin whose tips lie on a circle.

**(a)** What is the length of each vector and the angular spacing between adjacent vectors? Sketch for $N = 5$ and $N = 6$.

**(b)** Extend this concept to find the roots of $z^N = -1$ and sketch for $N = 5$ and $N = 6$.

[**Hints and Suggestions:** In part (b), note that $z^N = -1 = e^{j\pi}e^{j2k\pi} = e^{j(2k+1)\pi}$.]

**2.18. (Digital Frequency)** Set up an expression for each signal, using a digital frequency $|F| < 0.5$, and another expression using a digital frequency in the range $4 < F < 5$.

**(a)** $x[n] = \cos(\frac{4n\pi}{3})$ **(b)** $x[n] = \sin(\frac{4n\pi}{3}) + 3\sin(\frac{8n\pi}{3})$

[**Hints and Suggestions:** First find the digital frequency of each component in the principal range $(-0.5 < F \le 0.5)$. Then, add 4 or 5 as appropriate to bring each frequency into the required range.]

**2.19.** **(Digital Sinusoids)** Find the period $N$ of each signal if periodic. Express each signal using a digital frequency in the principal range ($|F| < 0.5$) and in the range $3 \le F \le 4$.

**(a)** $x[n] = \cos(\frac{7n\pi}{3})$ **(b)** $x[n] = \cos(\frac{7n\pi}{3}) + \sin(0.5n\pi)$ **(c)** $x[n] = \cos(n)$

**2.20.** **(Sampling and Aliasing)** Each of the following sinusoids is sampled at $S = 100$ Hz. Determine if aliasing has occurred and set up an expression for each sampled signal using a digital frequency in the principal range ($|F| < 0.5$).

**(a)** $x(t) = \cos(320\pi t + \frac{\pi}{4})$ **(b)** $x(t) = \cos(140\pi t - \frac{\pi}{4})$ **(c)** $x(t) = \sin(60\pi t)$

[**Hints and Suggestions:** Find the frequency $f_0$. If $S > 2f_0$ there is no aliasing and $F < 0.5$. Otherwise, bring $F$ into the principal range to write the expression for the sampled signal.]

**2.21.** **(Aliasing and Signal Reconstruction)** The signal $x(t) = \cos(320\pi t + \frac{\pi}{4})$ is sampled at 100 Hz, and the sampled signal $x[n]$ is reconstructed at 200 Hz to recover the analog signal $x_r(t)$.

**(a)** Has aliasing occurred? What is the period $N$ and the digital frequency $F$ of $x[n]$?
**(b)** How many full periods of $x(t)$ are required to generate one period of $x[n]$?
**(c)** What is the analog frequency of the recovered signal $x_r(t)$?
**(d)** Write expressions for $x[n]$ (using $|F| < 0.5$) and for $x_r(t)$.

[**Hints and Suggestions:** For part (b), if the digital frequency is expressed as $F = k/N$ where $N$ is the period and $k$ is an integer, it takes $k$ full cycles of the analog sinusoid to get $N$ samples of the sampled signal. In part (c), the frequency of the reconstructed signal is found from the aliased frequency in the principal range.]

**2.22.** **(Digital Pitch Shifting)** One way to accomplish *pitch shifting* is to play back (or reconstruct) a sampled signal at a *different* sampling rate. Let the analog signal $x(t) = \sin(15800\pi t + 0.25\pi)$ be sampled at a sampling rate of 8 kHz.

**(a)** Find its sampled representation with digital frequency $|F| < 0.5$.
**(b)** What frequencies are heard if the signal is reconstructed at a rate of 4 kHz?
**(c)** What frequencies are heard if the signal is reconstructed at a rate of 8 kHz?
**(d)** What frequencies are heard if the signal is reconstructed at a rate of 20 kHz?

[**Hints and Suggestions:** The frequency of the reconstructed signal is found from the aliased digital frequency in the principal range and the appropriate reconstruction rate.]

**2.23.** **(Discrete-Time Chirp Signals)** Consider the signal $x(t) = \cos[\phi(t)]$, where $\phi(t) = \alpha t^2$. Show that its instantaneous frequency $f_i(t) = \frac{1}{2\pi}\phi'(t)$ varies linearly with time.

**(a)** Choose $\alpha$ such that the frequency varies from 0 Hz to 2 Hz in 10 seconds, and generate the sampled signal $x[n]$ from $x(t)$, using a sampling rate of $S = 4$ Hz.
**(b)** It is claimed that, unlike $x(t)$, the signal $x[n]$ is periodic. Verify this claim, using the condition for periodicity ($x[n] = x[n+N]$), and determine the period $N$ of $x[n]$.
**(c)** The signal $y[n] = \cos(\pi F_0 n^2/M)$, $n = 0, 1, \ldots, M-1$, describes an $M$-sample chirp whose digital frequency varies linearly from 0 to $F_0$. What is the period of $y[n]$ if $F_0 = 0.25$ and $M = 8$?

[**Hints and Suggestions:** In part (b), if $x[n] = \cos(\beta n^2)$, periodicity requires $x[n] = x[n+N]$ or $\cos(\beta n^2) = \cos[\beta(n^2 + 2nN + N^2)]$. Thus $2nN\beta = 2m\pi$ and $N^2\beta = 2k\pi$ where $m$ and $k$ are integers. Satisfy these conditions for the smallest integer $N$.]

**2.24.** **(Time Constant)** For exponentially decaying discrete signals, the **time constant** is a measure of how fast a signal decays. The 60-dB time constant describes the (integer) number of samples it takes for the signal level to decay by a factor of 1000 (or $20 \log 1000 = 60$ dB).

**(a)** Let $x[n] = (0.5)^n u[n]$. Compute its 60-dB time constant and 40-dB time constant.
**(b)** Compute the time constant in seconds if the discrete-time signal is derived from an analog signal sampled at 1 kHz.

**2.25.** **(Signal Delay)** The delay $D$ of a discrete-time energy signal $x[n]$ is defined by

$$D = \frac{\displaystyle\sum_{k=-\infty}^{\infty} kx^2[k]}{\displaystyle\sum_{k=-\infty}^{\infty} x^2[k]}$$

**(a)** Verify that the delay of the symmetric sequence $x[n] = \{4, 3, 2, 1, \overset{\Downarrow}{0}, 1, 2, 3, 4\}$ is zero.
**(b)** Compute the delay of the signals $g[n] = x[n-1]$ and $h[n] = x[n-2]$.
**(c)** What is the delay of the signal $y[n] = 1.5(0.5)^n u[n] - 2\delta[n]$?

[**Hints and Suggestions:** For part (c), compute the summations required in the expression for the delay by using tables and the fact that $y[n] = -0.5$ for $n = 0$ and $y[n] = 1.5(0.5)^n$ for $n \geq 1$.]

**2.26.** **(Periodicity)** It is claimed that the sum of an absolutely summable signal $x[n]$ and its shifted (by multiples of $N$) replicas is a periodic signal $x_p[n]$ with period $N$. Verify this claim by sketching the following and, for each case, compute the average power in the resulting periodic signal $x_p[n]$ and compare the sum and energy of one period of $x_p[n]$ with the sum and energy of $x[n]$.

**(a)** the sum of $x[n] = \text{tri}(n/3)$ and its replicas shifted by $N = 7$
**(b)** the sum of $x[n] = \text{tri}(n/3)$ and its replicas shifted by $N = 6$
**(c)** the sum of $x[n] = \text{tri}(n/3)$ and its replicas shifted by $N = 5$
**(d)** the sum of $x[n] = \text{tri}(n/3)$ and its replicas shifted by $N = 4$
**(e)** the sum of $x[n] = \text{tri}(n/3)$ and its replicas shifted by $N = 3$

**2.27.** **(Periodic Extension)** The sum of an absolutely summable signal $x[n]$ and its shifted (by multiples of $N$) replicas is called the *periodic extension* of $x[n]$ with period $N$.

**(a)** Show that one period of the periodic extension of the signal $x[n] = \alpha^n u[n]$ with period $N$ is

$$y[n] = \frac{\alpha^n}{1 - \alpha^N}, \quad 0 \leq n \leq N - 1$$

**(b)** How does the sum of one period of the periodic extension $y[n]$ compare with the sum of $x[n]$?
**(c)** With $\alpha = 0.5$ and $N = 3$, compute the signal energy in $x[n]$ and the average power in $y[n]$.

[**Hints and Suggestions:** For one period ($n = 0$ to $n = N - 1$), only $x[n]$ and the tails of the replicas to its left contribute. So, find the sum of $x[n + kN] = \alpha^{n+kN}$ only from $k = 0$ to $k = \infty$.]

**2.28. (Signal Norms)** Norms provide a measure of the *size* of a signal. The $p$**-norm**, or **Hölder norm,** $\|x\|_p$ for discrete signals is defined by $\|x\|_p = (\sum |x|^p)^{1/p}$, where $0 < p < \infty$ is a positive integer. For $p = \infty$, we also define $\|x\|_\infty$ as the peak absolute value $|x|_{\max}$.

**(a)** Let $x[n] = \{3,\ -j4,\ 3 + j4\}$. Find $\|x\|_1$, $\|x\|_2$, and $\|x\|_\infty$.
**(b)** What is the significance of each of these norms?

## Computation and Design

**2.29. (Discrete Signals)** For each part, plot the signals $x[n]$ and $y[n]$ over $-10 \le n \le 10$ and compare.

**(a)** $x[n] = u[n+4] - u[n-4] + 2\delta[n+6] - \delta[n-3]$ $\qquad y[n] = x[-n-4]$
**(b)** $x[n] = r[n+6] - r[n+3] - r[n-3] + r[n-6]$ $\qquad y[n] = x[n-4]$
**(c)** $x[n] = \text{rect}(\frac{n}{10}) - \text{rect}(\frac{n-3}{6})$ $\qquad y[n] = x[n+4]$
**(d)** $x[n] = 6\,\text{tri}(\frac{n}{6}) - 3\,\text{tri}(\frac{n}{3})$ $\qquad y[n] = x[-n+4]$

**2.30. (Signal Interpolation)** Let $h[n] = \sin(n\pi/3),\ 0 \le n \le 10$. Plot the signal $h[n]$. Use this to generate and plot the *zero-interpolated*, *step-interpolated*, and *linearly interpolated* signals assuming interpolation by 3.

**2.31. (Discrete Exponentials)** A causal discrete exponential may be expressed as $x[n] = \alpha^n u[n]$, where the nature of $\alpha$ dictates the form of $x[n]$. Plot the following over $0 \le n \le 40$ and comment on the nature of each plot.

**(a)** The signal $x[n]$ for $\alpha = 1.2$, $\alpha = 1$, and $\alpha = 0.8$.
**(b)** The signal $x[n]$ for $\alpha = -1.2$, $\alpha = -1$, and $\alpha = -0.8$.
**(c)** The real part and imaginary parts of $x[n]$ for $\alpha = Ae^{j\pi/4}$, with $A = 1.2$, $A = 1$, and $A = 0.8$.
**(d)** The magnitude and phase of $x[n]$ for $\alpha = Ae^{j\pi/4}$, with $A = 1.2$, $A = 1$, and $A = 0.8$.

**2.32. (Discrete-Time Sinusoids)** Which of the following signals are periodic and with what period? Plot each signal over $-10 \le n \le 30$. Do the plots confirm your expectations?

**(a)** $x[n] = 2\cos(\frac{n\pi}{2}) + 5\sin(\frac{n\pi}{5})$ $\qquad$ **(b)** $x[n] = 2\cos(\frac{n\pi}{2})\sin(\frac{n\pi}{3})$
**(c)** $x[n] = \cos(0.5n)$ $\qquad$ **(d)** $x[n] = 5\sin(\frac{n\pi}{8} + \frac{\pi}{4}) - 5\cos(\frac{n\pi}{8} - \frac{\pi}{4})$

**2.33. (Complex-Valued Signals)** A complex-valued signal $x[n]$ requires *two* plots for a complete description in one of two forms—the *magnitude* and *phase* versus $n$ or the *real part* versus $n$ and *imaginary part* versus $n$.

**(a)** Let $x[n] = \{\overset{\Downarrow}{2},\ 1 + j,\ -j2,\ 2 - j2,\ -4\}$. Sketch each form for $x[n]$ by hand.
**(b)** Let $x[n] = e^{-j0.3n\pi}$. Use MATLAB to plot each form over $-30 \le n \le 30$. Is $x[n]$ periodic? If so, can you identify its period from the MATLAB plots? From which form, and how?

**2.34. (Complex Exponentials)** Let $x[n] = 5\sqrt{2}e^{j(\frac{n\pi}{9} - \frac{\pi}{4})}$. Plot the following signals and, for each case, derive analytic expressions for the signals plotted and compare with your plots. Is the signal $x[n]$ periodic? What is the period $N$? Which plots allow you determine the period of $x[n]$?

**(a)** the real part and imaginary part of $x[n]$ over $-20 \le n \le 20$
**(b)** the magnitude and phase of $x[n]$ over $-20 \le n \le 20$

**(c)** the sum of the real and imaginary parts over $-20 \leq n \leq 20$
**(d)** the difference of the real and imaginary parts over $-20 \leq n \leq 20$

2.35. **(Complex Exponentials)** Let $x[n] = (\sqrt{j})^n + (\sqrt{j})^{-n}$. Plot the following signals and, for each case, derive analytic expressions for the sequences plotted and compare with your plots. Is the signal $x[n]$ periodic? What is the period $N$? Which plots allow you determine the period of $x[n]$?

**(a)** the real part and imaginary part of $x[n]$ over $-20 \leq n \leq 20$
**(b)** the magnitude and phase of $x[n]$ over $-20 \leq n \leq 20$

2.36. **(Discrete-Time Chirp Signals)** An $N$-sample *chirp* signal $x[n]$ whose digital frequency *varies linearly* from $F_0$ to $F_1$ is described by

$$x[n] = \cos\left[2\pi\left(F_0 n + \frac{F_1 - F_0}{2N} n^2\right)\right], \qquad n = 0, 1, \ldots, N-1$$

**(a)** Generate and plot 800 samples of a chirp signal **x** whose digital frequency varies from $F = 0$ to $F = 0.5$. Using the MATLAB based routine **timefreq** (from the author's website), observe how the frequency of **x** varies linearly with time.
**(b)** Generate and plot 800 samples of a chirp signal **x** whose digital frequency varies from $F = 0$ to $F = 1$. Is the frequency always increasing? If not, what is the likely explanation?

2.37. **(Chirp Signals)** It is claimed that the chirp signal $x[n] = \cos(\pi n^2/6)$ is periodic (unlike the analog chirp signal $x(t) = \cos(\pi t^2/6)$). Plot $x[n]$ over $0 \leq n \leq 20$. Does $x[n]$ appear periodic? If so, can you identify the period $N$? Justify your results by trying to find an integer $N$ such that $x[n] = x[n + N]$ (the basis for periodicity).

2.38. **(Signal Averaging)** Extraction of signals from noise is an important signal-processing application. Signal averaging relies on averaging the results of many runs. The noise tends to average out to zero, and the signal quality or **signal-to-noise ratio** (SNR) improves.

**(a)** Generate samples of the sinusoid $x(t) = \sin(800\pi t)$ sampled at $S = 8192$ Hz for 2 seconds. The sampling rate is chosen so that you may also listen to the signal if your machine allows.
**(b)** Create a noisy signal $s[n]$ by adding $x[n]$ to samples of uniformly distributed noise such that $s[n]$ has an SNR of 10 dB. Compare the noisy signal with the original and compute the actual SNR of the noisy signal.
**(c)** Sum the signal $s[n]$ 64 times and average the result to obtain the signal $s_a[n]$. Compare the averaged signal $s_a[n]$, the noisy signal $s[n]$, and the original signal $x[n]$. Compute the SNR of the averaged signal $x_a[n]$. Is there an improvement in the SNR? Do you notice any (visual and audible) improvement? Should you?
**(d)** Create the averaged result $x_b[n]$ of 64 *different* noisy signals and compare the averaged signal $x_b[n]$ with the original signal $x[n]$. Compute the SNR of the averaged signal $x_b[n]$. Is there an improvement in the SNR? Do you notice any (visual and/or audible) improvement? Explain how the signal $x_b[n]$ differs from $x_a[n]$.
**(e)** The reduction in SNR is a function of the noise distribution. Generate averaged signals, using different noise distributions (such as Gaussian noise) and comment on the results.

2.39. **(The Central Limit Theorem)** The central limit theorem asserts that the sum of independent noise distributions tends to a Gaussian distribution as the number $N$ of distributions in the sum increases. In fact, one way to generate a random signal with a Gaussian distribution is to add many (typically 6 to 12) uniformly distributed signals.

**(a)** Generate the sum of uniformly distributed random signals using $N = 2$, $N = 6$, and $N = 12$ and plot the histograms of each sum. Does the histogram begin to take on a Gaussian shape as $N$ increases? Comment on the shape of the histogram for $N = 2$.

**(b)** Generate the sum of random signals with different distributions using $N = 6$ and $N = 12$. Does the central limit theorem appear to hold even when the distributions are not identical (as long as you select a large enough $N$)? Comment on the physical significance of this result.

**2.40. (Music Synthesis I: Raga Malkauns)** A musical composition is a combination of *notes*, or signals, at various frequencies. An *octave* covers a range of frequencies from $f_0$ to $2f_0$. In the western musical scale, there are 12 notes per octave, *logarithmically equispaced*. The frequencies of the notes from $f_0$ to $2f_0$ correspond to

$$f = 2^{k/12} f_0 \quad (k = 0, 1, 2, \ldots, 11)$$

The 12 notes are as follows (the $\sharp$ and $\flat$ stand for *sharp* and *flat*, and each pair of notes in parentheses has the same frequency):

A (A$^\sharp$ or B$^\flat$) B C (C$^\sharp$ or D$^\flat$) D (D$^\sharp$ or E$^\flat$) E F (F$^\sharp$ or G$^\flat$) G (G$^\sharp$ or A$^\flat$)

To synthesize the scale of *raga malkauns* in MATLAB, we start with a frequency $f_0$ corresponding to the first note D and go up in frequency to get the notes in the ascending scale; when we reach the note D, which is an octave higher, we go down in frequency to get the notes in the descending scale. Here is a MATLAB code fragment.

```
>> f0=340; d=f0;                          % Pick a frequency and the note D
>> f=f0*(2 ^ (3/12)); g=f0*(2 ^ (5/12));  % The notes F and G
>> bf=f0*(2 ^ (8/12)) ; c=f0*(2 ^ (10/12)); % The notes B(flat) and C
>> d2=2*d;                                % The note D (an octave higher)
```

Generate sampled sinusoids at these frequencies, using an appropriate sampling rate (say, 8192 Hz); concatenate them, assuming silent passages between each note; and play the resulting signal, using the MATLAB command **sound**. Use the following MATLAB code fragment as a guide:

```
>> ts=1/8192;                          % Sampling interval
>> t=0:ts:0.4;                         % Time for each note (0.4 s)
>> s1=0*(0:ts:0.1);                    % Silent period (0.1 s)
>> s2=0*(0:ts:0.05);                   % Shorter silent period (0.05 s)
>> tl=0:ts:1;                          % Time for last note of each scale
>> d1=sin(2*pi*d*t);                   % Start generating the notes
>> f1=sin(2*pi*f*t); g1=sin(2*pi*g*t);
>> bf1=sin(2*pi*bf*t); c1=sin(2*pi*c*t);
>> dl1=sin(2*pi*d2*tl); dl2=sin(2*pi*d*tl);
>> asc=[d1 s1 f1 s1 g1 s1 bf1 s1 c1 s2 dl1];   % Create ascending scale
>> dsc=[c1 s1 bf1 s1 g1 s1 f1 s1 dl2];         % Create descending scale
>> y=[asc s1 dsc s1]; sound(y)                % Malkauns scale (y)
```

**Raga Malkauns:** In Indian classical music, a *raga* is a musical composition based on an ascending and descending scale. The notes and their order form the musical alphabet and grammar from which the performer constructs musical passages, using only the notes allowed. The performance of a raga can last from a few minutes to an hour or more! *Raga malkauns* is a pentatonic raga (with five notes) and the following scales:

Ascending:
D F G B$^\flat$ C D

Descending:
C B$^\flat$ G F D

The final note in each scale is held twice as long as the rest.

**2.41. (Music Synthesis II)** The raw scale of *raga malkauns* will sound pretty dry! The reason for this is the manner in which the sound from a musical instrument is generated. Musical instruments produce sounds by the vibrations of a string (in string instruments) or a column of air (in woodwind instruments). Each instrument has its characteristic sound. In a guitar, for example, the strings are plucked, held, and then released to sound the notes. Once plucked, the sound dies out and decays. Furthermore, the notes are never pure but contain overtones (harmonics). For a realistic sound, we must include the overtones and the attack, sustain, and release (decay) characteristics. The sound signal may be considered to have the form $x(t) = \alpha(t)\cos(2\pi f_0 t + \theta)$, where $f_0$ is the pitch and $\alpha(t)$ is the envelope that describes the attack-sustain-release characteristics of the instrument played. A crude representation of some envelopes is shown in Figure P2.41 (the piecewise linear approximations will work just as well for our purposes). Wood-wind instruments have a much longer sustain time and a much shorter release time than do plucked-string and keyboard instruments. Experiment with the scale of *raga malkauns* (Problem 2.40) and try to produce a guitar-like sound, using the appropriate envelope shape.

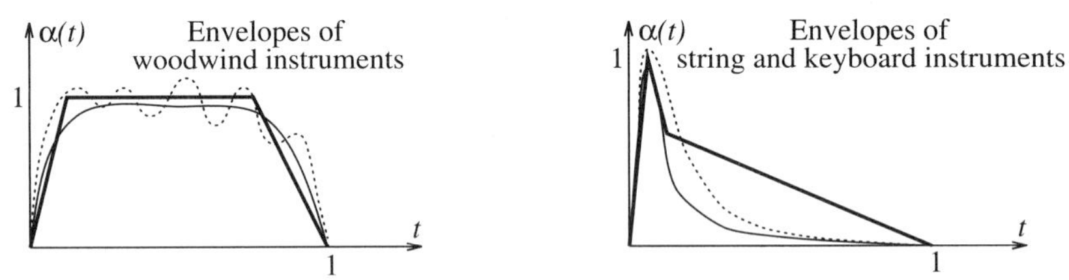


**FIGURE P.2.41** Envelopes and their piecewise linear approximations (dark) for Problem 2.41

**2.42.** **(Music Synthesis III)** Synthesize the following notes, using a woodwind envelope, and synthesize the same notes using a plucked-string envelope.

$$F^\sharp(0.3) \quad D(0.4) \quad E(0.4) \quad A(1) \quad A(0.4) \quad E(0.4) \quad F^\sharp(0.3) \quad D(1)$$

All the notes cover one octave, and the numbers in parentheses give a rough indication of their relative duration. Can you identify the music? (It is *Big Ben*.)

**2.43.** **(Music Synthesis IV)** Synthesize the first bar of *Pictures at an Exhibition* by Mussorgsky, which has the following notes:

$$A(3) \quad G(3) \quad C(3) \quad D(2) \quad G^\star(1) \quad E(3) \quad D(2) \quad G^\star(1) \quad E(3) \quad C(3) \quad D(3) \quad A(3) \quad G(3)$$

All the notes cover one octave except the note $G^\star$, which is an octave above G. The numbers in parentheses give a rough indication of the relative duration of the notes (for more details, you may want to listen to an actual recording). Assume that a keyboard instrument (such as a piano) is played.

**2.44.** **(DTMF Tones)** In dual-tone multifrequency (DTMF) or touch-tone telephone dialing, each number is represented by a dual-frequency tone. The frequencies for each digit are listed in Chapter 7.

**(a)** Generate DTMF tones corresponding to the telephone number 487-2550, by sampling the sum of two sinusoids at the required frequencies at $S = 8192$ Hz for each digit. Concatenate the signals by putting 50 zeros between each signal (to represent silence) and listen to the signal using the MATLAB command **sound**.

**(b)** Write a MATLAB program that generates DTMF signals corresponding to a vector input representing the digits in a phone number. Use a sampling frequency of $S = 8192$ Hz.

CHAPTER 3

# Time-Domain Analysis

## 3.0 Scope and Overview

Systems that process discrete-time signals are called **discrete-time** systems or **digital filters**. Their mathematical description relies heavily on how they respond to arbitrary or specific signals. This chapter starts with the classification of discrete systems and introduces the important concepts of linearity and time invariance. It presents the analysis of discrete-time systems described by difference equations. It concludes with the all-important concept of the impulse response and the process of convolution that forms an important method for finding the response of *linear, time-invariant* systems.

### 3.0.1 Goals and Learning Objectives

The goals of this chapter are to provide the mathematical tools necessary for the classification and time-domain analysis of discrete-time systems and digital filters. After going through this chapter, the reader should:

1. Understand linearity, superposition, time invariance, and causality.
2. Know how to identify LTI (linear time invariant) systems.
3. Understand the terminology and classification of digital filters.
4. Know how to set up the realization of recursive and nonrecursive filters.
5. Know how to solve a difference equation for its natural and forced response.
6. Know how to solve a difference equation for its zero-state and zero-input response.
7. Know how to find the impulse response of an LTI system described by its difference equation.
8. Know how to convert between the system difference equation and its impulse response.
9. Know how to find the convolution of finite-length sequences.
10. Know how to use the defining relation to find the convolution.
11. Understand the properties of convolution and know how to use them for problem solving.

12. Know how to find the impulse response of systems in cascade and parallel.
13. Understand the concept of stability.
14. Know how to determine stability from the system difference equation or impulse response.
15. Understand periodic convolution and know how to find the periodic convolution of two signals.
16. Know how to find the cross-correlation and autocorrelation.
17. Understand the connection between convolution and transform methods.

# 3.1 Discrete-Time Systems

In the time domain, many discrete-time systems can be modeled by **difference equations** relating the input and output. Difference equations typically involve input and output signals and their shifted versions. For example, the system described by $y[n] = \alpha y[n-1]+\beta x[n]$ produces the present output $y[n]$ as the sum of the previous output $y[n-1]$ and the present input $x[n]$. The quantities $\alpha$ and $\beta$ are called *coefficients*. Linear, time-invariant (LTI) systems are characterized by constant coefficients. Such systems have been studied extensively and their response can be obtained by well-established mathematical methods.

## 3.1.1 Linearity and Superposition

An operator allows us to transform one function to another. If an operator is represented by the symbol $\mathcal{O}$, the equation

$$\mathcal{O}\{x[n]\} = y[n] \tag{3.1}$$

implies that if the function $x[n]$ is treated exactly as the operator $\mathcal{O}$ tells us, we obtain the function $y[n]$. The operator $z^m$ describes an *advance* of $m$ units and transforms $x[n]$ to $x[n+m]$. We express this transformation in operator notation as $z^m\{x[n]\} = x[n+m]$. The operator $z^{-m}$ describes a *delay* of $m$ units and transforms $x[n]$ to $x[n-m]$. In operator notation, we have $z^{-m}\{x[n]\} = x[n-m]$. An operation may describe several steps. For example, the operation $\mathcal{O}\{\ \} = 4z^{-3}\{\ \} + 6$ says that to get $y[n]$, we must delay $x[n]$ by 3 units, multiply the result by 4, and then add 6 to finally obtain $4z^{-3}\{x[n]\} + 6 = 4x[n-3] + 6 = y[n]$.

If an operation on the sum of two functions is equivalent to the sum of operations applied to each separately, the operator is said to be **additive**. In other words,

$$\mathcal{O}\{x_1[n] + x_2[n]\} = \mathcal{O}\{x_1[n]\} + \mathcal{O}\{x_2[n]\} \quad \text{(for an additive operation)} \tag{3.2}$$

If an operation on $Kx[n]$ is equivalent to $K$ times the linear operation on $x[n]$ where $K$ is a scalar, the operator is said to be **homogeneous** or **scalable**. In other words,

$$\mathcal{O}\{Kx[n]\} = K\mathcal{O}\{x[n]\} \quad \text{(for a homogeneous or scalable operation)} \tag{3.3}$$

Together, the two results describe the **principle of superposition**. An operator $\mathcal{O}$ is termed a **linear operator** if it obeys superposition and is therefore both homogeneous (scalable) and additive:

$$\mathcal{O}\{Ax_1[n] + Bx_2[n]\} = A\mathcal{O}\{x_1[n]\} + B\mathcal{O}\{x_2[n]\} \quad \text{(for a linear operation)} \tag{3.4}$$

If a system fails the test for either additivity or homogeneity, it is termed nonlinear. It must pass both tests in order to be termed linear. However, in many instances, it suffices to test only for homogeneity or additivity to confirm the linearity of an operation (even though one does not imply the other). An important concept that forms the basis for the study of linear systems is that *the superposition of linear operators is also linear.*

**Example 3.1**

### Testing for Linear Operators

**(a)** Consider the operator $\mathcal{O}\{\ \} = C\{\ \} + D$.

$$\text{By the homogeneity test: } \mathcal{O}\{Ax[n]\} = ACx[n] + D, \quad \text{but}$$
$$A\mathcal{O}\{x[n]\} = A(Cx[n] + D) = ACx[n] + AD$$

The two differ, so the operation is nonlinear (it is linear only if $D = 0$).

**(b)** Consider the squaring operator $\mathcal{O}\{\ \} = \{\ \}^2$, which transforms $x[n]$ to $x^2[n]$.

$$\text{By the homogeneity test: } A\mathcal{O}\{x[n]\} = Ax^2[n], \quad \text{but}$$
$$\mathcal{O}\{Ax[n]\} = (Ax[n])^2 = A^2x^2[n]$$

The two are not equal, and the squaring operator is nonlinear.

# Drill Problem 3.1

Which of the following operations are linear?

**(a)** $\mathcal{O}\{\ \} = \cos\{\ \}$ **(b)** $\mathcal{O}\{\ \} = \log\{\ \}$ **(c)** $\mathcal{O}\{x[n]\} = \alpha^n\{x[n]\}$

**Answers:** Only (c) is linear.

## 3.1.2 Time Invariance

Time invariance implies that the shape of the response $y[n]$ depends only on the shape of the input $x[n]$ and not on the time when it is applied. If the input is shifted to $x[n - n_0]$, the response equals $y[n - n_0]$ and is shifted by the same amount. In other words, the system does not change with time. Formally, if the operator $\mathcal{O}$ transforms the input $x[n]$ to the output $y[n]$ such that $\mathcal{O}\{x[n]\} = y[n]$, time invariance means

$$\mathcal{O}\{x[n - n_0]\} = y[n - n_0] \quad \text{(for time invariance)} \tag{3.5}$$

In other words, if the input is delayed by $n_0$ units, the output is also delayed by the same amount and is simply a shifted replica of the original output.

**Example 3.2**

### Linearity and Time Invariance of Operators

**(a)** $y[n] = x[n]x[n-1]$ is nonlinear but time invariant.

The operation is $\mathcal{O}\{\ \} = (\{\ \})(z^{-1}\{\ \})$. We find that

$$\begin{aligned} A\mathcal{O}\{x[n]\} &= A(x[n]x[n-1]), \quad \text{but} \\ \mathcal{O}\{Ax[n]\} &= (Ax[n])(Ax[n-1]) \text{(the two are not equal)} \end{aligned}$$

$$\begin{aligned} \mathcal{O}\{x[n-n_0]\} &= x[n-n_0]x[n-n_0-1], \text{ and } y[n-n_0] \\ &= x[n-n_0]x[n-n_0-1] \quad \text{(the two are equal)} \end{aligned}$$

**(b)** $y[n] = nx[n]$ is linear but time varying.

The operation is $\mathcal{O}\{\ \} = n\{\ \}$. We find that

$$A\mathcal{O}\{x[n]\} = A(nx[n]), \text{ and } \mathcal{O}\{Ax[n]\} = n(Ax[n]) \text{ (the two are equal)}$$

$$\begin{aligned} \mathcal{O}\{x[n-n_0]\} &= n(x[n-n_0]), \quad \text{but} \\ y[n-n_0] &= (n-n_0)x[n-n_0] \quad \text{(the two are not equal)} \end{aligned}$$

**(c)** $y[n] = x[2n]$ is linear but time varying.

The operation $n \Rightarrow 2n$ reveals that

$$A\mathcal{O}\{x[n]\} = A(x[2n]), \text{ and } \mathcal{O}\{Ax[n]\} = (Ax[2n]) \text{ (the two are equal)}$$

$$\begin{aligned} \mathcal{O}\{x[n-n_0]\} &= x[2n-n_0], \quad \text{but} \\ y[n-n_0] &= x[2(n-n_0)] \quad \text{(the two are not equal)} \end{aligned}$$

**(d)** $y[n] = x[n-2]$ is linear and time invariant.

The operation $n \Rightarrow n-2$ reveals that

$$A\mathcal{O}\{x[n]\} = A(x[n-2]), \text{ and } \mathcal{O}\{Ax[n]\} = (Ax[n-2]) \text{ (the two are equal)}$$

$$\mathcal{O}\{x[n-n_0]\} = x[n-n_0-2], \text{ and } y[n-n_0] = x[n-n_0-2] \quad \text{(the two are equal)}$$

**(e)** $y[n] = 2^{x[n]}x[n]$ is nonlinear but time invariant.

The operation is $\mathcal{O}\{\ \} = (2)^{\{\ \}}\{\ \}$ and reveals that

$$A\mathcal{O}\{x[n]\} = A(2)^{x[n]}x[n], \text{ but } \mathcal{O}\{Ax[n]\} = (2)^{Ax[n]}(Ax[n]) \quad \text{(the two are not equal)}$$

$$\begin{aligned} \mathcal{O}\{x[n-n_0]\} &= (2)^{x[n-n_0]}x[n-n_0], \quad \text{and} \\ y[n-n_0] &= (2)^{x[n-n_0]}x[n-n_0] \quad \text{(the two are equal)} \end{aligned}$$

# Drill Problem 3.2

Which of the following operations are time-invariant?

**(a)** $\mathcal{O}\{x[n]\} = \cos\{x[n]\}$ **(b)** $\mathcal{O}\{x[n]\} = C\{x[n]\}$ **(c)** $\mathcal{O}\{x[n]\} = C^n\{x[n]\}$

**Answers:** (a) and (b) are time-invariant.

### 3.1.3 LTI Systems

Systems that are both linear and time-invariant are termed LTI (linear, time invariant). We may check for linearity or time-invariance by applying formal tests to the system equation as a whole or by looking at its individual operations. A difference equation is LTI if all its coefficients are constant (and no constant terms are present). Any nonlinear or time-varying behavior is recognized (by generalizing the results of the previous example) as follows:

1. If a constant term is present or a *term* contains products of the input and/or output, the system equation is nonlinear.
2. If a *coefficient* is an explicit function of $n$ or a scaled input or output (such as $y[2n]$) is present, the system equation is time varying.

**LTI Systems: What Makes a Difference Equation LTI or Nonlinear or Time Varying?**

It is **LTI** if all coefficients are constant and there are *no constant terms*.
It is **nonlinear** if a term is constant or a *nonlinear function* of $x[n]$ or $y[n]$.
It is **time varying** if a *coefficient* is an *explicit* function of $n$ or an input or output is scaled (*e.g.*, $y[2n]$).

**Example 3.3**

**Linearity and Time Invariance of Systems**

We check the following systems for linearity and time invariance.

**(a)** $y[n] - 2y[n-1] = 4x[n]$ (this is LTI)
**(b)** $y[n] - 2ny[n-1] = x[n]$ (this is linear but time varying)
**(c)** $y[n] + 2y^2[n] = 2x[n] - x[n-1]$ (this is nonlinear but time invariant)
**(d)** $y[n] - 2y[n-1] = (2)^{x[n]}x[n]$ (this is nonlinear but time invariant)
**(e)** $y[n] - 4y[n]y[2n] = x[n]$ (this is nonlinear and time varying)

# Drill Problem 3.3

What can you say about the linearity (L) and time-invariance (TI) of the following?
**(a)** $y[n] + 2y[n-1] = x[n]$ **(b)** $y[n] + 2^n y[n-1] = x[n]$
**(c)** $y^2[n] + 2^n y[n-1] = x[n]$

**Answers: (a)** L and TI **(b)** L but not TI **(c)** Not L, not TI

### 3.1.4 Causality and Memory

In many practical situations, we deal with systems whose inputs and outputs are right-sided signals. In a **causal system**, the present response $y[n]$ cannot depend on future values of the input, such as $x[n+2]$. Systems whose present response requires knowledge of future values of the input are termed **noncausal**. Consider an LTI system whose input and output are assumed to be right-sided signals. If such a system is described by the difference equation

$$y[n] + A_1 y[n-1] + \cdots + A_N y[n-N] = B_0 x[n+K]$$

it is causal as long as $K \le 0$. On the other hand, if the system is described by

$$y[n+L] + A_1 y[n+L-1] + \cdots + A_L y[n] = B_0 x[n+K]$$

it is causal as long as $K \le L$. The reason is that by time invariance, this system may also be described by

$$y[n] + A_1 y[n-1] + \cdots + A_L y[n-L] = B_0 x[n+K-L]$$

For causality, we require $K - L \le 0$ or $K \le L$. It is often easier to check for causality by examining the **operational transfer function** $H(z)$ derived from the difference equation. The general form of such a transfer function may be expressed as a ratio of polynomials in $z$

$$H(z) = \frac{B_0 z^P + B_1 z^{P-1} + \cdots + B_{P-1} z + B_P}{A_0 z^Q + A_1 z^{Q-1} + \cdots + A_{Q-1} z + A_Q} \qquad (3.6)$$

Assuming a right-sided input and output, this system is causal if $P \le Q$.

# Drill Problem 3.4

**(a)** Which of the two systems are causal?

$$y[n] = 2^{n+1} x[n] \qquad\qquad y[n] = 2^n x[n+1]$$

**(b)** Find the operational transfer function of each system and determine if it is causal.

$$y[n] - 2y[n-1] = x[n+1] \qquad\qquad y[n+1] - 2y[n-2] = x[n+1]$$

**Answers:** (a) The first is causal. (b) The second is causal.

### Instantaneous and Dynamic Systems

If the response of a system at $n = n_0$ depends only on the input at $n = n_0$ and not at any other times (past or future), the system is called **instantaneous** or **static**. The system equation of an instantaneous system has the form $y[n+\alpha] = Kx[n+\alpha]$. Note that the arguments of the input and output are *identical*. The response of a dynamic system depends on past (and/or future) inputs. Dynamic systems are usually described by (but not limited to) difference equations. The system $y[n] + 0.5y[n-1] = x[n]$ (a difference equation) is dynamic. The system $y[n] = 3x[n-2]$ is also dynamic (because the arguments of the input and output are different), but the system $y[n-2] = 3x[n-2]$ is instantaneous.

### What Makes a System Equation Noncausal or Dynamic?

**Noncausal:** If the numerator degree of the operational transfer function **exceeds** the denominator degree.
**Dynamic:** If the system equation has a form different from $y[n+\alpha] = Kx[n+\alpha]$.

**Example 3.4**

### Causal and Dynamic Systems

**(a)** $y[n] = x[n+2]$ is noncausal (to find $y[0]$, we need $x[2]$) and dynamic ($y[n_0]$ does not depend on $x[n_0]$ but on $x[n_0+2]$).
**(b)** $y[n+4] + y[n+3] = x[n+2]$. By time invariance, $y[n] + y[n-1] = x[n-2]$. So it is causal and dynamic.
**(c)** $y[n] = 2x[\alpha n]$ is causal and instantaneous for $\alpha = 1$, causal and dynamic for $\alpha < 1$, and noncausal and dynamic for $\alpha > 1$. It is also time varying if $\alpha \neq 1$.
**(d)** $y[n] = 2(n+1)x[n]$ is causal, instantaneous and time varying.

# Drill Problem 3.5

Which of the following systems are instantaneous?
**(a)** $y[n] = 2^{n+1}x[n]$ **(b)** $y[n] = 2^n x[n+1]$ **(c)** $y[n] - 2y[n-1] = x[n+1]$

**Answers:** Only (a) is instantaneous.

# 3.2 Digital Filters

A discrete-time system is also referred to as a **digital filter**. An important formulation for digital filters is based on difference equations. The general form of an $N$th-order difference equation may be written as

$$\begin{aligned} y[n] + A_1 y[n-1] + \cdots + A_N y[n-N] \\ = B_0 x[n] + B_1 x[n-1] + \cdots + B_M x[n-M] \end{aligned} \tag{3.7}$$

The **order** $N$ describes the output term with the largest delay. It is customary to normalize the leading coefficient to unity. The coefficients $A_k$ and $B_k$ are constant for an LTI digital filter. The response depends on the applied input and initial conditions that describe its state just before the input is applied. Systems with zero initial conditions are said to be *relaxed*. In general, if an input $x[n]$ to a relaxed LTI system undergoes a linear operation, the output $y[n]$ undergoes the same linear operation. Often, an arbitrary function may be decomposed into its simpler constituents, the response due to each analyzed separately and more effectively, and the total response found using superposition. This approach is the key to several methods of filter analysis described in subsequent chapters. An arbitrary signal $x[n]$ can be expressed as a weighted sum of shifted impulses. If the input to a discrete-time system is the unit impulse $\delta[n]$, the resulting output is called the **impulse response** and denoted $h[n]$. The impulse response is the basis for finding the filter response by a method called convolution.

### 3.2.1 Digital Filter Terminology

Consider a digital filter described by the equation

$$y[n] = B_0 x[n] + B_1 x[n-1] + \cdots + B_M x[n-M] \quad \text{(nonrecursive filter)} \tag{3.8}$$

Its present response depends only on the input terms and shows no dependence (recursion) on past values of the response. It is called a **nonrecursive filter** (or a **moving average filter**) because its response is just a weighted sum (moving average) of the input terms.

Now consider a digital filter described by the difference equation:

$$y[n] + A_1 y[n-1] + \cdots + A_N y[n-N] = B_0 x[n] \quad \text{(recursive or AR filter)} \tag{3.9}$$

This describes a **recursive filter** of order $N$ whose present output depends on its own past values $y[n-k]$ and on the present value of the input. It is also called an **infinite impulse response** (IIR) **filter**, because its impulse response $h[n]$ (the response to a unit impulse input) is usually of infinite duration. It is also called an AR (**autoregressive**) filter, because its output depends (regresses) on its own previous values.

Finally, consider the most general formulation described by the difference equation:

$$\begin{aligned} y[n] &+ A_1 y[n-1] + \cdots + A_N y[n-N] \\ &= B_0 x[n] + B_1 x[n-1] + \cdots + B_M x[n-M] \quad \text{(recursive or ARMA filter)} \end{aligned} \tag{3.10}$$

This is also a recursive filter. It is called an **autoregressive**, **moving average** (ARMA) filter because its present output depends not only on its own past values $y[n-k]$ but also on the past and present values of the input.

---

**The Terminology of Digital Filters**

**Nonrecursive or FIR:** $y[n] = B_0 x[n] + B_1 x[n-1] + \cdots + B_M x[n-M]$
**Recursive, IIR, or AR:** $y[n] + A_1 y[n-1] + \cdots + A_N y[n-N] = B_0 x[n]$
**Recursive or ARMA:** $y[n] + A_1 y[n-1] + \cdots + A_N y[n-N] = B_0 x[n] + B_1 x[n-1] + \cdots + B_M x[n-M]$

---

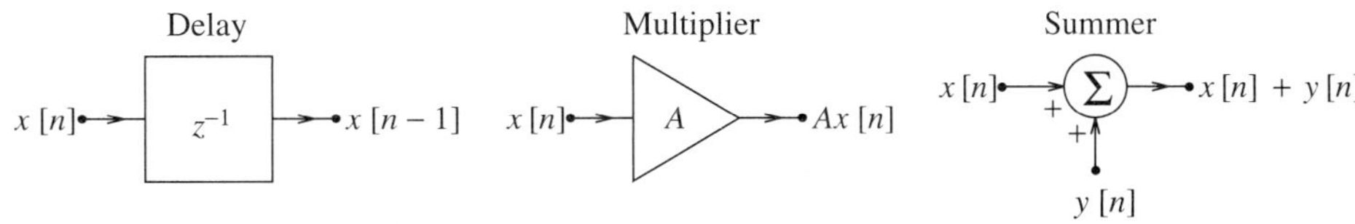


**FIGURE 3.1** The building blocks for digital filter realization include delay elements (that can be cascaded), scalar multipliers, and summers

### 3.2.2 Digital Filter Realization

Digital filters described by linear difference equations with constant coefficients may be *realized* by using elements corresponding to the operations of *scaling* (or scalar multiplication), *shift* (or delay), and *summing* (or addition) that naturally occur in such equations. These elements describe the **gain** (scalar multiplier), **delay**, and **summer** (or **adder**), represented symbolically in Figure 3.1.

Delay elements in cascade result in an output delayed by the sum of the individual delays. The operational notation for a delay of $k$ units is $z^{-k}$. A nonrecursive filter described by

$$y[n] = B_0 x[n] + B_1 x[n-1] + \cdots + B_N x[n-N] \tag{3.11}$$

can be realized using a **feed-forward** structure with $N$ delay elements, and a recursive filter of the form

$$y[n] = -A_1 y[n-1] - \cdots - A_N y[n-N] + x[n] \tag{3.12}$$

requires a **feedback** structure (because the output depends on its own past values). Each realization is shown in Figure 3.2 and requires $N$ delay elements.

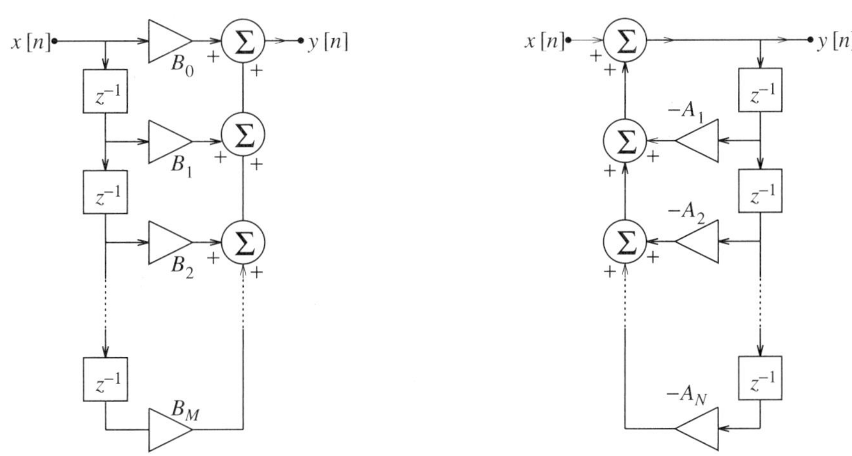


**FIGURE 3.2** The realization of a nonrecursive digital filter (left) shows only feed-forward paths. The realization of a recursive digital filter (right) includes feedback paths

# Drill Problem 3.6

Sketch the realizations of the following systems

**(a)** $y[n] - 0.6y[n-1] = x[n]$ **(b)** $y[n] - \frac{1}{6}y[n-1] - \frac{1}{6}y[n-2] = 4x[n]$

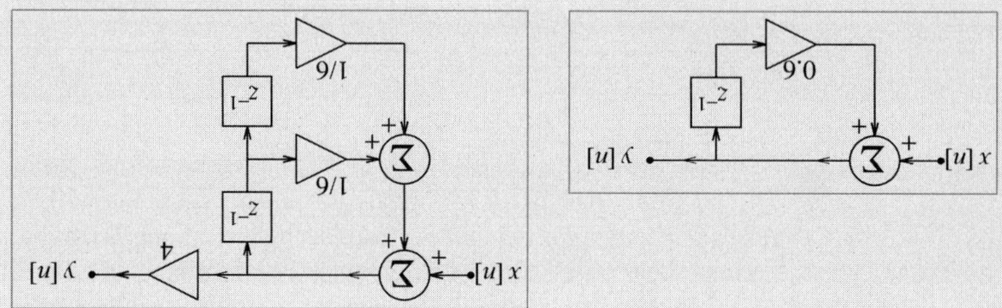


**Answer:**

The general form described by

$$y[n] = -A_1 y[n-1] - \cdots - A_N y[n-N] + B_0 x[n] + B_1 x[n-1] + \cdots + B_N x[n-N] \quad (3.13)$$

requires both feed-forward and feedback and may be realized using $2N$ delay elements, as shown in Figure 3.3. This describes a **direct form I** realization.

However, since LTI systems may be cascaded in any order (as we shall learn soon), we can switch the feedback and feedforward sections to obtain a **canonical realization** with only $N$ delays, as also shown in Figure 3.3. It is also called a **direct form II** realization. Other forms that also use only $N$ elements are also possible. We discuss various aspects of digital filter realization in more detail in subsequent chapters.

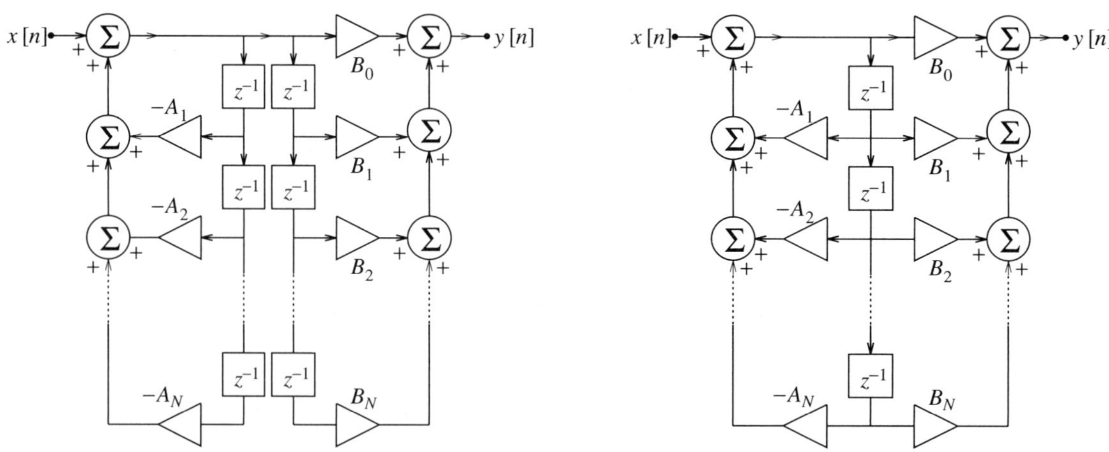


**FIGURE 3.3** The direct (direct form I) realization of an *N*th-order difference equation (left) requires 2*N* delay elements. The canonical (direct form II) realization the same difference equation requires only *N* delay elements

# Drill Problem 3.7

What is the difference equation of the digital filter whose realization is shown?

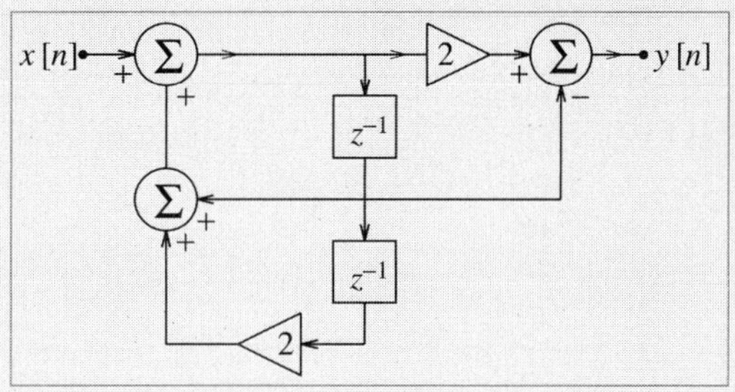


**Answer:** $y[n] - y[n-1] - 2y[n-2] = 2x[n] - x[n-1]$

# 3.3 Response of Digital Filters

A digital filter processes discrete signals and yields a discrete output in response to a discrete input. Its response depends not only on the applied input but also on the initial conditions that describe its state just prior to the application of the input. Systems with zero initial conditions are said to be *relaxed*. Digital filters may be analyzed in the time domain using any of the following models:

- **The difference equation representation** applies to linear, nonlinear, and time-varying systems. For LTI systems, it allows computation of the response using superposition even if initial conditions are present.
- **The impulse response representation** describes a *relaxed* LTI system by its impulse response $h[n]$. The output $y[n]$ appears explicitly in the governing relation called the *convolution sum*. It also allows us to relate time-domain and transformed-domain methods of system analysis.
- **The state variable representation** describes an $n$th-order system by $n$ simultaneous first-order difference equations called **state equations** in terms of $n$ state variables. It is useful for complex or nonlinear systems and those with multiple inputs and outputs. For LTI systems, state equations can be solved using matrix methods. The state variable form is also readily amenable to numerical solution. We do not pursue this method in this book.

## 3.3.1 Response of Nonrecursive Filters

The system equation of a nonrecursive filter is

$$y[n] = B_0x[n] + B_1x[n-1] + \cdots + B_Mx[n-M] \quad \text{(FIR filter)}$$

Since the output $y[n]$ depends only on the input $x[n]$ and its shifted versions, the response is simply a weighted sum of the input terms exactly as described by its system equation.

**Example 3.5**

**Response of Nonrecursive Filters**

Consider an FIR filter described by $y[n] = 2x[n] - 3x[n-2]$. Find its output $y[n]$ if the input to the system is $x[n] = (0.5)^n u[n]$ and compute $y[n]$ for $n = 1$ and $n = 2$.
We find

$$y[n] = 2x[n] - 3x[n-2] = 2(0.5)^n u[n] - 3(0.5)^{n-2} u[n-2]$$

This gives

$$y[1] = 2(0.5)^1 = 1 \text{ and } y[2] = 2(0.5)^2 - 3(0.5)^0 = 0.5 - 3 = -2.5$$

### 3.3.2 Response of Recursive Filters

The difference equation of a recursive digital filter is

$$y[n] + A_1 y[n-1] + \cdots + A_N y[n-N] = B_0 x[n] + B_1 x[n-1] + \cdots + B_M x[n-M]$$

The response shows dependence on its past values as well as values of the input. The output $y[n]$ requires the prior values of $y[n-1],\ y[n-2],\ \ldots,\ y[n-N]$. Once known, we can use $y[n]$ and the other previously known values to compute $y[n+1]$ and continue to use recursion to successively compute the values of the output as far as desired. Consider the second-order difference equation

$$y[n] + A_1 y[n-1] + A_2 y[n-2] = B_0 x[n] + B_1 x[n-1]$$

To find $y[n]$, we rewrite this equation as follows

$$y[n] = -A_1 y[n-1] - A_2 y[n-2] + B_0 x[n] + B_1 x[n-1]$$

We see that $y[n]$ can be found from its past values $y[n-1]$ and $y[n-2]$. To start the recursion at $n = 0$, we must be given values of $y[-1]$ and $y[-2]$, and once known, values of $y[n],\ n \geq 0$ may be computed successively as far as desired. We see that this method requires initial conditions to get the recursion started. In general, the response $y[n],\ n \geq 0$ of the $N$th-order difference equation requires the $N$ consecutive initial conditions $y[-1], y[-2], \ldots, y[-N]$. The recursive approach is effective and can be used even for nonlinear or time varying systems. Its main disadvantage is that a general closed form solution for the output is not always easy to discern.

**Example 3.6**

**System Response Using Recursion**

**(a)** Consider a system described by $y[n] = a_1 y[n-1] + b_0 u[n]$. Let the initial condition be $y[-1] = 0$. We then successively compute

$$\begin{aligned} y[0] &= a_1 y[-1] + b_0 u[0] = b_0 \\ y[1] &= a_1 y[0] + b_0 u[1] = a_1 b_0 + b_0 = b_0[1 + a_1] \\ y[2] &= a_1 y[1] + b_0 u[2] = a_1[a_1 b_0 + b_0] + b_0 = b_0\left[1 + a_1 + a_1^2\right] \end{aligned}$$

The form of $y[n]$ may be discerned as

$$y[n] = b_0\left[1 + a_1 + a_1^2 + \cdots + a_1^{n-1} + a_1^n\right]$$

Using the closed form for the geometric sequence results in

$$y[n] = \frac{b_0(1 - a_1^{n+1})}{1 - a_1}$$

If the coefficients appear as numerical values, the general form may not be easy to discern.

**(b)** Consider a system described by $y[n] = a_1 y[n-1] + b_0 n u[n]$. Let the initial condition be $y[-1] = 0$. We then successively compute

$$\begin{aligned}
y[0] &= a_1 y[-1] = 0 \\
y[1] &= a_1 y[0] + b_0 u[1] = b_0 \\
y[2] &= a_1 y[1] + 2b_0 u[2] = a_1 b_0 + 2b_0 \\
y[3] &= a_1 y[2] + 3b_0 u[3] = a_1[a_1 b_0 + 2b_0] + 3b_0 = a_1^2 b_0 + 2a_1 b_0 + 3b_0
\end{aligned}$$

The general form is thus $y[n] = b_0 a_1^{n-1} + 2b_0 a_1^{n-2} + 3b_0 a_1^{n-3} + (n-1)b_0 a_1 + n b_0$. We can find a more compact form for this, but not without some effort. By adding and subtracting $b_0 a_1^{n-1}$ and factoring out $a_1^n$, we obtain

$$y[n] = a_1^n - b_0 a_1^{n-1} + b_0 a_1^n\left[a_1^{-1} + 2a_1^{-2} + 3a_1^{-3} + \cdots + n a_1^{-n}\right]$$

Using the closed form for the sum $\sum kx^k$ from $k = 1$ to $k = N$ (with $x = a^{-1}$), we get

$$y[n] = a_1^n - b_0 a_1^{n-1} + b_0 a_1^n \frac{a^{-1}[1 - (n+1)a^{-n} + n a^{-(n+1)}]}{(1 - a^{-1})^2}$$

What a chore! More elegant ways of solving difference equations are described later in this chapter.

# Drill Problem 3.8

**(a)** Let $y[n] - y[n-1] - 2y[n-2] = u[n]$. Use recursion to compute $y[3]$ if $y[-1] = 2,\ y[-2] = 0$.

**(b)** Let $y[n] - 0.8y[n-1] = x[n]$. Use recursion to find the general form of $y[n]$ if $x[n] = \delta[n]$ and $y[-1] = 0$.

**Answers:** **(a)** 32 **(b)** $(0.8)^n$, $n \geq 0$ or $(0.8)^n u[n]$

## 3.4 Solving Difference Equations

For an LTI system governed by a linear constant-coefficient difference equation, a formal way of computing the output is by the **method of undetermined coefficients**. This method yields the total response $y[n]$ as the sum of the **forced response** $y_F[n]$ and the **natural response** $y_N[n]$. Consider the $N$th-order difference equation with the single unscaled input $x[n]$

$$y[n] + A_1 y[n-1] + A_2 y[n-2] + \cdots + A_N y[n-N] = x[n] \qquad (3.14)$$

with initial conditions $y[-1], y[-2], y[-3], \ldots, y[-N]$.

Its **forced response** arises due to the interaction of the system with the input, has the same form as the input, and satisfies the given difference equation. Table 3.1 summarizes the forced response for various types of inputs. The constants in the forced response are found by satisfying the given differential equation. For linear combinations of inputs, we find the forced response due to each component and add the results. The form of the natural response depends only on the system details and is independent of the nature of the input. The forced response arises due to the interaction of the system with the input and thus depends on both the input and the system details.

The **characteristic equation** is defined by the polynomial equation

$$1 + A_1 z^{-1} + A_2 z^{-2} + \cdots + A_N z^{-N} = z^N + A_1 z^{N-1} + \cdots + A_N = 0 \qquad (3.15)$$

This equation has $N$ roots, $z_1, z_2, \ldots, z_N$. The natural response is a linear combination of $N$ discrete-time exponentials of the form

$$y_N[n] = K_1 z_1^n + K_2 z_2^n + \cdots + K_N z_N^n \qquad (3.16)$$

This form must be modified for multiple roots. Since complex roots occur in conjugate pairs, their associated constants also form conjugate pairs to ensure that $y_N[n]$ is real. Algebraic details lead to the preferred form with two real constants. Table 3.2 summarizes the preferred forms for multiple or complex roots. The **total response** (sum of the forced and natural response) satisfies the given initial conditions and is used to find the undetermined constants.

**TABLE 3.1 ➤** Form of the Forced Response for Discrete LTI Systems

| Entry | Forcing Function (RHS) | Form of Forced Response |
|---|---|---|
| 1 | $C_0$ (constant) | $C_1$ (another constant) |
| 2 | $\alpha^n$ (see note above) | $C\alpha^n$ |
| 3 | $\cos(n\Omega + \beta)$ | $C_1\cos(n\Omega) + C_2\sin(n\Omega)$ or $C\cos(n\Omega + \phi)$ |
| 4 | $\alpha^n\cos(n\Omega + \beta)$ (see note above) | $\alpha^n[C_1\cos(n\Omega) + C_2\sin(n\Omega)]$ |
| 5 | $n$ | $C_0 + C_1 n$ |
| 6 | $n^p$ | $C_0 + C_1 n + C_2 n^2 + \cdots + C_p n^p$ |
| 7 | $n\alpha^n$ (see note above) | $\alpha^n(C_0 + C_1 n)$ |
| 8 | $n^p\alpha^n$ (see note above) | $\alpha^n(C_0 + C_1 n + C_2 n^2 + \cdots + C_p n^p)$ |
| 9 | $n\cos(n\Omega + \beta)$ | $(C_1 + C_2 n)\cos(n\Omega) + (C_3 + C_4 n)\sin(n\Omega)$ |

*NOTE:* If the right-hand side (RHS) is $\alpha^n$, where $\alpha$ is also a root of the characteristic equation repeated $p$ times, the forced response form must be multiplied by $n^p$.

TABLE 3.2 ➤ Form of the Natural Response for Discrete LTI Systems

| Entry | Root of Characteristic Equation | Form of Natural Response |
|---|---|---|
| 1 | Real and distinct: $r$ | $Kr^n$ |
| 2 | Complex conjugate: $re^{j\Omega}$ | $r^n[K_1\cos(n\Omega)+K_2\sin(n\Omega)]$ |
| 3 | Real, repeated: $r^{p+1}$ | $r^n(K_0+K_1n+K_2n^2+\cdots+K_pn^p)$ |
| 4 | Complex, repeated: $(re^{j\Omega})^{p+1}$ | $r^n\cos(n\Omega)(A_0+A_1n+A_2n^2+\cdots+A_pn^p)+$ $r^n\sin(n\Omega)(B_0+B_1n+B_2n^2+\cdots+B_pn^p)$ |

For stable systems, the natural response is also called the **transient response**, since it decays to zero with time. For a sinusoidal input, the forced response is a sinusoid at the input frequency and is termed the **steady-state response**.

### Response of LTI Systems Described by Difference Equations

**Total Response = Natural Response + Forced Response**

The roots of the characteristic equation determine *only the form* of the natural response.
The input terms (RHS) of the difference equation *completely determine* the forced response.
Initial conditions *satisfy the total response* to yield the constants in the natural response.

**Example 3.7**

### Forced and Natural Response

**(a)** Consider the system shown in Figure E3.7a.

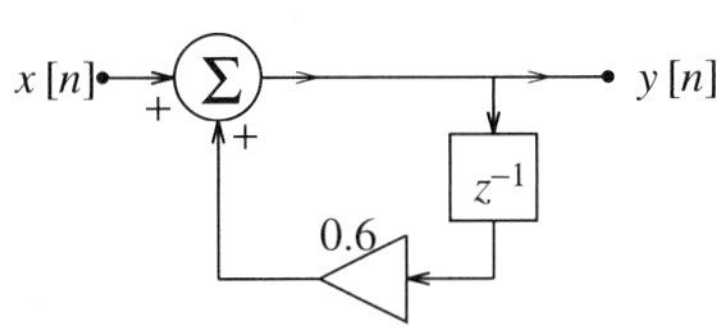


FIGURE E 3.7.a The system for Example 3.7(a)

Find its response if $x[n]=(0.4)^n,\ \ n\ge 0$ and the initial condition is $y[-1]=10$.

The difference equation describing this system is $y[n]-0.6y[n-1]=x[n]=(0.4)^n$, $n\ge 0$.
Its characteristic equation is $1-0.6z^{-1}=0$ or $z-0.6=0$.
Its root $z=0.6$ gives the form of the natural response $y_N[n]=K(0.6)^n$.
Since $x[n]=(0.4)^n$, the forced response is

$$y_F[n]=C(0.4)^n$$

We find $C$ by substituting for $y_F[n]$ into the difference equation:

$$y_F[n]-0.6y_F[n-1]=(0.4)^n=C(0.4)^n-0.6C(0.4)^{n-1}$$

Cancel out $(0.4)^n$ from both sides and solve for $C$ to get

$$C-1.5C=1 \text{ or } C=-2$$

Thus, $y_F[n] = -2(0.4)^n$. The total response is

$$y[n] = y_N[n] + y_F[n] = -2(0.4)^n + K(0.6)^n$$

We use the initial condition $y[-1] = 10$ on the *total response* to find $K$:

$$y[-1] = 10 = -5 + \frac{K}{0.6} \quad \text{and } K = 9$$

Thus,

$$y[n] = -2(0.4)^n + 9(0.6)^n, \quad n \geq 0$$

**(b)** Consider the difference equation $y[n] - 0.5y[n-1] = 5\cos(0.5n\pi), \quad n \geq 0$ with $y[-1] = 4$.

Its characteristic equation is $1 - 0.5z^{-1} = 0$ or $z - 0.5 = 0$.
Its root $z = 0.5$ gives the form of the natural response $y_N[n] = K(0.5)^n$.
Since $x[n] = 5\cos(0.5n\pi)$, the forced response is

$$y_F[n] = A\cos(0.5n\pi) + B\sin(0.5n\pi)$$

We find $y_F[n-1] = A\cos[0.5(n-1)\pi] + B\sin[0.5(n-1)\pi] = A\sin(0.5n\pi) - B\cos(0.5n\pi)$. Then,

$$\begin{aligned} y_F[n] - 0.5y_F[n-1] &= (A + 0.5B)\cos(0.5n\pi) - (0.5A - B)\sin(0.5n\pi) \\ &= 5\cos(0.5n\pi) \end{aligned}$$

Equate the coefficients of the cosine and sine terms to get

$$(A + 0.5B) = 5, \quad (0.5A - B) = 0 \text{ or } A = 4, \ B = 2$$

Thus,

$$y_F[n] = 4\cos(0.5n\pi) + 2\sin(0.5n\pi)$$

The total response is

$$y[n] = K(0.5)^n + 4\cos(0.5n\pi) + 2\sin(0.5n\pi)$$

With $y[-1] = 4$, we find $y[-1] = 4 = 2K - 2$ or $K = 3$, and thus
$y[n] = 3(0.5)^n + 4\cos(0.5n\pi) + 2\sin(0.5n\pi), \ n \geq 0$.
The steady-state response is $4\cos(0.5n\pi) + 2\sin(0.5n\pi)$, and the transient response is $3(0.5)^n$.
The forced response may also be assumed as $y_F[n] = C\cos(0.5n\pi + \theta)$. Then,

$$y_F[n] - 0.5y_F[n-1] = C\cos(0.5n\pi + \theta) - 0.5C\cos[(0.5(n-1)\pi + \theta] = 5\cos(0.5n\pi)$$

The constants $C$ and $\theta$ may be found by using complex number algebra as

$$Ce^{j(0.5n\pi+\theta)} - 0.5Ce^{j(0.5n\pi-0.5\pi+\theta)} = 5e^{j0.5n\pi}$$

This simplifies to

$$Ce^{j\theta} - 0.5Ce^{j(-0.5\pi+\theta)} = 5 \text{ or } Ce^{j\theta}(1 - 0.5e^{-j0.5\pi}) = 5$$

This yields:

$$Ce^{j\theta} = \frac{5}{1 - 0.5e^{-j0.5\pi}} = 4.47e^{-26.6^\circ} \text{ or } C = 4.47 \text{ and } \theta = -26.6^\circ$$

So, the forced response is

$$y_F[n] = C\cos(0.5n\pi + \theta) = 4.47\cos(0.5n\pi - 26.6^\circ)$$

**(c)** Consider the difference equation $y[n] - 0.5y[n-1] = 3(0.5)^n, \quad n \geq 0$ with $y[-1] = 2$.

Its characteristic equation is $1 - 0.5z^{-1} = 0$ or $z - 0.5 = 0$.
Its root, $z = 0.5$, gives the form of the natural response $y_N[n] = K(0.5)^n$.
Since $x[n] = (0.5)^n$ has the same form as the natural response, the forced response is

$$y_F[n] = Cn(0.5)^n$$

We find $C$ by substituting for $y_F[n]$ into the difference equation:

$$y_F[n] - 0.5y_F[n-1] = 3(0.5)^n = Cn(0.5)^n - 0.5C(n-1)(0.5)^{n-1}$$

Cancel out $(0.5)^n$ from both sides and solve for $C$ to get $Cn - C(n-1) = 3$, or $C = 3$. Thus, $y_F[n] = 3n(0.5)^n$. The total response is

$$y[n] = y_N[n] + y_F[n] = K(0.5)^n + 3n(0.5)^n$$

We use the initial condition $y[-1] = 2$ on the *total response* to find $K$:

$$y[-1] = 2 = 2K - 6 \text{ and } K = 4$$

Thus,

$$y[n] = 4(0.5)^n + 3n(0.5)^n = (4 + 3n)(0.5)^n, \quad n \geq 0$$

**(d) A Second-Order System**
Consider the system shown in Figure E.3.7d.

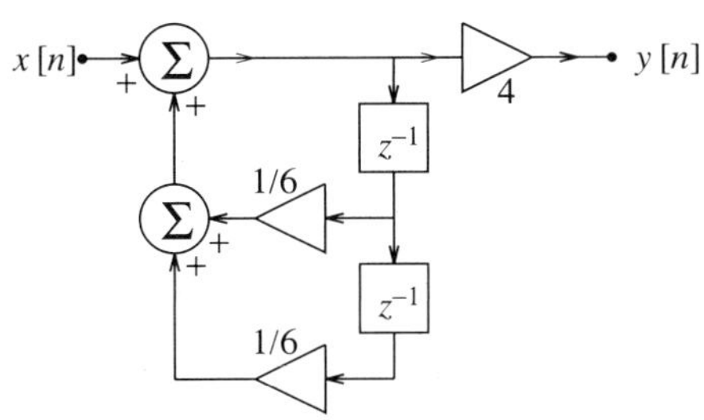


**FIGURE E 3.7.d** The system for Example 3.7(d)

Find the forced and natural response of this system if $x[n] = u[n]$ and $y[-1] = 0$, $y[-2] = 12$.
Comparison with the generic realization reveals that the system difference equation is

$$y[n] - \frac{1}{6}y[n-1] - \frac{1}{6}y[n-2] = 4x[n] = 4u[n]$$

Its characteristic equation is $1 - \frac{1}{6}z^{-1} - \frac{1}{6}z^{-2} = 0$ or $z^2 - \frac{1}{6}z - \frac{1}{6} = 0$. Its roots are $z_1 = \frac{1}{2}$ and $z_2 = -\frac{1}{3}$.
The natural response is thus

$$y_N[n] = K_1(z_1)^n + K_2(z_2)^n = K_1\left(\frac{1}{2}\right)^n + K_2\left(-\frac{1}{3}\right)^n$$

Since the forcing function is $4u[n]$ (a constant for $n \geq 0$), the forced response $y_F[n]$ is constant. Let $y_F[n] = C$. Then $y_F[n-1] = C$, $y_F[n-2] = C$, and

$$y_F[n] - \frac{1}{6}y_F[n-1] - \frac{1}{6}y_F[n-2] = C - \frac{1}{6}C - \frac{1}{6}C = 4$$

This yields $C = 6$.

Thus, $y_F[n] = 6$. The total response $y[n]$ is

$$y[n] = y_N[n] + y_F[n] = K_1\left(\frac{1}{2}\right)^n + K_2\left(-\frac{1}{3}\right)^n + 6$$

To find the constants $K_1$ and $K_2$, we use the initial conditions on the *total response* to obtain $y[-1] = 0 = 2K_1 - 3K_2 + 6$, and $y[-2] = 12 = 4K_1 + 9K_2 + 6$. We find $K_1 = -1.2$ and $K_2 = 1.2$.

Thus,

$$y[n] = -1.2\left(\frac{1}{2}\right)^n + 1.2\left(-\frac{1}{3}\right)^n + 6, \; n \geq 0$$

Its transient response is $-1.2(\frac{1}{2})^n + 1.2(-\frac{1}{3})^n$. Its steady-state response is a constant that equals 6.

---

## Drill Problem 3.9

**(a)** Let $y[n] - 0.8y[n-1] = 2$ with $\frac{1}{2}y[-1] = 5$. Solve for $y[n]$, $n \geq 0$.
**(b)** Let $y[n] + 0.8y[n-1] = 2(0.8)^n$ with $y[-1] = 10$. Solve for $y[n]$, $n \geq 0$.
**(c)** Let $y[n] - 0.8y[n-1] = 2(0.8)^n$ with $y[-1] = 5$. Solve for $y[n]$, $n \geq 0$.
**(d)** Let $y[n] - 0.8y[n-1] = 2(-0.8)^n + 2(0.4)^n$ with $y[-1] = -5$. Solve for $y[n]$, $n \geq 0$.

**Answers: (a)** $10 - 4(0.8)^n$, **(b)** $(0.8)^n - 7(-0.8)^n$, **(c)** $(2n + 6)(0.8)^n$, **(d)** $(0.8)^n + (-0.8)^n - 2(0.4)^n$

## 3.5 Zero-Input Response and Zero-State Response

A linear system is one for which superposition applies and implies that the system is relaxed (with zero initial conditions) and that the system equation involves only linear operators. However, we can use superposition even for a system with nonzero initial conditions that is otherwise linear. We treat it as a multiple-input system by including the initial conditions as additional inputs. The output then equals the superposition of the outputs due to each input acting alone, and any *changes* in the input are related linearly to *changes* in the response. As a result, its response can be written as the sum of a *zero-input response* (due to the initial conditions alone) and the *zero-state response* (due to the input alone). The zero-input response obeys superposition, as does the zero-state response.

It is often more convenient to describe the response $y[n]$ of an LTI system as the sum of its zero-state response (ZSR) $y_{zs}[n]$ (assuming zero initial conditions) and zero-input response (ZIR) $y_{zi}[n]$ (assuming zero input). Each component is found by solving the appropriate difference equation. Note that the natural and forced components $y_N[n]$ and $y_F[n]$ do not, in general, correspond to the zero-input and zero-state response, respectively, even though each pair adds up to the total response.

Consider the general difference equation described by

$$y[n] + A_1 y[n-1] + \cdots + A_N y[n-N] = B_0 x[n] + B_1 x[n-1] + \cdots + B_M x[n-M] \tag{3.17}$$

Its zero-input response $y_{zi}[n]$ is found by setting the right-hand side to zero and computing the response using the given initial conditions. Its zero-input response is found by superposition. We start with the single input system

$$y_0[n] + A_1 y_0[n-1] + A_2 y_0[n-2] + \cdots + A_N y_0[n-N] = x[n] \tag{3.18}$$

If its zero-state response is given by $y_{zs0}[n]$, the zero-state response of the original system is

$$y_{zs}[n] = B_0 y_{zs0}[n] + B_1 y_{zs0}[n-1] + \cdots + B_M y_{zs0}[n-M]$$

The total response $y[n]$ is found by adding the zero-input response and superposed zero-state response

$$y[n] = \underbrace{y_{zi}[n]}_{\text{ZIR}} + \underbrace{B_0 y_{zs0}[n] + B_1 y_{zs0}[n-1] + \cdots + B_M y_{zs0}[n-M]}_{\text{superposed ZSR}} \tag{3.19}$$

Note that the total zero-state response is found by superposition and the ZIR is included just once.

### Solving the General Difference Equation

Let $y[n]+A_1 y[n-1]+\cdots+A_N y[n-N] = B_0 x[n]+B_1 x[n-1]+\cdots+B_M x[n-M]$

1. Find its ZIR $y_{zi}[n]$ using the given initial conditions.
2. Start with the single-input system $y_0[n]$ for $y[n] + A_1 y[n-1] + \cdots + A_N y[n-N] = x[n]$.
3. Find its ZSR as $y_{zs0}[n]$ using zero initial conditions.
4. By superposition, get

$$y[n] = \underbrace{y_{zi}[n]}_{\text{ZIR}} + \underbrace{B_0 y_{zs0}[n] + B_1 y_{zs0}[n-1] + \cdots + B_M y_{zs0}[n-M]}_{\text{superposed ZSR}}$$

**Example 3.8** **Zero-Input and Zero-State Response for the Single-Input Case**

**(a) A First-Order System**

Consider the difference equation $y[n] - 0.6y[n-1] = (0.4)^n,\ n \geq 0$, with $y[-1] = 10$.

Its characteristic equation is $1 - 0.6z^{-1} = 0$ or $z - 0.6 = 0$.
Its root $z = 0.6$ gives the form of the natural response $y_N[n] = K(0.6)^n$.
Since $x[n] = (0.4)^n$, the forced response is

$$y_F[n] = C(0.4)^n$$

We find $C$ by substituting for $y_F[n]$ into the difference equation

$$y_F[n] - 0.6y_F[n-1] = (0.4)^n = C(0.4)^n - 0.6C(0.4)^{n-1}$$

Cancel out $(0.4)^n$ from both sides and solve for $C$ to get $C - 1.5C = 1$ or $C = -2$. Thus,

$$y_F[n] = -2(0.4)^n$$

The total response (subject to initial conditions) is

$$y[n] = y_F[n] + y_N[n] = -2(0.4)^n + K(0.6)^n$$

Note that the value of $K$ is different for different initial conditions.

1. Its ZSR is found from the form of the total response is $y_{zs}[n] = -2(0.4)^n + K_0(0.6)^n$, assuming zero initial conditions:

$$y_{zs}[-1] = 0 = -5 + \frac{K_0}{0.6} \qquad K_0 = 3 \qquad y_{zs}[n] = -2(0.4)^n + 3(0.6)^n, \quad n \geq 0$$

2. Its ZIR is found from the natural response $y_{zi}[n] = K_1(0.6)^n$, where $K_1$ is found from the given initial conditions:

$$y_{zi}[-1] = 10 = \frac{K_1}{0.6} \qquad K_1 = 6 \qquad y_{zi}[n] = 6(0.6)^n, \quad n \geq 0$$

3. The total response is

$$y[n] = y_{zi}[n] + y_{zs}[n] = -2(0.4)^n + 9(0.6)^n,\ n \geq 0$$

**(b) A Second-Order System**

Let $y[n] - \frac{1}{6}y[n-1] - \frac{1}{6}y[n-2] = 4,\ n \geq 0$, with $y[-1] = 0$ and $y[-2] = 12$.
Its characteristic equation is $1 - \frac{1}{6}z^{-1} - \frac{1}{6}z^{-2} = 0$ or $z^2 - \frac{1}{6}z - \frac{1}{6} = 0$. Its roots are $z_1 = \frac{1}{2}$ and $z_2 = -\frac{1}{3}$.

Since the forcing function is a constant for $n \geq 0$, the forced response $y_F[n]$ is constant.

Let $y_F[n] = C$. Then $y_F[n-1] = C$, $y_F[n-2] = C$, and

$$y_F[n] - \frac{1}{6}y_F[n-1] - \frac{1}{6}y_F[n-2] = C - \frac{1}{6}C - \frac{1}{6}C = 4$$

This yields $C = 6$ to give the forced response $y_F[n] = 6$.

1. The ZIR has the form of the natural response because the forced response is zero. Thus, we write,

$$y_{zi}[n] = A_1\left(\frac{1}{2}\right)^n + A_2\left(-\frac{1}{3}\right)^n$$

To find the constants, we use the given initial conditions $y[-1] = 0$ and $y[-2] = 12$:

$$0 = A_1\left(\frac{1}{2}\right)^{-1} + A_2\left(-\frac{1}{3}\right)^{-1} = 2A_1 - 3A_2$$

$$12 = A_1\left(\frac{1}{2}\right)^{-2} + A_2\left(-\frac{1}{3}\right)^{-2} = 4A_1 + 9A_2$$

Thus, $A_1 = 1.2,\ A_2 = 0.8$, and

$$y_{zi}[n] = 1.2\left(\frac{1}{2}\right)^n + 0.8\left(-\frac{1}{3}\right)^n, \quad n \geq 0$$

2. The ZSR has the same form as the total response but the initial conditions are assumed zero. Since the forced response is $y_F[n] = 6$, we have

$$y_{zs}[n] = B_1\left(\frac{1}{2}\right)^n + B_2\left(-\frac{1}{3}\right)^n + 6$$

To find the constants, we assume zero initial conditions, $y[-1]=0$ and $y[-2]=0$, to get

$$y[-1] = 0 = 2B_1 - 3B_2 + 6 \qquad y[-2] = 0 = 4B_1 + 9B_2 + 6$$

We find $B_1 = -2.4$ and $B_2 = 0.4$, and thus

$$y_{zs}[n] = -2.4\left(\frac{1}{2}\right)^n + 0.4\left(-\frac{1}{3}\right)^n + 6, \quad n \geq 0$$

3. The total response is

$$y[n] = y_{zi}[n] + y_{zs}[n] = -1.2\left(\frac{1}{2}\right)^n + 1.2\left(-\frac{1}{3}\right)^n + 6, \quad n \geq 0$$

**(c) Linearity and Superposition of the ZSR and ZIR**

An IIR filter is described by $y[n] - y[n-1] - 2y[n-2] = x[n]$, with $x[n] = 6u[n]$ and initial conditions $y[-1] = -1,\ y[-2] = 4$.

1. Find the zero-input response, zero-state response, and total response.
2. How does the total response change if $y[-1] = -1,\ y[-2] = 4$ as given, but $x[n] = 12u[n]$?
3. How does the total response change if $x[n] = 6u[n]$ as given, but $y[-1] = -2,\ y[-2] = 8$?

**Solution:**

1. We find the characteristic equation as $(1 - z^{-1} - 2z^{-2}) = 0$ or $(z^2 - z - 2) = 0$. The roots of the characteristic equation are $z_1 = -1$ and $z_2 = 2$. The form of the *natural* response is $y_N[n] = A(-1)^n + B(2)^n$. Since the input $x[n]$ is constant for $n \geq 0$, the form of the *forced response* is also constant.

So, choose $y_F[n] = C$ in the system equation and evaluate $C$:

$$y_F[n] - y_F[n-1] - 2y_F[n-2] = C - C - 2C = 6 \quad C = -3 \quad y_F[n] = -3$$

For the ZSR, we use the *form* of the total response and zero initial conditions:

$$y_{zs}[n] = y_F[n] + y_N[n] = -3 + A_0(-1)^n + B_0(2)^n, \quad y[-1] = y[-2] = 0$$

We obtain $y_{zs}[-1] = 0 = -3 - A_0 + 0.5B_0$ and $y_{zs}[-2] = 0 = -3 + A_0 + 0.25B_0$

Thus, $A_0 = 1$, $B_0 = 8$, and $y_{zs}[n] = -3 + (-1)^n + 8(2)^n, \; n \geq 0$.

For the ZIR, we use the *form* of the *natural* response and the *given* initial conditions:

$$y_{zi}[n] = y_N[n] = A_1(-1)^n + B_1(2)^n \qquad y[-1] = -1 \qquad y[-2] = 4$$

This gives $y_{zi}[-1] = -1 = -A_1 + 0.5B_1$, and $y_{zi}[-2] = 4 = A_1 + 0.25B_1$.

Thus, $A_1 = 3$, $B_1 = 4$, and $y_{zi}[n] = 3(-1)^n + 4(2)^n, \; n \geq 0$.

The total response is the sum of the zero-input and zero-state response:

$$y[n] = y_{zi}[n] + y_{zs}[n] = -3 + 4(-1)^n + 12(2)^n, \quad n \geq 0$$

2. If $x[n] = 12u[n]$, the input is now doubled and the zero-state response doubles to $y_{zs}[n] = -6 + 2(-1)^n + 16(2)^n$. We add the original zero-input response to get

$$y[n] = -6 + 2(-1)^n + 16(2)^n + 3(-1)^n + 4(2)^n = -6 + 5(-1)^n + 20(2)^n$$

3. If $y[-1] = -2$ and $y[-2] = 8$, the initial conditions are doubled and the zero-input response doubles to $y_{zi}[n] = 6(-1)^n + 8(2)^n$. We add the original zero-state response to get

$$y[n] = 6(-1)^n + 8(2)^n - 3 + (-1)^n + 8(2)^n = -3 + 7(-1)^n + 16(2)^n$$

# Drill Problem 3.10

**(a)** Let $y[n] - 0.8y[n-1] = 2$. Find its zero-state response.
**(b)** Let $y[n] + 0.8y[n-1] = x[n]$ with $y[-1] = -5$. Find its zero-input response.
**(c)** Let $y[n] - 0.4y[n-1] = (0.8)^n$ with $y[-1] = 10$. Find its zero-state and zero-input response.

**Answers: (a)** $10 - 8(0.8)^n$ **(b)** $4(-0.8)^n$ **(c)** $2(0.8)^n - (0.4)^n$, $4(0.4)^n$

**Example 3.9**

## Response of a General System

Consider the recursive digital filter whose realization is shown in Figure E3.9.

What is the response of this system if $x[n] = 6u[n]$ and $y[-1] = -1, \; y[-2] = 4$?

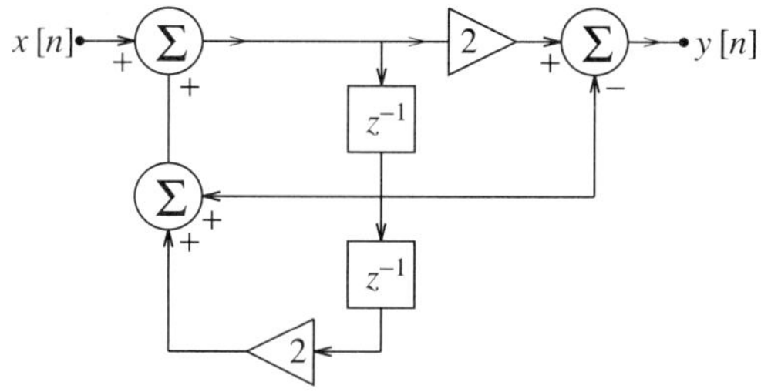


**FIGURE E.3.9** The digital filter for Example 3.9

**Solution:** Comparison with the generic realization reveals that the system difference equation is

$$y[n] - y[n-1] - 2y[n-2] = 2x[n] - x[n-1]$$

From the previous example, the ZSR of $y[n] - y[n-1] - 2y[n-2] = x[n]$ is $y_0[n] = [-3 + (-1)^n + 8(2)^n]u[n]$.

The ZSR for the input $2x[n] - x[n-1]$ is thus

$$\begin{aligned} y_{zs}[n] = 2y_0[n] - y_0[n-1] = &[-6 + 2(-1)^n + 16(2)^n]u[n] \\ &- [3 + (-1)^{n-1} + 8(2)^{n-1}]u[n-1] \end{aligned}$$

From the previous example, the ZIR of $y[n] - y[n-1] - 2y[n-2] = x[n]$ is $y_{zi}[n] = [3(-1)^n + 4(2)^n]u[n]$.

The total response is $y[n] = y_{zi}[n] + y_{zs}[n]$:

$$\begin{aligned} y[n] = &[3(-1)^n + 4(2)^n]u[n] + [-6 + 2(-1)^n + 16(2)^n]u[n] \\ &-[3 + (-1)^{n-1} + 8(2)^{n-1}]u[n-1] \end{aligned}$$

# Drill Problem 3.11

**(a)** Let $y[n] - 0.8y[n-1] = x[n]$ with $x[n] = 2(0.4)^n$ and $y[-1] = -10$. Find its ZSR and ZIR.

**(b)** Let $y[n] - 0.8y[n-1] = 2x[n]$ with $x[n] = 2(0.4)^n$ and $y[-1] = -10$. Find $y[n]$.

**(c)** Let $y[n] - 0.8y[n-1] = 2x[n] - x[n-1]$ with $x[n] = 2(0.4)^n$ and $y[-1] = -10$. Find $y[n]$.

**Answers:** **(a)** $4(0.8)^n - 2(0.4)^n$, $-8(0.8)^n$ **(b)** $-4(0.4)^n$ **(c)** $(0.4)^n - 5(0.8)^n$ (simplified)

## 3.6 The Impulse Response

The impulse response $h[n]$ of a *relaxed* LTI system is simply the response to a unit impulse input $\delta[n]$, as illustrated in Figure 3.4. The impulse response provides us with a powerful method for finding the zero-state response of LTI systems to arbitrary inputs using

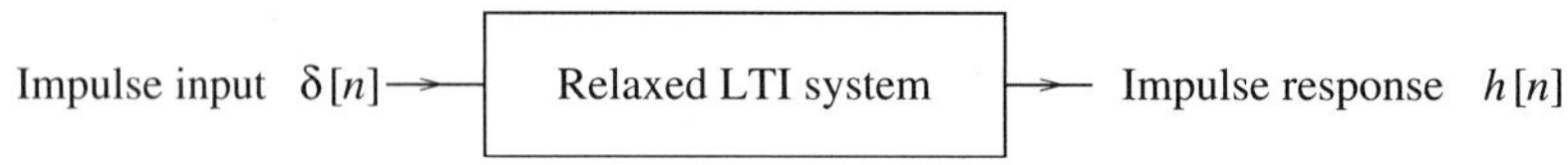


**FIGURE 3.4** The impulse response $h[n]$ of a system is the output if the input is the unit impulse $\delta[n]$. It is defined only for relaxed systems (whose initial conditions are zero) and useful primarily for LTI systems

superposition. The impulse response and the step response are often used to assess the time-domain performance of digital filters. The length of the impulse-response sequence is also used as a means of classification of digital filters. A filter whose impulse response is of finite length is called a **finite impulse response** (FIR) **filter**. A filter whose impulse response is of infinite length is called an **infinite impulse response** (IIR) **filter**.

### 3.6.1 Impulse Response of Nonrecursive Filters

For a nonrecursive filter of length $M + 1$ described by

$$y[n] = B_0x[n] + B_1x[n-1] + \cdots + B_Mx[n-M] \tag{3.20}$$

the impulse response $h[n]$ (with $x[n] = \delta[n]$) is an $M+1$ term sequence of the input terms, which may be written as

$$h[n] = B_0\delta[n] + B_1\delta[n-1] + \cdots + B_M\delta[n-M] \quad \text{or} \quad h[n] = \{\overset{\Downarrow}{B_0},\ B_1, \ldots, B_M\} \tag{3.21}$$

This also describes an FIR filter. The impulse response samples $B_k$ are called the **filter coefficients**.

# Drill Problem 3.12

**(a)** Let $y[n] = 2x[n+1] + 3x[n] - x[n-2]$. Write its impulse response as a sequence.
**(b)** Let $y[n] = x[n] - 2x[n-1] + 4x[n-3]$. Write its impulse response as a sum of impulses.

**Answers: (a)** $h[n] = \{2, \underset{\Uparrow}{3}, 0, -1\}$ **(b)** $h[n] = \delta[n] - 2\delta[n-1] + 4\delta[n-3]$

### 3.6.2 Impulse Response of Recursive Filters

Recursion provides a simple means of obtaining as many terms of the impulse response $h[n]$ of a *relaxed* recursive filter as we please, although we may not always be able to discern a closed form from the results. Remember that we must use zero initial conditions.

**Example 3.10**

### Impulse Response by Recursion

Find the impulse response of the system described by $y[n] - \alpha y[n-1] = x[n]$.

**Solution:** We find $h[n]$ as the solution to $h[n] = \alpha h[n-1] + \delta[n]$ subject to the initial condition $y[-1] = 0$. By recursion, we obtain

$$h[0] = \alpha h[-1] + \delta[0] = 1 \qquad h[2] = \alpha h[1] = \alpha^2$$

$$h[1] = \alpha h[0] = \alpha \qquad h[3] = \alpha h[2] = \alpha^3$$

The general form of $h[n]$ is easy to discern as $h[n] = \alpha^n u[n]$.

# Drill Problem 3.13

**(a)** Let $y[n] + 0.8y[n-1] = x[n]$. Find its impulse response by recursion.
**(b)** Let $y[n] + 0.8y[n-1] = 3x[n]$. Find its impulse response by recursion.
**(c)** Let $y[n] - \alpha y[n-1] = Cx[n]$. Find its impulse response by recursion.

**Answers:** **(a)** $(-0.8)^n$ **(b)** $3(-0.8)^n$ **(c)** $C\alpha^n$

## 3.6.3 General Method for Finding the Impulse Response

Consider the $N$th-order difference equation with a single input:

$$y[n] + A_1 y[n-1] + A_2 y[n-2] + \cdots + A_N y[n-N] = x[n] \tag{3.22}$$

To find its impulse response, we solve the difference equation

$$h[n] + A_1 h[n-1] + A_2 h[n-2] + \cdots + A_N h[n-N] = \delta[n] \quad \text{(zero initial conditions)} \tag{3.23}$$

Since the input $\delta[n]$ is zero for $n > 0$, we must apparently assume a forced response that is zero and thus solve for the natural response using initial conditions (leading to a trivial result). The trick is to use at least one nonzero initial condition, which we must find by recursion. By recursion, we find $h[0] = 1$. Since $\delta[n] = 0,\ n > 0$, the impulse response is found as the *natural response* of the homogeneous equation

$$h[n] + A_1 h[n-1] + A_2 h[n-2] + \cdots + A_N h[n-N] = 0, \quad h[0] = 1 \tag{3.24}$$

subject to the nonzero initial condition $h[0] = 1$. All the other initial conditions are assumed to be zero ($h[-1] = 0$ for a second-order system, $h[-1] = h[-2] = 0$ for a third-order system, and so on).

To find the impulse response of the general system described by

$$\begin{aligned} y[n] + A_1 y[n-1] + A_2 y[n-2] + \cdots + A_N y[n-N] \\ = B_0 x[n] + B_1 x[n-1] + \cdots + B_M x[n-M] \end{aligned} \tag{3.25}$$

we use linearity and superposition as follows:

1. Find the impulse response $h_0[n]$ of the single-input system

$$y_0[n] + A_1 y_0[n-1] + A_2 y_0[n-2] + \cdots + A_N y_0[n-N] = x[n] \qquad (3.26)$$

by solving the homogeneous equation

$$h_0[n] + A_1 h_0[n-1] + \cdots + A_N h_0[n-N] = 0, \quad h_0[0] = 1 \quad \text{(all other conditions zero)} \qquad (3.27)$$

2. Then, invoke superposition to find the actual impulse response $h[n]$ as

$$h[n] = B_0 h_0[n] + B_1 h_0[n-1] + \cdots + B_M h_0[n-M] \qquad (3.28)$$

### Impulse Response of an LTI System Described by a Difference Equation

- If $y[n] + A_1 y[n-1] + \cdots + A_N y[n-N] = x[n]$,
- Find $h[n]$ from $h[n] + A_1 h[n-1] + \cdots + A_N h[n-N] = 0$ with just $h[0] = 1$ (*all others zero*)
- If $y[n]+A_1 y[n-1]+\cdots+A_N y[n-N] = B_0 x[n]+B_1 x[n-1]+\cdots+B_M x[n-M]$,
- Find $h_0[n]$ from $h_0[n] + A_1 h_0[n-1] + \cdots + A_N h_0[n-N] = 0$ with just $h_0[0] = 1$ (*all others zero*)
- Find $h[n]$ (using superposition) as $h[n] = B_0 h_0[n]+B_1 h_0[n-1]+\cdots+B_M h_0[n-M]$

**Example 3.11**

### Impulse Response Computation for the Single-Input Case

**(a) A First-Order System**

Consider the difference equation $y[n] - 0.6y[n-1] = x[n]$.

Its impulse response is found by solving $h[n] - 0.6h[n-1] = 0$, $h[0] = 1$.
Its natural response is

$$h[n] = K(0.6)^n$$

With $h[0] = 1$, we find $K = 1$, and thus

$$h[n] = (0.6)^n u[n]$$

**(b) A Second-Order System**

Let $y[n] - \frac{1}{6}y[n-1] - \frac{1}{6}y[n-2] = x[n]$.

Its impulse response is found by solving $h[n] - \frac{1}{6}h[n-1] - \frac{1}{6}h[n-2] = 0$, $h[0] = 1$, $h[-1] = 0$.

Its characteristic equation is $1 - \frac{1}{6}z^{-1} - \frac{1}{6}z^{-2} = 0$ or $z^2 - \frac{1}{6}z - \frac{1}{6} = 0$.

Its roots, $z_1 = \frac{1}{2}$ and $z_2 = -\frac{1}{3}$, give the natural response

$$h[n] = K_1\left(\frac{1}{2}\right)^n + K_2\left(-\frac{1}{3}\right)^n$$

With $h[0] = 1$ and $h[-1] = 0$, we find $1 = K_1 + K_2$ and $0 = 2K_1 - 3K_2$. Solving for the constants, we obtain $K_1 = 0.6$ and $K_2 = 0.4$ Thus,

$$h[n] = \left[0.6\left(\frac{1}{2}\right)^n + 0.4\left(-\frac{1}{3}\right)^n\right]u[n]$$

# Drill Problem 3.14

**(a)** Let $y[n] - 0.9y[n-1] = x[n]$. Find its impulse response $h[n]$.
**(b)** Let $y[n] - 1.2y[n-1] + 0.32y[n-2] = x[n]$. Find its impulse response $h[n]$.

**Answers: (a)** $(0.9)^n$ **(b)** $2(0.8)^n - (0.4)^n$

**Example 3.12**

### Impulse Response for the General Case

**(a)** Find the impulse response of $y[n] - 0.6y[n-1] = 4x[n]$ and $y[n] - 0.6y[n-1] = 3x[n+1] - x[n]$.

We start with the single-input system

$$y_0[n] - 0.6y_0[n-1] = x[n]$$

Its impulse response $h_0[n]$ was found in the previous example as $h_0[n] = (0.6)^n u[n]$. Then, for the first system,

$$h[n] = 4h_0[n] = 4(0.6)^n u[n]$$

For the second system,

$$h[n] = 3h_0[n+1] - h_0[n] = 3(0.6)^{n+1}u[n+1] - (0.6)^n u[n]$$

This may also be expressed as

$$h[n] = 3\delta[n+1] + 0.8(0.6)^n u[n]$$

*Comment:* The general approach can be used for causal or noncausal systems.

**(b)** Let $y[n] - \frac{1}{6}y[n-1] - \frac{1}{6}y[n-2] = 2x[n] - 6x[n-1]$.
To find $h[n]$, start with the single-input system

$$y[n] - \frac{1}{6}y[n-1] - \frac{1}{6}y[n-2] = x[n]$$

Its impulse response $h_0[n]$ was found in the previous example as

$$h_0[n] = \left[0.6\left(\frac{1}{2}\right)^n + 0.4\left(-\frac{1}{3}\right)^n\right]u[n]$$

The impulse response of the given system is $h[n] = 2h_0[n] - 6h_0[n-1]$. This gives

$$h[n] = [1.2\left(\frac{1}{2}\right)^n + 0.8\left(-\frac{1}{3}\right)^n]u[n] - \left[3.6\left(\frac{1}{2}\right)^{n-1} + 2.4\left(-\frac{1}{3}\right)^{n-1}\right]u[n-1]$$

*Comment:* This may be simplified to $h[n] = [-6(\frac{1}{2})^n + 8(-\frac{1}{3})^n]u[n]$.

# Drill Problem 3.15

**(a)** Let $y[n] - 0.5y[n-1] = 2x[n] + x[n-1]$. Find its impulse response $h[n]$.
**(b)** Let $y[n] - 1.2y[n-1] + 0.32y[n-2] = x[n] + 2x[n-1]$. Find its impulse response $h[n]$.

**Answers: (a)** $2(0.5)^n u[n] + (0.5)^{n-1}u[n-1] = 4(0.5)^n u[n] - 2\delta[n]$
**(b)** $7(0.8)^n - 6(0.4)^n$ (simplified)

## 3.6.4 Impulse Response of Anti-Causal Systems

So far, we have focused on the response of systems described by difference equations to causal inputs. However, by specifying an anti-causal input and appropriate initial conditions, the same difference equation can be solved backward in time for $n < 0$ to generate an anti-causal response. For example, to find the causal impulse response $h[n]$, we assume that $h[n] = 0,\ n < 0$; but to find the anti-causal impulse response $h_A[n]$ of the same system, we would assume that $h[n] = 0,\ n > 0$. This means that the same system can be described by two different impulse response functions. How we distinguish between them is easily handled using the $z$-transform (described in the next chapter).

**Example 3.13**

### Causal and Anti-Causal Impulse Response

**(a)** Find the causal impulse response of the first-order system $y[n] - 0.4y[n-1] = x[n]$.

For the causal impulse response, we assume $h[n] = 0,\ n < 0$, and solve for $h[n],\ n > 0$, by recursion from $h[n] = 0.4h[n-1] + \delta[n]$. With $h[0] = 0.4h[-1] + \delta[0] = 1$ and $\delta[n] = 0,\ n \neq 0$, we find

$$h[1] = 0.4h[0] = 0.4 \qquad h[2] = 0.4h[1] = (0.4)^2$$

$$h[3] = 0.4h[2] = (0.4)^3 \quad \text{(etc.)}$$

The general form is easily discerned as $h[n] = (0.4)^n$ and is valid for $n \geq 0$.
*Comment:* The causal impulse response of $y[n] - \alpha y[n-1] = x[n]$ is $h[n] = \alpha^n u[n]$.

**(b)** Find the anti-causal impulse response of the first-order system $y[n] - 0.4y[n-1] = x[n]$.

For the anti-causal impulse response, we assume $h[n] = 0,\ n \geq 0$, and solve for $h[n],\ n < 0$, by recursion from $h[n-1] = 2.5(h[n] - \delta[n])$.

With $h[-1] = 2.5(h[0] - \delta[0]) = -2.5$, and $\delta[n] = 0,\ n \neq 0$, we find

$$h[-2] = 2.5h[-1] = -(2.5)^2$$
$$h[-3] = 2.5h[-2] = -(2.5)^3$$
$$h[-4] = 2.5h[-3] = -(2.5)^4 \quad \text{(etc.)}$$

The general form is easily discerned as $h[n] = -(2.5)^{-n} = -(0.4)^n$ and is valid for $n \leq -1$.

*Comment:* The anti-causal impulse response of $y[n] - \alpha y[n-1] = x[n]$ is $h[n] = -\alpha^n u[-n-1]$.

# Drill Problem 3.16

Let $y[n] - 0.5y[n-1] = 2x[n]$. Find its causal impulse response and anti-causal impulse response.

**Answers:** $h_c[n] = 2(0.5)^n u[n]$, $h_{ac}[n] = -2(0.5)^n u[-n-1]$

# 3.7 System Representation in Various Forms

An LTI system may be described by a difference equation, impulse response, or input-output data. All three are related and, given one form, we should be able to access the others. We have already studied how to obtain the impulse response from a difference equation. Here we shall describe how to represent a nonrecursive filter as a recursive filter, and how to obtain the system difference equation from its impulse response or from input-output data.

## 3.7.1 Recursive Forms for Nonrecursive Digital Filters

The terms *FIR* and *nonrecursive* are synonymous. A nonrecursive filter always has a finite impulse response. The terms *IIR* and *recursive* are often, but not always, synonymous. Not all recursive filters have an impulse response of infinite length. In fact, nonrecursive filters can always be implemented in recursive form if desired. A recursive filter may also be approximated by a nonrecursive filter of the form $y[n] = B_0 x[n] + B_1 x[n-1] + \cdots + B_M x[n-M]$ by truncating its impulse response sequence. The approximation improves as $M$ becomes large.

**Example 3.14** **Recursive Forms for Nonrecursive Filters**

**(a)** Consider the recursive system $y[n] = y[n-1] + x[n] - x[n-3]$.

If $x[n]$ equals $\delta[n]$ and $y[-1] = 0$, we successively obtain

$$y[0] = y[-1] + \delta[0] - \delta[-3] = 1 \qquad y[3] = y[2] + \delta[3] - \delta[0] = 1 - 1 = 0$$
$$y[1] = y[0] + \delta[1] - \delta[-2] = 1 \qquad y[4] = y[3] + \delta[4] - \delta[1] = 0$$
$$y[2] = y[1] + \delta[2] - \delta[-1] = 1 \qquad y[5] = y[4] + \delta[5] - \delta[2] = 0$$

The impulse response of this "recursive" filter is zero after the first three values and has a finite length. It is actually a nonrecursive (FIR) filter in disguise!

**(b)** Consider the nonrecursive filter $y[n] = x[n] + x[n-1] + x[n-2]$.

Its impulse response is

$$h[n] = \delta[n] + \delta[n-1] + \delta[n-2]$$

To cast this filter in recursive form, we compute
$y[n-1] = x[n-1] + x[n-2] + x[n-3]$.
Upon subtraction from the original equation, we obtain the recursive form

$$y[n] - y[n-1] = x[n] - x[n-3]$$

This describes a recursive formulation for the given nonrecursive, FIR filter.

## Drill Problem 3.17

Consider the nonrecursive filter $y[n] = x[n]+x[n-1]+x[n-2]$. What recursive filter do you obtain by computing $y[n]-y[n-2]$. Does the impulse response of the recursive filter match the impulse response of the nonrecursive filter?

**Answers:** $y[n] - y[n-2] = x[n] + x[n-1] - x[n-3] - x[n-4]$,
(the impulse responses match)

### 3.7.2 Difference Equations from the Impulse Response

In the time domain, the process of finding the difference equation from its impulse response is tedious. It is much easier implemented by other methods (such as the $z$-transform). The central idea is that the terms in the impulse response are an indication of the natural response (and the roots of the characteristic equation) from which the difference equation may be reconstructed if we can describe the combination of the impulse response and its delayed versions by a sum of impulses. The process is best illustrated by some examples.

**Example 3.15** **Difference Equations from the Impulse Response**

**(a)** Let $h[n] = u[n]$. Then $h[n-1] = u[n-1]$, and
$h[n] - h[n-1] = u[n] - u[n-1] = \delta[n]$.

The difference equation corresponding to $h[n] - h[n-1] = \delta[n]$ is simply

$$y[n] - y[n-1] = x[n]$$

**(b)** Let $h[n] = 3(0.6)^n u[n]$.

This suggests a difference equation whose left-hand side is $y[n] - 0.6y[n-1]$. We then set up $h[n] - 0.6h[n-1] = 3(0.6)^n u[n] - 1.8(0.6)^{n-1}u[n-1]$. This simplifies to

$$\begin{aligned} h[n] - 0.6h[n-1] &= 3(0.6)^n u[n] - 3(0.6)^n u[n-1] \\ &= 3(0.6)^n (u[n] - u[n-1]) = 3(0.6)^n \delta[n] = 3\delta[n] \end{aligned}$$

The difference equation corresponding to $h[n] - 0.6h[n-1] = 3\delta[n]$ is $y[n] - 0.6y[n-1] = 3x[n]$.

**(c)** Let $h[n] = 2(-0.5)^n u[n] + (0.5)^n u[n]$.

This suggests a characteristic equation $(z - 0.5)(z + 0.5)$. The left-hand side of the difference equation is thus $y[n] - 0.25y[n-2]$. We now compute

$$\begin{aligned} h[n] - 0.25h[n-2] = {} & 2(-0.5)^n u[n] + (0.5)^n u[n] \\ & - 0.25(2(-0.5)^{n-2}u[n-2] + (0.5)^{n-2}u[n-2]) \end{aligned}$$

This simplifies to

$$h[n] - 0.25h[n-2] = [2(-0.5)^n + (0.5)^n](u[n] - u[n-2])$$

Since $u[n] - u[n-2]$ has just two samples (at $n = 0$ and $n = 1$), it equals $\delta[n] + \delta[n-1]$, and we get

$$h[n] - 0.25h[n-2] = [2(-0.5)^n + (0.5)^n](\delta[n] + \delta[n-1])$$

This simplifies further to $h[n] - 0.25h[n-2] = 3\delta[n] - 0.5\delta[n-1]$. From this result, the difference equation is

$$y[n] - 0.25y[n-2] = 3x[n] - 0.5x[n-1]$$

# Drill Problem 3.18

**(a)** Set up the difference equation corresponding to the impulse response $h[n] = 2(-0.5)^n u[n]$.

**(b)** Set up the difference equation corresponding to the impulse response $h[n] = (0.5)^n u[n] + \delta[n]$.

**Answers: (a)** $y[n] + 0.5y[n-1] = 2x[n]$
**(b)** $y[n] - 0.5y[n-1] = 2x[n] - 0.5x[n-1]$

## 3.7.3 Difference Equations from Input-Output Data

The difference equation of LTI systems may also be obtained from input-output data. The response of the system described by $y[n] = 3x[n] + 2x[n-1]$ to $x[n] = \delta[n]$ is $y[n] = 3\delta[n] + 2\delta[n-1]$. Turning things around, the input $\delta[n]$ and output $3\delta[n] + 2\delta[n-1]$ then corresponds to the difference equation $y[n] = 3x[n] + 2x[n-1]$. Note how the coefficients of the input match the output data (and vice versa). Similarly, if the input to an LTI system is $x[n] = \{\overset{\Downarrow}{1},\ 2,\ 3\}$ and the output is $y[n] = \{\overset{\Downarrow}{3},\ 3\}$, the system equation may be written as $y[n] + 2y[n-1] + 3y[n-2] = 3x[n] + 3x[n-1]$.

## Drill Problem 3.19

**(a)** The impulse response of an LTI system is $h[n] = \{\overset{\Downarrow}{1},\ 2,\ -1\}$. What is the system equation?

**(b)** The input $\{\overset{\Downarrow}{2},\ 4\}$ to an LTI system gives the output $\{\overset{\Downarrow}{3},\ -1\}$. What is the system equation?

**Answers: (a)** $y[n] = x[n] + 2x[n-1] - x[n-2]$
**(b)** $2y[n] + 4y[n-1] = 3x[n] - x[n-1]$

# 3.8 Application-Oriented Examples

In this section, we explore various practical applications of digital filters, such as signal smoothing using averaging filters, echo cancellation using inverse filters, special audio effects using echo and reverb, and wave-table synthesis of musical tones for synthesizers.

### 3.8.1 Moving Average Filters

A **moving average** or **running average** filter is an FIR filter that replaces a signal value by an average of its neighboring values. The averaging process blurs the sharp details of the signal and results in an output that is a smoother version of the input. Moving average filtering is a simple way of smoothing a noisy signal and improving its signal-to-noise ratio. A causal $L$-point moving average (or averaging) filter replaces a signal value by the average of its past $L$ samples and is defined by

$$y[n] = \frac{1}{L}[x[n] + x[n-1] + x[n-2] + \cdots + x[n-(L-1)]]$$

$$= \frac{1}{L}\sum_{k=0}^{L-1} x[n-k] \tag{3.29}$$

This is an $L$-point FIR filter whose impulse response is

$$h[n] = \frac{1}{L}\{\underbrace{\overset{\Downarrow}{1},\ 1,\ 1,\ \ldots,\ 1,\ 1}_{L \text{ samples}}\} \tag{3.30}$$

Figure 3.5 shows a noisy sinusoid and the output when it is passed through two moving average filters of different lengths. We see that the output of the 10-point filter is indeed a smoother version of the noisy input. Note that the output is a delayed version of the input and shows a start-up transient for about 10 samples before the filter output settles down. This is typical of all filters. A filter of longer length will result in a longer transient. It should also produce better smoothing. While that may be generally true, we see that the 50-point averaging

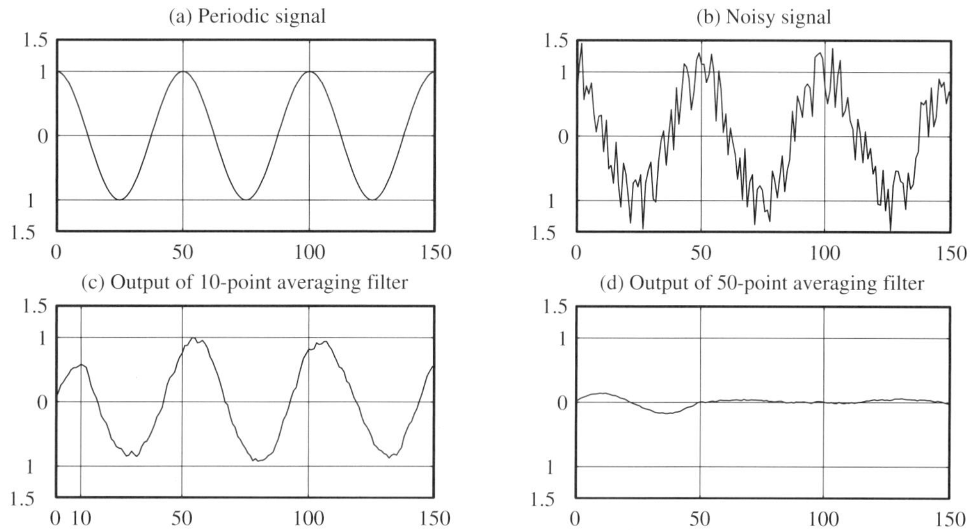


**FIGURE 3.5** The response of two averaging filters to a noisy sinusoid. The 10-point averaging filter results in a smoothing of the signal as expected. Since the sinusoid has a period of 50 samples, the 50-point averaging filter yields zero output (average of 50 successive samples) for the sinusoid and all we see is just the start-up transient and the 50-point average of the noise

filter produces an output that is essentially zero after the initial transient. The reason is that the sinusoid also has a period of 50 samples and its 50-point running average is thus zero! This suggests that while averaging filters of longer lengths usually do a better job of signal smoothing, the peculiarities of a given signal may not always result in a useful output.

**Finding the Inverse of an LTI System? Try Switching the Input and Output**

**System:** $y[n] + 2y[n-1] = 3x[n] + 4x[n-1]$
**Inverse system:** $3y[n] + 4y[n-1] = x[n] + 2x[n-1]$

How to find the inverse from the impulse response $h[n]$? Find the difference equation first.

## 3.8.2 Inverse Systems

Inverse systems are quite important in practical applications. For example, a measurement system (such as a transducer) invariably affects (distorts) the signal being measured. To undo the effects of the distortion requires a system that acts as the inverse of the measurement system. If an input $x[n]$ to a system results in an output $y[n]$, then its inverse is a system that recovers the signal $x[n]$ in response to the input $y[n]$, as illustrated in Figure 3.6. For invertible LTI systems described by difference equations, finding the inverse system is as easy as switching the input and output variables, as illustrated in the following example.

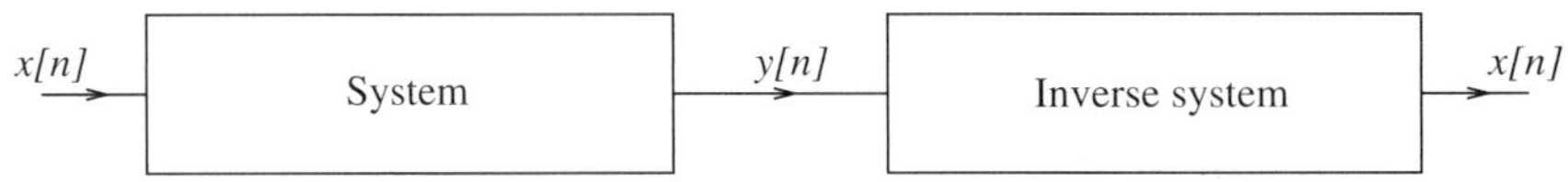


**FIGURE 3.6** A system and its inverse. An input $x[n]$ to the system results in an output $y[n]$. The inverse system is then used to recover the original signal $x[n]$

**Example 3.16**

### Inverse Systems

**(a)** Refer to the interconnected system shown in Figure E3.16a(1). Find the difference equation of the inverse system, sketch a realization of each system, and find the output of each system.

**FIGURE E.3.16.a(1)** The interconnected system for Example 3.16(a)

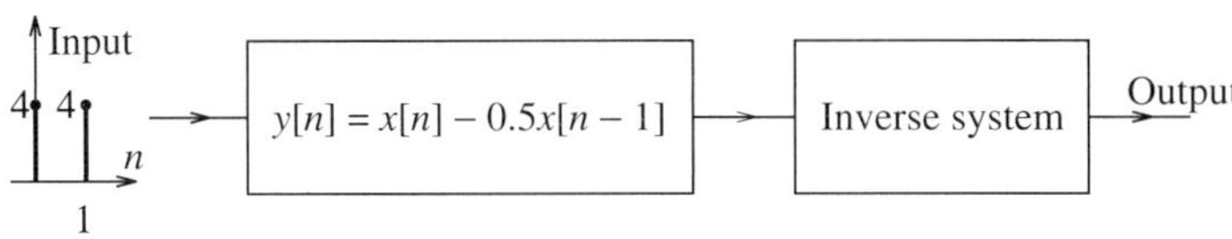


The original system is described by $y[n] = x[n] - 0.5x[n-1]$. By switching the input and output, the inverse system is described by $y[n] - 0.5y[n-1] = x[n]$. The realization of each system is shown in Figure E3.16a(2). Are they related? Yes. If you flip the realization of the echo system end-on-end *and change the sign of the feedback signal*, you get the inverse realization.

**FIGURE E.3.16.a(2)** Realization of the system and its inverse for Example 3.16(a)

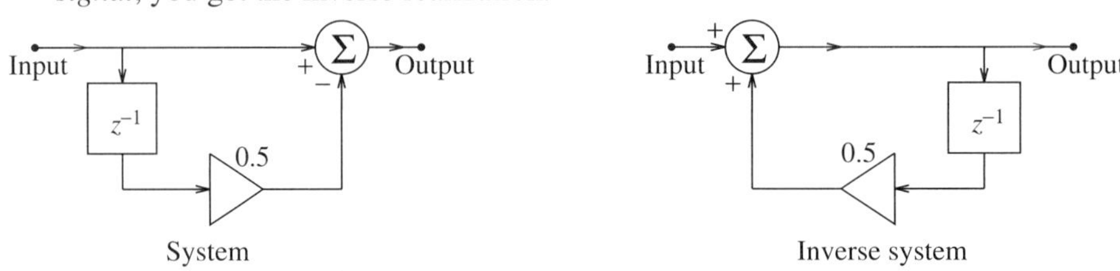


The response $g[n]$ of the first system is simply

$$g[n] = (4\delta[n]+4\delta[n-1]) - (2\delta[n-1]+2\delta[n-2]) = 4\delta[n]+2\delta[n-1]) - 2\delta[n-2])$$

If we let the output of the second system be $y_0[n]$, we have

$$y_0[n] = 0.5y_0[n-1] + 4\delta[n] + 2\delta[n-1]) - 2\delta[n-2])$$

Recursive solution gives

$$y_0[0] = 0.5y_0[-1] + 4\delta[0] = 4$$
$$y_0[1] = 0.5y_0[0] + 2\delta[0] = 4$$
$$y_0[2] = 0.5y_0[1] - 2\delta[0] = 0$$

All subsequent values of $y_0[n]$ are zero since the input terms are zero for $n > 2$. The output is thus $y_0[n] = \{\overset{\Downarrow}{4}, 4\}$, the same as the input to the overall system.

**(b)** Consider the FIR filter $y[n] = x[n] + 2x[n-1]$.

Its inverse is found by switching the input and output as $y[n] + 2y[n-1] = x[n]$. The inverse of an FIR filter always results in an IIR filter (while the converse is not always true).

# Drill Problem 3.20

**(a)** Consider the IIR filter $y[n] + 2y[n-1] = x[n]$. What is the impulse response of the inverse system? Is the inverse system FIR or IIR?
**(b)** Consider the IIR filter $y[n] + 2y[n-1] = x[n] + x[n-1]$. Is the inverse system FIR or IIR?
**(c)** Consider the FIR filter whose impulse response is $h[n] = \{\overset{\Downarrow}{1}, \ -0.4\}$. What is the impulse response of the inverse system? Is the inverse system FIR or IIR?

**Answers: (a)** $h[n] = \{\underset{\Uparrow}{1}, \ 2\}$, FIR **(b)** $y[n] + y[n-1] = x[n] + 2x[n-1]$, IIR
**(c)** $(0.4)^n u[n]$, IIR

## Invertible Systems

Not all systems have an inverse. For a system to have an inverse (or be **invertible**) distinct inputs must lead to distinct outputs. If a system produces an identical output for two different inputs, it does not have an inverse. For example, the system described by $y[n] = \cos(x[n])$ is not invertible because different inputs (such as $x[n] + 2k\pi, \ k = 0, \pm 1, \pm 2, \ldots$) yield an identical input.

**Example 3.17**

### Invertibility

**(a)** The nonlinear system $y[n] = x^2[n]$ does not have an inverse. Two inputs, differing in sign, yield the same output. If we try to recover $x[n]$ as $\sqrt{y[n]}$, we run into a sign ambiguity.

**(b)** The linear (but time-varying) decimating system $y[n] = x[2n]$ does not have an inverse. Two inputs, which differ in the samples discarded (for example, the signals $\{1, 2, 4, 5\}$ and $\{1, 3, 4, 8\}$ yield the same output $\{1, \ 4\}$). If we try to recover the original signal by interpolation, we cannot uniquely identify the original signal.

**(c)** The linear (but time-varying) interpolating system $y[n] = x^{\uparrow}[n/2]$ does have an inverse. Its inverse is a decimating system that discards the very samples inserted during interpolation and thus recovers the original signal exactly.

# Drill Problem 3.21

**(a)** Is the system $y[n] = e^{x[n]}$ invertible? If so, what is the inverse system?
**(b)** Is the system described by the impulse response $h[n] = (0.5)^n u[n]$ invertible? If so, what is the impulse response of the inverse system?

**Answers: (a)** Yes, $y[n] = \ln(x[n])$ **(b)** Yes, $h[n] = \delta[n] - 0.5\delta[n-1]$

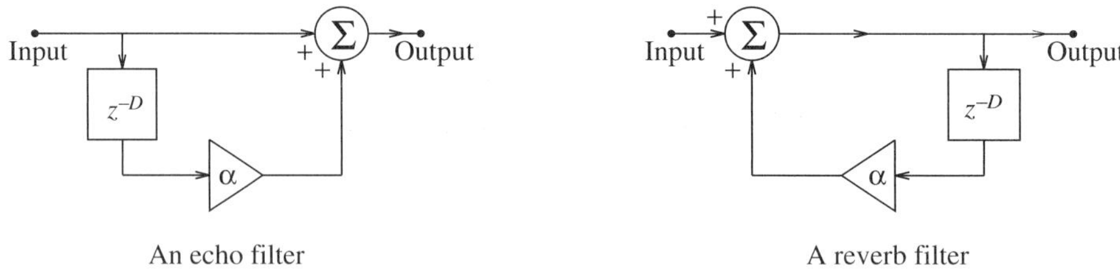


**FIGURE 3.7** A simple realization of first-order echo and reverb filters

### 3.8.3 Echo and Reverb

The digital processing of audio signals often involves digital filters to create various special effects such as echo and reverb, which are typical of modern-day studio sounds. An echo filter has the form

$$y[n] = x[n] + \alpha x[n-D] \quad \text{(an echo filter)} \tag{3.31}$$

This describes an FIR filter whose output $y[n]$ equals the input $x[n]$ and its delayed (by $D$ samples) and attenuated (by $\alpha$) replica of $x[n]$ (the echo term). Its realization is sketched in Figure 3.7. The $D$-sample delay is implemented by a cascade of $D$ delay elements and represented by the block marked $z^{-D}$. This filter is also called a **comb filter** (for reasons to be explained in later chapters).

Reverberations are due to multiple echoes (from the walls and other structures in a concert hall, for example). For simplicity, if we assume that the signal suffers the same delay $D$ and the same attenuation $\alpha$ in each round-trip to the source, we may describe the action of reverb by

$$y[n] = x[n] + \alpha x[n-D] + \alpha^2 x[n-2D] + \alpha^3 x[n-3D] + \cdots \tag{3.32}$$

If we delay both sides by $D$ units, we get

$$\alpha y[n-D] = \alpha x[n-D] + \alpha^2 x[n-2D] + \alpha^3 x[n-3D] + \alpha^4 x[n-4D] + \cdots \tag{3.33}$$

Subtracting the second equation from the first, we obtain a compact form for a reverb filter:

$$y[n] - \alpha y[n-D] = x[n] \quad \text{(a reverb filter)} \tag{3.34}$$

This is an IIR filter whose realization is also sketched in Figure 3.7. Its form is reminiscent of the inverse of the echo system $y[n] + \alpha y[n-D] = x[n]$, but with $\alpha$ replaced by $-\alpha$.

In concept, it should be easy to tailor the simple reverb filter to simulate realistic effects by including more terms with different delays and attenuation. In practice, however, this is no easy task, and the filter designs used by commercial vendors in their applications are often proprietary.

### 3.8.4 Periodic Sequences and Wave-Table Synthesis

Electronic music synthesizers typically possess *tone banks* of stored tone sequences that are replicated and used in their original or delayed forms or combined to synthesize new sounds and generate various musical effects. The periodic versions of such sequences can be generated by filters that have a recursive form developed from a nonrecursive filter whose impulse response corresponds to one period of the periodic signal. The difference equation of such a recursive filter is given by

$$y[n] - y[n - N] = x_1[n] \quad (N\text{-sample periodic signal generator}) \qquad (3.35)$$

where $x_1[n]$ corresponds to one period ($N$ samples) of the signal $x[n]$. This form actually describes a reverb system with no attenuation whose delay equals the period $N$. Hardware implementation often uses a circular buffer or *wave-table* (in which one period of the signal is stored), and cycling over it generates the periodic signal. The same wave-table can also be used to change the frequency (or period) of the signal (to double the frequency for example, we would cycle over alternate samples) or for storing a new signal.

**Example 3.18**

### Generating Periodic Signals Using Recursive Filters

Suppose we wish to generate the periodic signal $x[n]$ described by

$$x[n] = \{2, 3, 1, 6, 2, 3, 1, 6, 2, 3, 1, 6, 2, 3, 1, 6, \ldots\}$$

**Solution:** The impulse response sequence of a nonrecursive filter that generates this signal is simply

$$h[n] = \{2, 3, 1, 6, 2, 3, 1, 6, 2, 3, 1, 6, 2, 3, 1, 6, \ldots\}$$

If we delay this sequence by one period (four samples), we obtain

$$h[n-4] = \{0, 0, 0, 0, 2, 3, 1, 6, 2, 3, 1, 6, 2, 3, 1, 6, 2, 3, 1, 6, \ldots\}$$

Subtracting the delayed sequence from the original gives us $h[n] - h[n-4] = \{2, 3, 1, 6\}$. Its recursive form is $y[n] - y[n-4] = x_1[n]$, where $x_1[n] = \{2, 3, 1, 6\}$ describes one period of $x[n]$.

# Drill Problem 3.22

**(a)** What is the system equation of a filter whose impulse response is periodic with first period $\{\overset{\Downarrow}{1}, 2\}$.

**(b)** Find the response of the system $y[n] - y[n-3] = 2\delta[n] + 3\delta[n-2]$.

**Answers: (a)** $y[n] - y[n-2] = x[n] + 2x[n-1]$
**(b)** Periodic ($N = 3$) with period $y_1[n] = \{\overset{\Downarrow}{2}, 0, 3\}$

## 3.8.5 How Difference Equations Arise

We conclude this section with some examples of difference equations, which arise in many ways in various fields ranging from mathematics and engineering to economics and biology.

1. $y[n] = y[n-1] + n, \ y[-1] = 1$
   This difference equation describes the number of regions $y[n]$ into which $n$ lines divide a plane if no two lines are parallel and no three lines intersect.

2. $y[n+1] = (n+1)(y[n]+1), \ y[0] = 0$
   This difference equation describes the number of multiplications $y[n]$ required to compute the determinant of an $n \times n$ matrix using cofactors.

3. $y[n+2] = y[n+1] + y[n], \ y[0] = 0, \ y[1] = 1$
   This difference equation generates the Fibonacci sequence $\{y[n] = 0, \ 1, \ 1, \ 2, \ 3, \ 5, \ \ldots\}$, where each number is the sum of the previous two.

4. $y[n+2] - 2xy[n+1] + y[n] = 0, \ y[0] = 1, \ y[1] = x$
   This difference equation generates the **Chebyshev polynomials** $T_n(x) = y[n]$ *of the first kind*. We find that $T_2(x) = y[2] = 2x^2 - 1, \ T_3(x) = y[3] = 4x^3 - 3x$, etc. Similar difference equations called *recurrence relations* form the basis for generating other polynomial sets.

5. $y[n+1] = \alpha y[n](1 - y[n])$
   This difference equation (called a *logistic equation* in biology) is used to model the growth of populations that reproduce at discrete intervals.

6. $y[n+1] = (1+\alpha)y[n] + d[n]$
   This difference equation describes the bank balance $y[n]$ at the beginning of the $n$th-compounding period (day, month, etc.) if the percent interest rate is $\alpha$ per compounding period and $d[n]$ is the amount deposited in that period.

## 3.9 Discrete Convolution

Discrete-time convolution is a method of finding the *zero-state response of relaxed* linear time-invariant (LTI) systems. It is based on the concepts of linearity and time invariance and assumes that the system information is known in terms of its impulse response $h[n]$. In other words, if the input is $\delta[n]$, a unit sample at the origin $n = 0$, the system response is $h[n]$. Now, if the input is $x[0]\delta[n]$ (a scaled impulse at the origin) the response is $x[0]h[n]$ (by linearity). Similarly, if the input is the shifted impulse $x[1]\delta[n-1]$ at $n = 1$, the response is $x[1]h[n-1]$ (by time invariance). The response to the shifted impulse $x[k]\delta[n-k]$ at $n = k$ is $x[k]h[n-k]$ (by linearity and time invariance). Since an arbitrary input $x[n]$ is simply a sequence of samples, it can be described by a sum of scaled and shifted impulses:

$$x[n] = \sum_{k=-\infty}^{\infty} x[k]\delta[n-k] \tag{3.36}$$

By superposition, the response to $x[n]$ is the sum of scaled and shifted versions of the impulse response:

$$y[n] = \sum_{k=-\infty}^{\infty} x[k]h[n-k] = x[n] * h[n] \tag{3.37}$$

This defines the convolution operation and is also called **linear convolution** or the **convolution sum**, and denoted by $x[n] * h[n]$ in this book. The order in which we perform the operation does not matter, and we can interchange the arguments of $x$ and $h$ without affecting the result.

$$y[n] = \sum_{k=-\infty}^{\infty} x[k]h[n-k] = \sum_{k=-\infty}^{\infty} x[n-k]h[k] \quad \text{or} \quad y[n] = x[n] * h[n] = h[n] * x[n] \tag{3.38}$$

### 3.9.1 Analytical Evaluation of Discrete Convolution

If $x[n]$ and $h[n]$ are described by simple enough analytical expressions, the convolution sum can be implemented quite readily to obtain closed-form results. While evaluating the convolution sum, it is useful to keep in mind that $x[k]$ and $h[n-k]$ are functions of the summation variable $k$. For causal signals of the form $x[n]u[n]$ and $h[n]u[n]$, the summation involves step functions of the form $u[k]$ and $u[n-k]$. Since $u[k] = 0,\ k < 0$ and $u[n-k] = 0,\ k > n$, these can be used to simplify the lower and upper summation limits to $k = 0$ and $k = n$, respectively.

**Example 3.19**

**Analytical Evaluation of Discrete Convolution**

**(a)** Let $x[n] = h[n] = u[n]$. Then $x[k] = u[k]$ and $h[n-k] = u[n-k]$.

The lower limit on the convolution sum simplifies to $k = 0$ (because $u[k] = 0,$ $k < 0$), the upper limit to $k = n$ (because $u[n-k] = 0,\ k > n$), and we get

$$y[n] = \sum_{k=-\infty}^{\infty} u[k]u[n-k] = \sum_{k=0}^{n} 1 = (n+1)u[n] = r[n+1]$$

Note that $(n+1)u[n]$ also equals $r[n+1]$, and thus $u[n] * u[n] = r[n+1]$.

**(b)** Let $x[n] = (0.8)^n u[n]$ and $h[n] = (0.4)^n u[n]$. Then, $y[n] = x[k] * h[n]$ is evaluated from

$$\underbrace{\sum_{k=-\infty}^{\infty} (0.8)^k u[k](0.4)^{n-k} u[n-k]}_{\text{enter } x[k] \text{ and } h[n-k]} = \underbrace{\sum_{k=0}^{n} (0.8)^k (0.4)^{n-k}}_{u[k],\ u[n-k] \text{ modify limits}} = \underbrace{(0.4)^n \sum_{k=0}^{n} 2^k}_{\text{pull out terms in } n}$$

$$= \underbrace{(0.4)^n \frac{1-2^{n+1}}{1-2}}_{\text{evaluate sum}}$$

This simplifies to $y[n] = (0.4)^n(2^{n+1} - 1)u[n] = [2(0.8)^n - (0.4)^n]u[n]$.

**(c)** Let $x[n] = h[n] = a^n u[n]$. Then $x[k] = a^k u[k]$ and $h[n-k] = a^{n-k}u[n-k]$.

The lower limit on the convolution sum simplifies to $k = 0$ (because $u[k] = 0,$ $k < 0$), the upper limit to $k = n$ (because $u[n-k] = 0,\ k > n$), and we get

$$y[n] = \sum_{k=-\infty}^{\infty} a^k a^{n-k} u[k]u[n-k] = \sum_{k=0}^{n} a^k a^{n-k} = a^n \sum_{k=0}^{n} 1 = (n+1)a^n u[n]$$

The argument of the step function $u[n]$ is based on the fact that the upper limit on the summation must exceed or equal the lower limit (*i.e.*, $n \geq 0$).

**(d)** Let $x[n] = u[n-1]$ and $h[n] = \alpha^n u[n-1]$.

Then,

$$u[n-1] * \alpha^n u[n-1] = \sum_{k=-\infty}^{\infty} \alpha^k u[k-1]u[n-1-k] = \sum_{k=1}^{n-1} \alpha^k = \frac{(\alpha - \alpha^n)}{1-\alpha} u[n-2]$$

Here, we used the closed form result for the finite summation. The argument of the step function $u[n-2]$ is dictated by the fact that the upper limit on the summation must exceed or equal the lower limit (*i.e.*, $n-1 \geq 1$ or $n \geq 2$).

**(e)** Let $x[n] = nu[n]$ and $h[n] = a^{-n}u[n-1]$, $a < 1$.

With $h[n-k] = a^{-(n-k)}u[n-1-k]$ and $x[k] = ku[k]$, the lower and upper limits on the convolution sum become $k = 0$ and $k = n-1$. Then

$$y[n] = \sum_{k=0}^{n-1} ka^{-(n-k)} = a^{-n}\sum_{k=0}^{n-1} ka^k = \frac{a^{-n+1}}{(1-a)^2}[1 - na^{n-1} + (n-1)a^n]u[n-1]$$

Here, we used known results for the finite summation to generate the closed-form solution.

## Drill Problem 3.23

**(a)** Let $x[n] = (0.8)^n u[n]$ and $h[n] = (0.4)^n u[n-1]$. Find their convolution.
**(b)** Let $x[n] = (0.8)^n u[n-1]$ and $h[n] = (0.4)^n u[n]$. Find their convolution.
**(c)** Let $x[n] = (0.8)^n u[n-1]$ and $h[n] = (0.4)^n u[n-1]$. Find their convolution.

**Answers:** **(a)** $[(0.8)^n - (0.4)^n]u[n-1]$ **(b)** $2[(0.8)^n - (0.4)^n]u[n-1]$
**(c)** $[(0.8)^n - 2(0.4)^n]u[n-2]$

# 3.10 Convolution Properties

Many of the properties of discrete convolution are based on linearity and time invariance. For example, if $x[n]$ (or $h[n]$) is shifted by $n_0$, so is $y[n]$. Thus, if $y[n] = x[n] * h[n]$, then

$$x[n-n_0] * h[n] = x[n] * h[n-n_0] = y[n-n_0] \tag{3.39}$$

The sum of the samples in $x[n]$, $h[n]$, and $y[n]$ are related by

$$\sum_{n=-\infty}^{\infty} y[n] = \left(\sum_{n=-\infty}^{\infty} x[n]\right)\left(\sum_{n=-\infty}^{\infty} h[n]\right) \tag{3.40}$$

For causal systems ($h[n] = 0,\ n < 0$) and causal signals ($x[n] = 0,\ n < 0$), $y[n]$ is also causal. Thus,

$$y[n] = x[n] * h[n] = h[n] * x[n] = \sum_{k=0}^{n} x[k]h[n-k] = \sum_{k=0}^{n} h[k]x[n-k] \qquad (3.41)$$

An extension of this result is that the convolution of two left-sided signals is also left-sided and the convolution of two right-sided signals is also right-sided.

**Example 3.20**

### Properties of Convolution

**(a)** Here are two useful convolution results that are readily found from the defining relation:

$$\delta[n] * x[n] = x[n] \qquad \delta[n] * \delta[n] = \delta[n]$$

**(b)** We find $y[n] = u[n] * x[n]$. Since the step response is the running sum of the impulse response, the convolution of a signal $x[n]$ with a unit step is the running sum of the signal $x[n]$:

$$x[n] * u[n] = \sum_{k=-\infty}^{n} x[k]$$

**(c)** We find $y[n] = \mathrm{rect}(n/2N) * \mathrm{rect}(n/2N)$ where $\mathrm{rect}(n/2N) = u[n+N] - u[n-N-1]$.
The convolution contains four terms:

$$\begin{aligned} y[n] = {} & u[n+N] * u[n+N] - u[n+N] * u[n-N-1] \\ & - u[n-N-1] * u[n+N] + u[n-N-1] * u[n-N-1] \end{aligned}$$

Using the result $u[n] * u[n] = r[n+1]$ and the shifting property, we obtain

$$y[n] = r[n+2N+1] - 2r[n] + r[n-2N-1] = (2N+1)\mathrm{tri}\left(\frac{n}{2N+1}\right)$$

The convolution of two rect functions (with identical arguments) is thus a tri function.

# Drill Problem 3.24

**(a)** Let $x[n] = (0.8)^{n+1}u[n+1]$ and $h[n] = (0.4)^{n-1}u[n-2]$. Find their convolution.
**(b)** Let $x[n] = (0.8)^{n}u[n]$ and $h[n] = (0.4)^{n-1}u[n-2]$. Find their convolution.

**Answers: (a)** $[(0.8)^n - (0.4)^n]u[n-1]$ **(b)** $[(0.8)^{n-1} - (0.4)^{n-1}]u[n-2]$

# 3.11 Convolution of Finite Sequences

In practice, we often deal with sequences of finite length, and their convolution may be found by several methods. The convolution $y[n]$ of two finite-length sequences $x[n]$ and $h[n]$ is also of finite length and is subject to the following rules, which serve as useful consistency checks:

1. The starting index of $y[n]$ equals the sum of the starting indices of $x[n]$ and $h[n]$.
2. The ending index of $y[n]$ equals the sum of the ending indices of $x[n]$ and $h[n]$.
3. The length $L_y$ of $y[n]$ is related to the lengths $L_x$ and $L_h$ of $x[n]$ and $h[n]$ by $L_y = L_x + L_h - 1$.

## 3.11.1 The Sum-by-Column Method

This method is based on the idea that the convolution $y[n]$ equals the sum of the (shifted) impulse responses due to each of the impulses that make up the input $x[n]$. To find the convolution, we set up a row of index values beginning with the starting index of the convolution and $h[n]$ and $x[n]$ below it. We regard $x[n]$ as a sequence of weighted shifted impulses. Each element (impulse) of $x[n]$ generates a shifted impulse response (product with $h[n]$), starting at its index (to indicate the shift). Summing the response (by columns) gives the discrete convolution. *Note that none of the sequences is flipped.* It is better (if only to save paper) to let $x[n]$ be the shorter sequence. The starting index (and the marker location corresponding to $n = 0$) for the convolution $y[n]$ is found from the starting indices of $x[n]$ and $h[n]$.

**Example 3.21**

### Convolution of Finite-Length Signals

**(a)** An FIR (finite impulse response) filter has an impulse response given by $h[n] = \{\overset{\Downarrow}{1}, 2, 2, 3\}$. Find its response $y[n]$ to the input $x[n] = \{\overset{\Downarrow}{2}, -1, 3\}$. Assume that both $x[n]$ and $h[n]$ start at $n = 0$. The paper-and-pencil method expresses the input as $x[n] = 2\delta[n] - \delta[n-1] + 3\delta[n-2]$ and tabulates the response to each impulse and the total response as follows:

TABLE E 3.21.a ➤

| | | | | | | | | |
|---|---|---|---|---|---|---|---|---|
| $h[n]$ | | = | 1 | 2 | 2 | 3 | | |
| $x[n]$ | | = | 2 | −1 | 3 | | | |
| Input | Response | | | | | | | |
| $2\delta[n]$ | $2h[n]$ | = | 2 | 4 | 4 | 6 | | |
| $-\delta[n-1]$ | $-h[n-1]$ | = | | −1 | −2 | −2 | −3 | |
| $3\delta[n-2]$ | $3h[n-2]$ | = | | | 3 | 6 | 6 | 9 |
| Sum = $x[n]$ | Sum = $y[n]$ | = | 2 | 3 | 5 | 10 | 3 | 9 |

So,

$$y[n] = \{\overset{\Downarrow}{2}, 3, 5, 10, 3, 9\} = 2\delta[n] + 3\delta[n-1] + 5\delta[n-2] + 10\delta[n-3] + 3\delta[n-4] + 9\delta[n-5]$$

The convolution process is illustrated graphically in Figure E3.21a.

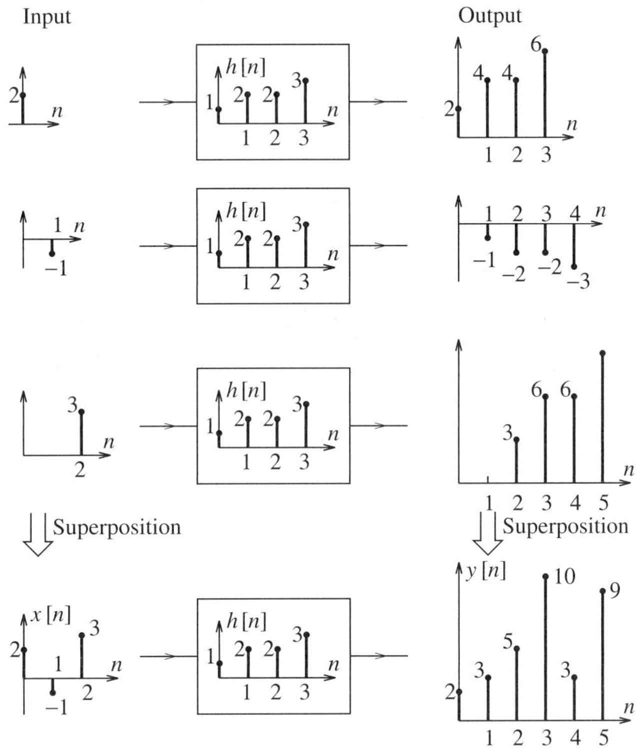


**FIGURE E 3.21.a** The discrete convolution for Example 3.21(a)

**(b)** Let $h[n] = \{2,\ 5,\ \overset{\Downarrow}{0},\ 4\}$ and $x[n] = \{4,\ \overset{\Downarrow}{1},\ 3\}$.

We note that the convolution starts at $n = -3$ and use this to set up the index array and generate the convolution as follows:

**TABLE E 3.21.b ➤**

| $n$ | $-3$ | $-2$ | $-1$ | 0 | 1 | 2 |
|---|---|---|---|---|---|---|
| $h[n]$ | 2 | 5 | 0 | 4 | | |
| $x[n]$ | 4 | 1 | 3 | | | |
| | 8 | 20 | 0 | 16 | | |
| | | 2 | 5 | 0 | 4 | |
| | | | 6 | 15 | 0 | 12 |
| $y[n]$ | 8 | 22 | 11 | $\overset{\Downarrow}{31}$ | 4 | 12 |

The marker is placed by noting that the convolution starts at $n = -3$, and we get

$$y[n] = \{8,\ 22,\ 11,\ \overset{\Downarrow}{31},\ 4,\ 12\}$$

**(c) Response of a Moving-Average Filter**

Let $x[n] = \{\overset{\Downarrow}{2},\ 4,\ 6,\ 8,\ 10,\ 12,\ \ldots\}$.

What system will result in the response $y[n] = \{\overset{\Downarrow}{1},\ 3,\ 5,\ 7,\ 9,\ 11,\ \ldots\}$?

At each instant, the response is the average of the input and its previous value. This system describes an averaging or moving average filter. Its difference equation is simply

$$y[n] = 0.5(x[n] + x[n-1])$$

Its impulse response is thus

$$h[n] = 0.5\{\delta[n] + \delta[n-1]\} \text{ or } h[n] = \{\overset{\Downarrow}{0.5},\ 0.5\}$$

Using discrete convolution, we find the response as follows:

TABLE E 3.21.c ➤

| | | | | | | | | |
|---|---|---|---|---|---|---|---|---|
| $x$: | 2 | 4 | 6 | 8 | 10 | 12 | ... | |
| $h$: | $\frac{1}{2}$ | $\frac{1}{2}$ | | | | | | |
| | 1 | 2 | 3 | 4 | 5 | 6 | ... | |
| | | 1 | 2 | 3 | 4 | 5 | 6 | ... |
| $y$: | 1 | 3 | 5 | 7 | 9 | 11 | ... | |

This result is indeed the averaging operation we expected.

# Drill Problem 3.25

**(a)** Let $x[n] = \{\overset{\Downarrow}{1},\ 4,\ 0,\ 2\}$ and $h[n] = \{\overset{\Downarrow}{1},\ 2,\ 1\}$. Find their convolution.

**(b)** Let $x[n] = \{1,\ 4,\ \overset{\Downarrow}{1},\ 3\}$ and $h[n] = \{2,\ \overset{\Downarrow}{1},\ -1\}$. Find their convolution.

**Answers: (a)** $\{\overset{\Uparrow}{1},\ 6,\ 9,\ 6,\ 4,\ 2\}$ **(b)** $\{2,\ 9,\ 5,\ \overset{\Uparrow}{3},\ 2,\ -3\}$

## 3.11.2 The Flip, Shift, Multiply, and Sum Concept

The convolution sum may also be interpreted as follows. We flip $x[n]$ to generate $x[-n]$ and shift the flipped signal $x[-n]$ to line up its last element with the first element of $h[n]$. We then successively shift $x[-n]$ (to the right) past $h[n]$, one index at a time, and find the convolution at each index as the sum of the pointwise products of the overlapping samples. One method of computing $y[n]$ is to list the values of the flipped function on a strip of paper and slide it along the stationary function, to better visualize the process. This technique has prompted the name **sliding strip method**. We simulate this method by showing the successive positions of the stationary and flipped sequence along with the resulting products, the convolution sum, and the actual convolution.

**Example 3.22**

### Convolution by the Sliding Strip Method

Find the convolution of $h[n] = \{\overset{\Downarrow}{2},\ 5,\ 0,\ 4\}$ and $x[n] = \{\overset{\Downarrow}{4},\ 1,\ 3\}$.

Since both sequences start at $n = 0$, the flipped sequence is $x[-k] = \{3,\ 1,\ \overset{\Downarrow}{4}\}$.

We line up the flipped sequence below $h[n]$ to begin overlap and shift it successively, summing the product sequence as we go, to obtain the discrete convolution. The results are computed in Figure E3.22.

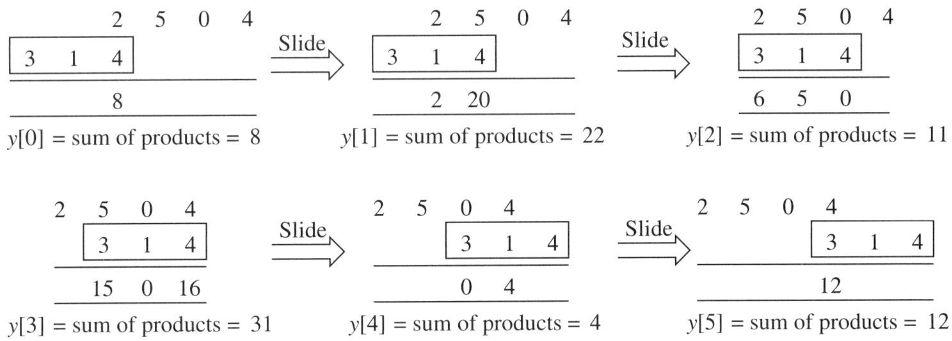


**FIGURE E.3.22** The discrete signals for Example 3.22 and their convolution

The discrete convolution is

$$y[n] = \{\overset{\Downarrow}{8},\ 22, 11, 31, 4, 12\}$$

### 3.11.3 Discrete Convolution, Multiplication, and Zero Insertion

The discrete convolution of two *finite-length* sequences $x[n]$ and $h[n]$ is entirely equivalent to the multiplication of two polynomials whose coefficients are described by the arrays $x[n]$ and $h[n]$ (in ascending or descending order). The convolution sequence corresponds to the coefficients of the product polynomial. Based on this result, if we insert $N$ zeros between each pair of adjacent samples of each sequence to be convolved, their convolution corresponds to the original convolution sequence with $N$ zeros inserted between each pair of its adjacent samples.

If we append zeros to one of the convolved sequences, the convolution result also will show as many appended zeros at the corresponding location. For example, leading zeros appended to a sequence will appear as leading zeros in the convolution result. Similarly, trailing zeros appended to a sequence will show up as trailing zeros in the convolution.

**Convolution of Finite-Length Signals Corresponds to Polynomial Multiplication**

**Example:** $\{\overset{\Downarrow}{1},\ 1,\ 3\} * \{\overset{\Downarrow}{1},\ 0,\ 2\} = \{\overset{\Downarrow}{1},\ 1,\ 5,\ 2,\ 6\}$ $\quad (x^2 + x + 3)(x^2 + 2) = x^4 + x^3 + 5x^2 + 2x + 6$

**Zero Insertion of Both Sequences Leads to Zero Insertion of the Convolution**

**Example:** If $\{\overset{\Downarrow}{1},\ 2\} * \{\overset{\Downarrow}{3},\ 1,\ 4\} = \{\overset{\Downarrow}{3},\ 7,\ 6,\ 8\}$

Then, $\{\overset{\Downarrow}{1},\ 0,\ 0,\ 2\} * \{\overset{\Downarrow}{3},\ 0,\ 0,\ 1,\ 0,\ 0,\ 4\} = \{\overset{\Downarrow}{3},\ 0,\ 0,\ 7,\ 0,\ 0,\ 6,\ 0,\ 0,\ 8\}$

**Zero-Padding of One Sequence Leads to Zero-Padding of the Convolution**

**Example:** If $x[n] * h[n] = y[n]$ then $\{0,\ 0, x[n],\ 0,\ 0\} * \{h[n],\ 0\} = \{0,\ 0,\ y[n],\ 0,\ 0,\ 0\}$.

**Example 3.23** **Polynomial Multiplication, Zero Insertion, Zero-Padding**

**(a)** Let $h[n] = \{\overset{\Downarrow}{2}, 5, 0, 4\}$ and $x[n] = \{\overset{\Downarrow}{4}, 1, 3\}$.

To find their convolution, we set up the polynomials:

$$h(z) = 2z^3 + 5z^2 + 0z + 4 \qquad x(z) = 4z^2 + 1z + 3$$

Their product is $y(z) = 8z^5 + 22z^4 + 11z^3 + 31z^2 + 4z + 12$.

The convolution is thus $y[n] = \{\overset{\Downarrow}{8}, 22, 11, 31, 4, 12\}$.

**(b) Zero-Insertion**

Zero insertion of each convolved sequence gives

$$h_1[n] = \{\overset{\Downarrow}{2}, 0, 5, 0, 0, 0, 4\} \qquad x_1[n] = \{\overset{\Downarrow}{4}, 0, 1, 0, 3\}$$

To find their convolution, we set up the polynomials:

$$h_1(z) = 2z^6 + 5z^4 + 0z^2 + 4 \qquad x_1(z) = 4z^4 + 1z^2 + 3$$

Their product is $y_1(z) = 8z^{10} + 22z^8 + 11z^6 + 31z^4 + 4z^2 + 12$.

The convolution is then $y_1[n] = \{\overset{\Downarrow}{8}, 0, 22, 0, 11, 0, 31, 0, 4, 0, 12\}$.
This result is just $y[n]$ with zeros inserted between adjacent samples.

**(c) Zero-Padding**

If we pad the first sequence by two zeros and the second by one zero, we get

$$h_2[n] = \{\overset{\Downarrow}{2}, 5, 0, 4, 0, 0\} \qquad x_2[n] = \{\overset{\Downarrow}{4}, 1, 3, 0\}$$

To find their convolution, we set up the polynomials:

$$h_2(z) = 2z^5 + 5z^4 + 4z^2 \qquad x_2(z) = 4z^3 + 1z^2 + 3z$$

Their product is $y_2(z) = 8z^8 + 22z^7 + 11z^6 + 31z^5 + 4z^4 + 12z^3$.

The convolution is then $y_1[n] = \{\overset{\Downarrow}{8}, 22, 11, 31, 4, 12, 0, 0, 0\}$.

This result is just $y[n]$ with three zeros appended at the end.

# Drill Problem 3.26

Use the result $\{\overset{\Downarrow}{1}, 0, 2\} * \{\overset{\Downarrow}{1}, 3\} = \{\overset{\Downarrow}{1}, 3, 2, 6\}$ to answer the following.

**(a)** Let $x[n] = \{\overset{\Downarrow}{1}, 0, 0, 0, 2\}$ and $h[n] = \{\overset{\Downarrow}{1}, 0, 3\}$. Find their convolution.

**(b)** Let $x[n] = \{\overset{\Downarrow}{1}, 0, 2, 0\}$ and $h[n] = \{\overset{\Downarrow}{1}, 3, 0, 0\}$. Find their convolution.

**(c)** Let $x[n] = \{0, 0, \overset{\Downarrow}{1}, 0, 2\}$ and $h[n] = \{\overset{\Downarrow}{1}, 3, 0\}$. Find their convolution.

**Answers:** **(a)** $\{\overset{\Downarrow}{1}, 0, 3, 0, 2, 0, 6\}$ **(b)** $\{\overset{\Downarrow}{1}, 3, 2, 6, 0, 0, 0\}$ **(c)** $\{0, 0, \overset{\Downarrow}{1}, 3, 2, 6, 0\}$

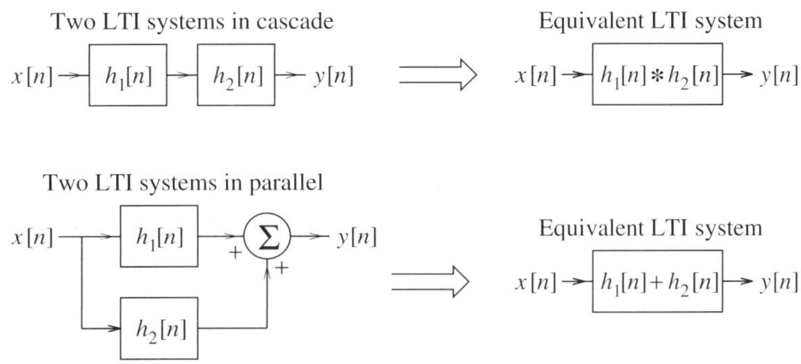


**FIGURE 3.8** Cascaded and parallel systems and their equivalents. The equivalent impulse response of systems in cascade is the convolution of the individual impulse responses. The equivalent impulse response of systems in parallel is the sum of the individual impulse responses

## 3.11.4 Impulse Response of LTI Systems in Cascade and Parallel

Consider the ideal cascade of two LTI systems shown in Figure 3.8. The response of the first system is $y_1[n] = x[n] * h_1[n]$. The response $y[n]$ of the second system is

$$y[n] = y_1[n] * h_2[n] = (x[n] * h_1[n]) * h_2[n] = x[n] * (h_1[n] * h_2[n]) \qquad (3.42)$$

If we wish to replace the cascaded system by an equivalent LTI system with impulse response $h[n]$ such that $y[n] = x[n] * h[n]$, it follows that $h[n] = h_1[n] * h_2[n]$. Generalizing this result, the impulse response $h[n]$ of $N$ ideally cascaded LTI systems is simply the convolution of the $N$ individual impulse responses

$$h[n] = h_1[n] * h_2[n] * \cdots * h_N[n] \quad \text{(for a cascade combination)} \qquad (3.43)$$

If the $h_k[n]$ are energy signals, the order of cascading is unimportant.
The overall impulse response of LTI systems in parallel equals the sum of the individual impulse responses, as shown in Figure 3.8:

$$h_P[n] = h_1[n] + h_2[n] + \cdots + h_N[n] \quad \text{(for a parallel combination)} \qquad (3.44)$$

---

### Impulse Response of *N* Interconnected Discrete LTI Systems

**In Cascade:** *Convolve* the impulse responses: $h_C[n] = h_1[n] * h_2[n] * \cdots * h_N[n]$
**In Parallel:** *Add* the impulse responses: $h_P[n] = h_1[n] + h_2[n] + \cdots + h_N[n]$

---

**Example 3.24**

### Interconnected Systems

Consider the interconnected system of Figure E3.24. Find its overall impulse response and the output. Comment on the results.

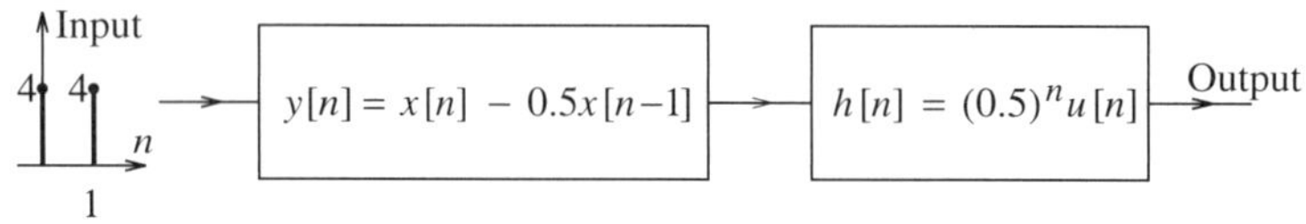


**FIGURE E.3.24** The interconnected system of Example 3.24

**Solution:** The impulse response of the first system is $h_1[n] = \delta[n] - 0.5\delta[n-1]$. The overall impulse response $h_C[n]$ is given by the convolution:

$$h_C[n] = (\delta[n] - 0.5\delta[n-1]) * (0.5)^n u[n] = (0.5)^n u[n] - 0.5(0.5)^{n-1} u[n-1]$$

This simplifies to

$$h_C[n] = (0.5)^n (u[n] - u[n-1]) = (0.5)^n \delta[n] = \delta[n]$$

What this means is that the overall system output equals the applied input. The second system thus acts as the inverse of the first.

## Drill Problem 3.27

**(a)** Find the impulse response of the cascade of two identical filters, each with $h[n] = \{\overset{\Downarrow}{1}, 1, 3\}$.

**(b)** The impulse response of two filters is $h_1[n] = \{\overset{\Downarrow}{1}, 0, 2\}$ and $h_2[n] = \{4, \overset{\Downarrow}{1}, 3\}$. Find the impulse response of their parallel combination.

**(c)** Two filters are described by $y[n] - 0.4y[n-1] = x[n]$ and $h_2[n] = 2(0.4)^n u[n]$. Find the impulse response of their parallel combination and cascaded combination.

**Answers:** **(a)** $\{\overset{\Downarrow}{1}, 2, 7, 6, 9\}$ **(b)** $\{4, \overset{\Downarrow}{2}, 3, 2\}$
**(c)** $h_P[n] = 3(0.4)^n u[n]$, $h_C[n] = 2(n+1)(0.4)^n u[n]$

# 3.12 Stability and Causality of LTI Systems

System stability is an important practical constraint in filter design and is defined in various ways. Here we introduce the concept of **bounded-input**, **bounded-output** (BIBO) **stability** that requires *every* bounded input to produce a bounded output.

### 3.12.1 Stability of FIR Filters

The system equation of an FIR filter describes the output as a weighted sum of shifted inputs. If the input remains bounded, the weighted sum of the inputs is also bounded. In other words, FIR *filters are always stable*. This can be a huge design advantage.

### 3.12.2 Stability of LTI Systems Described by Difference Equations

For an LTI system described by the difference equation

$$\begin{aligned} y[n] + A_1 y[n-1] + \cdots + A_N y[n-N] \\ = B_0 x[n] + B_1 x[n-1] + \cdots + B_M x[n-M] \end{aligned} \tag{3.45}$$

the conditions for BIBO stability involve the roots of the characteristic equation. A necessary and sufficient condition for BIBO stability of such an LTI system is that *every root of its characteristic equation must have a magnitude less than unity*. This criterion is based on the results of Tables 3.1 and 3.2. Root magnitudes less than unity ensure that the natural (and zero-input) response always decays with time (see Table 3.2), and the forced (and zero-state) response always remains bounded for *every* bounded input. Roots with magnitudes that equal unity make the system unstable. Simple (non-repeated) roots with unit magnitude produce a constant (or sinusoidal) natural response that is bounded; but if the input is also a constant (or sinusoid at the same frequency), the forced response is a ramp or growing sinusoid (see Table 3.1) and hence unbounded. Repeated roots with unit magnitude result in a natural response that is itself a growing sinusoid or polynomial and thus unbounded. In the next chapter, we shall see that the stability condition is equivalent to having an LTI system whose impulse response $h[n]$ is *absolutely summable*. The stability of nonlinear or time-varying systems usually must be checked by other means.

### 3.12.3 Stability of LTI Systems Described by the Impulse Response

For systems described by their impulse response, it turns out that BIBO stability requires that the impulse response $h[n]$ be absolutely summable. Here is why. If $x[n]$ is bounded such that $|x[n]| < M$, so too is its shifted version $x[n-k]$. The convolution sum then yields the following inequality:

$$|y[n]| < \sum_{k=-\infty}^{\infty} |h[k]||x[n-k]| < M \sum_{k=-\infty}^{\infty} |h[k]| \tag{3.46}$$

If the output is to remain bounded ($|y[n]| < \infty$), then

$$\sum_{k=-\infty}^{\infty} |h[k]| < \infty \qquad \text{(for a stable LTI system)} \tag{3.47}$$

In other words, $h[n]$ must be absolutely summable. This is both a necessary and sufficient condition. The stability of nonlinear systems must be investigated by other means.

### 3.12.4 Causality

In analogy with analog systems, causality of discrete-time systems implies a non-anticipating system with an impulse response $h[n] = 0,\ n < 0$. This ensures that an input $x[n]u[n-n_0]$. Starting at $n = n_0$ results in a response $y[n]$ also starting at $n = n_0$ (and not earlier). This follows from the convolution sum:

$$y[n] = \sum_{k=-\infty}^{\infty} x[k]u(k-n_0]h[n-k]u[n-k] = \sum_{n_0}^{n} x[k]h[n-k] \tag{3.48}$$

### Stability and Causality of Discrete LTI Systems

**Stability:** Every root $r$ of the characteristic equation must have magnitude $|r|$ *less than* unity. The impulse response $h[n]$ must be *absolutely summable* with

$$\sum_{k=-\infty}^{\infty} |h[k]| < \infty$$

**Note:** FIR *filters are always stable.*
**Causality:** The impulse response $h[n]$ must be zero for negative indices ($h[n]=0,\ n<0$).

**Example 3.25**

### Concepts Based on Stability and Causality

**(a)** The system $y[n] - \frac{1}{6}y[n-1] - \frac{1}{6}y[n-2] = x[n]$ is stable since the roots of its characteristic equation $z^2 - \frac{1}{6}z - \frac{1}{6} = 0$ are $z_1 = \frac{1}{2}$ and $z_2 = -\frac{1}{3}$ and their magnitudes are less than 1.

**(b)** The system $y[n] - y[n-1] = x[n]$ is unstable. The root of its characteristic equation $z - 1 = 0$ is $z = 1$ gives the natural response $y_N = Ku[n]$, which is actually bounded. However, for an input $x[n] = u[n]$, the forced response will have the form $Cnu[n]$, which becomes unbounded.

**(c)** The system $y[n] - 2y[n-1] + y[n-2] = x[n]$ is unstable. The roots of its characteristic equation $z^2 - 2z + 1 = 0$ are equal and produce the unbounded natural response $y_N[n] = Au[n] + Bnu[n]$.

**(d)** The system $y[n] - \frac{1}{2}y[n-1] = nx[n]$ is linear, time varying, and unstable. The (bounded) step input $x[n] = u[n]$ results in a response that includes the ramp $nu[n]$, which becomes unbounded.

**(e)** The system $y[n] = x[n] - 2x[n-1]$ is stable because it describes an FIR filter.

**(f)** The FIR filter described by $y[n] = x[n+1] - x[n]$ has the impulse response $h[n] = \{1, -\overset{\Downarrow}{1}\}$. It is a stable system, since $\sum |h[n]| = |1| + |-1| = 2$. It is also noncausal because $h[n] = \delta[n+1] - \delta[n]$ is not zero for $n < 0$. We emphasize that FIR filters are always stable because $\sum |h[n]|$ is the absolute sum of a finite sequence and is thus always finite.

**(g)** A filter described by $h[n] = (-0.5)^n u[n]$ is causal. It describes a system with the difference equation $y[n] = x[n] + ay[n-1]$. It is also stable because $\sum |h[n]|$ is finite. In fact, we find that

$$\sum_{n=-\infty}^{\infty} |h[n]| = \sum_{n=0}^{\infty} (-0.5)^n = \frac{1}{1-0.5} = 2$$

**(h)** A filter described by the difference equation $y[n] - 0.5y[n-1] = nx[n]$ is causal but time varying. It is also unstable. If we apply a step input $u[n]$ (bounded input), then $y[n] = nu[n] + 0.5y[n-1]$. The term $nu[n]$ grows without bound and makes this system unstable. We caution you that this approach is not a formal way of checking for the stability of time-varying systems.

## Drill Problem 3.28

**(a)** Is the filter described by $h[n] = \{2, \overset{\Downarrow}{1}, 1, 3\}$ causal? Is it stable?
**(b)** Is the filter described by $h[n] = 2^n u[n+2]$ causal? Is it stable?
**(c)** Is the filter described by $y[n] + 0.5y[n-1] = 4u[n]$ causal? Is it stable?
**(d)** Is the filter described by $y[n] + 1.5y[n-1] + 0.5y[n-2] = u[n]$ causal? Is it stable?

**Answers: (a)** Noncausal, stable **(b)** Noncausal, unstable **(c)** Causal, stable **(d)** Causal, unstable

# 3.13 System Response to Periodic Inputs

In analogy with analog systems, the response of a discrete-time system to a periodic input with period $N$ is also periodic with the same period $N$. A simple example demonstrates this concept.

**Example 3.26**

### Response to Periodic Inputs

**(a)** Let $x[n] = \{\overset{\Downarrow}{1}, 2, -3, 1, 2, -3, 1, 2, -3, \ldots\}$ and $h[n] = \{\overset{\Downarrow}{1}, 1\}$. The convolution $y[n] = x[n] * h[n]$, using the sum-by-column method, is

TABLE E 3.26.a ➤

| Index $n$ | 0 | 1 | 2 | 3 | 4 | 5 | 6 | 7 | 8 | 9 | 10 |
|---|---|---|---|---|---|---|---|---|---|---|---|
| $x[n]$ | 1 | 2 | −3 | 1 | 2 | −3 | 1 | 2 | −3 | 1 | ... |
| $h[n]$ | 1 | 1 | | | | | | | | | |
| | 1 | 2 | −3 | 1 | 2 | −3 | 1 | 2 | −3 | 1 | ... |
| | | 1 | 2 | −3 | 1 | 2 | −3 | 1 | 2 | −3 | ... |
| $y[n]$ | 1 | 3 | −1 | −2 | 3 | −1 | −2 | 3 | −1 | −2 | ... |

The convolution $y[n]$ is periodic with period $N = 3$, except for start-up effects (which last for one period). One period of the convolution is $y[n] = \{\overset{\Downarrow}{-2}, 3, -1\}$.

**(b)** Let $x[n] = \{\overset{\Downarrow}{1}, 2, -3, 1, 2, -3, 1, 2, -3, \ldots\}$ and $h[n] = \{\overset{\Downarrow}{1}, 1, 1\}$.
The convolution $y[n] = x[n] * h[n]$, using the sum-by-column method, is found as follows:

**TABLE E 3.26.b ➤**

| Index $n$ | 0 | 1 | 2 | 3 | 4 | 5 | 6 | 7 | 8 | 9 | 10 |
|---|---|---|---|---|---|---|---|---|---|---|---|
| $x[n]$ | 1 | 2 | −3 | 1 | 2 | −3 | 1 | 2 | −3 | 1 | ... |
| $h[n]$ | 1 | 1 | 1 | | | | | | | | |
| | 1 | 2 | −3 | 1 | 2 | −3 | 1 | 2 | −3 | 1 | ... |
| | | 1 | 2 | −3 | 1 | 2 | −3 | 1 | 2 | −3 | ... |
| | | | 1 | 2 | −3 | 1 | 2 | −3 | 1 | 2 | ... |
| $y[n]$ | 1 | 3 | 0 | 0 | 0 | 0 | 0 | 0 | 0 | 0 | ... |

Except for start-up effects, the convolution is *zero*. The system $h[n] = \{\overset{\Downarrow}{1}, 1, 1\}$ is a moving average filter. It extracts the 3-point running sum, which is always zero for the given periodic signal $x[n]$.

---

One way to find the system response to periodic inputs is to find the response to one period of the input and then use superposition. If we add an absolutely summable signal (or an energy signal) $x[n]$ and its infinitely many replicas shifted by multiples of $N$, we obtain a *periodic signal* with period $N$, which is called the periodic extension of $x[n]$:

$$x_{\text{pe}}[n] = \sum_{k=-\infty}^{\infty} x[n + kN] \tag{3.49}$$

For finite-length sequences, an equivalent way of finding one period of the periodic extension is to wrap around $N$-sample sections of $x[n]$ and add them all up. If $x[n]$ is shorter than $N$, we obtain one period of its periodic extension simply by padding $x[n]$ with enough zeros to increase its length to $N$.

**Example 3.27**

### Periodic Extension

**(a)** The periodic extension of $x[n] = \{\overset{\Downarrow}{1}, 5, 2, 0, 4, 3, 6, 7\}$ with period $N = 3$ is found by wrapping around blocks of 3 samples and finding the sum to give

$$\{\overset{\Downarrow}{1}, 5, 2, 0, 4, 3, 6, 7\} \Longrightarrow \text{wrap around} \Longrightarrow \left\{\begin{matrix} \overset{\Downarrow}{1} & 5 & 2 \\ 0 & 4 & 3 \\ 6 & 7 & \end{matrix}\right\} \Longrightarrow \text{sum} \Longrightarrow \{\overset{\Downarrow}{7}, 16, 5\}$$

In other words, if we add $x[n]$ to its shifted versions $x[n + kN]$ where $N = 3$ and $k = \pm 1, \pm 2, \pm 3, \ldots$, we get a periodic signal whose first period is $\{\overset{\Downarrow}{7}, 16, 5\}$.

**(b)** The periodic extension of the signal $x[n] = \alpha^n u[n]$ with period $N$ is given by

$$x_{\text{pe}}[n] = \sum_{k=-\infty}^{\infty} x[n + kN] = \sum_{k=0}^{\infty} \alpha^{n+kN} = \alpha^n \sum_{k=0}^{\infty} (\alpha^N)^k = \frac{\alpha^n}{1 - \alpha^N}, \quad 0 \le n \le N - 1$$

---

The methods for finding the response of a discrete-time system to periodic inputs rely on the concepts of periodic extension and wraparound. One approach is to find the output for one period of the input (using regular convolution) and find one period of the periodic output by superposition (using periodic extension). Another approach is to first find one period of the periodic extension of the impulse response, then find its regular convolution with one period of the input, and finally, wrap around the regular convolution to generate one period of the periodic output.

**Example 3.28**

### System Response to Periodic Inputs

**(a)** Let $x[n] = \{\overset{\Downarrow}{1}, 2, -3\}$ describe one period of a periodic input with period $N = 3$ to a system whose impulse response is $h[n] = \{\overset{\Downarrow}{1}, 1\}$. The response $y[n]$ is also periodic with period $N = 3$. To find $y[n]$ for one period, we find the regular convolution $y_1[n]$ of $h[n]$ and one period of $x[n]$ to give

$$y_1[n] = \{\overset{\Downarrow}{1}, 1\} * \{\overset{\Downarrow}{1}, 2, -3\} = \{\overset{\Downarrow}{1}, 3, -1, -3\}$$

We then wrap around $y_1[n]$ past three samples to obtain one period of $y[n]$ as

$$\{\overset{\Downarrow}{1}, 3, -1, -3\} \Longrightarrow \text{wrap around} \Longrightarrow \left\{\begin{array}{ccc} \overset{\Downarrow}{1} & 3 & -1 \\ -3 & & \end{array}\right\} \Longrightarrow \text{sum} \Longrightarrow \{\overset{\Downarrow}{-2}, 3, -1\}$$

This is identical to the result obtained in the previous example.

**(b)** We find the response $y_p[n]$ of a moving average filter described by $h[n] = \{\overset{\Downarrow}{2}, 1, 1, 3, 1\}$ to a periodic signal whose one period is $x_p[n] = \{\overset{\Downarrow}{2}, 1, 3\}$, with $N = 3$, using two methods.

1. **Method 1**
   We find the regular convolution $y[n] = x_p[n] * h[n]$ to obtain

$$y[n] = \{\overset{\Downarrow}{2}, 1, 3\} * \{\overset{\Downarrow}{2}, 1, 1, 3, 1\} = \{\overset{\Downarrow}{4}, 4, 9, 10, 8, 10, 3\}$$

   To find $y_p[n]$, values past $N = 3$ are wrapped around and summed to give

$$\{\overset{\Downarrow}{4}, 4, 9, 10, 8, 10, 3\} \Longrightarrow \text{wrap around} \Longrightarrow \left\{\begin{array}{ccc} \overset{\Downarrow}{4} & 4 & 9 \\ 10 & 8 & 10 \\ 3 & & \end{array}\right\} \Longrightarrow \text{sum}$$

$$\Longrightarrow \{\overset{\Downarrow}{17}, 12, 19\}$$

2. **Method 2**
   We first create the periodic extension of $h[n]$, with $N = 3$, (by wraparound) to get $h_p[n] = \{\overset{\Downarrow}{5}, 2, 1\}$. The regular convolution of $h_p[n]$ and one period of $x[n]$ gives

$$y[n] = \{\overset{\Downarrow}{2}, 1\ , 3\} * \{\overset{\Downarrow}{5}, 2, 1\} = \{\overset{\Downarrow}{10}, 9, 19, 7, 3\}$$

   This result is wrapped around past $N = 3$ to give $y_p[n] = \{\overset{\Downarrow}{17}, 12, 19\}$, as before.

**(c)** We find the response of the system $y[n] - 0.5y[n-1] = x[n]$ to $x[n] = \{\ldots, \overset{\Downarrow}{7}, 0, 0, 7, 0, 0, \ldots\}$.

The impulse response of the system is $h[n] = (0.5)^n u[n]$. The input is periodic with period $N = 3$ and first period $x_1 = \{\overset{\Downarrow}{7}, 0, 0\}$. The response due to this one period is just $y_1[n] = 7(0.5)^n u[n]$. The complete periodic output is the periodic extension of $y_1[n]$ with period $N = 3$. Using the result that the periodic extension of $\alpha^n u[n]$ with period $N$ is given by $\frac{\alpha^n}{1-\alpha^N}$, $0 \le n \le N-1$, we find one period of the periodic output as

$$y_p[n] = \frac{(0.5)^n}{1-(0.5)^3} = 0 \le n \le 2\{\overset{\Downarrow}{8}, 4, 2\}$$

## Drill Problem 3.29

**(a)** A filter is described by $h[n] = \{\overset{\Downarrow}{2}, 1, 1, 3, 2\}$. Find one period of its periodic output if the input is periodic with first period $x_1[n] = \{\overset{\Downarrow}{1}, 0, 0, 2\}$.
**(b)** A filter is described by $y[n] - 0.5y[n-1] = 7x[n]$. Find one period of its periodic output if the input is periodic with first period $x_1[n] = \{\overset{\Downarrow}{1}, 0, 0\}$.
**(c)** A filter is described by $y[n] - 0.5y[n-1] = 7x[n]$? Find one period of its periodic output if the input is periodic with first period $x_1[n] = \{\overset{\Downarrow}{1}, 0, 1\}$.

**Answers: (a)** $\{\overset{\Downarrow}{6}, 3, 7, 11\}$ **(b)** $\{\overset{\Downarrow}{8}, 4, 2\}$ **(c)** $\{\overset{\Downarrow}{12}, 6, 10\}$

## 3.14 Periodic or Circular Convolution

The regular convolution of two signals, both of which are periodic, does not exist. For this reason, we resort to **periodic convolution** by using averages. If *both* $x_p[n]$ and $h_p[n]$ are periodic with identical period $N$, their periodic convolution generates a convolution result $y_p[n]$ that is also periodic with the same period $N$. The periodic convolution or **circular convolution** or **cyclic convolution** $y_p[n]$ of $x_p[n]$ and $h_p[n]$ is denoted $y_p[n] = x_p[n] \circledast h_p[n]$ and, over one period ($n = 0, 1, \ldots, N-1$), it is defined by

$$y_p[n] = x_p[n] \circledast h_p[n] = h_p[n] \circledast x_p[n] = \sum_{k=0}^{N-1} x_p[k]h_p[n-k] = \sum_{k=0}^{N-1} h_p[k]x_p[n-k] \tag{3.50}$$

An averaging factor of $1/N$ is sometimes included with the summation. Periodic convolution can be implemented using wraparound. We find the *linear* convolution of one period of $x_p[n]$ and $h_p[n]$, which will have $(2N-1)$ samples. We then extend its length to $2N$ (by appending a zero), slice it in two halves (of length $N$ each), line up the second half with the first, and add the two halves to get the periodic convolution.

---

**Periodic Convolution of Periodic Discrete-Time Signals with Identical Period *N***

1. Find the regular convolution of their one-period segments (this will have length $2N - 1$).
2. Append a trailing zero. Wrap around the last $N$ samples and add to the first $N$ samples.

---

***Example 3.29***

**Periodic Convolution**

**(a)** Find the periodic convolution of $x_p[n] = \{\overset{\Downarrow}{1}, 0, 1, 1\}$ and $h_p[n] = \{\overset{\Downarrow}{1}, 2, 3, 1\}$.

The period is $N = 4$. First, we find the linear convolution $y[n]$.

| Index $n$ | 0 | 1 | 2 | 3 | 4 | 5 | 6 |
|---|---|---|---|---|---|---|---|
| $h_p[n]$ | 1 | 2 | 3 | 1 | | | |
| $x_p[n]$ | 1 | 0 | 1 | 1 | | | |
| | 1 | 2 | 3 | 1 | | | |
| | | 0 | 0 | 0 | 0 | | |
| | | | 1 | 2 | 3 | 1 | |
| | | | | 1 | 2 | 3 | 1 |
| $y[n]$ | 1 | 2 | 4 | 4 | 5 | 4 | 1 |

Then, we append a zero, wrap around the last four samples, and add.

| Index $n$ | 0 | 1 | 2 | 3 |
|---|---|---|---|---|
| First half of $y[n]$ | 1 | 2 | 4 | 4 |
| Wrapped around half of $y[n]$ | 5 | 4 | 1 | 0 |
| Periodic convolution $y_p[n]$ | 6 | 6 | 5 | 4 |

**(b)** Find the periodic convolution of $x_p[n] = \{\overset{\Downarrow}{1}, 2, 3\}$ and $h_p[n] = \{\overset{\Downarrow}{1}, 0, 2\}$, with period $N = 3$.

The regular convolution is easily found to be $y_R[n] = \{\overset{\Downarrow}{1}, 2, 5, 4, 6\}$. Appending a zero and wrapping around the last three samples gives $y_p[n] = \{\overset{\Downarrow}{5}, 8, 5\}$.

---

# Drill Problem 3.30

Find the periodic convolution of two identical signals whose first period is given by $x_1[n] = \{\overset{\Downarrow}{1}, 2, 0, 2\}$.

**Answer:** $\{\overset{\Downarrow}{9}, 4, 8, 4\}$

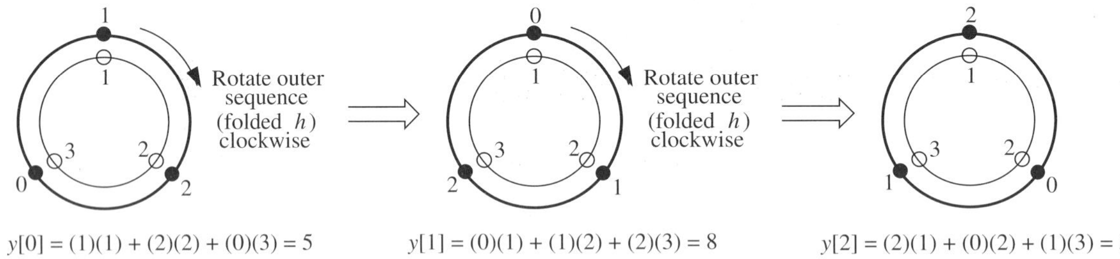


**FIGURE 3.9** The cyclic method of circular (periodic) convolution. The time reversed sequence is rotated one index at a time and the convolution result found by summing the products. After we come full circle, the convolution results replicate

## 3.14.1 Periodic Convolution by the Cyclic Method

To find the periodic convolution, we shift the flipped signal $x_p[-n]$ past $h_p[n]$—one index at a time—and find the convolution at each index as the sum of the pointwise product of their samples but only over a one-period window $(0, N-1)$. Values of $x_p[n]$ and $h_p[n]$ outside the range $(0, N-1)$ are generated by **periodic extension**. One way to visualize the process is to line up $x[k]$ clockwise around a circle and $h[k]$ counterclockwise (to indicated that it is flipped) on a concentric circle positioned to start the convolution, as shown in Figure 3.9.

Shifting the flipped sequence turns it clockwise. At each turn, the convolution equals the sum of the pairwise products. This approach clearly brings out the cyclic nature of periodic convolution.

## 3.14.2 Periodic Convolution by the Circulant Matrix

Periodic convolution may also be expressed as a matrix multiplication. We set up an $N \times N$ matrix whose columns equal $x[n]$ and its cyclically shifted versions (or whose rows equal successively shifted versions of the *first* period of the flipped signal $x[-n]$). This is called the **circulant matrix** or **convolution matrix**. An $N \times N$ circulant matrix $\mathbf{C}_x$ for $x[n]$ has the general form

$$\mathbf{C}_x = \begin{bmatrix} x[0] & x[N-1] & \ldots\ x[2] & x[1] \\ x[1] & x[0]\ldots & & x[2] \\ x[2] & x[1] & \ldots & x[3] \\ \vdots & \vdots & & \vdots \\ x[N-2] & & \ldots\ x[0] & x[N-1] \\ x[N-1] & x[N-2] & \ldots\ x[1] & x[0] \end{bmatrix} \tag{3.51}$$

Note that each diagonal of the circulant matrix has equal values. Such a *constant diagonal matrix* is also called a **Toeplitz matrix**. Its matrix product with an $N \times 1$ column matrix $\mathbf{h}$ describing $h[n]$ yields the periodic convolution $\mathbf{y} = \mathbf{Ch}$ as an $N \times 1$ column matrix.

**Example 3.30**

**Periodic Convolution by the Circulant Matrix**

Consider $x[n] = \{\overset{\Downarrow}{1}, 0, 2\}$ and $h[n] = \{\overset{\Downarrow}{1}, 2, 3\}$, described over one period $(N = 3)$.

**(a)** The circulant matrix $\mathbf{C}_x$ and periodic convolution $y_1[n]$ are given by

$$\mathbf{C}_x = \begin{bmatrix} 1 & 2 & 0 \\ 0 & 1 & 2 \\ 2 & 0 & 1 \end{bmatrix} \qquad \mathbf{h} = \begin{bmatrix} 1 \\ 2 \\ 3 \end{bmatrix} \qquad y_1[n] = \begin{bmatrix} 1 & 2 & 0 \\ 0 & 1 & 2 \\ 2 & 0 & 1 \end{bmatrix} \begin{bmatrix} 1 \\ 2 \\ 3 \end{bmatrix} = \begin{bmatrix} 5 \\ 8 \\ 5 \end{bmatrix}$$

*Comment:* Though not required, normalization by $N = 3$ gives $y_{p1}[n] = \frac{y_1[n]}{3} = \{\overset{\Downarrow}{\tfrac{5}{3}}, \tfrac{8}{3}, \tfrac{5}{3}\}$.

**(b)** The periodic convolution $y_2[n]$ of $x[n]$ and $h[n]$ over a two-period window yields

$$\mathbf{C}_2 = \begin{bmatrix} 1 & 2 & 0 & 1 & 2 & 0 \\ 0 & 1 & 2 & 0 & 1 & 2 \\ 2 & 0 & 1 & 2 & 0 & 1 \\ 1 & 2 & 0 & 1 & 2 & 0 \\ 0 & 1 & 2 & 0 & 1 & 2 \\ 2 & 0 & 1 & 2 & 0 & 1 \end{bmatrix} \qquad \mathbf{h}_2 = \begin{bmatrix} 1 \\ 2 \\ 3 \\ 1 \\ 2 \\ 3 \end{bmatrix} \qquad y_2[n] = \begin{bmatrix} 10 \\ 16 \\ 10 \\ 10 \\ 16 \\ 10 \end{bmatrix}$$

We see that $y_2[n]$ has double the length (and values) of $y_1[n]$, but it is still periodic with $N = 3$.

*Comment:* Normalization by a two-period window width (6 samples) gives

$$y_{p2}[n] = \frac{y_2[n]}{6} = \left\{ \overset{\Downarrow}{\frac{5}{3}}, \frac{8}{3}, \frac{5}{3}, \frac{5}{3}, \frac{8}{3}, \frac{5}{3} \right\}$$

Note that one period ($N = 3$) of $y_{p2}[n]$ is identical to the normalized result $y_{p1}[n]$ of part (a).

### 3.14.3 Regular Convolution from Periodic Convolution

The linear convolution of $x[n]$ (with length $N_x$) and $h[n]$ (with length $N_h$) may also be found using the *periodic* convolution of two zero-padded signals $x_z[n]$ and $h_z[n]$ (each of length $N_y = N_x + N_h - 1$). The *regular* convolution of the original, unpadded sequences equals the *periodic* convolution of the zero-padded sequences.

**Example 3.31**

**Regular Convolution by the Circulant Matrix**

Let $x[n] = \{\overset{\Downarrow}{2}, 5, 0, 4\}$ and $h[n] = \{\overset{\Downarrow}{4}, 1, 3\}$.

Their regular convolution has $S = M + N - 1 = 6$ samples. Using trailing zeros, we create the padded sequences

$$x_{\text{zp}}[n] = \{\overset{\Downarrow}{2}, 5, 0, 4, 0, 0\} \qquad h_{\text{zp}}[n] = \{\overset{\Downarrow}{4}, 1, 3, 0, 0, 0\}$$

The periodic convolution $x_{\text{zp}}[n] \circledast h_{\text{zp}}[n]$, using the circulant matrix, equals

$$\mathbf{C}_{\text{xzp}} = \begin{bmatrix} 2 & 0 & 0 & 4 & 0 & 5 \\ 5 & 2 & 0 & 0 & 4 & 0 \\ 0 & 5 & 2 & 0 & 0 & 4 \\ 4 & 0 & 5 & 2 & 0 & 0 \\ 0 & 4 & 0 & 5 & 2 & 0 \\ 0 & 0 & 4 & 0 & 5 & 2 \end{bmatrix} \qquad \mathbf{h}_{\text{zp}} = \begin{bmatrix} 4 \\ 1 \\ 3 \\ 0 \\ 0 \\ 0 \end{bmatrix} \qquad y_p[n] = \begin{bmatrix} 8 \\ 22 \\ 11 \\ 31 \\ 4 \\ 12 \end{bmatrix}$$

This is identical to the regular convolution $y[n] = x[n]*h[n]$ obtained previously by several other methods in previous examples.

---

## Drill Problem 3.31

Let $x[n] = \{\overset{\Downarrow}{1}, 2, 0, 2, 2\}$ and $h[n] = \{\overset{\Downarrow}{3}, 2\}$.

**(a)** How many zeros must be appended to $x[n]$ and $h[n]$ in order to generate their regular convolution from the periodic convolution of the zero-padded sequences.
**(b)** What is the regular convolution of the zero-padded sequences?
**(c)** What is the regular convolution of the original sequences?

**Answers: (a)** 1, 4 **(b)** $\{\overset{\Downarrow}{3}, 8, 4, 6, 10, 4, 0, 0, 0, 0, 0\}$ **(c)** $\{\overset{\Downarrow}{3}, 8, 4, 6, 10, 4\}$

# 3.15 Deconvolution

Given the system impulse response $h[n]$, the response $y[n]$ of the system to an input $x[n]$ is simply the convolution of $x[n]$ and $h[n]$. Given $x[n]$ and $y[n]$ instead, how do we find $h[n]$? This situation arises very often in practice and is referred to as **deconvolution** or **system identification**.

For discrete-time systems, we have a partial solution to this problem. Since discrete convolution may be thought of as polynomial multiplication, *discrete* deconvolution may be regarded as polynomial division. One approach to discrete deconvolution is to use the idea of *long division*, a familiar process, illustrated in the following section.

## 3.15.1 Deconvolution by Recursion

Deconvolution also may be recast as a recursive algorithm. The convolution

$$y[n] = x[n] * h[n] = \sum_{k=0}^{n} h[k]x[n-k] \tag{3.52}$$

when evaluated at $n = 0$, provides the seed value $h[0]$ as

$$y[0] = x[0]h[0] \qquad h[0] = \frac{y[0]}{x[0]} \tag{3.53}$$

We now separate the term containing $h[n]$ in the convolution relation:

$$y[n] = \sum_{k=0}^{n} h[k]x[n-k] = h[n]x[0] + \sum_{k=0}^{n-1} h[k]x[n-k] \tag{3.54}$$

and evaluate $h[n]$ for successive values of $n > 0$ from

$$h[n] = \frac{1}{x[0]}\left[y[n] - \sum_{k=0}^{n-1} h[k]x[n-k]\right], \quad n > 0 \qquad (3.55)$$

If all goes well, we need to evaluate $h[n]$ only at $M - N + 1$ points, where $M$ and $N$ are the lengths of $y[n]$ and $x[n]$, respectively.

Naturally, problems arise if a remainder is involved. This may well happen in the presence of noise, which could modify the values in the output sequence even slightly. In other words, the approach is quite susceptible to noise or roundoff error and not very practical.

**Example 3.32**

### Deconvolution

**(a) Deconvolution by Polynomial Division**

Consider $x[n] = \{\overset{\Downarrow}{2}, 5, 0, 4\}$ and $y[n] = \{\overset{\Downarrow}{8}, 22, 11, 31, 4, 12\}$.

We regard these as being the coefficients, in descending order, of the polynomials

$$x(w) = 2w^3 + 5w^2 + 0w + 4 \qquad y(w) = 8w^5 + 22w^4 + 11w^3 + 31w^2 + 4w + 12$$

The polynomial $h(w)$ may be deconvolved out of $x(w)$ and $y(w)$ by performing the division $y(w)/x(w)$:

$$\begin{array}{r|l}
 & 4w^2 + w + 3 \\
2w^3 + 5w^2 + 0w + 4 & 8w^5 + 22w^4 + 11w^3 + 31w^2 + 4w + 12 \\
 & 8w^5 + 20w^4 + 0w^3 + 16w^2 \\
 & \quad 2w^4 + 11w^3 + 15w^2 + 4w + 12 \\
 & \quad 2w^4 + 5w^3 + 0w^2 + 4w \\
 & \qquad 6w^3 + 15w^2 + 0w + 12 \\
 & \qquad 6w^3 + 15w^2 + 0w + 12 \\
 & \qquad\qquad 0
\end{array}$$

The coefficients of the quotient polynomial describe the sequence $h[n] = \{\overset{\Downarrow}{4}, 1, 3\}$.

**(b) Deconvolution by Recursion**

Let $x[n] = \{\overset{\Downarrow}{2}, 5, 0, 4\}$ and $y[n] = \{\overset{\Downarrow}{8}, 22, 11, 31, 4, 12\}$.

We note that $x[n]$ is of length $N = 4$ and $y[n]$ is of length $M = 4$. So, we need only $M - N + 1 = 6 - 4 + 1 = 3$ recursive evaluations to obtain $h[n]$. We compute

$$h[0] = \frac{y[0]}{x[0]} = 4$$

$$h[1] = \frac{1}{x[0]}\left[y[1] - \sum_{k=0}^{0} h[k]x[1-k]\right] = \frac{y[1] - h[0]x[1]}{x[0]} = 1$$

$$h[2] = \frac{1}{x[0]}\left[y[2] - \sum_{k=0}^{1} h[k]x[2-k]\right] = \frac{y[2] - h[0]x[2] - h[1]x[1]}{x[0]} = 3$$

As before, $h[n] = \{4, 1, 3\}$.

## Drill Problem 3.32

The input $x[n] = \{\overset{\Downarrow}{1},\ 2\}$ to an LTI system produces the output $y[n] = \{\overset{\Downarrow}{2},\ 3,\ 1,\ 6\}$. Use deconvolution to find the impulse response $h[n]$.

**Answer:** $\{\underset{\Uparrow}{2},\ -1,\ 3\}$

# 3.16 Discrete Correlation

Correlation is a measure of similarity between two signals and is found using a process similar to convolution. The discrete **cross-correlation** (denoted $**$) of $x[n]$ and $h[n]$ is defined by

$$r_{xh}[n] = x[n] ** h[n] = \sum_{k=-\infty}^{\infty} x[k]h[k-n] = \sum_{k=-\infty}^{\infty} x[k+n]h[k] \tag{3.56}$$

$$r_{hx}[n] = h[n] ** x[n] = \sum_{k=-\infty}^{\infty} h[k]x[k-n] = \sum_{k=-\infty}^{\infty} h[k+n]x[k] \tag{3.57}$$

Some authors prefer to switch the definitions of $r_{xh}[n]$ and $r_{hx}[n]$.

To find $r_{xh}[n]$, we line up the last element of $h[n]$ with the first element of $x[n]$ and start shifting $h[n]$ past $x[n]$, one index at a time. We sum the pointwise product of the overlapping values to generate the correlation at each index. This is equivalent to performing the *convolution* of $x[n]$ and the *flipped* signal $h[-n]$. The starting index of the correlation equals the sum of the starting indices of $x[n]$ and $h[-n]$. Similarly, $r_{hx}[n]$ equals the convolution of $x[-n]$ and $h[n]$, and its starting index equals the sum of the starting indices of $x[-n]$ and $h[n]$. However, $r_{xh}[n]$ does not equal $r_{hx}[n]$. The two are flipped (time reversed) versions of each other and related by $r_{xh}[n] = r_{hx}[-n]$.

**Correlation Is the Convolution of One Signal with a Flipped Version of the Other**

$r_{xh}[n] = x[n] ** h[n] = x[n] * h[-n]$ $\qquad$ $r_{hx}[n] = h[n] ** x[n] = h[n] * x[-n]$

**Correlation length:** $N_x + N_h - 1$ $\qquad$ **Correlation sum:** $\sum r[n] = (\sum x[n])(\sum h[n])$

**Example 3.33**

## Discrete Autocorrelation and Cross-Correlation

**(a)** Let $x[n] = \{2, \overset{\Downarrow}{5}, 0, 4\}$ and $h[n] = \{\overset{\Downarrow}{3}, 1, 4\}$.

To find $r_{xh}[n]$, we compute the convolution of $x[n]$ and $h[-n] = \{4, 1, \overset{\Downarrow}{3}\}$. The starting index of $r_{xh}[n]$ is $n = -3$. We use this to set up the index array and generate the result (using the sum-by-column method for convolution) as follows:

| $n$ | $-3$ | $-2$ | $-1$ | $0$ | $1$ | $2$ |
|---|---|---|---|---|---|---|
| $x[n]$ | 2 | 5 | 0 | 4 | | |
| $h[-n]$ | 4 | 1 | 3 | | | |
| | 8 | 20 | 0 | 16 | | |
| | | 2 | 5 | 0 | 4 | |
| | | | 6 | 15 | 0 | 12 |
| $r_{xh}[n]$ | 8 | 22 | 11 | $\overset{\Downarrow}{31}$ | 4 | 12 |

So, $r_{xh}[n] = \{8, 22, 11, \overset{\Downarrow}{31}, 4, 12\}$

**(b)** Let $x[n] = \{2, \overset{\Downarrow}{5}, 0, 4\}$ and $h[n] = \{\overset{\Downarrow}{3}, 1, 4\}$.

To find $r_{hx}[n]$, we compute the convolution of $x[-n] = \{4, 0, \overset{\Downarrow}{5}, 2\}$ and $h[n]$. The starting index of $r_{hx}[n]$ is $n = -2$. We use this to set up the index array and generate the result (using the sum-by-column method for convolution) as follows:

| $n$ | $-2$ | $-1$ | $0$ | $1$ | $2$ | $3$ |
|---|---|---|---|---|---|---|
| $x[-n]$ | 4 | 0 | 5 | 2 | | |
| $h[n]$ | 3 | 1 | 4 | | | |
| | 12 | 0 | 15 | 6 | | |
| | | 4 | 0 | 5 | 2 | |
| | | | 16 | 0 | 20 | 8 |
| $r_{hx}[n]$ | 12 | 4 | $\overset{\Downarrow}{31}$ | 11 | 22 | 8 |

So, $r_{xh}[n] = \{12, 4, \overset{\Downarrow}{31}, 11, 22, 8\}$.
(*Note*: $r_{xh}[n]$ and $r_{xh}[n]$ are flipped versions of each other with $r_{xh}[n] = r_{xh}[-n]$.)

**(c)** Let $x[n] = \{\overset{\Downarrow}{3}, 1, -4\}$.

To find $r_{xx}[n]$, we compute the convolution of $x[n]$ and $x[-n] = \{-4, 1, \overset{\Downarrow}{3}\}$. The starting index of $r_{xx}[n]$ is $n = -2$. We use this to set up the index array and generate the result (using the sum-by-column method for convolution) as follows:

| $n$ | $-2$ | $-1$ | $0$ | $1$ | $2$ |
|---|---|---|---|---|---|
| $x[n]$ | 3 | 1 | $-4$ | | |
| $x[-n]$ | $-4$ | 1 | 3 | | |
| | $-12$ | $-4$ | 16 | | |
| | | 3 | 1 | $-4$ | |
| | | | 9 | 3 | $-12$ |
| $r_{xx}[n]$ | $-12$ | $-1$ | $\overset{\Downarrow}{26}$ | $-1$ | $-12$ |

So, $r_{xx}[n] = \{-12, -1, \overset{\Downarrow}{26}, -1, -12\}$. Note that $r_{xx}[n]$ is even symmetric about the origin $n = 0$.

# Drill Problem 3.33

Let $x[n] = \{\overset{\Downarrow}{2},\ 1,\ 3\}$ and $h[n] = \{3, \overset{\Downarrow}{-2},\ 1\}$. Find $r_{xh}[n]$, $r_{hx}[n]$, $r_{xx}[n]$ and $r_{hh}[n]$

**Answers:** $\{2, \overset{\Downarrow}{-3}, 7, -3, 9\}$, $\{9, -3, 7, \overset{\Downarrow}{-3}, 2\}$, $\{6, 5, \overset{\Downarrow}{14}, 5, 6\}$, $\{3, -8, \overset{\Downarrow}{14}, -8, 3\}$

## 3.16.1 Autocorrelation

The correlation $r_{xx}[n]$ of a signal $x[n]$ with itself is called the *autocorrelation*. In terms of the convolution operation, we may write

$$r_{xx}[n] = x[n] ** x[n] = x[n] * x[-n] \tag{3.58}$$

The autocorrelation is an even symmetric function with

$$r_{xx}[n] = r_{xx}[-n] \tag{3.59}$$

The autocorrelation has a maximum at $n = 0$ and satisfies the inequality

$$|r_{xx}[n]| \le r_{xx}[0] \tag{3.60}$$

In other words, the absolute value of the autocorrelation at any index can never exceed its value at the origin. Correlation is an effective method of detecting signals buried in noise. Noise is essentially uncorrelated with the signal. This means that, if we correlate a noisy signal with itself, the correlation will be due only to the signal (if present) and will exhibit a sharp peak at $n = 0$.

**Example 3.34**

### Discrete Autocorrelation and Cross-Correlation

**(a)** Let $x[n] = (0.5)^n u[n]$ and $h[n] = (0.4)^n u[n]$.

We compute the cross-correlation $r_{xh}[n]$ as follows:

$$r_{xh}[n] = \sum_{k=-\infty}^{\infty} x[k]h[k-n] = \sum_{k=-\infty}^{\infty} (0.5)^k (0.4)^{k-n} u[k]u[k-n]$$

This summation requires evaluation over two ranges of $n$. If $n < 0$, the shifted step $u[k-n]$ is nonzero for some $k < 0$. But since $u[k] = 0$, $k < 0$, the lower limit on the summation reduces to $k = 0$ and we get

$$(n < 0) \quad r_{xh}[n] = \sum_{k=0}^{\infty} (0.5)^k (0.4)^{k-n} = (0.4)^{-n} \sum_{k=0}^{\infty} (0.2)^k$$

$$= \frac{(0.4)^{-n}}{1 - 0.2} = 1.25(0.4)^{-n} u[-n-1]$$

If $n \ge 0$, the shifted step $u[k-n]$ is zero for $k < n$, the lower limit on the summation reduces to $k = n$ and we obtain

$$(n \geq 0) \quad r_{xh}[n] = \sum_{k=n}^{\infty} (0.5)^k (0.4)^{k-n}$$

With the change of variable $m = k - n$, we get

$$(n \geq 0) \quad r_{xh}[n] = \sum_{m=0}^{\infty} (0.5)^{m+n} (0.4)^m = (0.5)^n \sum_{m=0}^{\infty} (0.2)^m = \frac{(0.5)^n}{1 - 0.2} = 1.25(0.5)^n u[n]$$

So, $r_{xh}[n] = 1.25(0.4)^{-n} u[-n-1] + 1.25(0.5)^n u[n]$.

**(b)** Let $x[n] = a^n u[n]$, $|a| < 1$.

To compute $r_{xx}[n]$ which is even symmetric, we need compute the result only for $n \geq 0$ and create its even extension. Following the previous part, we have.

$$(n \geq 0) \quad r_{xx}[n] = \sum_{k=-\infty}^{\infty} x[k]x[k-n] = \sum_{k=n}^{\infty} a^k a^{k-n} = \sum_{m=0}^{\infty} a^{m+n} a^m$$

$$= a^n \sum_{m=0}^{\infty} a^{2m} = \frac{a^n}{1 - a^2} u[n]$$

The even extension of this result gives $r_{xx}[n] = \frac{a^{|n|}}{1-a^2}$.

**(c)** Let $x[n] = a^n u[n]$, $|a| < 1$, and $y[n] = \text{rect}(n/2N)$.

To find $r_{xy}[n]$, we shift $y[k]$ and sum the products over different ranges. Since $y[k - n]$ shifts the pulse to the right over the limits $(-N + n, N + n)$, the correlation $r_{xy}[n]$ equals zero until $n = -N$. We then obtain

$$-N \leq n \leq N - 1 \text{ (partial overlap):} \quad r_{xy}[n] = \sum_{k=-\infty}^{\infty} x[k]y[k-n] = \sum_{k=0}^{N+1} a^k$$

$$= \frac{1 - a^{N+n+1}}{1 - a}$$

$$n \geq N \text{ (total overlap):} \quad r_{xy}[n] = \sum_{k=-N+1}^{N+1} a^k = \sum_{m=0}^{2N} a^{m-N+1}$$

$$= a^{-N+1} \frac{1 - a^{2N+1}}{1 - a}$$

# Drill Problem 3.34

**(a)** Let $x[n] = (0.5)^n u[n]$ and $h[n] = u[n]$. Find their correlation.
**(b)** Let $x[n] = (0.5)^n u[n]$ and $h[n] = (0.5)^{-n} u[-n]$. Find their correlation.

**Answers:** **(a)** $2u[-n-1] + 2(0.5)^n u[n]$ **(b)** $(n+1)(0.5)^n u[n]$

### 3.16.2 Periodic Discrete Correlation

For periodic sequences with identical period $N$, the periodic discrete correlation is defined as

$$r_{xhp}[n] = x[n]\circledast\circledast h[n] = \sum_{k=0}^{N-1} x[k]h[k-n] \quad r_{hxp}[n] = h[n]\circledast\circledast x[n] = \sum_{k=0}^{N-1} h[k]x[k-n] \tag{3.61}$$

As with discrete periodic convolution, an averaging factor of $1/N$ is sometimes included in the summation. We can find one period of the periodic correlation $r_{xhp}[n]$ by first computing the linear correlation of one period segments and then wrapping around the result. We find that $r_{hxp}[n]$ is a circularly flipped (time reversed) version of $r_{xhp}[n]$ with $r_{hxp}[n] = r_{xhp}[-n]$. We also find that the periodic autocorrelation $r_{xxp}[n]$ or $r_{hhp}[n]$ always displays circular even symmetry. This means that the periodic extension of $r_{xxp}[n]$ or $r_{hhp}[n]$ is even symmetric about the origin $n = 0$. The periodic autocorrelation function also attains a maximum at $n = 0$.

**Example 3.35**

**Discrete Periodic Autocorrelation and Cross-Correlation**

Consider two periodic signals whose first period is given by $x_1[n] = \{\overset{\Downarrow}{2},\ 5,\ 0,\ 4\}$ and $h_1[n] = \{\overset{\Downarrow}{3},\ 1,\ -1,\ 2\}$.

**(a)** To find the periodic cross-correlation $r_{xhp}[n]$, we first evaluate the linear cross-correlation

$$r_{xh}[n] = x_1[n] * h_1[-n] = \{4,\ 8,\ -3,\ \overset{\Downarrow}{19},\ 11,\ 4,\ 12\}$$

Wraparound gives the periodic cross-correlation as $r_{xhp} = \{15,\ 12,\ 9,\ \overset{\Downarrow}{19}\}$.

We invoke periodicity and describe the result in terms of its first period as $r_{xhp} = \{\overset{\Downarrow}{19},\ 15,\ 12,\ 9\}$.

**(b)** To find the periodic cross-correlation $r_{hxp}[n]$, we first evaluate the linear cross-correlation

$$r_{hx}[n] = r_{xh}[-n] = \{12,\ 4,\ 11,\ \overset{\Downarrow}{19},\ -3,\ 8,\ 4\}$$

Wraparound gives the periodic cross-correlation as $r_{hxp} = \{9,\ 12,\ 15,\ \overset{\Downarrow}{19}\}$.

We rewrite the result in terms of its first period as $r_{hxp} = \{\overset{\Downarrow}{19},\ 9,\ 12,\ 15\}$.
Note that $r_{hxp}[n]$ is a circularly flipped version of $r_{xhp}[n]$ with $r_{hxp}[n] = r_{xhp}[-n]$

**(c)** To find the periodic autocorrelation $r_{xxp}[n]$, we first evaluate the linear autocorrelation

$$r_{xx}[n] = x_1[n] * x_1[-n] = \{8,\ 20,\ 10,\ \overset{\Downarrow}{45},\ 10,\ 20,\ 8\}$$

Wraparound gives the periodic autocorrelation as $r_{xxp} = \{18,\ 40,\ 18,\ \overset{\Downarrow}{45}\}$.

We rewrite the result in terms of its first period as $r_{hxp} = \{\overset{\Downarrow}{45},\ 18,\ 40,\ 18\}$.
This displays circular even symmetry (its periodic extension is even symmetric about the origin $n = 0$).

**(d)** To find the periodic autocorrelation $r_{hhp}[n]$, we first evaluate the linear autocorrelation

$$r_{hh}[n] = h_1[n] * h_1[-n] = \{6,\ -1,\ 0,\ \overset{\Downarrow}{15},\ 0,\ -1,\ 6\}$$

Wraparound gives the periodic autocorrelation as $r_{xxp} = \{6,\ -2,\ 6,\ \overset{\Downarrow}{15}\}$.

We rewrite the result in terms of its first period as $r_{hxp} = \{\overset{\Downarrow}{15},\ 6,\ -2,\ 6\}$.
This displays circular even symmetry (its periodic extension is even symmetric about $n = 0$).

# Drill Problem 3.35

Let $x_1[n] = \{\overset{\Downarrow}{2},\ -1,\ 0,\ 3\}$ and $h_1[n] = \{1,\ \overset{\Downarrow}{0},\ 3,\ 2\}$ describe one-period segments of two periodic signals. Find the first period of $r_{xxp}[n]$, $r_{hhp}[n]$, $r_{xhp}[n]$ and $r_{hxp}[n]$

**Answers:** $\{\overset{\Downarrow}{14},\ 4,\ -6,\ 4\}$ $\{\overset{\Downarrow}{14},\ 8,\ 6,\ 8\}$ $\{\overset{\Downarrow}{0},\ 8,\ 12,\ 4\}$ $\{\overset{\Downarrow}{0},\ 4,\ 12,\ 8\}$

### 3.16.3 Matched Filtering and Target Ranging

Correlation finds widespread use in applications such as target ranging and estimation of periodic signals buried in noise. For target ranging, a sampled interrogating signal $x[n]$ is transmitted toward the target. The signal reflected from the target is $s[n] = \alpha x[n-D]+p[n]$, a delayed (by $D$) and attenuated (by $\alpha$) version of $x[n]$, contaminated by noise $p[n]$. The reflected signal $s[n]$ is correlated with the interrogating signal $x[n]$. If the noise is uncorrelated with the signal $x[n]$, its correlation with $x[n]$ is essentially zero. The correlation of $x[n]$ and its delayed version $\alpha x[n - D]$ yield a result that attains a peak at $n = D$. It is thus quite easy to identify the index $D$ from the correlation peak (rather than from the reflected signal directly), even in the presence of noise. The (round-trip) delay index $D$ may then be related to the target distance $d$ by $d = 0.5vD/S$, where $v$ is the propagation velocity and $S$ is the rate at which the signal is sampled. The device that performs the correlation of the received signal $s[n]$ and $x[n]$ is called a **correlation receiver**. The correlation of $s[n]$ with $x[n]$ is equivalent to the convolution of $s[n]$ with $x[-n]$, a flipped version of the interrogating signal. This means that the impulse response of the correlation receiver is $h[n] = x[-n]$ and is matched to the transmitted signal. For this reason, such a receiver is also called a **matched filter**.

Figure 3.10 illustrates the concept of matched filtering. The transmitted signal is a rectangular pulse. The impulse response of the matched filter is its flipped version and is noncausal. In an ideal situation, the received signal is simply a delayed version of the transmitted signal and the output of the matched filter yields a peak whose location gives the delay. This can also be identified from the ideal received signal itself. In practice, the received signal is contaminated by noise and it is difficult to identify where the delayed pulse is located. The output of the matched filter, however, provides a clear indication of the delay even for low signal-to-noise ratios.

Correlation also finds application in pattern recognition. For example, if we need to establish whether an unknown pattern belongs to one of several known patterns or templates, it can be compared (correlated) with each template in turn. A match occurs if the

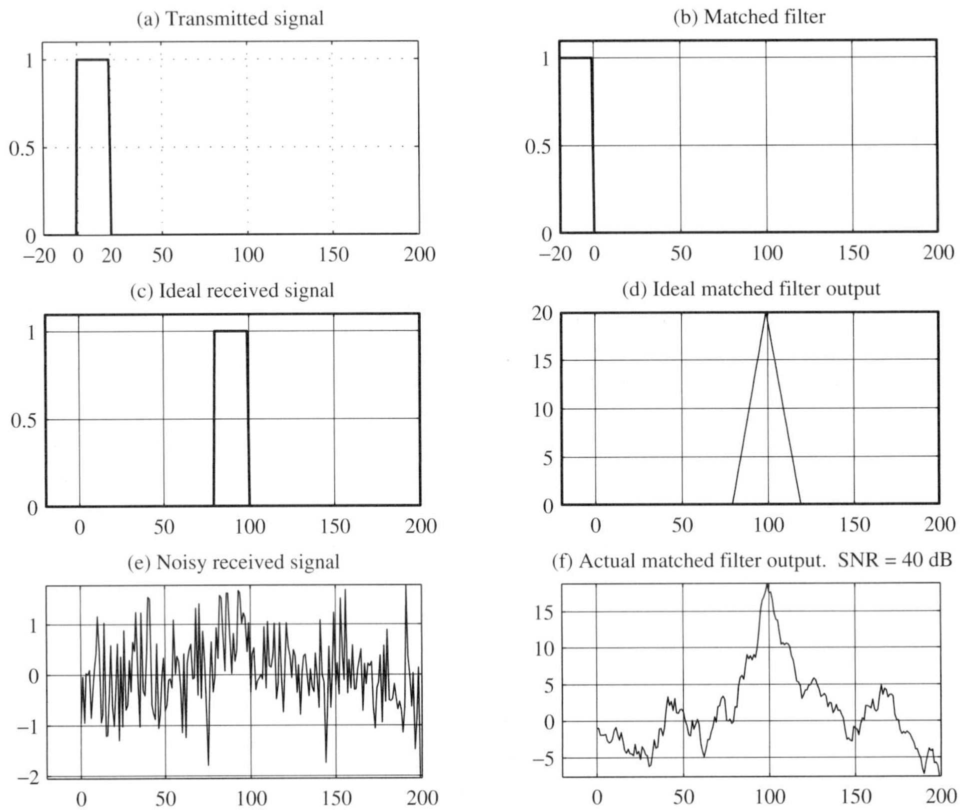


**FIGURE 3.10** The concept of matched filtering and target ranging. The output of the matched filter is the correlation of the transmitted and received signals. It exhibits a sharp peak even when the received signal is contaminated by noise. The time of the peak corresponds to the round-trip time. The target distance can be calculated from this value

autocorrelation of the template matches (or resembles) the cross-correlation of the template and the unknown pattern.

## Identifying Periodic Signals in Noise

Correlation methods may also be used to identify the presence of a periodic signal $x[n]$ buried in the noisy signal $s[n] = x[n] + p[n]$, where $p[n]$ is the noise component (presumably uncorrelated with the signal), and to extract the signal itself. The idea is to first identify the period of the signal from the *periodic* autocorrelation of the noisy signal. If the noisy signal contains a periodic component, the autocorrelation will show peaks at multiples of the period $N$. Once the period $N$ is established, we can recover $x[n]$ as the *periodic* cross-correlation of an impulse train $i[n] = \delta(n - kN)$, with period $N$, and the noisy signal $s[n]$. Since $i[n]$ is uncorrelated with the noise, the periodic cross-correlation of $i[n]$ and $s[n]$ yields (an amplitude scaled version of) the periodic signal $x[n]$.

Figure 3.11 illustrates the concept of identifying a periodic signal hidden in noise. A periodic sawtooth signal is contaminated by noise to yield a noisy signal. The peaks in the periodic autocorrelation of the noisy signal allows us to identify the period as $N = 20$. The periodic cross-correlation of the noisy signal with the impulse train $\delta[n - 20k]$ extracts the

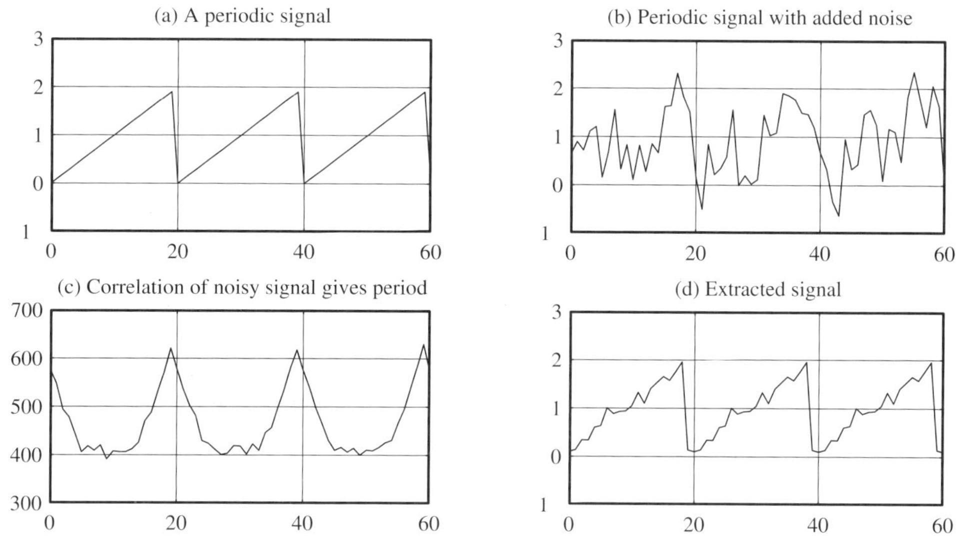


**FIGURE 3.11** Extraction of a periodic signal buried in noise. The periodic autocorrelation of a long stretch of the noisy signal reveals the period of the buried periodic signal in panel(c) as $N = 20$. The cross-correlation of the noisy signal with a periodic impulse train with period $N = 20$ extracts the buried periodic signal

signal from noise. Note that longer lengths of the noisy signal (compared to the period $N$) will improve the match between the recovered and buried periodic signal.

---

**How to Identify a Periodic Signal $x[n]$ Buried in a Noisy Signal $s[n]$**

1. Find the period $N$ of $x[n]$ from the *periodic* autocorrelation of the noisy signal $s[n]$.
2. Find the signal $x[n]$ as the *periodic* cross-correlation of $s[n]$ and an *impulse train* with period $N$.

---

# 3.17 Discrete Convolution and Transform Methods

Discrete-time convolution provides a connection between the time-domain and frequency-domain methods of system analysis for discrete-time signals. It forms the basis for every transform method described in this text, and its role in linking the time domain and the transformed domain is tied intimately to the concept of discrete eigensignals and eigenvalues. The everlasting exponential $z^n$ is an eigensignal of discrete-time linear systems. In this complex exponential $z^n$, the quantity $z$ has the general form $z = re^{j2\pi F}$. If the input to an LTI system is $z^n$, the output has the same form and is given by $Cz^n$, where $C$ is a (possibly complex) constant. Similarly, the everlasting discrete-time harmonic $z = e^{j2\pi F}$ (a special case with $r = 1$) is also an eigensignal of discrete-time systems.

### 3.17.1 The z-Transform

For an input $x[n] = r^n e^{j2\pi nF} = (re^{j2\pi F})^n = z^n$, where $z$ is complex, with magnitude $|z| = r$, the response may be written as

$$y[n] = x[n] * h[n] = \sum_{k=-\infty}^{\infty} z^{n-k}h[k] = z^n \sum_{k=-\infty}^{\infty} h[k]z^{-k} = x[n]H(z) \tag{3.62}$$

The response equals the input (eigensignal) modified by the system function $H(z)$, where

$$H(z) = \sum_{k=-\infty}^{\infty} h[k]z^{-k} \quad \text{(two-sided } z\text{-transform)} \tag{3.63}$$

The complex quantity $H(z)$ describes the **z-transform** of $h[n]$ and is not, in general, periodic in $z$. Denoting the $z$-transform of $x[n]$ and $y[n]$ by $X(z)$ and $Y(z)$, we write

$$Y(z) = \sum_{k=-\infty}^{\infty} y[k]z^{-k} = \sum_{k=-\infty}^{\infty} x[k]H(z)z^{-k} = H(z)X(z) \tag{3.64}$$

Convolution in the time domain thus corresponds to multiplication in the $z$-domain.

### 3.17.2 The Discrete-Time Fourier Transform

For the harmonic input $x[n] = e^{j2\pi nF}$, the response $y[n]$ equals

$$y[n] = \sum_{k=-\infty}^{\infty} e^{j2\pi(n-k)F}h[k] = e^{j2\pi nF} \sum_{k=-\infty}^{\infty} h[k]e^{-j2\pi kF} = x[n]H(F) \tag{3.65}$$

This is just the input modified by the system function $H(F)$, where

$$H(F) = \sum_{k=-\infty}^{\infty} h[k]e^{-j2\pi kF} \tag{3.66}$$

The quantity $H(F)$ describes the **discrete-time Fourier transform** (DTFT) or **discrete-time frequency response** or **spectrum** of $h[n]$. Any signal $x[n]$ may similarly be described by its DTFT $X(F)$. The response $y[n] = x[n]H[F]$ may then be transformed to its DTFT $Y[n]$ to give

$$Y(F) = \sum_{k=-\infty}^{\infty} y[k]e^{-j2\pi Fk} = \sum_{k=-\infty}^{\infty} x[k]H(F)e^{-j2\pi Fk} = H(F)X(F) \tag{3.67}$$

Once again, convolution in the time domain corresponds to multiplication in the frequency domain. Note that we obtain the DTFT of $h[n]$ from its $z$-transform $H(z)$ by letting $z = e^{j2\pi F}$ or $|z| = 1$ to give

$$H(F) = H(z)|_{z=\exp(j2\pi F)} = H(z)|_{|z|=1} \tag{3.68}$$

The DTFT is thus the $z$-transform evaluated on the unit circle $|z| = 1$. The system function $H(F)$ is also periodic in $F$ with a period of unity because $e^{-j2\pi kF} = e^{-j2\pi k(F+1)}$. This periodicity is a direct consequence of the discrete nature of $h[n]$.

# 3.18 Problems

**3.1.** **(Operators)** Which of the following describe linear operators?

**(a)** $\mathcal{O}\{\ \} = 4\{\ \}$ **(b)** $\mathcal{O}\{\ \} = 4\{\ \} + 3$ **(c)** $\mathcal{O}\{\ \} = \alpha^{\{\ \}}$

**3.2.** **(System Classification)** In each of the systems below, $x[n]$ is the input and $y[n]$ is the output. Check each system for linearity, shift invariance, memory, and causality.

**(a)** $y[n] - y[n-1] = x[n]$
**(b)** $y[n] + y[n+1] = nx[n]$
**(c)** $y[n] - y[n+1] = x[n+2]$
**(d)** $y[n+2] - y[n+1] = x[n]$
**(e)** $y[n+1] - x[n]y[n] = nx[n+2]$
**(f)** $y[n] + y[n-3] = x^2[n] + x[n+6]$
**(g)** $y[n] - 2^n y[n] = x[n]$
**(h)** $y[n] = x[n] + x[n-1] + x[n-2]$

**3.3.** **(System Classification)** Classify the following systems in terms of their linearity, time invariance, memory, causality, and stability.

**(a)** $y[n] = 3^n x[n]$
**(b)** $y[n] = e^{jn\pi} x[n]$
**(c)** $y[n] = \cos(0.5n\pi)x[n]$
**(d)** $y[n] = [1 + \cos(0.5n\pi)]x[n]$
**(e)** $y[n] = e^{x[n]}$
**(f)** $y[n] = x[n] + \cos[0.5(n+1)\pi]$

**3.4.** **(System Classification)** Classify the following systems in terms of their linearity, time invariance, memory, causality, and stability.

**(a)** $y[n] = x[n/3]$ (zero interpolation)
**(b)** $y[n] = \cos(n\pi)x[n]$ (modulation)
**(c)** $y[n] = [1 + \cos(n\pi)]x[n]$ (modulation)
**(d)** $y[n] = \cos(n\pi x[n])$ (frequency modulation)
**(e)** $y[n] = \cos(n\pi + x[n])$ (phase modulation)
**(f)** $y[n] = x[n] - x[n-1]$ (differencing operation)
**(g)** $y[n] = 0.5x[n] + 0.5x[n-1]$ (averaging operation)
**(h)** $y[n] = \frac{1}{N}\sum_{k=0}^{N-1} x[n-k]$ (moving average)
**(i)** $y[n] - \alpha y[n-1] = \alpha x[n],\ 0 < \alpha < 1$ (exponential averaging)
**(j)** $y[n] = 0.4(y[n-1] + 2) + x[n]$

**3.5.** **(Classification)** Classify each system in terms of its linearity, time invariance, memory, causality, and stability.

**(a)** the time-reversing system $y[n] = x[-n]$
**(b)** the decimating system $y[n] = x[2n]$
**(c)** the zero-interpolating system $y[n] = x[n/2]$
**(d)** the sign-inversion system $y[n] = \operatorname{sgn}\{x[n]\}$
**(e)** the rectifying system $y[n] = |x[n]|$

**3.6.** **(Classification)** Classify each system in terms of its linearity, time invariance, causality, and stability.

**(a)** $y[n] = \operatorname{round}\{x[n]\}$
**(b)** $y[n] = \operatorname{median}\{x[n+1], x[n], x[n-1]\}$
**(c)** $y[n] = x[n]\operatorname{sgn}(n)$
**(d)** $y[n] = x[n]\operatorname{sgn}\{x[n]\}$

**3.7.** **(Realization)** Find the difference equation for each system realization shown in Figure P3.7.

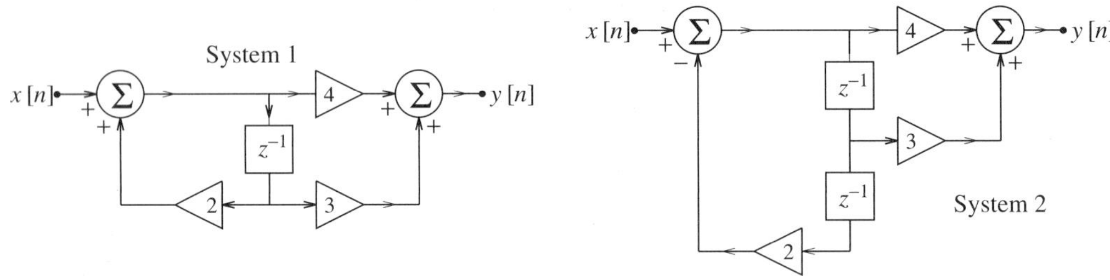


**FIGURE P.3.7** Filter realizations for Problem 3.7

[**Hints and Suggestions:** Compare with the generic first-order and second-order realizations to get the difference equations.]

**3.8. (Response by Recursion)** Use recursion to find the response $y[n]$ of the following systems for the first few values of $n$ and discern the general form for $y[n]$.

**(a)** $y[n] - ay[n-1] = \delta[n] \; y[-1] = 0$ **(b)** $y[n] - ay[n-1] = u[n] \; y[-1] = 1$
**(c)** $y[n] - ay[n-1] = u[n] \; y[-1] = 1$ **(d)** $y[n] - ay[n-1] = nu[n] \; y[-1] = 0$

[**Hints and Suggestions:** For parts (b) through (d), you may need to use a table of summations to simplify the results for the general form.]

**3.9. (Response by Recursion)** Let $y[n] + 4y[n-1] + 3y[n-2] = u[n-2]$ with $y[-1] = 0, \; y[-2] = 1$. Use recursion to compute $y[n]$ up to $n = 4$. Can you discern a general form for $y[n]$?

**3.10. (Forced Response)** Find the forced response of the following systems.

**(a)** $y[n] - 0.4y[n-1] = 3u[n]$ **(b)** $y[n] - 0.4y[n-1] = (0.5)^n$
**(c)** $y[n] + 0.4y[n-1] = (0.5)^n$ **(d)** $y[n] - 0.5y[n-1] = \cos(n\pi/2)$

[**Hints and Suggestions:** For part (a), $3u[n] = 3, \; n \geq 0$ and implies that the forced response (or its shifted version) is constant. So, choose $y_F[n] = C = y_F[n-1]$. For part (c), pick $y_F[n] = A\cos(0.5n\pi) + B\sin(0.5n\pi)$, expand terms like $\cos[0.5(n-1)\pi]$ using trigonometric identities, and compare the coefficients of $\cos(0.5n\pi)$ and $\sin(0.5n\pi)$ to generate two equations to solve for $A$ and $B$.]

**3.11. (Zero-State Response)** Find the zero-state response of the following systems.

**(a)** $y[n] - 0.5y[n-1] = 2u[n]$ **(b)** $y[n] - 0.4y[n-1] = (0.5)^n u[n]$
**(c)** $y[n] - 0.4y[n-1] = (0.4)^n u[n]$ **(d)** $y[n] - 0.5y[n-1] = \cos(n\pi/2)$

[**Hints and Suggestions:** Here, zero-state implies $y[-1] = 0$. Part (c) requires $y_F[n] = Cn(0.4)^n$ because the root of the characteristic equation is 0.4.]

**3.12. (Zero-State Response)** Consider the system $y[n] - 0.5y[n-1] = x[n]$. Find its zero-state response to the following inputs.

**(a)** $x[n] = u[n]$ **(b)** $x[n] = (0.5)^n u[n]$ **(c)** $x[n] = \cos(0.5n\pi)u[n]$
**(d)** $x[n] = (-1)^n u[n]$ **(e)** $x[n] = j^n u[n]$ **(f)** $x[n] = (\sqrt{j})^n u[n] + (\sqrt{j})^{-n} u[n]$

[**Hints and Suggestions:** For part (e), pick the forced response as $y_F[n] = C(j)^n$. This will give a complex response because the input is complex. For part (f), $x[n]$ simplifies to a sinusoid by using $j = e^{j\pi/2}$ and Euler's relation.]

**3.13.** **(Zero-State Response)** Find the zero-state response of the following systems.

**(a)** $y[n] - 1.1y[n-1] + 0.3y[n-2] = 2u[n]$
**(b)** $y[n] + 0.7y[n-1] + 0.1y[n-2] = (0.5)^n$
**(c)** $y[n] - 0.9y[n-1] + 0.2y[n-2] = (0.5)^n$
**(d)** $y[n] - 0.25y[n-2] = \cos(n\pi/2)$

[**Hints and Suggestions:** Zero-state implies $y[-1] = y[-2] = 0$. For part (b), use $y_F[n] = C(0.5)^n$, but for part (c), pick $y_F[n] = Cn(0.5)^n$ because one root of the characteristic equation is 0.5.]

**3.14.** **(System Response)** Let $y[n] - 0.5y[n-1] = x[n]$, with $y[-1] = -1$. Find the response of this system due to the following inputs for $n \geq 0$.

**(a)** $x[n] = 2u[n]$ **(b)** $x[n] = (0.25)^n u[n]$ **(c)** $x[n] = n(0.25)^n u[n]$
**(d)** $x[n] = (0.5)^n u[n]$ **(e)** $x[n] = n(0.5)^n u[n]$ **(f)** $x[n] = (0.5)^n \cos(0.5n\pi)u[n]$

[**Hints and Suggestions:** For part (c), pick $y_F[n] = (C + Dn)(0.5)^n$ (and compare coefficients of like powers of $n$ to solve for $C$ and $D$). For part (d), pick $y_F[n] = Cn(0.5)^n$ because the root of the characteristic equation is 0.5. Part(e) requires $y_F[n] = n(C + Dn)(0.5)^n$ for the same reason.]

**3.15.** **(System Response)** For the system realization shown in Figure P3.15, find the response to the following inputs and initial conditions.

**(a)** $x[n] = u[n]$ $y[-1] = 0$ **(b)** $x[n] = u[n]$ $y[-1] = 4$
**(c)** $x[n] = (0.5)^n u[n]$ $y[-1] = 0$ **(d)** $x[n] = (0.5)^n u[n]$ $y[-1] = 6$
**(e)** $x[n] = (-0.5)^n u[n]$ $y[-1] = 0$ **(f)** $x[n] = (-0.5)^n u[n]$ $y[-1] = -2$

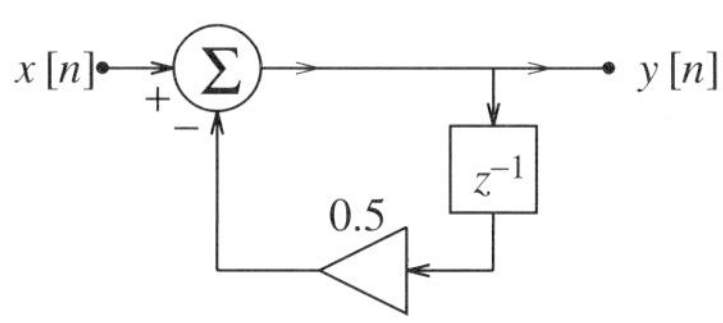


**FIGURE P.3.15** System realization for Problem 3.15

[**Hints and Suggestions:** For part (e), pick the forced response as $y_F[n] = Cn(-0.5)^n$.]

**3.16.** **(System Response)** Find the response of the following systems.

**(a)** $y[n] - 0.4y[n-1] = 2(0.5)^{n-1}u[n-1]$ $y[-1] = 0$
**(b)** $y[n] - 0.4y[n-1] = (0.4)^n u[n] + 2(0.5)^{n-1}u[n-1]$ $y[-1] = 2.5$
**(c)** $y[n] - 0.4y[n-1] = n(0.5)^n u[n] + 2(0.5)^{n-1}u[n-1]$ $y[-1] = 2.5$

[**Hints and Suggestions:** Start with $y[n] - 0.4y[n-1] = 2(0.5)^n$, $y[-1] = 0$ and find its zero-state response. Then use superposition and time invariance as required. For the input $(0.4)^n$ of part (b), assume $y_F[n] = Cn(0.4)^n$. For the input $n(0.5)^n$ of part (c), assume $y_F[n] = (A + Bn)(0.5)^n$.]

**3.17.** **(System Response)** Find the impulse response of the following filters.

**(a)** $y[n] = x[n] - x[n-1]$ (differencing operation)
**(b)** $y[n] = 0.5x[n] + 0.5x[n-1]$ (averaging operation)
**(c)** $y[n] = \frac{1}{N}\sum_{k=0}^{N-1} x[n-k]$, $N = 3$ (moving average)
**(d)** $y[n] = \frac{2}{N(N+1)}\sum_{k=0}^{N-1}(N-k)x[n-k]$, $N = 3$ (weighted moving average)
**(e)** $y[n] - \alpha y[n-1] = (1-\alpha)x[n]$, $N = 3$, $\alpha = \frac{N-1}{N+1}$ (exponential averaging)

**3.18.** **(System Response)** It is known that the response of the system $y[n] + \alpha y[n-1] = x[n]$, $\alpha \neq 0$, is given by $y[n] = [5 + 3(0.5)^n]u[n]$.

**(a)** Identify the natural response and forced response.
**(b)** Identify the values of $\alpha$ and $y[-1]$.
**(c)** Identify the zero-input response and zero-state response.
**(d)** Identify the input $x[n]$.

**3.19.** **(System Response)** It is known that the response of the system $y[n] + 0.5y[n-1] = x[n]$ is described by $y[n] = [5(0.5)^n + 3(-0.5)^n)]u[n]$.

**(a)** Identify the zero-input response and zero-state response.
**(b)** What is the zero-input response of the system $y[n] + 0.5y[n-1] = x[n]$ if $y[-1] = 10$?
**(c)** What is the response of the relaxed system $y[n] + 0.5y[n-1] = x[n-2]$?
**(d)** What is the response of the relaxed system $y[n] + 0.5y[n-1] = x[n-1] + 2x[n]$?

**3.20.** **(System Response)** It is known that the response of the system $y[n] + \alpha y[n-1] = x[n]$ is described by $y[n] = (5 + 2n)(0.5)^n u[n]$.

**(a)** Identify the zero-input response and zero-state response.
**(b)** What is the zero-input response of the system $y[n] + \alpha y[n-1] = x[n]$ if $y[-1] = 10$?
**(c)** What is the response of the relaxed system $y[n] + \alpha y[n-1] = x[n-1]$?
**(d)** What is the response of the relaxed system $y[n] + \alpha y[n-1] = 2x[n-1] + x[n]$?
**(e)** What is the complete response of the system $y[n] + \alpha y[n-1] = x[n] + 2x[n-1]$ if $y[-1] = 4$?

**3.21.** **(System Response)** Find the response of the following systems.

| | | |
|---|---|---|
| **(a)** $y[n] + 0.1y[n-1] - 0.3y[n-2] = 2u[n]$ | $y[-1] = 0$ | $y[-2] = 0$ |
| **(b)** $y[n] - 0.9y[n-1] + 0.2y[n-2] = (0.5)^n$ | $y[-1] = 1$ | $y[-2] = -4$ |
| **(c)** $y[n] + 0.7y[n-1] + 0.1y[n-2] = (0.5)^n$ | $y[-1] = 0$ | $y[-2] = 3$ |
| **(d)** $y[n] - 0.25y[n-2] = (0.4)^n$ | $y[-1] = 0$ | $y[-2] = 3$ |
| **(e)** $y[n] - 0.25y[n-2] = (0.5)^n$ | $y[-1] = 0$ | $y[-2] = 0$ |

[**Hints and Suggestions:** For parts (b) and (e), pick $y_F[n] = Cn(0.5)^n$ because one root of the characteristic equation is 0.5.]

**3.22.** **(System Response)** Sketch a realization for each system, assuming zero initial conditions. Then evaluate the complete response from the information given. Check your answer by computing the first few values by recursion.

| | | |
|---|---|---|
| **(a)** $y[n] - 0.4y[n-1] = x[n]$ | $x[n] = (0.5)^n u[n]$ | $y[-1] = 0$ |
| **(b)** $y[n] - 0.4y[n-1] = 2x[n] + x[n-1]$ | $x[n] = (0.5)^n u[n]$ | $y[-1] = 0$ |
| **(c)** $y[n] - 0.4y[n-1] = 2x[n] + x[n-1]$ | $x[n] = (0.5)^n u[n]$ | $y[-1] = 5$ |
| **(d)** $y[n] + 0.5y[n-1] = x[n] - x[n-1]$ | $x[n] = (0.5)^n u[n]$ | $y[-1] = 2$ |
| **(e)** $y[n] + 0.5y[n-1] = x[n] - x[n-1]$ | $x[n] = (-0.5)^n u[n]$ | $y[-1] = 0$ |

[**Hints and Suggestions:** For parts (b) through (c), use the results of part (a) plus linearity (superposition) and time invariance.]

**3.23.** **(System Response)** For each system, evaluate the natural, forced, and total response. Assume that $y[-1] = 0$, $y[-2] = 1$. Check your answer for the total response by computing its first few values by recursion.

**(a)** $y[n] + 4y[n-1] + 3y[n-2] = u[n]$
**(b)** $\{1 - 0.5z^{-1}\}y[n] = (0.5)^n \cos(0.5n\pi)u[n]$
**(c)** $y[n] + 4y[n-1] + 8y[n-2] = \cos(n\pi)u[n]$
**(d)** $\{(1 + 2z^{-1})^2\}y[n] = n(2)^n u[n]$
**(e)** $\{1 + \frac{3}{4}z^{-1} + \frac{1}{8}z^{-2}\}y[n] = (\frac{1}{3})^n u[n]$
**(f)** $\{1 + 0.5z^{-1} + 0.25z^{-2}\}y[n] = \cos(0.5n\pi)u[n]$

[**Hints and Suggestions:** For part (b), pick $y_F[n] = (0.5)^n[A\cos(0.5n\pi) + B\sin(0.5n\pi)]$, expand terms like $\cos[0.5(n-1)\pi]$ using trigonometric identities, and compare the coefficients of $\cos(0.5n\pi)$ and $\sin(0.5n\pi)$ to generate two equations to solve for $A$ and $B$. For part (d), pick $y_F[n] = (C + Dn)(2)^n$ and compare like powers of $n$ to solve for $C$ and $D$.]

**3.24. (System Response)** For each system, evaluate the zero-state, zero-input, and total response. Assume that $y[-1] = 0$, $y[-2] = 1$.

**(a)** $y[n] + 4y[n-1] + 4y[n-2] = 2^n u[n]$
**(b)** $\{z^2 + 4z + 4\}y[n] = 2^n u[n]$

[**Hints and Suggestions:** In part (b), $y[n+2] + 4y[n+1] + 4y[n] = (2)^n u[n]$. By time invariance, $y[n] + 4y[n-1] + 4y[n-2] = (2)^{n-2}u[n-2]$ and we shift the zero-state response of part (a) by two units ($n \to n-2$) and add to the zero-input response to get the result.]

**3.25. (System Response)** For each system, set up a difference equation and compute the zero-state, zero-input, and total response, assuming $x[n] = u[n]$ and $y[-1] = y[-2] = 1$.

**(a)** $\{1 - z^{-1} - 2z^{-2}\}y[n] = x[n]$
**(b)** $\{z^2 - z - 2\}y[n] = x[n]$
**(c)** $\{1 - \frac{3}{4}z^{-1} + \frac{1}{8}z^{-2}\}y[n] = x[n]$
**(d)** $\{1 - \frac{3}{4}z^{-1} + \frac{1}{8}z^{-2}\}y[n] = \{1 + z^{-1}\}x[n]$
**(e)** $\{1 - 0.25z^{-2}\}y[n] = x[n]$
**(f)** $\{z^2 - 0.25\}y[n] = \{2z^2 + 1\}x[n]$

[**Hints and Suggestions:** For part (b), use the result of part (a) and time-invariance to get the answer as $y_{zs}[n-2] + y_{zi}[n]$. For part (d), use the result of part (c) to get the answer as $y_{zi}[n] + y_{zs}[n] + y_{zs}[n-1]$. The answer for part (f) may be similarly obtained from part (e).]

**3.26. (Impulse Response by Recursion)** Find the impulse response $h[n]$ by recursion up to $n = 4$ for each of the following systems.

**(a)** $y[n] - y[n-1] = 2x[n]$
**(b)** $y[n] - 3y[n-1] + 6y[n-2] = x[n-1]$
**(c)** $y[n] - 2y[n-3] = x[n-1]$
**(d)** $y[n] - y[n-1] + 6y[n-2] = nx[n-1] + 2x[n-3]$

[**Hints and Suggestions:** For the impulse response, $x[n] = 1,\ n = 0$ and $x[n] = 0,\ n \neq 0$.]

**3.27. (Analytical Form for Impulse Response)** Classify each filter as recursive or FIR (nonrecursive) and causal or noncausal, and find an expression for its impulse response $h[n]$.

**(a)** $y[n] = x[n] + x[n-1] + x[n-2]$
**(b)** $y[n] = x[n+1] + x[n] + x[n-1]$
**(c)** $y[n] + 2y[n-1] = x[n]$
**(d)** $y[n] + 2y[n-1] = x[n-1]$
**(e)** $y[n] + 2y[n-1] = 2x[n] + 6x[n-1]$
**(f)** $y[n] + 2y[n-1] = x[n+1] + 4x[n] + 6x[n-1]$
**(g)** $\{1 + 4z^{-1} + 3z^{-2}\}y[n] = \{z^{-2}\}x[n]$
**(h)** $\{z^2 + 4z + 4\}y[n] = \{z + 3\}x[n]$
**(i)** $\{z^2 + 4z + 8\}y[n] = x[n]$
**(j)** $y[n] + 4y[n-1] + 4y[n-2] = x[n] - x[n+2]$

[**Hints and Suggestions:** To find the impulse response for the recursive filters, assume $y[0] = 1$ and (if required) $y[-1] = y[-2] = \cdots = 0$. If the right-hand side of the recursive filter equation is anything but $x[n]$, start with the single input $x[n]$ and then use superposition and time-invariance to get the result for the required input. The results for (d) through (f) can be found from the results of (c) in this way.]

3.28. **(Stability)** Investigate the causality and stability of the following right-sided systems.

**(a)** $y[n] = x[n-1] + x[n] + x[n+1]$
**(b)** $y[n] = x[n] + x[n-1] + x[n-2]$
**(c)** $y[n] - 2y[n-1] = x[n]$
**(d)** $y[n] - 0.2y[n-1] = x[n] - 2x[n+2]$
**(e)** $y[n] + y[n-1] + 0.5y[n-2] = x[n]$
**(f)** $y[n] - y[n-1] + y[n-2] = x[n] - x[n+1]$
**(g)** $y[n] - 2y[n-1] + y[n-2] = x[n] - x[n-3]$
**(h)** $y[n] - 3y[n-1] + 2y[n-2] = 2x[n+3]$

[**Hints and Suggestions:** Remember that FIR filters are always stable and for right-sided systems, every root of the caracteristic equation must have a magnitude (absolute value) less than 1.]

3.29. **(System Interconnections)** Two systems are said to be in cascade if the output of the first system acts as the input to the second. Find the response of the following cascaded systems if the input is a unit step and the systems are described as follows. In which instances does the response differ when the order of cascading is reversed? Can you use this result to justify that the order in which the systems are cascaded does not matter in finding the overall response if both systems are LTI?

**(a)** System 1: $y[n] = x[n] - x[n-1]$ System 2: $y[n] = 0.5y[n-1] + x[n]$
**(b)** System 1: $y[n] = 0.5y[n-1] + x[n]$ System 2: $y[n] = x[n] - x[n-1]$
**(c)** System 1: $y[n] = x^2[n]$ System 2: $y[n] = 0.5y[n-1] + x[n]$
**(d)** System 1: $y[n] = 0.5y[n-1] + x[n]$ System 2: $y[n] = x^2[n]$

3.30. **(Systems in Cascade and Parallel)** Consider the realization of Figure P3.30.

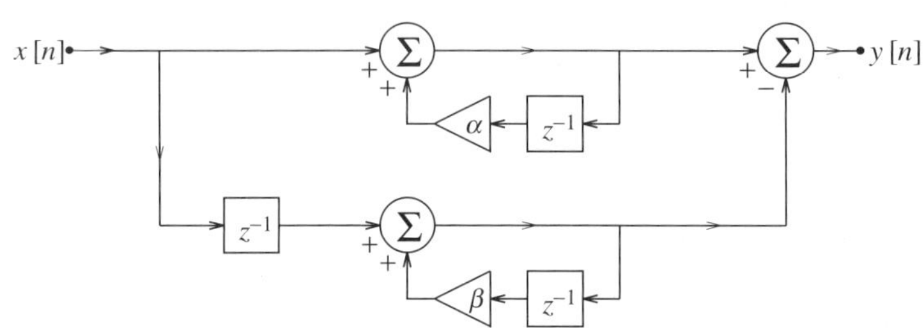


**FIGURE P.3.30** System realization for Problem 3.30

**(a)** Find its impulse response if $\alpha \neq \beta$. Is the overall system FIR or IIR?
**(b)** Find its difference equation and impulse response if $\alpha = \beta$. Is the overall system FIR or IIR?
**(c)** Find its difference equation and impulse response if $\alpha = \beta = 1$. What is the function of the overall system?

**3.31. (Difference Equations from Impulse Response)** Find the difference equations describing the following systems.

**(a)** $h[n] = \delta[n] + 2\delta[n-1]$ **(b)** $h[n] = \{2, \overset{\Downarrow}{3}, -1\}$

**(c)** $h[n] = (0.3)^n u[n]$ **(d)** $h[n] = (0.5)^n u[n] - (-0.5)^n u[n]$

[**Hints and Suggestions:** For part (c), the left-hand side of the difference equation is $y[n] - 0.3y[n-1]$. So, $h[n] - 0.3h[n-1]$ simplified to get impulses leads to the right-hand side. For part (d), start with the left-hand side as $y[n] - 0.25y[n-2]$.]

**3.32. (Difference Equations from Impulse Response)** A system is described by the impulse response $h[n] = (-1)^n u[n]$. Find the difference equation of this system. Then find the difference equation of the inverse system. Does the inverse system describe an FIR filter or IIR filter? What function does it perform?

**3.33. (Difference Equations)** For the filter realization shown in Figure P3.33, find the difference equation relating $y[n]$ and $x[n]$ if the impulse response of the filter is given by

**(a)** $h[n] = \delta[n] - \delta[n-1]$ **(b)** $h[n] = 0.5\delta[n] + 0.5\delta[n-1]$

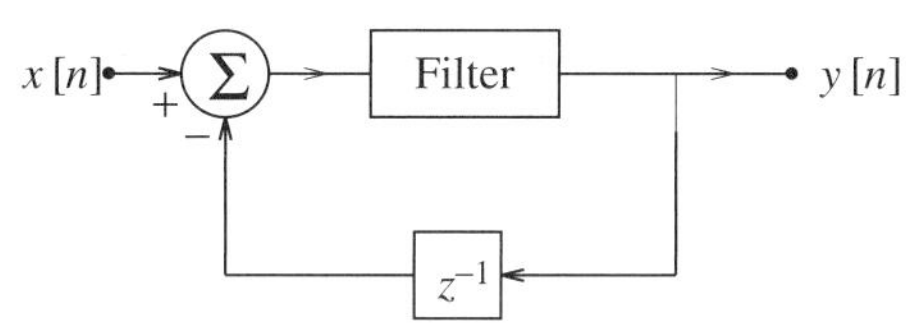


**FIGURE P.3.33** Filter realization for Problem 3.33

**3.34. (Difference Equations from Differential Equations)** This problem assumes some familiarity with analog theory. Consider an analog system described by

$$y''(t) + 3y'(t) + 2y(t) = 2u(t)$$

**(a)** Confirm that this describes a stable analog system.

**(b)** Convert this to a difference equation using the backward Euler algorithm and check the stability of the resulting digital filter.

**(c)** Convert this to a difference equation using the forward Euler algorithm and check the stability of the resulting digital filter.

**(d)** Which algorithm is better in terms of preserving stability? Can the results be generalized to any arbitrary analog system?

**3.35. (Inverse Systems)** Are the following systems invertible? If not, explain why; if invertible, find the inverse system.

**(a)** $y[n] = x[n] - x[n-1]$ (differencing operation)

**(b)** $y[n] = \frac{1}{3}(x[n] + x[n-1] + x[n-2])$ (moving average operation)

**(c)** $y[n] = 0.5x[n] + x[n-1] + 0.5x[n-2]$ (weighted moving average operation)

**(d)** $y[n] - \alpha y[n-1] = (1-\alpha)x[n], \quad 0 < \alpha < 1$ (exponential averaging operation)

**(e)** $y[n] = \cos(n\pi)x[n]$ (modulation)

**(f)** $y[n] = \cos(x[n])$

**(g)** $y[n] = e^{x[n]}$

[**Hints and Suggestions:** The inverse system is found by switching input and output and rearranging. Only one of these systems is not invertible.]

**3.36. (An Echo System and Its Inverse)** An echo system is described by

$$y[n] = x[n] + 0.5x[n - N]$$

Assume that the echo arrives after 1 ms and the sampling rate is 2 kHz.

**(a)** What is the value of $N$? Sketch a realization of this echo system.
**(b)** What is the impulse response and step response of this echo system?
**(c)** Find the difference equation of the inverse system. Then, sketch its realization and find its impulse response and step response.

**3.37. (Reverb)** A reverb filter is described by $y[n] = x[n] + 0.25y[n - N]$. Assume that the echoes arrive every millisecond and the sampling rate is 2 kHz.

**(a)** What is the value of $N$? Sketch a realization of this reverb filter.
**(b)** What is the impulse response and step response of this reverb filter?
**(c)** Find the difference equation of the inverse system. Then, sketch its realization and find its impulse response and step response.

**3.38. (Periodic Signal Generators)** Find the difference equation of a filter whose impulse response is a periodic sequence with first period $x[n] = \{\overset{\Downarrow}{1}, 2, 3, 4, 6, 7, 8\}$. Sketch a realization for this filter.

**3.39. (Recursive and IIR Filters)** The terms *recursive* and *IIR* are not always synonymous. A recursive filter could in fact have a finite impulse response. Use recursion to find the the impulse response $h[n]$ for each of the following recursive filters. Which filters (if any) describe IIR filters?

**(a)** $y[n] - y[n - 1] = x[n] - x[n - 2]$
**(b)** $y[n] - y[n - 1] = x[n] - x[n - 1] - 2x[n - 2] + 2x[n - 3]$

**3.40. (Recursive Forms of FIR Filters)** An FIR filter may always be recast in recursive form by the simple expedient of including identical factors on the left-hand and right-hand side of its difference equation in operational form. For example, the filter $y[n] = (1 - z^{-1})x[n]$ is FIR, but the identical filter $(1 + z^{-1})y[n] = (1 + z^{-1})(1 - z^{-1})x[n]$ has the difference equation $y[n] + y[n - 1] = x[n] - x[n - 2]$ and can be implemented recursively. Find two different recursive difference equations (with different orders) for each of the following filters.

**(a)** $y[n] = x[n] - x[n - 2]$ **(b)** $h[n] = \{1, \overset{\Downarrow}{2}, 1\}$

**3.41. (Nonrecursive Forms of IIR Filters)** An FIR filter may always be represented exactly in recursive form, but we can also approximate an IIR filter as an FIR filter by truncating its impulse response to $N$ terms. The larger the truncation index $N$, the better is the approximation. Consider the IIR filter described by $y[n] - 0.8y[n - 1] = x[n]$. Find its impulse response $h[n]$ and truncate it to three terms to obtain $h_3[n]$, the impulse response of the approximate FIR equivalent. Would you expect the greatest mismatch in the response of the two filters to identical inputs to occur for lower or higher values of $n$? Compare the step response of the two filters up to $n = 6$ to justify your expectations.

**3.42. (Nonlinear Systems)** One way to solve nonlinear difference equations is by recursion. Consider the nonlinear difference equation $y[n]y[n - 1] - 0.5y^2[n - 1] = 0.5Au[n]$.

**(a)** What makes this system nonlinear?
**(b)** Using $y[-1] = 2$, recursively obtain $y[0]$, $y[1]$, and $y[2]$.

**(c)** Use $A = 2$, $A = 4$, and $A = 9$ in the results of part (b) to confirm that this system finds the square root of $A$.

**(d)** Repeat parts (b) and (c) with $y[-1] = 1$ to check whether the choice of the initial condition affects system operation.

**3.43.** **(LTI Concepts and Stability)** Argue that neither of the following describes an LTI system. Then, explain how you might check for their stability and determine which of the systems are stable.

**(a)** $y[n] + 2y[n-1] = x[n] + x^2[n]$ **(b)** $y[n] - 0.5y[n-1] = nx[n] + x^2[n]$

**3.44.** **(Response of Causal and Noncausal Systems)** A difference equation may describe a causal or noncausal system depending on how the initial conditions are prescribed. Consider a first-order system governed by $y[n] + \alpha y[n-1] = x[n]$.

**(a)** With $y[n] = 0, n < 0$, this describes a causal system. Assume $y[-1] = 0$ and find the first few terms $y[0], y[1], \ldots$ of the impulse response and step response, using recursion, and establish the general form for $y[n]$.

**(b)** With $y[n] = 0, n > 0$, we have a noncausal system. Assume $y[0] = 0$ and rewrite the difference equation as $y[n-1] = \{-y[n] + x[n]\}/\alpha$ to find the first few terms $y[0], y[-1], y[-2], \ldots$ of the impulse response and step response, using recursion, and establish the general form for $y[n]$.

**3.45.** **(Time Reversal)** For each signal $x[n]$, sketch $g[k] = x[3-k]$ versus $k$ and $h[k] = x[2+k]$ versus $k$.

**(a)** $x[n] = \{\overset{\Downarrow}{1}, 2, 3, 4\}$ **(b)** $x[n] = \{3, 3, \overset{\Downarrow}{3}, 2, 2, 2\}$

[**Hints and Suggestions:** Note that $g[k]$ and $h[k]$ will be plotted against the index $k$.]

**3.46.** **(Closed-Form Convolution)** Find the convolution $y[n] = x[n] * h[n]$ for the following:

**(a)** $x[n] = u[n]$ $h[n] = u[n]$
**(b)** $x[n] = (0.8)^n u[n]$ $h[n] = (0.4)^n u[n]$
**(c)** $x[n] = (0.5)^n u[n]$ $h[n] = (0.5)^n \{u[n+3] - u[n-4]\}$
**(d)** $x[n] = \alpha^n u[n]$ $h[n] = \alpha^n u[n]$
**(e)** $x[n] = \alpha^n u[n]$ $h[n] = \beta^n u[n]$
**(f)** $x[n] = \alpha^n u[n]$ $h[n] = \text{rect}(n/2N)$

[**Hints and Suggestions:** The summations will be over the index $k$ and functions of $n$ should be pulled out before evaluating them using tables. For (a), (b), (d) and (e), summations will be from $k = 0$ to $k = n$. For part (c) and (f), use superposition. For (a) and (d), the sum $\Sigma(1)^k = \Sigma(1)$ from $k = 0$ to $k = n$ equals $n + 1$.]

**3.47.** **(Convolution with Impulses)** Find the convolution $y[n] = x[n] * h[n]$ of the following signals.

**(a)** $x[n] = \delta[n-1]$ $h[n] = \delta[n-1]$
**(b)** $x[n] = \cos(0.25n\pi)$ $h[n] = \delta[n] - \delta[n-1]$
**(c)** $x[n] = \cos(0.25n\pi)$ $h[n] = \delta[n] - 2\delta[n-1] + \delta[n-2]$
**(d)** $x[n] = (-1)^n$ $h[n] = \delta[n] + \delta[n-1]$

[**Hints and Suggestions:** Start with $\delta[n] * g[n] = g[n]$ and use linearity and time invariance.]

**3.48. (Convolution)** Find the convolution $y[n] = x[n] * h[n]$ for each pair of signals.

**(a)** $x[n] = (0.4)^{-n}u[n] \qquad h[n] = (0.5)^{-n}u[n]$
**(b)** $x[n] = \alpha^{-n}u[n] \qquad h[n] = \beta^{-n}u[n]$
**(c)** $x[n] = \alpha^{n}u[-n] \qquad h[n] = \beta^{n}u[-n]$
**(d)** $x[n] = \alpha^{-n}u[-n] \qquad h[n] = \beta^{-n}u[-n]$

[**Hints and Suggestions:** For parts (a) and (b) write the exponentials in the form $r^n$. For parts (c) and (d) find the convolution of $x[-n]$ and $h[-n]$ and flip the result to get $y[n]$.]

**3.49. (Convolution of Finite Sequences)** Find the convolution $y[n] = x[n] * h[n]$ for each of the following signal pairs. Use a marker to indicate the origin $n = 0$.

**(a)** $x[n] = \{\overset{\Downarrow}{1}, 2, 0, 1\} \qquad h[n] = \{\overset{\Downarrow}{2}, 2, 3\}$
**(b)** $x[n] = \{\overset{\Downarrow}{0}, 2, 4, 6\} \qquad h[n] = \{\overset{\Downarrow}{6}, 4, 2, 0\}$
**(c)** $x[n] = \{-3, -2, -\overset{\Downarrow}{1}, 0, 1\} \qquad h[n] = \{\overset{\Downarrow}{4}, 3, 2\}$
**(d)** $x[n] = \{3, 2, \overset{\Downarrow}{1}, 1, 2\} \qquad h[n] = \{4, \overset{\Downarrow}{2}, 3, 2\}$
**(e)** $x[n] = \{3, 0, 2, 0, \overset{\Downarrow}{1}, 0, 1, 0, 2\} \qquad h[n] = \{4, 0, \overset{\Downarrow}{2}, 0, 3, 0, 2\}$
**(f)** $x[n] = \{\overset{\Downarrow}{0}, 0, 0, 3, 1, 2\} \qquad h[n] = \{4, \overset{\Downarrow}{2}, 3, 2\}$

[**Hints and Suggestions:** Since the starting index of the convolution equals the sum of the starting indices of the sequences convolved, ignore markers during convolution and assign as the last step.]

**3.50. (Convolution of Symmetric Sequences)** The convolution of sequences that are symmetric about their midpoint is also endowed with symmetry (about its midpoint). Compute $y[n] = x[n] * h[n]$ for each pair of signals and use the results to establish the type of symmetry (about the midpoint) in the convolution if the convolved signals are both even symmetric (about their midpoint), both odd symmetric (about their midpoint), or one of each type.

**(a)** $x[n] = \{2,\ 1,\ 2\} \qquad h[n] = \{1,\ 0,\ 1\}$
**(b)** $x[n] = \{2,\ 1,\ 2\} \qquad h[n] = \{1,\ 1\}$
**(c)** $x[n] = \{2,\ 2\} \qquad h[n] = \{1,\ 1\}$
**(d)** $x[n] = \{2,\ 0,\ -2\} \qquad h[n] = \{1,\ 0,\ -1\}$
**(e)** $x[n] = \{2,\ 0,\ -2\} \qquad h[n] = \{1,\ -1\}$
**(f)** $x[n] = \{2,\ -2\} \qquad h[n] = \{1,\ -1\}$
**(g)** $x[n] = \{2,\ 1,\ 2\} \qquad h[n] = \{1,\ 0,\ -1\}$
**(h)** $x[n] = \{2,\ 1,\ 2\} \qquad h[n] = \{1,\ -1\}$
**(i)** $x[n] = \{2,\ 2\} \qquad h[n] = \{1,\ -1\}$

**3.51. (Properties)** Let $x[n] = h[n] = \{\overset{\Downarrow}{3}, 4, 2, 1\}$. Compute the following:

**(a)** $y[n] = x[n] * h[n]$ **(b)** $g[n] = x[-n] * h[-n]$
**(c)** $p[n] = x[n] * h[-n]$ **(d)** $f[n] = x[-n] * h[n]$
**(e)** $r[n] = x[n-1] * h[n+1]$ **(f)** $s[n] = x[n-1] * h[n+4]$

[**Hints and Suggestions:** The results for (b) and (d) can be found by flipping the results for (a) and (c) respectively. The result for (f) can be found by shifting the result for (e) (time-invariance).]

**3.52.** **(Properties)** Let $x[n] = h[n] = \{\overset{\Downarrow}{2}, 6, 0, 4\}$. Compute the following:

**(a)** $y[n] = x[2n] * h[2n]$
**(b)** Find $g[n] = x[n/2] * h[n/2]$, assuming zero interpolation.
**(c)** Find $p[n] = x[n/2] * h[n]$, assuming step interpolation where necessary.
**(d)** Find $r[n] = x[n] * h[n/2]$, assuming linear interpolation where necessary.

**3.53.** **(Application)** Consider a 2-point averaging filter whose present output equals the average of the present and previous input.

**(a)** Set up a difference equation for this system.
**(b)** What is the impulse response of this system?
**(c)** What is the response of this system to the sequence $\{\overset{\Downarrow}{1}, 2, 3, 4, 5\}$?
**(d)** Use convolution to show that the system performs the required averaging operation.

**3.54.** **(Step Response)** Given the impulse response $h[n]$, find the step response $s[n]$ of each system.

**(a)** $h[n] = (0.5)^n u[n]$
**(b)** $h[n] = (0.5)^n \cos(n\pi)u[n]$
**(c)** $h[n] = (0.5)^n \cos(n\pi + 0.5\pi)u[n]$
**(d)** $h[n] = (0.5)^n \cos(n\pi + 0.25\pi)u[n]$
**(e)** $h[n] = n(0.5)^n u[n]$
**(f)** $h[n] = n(0.5)^n \cos(n\pi)u[n]$

[**Hints and Suggestions:** Note that $s[n] = x[n] * h[n]$ where $x[n] = u[n]$. In part (b) and (f), note that $\cos(n\pi) = (-1)^n$. In part (d) expand $\cos(n\pi + 0.25\pi)$ and use the results of parts (b).]

**3.55.** **(Convolution and System Response)** Consider the system $y[n] - 0.5y[n-1] = x[n]$.

**(a)** What is the impulse response $h[n]$ of this system?
**(b)** Find its output if $x[n] = (0.5)^n u[n]$ by convolution.
**(c)** Find its output if $x[n] = (0.5)^n u[n]$ and $y[-1] = 0$ by solving the difference equation.
**(d)** Find its output if $x[n] = (0.5)^n u[n]$ and $y[-1] = 2$ by solving the difference equation.
**(e)** Are any of the outputs identical? Should they be? Explain.

[**Hints and Suggestions:** For part (e), remember that convolution finds the zero-state response.]

**3.56.** **(Convolution and Interpolation)** Let $x[n] = \{\overset{\Downarrow}{2}, 4, 6, 8\}$.

**(a)** Find the convolution $y[n] = x[n] * x[n]$.
**(b)** Find the convolution $y_1[n] = x[2n] * x[2n]$. Is $y_1[n]$ related to $y[n]$? Should it be? Explain.
**(c)** Find the convolution $y_2[n] = x[n/2] * x[n/2]$, assuming zero interpolation. Is $y_2[n]$ related to $y[n]$? Should it be? Explain.
**(d)** Find the convolution $y_3[n] = x[n/2] * x[n/2]$, assuming step interpolation. Is $y_3[n]$ related to $y[n]$? Should it be? Explain.
**(e)** Find the convolution $y_4[n] = x[n/2] * x[n/2]$, assuming linear interpolation. Is $y_4[n]$ related to $y[n]$? Should it be? Explain.

**3.57.** **(Linear Interpolation)** Consider a system that performs linear interpolation by a factor of $N$. One way to construct such a system (as shown) is to perform up-sampling by $N$ (zero interpolation between signal samples) and pass the up-sampled signal through a filter with impulse response $h[n]$ whose output $y[n]$ is the linearly interpolated signal.

$$x[n] \longrightarrow \boxed{\text{up-sample (zero interpolate) by } N} \longrightarrow \boxed{\text{filter}} \longrightarrow y[n]$$

**(a)** What should $h[n]$ be for linear interpolation by a factor of $N$?
**(b)** Let $x[n] = 4\text{tri}(0.25n)$. Find $y_1[n] = x[n/2]$ by linear interpolation.
**(c)** Find the system output $y[n]$ for $N = 2$. Does $y[n]$ equal $y_1[n]$?

3.58. **(Causality)** Argue that the impulse response $h[n]$ of a causal system must be zero for $n < 0$. Based on this result, if the input to a causal system starts at $n = n_0$, when does the response start?

3.59. **(Stability)** Investigate the causality and stability of the following systems.

**(a)** $h[n] = (2)^n u[n-1]$ **(b)** $y[n] = 2x[n+1] + 3x[n] - x[n-1]$

**(c)** $h[n] = (-0.5)^n u[n]$ **(d)** $h[n] = \{3, 2, \overset{\Downarrow}{1}, 1, 2\}$

**(e)** $h[n] = (0.5)^{-n} u[-n]$ **(f)** $h[n] = (0.5)^{|n|}$

[**Hints and Suggestions:** Only one of these is unstable. For part (e), note that summing $|h[n]|$ is equivalent to summing its flipped version.]

3.60. **(Numerical Convolution)** The convolution $y(t)$ of two analog signals $x(t)$ and $h(t)$ may be approximated by sampling each signal at intervals $t_s$ to obtain the signals $x[n]$ and $h[n]$ and flipping and shifting the samples of one function past the other in steps of $t_s$ (to line up the samples). At each instant $kt_s$, the convolution equals the sum of the product samples multiplied by $t_s$. This is equivalent to using the rectangular rule to approximate the area. If $x[n]$ and $h[n]$ are convolved using the sum-by-column method, the columns make up the product, and their sum multiplied by $t_s$ approximates $y(t)$ at $t = kt_s$.

**(a)** Let $x(t) = \text{rect}(t/2)$ and $h(t) = \text{rect}(t/2)$. Find $y(t) = x(t) * h(t)$ and compute $y(t)$ at intervals of $t_s = 0.5$ s.
**(b)** Sample $x(t)$ and $h(t)$ at intervals of $t_s = 0.5$ s to obtain $x[n]$ and $h[n]$. Compute $y[n] = x[n] * h[n]$ and the convolution estimate $y_R(nt_s) = t_s y[n]$. Do the values of $y_R(nt_s)$ match the exact result $y(t)$ at $t = nt_s$? If not, what are the likely sources of error?
**(c)** Argue that the trapezoidal rule for approximating the convolution is equivalent to subtracting half the sum of the two end samples of each column from the discrete convolution result and then multiplying by $t_s$. Use this rule to obtain the convolution estimate $y_T(nt_s)$. Do the values of $y_T(nt_s)$ match the exact result $y(t)$ at $t = nt_s$? If not, what are the likely sources of error?
**(d)** Obtain estimates based on the rectangular rule and trapezoidal rule for the convolution $y(t)$ of $x(t) = 2\,\text{tri}(t)$ and $h(t) = \text{rect}(t/2)$ by sampling the signals at intervals of $t_s = 0.5$ s. Which rule would you expect to yield a better approximation, and why?

3.61. **(Convolution)** Let $x[n] = \text{rect}(n/2)$ and $h[n] = \text{rect}(n/4)$.

**(a)** Find $f[n] = x[n] * x[n]$ and $g[n] = h[n] * h[n]$.
**(b)** Express these results as $f[n] = A\,\text{tri}(n/M)$ and $g[n] = B\,\text{tri}(n/K)$ by selecting appropriate values for the constants $A$, $M$, $B$ and $K$.
**(c)** Generalize the above results to show that $\text{rect}(n/2N) * \text{rect}(n/2N) = (2N+1)\text{tri}(\frac{n}{2N+1})$.

**3.62. (Impulse Response of Difference Algorithms)** Two systems to compute the forward difference and backward difference are described by

$$\text{Forward difference: } y_F[n] = x[n+1] - x[n]$$

$$\text{Backward difference: } y_B[n] = x[n] - x[n-1]$$

**(a)** What is the impulse response of each system?
**(b)** Which of these systems is stable? Which of these systems is causal?
**(c)** Find the impulse response of their parallel connection. Is the parallel system stable? Is it causal?
**(d)** What is the impulse response of their cascade? Is the cascaded system stable? Is it causal?

**3.63. (System Response)** Find the response of the following filters to the unit step $x[n] = u[n]$, and to the alternating unit step $x[n] = (-1)^n u[n]$, using convolution concepts.

**(a)** $h[n] = \delta[n] - \delta[n-1]$ (differencing operation)
**(b)** $h[n] = \{\overset{\Downarrow}{0.5}, 0.5\}$ (2-point average)
**(c)** $h[n] = \frac{1}{N}\sum_{k=0}^{N-1}\delta[n-k],\ N = 3$ (moving average)
**(d)** $h[n] = \frac{2}{N(N+1)}\sum_{k=0}^{N-1}(N-k)\delta[n-k],\ N = 3$ (weighted moving average)
**(e)** $y[n] + \frac{N-1}{N+1}y[n-1] = \frac{2}{N+1}x[n],\ N = 3$ (exponential average)

**3.64. (Convolution and Interpolation)** Consider the following system with $x[n] = \{\overset{\Downarrow}{0}, 3, 9, 12, 15, 18\}$.

$$x[n] \longrightarrow \boxed{\text{zero interpolate by } N} \longrightarrow \boxed{\text{filter}} \longrightarrow y[n]$$

**(a)** Find the response $y[n]$ if $N = 2$ and the filter impulse response is $h[n] = \{\overset{\Downarrow}{1}, 1\}$. Show that, except for end effects, the output describes a step interpolation between the samples of $x[n]$.
**(b)** Find the response $y[n]$ if $N = 3$ and the filter impulse response is $h[n] = \{\overset{\Downarrow}{1}, 1, 1\}$. Does the output describe a step interpolation between the samples of $x[n]$?
**(c)** Pick $N$ and $h[n]$ if the system is to perform step interpolation by 4.

**3.65. (Convolution and Interpolation)** Consider the following system with $x[n] = \{\overset{\Downarrow}{0}, 3, 9, 12, 15, 18\}$.

$$x[n] \longrightarrow \boxed{\text{zero interpolate by } N} \longrightarrow \boxed{\text{filter}} \longrightarrow y[n]$$

**(a)** Find the response $y[n]$ if $N = 2$ and the filter impulse response is $h[n] = \text{tri}(n/2)$. Show that, except for end effects, the output describes a linear interpolation between the samples of $x[n]$.
**(b)** Find the response $y[n]$ if $N = 3$ and the filter impulse response is $h[n] = \text{tri}(n/3)$. Does the output describe a linear interpolation between the samples of $x[n]$?
**(c)** Pick $N$ and $h[n]$ if the system is to perform linear interpolation by 4.

**3.66. (Interconnected Systems)** Consider two systems described by

$$h_1[n] = \delta[n] + \delta[n-1] \qquad h_2[n] = (0.5)^n u[n]$$

Find the response to the input $x[n] = (0.5)^n u[n]$ if

**(a)** The two systems are connected in parallel with $\alpha = 0.5$.
**(b)** The two systems are connected in parallel with $\alpha = -0.5$.
**(c)** The two systems are connected in cascade with $\alpha = 0.5$.
**(d)** The two systems are connected in cascade with $\alpha = -0.5$.

**3.67. (Systems in Cascade and Parallel)** Consider the realization of Figure P3.67.

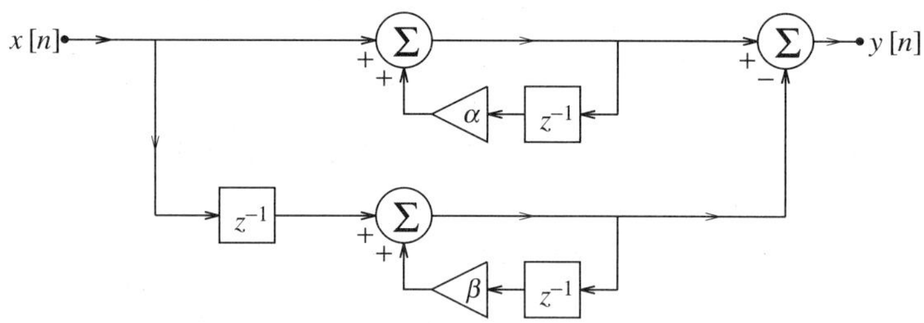


**FIGURE P.3.67** System realization for Problem 3.67

**(a)** Find its impulse response if $\alpha \neq \beta$. Is the overall system FIR or IIR?
**(b)** Find its impulse response if $\alpha = \beta$. Is the overall system FIR or IIR?
**(c)** Find its impulse response if $\alpha = \beta = 1$. What does the overall system represent?

**3.68. (Cascading)** The impulse response of two cascaded systems equals the convolution of their impulse responses. Does the step response $s_C[n]$ of two cascaded systems equal $s_1[n]*s_2[n]$, the convolution of their step responses? If not, how is $s_C[n]$ related to $s_1[n]$ and $s_2[n]$?

**3.69. (Cascading)** System 1 is a squaring circuit, and system 2 is an exponential averager described by $h[n] = (0.5)^n u[n]$. Find the output of each cascaded combination. Will their output be identical? Should it be? Explain.

**(a)** $2(0.5)^n u[n] \longrightarrow$ [system 1] $\longrightarrow$ [system 2] $\longrightarrow y[n]$

**(b)** $2(0.5)^n u[n] \longrightarrow$ [system 2] $\longrightarrow$ [system 1] $\longrightarrow y[n]$

**3.70. (Cascading)** System 1 is an IIR filter with the difference equation $y[n] = 0.5y[n-1] + x[n]$, and system 2 is a filter with impulse response $h[n] = \delta[n] - \delta[n-1]$. Find the output of each cascaded combination. Will their output be identical? Should it be? Explain.

**(a)** $2(0.5)^n u[n] \longrightarrow$ [system 1] $\longrightarrow$ [system 2] $\longrightarrow y[n]$

**(b)** $2(0.5)^n u[n] \longrightarrow$ [system 2] $\longrightarrow$ [system 1] $\longrightarrow y[n]$

**3.71. (Cascading)** System 1 is an IIR filter with the difference equation $y[n] = 0.5y[n-1] + x[n]$, and system 2 is a filter with impulse response $h[n] = \delta[n] - (0.5)^n u[n]$.

**(a)** Find the impulse response $h_P[n]$ of their parallel connection.
**(b)** Find the impulse response $h_{12}[n]$ of the cascade of system 1 and system 2.
**(c)** Find the impulse response $h_{21}[n]$ of the cascade of system 2 and system 1.
**(d)** Are $h_{12}[n]$ and $h_{21}[n]$ identical? Should they be? Explain.
**(e)** Find the impulse response $h_I[n]$ of a system whose parallel connection with $h_{12}[n]$ yields $h_P[n]$.

**3.72. (Cascading)** System 1 is a lowpass filter described by $y[n] = 0.5y[n-1] + x[n]$, and system 2 is described by $h[n] = \delta[n] - 0.5\delta[n-1]$.

**(a)** What is the output of the cascaded system to the input $x[n] = 2(0.5)^n u[n]$?
**(b)** What is the output of the cascaded system to the input $x[n] = \delta[n]$?
**(c)** How are the two systems related?

**3.73. (Convolution in Practice)** Often, the convolution of a long sequence $x[n]$ and a short sequence $h[n]$ is performed by breaking the long signal into shorter pieces, finding the convolution of each short piece with $h[n]$, and "gluing" the results together. Let $x[n] = \{1,\ 1,\ 2,\ 3,\ 5,\ 4,\ 3,\ 1\}$ and $h[n] = \{4,\ 3,\ 2,\ 1\}$.

**(a)** Split $x[n]$ into two equal sequences $x_1[n] = \{1,\ 1,\ 2,\ 3\}$ and $x_2[n] = \{5,\ 4,\ 3,\ 1\}$.
**(b)** Find the convolution $y_1[n] = h[n] * x_1[n]$.
**(c)** Find the convolution $y_2[n] = h[n] * x_2[n]$.
**(d)** Find the convolution $y[n] = h[n] * x[n]$.
**(e)** How can you find $y[n]$ from $y_1[n]$ and $y_2[n]$?

[**Hints and Suggestions:** For part (e), use superposition and add the shifted version of $y_2[n]$ to $y_1[n]$ to get $y[n]$. This forms the basis for the *overlap-add* method of convolution.]

**3.74. (Periodic Convolution)** Find the regular convolution $y[n] = x[n] * h[n]$ of one period of each pair of periodic signals. Then, use wraparound to compute the periodic convolution $y_p[n] = x[n] \circledast h[n]$. In each case, specify the *minimum* number of padding zeros we must use if we wish to find the regular convolution from the periodic convolution of the zero-padded signals.

**(a)** $x[n] = \{\overset{\Downarrow}{1},\ 2,\ 0,\ 1\}$ $\quad h[n] = \{\overset{\Downarrow}{2},\ 2,\ 3,\ 0\}$
**(b)** $x[n] = \{\overset{\Downarrow}{0},\ 2,\ 4,\ 6\}$ $\quad h[n] = \{\overset{\Downarrow}{6},\ 4,\ 2,\ 0\}$
**(c)** $x[n] = \{-3,\ -2,\ \overset{\Downarrow}{-1},\ 0,\ 1\}$ $\quad h[n] = \{\overset{\Downarrow}{4},\ 3,\ 2,\ 0,\ 0\}$
**(d)** $x[n] = \{3,\ 2,\ 1,\ \overset{\Downarrow}{1},\ 2\}$ $\quad h[n] = \{4,\ 2,\ 3,\ \overset{\Downarrow}{2},\ 0\}$

[**Hints and Suggestions:** First assign the marker for the regular convolution. After wraparound, this also corresponds to the marker for the periodic convolution.]

**3.75. (Periodic Convolution)** Find the periodic convolution $y_p[n] = x[n] \circledast h[n]$ for each pair of signals using the circulant matrix for $x[n]$.

**(a)** $x[n] = \{\overset{\Downarrow}{1},\ 2,\ 0,\ 1\}$ $\quad h[n] = \{\overset{\Downarrow}{2},\ 2,\ 3,\ 0\}$
**(b)** $x[n] = \{\overset{\Downarrow}{0},\ 2,\ 4,\ 6\}$ $\quad h[n] = \{\overset{\Downarrow}{6},\ 4,\ 2,\ 0\}$

**3.76. (Periodic Convolution)** Consider a system whose impulse response is $h[n] = (0.5)^n u[n]$. Show that one period of its periodic extension with period $N$ is given by

$$h_{\text{pe}}[n] = \frac{(0.5)^n}{1 - (0.5)^N},\quad 0 \le n \le N-1$$

Use this result to find the response of this system to the following periodic inputs.

**(a)** $x[n] = \cos(n\pi)$ $\quad$ **(b)** $x[n] = \{\overset{\Downarrow}{1},\ 1,\ 0,\ 0\}$, with $N = 4$
**(c)** $x[n] = \cos(0.5n\pi)$ $\quad$ **(d)** $x[n] = (0.5)^n,\ 0 \le n \le 3$, with $N = 4$

[**Hints and Suggestions:** In each case, compute $N$ samples of $h_{\text{pe}}[n]$ and then get the periodic convolution. For example, the period of $x[n]$ in part (a) is $N = 2$.]

**3.77.** **(Correlation)** For each pair of signals, compute the autocorrelation $r_{xx}[n]$, the autocorrelation $r_{hh}[n]$, the cross-correlation $r_{xh}[n]$, and the cross-correlation $r_{hx}[n]$. For each result, indicate the location of the origin $n = 0$ by a marker.

**(a)** $x[n] = \{\overset{\Downarrow}{1},\ 2,\ 0,\ 1\}$ $\qquad h[n] = \{\overset{\Downarrow}{2},\ 2,\ 3\}$

**(b)** $x[n] = \{\overset{\Downarrow}{0},\ 2,\ 4,\ 6\}$ $\qquad h[n] = \{\overset{\Downarrow}{6},\ 4,\ 2\}$

**(c)** $x[n] = \{-3,\ -2,\ \overset{\Downarrow}{-1},\ 2\}$ $\qquad h[n] = \{\overset{\Downarrow}{4},\ 3,\ 2\}$

**(d)** $x[n] = \{3,\ 2,\ \overset{\Downarrow}{1},\ 1,\ 2\}$ $\qquad h[n] = \{4,\ \overset{\Downarrow}{2},\ 3,\ 2\}$

[**Hints and Suggestions:** Use convolution to get the correlation results. For example, $r_{xh}[n] = x[n] * h[-n]$ and the marker for the result is based on $x[n]$ and $h[-n]$ (the sequences convolved).]

**3.78.** **(Correlation)** Let $x[n] = \text{rect}[(n-4)/2]$ and $h[n] = \text{rect}[n/4]$.

**(a)** Find the autocorrelation $r_{xx}[n]$.
**(b)** Find the autocorrelation $r_{hh}[n]$.
**(c)** Find the cross-correlation $r_{xh}[n]$.
**(d)** Find the cross-correlation $r_{hx}[n]$.
**(e)** How are the results of parts (c) and (d) related?

**3.79.** **(Correlation)** Find the correlation $r_{xh}[n]$ of the following signals.

**(a)** $x[n] = \alpha^n u[n]$ $\qquad h[n] = \alpha^n u[n]$
**(b)** $x[n] = n\alpha^n u[n]$ $\qquad h[n] = \alpha^n u[n]$
**(c)** $x[n] = \text{rect}(n/2N)$ $\qquad h[n] = \text{rect}(n/2N)$

[**Hints and Suggestions:** In parts (a) and (b), each correlation will cover two ranges. For $n \ge 0$, the signals overlap over $n \le k \le \infty$ and for $n < 0$, the overlap is for $0 \le k \le \infty$. For part (c), $x[n]$ and $h[n]$ are identical and even symmetric and their correlation equals their convolution.]

**3.80.** **(Periodic Correlation)** For each pair of periodic signals described for one period, compute the periodic autocorrelations $r_{pxx}[n]$ and $r_{phh}[n]$, and the periodic cross-correlations $r_{pxh}[n]$ and $r_{phx}[n]$. For each result, indicate the location of the origin $n = 0$ by a marker.

**(a)** $x[n] = \{\overset{\Downarrow}{1},\ 2,\ 0,\ 1\}$ $\qquad h[n] = \{\overset{\Downarrow}{2},\ 2,\ 3,\ 0\}$

**(b)** $x[n] = \{\overset{\Downarrow}{0},\ 2,\ 4,\ 6\}$ $\qquad h[n] = \{\overset{\Downarrow}{6},\ 4,\ 2,\ 0\}$

**(c)** $x[n] = \{-3,\ -2,\ \overset{\Downarrow}{-1},\ 2\}$ $\qquad h[n] = \{0,\ \overset{\Downarrow}{4},\ 3,\ 2\}$

**(d)** $x[n] = \{3,\ 2,\ \overset{\Downarrow}{1},\ 1,\ 2\}$ $\qquad h[n] = \{4,\ \overset{\Downarrow}{2},\ 3,\ 2,\ 0\}$

[**Hints and Suggestions:** First get the regular correlation (by regular convolution) and then use wraparound. For example, $r_{xh}[n] = x[n] * h[-n]$ and the marker for the result is based on $x[n]$ and $h[-n]$ (the sequences convolved). Then, use wraparound to get $r_{pxh}[n]$ (the marker may get wrapped around in some cases).]

**3.81.** **(Mean and Variance from Autocorrelation)** The mean value $m_x$ of a random signal $x[n]$ (with nonzero mean value) may be computed from its autocorrelation function $r_{xx}[n]$ as $m_x^2 = \lim_{|n|\to\infty} r_{xx}[n]$. The variance of $x[n]$ is then given by $\sigma_x^2 = r_{xx}(0) - m_x^2$. Find the mean, variance, and average power of a random signal whose autocorrelation function is $r_{xx}[n] = 10(\frac{1+2n^2}{2+5n^2})$.

## Computation and Design

**3.82.** **(Numerical Integration Algorithms)** Numerical integration algorithms approximate the area $y[n]$ from $y[n-1]$ or $y[n-2]$ (one or more time steps away). Consider the following integration algorithms.

**(a)** $y[n] = y[n-1] + t_s x[n]$ (rectangular rule)
**(b)** $y[n] = y[n-1] + \frac{t_s}{2}(x[n] + x[n-1])$ (trapezoidal rule)
**(c)** $y[n] = y[n-1] + \frac{t_s}{12}(5x[n] + 8x[n-1] - x[n-2])$ (Adams-Moulton rule)
**(d)** $y[n] = y[n-2] + \frac{t_s}{3}(x[n] + 4x[n-1] + x[n-2])$ (Simpson's rule)
**(e)** $y[n] = y[n-3] + \frac{3t_s}{8}(x[n] + 3x[n-1] + 3x[n-2] + x[n-3])$ (Simpson's three-eighths rule)

Use each of the rules to approximate the area of $x(t) = \text{sinc}(t)$, $0 \le t \le 3$, with $t_s = 0.1$ s and $t_s = 0.3$ s, and compare with the expected result of 0.53309323761827. How does the choice of the time step $t_s$ affect the results? Which algorithm yields the most accurate results?

**3.83.** **(System Response)** Use the MATLAB routine **filter** to obtain and plot the response of the filter described by $y[n] = 0.25(x[n] + x[n-1] + x[n-2] + x[n-3])$ to the following inputs and comment on your results.

**(a)** $x[n] = 1, \; 0 \le n \le 60$
**(b)** $x[n] = 0.1n, \; 0 \le n \le 60$
**(c)** $x[n] = \sin(0.1n\pi), \; 0 \le n \le 60$
**(d)** $x[n] = 0.1n + \sin(0.5n\pi), \; 0 \le n \le 60$
**(e)** $x[n] = \sum_{k=-\infty}^{\infty} \delta[n-5k], \; 0 \le n \le 60$
**(f)** $x[n] = \sum_{k=-\infty}^{\infty} \delta[n-4k], \; 0 \le n \le 60$

**3.84.** **(System Response)** Use the MATLAB routine **filter** to obtain and plot the response of the filter described by $y[n] - y[n-4] = 0.25(x[n] + x[n-1] + x[n-2] + x[n-3])$ to the following inputs and comment on your results.

**(a)** $x[n] = 1, \; 0 \le n \le 60$
**(b)** $x[n] = 0.1n, \; 0 \le n \le 60$
**(c)** $x[n] = \sin(0.1n\pi), \; 0 \le n \le 60$
**(d)** $x[n] = 0.1n + \sin(0.5n\pi), \; 0 \le n \le 60$
**(e)** $x[n] = \sum_{k=-\infty}^{\infty} \delta[n-5k], \; 0 \le n \le 60$
**(f)** $x[n] = \sum_{k=-\infty}^{\infty} \delta[n-4k], \; 0 \le n \le 60$

**3.85.** **(System Response)** Use MATLAB to obtain and plot the response of the following systems over the range $0 \le n \le 199$.

**(a)** $y[n] = x[n/3], \; x[n] = (0.9)^n u[n]$ (assume zero interpolation)
**(b)** $y[n] = \cos(0.2n\pi)x[n], \; x[n] = \cos(0.04n\pi)$ (modulation)
**(c)** $y[n] = [1 + \cos(0.2n\pi)]x[n], \; x[n] = \cos(0.04n\pi)$ (modulation)

**3.86.** **(System Response)** Use MATLAB to obtain and plot the response of the following filters, using direct commands (where possible) and also using the routine **filter**, and compare your results. Assume that the input is given by $x[n] = 0.1n + \sin(0.1n\pi),\ 0 \le n \le 60$. Comment on your results.

**(a)** $y[n] = \frac{1}{N}\sum_{k=0}^{N-1} x[n-k],\ N = 4$ (moving average)

**(b)** $y[n] = \frac{2}{N(N+1)}\sum_{k=0}^{N-1}(N-k)x[n-k],\ N = 4$ (weighted moving average)

**(c)** $y[n] - \alpha y[n-1] = (1-\alpha)x[n],\ N = 4,\ \alpha = \frac{N-1}{N+1}$ (exponential average)

**3.87.** **(System Response)** Use MATLAB to obtain and plot the response of the following filters, using direct commands and using the routine **filter**, and compare your results. Use an input that consists of the sum of the signal $x[n] = 0.1n + \sin(0.1n\pi),\ 0 \le n \le 60$ and uniformly distributed random noise with a mean of 0. Comment on your results.

**(a)** $y[n] = \frac{1}{N}\sum_{k=0}^{N-1} x[n-k],\ N = 4$ (moving average)

**(b)** $y[n] = \frac{2}{N(N+1)}\sum_{k=0}^{N-1}(N-k)x[n-k],\ N = 4$ (weighted moving average)

**(c)** $y[n] - \alpha y[n-1] = (1-\alpha)x[n],\ N = 4,\ \alpha = \frac{N-1}{N+1}$ (exponential averaging)

**3.88.** **(System Response)** Use the MATLAB routine **filter** to obtain and plot the response of the following FIR filters. Assume that $x[n] = \sin(n\pi/8),\ 0 \le n \le 60$. Comment on your results. From the results, can you describe the the function of these filters?

**(a)** $y[n] = x[n] - x[n-1]$ (first difference)
**(b)** $y[n] = x[n] - 2x[n-1] + x[n-2]$ (second difference)
**(c)** $y[n] = \frac{1}{3}(x[n] + x[n-1] + x[n-2])$ (moving average)
**(d)** $y[n] = 0.5x[n] + x[n-1] + 0.5x[n-2]$ (weighted average)

**3.89.** **(System Response in Symbolic Form)** Determine the response $y[n]$ of the following filters and plot over $0 \le n \le 30$.

**(a)** the step response of $y[n] - 0.5y[n-1] = x[n]$
**(b)** the impulse response of $y[n] - 0.5y[n-1] = x[n]$
**(c)** the zero-state response of $y[n] - 0.5y[n-1] = (0.5)^n u[n]$
**(d)** the complete response of $y[n] - 0.5y[n-1] = (0.5)^n u[n],\ y[-1] = -4$
**(e)** the complete response of $y[n] + y[n-1] + 0.5y[n-2] = (0.5)^n u[n]$, $y[-1] = -4,\ y[-2] = 3$

**3.90.** **(Inverse Systems and Echo Cancellation)** A signal $x(t)$ is passed through the echo-generating system $y(t) = x(t) + 0.9x(t-\tau) + 0.8x(t-2\tau)$, with $\tau = 93.75$ ms. The resulting echo signal $y(t)$ is sampled at $S = 8192$ Hz to obtain the sampled signal $y[n]$.

**(a)** The difference equation of a digital filter that generates the output $y[n]$ from $x[n]$ may be written as $y[n] = x[n] + 0.9x[n-N] + 0.8x[n-2N]$. What is the value of the index $N$?

**(b)** What is the difference equation of an echo-canceling filter (inverse filter) that could be used to recover the input signal $x[n]$?

**(c)** The echo signal is supplied on the author's website as **echosig**. Load this signal into MATLAB (using the command **load echosig**). Listen to this signal using the MATLAB command **sound**. Can you hear the echoes? Can you make out what is being said?

**(d)** Filter the echo signal using your inverse filter and listen to the filtered signal. Have you removed the echoes? Can you make out what is being said? Do you agree with what is being said?

3.91. **(Nonrecursive Forms of IIR Filters)** An FIR filter may always be exactly represented in recursive form, but we can only approximately represent an IIR filter by an FIR filter by truncating its impulse response to $N$ terms. The larger the truncation index $N$, the better is the approximation. Consider the IIR filter described by $y[n] - 0.8y[n-1] = x[n]$. Find its impulse response $h[n]$ and truncate it to 20 terms to obtain $h_A[n]$, the impulse response of the approximate FIR equivalent. Would you expect the greatest mismatch in the response of the two filters to identical inputs to occur for lower or higher values of $n$?

**(a)** Use the MATLAB routine **filter** to find and compare the step response of each filter up to $n = 15$. Are there any differences? Should there be? Repeat by extending the response to $n = 30$. Are there any differences? For how many terms does the response of the two systems stay identical, and why?

**(b)** Use the MATLAB routine **filter** to find and compare the response to $x[n] = 1,\ 0 \le n \le 10$ for each filter up to $n = 15$. Are there any differences? Should there be? Repeat by extending the response to $n = 30$. Are there any differences? For how many terms does the response of the two systems stay identical, and why?

3.92. **(Convolution of Symmetric Sequences)** The convolution of sequences that are symmetric about their midpoint is also endowed with symmetry (about its midpoint). Use the MATLAB command **conv** to find the convolution of the following sequences and establish the type of symmetry (about the midpoint) in the convolution.

**(a)** $x[n] = \sin(0.2n\pi), \quad -10 \le n \le 10 \qquad h[n] = \sin(0.2n\pi), \quad -10 \le n \le 10$

**(b)** $x[n] = \sin(0.2n\pi), \quad -10 \le n \le 10 \qquad h[n] = \cos(0.2n\pi), \quad -10 \le n \le 10$

**(c)** $x[n] = \cos(0.2n\pi), \quad -10 \le n \le 10 \qquad h[n] = \cos(0.2n\pi), \quad -10 \le n \le 10$

**(d)** $x[n] = \mathrm{sinc}(0.2n), \quad -10 \le n \le 10 \qquad h[n] = \mathrm{sinc}(0.2n), \quad -10 \le n \le 10$

3.93. **(Extracting Periodic Signals Buried in Noise)** Extraction of periodic signals buried in noise requires autocorrelation (to identify the period) and cross-correlation (to recover the signal itself).

**(a)** Generate the signal $x[n] = \sin(0.1n\pi),\ 0 \le n \le 499$. Add some uniform random noise (with a noise amplitude of 2 and a mean of 0) to obtain the noisy signal $s[n]$. Plot each signal. Can you identify any periodicity from the plot of $x[n]$? If so, what is the period $N$? Can you identify any periodicity from the plot of $s[n]$?

**(b)** Obtain the periodic autocorrelation $r_{px}[n]$ of $x[n]$ and plot. Can you identify any periodicity from the plot of $r_{px}[n]$? If so, what is the period $N$? Is it the same as the period of $x[n]$?

**(c)** Use the value of $N$ found above (or identify $N$ from $x[n]$ if not) to generate the 500-sample impulse train $i[n] = \sum \delta[n - kN],\ 0 \le n \le 499$. Find the *periodic* cross-correlation $y[n]$ of $s[n]$ and $i[n]$. Choose a normalizing factor that makes the peak value of $y[n]$ unity. How is the normalizing factor related to the signal length and the period $N$?

**(d)** Plot $y[n]$ and $x[n]$ on the same plot. Is $y[n]$ a close match to $x[n]$? Explain how you might improve the results.

CHAPTER 4

# z-Transform Analysis

## 4.0 Scope and Overview

This chapter deals with the $z$-transform as a method of system analysis in a transformed domain. Even though its genesis was outlined in the previous chapter, we develop the $z$-transform as an independent transformation method in order to keep the discussion self-contained. We concentrate on the operational properties of the $z$-transform and its applications in systems analysis. Connections with other transform methods and system analysis methods are explored in later chapters.

### 4.0.1 Goals and Learning Objectives

The goals of this chapter are to provide an introduction to the $z$-transform and its properties and the tools necessary for the analysis of discrete-time systems and digital filters using the $z$-transform. After going through this chapter the reader should:

1. Understand the definition of the two-sided $z$-transform and the region of convergence (ROC).
2. Understand the basic properties of the $z$-transform and how to use them to find the transform of one-sided or two-sided signals.
3. Understand the concepts of the transfer function, poles, and zeros.
4. Know how to sketch a system realization from its transfer function.
5. Know how to find the transfer function of systems in cascade and parallel.
6. Know how to convert between various forms of system representation.
7. Understand the concept of stability and its connection to the ROC.
8. Know how to find the inverse $z$-transform by long division.
9. Know how to find the inverse $z$-transform by partial fractions and table look-up.
10. Understand the definition of the one-sided $z$-transform and its properties.
11. Know how to use the shift property of the $z$-transform to solve difference equations.
12. Know how to use the $z$-transform for system analysis.
13. Know how to find the steady-state response of systems to harmonic inputs.
14. Be able to find initial and final values using the initial value and final value theorem.

# 4.1 The Two-Sided $z$-Transform

The **two-sided $z$-transform** $X(z)$ of a discrete signal $x[n]$ is defined as

$$X(z) = \sum_{m=-\infty}^{\infty} x[m]z^{-m} \quad \text{(two-sided } z\text{-transform)} \tag{4.1}$$

The relation between $x[n]$ and $X(z)$ is denoted symbolically by

$$x[n] \Leftarrow\boxed{\text{ZT}}\Rightarrow X(z) \tag{4.2}$$

Here, $x[n]$ and $X(z)$ form a transform pair, and the double arrow implies a one-to-one correspondence between the two.

## 4.1.1 What the $z$-Transform Reveals

The complex quantity $z$ generalizes the concept of digital frequency $F$ or $\Omega$ to the complex domain and is usually described in polar form as

$$z = Re^{j2\pi F} = Re^{j\Omega} \tag{4.3}$$

Values of the complex quantity $z$ may be displayed graphically in the **$z$-plane** in terms of its real and imaginary parts or in terms of its magnitude $R$ and angle $\Omega$.

The defining relation for the $z$-transform is a power series (Laurent series) in $z$. The term for any index $m$ is the product of the sample value $x[m]$ and $z^{-m}$.

For the sequence $x[n] = \{-7,\ 3,\ \overset{\Downarrow}{1},\ 4,\ -8,\ 5\}$, for example, the $z$-transform may be written as

$$X(z) = -7z^2 + 3z^1 + z^0 + 4z^{-1} - 8z^{-2} + 5z^{-3}$$

Comparing $x[n]$ and $X(z)$, we observe that the quantity $z^{-1}$ plays the role of a unit delay operator. The sample location $n = 2$ in $x[n]$, for example, corresponds to the term with $z^{-2}$ in $X(z)$. In concept, then, it is not hard to go back and forth between a sequence and its $z$-transform if all we are given is a finite number of samples.

# Drill Problem 4.1

**(a)** Let $x[n] = \{\overset{\Downarrow}{2},\ 1,\ 0,\ 4\}$. Find its $z$-transform $X(z)$.
**(b)** Let $x[n] = \{2,\ -3,\ \overset{\Downarrow}{1},\ 0,\ 4\}$. Find its $z$-transform $X(z)$.
**(c)** Let $X(z) = 3z^2 + z - 3z^{-1} + 5z^{-2}$. Find $x[n]$.

**Answers:** (a) $2 + z^{-1} + 4z^{-3}$ (b) $2z^2 - 3z + 1 + 4z^{-2}$ (c) $\{3,\ 1,\ \overset{\Downarrow}{0},\ -3,\ 5\}$

Since the defining relation for $X(z)$ describes a power series, it may not converge for all $z$. The values of $z$ for which it does converge define the **region of convergence** (ROC) for $X(z)$. Two completely different sequences may produce the same *two-sided* $z$-transform $X(z)$, but with different regions of convergence. It is important that we specify the ROC associated with each $X(z)$, especially when dealing with the two-sided $z$-transform.

**TABLE 4.1** ➤ A Short Table of $z$-Transform Pairs

| Entry | Signal | $z$-Transform | ROC |
|---|---|---|---|
| | | Finite Sequences | |
| 1 | $\delta[n]$ | $1$ | all $z$ |
| 2 | $u[n]-u[n-N]$ | $\dfrac{1-z^{-N}}{1-z^{-1}}$ | $z \ne 0$ |
| | | Causal Signals | |
| 3 | $u[n]$ | $\dfrac{z}{z-1}$ | $\lvert z\rvert > 1$ |
| 4 | $\alpha^n u[n]$ | $\dfrac{z}{z-\alpha}$ | $\lvert z\rvert > \lvert\alpha\rvert$ |
| 5 | $(-\alpha)^n u[n]$ | $\dfrac{z}{z+\alpha}$ | $\lvert z\rvert > \lvert\alpha\rvert$ |
| 6 | $nu[n]$ | $\dfrac{z}{(z-1)^2}$ | $\lvert z\rvert > 1$ |
| 7 | $n\alpha^n u[n]$ | $\dfrac{z\alpha}{(z-\alpha)^2}$ | $\lvert z\rvert > \lvert\alpha\rvert$ |
| 8 | $\cos(n\Omega)u[n]$ | $\dfrac{z^2-z\cos\Omega}{z^2-2z\cos\Omega+1}$ | $\lvert z\rvert > 1$ |
| 9 | $\sin(n\Omega)u[n]$ | $\dfrac{z\sin\Omega}{z^2-2z\cos\Omega+1}$ | $\lvert z\rvert > 1$ |
| 10 | $\alpha^n\cos(n\Omega)u[n]$ | $\dfrac{z^2-\alpha z\cos\Omega}{z^2-2\alpha z\cos\Omega+\alpha^2}$ | $\lvert z\rvert > \lvert\alpha\rvert$ |
| 11 | $\alpha^n\sin(n\Omega)u[n]$ | $\dfrac{\alpha z\sin\Omega}{z^2-2\alpha z\cos\Omega+\alpha^2}$ | $\lvert z\rvert > \lvert\alpha\rvert$ |
| | | Anti-Causal Signals | |
| 12 | $-u[-n-1]$ | $\dfrac{z}{z-1}$ | $\lvert z\rvert < 1$ |
| 13 | $-nu[-n-1]$ | $\dfrac{z}{(z-1)^2}$ | $\lvert z\rvert < 1$ |
| 14 | $-\alpha^n u[-n-1]$ | $\dfrac{z}{z-\alpha}$ | $\lvert z\rvert < \lvert\alpha\rvert$ |
| 15 | $-n\alpha^n u[-n-1]$ | $\dfrac{z\alpha}{(z-\alpha)^2}$ | $\lvert z\rvert < \lvert\alpha\rvert$ |

### 4.1.2 Some z-Transform Pairs using the Defining Relation

Table 4.1 lists the $z$-transforms of some useful signals. We provide some examples using the defining relation to find $z$-transforms. For finite-length sequences, the $z$-transform may be written as a polynomial in $z$. For sequences with a large number of terms, the polynomial form can get to be unwieldy unless we can find closed-form solutions.

**Example 4.1**

**The z-Transform from the Defining Relation**

**(a)** Let $x[n] = \delta[n]$. Its $z$-transform is $X(z) = 1$. The ROC is the entire $z$-plane.

**(b)** Let $x[n] = 2\delta[n+1] + \delta[n] - 5\delta[n-1] + 4\delta[n-2]$. This describes the sequence $x[n] = \{2, \overset{\Downarrow}{1}, -5, 4\}$. Its $z$-transform is evaluated as $X(z) = 2z + 1 - 5z^{-1} + 4z^{-2}$. No simplifications are possible. The ROC is the entire $z$-plane, except $z = 0$ and $z = \infty$ (or $0 < |z| < \infty$).

**(c)** Let $x[n] = u[n] - u[n-N]$. This represents a sequence of $N$ samples, and its $z$-transform may be written as

$$X(z) = 1 + z^{-1} + z^{-2} + \cdots + z^{-(N-1)}$$

Its ROC is $|z| > 0$ (the entire $z$-plane except $z = 0$). A closed-form result for $X(z)$ may be found using the defining relation as follows:

$$X(z) = \sum_{k=0}^{N-1} z^{-k} = \frac{1 - z^{-N}}{1 - z^{-1}}, \qquad z \neq 1$$

**(d)** Let $x[n] = u[n]$. We evaluate its $z$-transform using the defining relation as follows:

$$X(z) = \sum_{k=0}^{\infty} z^{-k} = \sum_{k=0}^{\infty} (z^{-1})^k = \frac{1}{1 - z^{-1}} = \frac{z}{z - 1}, \qquad \text{ROC: } |z| > 1$$

Its ROC is $|z| > 1$ and is based on the fact that the geometric series $\sum_{k=0}^{\infty} r^k$ converges only if $|r| < 1$.

**(e)** Let $x[n] = \alpha^n u[n]$. Using the defining relation, its $z$-transform and ROC are

$$X(z) = \sum_{k=0}^{\infty} \alpha^k z^{-k} = \sum_{k=0}^{\infty} \left(\frac{\alpha}{z}\right)^k = \frac{1}{1 - (\alpha/z)} = \frac{z}{z - \alpha}, \qquad \text{ROC: } |z| > |\alpha|$$

Its ROC ($|z| > \alpha$) is also based on the fact that the geometric series $\sum_{k=0}^{\infty} r^k$ converges only if $|r| < 1$.

# Drill Problem 4.2

**(a)** Let $x[n] = (0.5)^n u[n]$. Find its $z$-transform $X(z)$ and its ROC.
**(b)** Let $y[n] = (-0.5)^n u[n]$. Find its $z$-transform $Y(z)$ and its ROC.
**(c)** Let $g[n] = -(0.5)^n u[-n-1]$. Find its $z$-transform $G(z)$ and its ROC.

**Answers: (a)** $X(z) = \frac{z}{z-0.5}$, $|z| > 0.5$ **(b)** $X(z) = \frac{z}{z+0.5}$, $|z| > 0.5$
**(c)** $G(z) = \frac{z}{z-0.5}$, $|z| < 0.5$

### 4.1.3 More on the ROC

For a finite sequence $x[n]$, the $z$-transform $X(z)$ is a polynomial in $z$ or $z^{-1}$ and converges (is finite) for all $z$, except $z = 0$, if $X(z)$ contains terms of the form $z^{-k}$ (or $x[n]$ is nonzero for $n > 0$), and/or $z = \infty$ if $X(z)$ contains terms of the form $z^k$ (or $x[n]$ is nonzero for $n < 0$). Thus, the ROC for finite sequences is the entire $z$-plane, except perhaps for $z = 0$ and/or $z = \infty$, as applicable.

In general, if $X(z)$ is a rational function in $z$, as is often the case, its ROC actually depends on the one- or two-sidedness of $x[n]$, as illustrated in Figure 4.1.

The ROC *excludes all pole locations* (denominator roots) where $X(z)$ becomes infinite. As a result, the ROC of right-sided signals is $|z| > |p|_{max}$ and lies exterior to a circle of radius $|p|_{max}$, the magnitude of the largest pole. The ROC of causal signals, with $x[n] = 0,\ n < 0$,

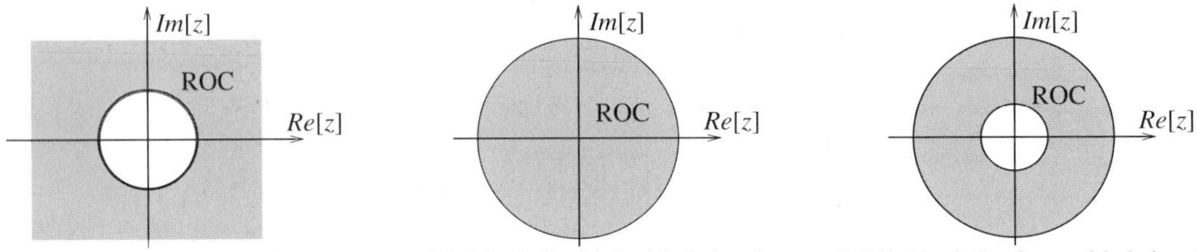


ROC (shaded) of right-sided signals ROC (shaded) of left-sided signals ROC (shaded) of two-sided signals

**FIGURE 4.1** The ROC (shown shaded) of the $z$-transform for various sequences. For a right-sided signal or causal signal, the ROC lies outside a circle of finite radius. For a left-sided signal or anti-causal signal, the ROC lies inside a circle of finite radius. For a two-sided signal, the ROC is an annular region

excludes the location $z = 0$ and is given by $0 > |z| > |p|_{\max}$. Note that $z = \infty$ is included in the ROC of causal signals. Thus, a necessary condition for the causality of $x[n]$ is that $X(z)$ remain finite at $z = \infty$ (with $\lim_{z\to\infty} \neq \infty$). Similarly, the ROC of a left-sided signal $x[n]$ is $|z| < |p|_{\min}$ and lies interior to a circle of radius $|p|_{\min}$, the smallest pole magnitude of $X(z)$. Finally, the ROC of a two-sided signal $x[n]$ is $|p|_{\min} < |z| < |p|_{\max}$, an annulus whose radii correspond to the smallest and largest pole locations in $X(z)$. We use *inequalities* of the form $|z| < |\alpha|$ (and not $|z| \leq |\alpha|$), for example, because $X(z)$ may not converge at the boundary $|z| = |\alpha|$.

### The ROC of the $z$-Transform $X(z)$ Determines the Nature of the Signal $x[n]$

**Finite-Length** $x[n]$: ROC of $X(z)$ is all the $z$-plane, except perhaps for $z = 0$ and/or $z = \infty$.

**Right-Sided** $x[n]$: ROC of $X(z)$ is outside a circle whose radius is the largest pole magnitude.

**Left-Sided** $x[n]$: ROC of $X(z)$ is inside a circle whose radius is the smallest pole magnitude.

**Two-Sided** $x[n]$: ROC of $X(z)$ is an annulus bounded by the largest and smallest pole radius.

## Why We Must Specify the ROC

Consider the signal $y[n] = -\alpha^n u[-n-1] = -1,\ n = -1, -2, \ldots$. The two-sided $z$-transform of $y[n]$, using the change of variable $m = -k$, can be written as

$$Y(z) = \sum_{k=-\infty}^{-1} -\alpha^k z^{-k} \underbrace{=}_{m=-k} -\sum_{m=1}^{\infty} \left(\frac{z}{\alpha}\right)^m = -\frac{z/\alpha}{1-(z/\alpha)} = \frac{z}{z-\alpha}, \qquad \text{ROC: } |z| < |\alpha|$$

The ROC of $Y(z)$ is $|z| < |\alpha|$. Recall that the $z$-transform of $x[n] = \alpha^n u[n]$ is $X(z) = \dfrac{z}{z-\alpha}$. This is identical to $Y(z)$ but the ROC of $X(z)$ is $|z| > |\alpha|$. So, we have a situation where two entirely different signals may possess an identical $z$-transform and the only way to distinguish between the them is by their ROC. In other words, we cannot uniquely identify a signal from its transform alone. We must also specify the ROC. In this book, we shall assume a right-sided signal if no ROC is specified.

**Example 4.2**

### Identifying the ROC

**(a)** Let $x[n] = \{4, \ -3, \ \overset{\Downarrow}{2}, \ 6\}$. The ROC of $X(z)$ is $0 < |z| < \infty$ and *excludes* $z = 0$ and $z = \infty$ because $x[n]$ is nonzero for $n < 0$ and $n > 0$.

**(b)** Let $X(z) = \dfrac{z}{z-2} + \dfrac{z}{z+3}$.
Its ROC depends on the nature of $x[n]$.

If $x[n]$ is assumed right-sided, the ROC is $|z| > 3$ (because $|p|_{\max} = 3$).
If $x[n]$ is assumed left-sided, the ROC is $|z| < 2$ (because $|p|_{\min} = 2$).
If $x[n]$ is assumed two-sided, the ROC is $2 < |z| < 3$.

The region $|z| < 2$ and $|z| > 3$ does not correspond to a valid region of convergence because we must find a region that is common to both terms.

# Drill Problem 4.3

**(a)** Let $X(z) = \dfrac{z+0.5}{z}$. What is its ROC?

**(b)** Let $Y(z) = \dfrac{z+1}{(z-0.1)(z+0.5)}$. What is its ROC if $y[n]$ is right-sided?

**(c)** Let $G(z) = \dfrac{z}{z+2} + \dfrac{z}{z-1}$. What is its ROC if $g[n]$ is two-sided?

**(d)** Let $H(z) = \dfrac{(z+3)}{(z-2)(z+1)}$. What is its ROC if $h[n]$ is left-sided?

**Answers: (a)** $|z| \neq 0$ **(b)** $|z| > 0.5$ **(c)** $1 < |z| < 2$ **(d)** $|z| < 1$

## 4.2 Properties of the Two-Sided $z$-Transform

The $z$-transform is a linear operation and obeys superposition. The properties of the $z$-transform listed in Table 4.2 are based on the linear nature of the $z$-transform operation.

### The Time-Shift Property

To prove the time-shift property of the two-sided $z$-transform, we use a change of variables. We start with the pair $x[n] \Leftarrow\boxed{\text{ZT}}\Rightarrow X(z)$. If $y[n] = x[n-N]$, its $z$-transform is

$$Y(z) = \sum_{k=-\infty}^{\infty} x[k-N]z^{-k} \tag{4.4}$$

With the change of variable $m = k - N$, the new summation index $m$ still ranges from $-\infty$ to $\infty$ (since $N$ is finite), and we obtain

$$Y(z) = \sum_{m=-\infty}^{\infty} x[m]z^{-(m+N)} \underbrace{=}_{m=k-N} z^{-N} \sum_{m=-\infty}^{\infty} x[m]z^{-m} = z^{-N}X(z) \tag{4.5}$$

**TABLE 4.2** ➤ Properties of the Two-Sided $z$-Transform

| Entry | Property | Signal | $z$-Transform |
|---|---|---|---|
| 1 | Shifting | $x[n-N]$ | $z^{-N}X(z)$ |
| 2 | Time-reversal | $x[-n]$ | $X(\frac{1}{z})$ |
| 3 | Anti-causal | $x[-n]u[-n-1]$ | $X(\frac{1}{z}) - x[0]$ (for causal $x[n]$) |
| 4 | Scaling | $\alpha^n x[n]$ | $X(\frac{z}{\alpha})$ |
| 5 | Times-$n$ | $nx[n]$ | $-z\frac{dX(z)}{dz}$ |
| 6 | Times-cos | $\cos(n\Omega)x[n]$ | $0.5[X(ze^{j\Omega}) + X(ze^{-j\Omega})]$ |
| 7 | Times-sin | $\sin(n\Omega)x[n]$ | $j0.5[X(ze^{j\Omega}) - X(ze^{-j\Omega})]$ |
| 8 | Convolution | $x[n] * h[n]$ | $X(z)H(z)$ |

The factor $z^{-N}$ with $X(z)$ induces a right shift of $N$ in $x[n]$.

# Drill Problem 4.4

**(a)** Let $X(z) = 2 + 5z^{-1}$. Find the $z$-transform of $y[n] = x[n-3]$.
**(b)** Use the result $(0.5)^n u[n] \Leftarrow\boxed{\text{ZT}}\Rightarrow \frac{z}{z-0.5}$ to find $g[n]$ if $G(z) = \frac{1}{z-0.5}$.

**Answers: (a)** $Y(z) = 2z^{-3} + 5z^{-4}$ **(b)** $g[n] = (0.5)^{n-1}u[n-1]$

## The Times-$n$ Property

The times-$n$ property is established by taking derivatives, to yield

$$X(z) = \sum_{k=-\infty}^{\infty} x[k]z^{-k} \qquad \frac{dX(z)}{dz} = \sum_{k=-\infty}^{\infty} \frac{d}{dz}\left[x[k]z^{-k}\right] = \sum_{k=-\infty}^{\infty} -kx[k]z^{-(k+1)} \tag{4.6}$$

Multiplying both sides by $-z$, we obtain

$$-z\frac{dX(z)}{dz} = \sum_{k=-\infty}^{\infty} kx[k]z^{-k} \tag{4.7}$$

This represents the transform of $nx[n]$.

# Drill Problem 4.5

**(a)** Let $X(z) = 2 + 5z^{-1} - 4z^{-2}$. Find the $z$-transform of $y[n] = nx[n]$.
**(b)** Let $G(z) = \frac{z}{z-0.5}$, ROC : $|z| > 0.5$. Find the $z$-transform of $h[n] = ng[n]$ and its ROC.

**Answers: (a)** $Y(z) = 5z^{-1} - 8z^{-2}$ **(b)** $H(z) = \frac{0.5z}{(z-0.5)^2}$, $|z| > 0.5$

### The Scaling Property

The scaling property follows from the transform of $y[n] = \alpha^n x[n]$, to yield

$$Y(z) = \sum_{k=-\infty}^{\infty} \alpha^k x[k] z^{-k} = \sum_{k=-\infty}^{\infty} x[k]\left(\frac{z}{\alpha}\right)^{-k} = X\left(\frac{z}{\alpha}\right) \tag{4.8}$$

It is important to realize that the scaling property also leads to a scaling of the ROC. If the ROC of $X(z)$ is $|z| > |K|$, the scaling property changes the ROC of $Y(z)$ to $|z| > |K\alpha|$. In particular, if $\alpha = -1$, we obtain the useful result $(-1)^n x[n] \Leftarrow\boxed{\text{ZT}}\Rightarrow X(-z)$. This result says that if we change the sign of alternating (odd indexed) samples of $x[n]$ to get $y[n]$, its $z$-transform $Y(z)$ is given by $Y(z) = X(-z)$ and has the same ROC.

# Drill Problem 4.6

**(a)** Let $X(z) = 2 - 3z^{-2}$. Find the $z$-transform of $y[n] = (2)^n x[n]$ and its ROC.
**(b)** Let $G(z) = \frac{z}{z-0.5}$, $|z| > 0.5$. Find the $z$-transform of $h[n] = (-0.5)^n g[n]$ and its ROC.
**(c)** Let $F(z) = 2 + 5z^{-1} - 4z^{-2}$. Find the $z$-transform of $p[n] = f[-n]$ and its ROC.

**Answers: (a)** $2 - 12z^{-2}$, $z \neq 0$ **(b)** $\frac{z}{z+0.25}$, $|z| > 0.25$ **(c)** $2 + 5z - 4z^2$, $z \neq \infty$

### The Times-Cos and Times-Sin Properties

If $x[n]$ is multiplied by $e^{jn\Omega}$ or $(e^{j\Omega})^n$, we then obtain the pair $e^{jn\Omega}x[n] \Leftarrow\boxed{\text{ZT}}\Rightarrow X(ze^{-j\Omega})$. An extension of this result, using Euler's relation, leads to the times-cos and times-sin properties

$$\cos(n\Omega)x[n] = 0.5x[n](e^{jn\Omega} + e^{-jn\Omega}) \Leftarrow\boxed{\text{ZT}}\Rightarrow 0.5[X(ze^{j\Omega}) + X(ze^{-j\Omega})] \tag{4.9}$$

$$\sin(n\Omega)x[n] = -j0.5x[n](e^{jn\Omega} - e^{-jn\Omega}) \Leftarrow\boxed{\text{ZT}}\Rightarrow j0.5[X(ze^{j\Omega}) - X(ze^{-j\Omega})] \tag{4.10}$$

The ROC is not affected by the times-cos and time-sin properties.

# Drill Problem 4.7

**(a)** Let $X(z) = \frac{z}{z-1}$, $|z| > 1$ and $y[n] = \cos(0.5n\pi)x[n]$. Find $Y(z)$ using the times-cos property.
**(b)** Let $G(z) = \frac{z}{z-0.5}$, $|z| > 0.5$ and $h[n] = \sin(0.5n\pi)g[n]$. Find $H(z)$ using the times-sin property.
**(c)** Let $P(z) = \frac{z^2}{z^2+0.25}$, $|z| > 0.5$ and $q[n] = \cos(0.5n\pi)p[n]$. Find $Q(z)$ using the times-cos property.
**(d)** Let $V(z) = \frac{z^2}{z^2+0.25}$, $|z| > 0.5$ and $s[n] = \cos(0.25n\pi)v[n]$. Find $S(z)$ using the times-cos property.

**Answers: (a)** $\frac{z^2}{z^2+1}$ **(b)** $\frac{0.5z}{z^2+0.25}$ **(c)** $\frac{z^2}{z^2-0.25}$ **(d)** $\frac{z^4}{z^4+0.0625}$

### The Convolution Property

The convolution property is based on the fact that multiplication in the time domain corresponds to convolution in any transformed domain. The $z$-transforms of sequences are polynomials, and multiplication of two polynomials corresponds to the convolution of their coefficient sequences. This property finds extensive use in the analysis of systems in the transformed domain.

### The Time-Reversal Property

With $x[n] \Leftarrow\boxed{\text{ZT}}\Rightarrow X(z)$ and $y[n] = x[-n]$, the change of variable $m \to -k$ in the defining relation gives

$$Y(z) = \sum_{k=-\infty}^{\infty} x[-k]z^{-k} \underbrace{=}_{m=-k} \sum_{m=-\infty}^{\infty} x[m]z^{m} = \sum_{m=-\infty}^{\infty} x[m](1/z)^{-m} = X(1/z) \tag{4.11}$$

If the ROC of $x[n]$ is $|z| > |\alpha|$, the ROC of the time-reversed signal $x[-n]$ becomes $|1/z| > |\alpha|$ or $|z| < 1/|\alpha|$.

# Drill Problem 4.8

**(a)** Let $X(z) = 2 + 3z^{-1}$, $z \neq 0$. Find the $z$-transform of $y[n] = x[-n]$ and its ROC.
**(b)** Let $G(z) = \frac{z}{z-0.5}$, $|z| > 0.5$. Find the $z$-transform of $h[n] = g[-n]$ and its ROC.

**Answers: (a)** $Y(z) = 3z + 2$, $|z| \neq \infty$ **(b)** $H(z) = \frac{1}{1-0.5z}$, $|z| < 2$

### The Reversal Property and Symmetric Signals

The reversal property is useful in checking for signal symmetry from its $z$-transform. For a signal $x[n]$ that has even symmetry about the origin $n = 0$, we have $x[n] = x[-n]$, and thus $X(z) = X(1/z)$. Similarly, if $x[n]$ has odd symmetry about $n = 0$, we have $x[n] = -x[-n]$, and thus $X(z) = -X(1/z)$. If $x[n]$ is symmetric about its midpoint and the center of symmetry is not $n = 0$, we observe that $X(z) = z^M X(1/z)$ for even symmetry and $X(z) = -z^M X(1/z)$ for odd symmetry. The factor $z^M$, where $M$ is an integer, accounts for the shift of the center of symmetry from the origin.

**A Property of the z-Transform of Symmetric Sequences**

**Even Symmetry:** $x[n] = x[-n]$ and $X(z) = z^M X(1/z)$
**Odd Symmetry:** $x[n] = -x[-n]$ and $X(z) = -z^M X(1/z)$

# Drill Problem 4.9

**(a)** Let $x[n] = \{\overset{\Downarrow}{0},\ 1,\ 5,\ 1\}$. Show that $X(z) = z^M X(1/z)$ and find $M$.

**(b)** Let $y[n] = \{2,\ \overset{\Downarrow}{2}\}$. Show that $Y(z) = z^M Y(1/z)$ and find $M$.

**(c)** Let $g[n] = \{3,\ \overset{\Downarrow}{0},\ -3\}$. Show that $G(z) = -z^M G(1/z)$ and find $M$.

**Answers: (a)** $M = -4$ **(b)** $M = 1$ **(c)** $M = 0$

## The Reversal Property and Anti-Causal Signals

The reversal property is also useful in finding the transform of anti-causal signals. From the causal signal $x[n]u[n] \Leftarrow\boxed{\text{ZT}}\Rightarrow X(z)$ (with ROC $|z| > |\alpha|$), we find the transform of $x[-n]u[-n]$ as $X(1/z)$ (whose ROC is $|z| < 1/|\alpha|$). The anti-causal signal $y[n] = x[-n]u[-n-1]$ (which excludes the sample at $n = 0$) can then be written as $y[n] = x[-n]u[-n] - x[0]\delta[n]$, as illustrated in Figure 4.2.

With $x[n]u[n] \Leftarrow\boxed{\text{ZT}}\Rightarrow X(z),\ |z| > |\alpha|$, the $z$-transform of $y[n] = x[-n]u[-n-1]$ gives

$$y[n] = x[-n]u[-n-1] \Leftarrow\boxed{\text{ZT}}\Rightarrow Y(z) = X(1/z) - x[0], \quad |z| < 1/|\alpha| \tag{4.12}$$

### How to Find the *z*-Transform of an Anti-Causal Signal from its Causal Version

If $x[n]u[n] \Leftarrow\boxed{\text{ZT}}\Rightarrow X(z),\ |z| > |\alpha|$, then $x[-n]u[-n-1] \Leftarrow\boxed{\text{ZT}}\Rightarrow X(1/z) - x[0],\ |z| < 1/|\alpha|$.

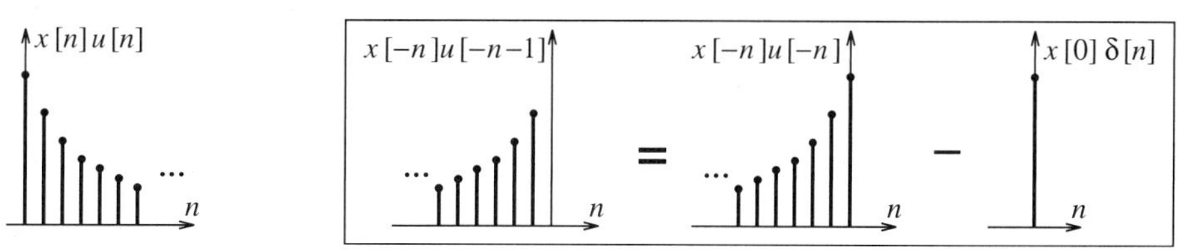


**FIGURE 4.2** To create an anti-causal signal from its causal version $x[n]u[n]$, we must flip it (to obtain $x[n]u[n]$) and subtract the impulse at the origin (whose sample value equals $x[0]$). This allows us to find the $z$-transform of the anti-causal signal from the $z$-transform of the causal version

# Drill Problem 4.10

**(a)** Let $x[n]u[n] \Leftarrow\boxed{\text{ZT}}\Rightarrow 2 + 5z^{-1}$, $z \neq 0$. Find the $z$-transform of $y[n] = x[-n]$ $u[-n-1]$ and its ROC.

**(b)** Let $f[n]u[n] \Leftarrow\boxed{\text{ZT}}\Rightarrow \frac{z+1}{z+0.5}$, $|z| > 0.5$. Find the $z$-transform of $g[n] = f[-n]$ $u[-n-1]$ and its ROC.

**Answers:** **(a)** $X(z) = 5z$, $|z| \neq \infty$ **(b)** $G(z) = \frac{0.5z}{0.5z+1}$, $|z| < 2$

**Example 4.3**

## z-Transforms using Properties

**(a)** Using the times-$n$ property, the $z$-transform of $y[n] = nu[n]$ is

$$Y(z) = -z\frac{d}{dz}\left[\frac{z}{z-1}\right] = -z\left[\frac{-z}{(z-1)^2} + \frac{1}{z-1}\right] = \frac{z}{(z-1)^2}$$

**(b)** With $x[n] = \alpha^n nu[n]$, we use scaling to obtain the $z$-transform:

$$X(z) = \frac{z/\alpha}{[(z/\alpha)-1]^2} = \frac{z\alpha}{(z-\alpha)^2}$$

**(c)** We find the transform of the $N$-sample exponential pulse $x[n] = \alpha^n(u[n] - u[n-N])$. We let $y[n] = u[n] - u[n-N]$. Its $z$-transform is

$$Y(z) = \frac{1-z^{-N}}{1-z^{-1}}, \quad |z| \neq 1$$

Then, the $z$-transform of $x[n] = \alpha^n y[n]$ becomes

$$X[z] = \frac{1-(z/\alpha)^{-N}}{1-(z/\alpha)^{-1}}, \quad z \neq \alpha$$

**(d)** The $z$-transforms of $x[n] = \cos(n\Omega)u[n]$ and $y[n] = \sin(n\Omega)u[n]$ are found using the times-cos and times-sin properties:

$$X(z) = 0.5\left[\frac{ze^{j\Omega}}{ze^{j\Omega}-1} + \frac{ze^{-j\Omega}}{ze^{-j\Omega}-1}\right] = \frac{z^2 - z\cos\Omega}{z^2 - 2z\cos\Omega + 1}$$

$$Y(z) = j0.5\left[\frac{ze^{j\Omega}}{ze^{j\Omega}-1} - \frac{ze^{-j\Omega}}{ze^{-j\Omega}-1}\right] = \frac{z\sin\Omega}{z^2 - 2z\cos\Omega + 1}$$

**(e)** The $z$-transforms of $f[n] = \alpha^n\cos(n\Omega)u[n]$ and $g[n] = \alpha^n\sin(n\Omega)u[n]$ follow from the results of part (d) and the scaling property:

$$F(z) = \frac{(z/\alpha)^2 - (z/\alpha)\cos\Omega}{(z/\alpha)^2 - 2(z/\alpha)\cos\Omega + 1} = \frac{z^2 - \alpha z\cos\Omega}{z^2 - 2\alpha z\cos\Omega + \alpha^2}$$

$$G(z) = \frac{(z/\alpha)\sin\Omega}{(z/\alpha)^2 - 2(z/\alpha)\cos\Omega + 1} = \frac{\alpha z\sin\Omega}{z^2 - 2\alpha z\cos\Omega + \alpha^2}$$

**(f)** We use the reversal property to find the transform of $y[n] = \alpha^{-n}u[-n-1]$. We start with the transform pair $x[n] = \alpha^n u[n] \Leftarrow\boxed{\text{ZT}}\Rightarrow \frac{z}{z-\alpha}$, ROC: $|z| > |\alpha|$. With

$x[0] = 1$, we find

$$y[n] = \alpha^{-n}u[-n-1] \Leftarrow\boxed{\text{ZT}}\Rightarrow X(1/z) - x[0] = \frac{1/z}{1/z - \alpha} - 1$$

$$= \frac{\alpha z}{1-\alpha z}, \qquad \text{ROC: } |z| < \frac{1}{|\alpha|}$$

If we replace $\alpha$ by $1/\alpha$ and change the sign of the result, we get

$$-\alpha^n u[-n-1] \Leftarrow\boxed{\text{ZT}}\Rightarrow \frac{z}{z-\alpha}, \qquad \text{ROC: } |z| < |\alpha|$$

This is listed as a standard transform pair in tables of $z$-transforms.

**(g)** We use the reversal property to find the transform of $x[n] = \alpha^{|n|}$, $|\alpha| < 1$ (a two-sided decaying exponential). We write this as $x[n] = \alpha^n u[n] + \alpha^{-n}u[n] - \delta[n]$ (a one-sided decaying exponential and its time-reversed version, less the extra sample included at the origin), as illustrated in Figure E4.3g.

**FIGURE E 4.3.g** The signal for Example 4.3(g).

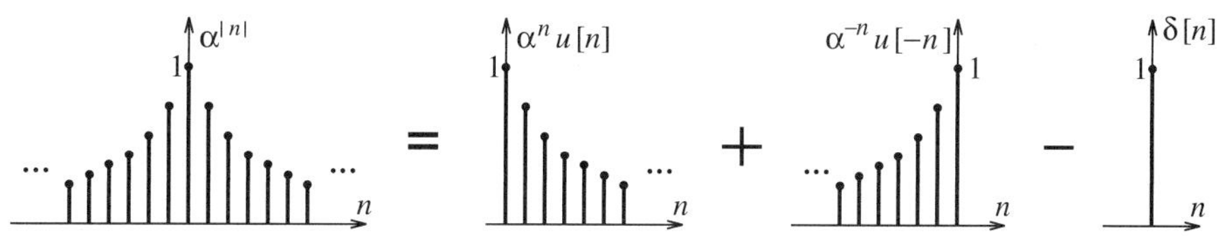


Its $z$-transform then becomes

$$X(z) = \frac{z}{z-\alpha} + \frac{1/z}{(1/z)-\alpha} - 1 = \frac{z}{z-\alpha} - \frac{z}{z-(1/\alpha)}, \qquad \text{ROC: } |\alpha| < |z| < \frac{1}{|\alpha|}$$

Note that the ROC is an annulus that corresponds to a two-sided sequence and describes a valid region only if $|\alpha| < 1$.

## 4.3 Poles, Zeros, and the $z$-Plane

The $z$-transform of many signals is a *rational* function of the form

$$X(z) = \frac{N(z)}{D(z)} = \frac{B_M z^M + B_{M-1}z^{M-1} + \cdots + B_2 z^2 + B_1 z + B_0}{A_N z^N + A_{N-1}z^{N-1} + \cdots + A_2 z^2 + A_1 z + A_0} \tag{4.13}$$

Denoting the roots of $N(z)$ by $z_i$, $i = 1, 2, \ldots, M$ and the roots of $D(z)$ by $p_k$, $k = 1, 2, \ldots, N$, we may also express $X(z)$ in factored form as

$$X(z) = K\frac{N(z)}{D(z)} = K\frac{(z-z_1)(z-z_2)\cdots(z-z_M)}{(z-p_1)(z-p_2)\cdots(z-p_N)} \tag{4.14}$$

The roots of $N(z)$ are termed **zeros** and the roots of $D(z)$ are termed **poles**. A plot of the poles (shown as $\times$) and zeros (shown as o) in the $z$-plane constitutes a **pole-zero plot** and provides a visual picture of the root locations. For multiple roots, we indicate their multiplicity next to the root location on the plot. Clearly, we can also find $X(z)$ in its entirety from a pole-zero plot of the root locations but only if the value of the multiplicative constant $K$ is also displayed on the plot.

**Example 4.4**

### Pole-Zero Plots

**(a)** Let $H(z) = \frac{2z(z+1)}{(z-\frac{1}{3})(z^2+\frac{1}{4})(z^2+4z+5)}$.

The numerator degree is 2. The two zeros are $z = 0$ and $z = -1$.

The denominator degree is 5. The five finite poles are at $z = \frac{1}{3}$, $z = \pm j\frac{1}{2}$, and $z = -2 \pm j$.

The multiplicative factor is $K = 2$. The pole-zero plot is shown in Figure E4.4(a).

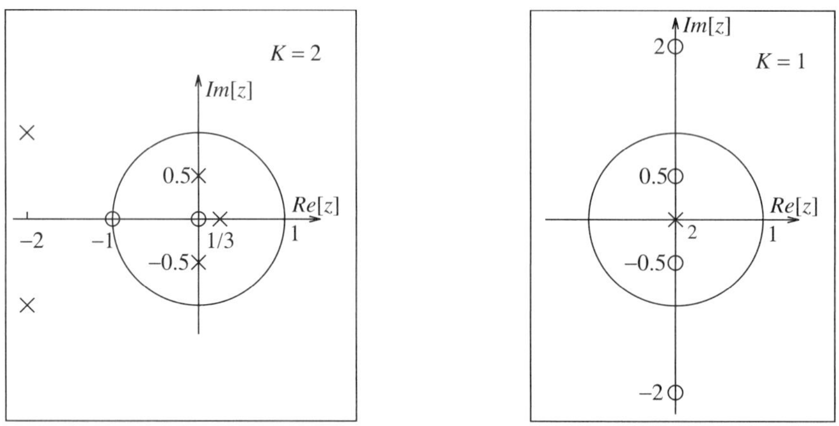


(a) (b)

**FIGURE E.4.4** Pole-zero plots for Example 4.4

**(b)** What is the $z$-transform corresponding to the pole-zero pattern of Figure E4.4(b)? Does it represent a symmetric signal?

If we let $X(z) = \frac{KN(z)}{D(z)}$, the four zeros correspond to the numerator $N(z)$ given by

$$N(z) = (z - j0.5)(z + j2)(z + j0.5)(z - j2) = z^4 + 4.25z^2 + 1$$

The two poles at the origin correspond to the denominator $D(z) = z^2$. With $K = 1$, the $z$-transform is given by

$$X(z) = K\frac{N(z)}{D(z)} = \frac{z^4 + 4.25z^2 + 1}{z^2} = z^2 + 4.25 + z^{-2}$$

Checking for symmetry, we find that $X(z) = X(1/z)$, and thus, $x[n]$ is even symmetric. In fact, $x[n] = \delta[n+2] + 4.25\delta[n] + \delta[n-2] = \{1, \overset{\Downarrow}{4.25}, 1\}$. We also note that each zero is paired with its reciprocal ($j0.5$ with $-j2$, and $-j0.5$ with $j2$) and is a characteristic of symmetric sequences.

# 4.4 The Transfer Function

The response $y[n]$ of a system with impulse response $h[n]$ to an arbitrary input $x[n]$ is given by the convolution $y[n] = x[n] * h[n]$. Since the convolution operation transforms to a product, we have

$$Y(z) = X(z)H(z) \quad \text{or} \quad H(z) = \frac{Y(z)}{X(z)} \tag{4.15}$$

The time-domain and $z$-domain equivalence of these operations is illustrated in Figure 4.3.

The transfer function is defined only for *relaxed LTI systems,* either as the ratio of the output $Y(z)$ and input $X(z)$ or as the $z$-transform of the system impulse response $h[n]$.

A relaxed LTI system is also described by the difference equation:

$$y[n]+A_1y[n-1]+\cdots+A_Ny[n-N] = B_0x[n]+B_1x[n-1]+\cdots+B_Mx[n-M] \tag{4.16}$$

Its $z$-transform results in the transfer function:

$$H(z) = \frac{Y(z)}{X(z)} = \frac{B_0 + B_1z^{-1} + \cdots + B_Mz^{-M}}{1 + A_1z^{-1} + \cdots + A_Nz^{-N}} \tag{4.17}$$

The transfer function is thus a ratio of polynomials in $z$. An LTI system may be described by its transfer function, its impulse response, its difference equation, or its pole-zero plot. Given one form, it is possible to obtain any of the other forms.

# Drill Problem 4.11

**(a)** Find the transfer function $H(z)$ of the digital filter described by $y[n] - 0.4y[n-1] = 2x[n]$.
**(b)** Find the difference equation of the digital filter described by $H(z) = \frac{z-1}{z+0.5}$.
**(c)** Find the difference equation of the digital filter described by $h[n] = (0.5)^n u[n] - \delta[n]$.

**Answers: (a)** $\frac{2z}{z-0.4}$ **(b)** $y[n] + 0.5y[n-1] = x[n] - x[n-1]$
**(c)** $y[n] - 0.5y[n-1] = 0.5x[n-1]$

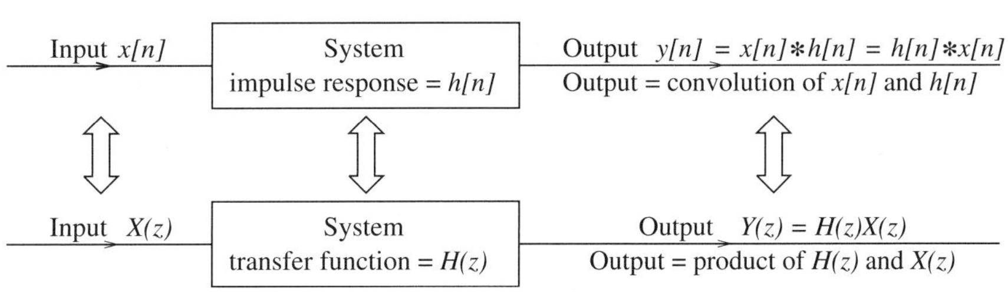


**FIGURE 4.3** System description in the time domain and *z*-domain. In the time domain, the system output is found by the convolution of $x[n]$ and $h[n]$. In the *z*-domain, the transformed output $Y(z)$ is found by the product of $X(z)$ and $H(z)$. Convolution in one domain transforms to multiplication in the other

The poles of a transfer function $H(z)$ are called **natural modes** or **natural frequencies**. The poles of $Y(z) = H(z)X(z)$ determine the form of the system response. Clearly, the natural frequencies in $H(z)$ will always appear in the system response, unless they are canceled by any corresponding zeros in $X(z)$. The zeros of $H(z)$ may be regarded as the (complex) frequencies that are blocked by the system.

## 4.5 Interconnected Systems

The $z$-transform is well suited to the study of interconnected LTI systems. Figure 4.4 shows the interconnection of two *relaxed* systems in cascade and in parallel.

The overall transfer function of a **cascaded system** is the product of the individual transfer functions. For $n$ systems in cascade, the overall impulse response $h_C[n]$ is the convolution of the individual impulse responses $h_1[n]$, $h_2[n]$, .... Since the convolution operation transforms to a product, we have

$$H_C(z) = H_1(z)H_2(z)\cdots H_n(z) \quad \text{(for } n \text{ systems in cascade)} \tag{4.18}$$

We can also factor a given transfer function $H(z)$ into the product of first-order and second-order transfer functions and realize $H(z)$ in cascaded form.

For systems in **parallel** the overall transfer function is the sum of the individual transfer functions. For $n$ systems in parallel,

$$H_P(z) = H_1(z) + H_2(z) + \cdots + H_n(z) \quad \text{(for } n \text{ systems in parallel)} \tag{4.19}$$

We can also use partial fractions to express a given transfer function $H(z)$ as the sum of first-order and/or second-order subsystems and realize $H(z)$ as a parallel combination.

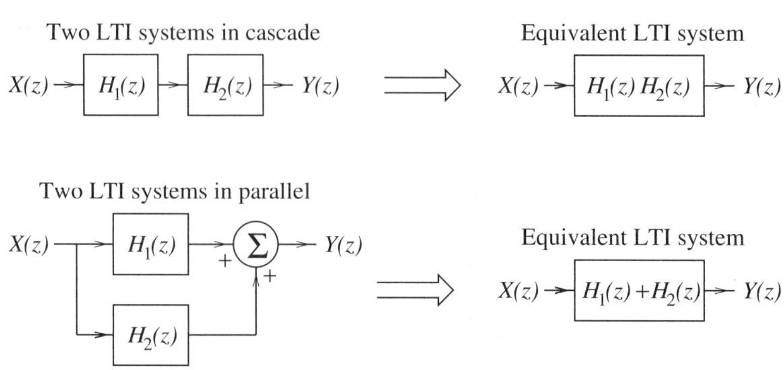


**FIGURE 4.4** The equivalent transfer function of systems in cascade is the product of the individual transfer functions. The equivalent transfer function of systems in parallel is the sum of the individual transfer functions

---

### Overall Impulse Response and Transfer Function of Systems in Cascade and Parallel

**Cascade:** Convolve individual impulse responses. Multiply individual transfer functions.
**Parallel:** Add individual impulse responses. Add individual transfer functions.

---

*Example 4.5*

### Systems in Cascade and Parallel

**(a)** Two digital filters are described by $h_1[n] = \alpha^n u[n]$ and $h_2[n] = (-\alpha)^n u[n]$. The transfer function of their cascade is $H_C(z)$ and of their parallel combination is $H_P(z)$. How are $H_C(z)$ and $H_P(z)$ related?

The transfer functions of the two filters are $H_1(z) = \dfrac{z}{z-\alpha}$ and $H_2(z) = \dfrac{z}{z+\alpha}$. Thus,

$$H_C(z) = H_1(z)H_2(z) = \frac{z^2}{z^2-\alpha^2} \qquad H_P(z) = H_1(z) + H_2(z) = \frac{2z^2}{z^2-\alpha^2}$$

So, $H_P(z) = 2H_C(z)$.

**(b)** Is the cascade or parallel combination of two *linear-phase* filters also linear phase? Explain.

Linear-phase filters are described by symmetric impulse response sequences.

The impulse response of their cascade is also symmetric because it is the convolution of two symmetric sequences. So, the cascade of two linear-phase filters is *always* linear phase.

The impulse response of their parallel combination is the sum of their impulse responses. Since the sum of symmetric sequences is not always symmetric (unless both are odd symmetric or both are even symmetric), the parallel combination of two linear-phase filters is not always linear phase.

---

# Drill Problem 4.12

**(a)** Find the transfer function of the parallel connection of two filters described by $h_1[n] = \{\overset{\Downarrow}{1}, 3\}$ and $h_2[n] = \{2, \overset{\Downarrow}{-1}\}$.

**(b)** The transfer function of the cascade of two filters is $H_C(z) = 1$. If the impulse response of one filter is $h_1[n] = 2(0.5)^n u[n]$, find the impulse response of the second.

**(c)** Two filters are described by $y[n] - 0.4y[n-1] = x[n]$ and $h_2[n] = 2(0.4)^n u[n]$. Find the transfer function of their parallel combination and cascaded combination.

**Answers:** **(a)** $2z + 3z^{-1}$ **(b)** $\{\underset{\Uparrow}{0.5}, -0.25\}$ **(c)** $H_P(z) = \frac{3z}{z-0.4}$, $H_C(z) = \frac{2z^2}{(z-0.4)^2}$

## 4.6 Transfer Function Realization

The realization of digital filters described by transfer functions parallels the realization based on difference equations. The nonrecursive and recursive filters described by

$$H_N(z) = B_0 + B_1 z^{-1} + \cdots + B_M z^{-M}$$
$$y[n] = B_0 x[n] + B_1 x[n-1] + \cdots + B_M x[n-M] \tag{4.20}$$
$$H_R(z) = \frac{1}{1 + A_1 z^{-1} + \cdots + A_N z^{-N}}$$
$$y[n] = -A_1 y[n-1] - \cdots - A_N y[n-N] + x[n] \tag{4.21}$$

can be realized using the **feed-forward** (nonrecursive) structure and **feedback** (recursive) structure, as shown in Figure 4.5.

Now, consider the general difference equation

$$y[n] = -A_1 y[n-1] - \cdots - A_N y[n-N] + B_0 x[n] + B_1 x[n-1] + \cdots + B_N x[n-N] \tag{4.22}$$

We choose $M = N$ with no loss of generality, since some of the coefficients $B_k$ always may be set to zero. The transfer function (with $M = N$) then becomes

$$H(z) = \frac{B_0 + B_1 z^{-1} + \cdots + B_N z^{-N}}{1 + A_1 z^{-1} + A_2 z^{-2} + \cdots + A_N z^{-N}} = H_N(z) H_R(z) \tag{4.23}$$

The transfer function $H(z) = H_N(z)H_R(z)$ is the product of the transfer functions of a recursive and a nonrecursive system. Its realization is thus a cascade of the realizations for the recursive and nonrecursive portions, as shown in Figure 4.6(a). This form describes a **direct form I** realization. It uses $2N$ delay elements to realize an $N$th-order difference equation and is therefore not very efficient.

Since LTI systems can be cascaded in any order, we can switch the recursive and nonrecursive parts to get the structure of Figure 4.6(b). This structure suggests that each

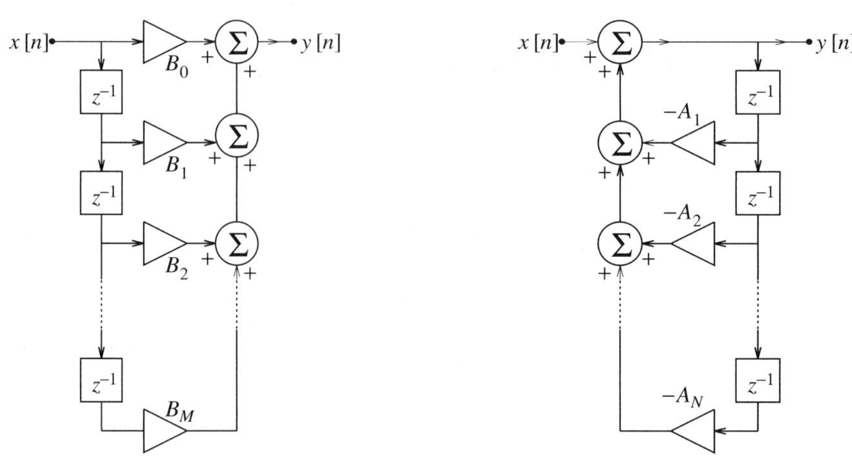


**FIGURE 4.5** Realization of a nonrecursive (left) and recursive (right) digital filter

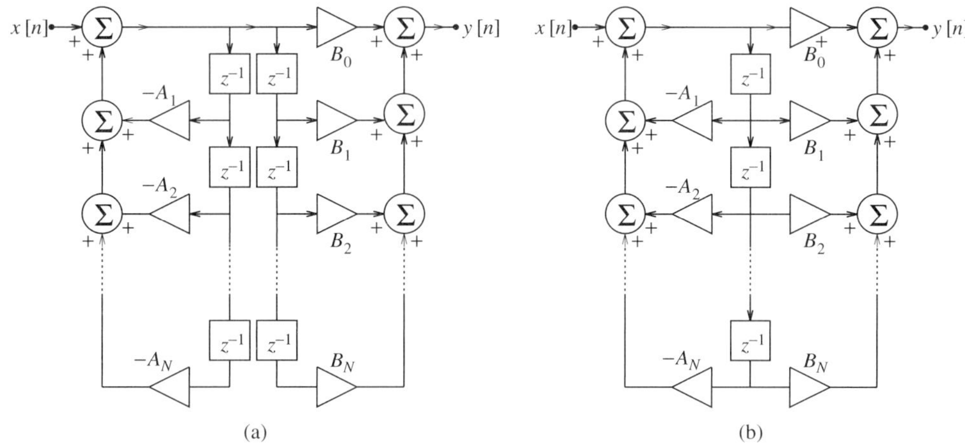


**FIGURE 4.6** (a) Direct form I and (b) canonical or direct form II realizations of a digital filter

pair of feed-forward and feedback signals can be obtained from a single delay element instead of two. This allows us to use only $N$ (rather than $2N$) delay elements and results in the **direct form II**, or **canonic**, realization. The term *canonic* implies a realization with the minimum number of delay elements.

If $M$ and $N$ are not equal, some of the coefficients ($A_k$ or $B_k$) will equal zero and will result in missing signal paths corresponding to these coefficients in the filter realization.

## 4.6.1 Transposed Realization

The direct form II also yields a transposed realization if we turn the realization around (and reverse the input and input), replace summing junctions by nodes (and vice versa), and reverse the direction of signal flow. Such a realization is developed in Figure 4.7.

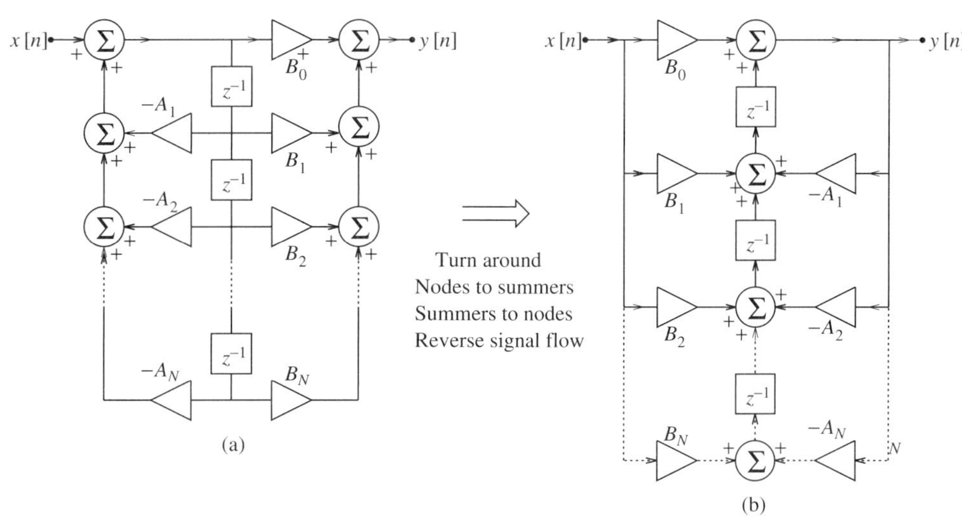


**FIGURE 4.7** (a) Direct form II and (b) transposed realizations of a digital filter

**Example 4.6**

### Direct Form II and Transposed Realizations

Consider a system described by $2y[n] - y[n-2] - 4y[n-3] = 3x[n-2]$.

Its transfer function is

$$H(z) = \frac{3z^{-2}}{2 - z^{-2} - 4z^{-3}} = \frac{1.5z}{z^3 - 0.5z - 2}$$

This is a third-order system. To sketch its direct form II and transposed realizations, we compare $H(z)$ with the generic third-order transfer function to get

$$H(z) = \frac{B_0 z^3 + B_1 z^2 + B_2 z + B_3}{z^3 + A_1 z^2 + A_2 z + A_3}$$

The nonzero constants are $B_2 = 1.5$, $A_2 = -0.5$, and $A_3 = -2$. Using these, we obtain the direct form II and transposed realizations shown in Figure E4.6.

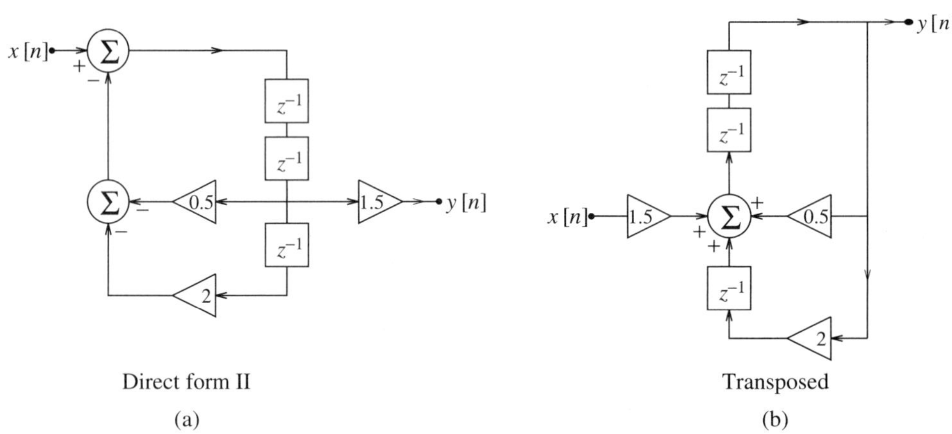


**FIGURE E.4.6** Direct form II and (b) transposed realizations of the system for Example 4.6

## 4.6.2 Cascaded and Parallel Realization

The overall transfer function of a **cascaded system** is the product of the individual transfer functions.

$$H_C(z) = H_1(z)H_2(z)\cdots H_n(z) \quad \text{(for } n \text{ systems in cascade)} \tag{4.24}$$

As a consequence, we can factor a given transfer function $H(z)$ into the product of several transfer functions to realize $H(z)$ in cascaded form. Typically, an $N$th-order transfer function is realized as a cascade of second-order sections (with an additional first-order section if $N$ is odd).

For systems in **parallel**, the overall transfer function is the sum of the individual transfer functions. For $n$ systems in parallel,

$$H_P(z) = H_1(z) + H_2(z) + \cdots + H_n(z) \quad \text{(for } n \text{ systems in parallel)} \tag{4.25}$$

As a result, we may use partial fractions to express a given transfer function $H(z)$ as the sum of first-order and/or second-order subsystems and realize $H(z)$ as a parallel combination.

**Example 4.7**

### System Realization

**(a)** Find a cascaded realization for $H(z) = \dfrac{z^2(6z-2)}{(z-1)(z^2-\frac{1}{6}z-\frac{1}{6})}$.

This system may be realized as a cascade $H(z) = H_1(z)H_2(z)$, as shown in Figure E4.7a, where

$$H_1(z) = \frac{z^2}{z^2 - \frac{1}{6}z - \frac{1}{6}} \qquad H_2(z) = \frac{6z-2}{z-1}$$

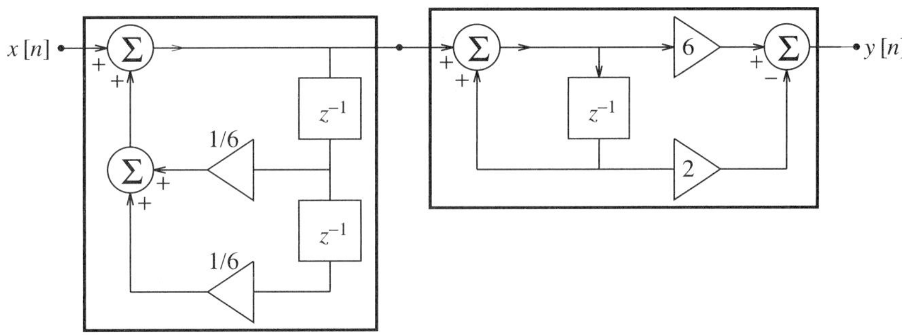


**FIGURE E 4.7.a** Cascade realization of the system for Example 4.7(a)

**(b)** Find a parallel realization for $H(z) = \dfrac{z^2}{(z-1)(z-0.5)}$.

Using partial fractions, we find

$$H(z) = \frac{2z}{z-1} - \frac{z}{z-0.5} = H_1(z) - H_2(z)$$

The two subsystems $H_1(z)$ and $H_2(z)$ may now be used to obtain the parallel realization, as shown in Figure E4.7b.

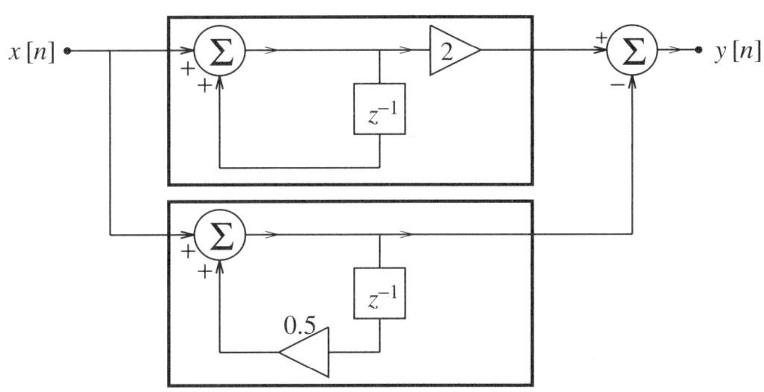


**FIGURE E 4.7.b** Parallel realization of the system for Example 4.7(b)

# 4.7 Causality and Stability of LTI Systems

In the time domain, a causal system requires a causal impulse response $h[n]$ with $h[n] = 0$, $n < 0$. If $H(z)$ describes the transfer function of this system, the number of zeros cannot exceed the number of poles. In other words, the degree of the numerator polynomial in $H(z)$

cannot exceed the degree of the denominator polynomial. This means that the transfer function of a causal system must be *proper* and its ROC must be outside a circle of finite radius.

For an LTI system to be bounded-input, bounded-output (BIBO) stable, every bounded input must result in a bounded output. In the time domain, BIBO stability of an LTI system requires an absolutely summable impulse response $h[n]$. For a causal system, this is equivalent to requiring the poles of the transfer function $H(z)$ to lie entirely *within* the unit circle in the $z$-plane. This equivalence stems from the following observations:

- **Poles outside the unit circle** ($|z| > 1$) lead to exponential growth even if the input is bounded.
  **Example:** $H(z) = \dfrac{z}{z-3}$ results in the growing exponential $(3)^n u[n]$.
- **Multiple poles on the unit circle** always result in polynomial growth.
  **Example:** $H(z) = \dfrac{1}{z(z-1)^2}$ produces a ramp function in $h[n]$.
- **Simple (non-repeated) poles on the unit circle** can also lead to an unbounded response.
  **Example:** A simple pole at $z = 1$ leads to $H(z)$ with a factor $\dfrac{z}{z-1}$. If $X(z)$ also contains a pole at $z = 1$, the response $Y(z)$ will contain the term $\dfrac{z}{(z-1)^2}$ and exhibit polynomial growth.

None of these types of time-domain terms is absolutely summable, and their presence leads to system instability. Formally, for BIBO stability, all of the poles of $H(z)$ must *lie inside (and exclude)* the unit circle $|z| = 1$. This is both a necessary and sufficient condition for the stability of causal systems.

If a system has simple (non-repeated) poles on the unit circle, it is sometimes called **marginally stable**. If a system has all of its poles and zeros inside the unit circle, it is called a **minimum-phase system**.

### 4.7.1 Stability and the ROC

For any LTI system, causal or otherwise, to be stable, *the ROC must include the unit circle.* The various situations are illustrated in Figure 4.8.

The stability of a causal system requires all of the poles to lie inside the unit circle. Thus, the ROC includes the unit circle. The stability of an anti-causal system requires all of the poles to lie outside the unit circle. Thus, the ROC once again includes the unit circle. Similarly, the ROC of a stable system with a two-sided impulse response is an annulus that includes the unit circle, and all of its poles lie *outside* this annulus. The poles inside the inner

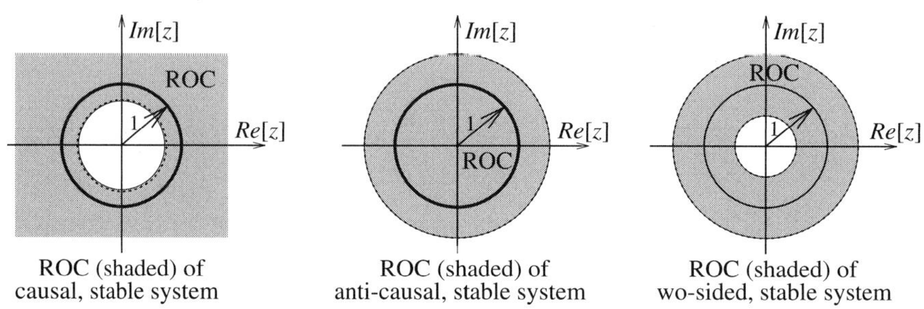


**FIGURE 4.8** The ROC of stable systems (shown shaded) always includes the unit circle

circle of the annulus contribute to the causal portion of the impulse response, while poles outside the outer circle of the annulus make up the anti-causal portion of the impulse response.

**The ROC of Stable LTI Systems Always Includes the Unit Circle**

**Stable, Causal System:** All the poles must lie inside the unit circle.
**Stable, Anti-Causal System:** All the poles must lie outside the unit circle.
**Stability from Impulse Response:** $h[n]$ must be *absolutely summable* ($\sum |h[k]| < \infty$).

***Example 4.8***

**Stability of a Recursive Filter**

**(a)** Let $H(z) = \dfrac{z}{z-\alpha}$.

If the ROC is $|z| > |\alpha|$, its impulse response is $h[n] = \alpha^n u[n]$, and the system is causal.

For stability, we require $|\alpha| < 1$ (for the ROC to include the unit circle).

**(b)** Let $H(z) = \dfrac{z}{z-\alpha}$, as before.

If the ROC is $|z| < |\alpha|$, its impulse response is $h[n] = -\alpha^n u[-n-1]$, and the system is anti-causal.

For stability, we require $|\alpha| > 1$ (for the ROC to include the unit circle).

# Drill Problem 4.13

**(a)** Is the filter described by $H(z) = \dfrac{2z+1}{(z-0.5)(z+0.5)}$, $|z| > 0.5$ stable? Causal?

**(b)** Is the filter described by $H(z) = \dfrac{2z+1}{(z-1.5)(z+0.5)}$, $|z| > 1.5$ stable? Causal?

**(c)** Is the filter described by $H(z) = \dfrac{2z+1}{(z-1.5)(z+0.5)}$, $0.5 < |z| < 1.5$ stable? Causal?

**Answers: (a)** Stable, causal **(b)** Unstable, causal **(c)** Stable, two-sided

## 4.7.2 Inverse Systems

The **inverse system** corresponding to a transfer function $H(z)$ is denoted by $H^{-1}(z)$, and defined as

$$H^{-1}(z) = H_I(z) = \frac{1}{H(z)} \tag{4.26}$$

The cascade of a system and its inverse has a transfer function of unity:

$$H_C(z) = H(z)H^{-1}(z) = 1 \qquad h_C[n] = \delta[n] \tag{4.27}$$

This cascaded system is called an **identity system**, and its impulse response equals $h_C[n] = \delta[n]$. The inverse system can be used to undo the effect of the original system. We can also describe $h_C[n]$ by the convolution $h[n] * h_I[n]$. It is far easier to find the inverse of a system in the transformed domain.

**Example 4.9**

### Inverse Systems

**(a)** Consider a system with the difference equation $y[n] + \alpha y[n-1] = x[n] + \beta x[n-1]$.

To find the inverse system, we evaluate $H(z)$ and take its reciprocal. Thus,

$$H(z) = \frac{1 + \beta z^{-1}}{1 + \alpha z^{-1}} \qquad H_I(z) = \frac{1}{H(z)} = \frac{1 + \alpha z^{-1}}{1 + \beta z^{-1}}$$

The difference equation of the inverse system is $y[n] + \beta y[n-1] = x[n] + \alpha x[n-1]$. Note that, in general, the inverse of an IIR filter is also an IIR filter.

**(b)** Consider an FIR filter whose system equation is $y[n] = x[n] + 2x[n-1] + 3x[n-2]$.

To find the inverse system, we evaluate $H(z)$ and take its reciprocal. Thus,

$$H(z) = 1 + 2z^{-1} + 3z^{-2} \qquad H_I(z) = \frac{1}{H(z)} = \frac{1}{1 + 2z^{-1} + 3z^{-2}}$$

The difference equation of the inverse system is $y[n] + 2y[n-1] + 3y[n-2] = x[n]$. Note that the inverse of an FIR filter is an IIR filter.

# Drill Problem 4.14

**(a)** The inverse of an FIR filter is always an IIR filter. True or false? If false, give a counterexample.

**(b)** The inverse of an IIR filter is always an FIR filter. True or false? If false, give a counterexample.

**Answers: (a)** True **(b)** False. For example, the inverse of $H(z) = \frac{z-0.4}{z-0.5}$ is also IIR.

## 4.8 The Inverse z-Transform

The formal inversion relation that yields $x[n]$ from $X(z)$ actually involves complex integration and is described by

$$x[n] = \frac{1}{j2\pi} \oint_\Gamma X(z) z^{n-1}\, dz \tag{4.28}$$

Here, $\Gamma$ describes a clockwise contour of integration (such as the unit circle) that encloses the origin. Evaluation of this integral requires knowledge of complex variable theory. In this text, we pursue simpler alternatives, which include long division and partial fraction expansion.

### 4.8.1 Inverse z-Transform of Finite Sequences

For finite-length sequences, $X(z)$ has a polynomial form that immediately reveals the required sequence $x[n]$. The ROC can also be discerned from the polynomial form of $X(z)$.

**Example 4.10**

**Inverse Transform of Sequences**

**(a)** Let $X(z) = 3z^{-1} + 5z^{-3} + 2z^{-4}$.

This transform corresponds to a causal sequence. We recognize $x[n]$ as a sum of shifted impulses given by

$$x[n] = 3\delta[n-1] + 5\delta[n-3] + 2\delta[n-4]$$

This sequence can also be written as $x[n] = \{\overset{\Downarrow}{0},\ 3,\ 0,\ 5,\ 2\}$.

**(b)** Let $X(z) = 2z^2 - 5z + 5z^{-1} - 2z^{-2}$.

This transform corresponds to a noncausal sequence. Its inverse transform is written, by inspection, as

$$x[n] = \{2,\ -5,\ \overset{\Downarrow}{0},\ 5,\ -2\}$$

*Comment:* Since $X(z) = -X(1/z)$, $x[n]$ should possess odd symmetry. It does.

### 4.8.2 Inverse z-Transform by Long Division

A second method requires that we set up $X(z) = \frac{N(z)}{D(z)}$ as a rational function (a ratio of polynomials) in powers of $z$ along with its ROC. For a right-sided signal (whose ROC is $|z| > |\alpha|$), we arrange $N(z)$ and $D(z)$ in *decreasing powers of* $z$ and use long division to obtain a power series in powers of $z^{-1}$, whose inverse transform corresponds to the right-sided sequence.

For a left-sided signal (whose ROC is $|z| < |\alpha|$), we arrange $N(z)$ and $D(z)$ in *increasing powers of* $z$ and use long division to obtain a power series in powers of $z$, whose inverse transform corresponds to the left-sided sequence.

This approach, however, becomes cumbersome if more than just the first few terms of $x[n]$ are required. It is not often that the first few terms of the resulting sequence allow its general nature or form to be discerned. If we regard the rational $z$-transform as a transfer function $H(z)$, the method of long division is simply equivalent to finding the first few terms of its impulse response recursively from its difference equation.

**Finding Inverse Transforms of $X(z) = N(z)/D(z)$ by Long Division**

**Right-Sided:** Put $N(z)$, $D(z)$ in *descending powers of* $z$. Obtain a power series in powers of $z^{-1}$.

**Left-Sided:** Put $N(z)$, $D(z)$ in *ascending powers of* $z$. Obtain a power series in powers of $z$.

**Example 4.11**

### Inverse Transforms by Long Division

**(a)** We find the right-sided inverse of $H(z) = \dfrac{z-4}{1-z+z^2}$.

We arrange the polynomials in descending powers of $z$ and use long division to get

$$
\begin{array}{r l}
 & z^{-1} - 3z^{-2} - 4z^{-3} \cdots \\
z^2 - z + 1 \,\big) & z - 4 \\
 & \underline{z - 1 + z^{-1}} \\
 & -3 - z^{-1} \\
 & \underline{-3 + 3z^{-1} - 3z^{-2}} \\
 & -4z^{-1} + 3z^{-2} \\
 & \underline{-4z^{-1} + 4z^{-2} - 4z^{-3}} \\
 & -z^{-2} + 4z^{-3} \\
 & \cdots
\end{array}
$$

This leads to $H(z) = z^{-1} - 3z^{-2} - 4z^{-3} + \cdots$. The sequence $h[n]$ can be written as

$$h[n] = \delta(n-1) - 3\delta[n-2] - 4\delta[n-3] + \cdots \quad \text{or} \quad h[n] = \{\overset{\Downarrow}{0}, 1, -3, -4, \ldots\}$$

**(b)** We could also have found the inverse by setting up the the difference equation corresponding to $H(z) = Y(z)/X(z)$, to give

$$y[n] - y[n-1] + y[n-2] = x[n-1] - 4x[n-2]$$

With $x[n] = \delta[n]$, its impulse response $h[n]$ is

$$h[n] - h[n-1] + h[n-2] = \delta[n] - 4\delta[n-2]$$

With $h[-1] = h[-2] = 0$, (a relaxed system), we recursively obtain the first few values of $h[n]$ as

$$n = 0: \quad h[0] = h[-1] - h[-2] + \delta[-1] - 4\delta[-2] = 0 - 0 + 0 - 0 = 0$$

$$n = 1: \quad h[1] = h[0] - h[-1] + \delta[0] - 4\delta[-1] = 0 - 0 + 1 - 0 = 1$$

$$n = 2: \quad h[2] = h[1] - h[0] + \delta[1] - 4\delta[0] = 1 - 0 + 0 - 4 = -3$$

$$n = 3: \quad h[3] = h[2] - h[1] + \delta[2] - 4\delta[1] = -3 - 1 + 0 + 0 = -4$$

These are identical to the values obtained using long division in part (a).

**(c)** We find the left-sided inverse of $H(z) = \dfrac{z-4}{1-z+z^2}$.

We arrange the polynomials in ascending powers of $z$ and use long division to obtain

$$
\begin{array}{r l}
 & -4 - 3z + z^2 \cdots \\
1 - z + z^2 \,\big) & -4 + z \\
 & \underline{-4 + 4z - 4z^2} \\
 & -3z + 4z^2 \\
 & \underline{-3z + 3z^2 - 3z^3} \\
 & z^2 + 3z^3 \\
 & \underline{z^2 - z^3 + z^4} \\
 & 4z^3 - z^4 \\
 & \cdots
\end{array}
$$

Thus, $H(z) = -4 - 3z + z^2 + \cdots$. The sequence $h[n]$ can then be written as

$$h[n] = -4\delta[n] - 3\delta[n+1] + \delta[n+2] + \cdots \quad \text{or} \quad h[n] = \{\ldots,\ 1,\ -3,\ -\overset{\Downarrow}{4}\}$$

**(d)** We could also have found the inverse by setting up the difference equation in the form

$$h[n-2] = h[n-1] + \delta[n-1] - 4\delta[n-2]$$

With $h[1] = h[2] = 0$, we can generate $h[0], h[-1], h[-2], \ldots$, recursively, to obtain the same result as in part (c).

# Drill Problem 4.15

**(a)** Let $H(z) = \dfrac{z-4}{z^2+1}$, $|z| > 1$. Find the first few terms of $h[n]$ by long division.

**(b)** Let $H(z) = \dfrac{z-4}{z^2+4}$, $|z| < 2$. Find the first few terms of $h[n]$ by long division.

**Answers: (a)** $\{\overset{\Downarrow}{0},\ 1,\ -4,\ -1,\ 4, \ldots\}$ **(b)** $\{\ldots,\ -0.06250,\ 0.25,\ 0.25,\ \overset{\Downarrow}{-1}\}$

## 4.8.3 Inverse *z*-Transform from Partial Fractions

A much more useful method for inversion of the $z$-transform relies on its *partial fraction expansion* into terms whose inverse transform can be identified using a table of transform pairs. Since the $z$-transform of standard sequences in Table 4.1 involves the factor $z$ in the numerator, it is more convenient to perform the partial fraction expansion for $Y(z) = X(z)/z$ rather than for $X(z)$. We then multiply through by $z$ to obtain terms describing $X(z)$ in a form ready for inversion. For partial fraction expansion of $X(z)$, the order of its numerator must not exceed the order of its denominator. The constants in the partial fraction expansion are often called **residues**. The form of the expansion depends on the nature of the poles (denominator roots) of $Y(z)$, as summarized in the following review panel. The inverse transform of the various partial fraction forms is listed in Table 4.3.

**TABLE 4.3** ➤ Inverse *z*-Transform of Partial Fraction Expansion (PFE) Terms

| Entry | PFE Term $X(z)$ | Causal Signal $x[n],\ n \ge 0$ |
|---|---|---|
| 1 | $\dfrac{z}{z-\alpha}$ | $\alpha^n$ |
| 2 | $\dfrac{z}{(z-\alpha)^2}$ | $n\alpha^{(n-1)}$ |
| 3 | $\dfrac{z}{(z-\alpha)^{N+1}} \quad (N > 1)$ | $\dfrac{n(n-1)\cdots(n-N+1)}{N!}\alpha^{(n-N)}$ |
| 4 | $\dfrac{z\bar{K}}{z-\alpha e^{j\Omega}} + \dfrac{z\bar{K}^*}{z-\alpha e^{-j\Omega}}$ | $2K\alpha^n\cos(n\Omega+\phi) = 2\alpha^n[C\cos(n\Omega) - D\sin(n\Omega)]$ |
| 5 | $\dfrac{z\bar{K}}{(z-\alpha e^{j\Omega})^2} + \dfrac{z\bar{K}^*}{(z-\alpha e^{-j\Omega})^2}$ | $2Kn\alpha^{n-1}\cos[(n-1)\Omega+\phi]$ |
| 6 | $\dfrac{z\bar{K}}{(z-\alpha e^{j\Omega})^{N+1}} + \dfrac{z\bar{K}^*}{(z-\alpha e^{-j\Omega})^{N+1}}$ | $2K\dfrac{n(n-1)\cdots(n-N+1)}{N!}\alpha^{(n-N)}\cos[(n-N)\Omega+\phi]$ |

*NOTE 1:* Where applicable, $\bar{K} = Ke^{j\phi} = C + jD$

*NOTE 2:* For anti-causal sequences, we get the signal $-x[n]u[-n-1]$, where $x[n]$ is as listed.

### Non-Repeated Real Roots

If $Y(z)$ contains only real and distinct poles, we express it in the form:

$$Y(z) = \frac{P(z)}{(z+p_1)(z+p_2)\cdots(z+p_N)} = \frac{K_1}{z+p_1} + \frac{K_2}{z+p_2} + \cdots + \frac{K_N}{z+p_N} \tag{4.29}$$

To find the $m$th coefficient $K_m$, we multiply both sides by $(z+p_m)$ to get

$$(z+p_m)Y(z) = K_1\frac{(z+p_m)}{(z+p_1)} + \cdots + K_m + \cdots + K_N\frac{(z+p_m)}{(z+p_N)} \tag{4.30}$$

With both sides evaluated at $z = -p_m$, we obtain $K_m$ as

$$K_m = (z+p_m)Y(z)|_{z=-p_m} \tag{4.31}$$

**Example 4.12**

### Inverse Transform of Right-Sided Signals

**(a) Non-Repeated Roots**

We find the causal inverse of $X(z) = \dfrac{z}{(z-0.25)(z-0.5)}$.

We first form $Y(z) = \frac{X(z)}{z}$, and expand $Y(z)$ into partial fractions, to obtain

$$Y(z) = \frac{X(z)}{z} = \frac{1}{(z-0.25)(z-0.5)} = \frac{-4}{z-0.25} + \frac{4}{z-0.5}$$

Multiplying through by $z$, we get

$$X(z) = zY(z) = \frac{-4z}{z-0.25} + \frac{4z}{z-0.5} \qquad x[n] = -4(0.25)^n u[n] + 4(0.5)^n u[n]$$

**(b) Non-Repeated Roots**

We find the causal inverse of $X(z) = \dfrac{1}{(z-0.25)(z-0.5)}$.

We first form $Y(z) = \frac{X(z)}{z}$, and expand $Y(z)$ into partial fractions, to obtain

$$Y(z) = \frac{X(z)}{z} = \frac{1}{z(z-0.25)(z-0.5)} = \frac{8}{z} - \frac{16}{z-0.25} + \frac{8}{z-0.5}$$

Multiplying through by $z$, we get

$$X(z) = zY(z) = 8 - \frac{16z}{z-0.25} + \frac{8z}{z-0.5} \qquad x[n] = 8\delta[n] - 16(0.25)^n u[n] + 8(0.5)^n u[n]$$

*Comment:* An alternate (but non-standard) approach is to expand $X(z)$ itself as

$$X(z) = \frac{-4}{z-0.25} + \frac{4}{z-0.5} \qquad x[n] = -4(0.25)^{n-1}u[n-1] + 4(0.5)^{n-1}u[n-1]$$

Its inverse requires the shifting property. This form is functionally equivalent to the previous case. For example, we find that $x[0] = 0,\ x[1] = 0,\ x[2] = 1$, and $x[3] = 0.25$, as before.

# Drill Problem 4.16

Let $H(z) = \frac{N(z)}{(z-0.5)(z-0.4)}$ and $h[n] = A\delta[n] + B(0.5)^n u[n] + C(0.4)^n u[n]$. Find $A$, $B$ and $C$ if

**(a)** $N(z) = z^2$
**(b)** $N(z) = z$
**(c)** $N(z) = 1$

**Answers: (a)** 0, 5, −4 **(b)** 0, 10, −10 **(c)** 5, 20, −25

## Quadratic Forms and Complex Roots

If $Y(z)$ has a quadratic denominator with complex conjugate roots, its inverse transform may be found by comparing the coefficients of the powers of $z$ in its numerator to a combination of the generic forms:

$$A\alpha^n \cos(n\Omega)u[n] \Leftarrow\boxed{\text{ZT}}\Rightarrow \frac{A(z^2 - z\alpha\cos\Omega)}{z^2 - 2z\alpha\cos\Omega + \alpha^2}$$

$$B\alpha^n \sin(n\Omega)u[n] \Leftarrow\boxed{\text{ZT}}\Rightarrow \frac{Bz\alpha\sin\Omega}{z^2 - 2z\alpha\cos\Omega + \alpha^2}$$

If $Y(z)$ has both distinct real and complex conjugate poles, its partial fraction expansion may be written as

$$Y(z) = \frac{K_1}{z+p_1} + \frac{K_2}{z+p_2} + \cdots + \frac{A_1}{z+r_1} + \frac{A_1^*}{z+r_1^*} + \frac{A_2}{z+r_2} + \frac{A_2^*}{z+r_2^*} + \cdots \quad (4.32)$$

For a real root, the residue (coefficient) will also be real. For each pair of complex conjugate roots, the residues will also be complex conjugates. The inverse transform of each complex conjugate pair may be simplified to the form listed in Table 4.3.

**Example 4.13**

### Inverse Transform of Right-Sided Signals

**(a) Inverse Transform of Quadratic Forms**

We find the causal inverse of $X(z) = \frac{z}{z^2+4}$.

The numerator suggests the generic form $x[n] = B\alpha^n \sin(n\Omega)u[n]$, because

$$B\alpha^n \sin(n\Omega)u[n] \Leftarrow\boxed{\text{ZT}}\Rightarrow \frac{Bz\alpha\sin\Omega}{z^2 - 2z\alpha\cos\Omega + \alpha^2}$$

Comparing denominators, we find that $\alpha^2 = 4$ and $2\alpha\cos\Omega = 0$. Thus, $\alpha = \pm 2$. If we pick $\alpha = 2$, we get $4\cos\Omega = 0$ or $\cos\Omega = 0$ and thus $\Omega = \pi/2$. Finally, comparing numerators, $Bz\alpha\sin\Omega = z$ or $B = 0.5$. Thus,

$$x[n] = B\alpha^n \sin(n\Omega)u[n] = 0.5(2)^n \sin(n\pi/2)u[n]$$

**(b) Non-Repeated Complex Roots**

We find the causal inverse of $X(z) = \dfrac{z^2 - 3z}{(z-2)(z^2-2z+2)}$.

We set up the partial fraction expansion for $Y(z) = \frac{X(z)}{z}$ as

$$Y(z) = \frac{X(z)}{z} = \frac{z-3}{(z-2)(z-1-j)(z-1+j)}$$
$$= \frac{A}{z-2} + \frac{\bar{K}}{z-\sqrt{2}e^{j\pi/4}} + \frac{\bar{K}^*}{z-\sqrt{2}e^{-j\pi/4}}$$

We evaluate the constants $A$ and $\bar{K}$, to give

$$A = \left.\frac{z-3}{z^2-2z+2}\right|_{z=2} = -0.5$$

$$\bar{K} = \left.\frac{z-3}{(z-2)(z-1+j)}\right|_{z=1+j} = 0.7906e^{-j71.56^\circ} = 0.25 - j0.75$$

Multiplying through by $z$, we get

$$X(z) = zY(z) = \frac{-0.5z}{z-2} + \frac{z\bar{K}}{z-\sqrt{2}e^{j\pi/4}} + \frac{z\bar{K}^*}{z-\sqrt{2}e^{-j\pi/4}}$$

The inverse of the first term is easy. For the remaining pair, we use entry 4 of Table 4.3 for inversion of partial fraction forms with $\alpha = \sqrt{2}$, $\Omega = \frac{\pi}{4}$, $K = 0.7906$, and $\phi = -71.56^\circ$ to give

$$x[n] = -0.5(2)^n u[n] + 2(0.7906)(\sqrt{2})^n \cos\left(\frac{n\pi}{4} - 71.56^\circ\right) u[n]$$

With $C = 0.25$ and $D = -0.75$, this may also be expressed in the alternate form:

$$x[n] = -0.5(2)^n u[n] + 2(\sqrt{2})^n \left[0.25\cos\left(\frac{n\pi}{4}\right) + 0.75\sin\left(\frac{n\pi}{4}\right)\right] u[n]$$

**(c) Inverse Transform of Quadratic Forms**

Let $X(z) = \dfrac{z^2+z}{z^2-2z+4}$.

The quadratic numerator suggests the form $x[n] = A\alpha^n \cos(n\Omega)u[n] + B\alpha^n \sin(n\Omega)u[n]$ because

$$A\alpha^n \cos(n\Omega)u[n] \Leftarrow\boxed{\text{ZT}}\Rightarrow \frac{A(z^2 - z\alpha\cos\Omega)}{z^2 - 2z\alpha\cos\Omega + \alpha^2}$$

$$B\alpha^n \sin(n\Omega)u[n] \Leftarrow\boxed{\text{ZT}}\Rightarrow \frac{Bz\alpha\sin\Omega}{z^2 - 2z\alpha\cos\Omega + \alpha^2}$$

Comparing denominators, we find $\alpha^2 = 4$ and $2\alpha\cos\Omega = 2$. Thus, $\alpha = \pm 2$. If we pick $\alpha = 2$, we get $\cos\Omega = 0.5$ or $\Omega = \pi/3$.

Now, $A(z^2 - z\alpha\cos\Omega) = A(z^2 - z)$ and $Bz\alpha\sin\Omega = Bz\sqrt{3}$. We express the numerator of $X(z)$ as a sum of these forms to get $z^2 + z = (z^2 - z) + 2z = (z^2 - z) + (2/\sqrt{3})(z\sqrt{3})$ (with $A = 1$ and $B = 2/\sqrt{3} = 1.1547$). Thus,

$$x[n] = A\alpha^n \cos(n\Omega)u[n] + B\alpha^n \sin(n\Omega)u[n]$$
$$= (2)^n \cos\left(\frac{n\pi}{3}\right) u[n] + 1.1547(2)^n \sin\left(\frac{n\pi}{3}\right) u[n]$$

The formal approach is to use partial fractions. With $z^2 - 2z + 4 = (z - 2e^{j\pi/3})(z - 2e^{-j\pi/3})$, we find

$$\frac{X(z)}{z} = \frac{z+1}{z^2 - 2z + 4} = \frac{\bar{K}}{z - 2e^{j\pi/3}} + \frac{\bar{K}^*}{z - 2e^{-j\pi/3}}$$

We find $\bar{K} = 0.7638e^{-j49.11^\circ} = 0.5 - j0.5774$ and entry 4 of Table 4.3 for inversion of partial fraction forms (with $\alpha = 2,\ \Omega = \frac{\pi}{3}$) gives

$$x[n] = 1.5275(2)^n \cos\left(\frac{n\pi}{3} - 49.11^\circ\right) u[n] = (2)^n \cos\left(\frac{n\pi}{3}\right) u[n] + 1.1547(2)^n \sin\left(\frac{n\pi}{3}\right) u[n]$$

The second form of this result is identical to what was found earlier.

# Drill Problem 4.17

Let $H(z) = \frac{N(z)}{z^2 - 0.8z + 0.64}$. Find its causal inverse $h[n]$

**(a)** by comparing with a standard transform pair if $N(z) = z$
**(b)** by comparing with standard transform pairs or by partial fractions if $N(z) = z^2$
**(c)** by partial fractions if $N(z) = 1$

**Answers:** **(a)** $1.44(0.8)^n \sin(\frac{n\pi}{3})u[n]$ **(b)** $(1.155(0.8)^n \cos(\frac{n\pi}{3} - \frac{\pi}{6})u[n]$
**(c)** $1.80(0.8)^n \cos(\frac{n\pi}{3} - \frac{5\pi}{6})u[n] + 1.5625\delta[n]$

## Repeated Factors

If the denominator of $Y(z)$ contains the repeated term $(z+r)^M$, the partial fraction expansion corresponding to the repeated terms has the form

$$Y(z) = (\text{other terms}) + \frac{A_0}{(z+r)^M} + \frac{A_1}{(z+r)^{M-1}} + \cdots + \frac{A_{M-1}}{z+r} \tag{4.33}$$

Observe that the constants $A_j$ ascend in index $j$ from 0 to $M-1$, whereas the denominators $(z+r)^k$ descend in power $k$ from $M$ to 1. Their evaluation requires $(z+r)^M Y(z)$ and its derivatives. We successively find

$$A_0 = (z+r)^M Y(z)\big|_{z=-r} \qquad A_2 = \frac{1}{2!}\frac{d^2}{dz^2}[(z+r)^M Y(z)]\bigg|_{z=-r}$$
$$A_1 = \frac{d}{dz}[(z+r)^M Y(z)]\bigg|_{z=-r} \qquad A_k = \frac{1}{k!}\frac{d^k}{dz^k}[(z+r)^M Y(z)]\bigg|_{z=-r} \tag{4.34}$$

Even though this process allows us to find the coefficients independently of each other, finding the constants involving derivatives can become tedious if the multiplicity $M$ of the roots exceeds 2 or 3. A simpler alternative to finding the constants that involve derivatives (or any constants) is to solve simultaneous equations. For example, if we wish to find $N$ constants, we evaluate the partial fraction expression at $N$ distinct values of $z$ (other than the pole locations) and solve the resulting $N$ simultaneous equations. Table 4.3 lists some transform pairs for inversion of the $z$-transform based on repeated linear factors.

**Example 4.14**

### Inverse Transform of Right-Sided Signals

**(a) Repeated Roots**

We find the inverse of $X(z) = \dfrac{z}{(z-1)^2(z-2)}$.

We obtain $Y(z) = \frac{X(z)}{z}$, and set up its partial fraction expansion as

$$Y(z) = \frac{X(z)}{z} = \frac{1}{(z-1)^2(z-2)} = \frac{A}{z-2} + \frac{K_0}{(z-1)^2} + \frac{K_1}{z-1}$$

The constants in the partial fraction expansion are

$$A = \left.\frac{1}{(z-1)^2}\right|_{z=2} = 1, \quad K_0 = \left.\frac{1}{z-2}\right|_{z=1} = -1, \quad K_1 = \left.\frac{d}{dz}\left[\frac{1}{z-2}\right]\right|_{z=1} = -1$$

Once $A$ and $K_0$ are known, an alternative way of finding $K_1$ without using derivatives is to evaluate the partial fraction expression at, say $z = 0$, to give

$$\frac{1}{(-1)^2(-2)} = -0.5 = \frac{A}{-2} + \frac{K_0}{(-1)^2} + \frac{K_1}{-1} = -0.5 - 1 - K_1 \qquad \text{or} \qquad K_1 = -1$$

Finally, multiplying $Y(z)$ by $z$, we get

$$X(z) = zY(z) = \frac{z}{z-2} - \frac{z}{(z-1)^2} - \frac{z}{z-1},$$

$$x[n] = (2)^n u[n] - nu[n] - u[n] = (2^n - n - 1)u[n]$$

The first few values $x[0] = 0$, $x[1] = 0$, $x[2] = 1$, $x[3] = 4$, and $x[4] = 11$ can be checked easily by long division.

**(b) Repeated Roots**

We find the inverse of $X(z) = \dfrac{1}{z(z-0.5)}$.

We obtain $Y(z) = \frac{X(z)}{z}$, and set up its partial fraction expansion as

$$Y(z) = \frac{X(z)}{z} = \frac{1}{z^2(z-0.5)} = \frac{A}{z-0.5} + \frac{K_0}{z^2} + \frac{K_1}{z}$$

The constants in the partial fraction expansion are

$$A = \left.\frac{1}{z^2}\right|_{z=0.5} = 4, \quad K_0 = \left.\frac{1}{z-0.5}\right|_{z=0} = -2, \quad K_1 = \left.\frac{d}{dz}\left[\frac{1}{z-0.5}\right]\right|_{z=0} = -4$$

Substituting into $Y(z)$ and multiplying through by $z$, we get

$$X(z) = zY(z) = \frac{4z}{z-0.5} - \frac{2}{z} - 4, \qquad x[n] = 4(0.5)^n u[n] - 2\delta[n-1] - 4\delta[n]$$

# Drill Problem 4.18

**(a)** Let $X(z) = \dfrac{4}{z(z-0.4)}$. Find its causal inverse transform $x[n]$.

**(b)** Let $H(z) = \dfrac{z-0.8}{(z-0.4)^2}$. Find its causal inverse transform $h[n]$.

**Answers: (a)** $25(0.4)^n u[n] - 25\delta[n] - 10\delta[n-1]$
**(b)** $5(0.4)^n u[n] - n(0.4)^{n-1}u[n] - 5\delta[n]$

## 4.8.4 The ROC and Inversion

We have so far been assuming right-sided sequences when no ROC is given. Only when the ROC is specified do we obtain a unique sequence from $X(z)$. Sometimes, the ROC may be specified indirectly by requiring the system to be stable, for example. Since the ROC of a stable system includes the unit circle, this gives us a clue to the type of inverse we require.

**Example 4.15**

**Inversion and the ROC**

**(a)** Find all possible inverse transforms of $X(z) = \dfrac{z}{(z-0.25)(z-0.5)}$.

The partial fraction expansion of $Y(z) = \dfrac{X(z)}{z}$ leads to $X(z) = \dfrac{-4z}{z-0.25} + \dfrac{4z}{z-0.5}$.

1. If the ROC is $|z| > 0.5$, $x[n]$ is causal and stable, and we obtain

$$x[n] = -4(0.25)^n u[n] + 4(0.5)^n u[n]$$

2. If the ROC is $|z| < 0.25$, $x[n]$ is anti-causal and unstable, and we obtain

$$x[n] = 4(0.25)^n u[-n-1] - 4(0.5)^n u[-n-1]$$

3. If the ROC is $0.25 < |z| < 0.5$, $x[n]$ is two-sided and unstable. This ROC is valid only if $\frac{-4z}{z-0.25}$ describes a causal sequence (ROC $|z| > 0.25$), and $\frac{4z}{z-0.5}$ describes an anti-causal sequence (ROC $|z| < 0.5$). With this in mind, we obtain

$$x[n] = \underbrace{-4(0.25)^n u[n]}_{\text{ROC: } |z|>0.25} \quad \underbrace{-4(0.5)^n u[-n-1]}_{\text{ROC: } |z|<0.5}$$

**(b)** Find the unique inverse transforms of the following, assuming each system is stable:

$$H_1(z) = \frac{z}{(z-0.4)(z+0.6)} \qquad H_2(z) = \frac{2.5z}{(z-0.5)(z+2)} \qquad H_3(z) = \frac{z}{(z-2)(z+3)}$$

Partial fraction expansion leads to

$$H_1(z) = \frac{z}{z-0.4} - \frac{z}{z+0.6} \qquad H_2(z) = \frac{z}{z-0.5} - \frac{z}{z+2} \qquad H_3(z) = \frac{z}{z-2} - \frac{z}{z+3}$$

To find the appropriate inverse, the key is to recognize that the ROC must include the unit circle. Looking at the pole locations, we see that

1. $H_1(z)$ is stable if its ROC is $|z| > 0.6$. Its inverse is causal, with

$$h_1[n] = (0.4)^n u[n] - (-0.6)^n u[n]$$

**2.** $H_2(z)$ is stable if its ROC is $0.5 < |z| < 2$. Its inverse is two-sided, with

$$h_2[n] = (0.5)^n u[n] + (-2)^n u[-n-1]$$

**3.** $H_3(z)$ is stable if its ROC is $|z| < 2$. Its inverse is anti-causal, with

$$h_3[n] = -(2)^n u[-n-1] + (-3)^n u[-n-1]$$

## Drill Problem 4.19

Let $H(z) = \frac{0.1z}{(z-0.5)(z-0.4)}$. Find its inverse transform $h[n]$ if the region of convergence (ROC) is

**(a)** $|z| > 0.5$
**(b)** $|z| < 0.4$
**(c)** $0.4 < |z| < 0.5$

**Answers: (a)** $[(0.5)^n - (0.4)^n]u[n]$ **(b)** $[(0.4)^n - (0.5)^n]u[-n-1]$
**(c)** $-(0.4)^n u[n] - (0.5)^n u[-n-1]$

# 4.9 The One-Sided *z*-Transform

The **one-sided *z*-transform** is useful particularly in the analysis of causal LTI systems. It is defined by

$$X(z) = \sum_{k=0}^{\infty} x[k]z^{-k} \quad \text{(one-sided } z\text{-transform)} \tag{4.35}$$

The lower limit of zero in the summation implies that the one-sided $z$-transform of an arbitrary signal $x[n]$ and its causal version $x[n]u[n]$ are identical. Most of the properties of the two-sided $z$-transform also apply to the one-sided version. However, the shifting property of the two-sided $z$-transform must be modified for use with right-sided (or causal) signals that are nonzero for $n < 0$. We also develop new properties, such as the initial value theorem and the final value theorem that are unique to the one-sided $z$-transform. These properties are summarized in Table 4.4.

**Example 4.16**

### Properties of the One-Sided *z*-Transform

**(a)** Find the $z$-transform of $x[n] = n(4)^{0.5n}u[n]$.

We rewrite this as $x[n] = n(2)^n u[n]$ to get $X(z) = \dfrac{2z}{(z-2)^2}$.

**(b)** Find the $z$-transform of $x[n] = (2)^{n+1}u[n-1]$.

We rewrite this as $x[n] = (2)^2(2)^{n-1}u[n-1]$ to get $X(z) = \dfrac{z^{-1}(4z)}{z-2} = \dfrac{4}{z-2}$.

**TABLE 4.4** ➤ Properties Unique to the One-Sided z-Transform

| Property | Signal | One-Sided z-Transform |
|---|---|---|
| Right shift | $y[n-1]$ | $z^{-1}Y(z)+y[-1]$ |
| | $y[n-2]$ | $z^{-2}Y(z)+z^{-1}y[-1]+y[-2]$ |
| | $y[n-N]$ | $z^{-N}Y(z)+z^{-(N-1)}y[-1]+z^{-(N-2)}y[-2]+\cdots+y[-N]$ |
| Left shift | $y[n+1]$ | $zY(z)-zy[0]$ |
| | $y[n+2]$ | $z^2Y(z)-z^2y[0]-zy[1]$ |
| | $y[n+N]$ | $z^NY(z)-z^Ny[0]-z^{N-1}y[1]-\cdots-zy[N-1]$ |
| Switched periodic | $x_p[n]u[n]$ | $\dfrac{X_1(z)}{1-z^{-N}}$ ($x_1[n]$ is the first period of $x_p[n]$) |
| **Initial Value Theorem:** | $x[0]=\lim_{z\to\infty}X(z)$ | |
| **Final Value Theorem:** | $\lim_{n\to\infty}x[n]=\lim_{z\to 1}(z-1)X(z)$ | |

**(c)** Let $x[n] \Leftarrow\boxed{\text{ZT}}\Rightarrow \frac{4z}{(z+0.5)^2} = X(z)$, with ROC: $|z| > 0.5$. Find the $z$-transform of the signals $h[n] = nx[n]$ and $y[n] = x[n] * x[n]$.

By the times-$n$ property, we have $H(z) = -zX'(z)$, which gives

$$H(z) = -z\left[\frac{-8z}{(z+0.5)^3} + \frac{4}{(z+0.5)^2}\right] = \frac{4z^2-2z}{(z+0.5)^3}$$

By the convolution property, $Y(z) = X^2(z) = \frac{16z^2}{(z+0.5)^4}$.

**(d)** Let $(4)^n u[n] \Leftarrow\boxed{\text{ZT}}\Rightarrow X(z)$. Find the signal corresponding to $F(z) = X^2(z)$ and $G(z) = X(2z)$.

By the convolution property, $f[n] = (4)^n u[n] * (4)^n u[n] = (n+1)(4)^n u[n]$.

By the scaling property, $G(z) = X(2z) = X(z/0.5)$ corresponds to the signal $g[n] = (0.5)^n x[n]$.

Thus, we have $g[n] = (2)^n u[n]$.

### 4.9.1 The Right-Shift Property of the One-Sided z-Transform

The one-sided $z$-transform of a sequence $y[n]$ and its causal version $y[n]u[n]$ are identical. A right shift of $y[n]$ brings samples for $n < 0$ into the range $n \geq 0$, as illustrated in Figure 4.9, and leads to the $z$-transforms

$$y[n-1] \Leftarrow\boxed{\text{ZT}}\Rightarrow z^{-1}Y(z)+y[-1] \qquad y[n-2] \Leftarrow\boxed{\text{ZT}}\Rightarrow z^{-2}Y(z)+z^{-1}y[-1]+y[-2] \tag{4.36}$$

These results generalize to

$$y[n-N] \Leftarrow\boxed{\text{ZT}}\Rightarrow z^{-N}Y(z)+z^{-(N-1)}y[-1]+z^{-(N-2)}y[-2]+\cdots+y[-N] \tag{4.37}$$

For a causal signal with $y[n] = 0,\ n < 0$, this result reduces to $y[n-N] \Leftarrow\boxed{\text{ZT}}\Rightarrow z^{-N}Y(z)$.

**Example 4.17**

**The Right-Shift Property**

**(a)** Using the right-shift property and superposition, we obtain the $z$-transform of the first difference of $x[n]$ as

$$y[n] = x[n]-x[n-1] \Leftarrow\boxed{\text{ZT}}\Rightarrow X(z)-z^{-1}X(z) = (1-z^{-1})X(z)$$

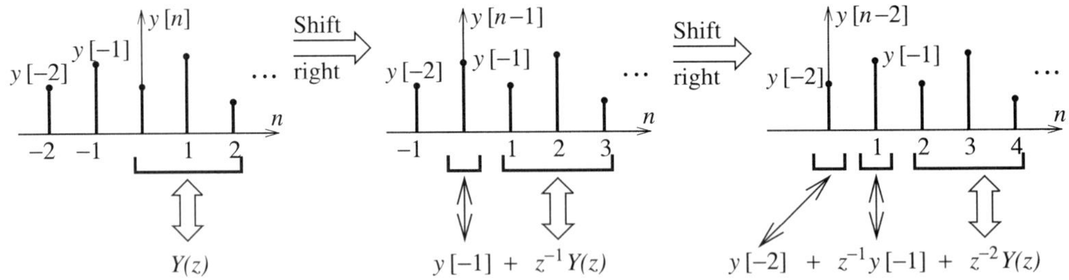


**FIGURE 4.9** Illustrating the right-shift property of the one-sided z-transform. A right shift brings samples from the left of the origin into the range $n \geq 0$. These samples now contribute to the z-transform of the shifted signal

**(b)** Consider the noncausal signal $x[n] = \alpha^n$. Its one-sided z-transform is identical to that of $\alpha^n u[n]$ and equals $X(z) = \dfrac{z}{z-\alpha}$. If $y[n] = x[n-1]$, the right-shift property, with $N = 1$, yields

$$Y(z) = z^{-1}X(z) + x[-1] = \frac{1}{z-\alpha} + \alpha^{-1}$$

The additional term $\alpha^{-1}$ arises because $x[n]$ is not causal.

# Drill Problem 4.20

**(a)** Let $y[n] - 0.5y[n-1] = (0.4)^n$ with $y[-1] = -2$. Use the right-shift property to find $Y(z)$.

**(b)** Let $y[n] + 0.5y[n-2] = \delta[n]$ with $y[-1] = -2,\ y[-2] = 2$. Use the right-shift property to find $Y(z)$.

**Answers: (a)** $\frac{0.4z}{(z-0.4)(z-0.5)}$ **(b)** $\frac{z}{z^2+0.5}$

## 4.9.2 The Left-Shift Property of the One-Sided z-Transform

A left shift of the signal $y[n]u[n]$ moves its samples for $n \geq 0$ into the range $n < 0$, and these samples no longer contribute to the z-transform of the causal portion, as illustrated in Figure 4.10.

This leads to the z-transforms

$$y[n+1] \Leftarrow\boxed{\text{ZT}}\Rightarrow zY(z) - zy[0] \qquad y[n+2] \Leftarrow\boxed{\text{ZT}}\Rightarrow z^2Y(z) - z^2y[0] - zy[1] \quad (4.38)$$

By successively shifting $y[n]u[n]$ to the left, we obtain the general relation

$$y[n+N] \Leftarrow\boxed{\text{ZT}}\Rightarrow z^NY(z) - z^Ny[0] - z^{N-1}y[1] - \cdots - zy[N-1] \quad (4.39)$$

The right-shift and left-shift properties of the one-sided z-transform form the basis for finding the response of causal LTI systems with nonzero initial conditions.

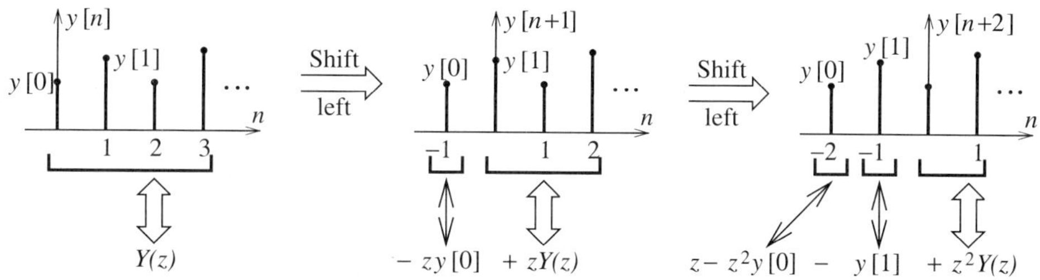


**FIGURE 4.10** Illustrating the left-shift property of the one-sided $z$-transform. A left shift moves samples from the causal region $n \ge 0$ to the left of the origin. These samples no longer contribute to the $z$-transform of the shifted signal

**Example 4.18**

**The Left-Shift Property**

**(a)** Consider the shifted step $u[n+1]$. Its one-sided $z$-transform should be identical to that of $u[n]$ since $u[n]$ and $u[n+1]$ are identical for $n \ge 0$.

With $u[n] \Leftarrow\boxed{\text{ZT}}\Rightarrow \frac{z}{z-1}$ and $u[0] = 1$, the left-shift property gives

$$u[n+1] \Leftarrow\boxed{\text{ZT}}\Rightarrow zU(z) - zu[0] = \frac{z^2}{z-1} - z = \frac{z}{z-1}$$

**(b)** With $y[n] = \alpha^n u[n] \Leftarrow\boxed{\text{ZT}}\Rightarrow \frac{z}{z-\alpha}$ and $y[0] = 1$, the left-shift property gives

$$y[n+1] = \alpha^{n+1}u[n+1] \Leftarrow\boxed{\text{ZT}}\Rightarrow z\left[\frac{z}{z-\alpha}\right] - z = \frac{\alpha z}{z-\alpha}$$

# Drill Problem 4.21

**(a)** Let $y[n+1] - 0.5y[n] = \delta[n]$ with $y[0] = 2$. Use the left-shift property to find $Y(z)$.

**(b)** Let $y[n+1] - 0.5y[n] = \delta[n]$ with $y[-1] = 2$. Use the left-shift property to find $Y(z)$.

**Answers: (a)** $\frac{2z+1}{z-0.5}$ **(b)** $\frac{z+1}{z-0.5}$

## 4.9.3 The Initial Value Theorem and Final Value Theorem

The initial value theorem and final value theorem apply only to the one-sided $z$-transform and the proper part $X(z)$ of a rational $z$-transform.

With $X(z)$ described by $x[0] + x[1]z^{-1} + x[2]z^{-2} + \cdots$, it should be obvious that only $x[0]$ survives as $z \to \infty$ and the initial value equals $x[0] = \lim_{z\to\infty} X(z)$.

To find the final value, we evaluate $(z-1)X(z)$ at $z = 1$. It yields meaningful results only when the poles of $(z-1)X(z)$ have magnitudes smaller than unity (lie within the unit circle in the $z$-plane). As a result:

1. $x[\infty] = 0$ if all poles of $X(z)$ lie within the unit circle (since $x[n]$ will then contain only exponentially damped terms).

2. $x[\infty]$ is constant if there is a single pole at $z = 1$ (since $x[n]$ will then include a step).

3. $x[\infty]$ is indeterminate if there are complex conjugate poles on the unit circle (since $x[n]$ will then include sinusoids). The final value theorem can yield absurd results if used in this case.

**Example 4.19**

**Initial and Final Value Theorems**

Let $X(z) = \frac{z(z-2)}{(z-1)(z-0.5)}$

$$\text{The initial value is } x[0] = \lim_{z\to\infty} X(z) = \lim_{z\to\infty} \frac{1 - 2z^{-1}}{(1 - z^{-1})(1 - 0.5z^{-1})} = 1$$

$$\text{The final value is } \lim_{n\to\infty} x[n] = \lim_{z\to 1}(z - 1)X(z) = \lim_{z\to 1} \frac{z(z - 2)}{z - 0.5} = -2$$

# Drill Problem 4.22

Find the initial value and final value (if it exists) of each causal signal described by

**(a)** $X(z) = \frac{z^2}{(z-1)(z-0.8)}$
**(b)** $X(z) = \frac{z^2}{z^2+1}$
**(c)** $X(z) = \frac{z}{(z-0.4)(z-0.5)}$

**Answers:** **(a)** 1, 5 **(b)** 1, undefined **(c)** 0, 0

### 4.9.4 The z-Transform of Switched Periodic Signals

Consider a causal signal $x[n] = x_p[n]u[n]$, where $x_p[n]$ is periodic with period $N$. If $x_1[n]$ describes the first period of $x[n]$ and has the $z$-transform $X_1(z)$, then the $z$-transform of $x[n]$ can be found as the superposition of the $z$-transform of the shifted versions of $x_1[n]$:

$$X(z) = X_1(z) + z^{-N}X_1(z) + z^{-2N}X_1(z) + \cdots = X_1(z)[1 + z^{-N} + z^{-2N} + \cdots] \quad (4.40)$$

Expressing the geometric series in closed form, we obtain

$$X(z) = \frac{1}{1 - z^{-N}}X_1(z) = \frac{z^N}{z^N - 1}X_1(z) \quad (4.41)$$

**Example 4.20**

**z-Transform of Switched Periodic Signals**

**(a)** Find the $z$-transform of a periodic signal whose first period is $x_1[n] = \{\overset{\Downarrow}{0}, 1, -2\}$.

The period of $x[n]$ is $N = 3$. We then find the $z$-transform of $x[n]$ as

$$X(z) = \frac{X_1(z)}{1 - z^{-N}} = \frac{z^{-1} - 2z^{-2}}{1 - z^{-3}}$$

**(b)** Find the $z$-transform of $x[n] = \sin(0.5n\pi)u[n]$.

The digital frequency of $x[n]$ is $F = \frac{1}{4}$. So $N = 4$. The first period of $x[n]$ is $x_1[n] = \{\overset{\Downarrow}{0},\ 1,\ 0,\ -1\}$. The $z$-transform of $x[n]$ is thus

$$X(z) = \frac{X_1(z)}{1 - z^{-N}} = \frac{z^{-1} - z^{-3}}{1 - z^{-4}} = \frac{z^{-1}}{1 + z^{-2}}$$

**(c)** Find the causal signal corresponding to $X(z) = \dfrac{2 + z^{-1}}{1 - z^{-3}}$.

Comparing with the $z$-transform for a switched periodic signal, we recognize $N = 3$ and $X_1(z) = 2 + z^{-1}$. Thus, the first period of $x[n]$ is $\{\overset{\Downarrow}{2},\ 1,\ 0\}$.

**(d)** Find the causal signal corresponding to $X(z) = \dfrac{z^{-1}}{1 + z^{-3}}$.

We first rewrite $X(z)$ as

$$X(z) = \frac{z^{-1}(1 - z^{-3})}{(1 + z^{-3})(1 - z^{-3})} = \frac{z^{-1} - z^{-4}}{1 - z^{-6}}$$

Comparing with the $z$-transform for a switched periodic signal, we recognize the period as $N = 6$, and $X_1(z) = z^{-1} - z^{-4}$. Thus, the first period of $x[n]$ is $\{\overset{\Downarrow}{0},\ 1,\ 0,\ 0,\ -1,\ 0\}$.

## Drill Problem 4.23

**(a)** One period of a periodic signal $x[n]$ is $\{\overset{\Downarrow}{1},\ 0,\ 0,\ 2,\ 0\}$. Find its $z$-transform $X(z)$.
**(b)** Let $X(z) = \frac{1-2z^{-2}}{1-z^{-4}}$. Find one period of $x[n]$.
**(c)** Let $X(z) = \frac{z}{z^3+1}$. Find one period of $x[n]$.

**Answers: (a)** $\frac{1+2z^{-3}}{1-z^{-5}}$ **(b)** $\{\overset{\Downarrow}{1},\ 0,\ -2,\ 0\}$ **(c)** $\{\overset{\Downarrow}{0},\ 1,\ 0,\ 0,\ -1\}$

# 4.10 The $z$-Transform and System Analysis

The one-sided $z$-transform serves as a useful tool for analyzing LTI systems described by difference equations or transfer functions. The key is (of course) that the solution methods are much simpler in the transformed domain because convolution transforms to a multiplication. Naturally, the time-domain response requires an inverse transformation—a penalty exacted by all methods in the transformed domain.

### 4.10.1 Systems Described by Difference Equations

For a system described by a difference equation, the solution is based on transformation of the difference equation using the shift property and incorporating the effect of initial conditions (if present), and subsequent inverse transformation using partial fractions to obtain the time-domain response. The response may be separated into its zero-state component (due only to the input) and zero-input component (due to the initial conditions) in the $z$-domain itself.

**Example 4.21**

### Solution of Difference Equations

**(a)** Solve the difference equation $y[n] - 0.5y[n-1] = 2(0.25)^n u[n]$ with $y[-1] = -2$.

Transformation using the right-shift property yields

$$Y(z) - 0.5(z^{-1}Y(z) + y[-1]) = \frac{2z}{z - 0.25} \qquad Y(z) = \frac{z(z+0.25)}{(z-0.25)(z-0.5)}$$

We use partial fractions to get

$$\frac{Y(z)}{z} = \frac{z+0.25}{(z-0.25)(z-0.5)} = \frac{-2}{z-0.25} + \frac{3}{z-0.5}$$

Multiplying through by $z$ and taking inverse transforms, we obtain

$$Y(z) = \frac{-2z}{z-0.25} + \frac{3z}{z-0.5} \qquad y[n] = [-2(0.25)^n + 3(0.5)^n]u[n]$$

**(b)** Solve the difference equation $y[n+1] - 0.5y[n] = 2(0.25)^{n+1}u[n+1]$ with $y[-1] = -2$.

We transform the difference equation using the *left-shift property*. The solution will require $y[0]$.

By recursion, with $n = -1$, we obtain $y[0] - 0.5y[-1] = 2$ or $y[0] = 2 + 0.5y[-1] = 2 - 1 = 1$.

Let $x[n] = (0.25)^n u[n]$. Then, by the left-shift property $x[n+1] \Leftarrow\boxed{\text{ZT}}\Rightarrow zX(z) - zx[0]$ (with $x[0] = 1$),

$$(0.25)^{n+1}u[n+1] \Leftarrow\boxed{\text{ZT}}\Rightarrow z\left[\frac{z}{z-0.25}\right] - z = \frac{0.25z}{z-0.25}$$

We now transform the difference equation using the left-shift property:

$$zY(z) - zy[0] - 0.5Y(z) = \frac{0.5z}{z-0.25} \qquad Y(z) = \frac{z(z+0.25)}{(z-0.25)(z-0.5)}$$

This is identical to the result of part (a), and thus $y[n] = -2(0.25)^n + 3(0.5)^n$, as before.

*Comment:* By time invariance, this represents the same system as in part (a).

**(c) Zero-Input and Zero-State Response**

Solve the difference equation $y[n] - 0.5y[n-1] = 2(0.25)^n u[n]$, with $y[-1] = -2$.

Upon transformation using the right-shift property, we obtain

$$Y(z) - 0.5(z^{-1}Y(z) + y[-1]) = \frac{2z}{z-0.25} \qquad (1 - 0.5z^{-1})Y(z) = \frac{2z}{z-0.25} - 1$$

1. **Zero-state response:** For the zero-state response, we assume zero initial conditions to obtain

$$(1 - 0.5z^{-1})Y_{zs}(z) = \frac{2z}{z-0.25} \qquad Y_{zs}(z) = \frac{2z^2}{(z-0.25)(z-0.5)}$$

Upon partial fraction expansion, we obtain

$$\frac{Y_{zs}(z)}{z} = \frac{2z}{(z-0.25)(z-0.5)} = -\frac{2}{z-0.25} + \frac{4}{z-0.5}$$

Multiplying through by $z$ and inverse transforming the result, we get

$$Y_{zs}(z) = \frac{-2z}{z-0.25} + \frac{4z}{z-0.5} \qquad y_{zs}[n] = -2(0.25)^n u[n] + 4(0.5)^n u[n]$$

2. **Zero-input response:** For the zero-input response, we assume zero input (the right-hand side) and use the right-shift property to get

$$Y_{zi}(z) - 0.5(z^{-1}Y_{zi}(z) + y[-1]) = 0 \qquad Y_{zi}(z) = \frac{-z}{z-0.5}$$

This easily is inverted to give $y_{zi}[n] = -(\frac{1}{2})^n u[n]$.

3. **Total response:** We find the total response as

$$y[n] = y_{zs}[n] + y_{zi}[n] = -2(0.25)^n u[n] + 3(0.5)^n u[n]$$

# Drill Problem 4.24

Find the response $y[n],\ n \ge 0$ of each filter using the $z$-transform.

**(a)** $y[n] - 0.8y[n-1] = 2u[n]$ with $y[-1] = 5$
**(b)** $y[n] + 0.8y[n-1] = 2(0.8)^n$ with $y[-1] = 10$
**(c)** $y[n] - 0.8y[n-1] = 2(0.8)^n$ with $y[-1] = 5$
**(d)** $y[n] - 0.8y[n-1] = 2(-0.8)^n + 2(0.4)^n$ with $y[-1] = -5$

**Answers:** **(a)** $10 - 4(0.8)^n$, **(b)** $(0.8)^n - 7(-0.8)^n$, **(c)** $(2n+6)(0.8)^n$, **(d)** $(0.8)^n + (-0.8)^n - 2(0.4)^n$

# Drill Problem 4.25

Use the $z$-transform to find:

**(a)** The ZSR and ZIR if $y[n] - 0.8y[n-1] = x[n]$ with $x[n] = 2(0.4)^n$ and $y[-1] = -10$
**(b)** The output $y[n]$ if $y[n] - 0.8y[n-1] = 2x[n]$ with $x[n] = 2(0.4)^n$ and $y[-1] = -10$
**(c)** The output $y[n]$ if $y[n] - 0.8y[n-1] = 2x[n] - x[n-1]$ with $x[n] = 2(0.4)^n$ and $y[-1] = -10$

**Answers:** **(a)** $4(0.8)^n - 2(0.4)^n$, $-8(0.8)^n$ **(b)** $-4(0.4)^n$ **(c)** $(0.4)^n - 5(0.8)^n$

# Drill Problem 4.26

Use the $z$-transform to find the impulse response $h[n]$ of the following filters.

**(a)** $y[n] - 0.9y[n-1] = x[n]$
**(b)** $y[n] - 0.5y[n-1] = 2x[n] + x[n-1]$
**(c)** $y[n] - 1.2y[n-1] + 0.32y[n-2] = x[n]$
**(d)** $y[n] - 1.2y[n-1] + 0.32y[n-2] = x[n] + 2x[n-1]$

**Answers: (a)** $(0.9)^n$ **(b)** $4(0.5)^n u[n] - 2\delta[n]$ **(c)** $2(0.8)^n - (0.4)^n$ **(d)** $7(0.8)^n - 6(0.4)^n$

## 4.10.2 Systems Described by the Transfer Function

The response $Y(z)$ of a relaxed LTI system equals the product $X(z)H(z)$ of the transformed input and the transfer function. It is often much easier to work with the transfer function description of a linear system. If we let $H(z) = N(z)/D(z)$, the zero-state response $Y(z)$ of a relaxed system to an input $X(z)$ may be expressed as $Y(z) = X(z)H(z) = X(z)N(z)/D(z)$. If the system is not relaxed, the initial conditions result in an additional contribution: the zero-input response $Y_{zi}(z)$, which may be written as $Y_{zi}(z) = N_{zi}(z)/D(z)$. To evaluate $Y_{zi}(z)$, we first set up the system difference equation and then use the shift property to transform it in the presence of initial conditions.

**System Analysis using the Transfer Function**

**Zero-State Response:** Evaluate $Y(z) = X(z)H(z)$ and take inverse transform.
**Zero-Input Response:** Find difference equation. Transform this using the shift property and initial conditions. Find the response in the $z$-domain and take inverse transform.

***Example 4.22***

**System Response from the Transfer Function**

**(a) A Relaxed System**

Let $H(z) = \dfrac{3z}{z-0.4}$.

To find the zero-state response of this system to $x[n] = (0.4)^n u[n]$, we first transform the input to $X(z) = \frac{z}{z-0.4}$. Then,

$$Y(z) = H(z)X(z) = \frac{3z^2}{(z-0.4)^2} \qquad y[n] = 3(n+1)(0.4)^n u[n]$$

**(b) Step Response**

Let $H(z) = \dfrac{4z}{z-0.5}$.

To find its step response, we let $x[n] = u[n]$. Then $X(z) = \frac{z}{z-1}$, and the output equals

$$Y(z) = H(z)X(z) = \left(\frac{4z}{z-0.5}\right)\left(\frac{z}{z-1}\right) = \frac{4z^2}{(z-1)(z-0.5)}$$

Using partial fraction expansion of $Y(z)/z$, we obtain

$$\frac{Y(z)}{z} = \frac{4z}{(z-1)(z-0.5)} = \frac{8}{z-1} - \frac{4}{z-0.5}$$

Thus,

$$Y(z) = \frac{8z}{z-1} - \frac{4z}{z-0.5} \qquad y[n] = 8u[n] - 4(0.5)^n u[n]$$

The first term in $y[n]$ is the *steady-state* response, which can be found much more easily, as described shortly.

**(c) A Second-Order System**

Let $H(z) = \dfrac{z^2}{z^2 - \frac{1}{6}z - \frac{1}{6}}$. Let the input be $x[n] = 4u[n]$ and the initial conditions be $y[-1] = 0,\ y[-2] = 12$.

1. **Zero-state and zero-input response:** The zero-state response is found directly from $H(z)$ as

$$Y_{zs}(z) = X(z)H(z) = \frac{4z^3}{(z^2 - \frac{1}{6}z - \frac{1}{6})(z-1)} = \frac{4z^3}{(z-\frac{1}{2})(z+\frac{1}{3})(z-1)}$$

Partial fractions of $Y_{zs}(z)/z$ and inverse transformation give

$$Y_{zs}(z) = \frac{-2.4z}{z-\frac{1}{2}} + \frac{0.4z}{z+\frac{1}{3}} + \frac{6z}{z-1}$$

$$y_{zs}[n] = -2.4\left(\frac{1}{2}\right)^n u[n] + 0.4\left(-\frac{1}{3}\right)^n u[n] + 6u[n]$$

To find the zero-input response, we first set up the difference equation. We start with

$$H(z) = \frac{Y(z)}{X(z)} = \frac{z^2}{z^2 - \frac{1}{6}z - \frac{1}{6}}, \text{ or } (z^2 - \tfrac{1}{6}z - \tfrac{1}{6})Y(z) = z^2X(z). \text{ This gives}$$

$$\left(1 - \frac{1}{6}z^{-1} - \frac{1}{6}z^{-2}\right)Y(z) = X(z) \qquad y[n] - \frac{1}{6}y[n-1] - \frac{1}{6}y[n-2] = x[n]$$

We now set the right-hand side to zero (for zero input) and transform this equation, using the *right-shift* property, to obtain the zero-input response from

$$Y_{zi}(z) - \frac{1}{6}(z^{-1}Y_{zi}(z) + y[-1]) - \frac{1}{6}(z^{-2}Y_{zi}(z) + z^{-1}y[-1] + y[-2]) = 0$$

With $y[-1] = 0$ and $y[-2] = 12$, this simplifies to

$$Y_{zi}(z) = \frac{2z^2}{z^2 - \frac{1}{6}z - \frac{1}{6}} = \frac{2z^2}{(z-\frac{1}{2})(z+\frac{1}{3})}$$

Partial fraction expansion of $Y_{zi}(z)/z$ and inverse transformation lead to

$$Y_{zi}(z) = \frac{1.2z}{z-\frac{1}{2}} + \frac{0.8z}{z+\frac{1}{3}} \qquad y_{zi}[n] = 1.2\left(\frac{1}{2}\right)^n u[n] + 0.8\left(-\frac{1}{3}\right)^n u[n]$$

Finally, we find the total response as

$$y[n] = y_{zs}[n] + y_{zi}[n] = -1.2\left(\frac{1}{2}\right)^n u[n] + 1.2\left(-\frac{1}{3}\right)^n u[n] + 6u[n]$$

2. **Natural and forced response:** By inspection, the natural and forced components of $y[n]$ are

$$y_N[n] = -1.2(\frac{1}{2})^n u[n] + 1.2(-\frac{1}{3})^n u[n] \qquad y_F[n] = 6u[n]$$

*Comment:* Alternatively, we could transform the system difference equation to obtain

$$Y(z) - \frac{1}{6}(z^{-1}Y(z) + y[-1]) - \frac{1}{6}(z^{-2}Y(z) + z^{-1}y[-1] + y[-2]) = \frac{4z}{z-1}$$

This simplifies to

$$Y(z) = \frac{z^2(6z-2)}{(z-1)(z^2 - \frac{1}{6}z - \frac{1}{6})} = \frac{-1.2z}{z-\frac{1}{2}} + \frac{1.2z}{z+\frac{1}{3}} + \frac{6z}{z-1}$$

The steady-state response corresponds to terms of the form $z/(z-1)$ (step functions). For this example, $Y_F(z) = \frac{6z}{z-1}$ and $y_F[n] = 6u[n]$. Since the poles of $(z-1)Y(z)$ lie within the unit circle, $y_F[n]$ can also be found by the final value theorem:

$$y_F[n] = \lim_{z\to 1}(z-1)Y(z) = \lim_{z\to 1}\frac{z^2(6z-2)}{z^2 - \frac{1}{6}z - \frac{1}{6}} = 6$$

# Drill Problem 4.27

**(a)** Let $H(z) = \dfrac{3z}{z-0.4}$. Find its step response $s[n]$.

**(b)** Let $H(z) = \dfrac{z}{z-0.5}$ with $y[-1] = 6$. Find its zero-input response.

**(c)** Let $H(z) = \dfrac{z-0.8}{z-0.4}$. with $x[n] = 3u[n]$ and $y[-1] = 5$. Find its ZIR and total response.

**Answers:** **(a)** $[5 - 2(0.4)^n]u[n]$ **(b)** $3(0.5)^n u[n]$ **(c)** $2(0.4)^n u[n]$, $[1 + 4(0.4)^n]u[n]$

## 4.10.3 Forced and Steady-State Response from the Transfer Function

In the time domain, the forced response is found by assuming that it has the same form as the input and then satisfying the difference equation describing the LTI system. If the LTI system is described by its transfer function $H(z)$, the forced response may also be found by evaluating $H(z)$ at the complex frequency $z_0$ of the input. For example, the input $x[n] = K\alpha^n \cos(n\Omega + \phi)$ has a complex frequency given by $z_0 = \alpha e^{j\Omega}$. Once $H(z_0) = H_0 e^{j\phi_0}$ is evaluated as a complex quantity, the forced response equals $y_F[n] = KH_0\alpha^n \cos(n\Omega + \phi + \phi_0)$.

For multiple inputs, we simply add the forced response due to each input. For dc and sinusoidal inputs (with $\alpha = 1$), the forced response is also called the **steady-state** response.

### Example 4.23 Finding the Forced Response

**(a)** Find the steady-state response of the a filter described by $H(z) = \dfrac{z}{z - 0.4}$ to the input $x[n] = \cos(0.6n\pi)$.

The complex input frequency is $z_0 = e^{j0.6\pi}$. We evaluate $H(z)$ at $z = z_0$ to give

$$H(z)|_{z=e^{j0.6\pi}} = \frac{e^{j0.6\pi}}{e^{j0.6\pi} - 0.4} = 0.843e^{-j18.7^\circ}$$

The steady-state response is thus $y_F[n] = 0.843\cos(0.6n\pi - 18.7^\circ)$.

**(b)** Find the forced response of the system $H(z) = \dfrac{z}{z - 0.4}$ to the input $x[n] = 5(0.6)^n$.

The complex input frequency is $z_0 = 0.6$. We evaluate $H(z)$ at $z = 0.6$ to give

$$H(z)|_{z=0.6} = \frac{0.6}{0.6 - 0.4} = 3$$

The forced response is thus $y_F[n] = (3)(5)(0.6)^n = 15(0.6)^n$.

**(c)** Find the forced response of the system $H(z) = \dfrac{3z^2}{z^2 - z + 1}$.

The input contains two components and is given by
$x[n] = x_1[n] + x_2[n] = (0.6)^n + 2(0.4)^n\cos(0.5n\pi - 100^\circ)$.

The forced response will be the sum of the forced component $y_1[n]$ due to $x_1[n]$ and $y_2[n]$ due to $x_2[n]$.

The complex input frequency of $x_1[n]$ is $z_0 = 0.6$. We evaluate $H(z)$ at $z = 0.6$ to give

$$H(z)|_{z=0.6} = \frac{3(0.36)}{0.36 - 0.6 + 1} = 1.4211$$

So, $y_1[n] = 1.4211(0.6)^n$.
The complex input frequency of $x_2[n]$ is $z_0 = 0.4e^{j0.5\pi} = j0.4$. We evaluate $H(z)$ at $z = j0.4$ to give

$$H(z)|_{z=j0.4} = \frac{3(-0.16)}{-0.16 - j0.4 + 1} = 0.5159e^{j154.54^\circ}$$

So, $y_2[n] = (0.5159)(2)(0.4)^n\cos(0.5n\pi - 100^\circ + 154.54^\circ) = 1.0318(0.4)^n$ $\cos(0.5n\pi + 54.54^\circ)$.

Finally, by superposition, the complete forced response is

$$y_F[n] = y_1[n] + y_2[n] = 1.4211(0.6)^n + 1.0318(0.4)^n\cos(0.5n\pi + 54.54^\circ)$$

## Drill Problem 4.28

Find the forced or steady-state response of the following systems

**(a)** $y[n] + 0.2y[n-2] = x[n] + 2x[n-1]$ with $x[n] = 2\cos(0.4n\pi + 60^\circ)$

**(b)** $H(z) = \dfrac{z}{z-0.5}$ with $x[n] = 3\cos(0.2n\pi) + 2\cos(0.5n\pi)$

**(c)** $H(z) = \dfrac{z+0.2}{z-0.2}$ with $x[n] = 2(0.4)^n$.

**Answers:** **(a)** $5.9\cos(0.4n\pi + 18.4^\circ)$
**(b)** $4.5\cos(0.2n\pi - 26.3^\circ) + 1.8\cos(0.5n\pi - 26.6^\circ)$ **(c)** $6(0.4)^n$

# 4.11 Problems

**4.1. (The z-Transform of Sequences)** Use the defining relation to find the $z$-transform and its region of convergence for the following:

**(a)** $x[n] = \{1,\ 2,\ \overset{\Downarrow}{3},\ 2,\ 1\}$ **(b)** $y[n] = \{-1,\ 2,\ \overset{\Downarrow}{0},\ -2,\ 1\}$

**(c)** $f[n] = \{\overset{\Downarrow}{1},\ 1,\ 1,\ 1\}$ **(d)** $g[n] = \{1,\ 1,\ -1,\ -\overset{\Downarrow}{1}\}$

**4.2. (z-Transforms)** Find the $z$-transforms and specify the ROC for the following:

**(a)** $x[n] = (2)^{n+2}u[n]$ **(b)** $y[n] = n(2)^{2n}u[n]$
**(c)** $f[n] = (2)^{n+2}u[n-1]$ **(d)** $g[n] = n(2)^{n+2}u[n-1]$
**(e)** $p[n] = (n+1)(2)^n u[n]$ **(f)** $q[n] = (n-1)(2)^{n+2}u[n]$

[**Hints and Suggestions:** For (a), $(2)^{n+2} = (2)^2(2)^n = 4(2)^n$. For (b), $(2)^{2n} = ((2^2)^n = (4)^n$. For (e), use superposition with $(n+1)(2)^n = n(2)^n + (2)^n$.]

**4.3. (The z-Transform of Sequences)** Find the $z$-transform and its ROC for the following:

**(a)** $x[n] = u[n+2] - u[n-2]$
**(b)** $y[n] = (0.5)^n(u[n+2] - u[n-2])$
**(c)** $f[n] = (0.5)^{|n|}(u[n+2] - u[n-2])$

[**Hints and Suggestions:** First write each signal as a sequence of sample values.]

**4.4. (z-Transforms)** Find the $z$-transforms and specify the ROC for the following:

**(a)** $x[n] = (0.5)^{2n}u[n]$ **(b)** $x[n] = n(0.5)^{2n}u[n]$ **(c)** $(0.5)^{-n}u[n]$
**(d)** $(0.5)^n u[-n]$ **(e)** $(0.5)^n u[-n-1]$ **(f)** $(0.5)^{-n}u[-n-1]$

[**Hints and Suggestions:** For (a) through (b), $(0.5)^{2n} = (0.5^2)^n = (0.25)^n$. For (c) and (f), $(0.5)^{-n} = (2)^n$. For (d), flip the results of (c).]

**4.5. (z-Transforms)** Find the $z$-transforms and specify the ROC for the following:

**(a)** $x[n] = \cos(\frac{n\pi}{4} - \frac{\pi}{4})u[n]$ **(b)** $y[n] = (0.5)^n\cos(\frac{n\pi}{4})u[n]$
**(c)** $f[n] = (0.5)^n\cos(\frac{n\pi}{4} - \frac{\pi}{4})u[n]$ **(d)** $g[n] = (\frac{1}{3})^n(u[n] - u[n-4])$
**(e)** $p[n] = n(0.5)^n\cos(\frac{n\pi}{4})u[n]$ **(f)** $q[n] = [(0.5)^n - (-0.5)^n]nu[n]$

[**Hints and Suggestions:** For (a), $\cos(0.25n\pi - 0.25\pi) = 0.7071\cos(0.25n\pi) + 0.7071\sin(0.25n\pi)$. For (b), start with the transform of $\cos(0.25n\pi)$ and use properties. For (c), start with the result for (a) and use the times-$\alpha^n$ property. For (e) start with the result for (b) and use the times-$n$ property.]

**4.6.** **(Two-Sided $z$-Transform)** Find the $z$-transform $X(z)$ and its ROC for the following:

**(a)** $x[n] = u[-n-1]$ **(b)** $y[n] = (0.5)^{-n}u[-n-1]$
**(c)** $f[n] = (0.5)^{|n|}$ **(d)** $g[n] = u[-n-1] + (\frac{1}{3})^n u[n]$
**(e)** $p[n] = (0.5)^{-n}u[-n-1] + (\frac{1}{3})^n u[n]$ **(f)** $q[n] = (0.5)^{|n|} + (-0.5)^{|n|}$

[**Hints and Suggestions:** For (a), start with the transform of $u[n-1]$ and use the reversal property. For (c), note that $(0.5)^{|n|} = (0.5)^n u[n] + (0.5)^{-n}u[-n] - \delta[n]$. For the ROC, note that (c) through (f) are two-sided signals.]

**4.7.** **(The ROC)** The transfer function of a system is $H(z)$. What can you say about the ROC of $H(z)$ for the following cases?

**(a)** $h[n]$ is a causal signal.
**(b)** The system is stable.
**(c)** The system is stable, and $h[n]$ is a causal signal.

**4.8.** **(Poles, Zeros, and the ROC)** The transfer function of a system is $H(z)$. What can you say about the poles and zeros of $H(z)$ for the following cases?

**(a)** The system is stable.
**(b)** The system is causal and stable.
**(c)** The system is an FIR filter with real coefficients.
**(d)** The system is a linear-phase FIR filter with real coefficients.
**(e)** The system is a causal, linear-phase FIR filter with real coefficients.

**4.9.** **($z$-Transforms and ROC)** Consider the signal $x[n] = \alpha^n u[n] + \beta^n u[-n-1]$. Find its $z$-transform $X(z)$. Will $X(z)$ represent a valid transform for the following cases?

**(a)** $\alpha > \beta$ **(b)** $\alpha < \beta$ **(c)** $\alpha = \beta$

**4.10.** **($z$-Transforms)** Find the $z$-transforms (if they exist) and specify their ROC.

**(a)** $x[n] = (2)^{-n}u[n] + 2^n u[-n]$ **(b)** $y[n] = (0.25)^n u[n] + 3^n u[-n]$
**(c)** $f[n] = (0.5)^n u[n] + 2^n u[-n-1]$ **(d)** $g[n] = (2)^n u[n] + (0.5)^n u[-n-1]$
**(e)** $p[n] = \cos(0.5n\pi)u[n]$ **(f)** $q[n] = \cos(0.5n\pi + 0.25\pi)u[n]$
**(g)** $s[n] = e^{jn\pi}u[n]$ **(h)** $t[n] = e^{jn\pi/2}u[n]$
**(i)** $v[n] = e^{jn\pi/4}u[n]$ **(j)** $w[n] = (\sqrt{j})^n u[n] + (\sqrt{j})^{-n}u[n]$

[**Hints and Suggestions:** Parts (a) through (c) are two-sided signals. Part (d) does not have a valid transform. For (f), $\cos(0.5n\pi + 0.25n\pi) = 0.7071[\cos(0.5n\pi) - \sin(0.5n\pi)]$. For part (j), use $j = e^{j\pi/2}$ and Euler's relation to express $w[n]$ as a sinusoid.]

**4.11.** **($z$-Transforms and ROC)** The causal signal $x[n] = \alpha^n u[n]$ has the transform $X(z)$ whose ROC is $|z| > \alpha$. Find the ROC of the $z$-transform of the following:

**(a)** $y[n] = x[n-5]$
**(b)** $p[n] = x[n+5]$
**(c)** $g[n] = x[-n]$
**(d)** $h[n] = (-1)^n x[n]$
**(e)** $p[n] = \alpha^n x[n]$

**4.12.** **($z$-Transforms and ROC)** The anti-causal signal $x[n] = -\alpha^n u[-n-1]$ has the transform $X(z)$ whose ROC is $|z| < \alpha$. Find the ROC of the $z$-transform of the following:

**(a)** $y[n] = x[n-5]$
**(b)** $p[n] = x[n+5]$
**(c)** $g[n] = x[-n]$
**(d)** $h[n] = (-1)^n x[n]$
**(e)** $r[n] = \alpha^n x[n]$

**4.13.** **($z$-Transforms)** Find the $z$-transform $X(z)$ of $x[n] = \alpha^{|n|}$ and specify the region of convergence of $X(z)$. Consider the cases $|\alpha| < 1$ and $|\alpha| > 1$ separately.

**4.14.** **(Properties)** Let $x[n] = nu[n]$. Find $X(z)$, using the following:

**(a)** The defining relation for the $z$-transform
**(b)** The times-$n$ property
**(c)** The convolution result $u[n] * u[n] = (n+1)u[n+1]$ and the shifting property
**(d)** The convolution result $u[n] * u[n] = (n+1)u[n]$ and superposition

**4.15.** **(Properties)** The $z$-transform of $x[n]$ is $X(z) = \dfrac{4z}{(z+0.5)^2}$, $|z| > 0.5$. Find the $z$-transform of the following using properties and specify the region of convergence.

**(a)** $y[n] = x[n-2]$ **(b)** $d[n] = (2)^n x[n]$ **(c)** $f[n] = nx[n]$
**(d)** $g[n] = (2)^n nx[n]$ **(e)** $h[n] = n^2 x[n]$ **(f)** $p[n] = [n-2]x[n]$
**(g)** $q[n] = x[-n]$ **(h)** $r[n] = x[n] - x[n-1]$ **(i)** $s[n] = x[n] * x[n]$

[**Hints and Suggestions:** For (d) through (f), use the results of (c).]

**4.16.** **(Properties)** The $z$-transform of $x[n] = (2)^n u[n]$ is $X(z)$. Use properties to find the time signal corresponding to the following:

**(a)** $Y(z) = X(2z)$ **(b)** $F(z) = X(1/z)$ **(c)** $G(z) = zX'(z)$
**(d)** $H(z) = \dfrac{zX(z)}{z-1}$ **(e)** $D(z) = \dfrac{zX(2z)}{z-1}$ **(f)** $P(z) = z^{-1}X(z)$
**(g)** $Q(z) = z^{-2}X(2z)$ **(h)** $R(z) = X^2(z)$ **(i)** $S(z) = X(-z)$

[**Hints and Suggestions:** Parts(d) through (e) require the summation property.]

**4.17.** **(Properties)** The $z$-transform of a signal $x[n]$ is $X(z) = \dfrac{4z}{(z+0.5)^2}$, $|z| > 0.5$. Find the $z$-transform and its ROC for the following.

**(a)** $y[n] = (-1)^n x[n]$ **(b)** $f[n] = x[2n]$
**(c)** $g[n] = (j)^n x[n]$ **(d)** $h[n] = x[n+1] + x[n-1]$

[**Hints and Suggestions:** For part (b), find $x[n]$ first and use it to get $f[n] = x[2n]$ and $F(z)$. For the rest, use properties.]

**4.18.** **(Properties)** The $z$-transform of the signal $x[n] = (2)^n u[n]$ is $X(z)$. Use properties to find the time signal corresponding to the following.

**(a)** $F(z) = X(-z)$ **(b)** $G(z) = X(1/z)$ **(c)** $H(z) = zX'(-z)$

**4.19.** **(Properties)** The $z$-transform of a causal signal $x[n]$ is $X(z) = \dfrac{z}{z-0.4}$.

**(a)** Let $x_e[n]$ be the even part of $x[n]$. Without computing $x[n]$ or $x_e[n]$, find $X_e(z)$ and its ROC.
**(b)** Confirm your answer by first computing $x_e[n]$ from $x[n]$ and then finding its $z$-transform.
**(c)** Can you find $X_e(z)$ if $x[n]$ represents an anti-causal signal? Explain.

**4.20.** **(Properties)** Find the $z$-transform of $x[n] = \text{rect}(n/2N) = u[n+N] - u[n-N-1]$. Use this result to evaluate the $z$-transform of $y[n] = \text{tri}(n/M)$ where $M = 2N+1$. [**Hints and Suggestions:** Recall that $\text{rect}(n/2N) * \text{rect}(n/2N) = M\text{tri}(n/M)$ where $M = 2N+1$ and use the convolution property of $z$-transforms.]

**4.21.** **(Poles and Zeros)** Make a rough sketch of the pole and zero locations of the $z$-transform of each of the signals shown in Figure P4.21.

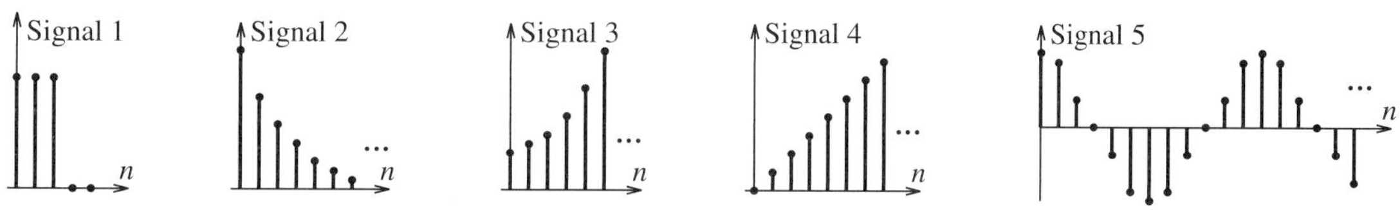


**FIGURE P.4.21** Figure for Problem 4.21

[**Hints and Suggestions:** Signal 1 has only three samples. Signals 2 and 3 appear to be exponentials. Signal 4 is a ramp. Signal 5 appears to be a sinusoid.]

**4.22.** **(Pole-Zero Patterns and Symmetry)** Plot the pole-zero patterns for each $X(z)$. Which of these correspond to symmetric sequences?

**(a)** $X(z) = \dfrac{z^2+z-1}{z}$ **(b)** $X(z) = \dfrac{z^4+2z^3+3z^2+2z+1}{z^2}$

**(c)** $X(z) = \dfrac{z^4-z^3+z-1}{z^2}$ **(d)** $X(z) = \dfrac{(z^2-1)(z^2+1)}{z^2}$

[**Hints and Suggestions:** If a sequence is symmetric about its midpoint, the zeros exhibit conjugate reciprocal symmetry. Also, $X(z) = \pm X(1/z)$ for symmetry about the origin.]

**4.23.** **(Realization)** Sketch the direct form I, direct form II, and transposed realization for each filter.

**(a)** $y[n] - \frac{1}{6}y[n-1] - \frac{1}{2}y[n-2] = 3x[n]$ **(b)** $H(z) = \dfrac{z-2}{z^2-0.25}$

**(c)** $y[n] - 3y[n-1] + 2y[n-2] = 2x[n-2]$ **(d)** $H(z) = \dfrac{2z^2+z-2}{z^2-1}$

[**Hints and Suggestions:** For each part, start with the generic second-order realization and delete any signal paths corresponding to missing coefficients.]

**4.24.** **(Realization)** Find the transfer function and difference equation for each system realization shown in Figure P4.24.

**FIGURE P.4.24** Filter realizations for Problem 4.24

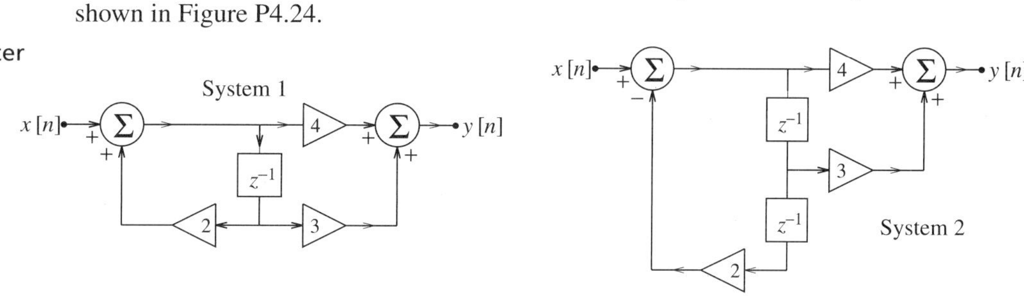


[**Hints and Suggestions:** Compare with the generic first-order and second-order realizations to get the difference equations and transfer functions.]

**4.25.** **(Realization)** Find the transfer function and difference equation for each digital filter realization shown in Figure P4.25.

**FIGURE P.4.25** Filter realizations for Problem 4.25

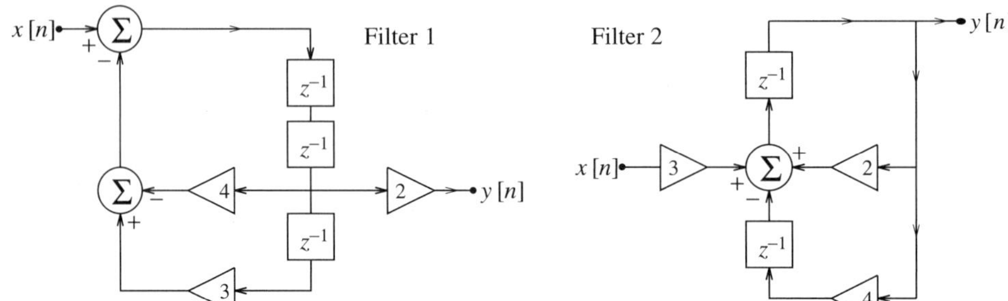


[**Hints and Suggestions:** For (a), compare with a generic third-order direct form II realization. For (b), compare with a generic second-order transposed realization.]

**4.26.** **(Inverse Systems)** Find the transfer function of the inverse systems for each of the following. Which inverse systems are causal? Which inverse systems are stable?

**(a)** $H(z) = \dfrac{z^2 + 0.1}{z^2 - 0.2}$ **(b)** $H(z) = \dfrac{z + 2}{z^2 + 0.25}$

**(c)** $y[n] - 0.5y[n-1] = x[n] + 2x[n-1]$ **(d)** $h[n] = n(2)^n u[n]$

[**Hints and Suggestions:** For parts (c) and (d), set up $H(z)$ and find $H_I(z) = 1/H(z)$.]

**4.27.** **(Causality and Stability)** How can you identify whether a system is a causal and/or stable system from the following information?

**(a)** Its impulse response $h[n]$
**(b)** Its transfer function $H(z)$ and its region of convergence
**(c)** Its system difference equation
**(d)** Its pole-zero plot

**4.28.** **(Switched Periodic Signals)** Find the $z$-transform of each switched periodic signal.

**(a)** $x[n] = \{\overset{\Downarrow}{2}, 1, 3, 0, \ldots\}, \quad N = 4$ **(b)** $y[n] = \cos(n\pi/2)u[n]$

**(c)** $f[n] = \{\overset{\Downarrow}{0}, 1, 1, 0, 0, \ldots\}, \quad N = 5$ **(d)** $g[n] = \cos(0.5n\pi + 0.25\pi)u[n]$

[**Hints and Suggestions:** For parts (b) and (d), the period is $N = 4$. Write out the sequence for the first period to find the transform of the periodic signal.]

**4.29.** **(Inverse Transforms of Polynomials)** Find the inverse $z$-transform of the following:

**(a)** $X(z) = 2 - z^{-1} + 3z^{-3}$ **(b)** $Y(z) = (2 + z^{-1})^3$
**(c)** $F(z) = (z - z^{-1})^2$ **(d)** $G(z) = (z - z^{-1})^2(2 + z)$

[**Hints and Suggestions:** For parts (b) through (d), expand the transform first.]

**4.30.** **(Inverse Transforms by Long Division)** Assuming right-sided signals, determine the ROC of the following $z$-transforms and compute the inverse transform by long division up to $n = 3$.

**(a)** $X(z) = \dfrac{(z+1)^2}{z^2 + 1}$ **(b)** $Y(z) = \dfrac{z + 1}{z^2 + 2}$

**(c)** $F(z) = \dfrac{1}{z^2 - 0.25}$ **(d)** $G(z) = \dfrac{1 - z^{-2}}{2 + z^{-1}}$

[**Hints and Suggestions:** Before dividing, set up the numerator and denominator as polynomials in descending powers of $z$ and obtain the quotient in powers of $z^{-1}$.]

**4.31. (Inverse Transforms by Partial Fractions)** Assuming right-sided signals, determine the ROC of the following $z$-transforms and compute the inverse transform using partial fractions.

**(a)** $X(z) = \dfrac{z}{(z+1)(z+2)}$ **(b)** $Y(z) = \dfrac{16}{(z-2)(z+2)}$

**(c)** $F(z) = \dfrac{3z^2}{(z^2-1.5z+0.5)(z-0.25)}$ **(d)** $G(z) = \dfrac{3z^3}{(z^2-1.5z+0.5)(z-0.25)}$

**(e)** $P(z) = \dfrac{3z^4}{(z^2-1.5z+0.5)(z-0.25)}$ **(f)** $Q(z) = \dfrac{4z}{(z+1)^2(z+3)}$

[**Hints and Suggestions:** For part (e), $P(z)$ is not proper. Use long division to get $P(z) = 3z + P_1(z)$ and use partial fractions for $P_1(z)$. Only part (f) has repeated roots.]

**4.32. (Inverse Transforms by Partial Fractions)** Assuming right-sided signals, determine the ROC of the following $z$-transforms and compute the inverse transform using partial fractions.

**(a)** $X(z) = \dfrac{z}{(z^2+z+0.25)(z+1)}$ **(b)** $Y(z) = \dfrac{z}{(z^2+z+0.25)(z+0.5)}$

**(c)** $F(z) = \dfrac{1}{(z^2+z+0.25)(z+1)}$ **(d)** $G(z) = \dfrac{z}{(z^2+z+0.5)(z+1)}$

**(e)** $P(z) = \dfrac{z^3}{(z^2-z+0.5)(z-1)}$ **(f)** $Q(z) = \dfrac{z^2}{(z^2+z+0.5)(z+1)}$

**(g)** $S(z) = \dfrac{2z}{(z^2-0.25)^2}$ **(h)** $T(z) = \dfrac{2}{(z^2-0.25)^2}$

**(i)** $v(z) = \dfrac{z}{(z^2+0.25)^2}$ **(j)** $w(z) = \dfrac{z^2}{(z^2+0.25)^2}$

[**Hints and Suggestions:** Parts (a) through (c) and (g) through (h) have real repeated roots. Parts (d) through (f) have complex roots. Parts (i) through (j) have complex repeated roots.]

**4.33. (Inverse Transforms by Long Division)** Assuming left-sided signals, determine the ROC of the following $z$-transforms and compute the inverse transform by long division for $n = -1, -2, -3$.

**(a)** $X(z) = \dfrac{z^2+4z}{z^2-z+2}$ **(b)** $Y(z) = \dfrac{z}{(z+1)^2}$

**(c)** $F(z) = \dfrac{z^2}{z^3+z-1}$ **(d)** $G(z) = \dfrac{z^3+1}{z^2+1}$

[**Hints and Suggestions:** Before dividing, set up the numerator and denominator as polynomials in ascending powers of $z$ and obtain the quotient in ascending powers of $z$.]

**4.34. (The ROC and Inverse Transforms)** Let $X(z) = \dfrac{z^2+5z}{z^2-2z-3}$. Which of the following describe a valid ROC for $X(z)$? For each valid ROC, find $x[n]$, using partial fractions.

**(a)** $|z| < 1$ **(b)** $|z| > 3$ **(c)** $1 < |z| < 3$ **(d)** $|z| < 1$ and $|z| > 3$

**4.35. (Inverse Transforms)** For each $X(z)$, find the signal $x[n]$ for each valid ROC.

**(a)** $X(z) = \dfrac{z}{(z+0.4)(z-0.6)}$ **(b)** $X(z) = \dfrac{3z^2}{z^2-1.5z+0.5}$

**4.36.** **(Inverse Transforms)** Consider the *stable* system described by $y[n]+\alpha y[n-1] = x[n]+x[n-1]$.

**(a)** Find its causal impulse response $h[n]$ and specify the range of $\alpha$ and the ROC of $H(z)$.
**(b)** Find its anti-causal impulse response $h[n]$ and specify the range of $\alpha$ and the ROC of $H(z)$.

**4.37.** **(Inverse Transforms)** Each $X(z)$ below represents the $z$-transform of a switched periodic signal $x_p[n]u[n]$. Find one period $x_1[n]$ of each signal. You may verify your results using inverse transformation by long division.

**(a)** $X(z) = \dfrac{1}{1-z^{-3}}$ **(b)** $X(z) = \dfrac{1}{1+z^{-3}}$

**(c)** $X(z) = \dfrac{1+2z^{-1}}{1-z^{-4}}$ **(d)** $X(z) = \dfrac{3+2z^{-1}}{1+z^{-4}}$

[**Hints and Suggestions:** Set up $X(z)$ in the form $\dfrac{X_1(z)}{1-z^{-N}}$ in order to identify the period $N$ and the sequence representing the first period $x_1[n]$. For part(b), for example, multiply the numerator and denominator by $(1-z^{-3})$.]

**4.38.** **(Inverse Transforms)** Let $H(z) = z^{-2}(z-0.5)(2z+4)(1-z^{-2})$.

**(a)** Find its inverse transform $h[n]$.
**(b)** Does $h[n]$ show symmetry about the origin?
**(c)** Does $h[n]$ describe a linear phase sequence?

[**Hints and Suggestions:** Expand $H(z)$ first.]

**4.39.** **(Inverse Transforms)** Let $H(z) = \dfrac{z}{(z-0.5)(z+2)}$.

**(a)** Find its impulse response $h[n]$ if it is known that this represents a stable system. Is this system causal?
**(b)** Find its impulse response $h[n]$ if it is known that this represents a causal system. Is this system stable?

**4.40.** **(Inverse Transforms)** Let $H(z) = \dfrac{z}{(z-0.5)(z+2)}$. Establish the ROC of $H(z)$, find its impulse response $h[n]$, and investigate its stability for the following:

**(a)** A causal $h[n]$ **(b)** An anti-causal $h[n]$ **(c)** A two-sided $h[n]$

**4.41.** **(Convolution)** Simplify each convolution using the $z$-transform. You may verify your results by using time-domain convolution.

**(a)** $y[n] = \{-\overset{\Downarrow}{1}, 2, 0, 3\} * \{2, 0, \overset{\Downarrow}{3}\}$ **(b)** $y[n] = \{-1, 2, \overset{\Downarrow}{0}, -2, 1\} * \{-1, 2, \overset{\Downarrow}{0}, -2, 1\}$

**(c)** $y[n] = (2)^n u[n] * (2)^n u[n]$ **(d)** $y[n] = (2)^n u[n] * (3)^n u[n]$

**4.42.** **(Periodic Signal Generators)** Find the transfer function $H(z)$ of a filter whose impulse response is a periodic sequence with first period $x[n] = \{\overset{\Downarrow}{1}, 2, 3, 4, 6, 7, 8\}$. Find the difference equation and sketch a realization for this filter.

**4.43.** **(Periodic Signal Generators)** It is required to design a filter whose impulse response is a pure cosine at a frequency of $F_0 = 0.25$ and unit amplitude.

**(a)** What is the impulse response of this filter?
**(b)** Find the transfer function $H(z)$ and the difference equation of this filter.
**(c)** Sketch a realization for this filter.
**(d)** Find the step response of this filter.

**4.44.** **(Initial Value and Final Value Theorems)** Assuming right-sided signals, find the initial and final signal values without using inverse transforms.

**(a)** $X(z) = \dfrac{2}{z^2 + \frac{1}{6}z - \frac{1}{6}}$ **(b)** $Y(z) = \dfrac{2z^2}{z^2 + z + 0.25}$

**(c)** $F(z) = \dfrac{2z}{z^2 + z - 1}$ **(d)** $G(z) = \dfrac{2z^2 + 0.25}{(z-1)(z+0.25)}$

**(e)** $P(z) = \dfrac{z + 0.25}{z^2 + 0.25}$ **(f)** $Q(z) = \dfrac{2z + 1}{z^2 - 0.5z - 0.5}$

[**Hints and Suggestions:** To find the initial value, set up each transform as the ratio of polynomials in $z^{-1}$. The final value is nonzero only if there is a single pole at $z = 1$ and all other poles are inside the unit circle. For part (c), the final value theorem does not hold.]

**4.45.** **(System Representation)** Find the transfer function and difference equation for the following causal systems. Investigate their stability, using each system representation.

**(a)** $h[n] = (2)^n u[n]$ **(b)** $h[n] = [1 - (\frac{1}{3})^n]u[n]$
**(c)** $h[n] = n(\frac{1}{3})^n u[n]$ **(d)** $h[n] = 0.5\delta[n]$
**(e)** $h[n] = \delta[n] - (-\frac{1}{3})^n u[n]$ **(f)** $h[n] = [(2)^n - (3)^n]u[n]$

[**Hints and Suggestions:** To find the difference equation, set up $H(z)$ as a ratio of polynomials in $z^{-1}$, equate with $Y(z)/X(z)$, cross multiply and find the inverse transform.]

**4.46.** **(System Representation)** Find the difference equation of the following causal systems. Investigate the stability of each system.

**(a)** $y[n] + 3y[n-1] + 2y[n-2] = 2x[n] + 3x[n-1]$
**(b)** $y[n] + 4y[n-1] + 4y[n-2] = 2x[n] + 3x[n-1]$
**(c)** $y[n] = 0.2x[n]$
**(d)** $y[n] = x[n] + x[n-1] + x[n-2]$

**4.47.** **(System Representation)** Set up the system difference equations of the following causal systems. Investigate the stability of each system.

**(a)** $H(z) = \dfrac{3}{z+2}$ **(b)** $H(z) = \dfrac{1 + 2z + z^2}{(1+z^2)(4+z^2)}$

**(c)** $H(z) = \dfrac{2}{1+z} - \dfrac{1}{2+z}$ **(d)** $H(z) = \dfrac{2z}{1+z} - \dfrac{1}{2+z}$

[**Hints and Suggestions:** Set up $H(z)$ as a ratio of polynomials in $z^{-1}$, equate with $Y(z)/X(z)$, cross multiply and find the inverse transform.]

**4.48.** **(Zero-State Response)** Find the zero-state response of the following systems, using the $z$-transform.

**(a)** $y[n] - 0.5y[n-1] = 2u[n]$ **(b)** $y[n] - 0.4y[n-1] = (0.5)^n u[n]$
**(c)** $y[n] - 0.4y[n-1] = (0.4)^n u[n]$ **(d)** $y[n] - 0.5y[n-1] = \cos(n\pi/2)$

[**Hints and Suggestions:** Use the $z$-transform to get $Y(z)$ assuming zero initial conditions. Then find the inverse transform by partial fractions.]

**4.49.** **(System Response)** Consider the system $y[n] - 0.5y[n-1] = x[n]$. Find its zero-state response to the following inputs, using the $z$-transform.

**(a)** $x[n] = u[n]$ **(b)** $x[n] = (0.5)^n u[n]$ **(c)** $x[n] = \cos(n\pi/2)u[n]$
**(d)** $x[n] = (-1)^n u[n]$ **(e)** $x[n] = (j)^n u[n]$ **(f)** $x[n] = (\sqrt{j})^n u[n] + (\sqrt{j})^{-n} u[n]$

[**Hints and Suggestions:** In part (e), $y[n]$ will be complex because the input is complex. In part (f), using $j = e^{j\pi/2}$ and Euler's relation, $x[n]$ simplifies to a sinusoid. Therefore, the *forced response* $y_F[n]$ is easy to find. Then, $y[n] = K(0.5)^n + y_F[n]$ with $y[-1] = 0$.]

**4.50.** **(Zero-State Response))** Find the zero-state response of the following systems, using the $z$-transform.

**(a)** $y[n] - 1.1y[n-1] + 0.3y[n-2] = 2u[n]$
**(b)** $y[n] - 0.9y[n-1] + 0.2y[n-2] = (0.5)^n$
**(c)** $y[n] + 0.7y[n-1] + 0.1y[n-2] = (0.5)^n$
**(d)** $y[n] - 0.25y[n-2] = \cos(n\pi/2)$

**4.51.** **(System Response))** Let $y[n] - 0.5y[n-1] = x[n]$, with $y[-1] = -1$. Find the response $y[n]$ of this system for the following inputs, using the $z$-transform.

**(a)** $x[n] = 2u[n]$ **(b)** $x[n] = (0.25)^n u[n]$ **(c)** $x[n] = n(0.25)^n u[n]$
**(d)** $x[n] = (0.5)^n u[n]$ **(e)** $x[n] = n(0.5)^n$ **(f)** $x[n] = (0.5)^n \cos(0.5n\pi)$

**4.52.** **(System Response)** Find the response $y[n]$ of the following systems, using the $z$-transform.

**(a)** $y[n] + 0.1y[n-1] - 0.3y[n-2] = 2u[n]$ $y[-1] = 0$ $y[-2] = 0$
**(b)** $y[n] - 0.9y[n-1] + 0.2y[n-2] = (0.5)^n$ $y[-1] = 1$ $y[-2] = -4$
**(c)** $y[n] + 0.7y[n-1] + 0.1y[n-2] = (0.5)^n$ $y[-1] = 0$ $y[-2] = 3$
**(d)** $y[n] - 0.25y[n-2] = (0.4)^n$ $y[-1] = 0$ $y[-2] = 3$
**(e)** $y[n] - 0.25y[n-2] = (0.5)^n$ $y[-1] = 0$ $y[-2] = 0$

**4.53.** **(System Response)** For each system, evaluate the response $y[n]$, using the $z$-transform.

**(a)** $y[n] - 0.4y[n-1] = x[n]$ $x[n] = (0.5)^n u[n]$ $y[-1] = 0$
**(b)** $y[n] - 0.4y[n-1] = 2x[n] + x[n-1]$ $x[n] = (0.5)^n u[n]$ $y[-1] = 0$
**(c)** $y[n] - 0.4y[n-1] = 2x[n] + x[n-1]$ $x[n] = (0.5)^n u[n]$ $y[-1] = 5$
**(d)** $y[n] + 0.5y[n-1] = x[n] - x[n-1]$ $x[n] = (0.5)^n u[n]$ $y[-1] = 2$
**(e)** $y[n] + 0.5y[n-1] = x[n] - x[n-1]$ $x[n] = (-0.5)^n u[n]$ $y[-1] = 0$

**4.54.** **(System Response)** Find the response $y[n]$ of the following systems, using the $z$-transform.

**(a)** $y[n] - 0.4y[n-1] = 2(0.5)^{n-1}u[n-1]$ $y[-1] = 2$
**(b)** $y[n] - 0.4y[n-1] = (0.4)^n u[n] + 2(0.5)^{n-1}u[n-1]$ $y[-1] = 2.5$
**(c)** $y[n] - 0.4y[n-1] = n(0.5)^n u[n] + 2(0.5)^{n-1}u[n-1]$ $y[-1] = 2.5$

**4.55.** **(System Response)** The transfer function of a system is $H(z) = \dfrac{2z(z-1)}{4 + 4z + z^2}$. Find its response $y[n]$ for the following inputs.

**(a)** $x[n] = \delta[n]$ **(b)** $x[n] = 2\delta[n] + \delta[n+1]$ **(c)** $x[n] = u[n]$
**(d)** $x[n] = (2)^n u[n]$ **(e)** $x[n] = nu[n]$ **(f)** $x[n] = \cos(\frac{n\pi}{2})u[n]$

**4.56.** **(System Analysis)** Find the impulse response $h[n]$ and the step response $s[n]$ of the causal digital filters described by

**(a)** $H(z) = \dfrac{4z}{z - 0.5}$ **(b)** $y[n] + 0.5y[n-1] = 6x[n]$

[**Hints and Suggestions:** Note that $y[-1] = 0$. Choose $x[n] = u[n]$ to compute the step response.]

**4.57.** **(System Analysis)** Find the zero-state response, zero-input response, and total response for each of the following systems, using the $z$-transform.

**(a)** $y[n] - \frac{1}{4}y[n-1] = (\frac{1}{3})^n u[n]$ $y[-1] = 8$

**(b)** $y[n] + 1.5y[n-1] + 0.5y[n-2] = (-0.5)^n u[n]$ $y[-1] = 2$ $y[-2] = -4$

**(c)** $y[n] + y[n-1] + 0.25y[n-2] = 4(0.5)^n u[n]$ $y[-1] = 6$ $y[-2] = -12$

**(d)** $y[n] - y[n-1] + 0.5y[n-2] = (0.5)^n u[n]$ $y[-1] = -1$ $y[-2] = -2$

**4.58.** **(Steady-State Response)** The transfer function of a system is $H(z) = \dfrac{2z(z-1)}{z^2 + 0.25}$. Find its *steady-state* response for the following inputs.

**(a)** $x[n] = 4u[n]$ **(b)** $x[n] = 4\cos(\frac{n\pi}{2} + \frac{\pi}{4})u[n]$

**(c)** $x[n] = \cos(\frac{n\pi}{2}) + \sin(\frac{n\pi}{2})$ **(d)** $x[n] = 4\cos(\frac{n\pi}{4}) + 4\sin(\frac{n\pi}{2})$

[**Hints and Suggestions:** For parts (c) through (d), add the forced response due to each component.]

**4.59.** **(Steady-State Response)** The filter $H(z) = A\dfrac{z - \alpha}{z - 0.5\alpha}$ is designed to have a steady-state response of unity if the input is $u[n]$ and a steady-state response of zero if the input is $\cos(n\pi)$. What are the values of $A$ and $\alpha$?

**4.60.** **(Steady-State Response)** The filter $H(z) = A\dfrac{z - \alpha}{z - 0.5\alpha}$ is designed to have a steady-state response of zero if the input is $u[n]$ and a steady-state response of unity if the input is $\cos(n\pi)$. What are the values of $A$ and $\alpha$?

**4.61.** **(System Response)** Find the response of the following filters to the unit step $x[n] = u[n]$, and to the alternating unit step $x[n] = (-1)^n u[n]$.

**(a)** $h[n] = \delta[n] - \delta[n-1]$ (differencing operation)

**(b)** $h[n] = \{\overset{\Downarrow}{0.5}, 0.5\}$ (2-point average)

**(c)** $h[n] = \frac{1}{N}\sum_{k=0}^{N-1} \delta[n-k]$, $N = 3$ (moving average)

**(d)** $h[n] = \frac{2}{N(N+1)}\sum_{k=0}^{N-1} (N-k)\delta[n-k]$, $N = 3$ (weighted moving average)

**(e)** $y[n] - \alpha y[n-1] = (1-\alpha)x[n]$, $\alpha = \frac{N-1}{N+1}$, $N = 3$ (exponential average)

**4.62.** **(Steady-State Response)** Consider the following DSP system:

$$x(t) \longrightarrow \boxed{\text{sampler}} \longrightarrow \boxed{\text{digital filter } H(z)} \longrightarrow \boxed{\text{ideal LPF}} \longrightarrow y(t)$$

The input is $x(t) = 2 + \cos(10\pi t) + \cos(20\pi t)$. The sampler is ideal and operates at a sampling rate of $S$ Hz. The digital filter is described by $H(z) = 0.1S\dfrac{z-1}{z-0.5}$. The ideal lowpass filter has a cutoff frequency of $0.5S$ Hz.

**(a)** What is the smallest value of $S$ that will prevent aliasing?
**(b)** Let $S = 40$ Hz and $H(z) = 1 + z^{-2} + z^{-4}$. What is the *steady-state* output $y(t)$?
**(c)** Let $S = 40$ Hz and $H(z) = \dfrac{z^2+1}{z^4+0.5}$. What is the *steady-state* output $y(t)$?

4.63. **(Response of Digital Filters)** Consider the averaging filter

$$y[n] = 0.5x[n] + x[n-1] + 0.5x[n-2]$$

**(a)** Find its impulse response $h[n]$ and its transfer function $H(z)$.
**(b)** Find its response $y[n]$ to the input $x[n] = \{\overset{\Downarrow}{2}, 4, 6, 8\}$.
**(c)** Find its response $y[n]$ to the input $x[n] = \cos(\frac{n\pi}{3})$.
**(d)** Find its response $y[n]$ to the input $x[n] = \cos(\frac{n\pi}{3}) + \sin(\frac{2n\pi}{3}) + \cos(\frac{n\pi}{2})$.

[**Hints and Suggestions:** For part (b), use convolution. For part (c), find the steady-state response. For part (d), add the steady state response due to each component.]

4.64. **(Transfer Function)** The input to a digital filter is $x[n] = \{\overset{\Downarrow}{1}, 0.5\}$, and the response is described by $y[n] = \delta[n+1] - 2\delta[n] - \delta[n-1]$.

**(a)** What is the filter transfer function $H(z)$?
**(b)** Does $H(z)$ describe an IIR filter or FIR filter?
**(c)** Is the filter stable? Is it causal?

4.65. **(System Analysis)** Consider a system whose impulse response is $h[n] = (0.5)^n u[n]$. Find its response to the following inputs.

**(a)** $x[n] = \delta[n]$ **(b)** $x[n] = u[n]$
**(c)** $x[n] = (0.25)^n u[n]$ **(d)** $x[n] = (0.5)^n u[n]$
**(e)** $x[n] = \cos(n\pi)$ **(f)** $x[n] = \cos(n\pi)u[n]$
**(g)** $x[n] = \cos(0.5n\pi)$ **(h)** $x[n] = \cos(0.5n\pi)u[n]$

[**Hints and Suggestions:** For parts (e) and (g), the output is the steady-state response $y_{ss}[n]$. For parts (f) and (h) use $y[n] = K(0.5)^n + y_{ss}[n]$ with $y[-1] = 0$ and find $K$.]

4.66. **(System Analysis)** Consider a system whose impulse response is $h[n] = n(0.5)^n u[n]$. What input $x[n]$ will result in each of the following steady-state outputs?

**(a)** $y[n] = \cos(0.5n\pi)$
**(b)** $y[n] = 2 + \cos(0.5n\pi)$
**(c)** $y[n] = \cos^2(0.25n\pi)$

[**Hints and Suggestions:** For (a), assume $x[n] = A\cos(0.5n\pi) + B\sin(0.5n\pi)$ and compare the resulting steady state response with $y[n]$. For (b), assume $x[n] = C + A\cos(0.5n\pi) + B\sin(0.5n\pi)$. For (c), note that $\cos^2\alpha = 0.5 + 0.5\cos 2\alpha$.]

4.67. **(System Response)** Consider the system $y[n] - 0.25y[n-2] = x[n]$. Find its response $y[n]$, using $z$-transforms, for the following inputs.

**(a)** $x[n] = 2\delta[n-1] + u[n]$ **(b)** $x[n] = 2 + \cos(0.5n\pi)$

[**Hints and Suggestions:** For part (a), find $Y(z)$ and its inverse transform by partial fractions. For part (b), add the steady state response for each component of the input.]

4.68. **(System Response)** The signal $x[n] = (0.5)^n u[n]$ is applied to a digital filter, and the response is $y[n]$. Find the filter transfer function and state whether it is an IIR or FIR filter and whether it is a linear-phase filter if the system output $y[n]$ is the following:

**(a)** $y[n] = \delta[n] + 0.5\delta[n-1]$
**(b)** $y[n] = \delta[n] - 2\delta[n-1]$
**(c)** $y[n] = (-0.5)^n u[n]$

**4.69.** **(Interconnected Systems)** Consider two systems whose impulse response is $h_1[n] = \delta[n] + \alpha\delta[n-1]$ and $h_2[n] = (0.5)^n u[n]$. Find the overall system transfer function and the response $y[n]$ of the overall system to the input $x[n] = (0.5)^n u[n]$, and to the input $x[n] = \cos(n\pi)$ if

**(a)** The two systems are connected in parallel with $\alpha = 0.5$.
**(b)** The two systems are connected in parallel with $\alpha = -0.5$.
**(c)** The two systems are connected in cascade with $\alpha = 0.5$.
**(d)** The two systems are connected in cascade with $\alpha = -0.5$.

**4.70.** **(Interconnected Systems)** The transfer function $H(z)$ of the cascade of two systems $H_1(z)$ and $H_2(z)$ is known to be $H(z) = \dfrac{z^2 + 0.25}{z^2 - 0.25}$. It is also known that the unit step response of the first system is $[2 - (0.5)^n]u[n]$. Determine $H_1(z)$ and $H_2(z)$.

**4.71.** **(Feedback Systems))** Consider the filter realization of Figure P4.71. Find the transfer function $H(z)$ of the overall system if the impulse response of the filter is given by

**(a)** $h[n] = \delta[n] - \delta[n-1]$. **(b)** $h[n] = 0.5\delta[n] + 0.5\delta[n-1]$.

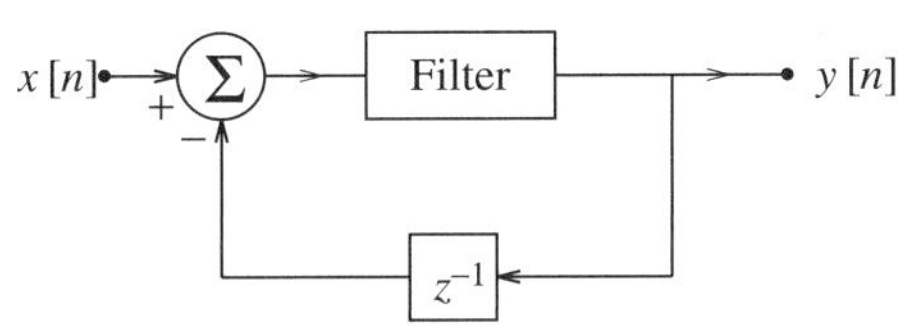


**FIGURE P.4.71** Filter realization for Problem 4.71

Find the difference equation relating $y[n]$ and $x[n]$ from $H(z)$ and investigate the stability of the overall system.

**4.72.** **(Systems in Cascade)** Consider the following system:

$$x[n] \longrightarrow \boxed{H_1(z)} \longrightarrow \boxed{H_2(z)} \longrightarrow \boxed{H_3(z)} \longrightarrow y[n]$$

It is known that $h_1[n] = 0.5(0.4)^n u[n]$, $H_2(z) = \dfrac{A(z+\alpha)}{z+\beta}$, and $h_3[n] = \delta[n] + 0.5\delta[n-1]$. Choose $A$, $\alpha$, and $\beta$ such that the overall system represents an identity system. [**Hints and Suggestions:** Set up $H_1(z)H_2(z)H_3(z) = 1$ to find the constants.]

**4.73.** **(Recursive and Non-Recursive Filters)** Consider two filters described by

$$(1)\ h[n] = \{\overset{\Downarrow}{1},\ 1,\ 1\} \qquad (2)\ y[n] - y[n-1] = x[n] - x[n-3]$$

**(a)** Find the transfer function of each filter.
**(b)** Find the response of each filter to the input $x[n] = \cos(n\pi)$.
**(c)** Are the two filters related in any way?

**4.74.** **(Feedback Compensation)** Feedback compensation is often used to stabilize unstable filters. It is required to stabilize the unstable filter $G(z) = \dfrac{6}{z - 1.2}$ by putting it in the forward path of a negative feedback system. The feedback block has the form $H(z) = \dfrac{\alpha}{z-\beta}$.

**(a)** What values of $\alpha$ and $\beta$ are required for the overall system to have two poles at $z = 0.4$ and $z = 0.6$? What is the overall transfer function and impulse response?

**(b)** What values of $\alpha$ and $\beta$ are required for the overall system to have both poles at $z = 0.6$? What is the overall transfer function and impulse response? How does the double pole affect the impulse response?

[**Hints and Suggestions:** For the negative feedback system, the overall transfer function is given by $T(z) = \frac{G(z)}{1+G(z)H(z)}$.]

**4.75.** **(Recursive Forms of FIR Filters)** An FIR filter may always be recast in recursive form by the simple expedient of including poles and zeros at identical locations. This is equivalent to multiplying the transfer function numerator and denominator by identical factors. For example, the filter $H(z) = 1 - z^{-1}$ is FIR but if we multiply the numerator and denominator by the identical term $1 + z^{-1}$, the new filter and its difference equation become

$$H_N(z) = \frac{(1-z^{-1})(1+z^{-1})}{1+z^{-1}} = \frac{1-z^{-2}}{1+z^{-1}} \qquad y[n] + y[n-1] = x[n] - x[n-2]$$

The difference equation can be implemented recursively. Find two different recursive difference equations (with different orders) for each of the following filters.

**(a)** $h[n] = \{1, \overset{\Downarrow}{2}, 1\}$

**(b)** $H(z) = \dfrac{z^2 - 2z + 1}{z^2}$

**(c)** $y[n] = x[n] - x[n-2]$

[**Hints and Suggestions:** Set up $H(z)$ and multiply both numerator and denominator by identical polynomials in $z$ (linear, quadratic etc). Use this to find the recursive difference equation.]

## Computation and Design

**4.76.** **(System Response in Symbolic Form)** The MATLAB based routine **sysresp2** (on the author's website) returns the system response in symbolic form. Obtain the response of the following filters and plot the response for $0 \le n \le 30$.

**(a)** The step response of $y[n] - 0.5y[n-1] = x[n]$.

**(b)** The impulse response of $y[n] - 0.5y[n-1] = x[n]$.

**(c)** The zero-state response of $y[n] - 0.5y[n-1] = (0.5)^n u[n]$.

**(d)** The complete response of $y[n] - 0.5y[n-1] = (0.5)^n u[n],\ y[-1] = -4$.

**(e)** The complete response of $y[n] + y[n-1] + 0.5y[n-2] = (0.5)^n u[n]$, $y[-1] = -4,\ y[-2] = 3$.

**4.77.** **(Steady-State Response in Symbolic Form)** The MATLAB based routine **ssresp** (on the author's website) yields a symbolic expression for the steady-state response to sinusoidal inputs. Find the steady-state response to the input $x[n] = 2\cos(0.2n\pi - \frac{\pi}{3})$ for each of the following systems and plot the results over $0 \le n \le 50$.

**(a)** $y[n] - 0.5y[n-1] = x[n]$

**(b)** $y[n] + y[n-1] + 0.5y[n-2] = 3x[n]$

CHAPTER 5

# Frequency Domain Analysis

## 5.0 Scope and Overview

This chapter develops the **discrete-time Fourier transform** (DTFT) as an analysis tool in the frequency domain for both discrete-time signals and discrete systems. It introduces the DTFT as a special case of the $z$-transform, develops the properties of the DTFT, describes the applications of the DTFT to system analysis and signal processing, and concludes with connections between the various methods of signal and system analysis covered thus far in the text.

### 5.0.1 Goals and Learning Objectives

The goals of this chapter are to provide an introduction to the DTFT and its properties and the techniques used for the frequency-domain analysis of discrete signals and systems. After going through this chapter the reader should:

1. Understand the definition of the DTFT and be able to find the DTFT of simple signals from the definition.
2. Understand the concept of magnitude spectrum, gain, and phase spectrum.
3. Understand the symmetry of the DTFT for real signals.
4. Know how to use the basic properties of the DTFT to find the DTFT of unfamiliar signals.
5. Know how to plot the magnitude and phase spectrum of a signal.
6. Understand the concept of the transfer function and frequency response of a system.
7. Understand Parseval's theorem and be able to use it to find signal energy.
8. Know how to find the DTFT of discrete periodic signals.
9. Understand the connection between the DTFT and the DFT.
10. Know how to use the DTFT for the analysis of relaxed systems.
11. Know how to use the DTFT to find the steady-state response of systems to harmonic inputs.

# 5.1 The DTFT from the $z$-Transform

The $z$-transform describes a discrete-time signal as a sum of weighted harmonics $z^{-m}$

$$X(z) = \sum_{m=-\infty}^{\infty} x[m]z^{-m} = \sum_{m=-\infty}^{\infty} x[m](Re^{j2\pi F})^{-m} \tag{5.1}$$

where the complex exponential $z = Re^{j2\pi F} = re^{j\Omega}$ includes a real weighting factor $R$. If we let $R = 1$, we obtain $z = e^{j2\pi F} = e^{j\Omega}$ and $z^{-m} = e^{-j2\pi mF} = e^{-jm\Omega}$. The expression for the $z$-transform then reduces to

$$X(F) = \sum_{m=-\infty}^{\infty} x[m]e^{-j2\pi mF} \qquad X(\Omega) = \sum_{m=-\infty}^{\infty} x[m]e^{-jm\Omega} \tag{5.2}$$

The quantity $X(F)$ (or $X(\Omega)$) is now a function of the frequency $F$ (or $\Omega$) alone and describes the **discrete-time Fourier transform** (DTFT) of $x[n]$ as a sum of weighted harmonics. The DTFT is a frequency-domain description of a discrete-time signal. The DTFT of $x[n]$ may be viewed as its $z$-transform $X(z)$ evaluated for $R = 1$ (along the unit circle in the $z$-plane). The DTFT is also called the **spectrum** and the DTFT $H(F)$ of the system impulse response is also referred to as the **frequency response** or the frequency domain **transfer function**.

Note that $X(F)$ is periodic in $F$ with *unit period* because $X(F) = X(F+1)$. We find

$$X(F+1) = \sum_{m=-\infty}^{\infty} x[m]e^{-j2\pi m(F+1)} = \sum_{m=-\infty}^{\infty} x[m]e^{-j2m\pi}e^{-j2\pi mF}$$
$$= \sum_{m=-\infty}^{\infty} x[m]e^{-j2\pi mF} = X(F)$$

The unit interval $-0.5 \le F \le 0.5$ (or $0 \le F \le 1$) defines the **principal period** or **central period**. Similarly, $X(\Omega)$ is periodic in $\Omega$ with period $2\pi$ and represents a scaled (stretched by $2\pi$) version of $X(F)$. The principal period of $X(\Omega)$ corresponds to the interval $-\pi \le \Omega \le \pi$ or $0 \le \Omega \le 2\pi$.

The inverse DTFT allows us to recover $x[n]$ from one period of its DTFT and is defined by

$$x[n] = \int_{-1/2}^{1/2} X(F)e^{j2\pi nF}\,dF \quad \text{(the } F\text{-form)}$$

$$x[n] = \frac{1}{2\pi}\int_{-\pi}^{\pi} X(\Omega)e^{jn\Omega}\,d\Omega \quad \text{(the } \Omega\text{-form)} \tag{5.3}$$

We will find it convenient to work with the $F$-form, especially while using the inverse transform relation, because it rids us of factors of $2\pi$ in many situations. The discrete signal $x[n]$ and its discrete-time Fourier transform $X(F)$ or $X(\Omega)$ form a unique transform pair, and their relationship is shown symbolically using a double arrow:

$$x[n] \Leftarrow \boxed{\text{DTFT}} \Rightarrow X(F) \quad \text{or} \quad x[n] \Leftarrow \boxed{\text{DTFT}} \Rightarrow X(\Omega) \tag{5.4}$$

**The DTFT is a Frequency-Domain Representation of Discrete-Time Signals**

| **Form:** | **DTFT:** | **Inverse DTFT:** |
|---|---|---|
| $F$-form | $X(F) = \sum_{m=-\infty}^{\infty} x[m]e^{-j2\pi mF}$ | $x[n] = \int_{-1/2}^{1/2} X(F)e^{j2\pi nF}\,dF$ |
| $\Omega$-form | $X(\Omega) = \sum_{m=-\infty}^{\infty} x[m]e^{-jm\Omega}$ | $x[n] = \frac{1}{2\pi}\int_{-\pi}^{\pi} X(\Omega)e^{jn\Omega}\,d\Omega$ |

## 5.1.1 Symmetry of the Spectrum for a Real Signal

The DTFT of a real signal is, in general, complex. A plot of the magnitude of the DTFT against frequency is called the **magnitude spectrum**. If $H(F)$ describes the DTFT of the impulse response $h[n]$ of a digital filter, its magnitude $|H(F)|$ is also called the **gain**. A plot of the phase of the DTFT against frequency is called the phase spectrum. The phase spectrum may be restricted to a 360° range (−180°, 180°). Sometimes, it is more convenient to *unwrap* the phase (by adding/subtracting multiples of 360°) to plot it as a *monotonic* function. The DTFT $X(F)$ of a signal $x[n]$ may be expressed in any of the following ways

$$X(F) = R(F) + jI(F) = |X(F)|e^{j\phi(F)} = |X(F)|\angle\phi(F) \quad \text{(the } F\text{-form)} \tag{5.5}$$

$$X(\Omega) = R(\Omega) + jI(\Omega) = |X(\Omega)|e^{j\phi(\Omega)} = |X(\Omega)|\angle\phi(\Omega) \quad \text{(the } \Omega\text{-form)} \tag{5.6}$$

For a real signal, the DTFT shows *conjugate symmetry* about $F = 0$ (or $\Omega = 0$) with

$$X(F) = X^*(-F) \quad |X(F)| = |X(-F)| \quad \phi(F) = -\phi(-F) \quad \text{(the } F\text{-form)} \tag{5.7}$$

$$X(\Omega) = X^*(-\Omega) \quad |X(\Omega)| = |X(-\Omega)| \quad \phi(\Omega) = -\phi(-\Omega) \quad \text{(the } \Omega\text{-form)} \tag{5.8}$$

Conjugate symmetry of $X(F)$ about the origin means that its magnitude spectrum displays *even symmetry* about the origin and its phase spectrum displays *odd symmetry* about the origin. It is easy to show that $X(F)$ also displays conjugate symmetry about $F = 0.5$. Since it is periodic, we find it convenient to plot just one period of $X(F)$ over the principal period $(-0.5 \le F \le 0.5)$ with conjugate symmetry about $F = 0$. We may even plot $X(F)$ over $(0 \le F \le 1)$ with conjugate symmetry about $F = 0.5$. These ideas are illustrated in Figure 5.1.

Similarly, $X(\Omega)$ shows conjugate symmetry about the origin $\Omega = 0$, and about $\Omega = \pi$ and may also be plotted only over its principal period $(-\pi \le \Omega \le \pi)$ (with conjugate symmetry about $\Omega = 0$) or over $(0 \le \Omega \le 2\pi)$ (with conjugate symmetry about $\Omega = \pi$). The principal period for each form is illustrated in Figure 5.2.

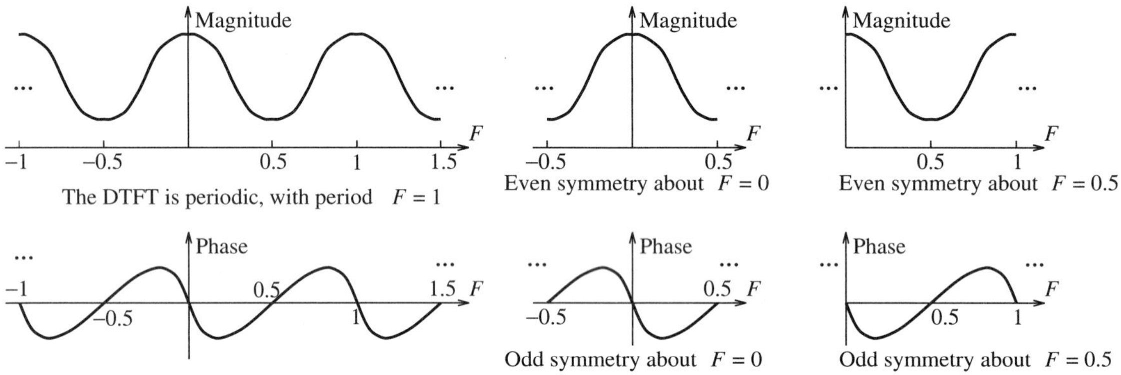


**FIGURE 5.1** Illustrating the symmetry in the DTFT spectrum of real signals. The magnitude shows even symmetry about $F = 0$ and about $F = 0.5$. The phase shows odd symmetry about $F = 0$ and about $F = 0.5$. Both are periodic, with unit period (in $F$)

**The DTFT is Always Periodic and Shows Conjugate Symmetry for Real Signals**

***F*-form:** $X(F)$ is periodic with unit period and conjugate symmetric about $F = 0$ and $F = 0.5$.

**Ω-form:** $X(\Omega)$ is periodic with period $2\pi$ and conjugate symmetric about $\Omega = 0$ and $\Omega = \pi$.

**Plotting:** It is sufficient to plot the DTFT over one period ($-0.5 \le F \le 0.5$ or $-\pi \le \Omega \le \pi$).

# Drill Problem 5.1

**(a)** If $X(F)$ evaluated at $F = -0.2$ equals $2e^{j\pi/3}$, find $X(F)|_{F=0.2}$, $X(F)|_{F=0.8}$.
**(b)** If $X(F)$ evaluated at $F = -0.2$ equals $2e^{j\pi/3}$, find $X(F)|_{F=3.2}$, $X(F)|_{F=5.8}$.

**Answers: (a)** $2e^{-j\pi/3}$, $2e^{j\pi/3}$ **(b)** $2e^{-j\pi/3}$, $2e^{j\pi/3}$

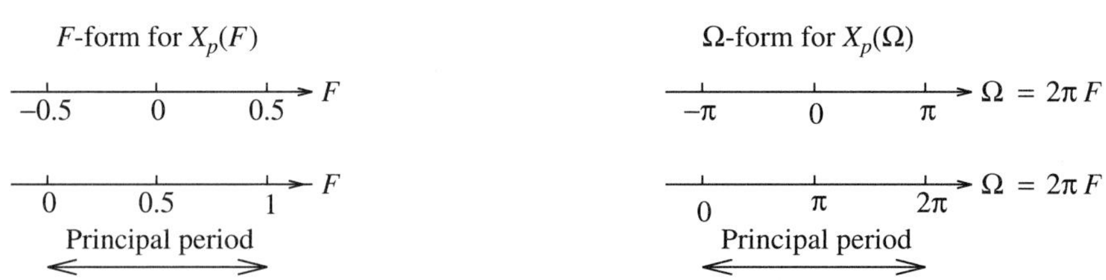


**FIGURE 5.2** Two ways of plotting the DTFT spectrum over one period. A convenient way is to plot over the principal period centered at the origin. An alternate way is to plot over the principal period that starts at the origin

If we know the spectrum of a real signal over the half-period $0 < F \le 0.5$ (or $0 < \Omega \le \pi$), we can use conjugate symmetry about the origin to obtain the spectrum for one full period and replicate this to generate the periodic spectrum. For this reason, the highest useful frequency present in the spectrum is $F = 0.5$ or $\Omega = \pi$. For sampled signals, this also corresponds to an analog frequency of $0.5S$ Hz (half the sampling frequency).

If a real signal $x[n]$ is even symmetric about $n = 0$, its DTFT $X(F)$ is always real and even symmetric in $F$ and has the form $X(F) = A(F)$. If a real signal $x[n]$ is odd symmetric, $X(F)$ is always imaginary and odd symmetric in $F$ and has the form $X(F) = \pm jA(F)$. A real symmetric signal is called a **linear-phase signal**. The real quantity $A(F)$ (which may not always be positive for all frequencies) is called the **amplitude spectrum**. For a linearphase signal, it is much more convenient to plot the amplitude (not magnitude) spectrum because its phase is then just zero or just $\pm 90^\circ$ (and uncomplicated by phase jumps of $180^\circ$ whenever $A(F)$ changes sign).

# Drill Problem 5.2

**(a)** Let $x[n] = \{\overset{\Downarrow}{3}, 1, 3, 4\}$. Is $X(F)$ periodic? Is $X(F)$ real? Is $X(F)$ conjugate symmetric?

**(b)** Let $x[n] = \{2, 1, \overset{\Downarrow}{3}, 1, 2\}$. Is $X(F)$ periodic? Is $X(F)$ real? Is $X(F)$ conjugate symmetric?

**(c)** Let $x[n] = \{2, -1, \overset{\Downarrow}{0}, 1, -2\}$. Is $X(F)$ periodic? Is $X(F)$ real? Is $X(F)$ conjugate symmetric?

**(d)** Let $x[n] = \{2, -j, \overset{\Downarrow}{0}, j, -2\}$. Is $X(F)$ periodic? Is $X(F)$ real? Is $X(F)$ conjugate symmetric?

**Answers:** **(a)** Yes, no, yes **(b)** Yes, yes, yes **(c)** Yes, no, yes **(d)** Yes, no, no

## 5.1.2 Some DTFT Pairs

The DTFT is a summation. It always exists if the summand $x[n]e^{j2\pi nF} = x[n]e^{jn\Omega}$ is absolutely integrable. Since $|e^{j2\pi nF}| = |e^{jn\Omega}| = 1$, the DTFT of absolutely summable signals always exists. The DTFT of sinusoids, steps, or constants (which are not absolutely summable) includes impulses (of the continuous kind). The DTFT of signals that grow exponentially or faster does not exist. A list of DTFT pairs appears in Table 5.1. Both the $F$-form and $\Omega$-form are shown. The transforms are identical (with $\Omega = 2\pi F$), except for the extra factors of $2\pi$ in the impulsive terms of the $\Omega$-form (in the transform of the constant, the step function, and the sinusoid). The reason for these extra factors is the scaling property of impulses that says $\delta(F) = \delta(\Omega/2\pi) = 2\pi\delta(\Omega)$.

**Differences between the $F$-Form and $\Omega$-Form of the DTFT**

**If the DTFT Contains No Impulses:** $H(F)$ and $H(\Omega)$ are related by $\Omega = 2\pi F$.

**If the DTFT Contains Impulses:** Replace $\delta(F)$ by $2\pi\delta(\Omega)$ (and $2\pi F$ by $\Omega$ elsewhere) to get $H(\Omega)$.

**TABLE 5.1** ➤ Some Useful DTFT Pairs

| Entry | Signal $x[n]$ | The $F$-Form: $X(F)$ | The $\Omega$-Form: $X(\Omega)$ |
|---|---|---|---|
| 1 | $\delta[n]$ | $1$ | $1$ |
| 2 | $\alpha^n u[n],\ \lvert\alpha\rvert < 1$ | $\dfrac{1}{1-\alpha e^{-j2\pi F}}$ | $\dfrac{1}{1-\alpha e^{-j\Omega}}$ |
| 3 | $n\alpha^n u[n],\ \lvert\alpha\rvert < 1$ | $\dfrac{\alpha e^{-j2\pi F}}{(1-\alpha e^{-j2\pi F})^2}$ | $\dfrac{\alpha e^{-j\Omega}}{(1-\alpha e^{-j\Omega})^2}$ |
| 4 | $(n+1)\alpha^n u[n],\ \lvert\alpha\rvert < 1$ | $\dfrac{1}{(1-\alpha e^{-j2\pi F})^2}$ | $\dfrac{1}{(1-\alpha e^{-j\Omega})^2}$ |
| 5 | $\alpha^{\lvert n\rvert},\ \lvert\alpha\rvert < 1$ | $\dfrac{1-\alpha^2}{1-2\alpha\cos(2\pi F)+\alpha^2}$ | $\dfrac{1-\alpha^2}{1-2\alpha\cos\Omega+\alpha^2}$ |
| 6 | $1$ | $\delta(F)$ | $2\pi\delta(\Omega)$ |
| 7 | $\cos(2n\pi F_0) = \cos(n\Omega_0)$ | $0.5[\delta(F+F_0)+\delta(F-F_0)]$ | $\pi[\delta(\Omega+\Omega_0)+\delta(\Omega-\Omega_0)]$ |
| 8 | $\sin(2n\pi F_0) = \sin(n\Omega_0)$ | $j0.5[\delta(F+F_0)-\delta(F-F_0)]$ | $j\pi[\delta(\Omega+\Omega_0)-\delta(\Omega-\Omega_0)]$ |
| 9 | $2F_C\,\mathrm{sinc}(2nF_C) = \dfrac{\sin(n\Omega_C)}{n\pi}$ | $\mathrm{rect}\left(\dfrac{F}{2F_C}\right)$ | $\mathrm{rect}\left(\dfrac{\Omega}{2\Omega_C}\right)$ |
| 10 | $u[n]$ | $0.5\delta(F)+\dfrac{1}{1-e^{-j2\pi F}}$ | $\pi\delta(\Omega)+\dfrac{1}{1-e^{-j\Omega}}$ |

*NOTE:* In all cases, we assume $\lvert\alpha\rvert < 1$.

**Example 5.1**

### DTFT from the Defining Relation

**(a)** The DTFT of $x[n] = \delta[n]$ follows immediately from the definition as

$$X(F) = \sum_{k=-\infty}^{\infty} x[k]e^{-j2\pi kF} = \sum_{k=-\infty}^{\infty} \delta[k]e^{-j2\pi kF} = 1$$

**(b)** The DTFT of the sequence $x[n] = \{1, 0, 3, -2\}$ also follows from the definition as

$$X(F) = \sum_{k=-\infty}^{\infty} x[k]e^{-j2\pi kF} = 1 + 3e^{-j4\pi F} - 2e^{-j6\pi F}$$

In the $\Omega$-form, we have

$$X(\Omega) = \sum_{k=-\infty}^{\infty} x[k]e^{-jk\Omega} = 1 + 3e^{-j2\Omega} - 2e^{-j3\Omega}$$

For finite sequences, the DTFT can be written just by inspection. Each term is the product of a sample value at index $n$ and the exponential $e^{-j2\pi nF}$ (or $e^{-jn\Omega}$).

**(c)** The DTFT of the exponential signal $x[n] = \alpha^n u[n]$ follows from the definition and the closed form for the resulting geometric series:

$$X(F) = \sum_{k=0}^{\infty} \alpha^k e^{-j2\pi kF} = \sum_{k=0}^{\infty} (\alpha e^{-j2\pi F})^k = \frac{1}{1-\alpha e^{-j2\pi F}}, \qquad |\alpha| < 1$$

The sum converges only if $|\alpha e^{-j2\pi F}| < 1$, or $|\alpha| < 1$ (since $|e^{-j2\pi F}| = 1$). In the $\Omega$-form,

$$X(\Omega) = \sum_{k=0}^{\infty} \alpha^k e^{-jk\Omega} = \sum_{k=0}^{\infty} (\alpha e^{-j\Omega})^k = \frac{1}{1-\alpha e^{-j\Omega}}, \qquad |\alpha| < 1$$

**(d)** The signal $x[n] = u[n]$ is a limiting form of $\alpha^n u[n]$ as $\alpha \to 1$ but must be handled with care, since $u[n]$ is not absolutely summable. In fact, $X(F)$ also includes an impulse (now an impulse train due to the periodic spectrum). Over the principal period,

$$X(F) = \frac{1}{1 - e^{-j2\pi F}} + 0.5\delta(F) \quad (F\text{-form}) \quad X(\Omega) = \frac{1}{1 - e^{-j\Omega}} + \pi\delta(\Omega) \quad (\Omega\text{-form})$$

# Drill Problem 5.3

**(a)** Let $x[n] = \{\overset{\Downarrow}{3}, 1, 3, 4\}$. Find $X(F)$ and compute $X(F)$ at $F = 0.2$.
**(b)** Let $x[n] = \{2, \overset{\Downarrow}{3}, 0, 1\}$. Find $X(F)$ and compute $X(F)$ at $F = 0$ and $F = 0.5$.
**(c)** Let $x[n] = 8(0.6)^n u[n]$. Find $X(F)$ and compute $X(F)$ at $F = 0, 0.25, 0.5$.

**Answers:** **(a)** $2.38e^{-j171^\circ}$ **(b)** 6, 2 **(c)** 20, $6.86e^{-j31^\circ}$, 5

## 5.1.3 Relating the $z$-Transform and DTFT

The DTFT describes a signal as a sum of weighted harmonics or complex exponentials. However, it cannot handle exponentially growing signals. The $z$-transform overcomes these shortcomings by using exponentially weighted harmonics in its definition. The $z$-transform may be viewed as a generalization of the DTFT to complex frequencies.

For absolutely summable signals, the DTFT is simply the $z$-transform with $z = e^{j2\pi F}$. The DTFT of signals that are not absolutely summable almost invariably contains impulses. However, for such signals the $z$-transform equals just the non-impulsive portion of the DTFT, with $z = e^{j2\pi F}$. In other words, for absolutely summable signals, we can always find their $z$-transform from the DTFT, but we cannot always find their DTFT from the $z$-transform.

**Relating the $z$-Transform and the DTFT**

**From $X(z)$ to DTFT:** If $x[n]$ is *absolutely summable*, simply replace $z$ by $e^{j2\pi F}$ (or $e^{j\Omega}$).
**From DTFT to $X(z)$:** Delete impulsive terms in DTFT and replace $e^{j2\pi F}$ (or $e^{j\Omega}$) by $z$.

**Example 5.2**

**The $z$-Transform and DTFT**

**(a)** The signal $x[n] = \alpha^n u[n]$, $|\alpha| < 1$ is absolutely summable. Its DTFT equals

$$X_p(F) = \frac{1}{1 - \alpha e^{-j2\pi F}}$$

We can find the $z$-transform of $x[n]$ from its DTFT as

$$X(z) = \frac{1}{1 - \alpha z^{-1}} = \frac{z}{z - \alpha}$$

We can also find the DTFT from the $z$-transform by reversing the steps.

**(b)** The signal $x[n] = u[n]$ is not absolutely summable. Its DTFT is

$$X_p(F) = \frac{1}{1 - e^{-j2\pi F}} + 0.5\delta(F)$$

We can find the $z$-transform of $u[n]$ as the impulsive part in the DTFT, with $e^{j2\pi F} = z$, to give $X(z) = \frac{1}{1-z^{-1}} = \frac{z}{z-1}$. However, we cannot recover the DTFT from its $z$-transform in this case.

---

# 5.2 Properties of the DTFT

The properties of the DTFT are summarized in Table 5.2. The proofs of most of the properties follow from the defining relations if we start with the basic transform pair $x[n] \Leftarrow \boxed{\text{DTFT}} \Rightarrow X(F)$.

## 5.2.1 Time Reversal

With $x[n] \Leftarrow \boxed{\text{DTFT}} \Rightarrow X(F)$, the DTFT of the signal $y[n] = x[-n]$ may be written (using a change of variable) as

$$Y(F) = \sum_{k=-\infty}^{\infty} x[-k]e^{-j2k\pi F} \underbrace{=}_{m=-k} \sum_{m=-\infty}^{\infty} x[m]e^{j2m\pi F} = X(-F) \qquad (5.9)$$

A flipping of $x[n]$ to $x[-n]$ results in a flipping of $X(F)$ to $X(-F)$. For real signals, $X(-F) = X^*(F)$ implying an identical magnitude spectrum and reversed phase.

# Drill Problem 5.4

**(a)** Let $X(F) = 4 - 2e^{-j4\pi F}$. Find the DTFT of $y[n] = x[-n]$ and compute at $F = 0.2$.

**(b)** For a signal $x[n]$, we find $X(F)|_{F=0.2} = 2e^{-j\pi/3}$. Compute the DTFT of $y[n] = x[-n]$ at $F = 0.2,\ 1.8$.

**Answers: (a)** $4 - 2e^{j4\pi F}$, $5.74e^{-j12°}$ **(b)** $2e^{j\pi/3}$, $2e^{-j\pi/3}$

## 5.2.2 Time Shift of $x[n]$

With $x[n] \Leftarrow \boxed{\text{DTFT}} \Rightarrow X(F)$, the DTFT of the signal $y[n] = x[n - m]$ may be written (using a change of variable) as

$$Y(F) = \sum_{k=-\infty}^{\infty} x[k-m]e^{-j2k\pi F} \underbrace{=}_{l=k-m} \sum_{l=-\infty}^{\infty} x[l]e^{-j2(l+m)\pi F} = X(F)e^{-j2\pi mF} \qquad (5.10)$$

**TABLE 5.2** ➤ Properties of the DTFT

| Property | DT Signal | Result ($F$-Form) | Result ($\Omega$-Form) |
|---|---|---|---|
| Reversal | $x[-n]$ | $X(-F) = X^*(F)$ | $X(-\Omega) = X^*(\Omega)$ |
| Time shift | $x[n-m]$ | $e^{-j2\pi mF}X(F)$ | $e^{-j\Omega m}X(\Omega)$ |
| Frequency shift | $e^{j2\pi nF_0}x[n]$ | $X(F-F_0)$ | $X(\Omega-\Omega_0)$ |
| Half-period shift | $(-1)^n x[n]$ | $X(F-0.5)$ | $X(\Omega-\pi)$ |
| Modulation | $\cos(2\pi nF_0)x[n]$ | $0.5[X(F+F_0)+X(F-F_0)]$ | $\pi[(\Omega+\Omega_0)+X(\Omega-\Omega_0)]$ |
| Convolution | $x[n] * y[n]$ | $X(F)Y(F)$ | $X(\Omega)Y(\Omega)$ |
| Product | $x[n]y[n]$ | $X(F) \circledast Y(F)$ | $\frac{1}{2\pi}[X(\Omega) \circledast Y(\Omega)]$ |
| Times-$n$ | $nx[n]$ | $\frac{j}{2\pi}\frac{dX(F)}{dF}$ | $j\frac{dX(\Omega)}{d\Omega}$ |
| Parseval's relation | $\sum_{k=-\infty}^{\infty} x^2[k] = \int_1 \lvert X(F)\rvert^2\,dF = \frac{1}{2\pi}\int_{2\pi} \lvert X(\Omega)\rvert^2\,d\Omega$ | | |
| Central ordinates | $x[0] = \int_1 X(F)\,dF = \frac{1}{2\pi}\int_{2\pi} X(\Omega)\,d\Omega \qquad X(0) = \sum_{n=-\infty}^{\infty} x[n]$ <br> $X(F)\rvert_{F=0.5} = X(\Omega)\rvert_{\Omega=\pi} = \sum_{n=-\infty}^{\infty} (-1)^n x[n]$ | | |

A time shift of $x[n]$ to $x[n-m]$ does not affect the magnitude spectrum. It augments the phase spectrum by $\theta(F) = -2\pi mF$ (or $\theta(\Omega) = -m\Omega$), which varies linearly with frequency.

# Drill Problem 5.5

**(a)** Let $X(F) = 4 - 2e^{-j4\pi F}$. If $y[n] = x[n-2]$, find $Y(F)$ and compute at $F = 0.2,\ 0.3$.
**(b)** If $g[n] = h[n-2]$, find the phase difference $\angle H(F) - \angle G(F)$ at $F = 0.2,\ 0.4$.

**Answers: (a)** $4e^{-j4\pi F} - 2e^{-j8\pi F}$, $5.74e^{-j132°}$, $5.74e^{j132°}$ **(b)** $144°$, $-72°$

### 5.2.3 Frequency Shift of $X(F)$

By duality, a frequency shift of $X(F)$ to $X(F - F_0)$ yields the signal $x[n]e^{j2\pi nF_0}$.

#### Half-Period Frequency Shift

If $X(F)$ is shifted by 0.5 to $X(F \pm 0.5)$, then $x[n]$ changes to $e^{\mp jn\pi} = (-1)^n x[n]$. Thus, samples of $x[n]$ at odd index values ($n = \pm 1, \pm 3, \pm 5, \ldots$) change sign.

# Drill Problem 5.6

**(a)** Let $X(F) = 4 - 2e^{-j4\pi F}$. If $y[n] = (-1)^n x[n]$, compute $Y(F)$ at $F = 0.2,\ 0.4$.
**(b)** Let $X(F) = 4 - 2e^{-j4\pi F}$. If $y[n] = (j)^n x[n]$, compute $Y(F)$ at $F = 0.2,\ 0.4$. [*Hint*: $j = e^{j\pi/2}$]

**Answers: (a)** $5.74e^{j12°}$, $3.88e^{-j29°}$ **(b)** $2.66e^{-j26°}$, $4.99e^{j22°}$

### 5.2.4 Modulation

Using the frequency-shift property and superposition gives the modulation property

$$\left(\frac{e^{j2\pi nF_0}+e^{-j2\pi nF_0}}{2}\right)x[n] = \cos(2\pi nF_0)x[n] \Leftarrow \boxed{\text{DTFT}} \Rightarrow \frac{X(F+F_0)+X(F-F_0)}{2} \tag{5.11}$$

Modulation results in a spreading of the original spectrum.

## Drill Problem 5.7

**(a)** The central period of $X(F)$ is defined by $X(F) = 1,\ |F| < 0.1$ and zero elsewhere. Consider the signal $y[n] = x[n]\cos(0.2n\pi)$. Sketch $Y(F)$ and evaluate at $F = 0,\ 0.1,\ 0.3,\ 0.4$.
**(b)** The central period of $X(F)$ is defined by $X(F) = 1,\ |F| < 0.1$ and zero elsewhere. Consider the signal $y[n] = x[n]\cos(0.5n\pi)$. Sketch $Y(F)$ and evaluate at $F = 0,\ 0.1,\ 0.3,\ 0.4$.
**(c)** The central period of $X(F)$ is defined by $X(F) = 1,\ |F| < 0.25$ and zero elsewhere. Consider the signal $y[n] = x[n]\cos(0.2n\pi)$. Sketch $Y(F)$ and evaluate at $F = 0,\ 0.1,\ 0.3,\ 0.4$.

**Answers: (a)** 0.5, 0.5, 0, 0 **(b)** 0, 0, 0.5, 0 **(c)** 1, 1, 0.5, 0

### 5.2.5 Convolution

The *regular* convolution of discrete-time signals results in the product of their DTFTs. This result follows from the fact that the DTFT may be regarded as a polynomial in powers of $e^{-j2\pi F}$ and discrete convolution corresponds to polynomial multiplication. If two discrete signals are multiplied together, the DTFT of the product corresponds to the *periodic* or circular convolution of the individual DTFTs. In other words, multiplication in one domain corresponds to convolution in the other.

## Drill Problem 5.8

**(a)** Let $X(F) = 4 - 2e^{-j4\pi F}$. If $y[n] = x[n] * x[n]$, compute $Y(F)$ at $F = 0.2,\ 0.4$.
**(b)** Let $X(F) = 4 - 2e^{-j4\pi F}$. If $y[n] = x[n-2] * x[-n]$, compute $Y(F)$ at $F = 0.2,\ 0.4$.
**(c)** The central period of $X(F)$ is defined by $X(F) = 1,\ |F| < 0.2$ and zero elsewhere. Consider the signal $y[n] = x^2[n]$. Sketch $Y(F)$ and evaluate at $F = 0,\ 0.1,\ 0.2,\ 0.3$.

**Answers: (a)** $32.94e^{j24^\circ}$, $15.06e^{-j59^\circ}$ **(b)** $32.94e^{-j144^\circ}$, $15.06e^{j72^\circ}$
**(c)** 0.4, 0.3, 0.2, 0.1

### 5.2.6 The Times-*n* Property:

With $x[n] \Leftarrow \boxed{\text{DTFT}} \Rightarrow X(F)$, differentiation of the defining DTFT relation gives

$$\frac{dX(F)}{dF} = \sum_{k=-\infty}^{\infty} (-j2k\pi)x[k]e^{-j2k\pi F} \tag{5.12}$$

The corresponding signal is $(-j2n\pi)x[n]$, and thus the DTFT of $y[n] = nx[n]$ is

$$Y(F) = \frac{j}{2\pi}\frac{dX(F)}{dF}$$

# Drill Problem 5.9

**(a)** Let $X(F) = 4 - 2e^{-j4\pi F}$. If $y[n] = nx[n]$, find $Y(F)$.
**(b)** Let $X(F) = 4 - 2e^{-j4\pi F}$. If $y[n] = nx[n-2]$, find $Y(F)$.
**(c)** Let $X(F) = 4 - 2e^{-j4\pi F}$. If $y[n] = (n-2)x[n]$, find $Y(F)$.
**(d)** Let $X(F) = \dfrac{1}{4 - 2e^{-j4\pi F}}$. If $y[n] = nx[n]$, find $Y(F)$.

**Answers: (a)** $-4e^{-j4\pi F}$ **(b)** $8e^{-j4\pi F} - 8e^{-j8\pi F}$ **(c)** $-8$ **(d)** $\dfrac{4e^{-j4\pi F}}{(4-2e^{-j4\pi F})^2}$

### 5.2.7 Parseval's Relation

The DTFT is an energy-conserving transform, and the signal energy may be found from either the signal $x[n]$ or from one period of its periodic magnitude spectrum $|X(F)|$ using Parseval's theorem

$$\sum_{k=-\infty}^{\infty} x^2[k] = \int_{-1/2}^{1/2} |X(F)|^2\, dF = \frac{1}{2\pi}\int_{-\pi}^{\pi} |X(\Omega)|^2\, d\Omega \quad \text{(Parseval's relation)} \tag{5.13}$$

# Drill Problem 5.10

**(a)** Let $X(F) = 5,\ |F| < 0.2$ and zero elsewhere in the central period. Find its total signal energy and its signal energy in the frequency range $|F| \le 0.15$.
**(b)** Let $X(F) = \dfrac{6}{4 - 2e^{-j2\pi F}}$. Find its signal energy.

**Answers: (a)** 10, 7.5 **(b)** 3

### 5.2.8 Central Ordinate Theorems

The DTFT obeys the *central ordinate* relations found by substituting $F = 0$ (or $\Omega = 0$) in the DTFT or $n = 0$ in the IDTFT.

$$x[0] = \int_{-1/2}^{1/2} X(F)\, dF = \frac{1}{2\pi}\int_{-\pi}^{\pi} X(\Omega)\, d\Omega \qquad X(0) = \sum_{n=-\infty}^{\infty} x[n] \quad \text{(central ordinates)} \tag{5.14}$$

With $F = 0.5$ (or $\Omega = \pi$), we also have the useful result

$$X(F)|_{F=0.5} = X(\Omega)|_{\Omega=\pi} = \sum_{n=-\infty}^{\infty} (-1)^n x[n] \tag{5.15}$$

For a filter whose impulse response is $h[n]$, the central ordinate theorems allow us to find the dc gain (at $F = 0$) and high-frequency gain (at $F = 0.5$) without having to first formally evaluate $H(F)$.

# Drill Problem 5.11

**(a)** Let $X(F) = 5,\ |F| < 0.2$ and zero elsewhere in the central period. Find the value of $x[n]$ at $n = 0$.
**(b)** Let $x[n] = 9(0.8)^n u[n]$. What is the value of $X(F)$ at $F = 0$ and $F = 0.5$.
**(c)** What is the dc gain and high-frequency gain of the filter described by $h[n] = \{\overset{\Downarrow}{1}, 2, 3, 4\}$.

**Answers:** **(a)** 2 **(b)** 45, 5 **(c)** 10, 2

**Example 5.3**

## Some DTFT Pairs using the Properties

**(a)** The DTFT of $x[n] = n\alpha^n u[n],\ |\alpha < 1|$ may be found using the times-$n$ property as

$$X(F) = \frac{j}{2\pi}\frac{d}{dF}\left[\frac{1}{1-\alpha e^{-j2\pi F}}\right] = \frac{\alpha e^{-j2\pi F}}{(1-\alpha e^{-j2\pi F})^2}$$

In the $\Omega$-form,

$$X(\Omega) = j\frac{d}{d\Omega}\left[\frac{1}{1-\alpha e^{-j\Omega}}\right] = \frac{\alpha e^{-j\Omega}}{(1-\alpha e^{-j\Omega})^2}$$

**(b)** The DTFT of the signal $x[n] = (n+1)\alpha^n u[n]$ may be found if we write $x[n] = n\alpha^n u[n] + \alpha^n u[n]$ and use superposition to give

$$X(F) = \frac{\alpha e^{-j2\pi F}}{(1-\alpha e^{-j2\pi F})^2} + \frac{1}{1-\alpha e^{-j2\pi F}} = \frac{1}{(1-\alpha e^{-j2\pi F})^2}$$

In the $\Omega$-form,

$$X(\Omega) = \frac{\alpha e^{-j\Omega}}{(1-\alpha e^{-j\Omega})^2} + \frac{1}{1-\alpha e^{-j\Omega}} = \frac{1}{(1-\alpha e^{-j\Omega})^2}$$

By the way, if we recognize that $x[n] = \alpha^n u[n] * \alpha^n u[n]$, we can also use the convolution property to obtain the same result.

**(c)** To find DTFT of the $N$-sample exponential pulse $x[n] = \alpha^n,\ 0 \le n < N$, express it as $x[n] = \alpha^n(u[n] - u[n-N]) = \alpha^n u[n] - \alpha^N \alpha^{n-N} u[n-N]$ and use the shifting property to get

$$X(F) = \frac{1}{1-\alpha e^{-j2\pi F}} - \alpha^N \frac{e^{-j2\pi FN}}{1-\alpha e^{-j2\pi F}} = \frac{1-(\alpha e^{-j2\pi F})^N}{1-\alpha e^{-j2\pi F}}$$

In the $\Omega$-form,

$$X(\Omega) = \frac{1}{1-\alpha e^{-j\Omega}} - \alpha^N \frac{e^{-j\Omega N}}{1-\alpha e^{-j\Omega}} = \frac{1-(\alpha e^{-j\Omega})^N}{1-\alpha e^{-j\Omega}}$$

**(d)** The DTFT of the two-sided decaying exponential $x[n] = \alpha^{|n|}$, $|\alpha| < 1$, may be found by rewriting this signal as $x[n] = \alpha^n u[n] + \alpha^{-n}u[-n] - \delta[n]$ and using the reversal property to give

$$X(F) = \frac{1}{1-\alpha e^{-j2\pi F}} + \frac{1}{1-\alpha e^{j2\pi F}} - 1$$

Simplification leads to the result

$$X(F) = \frac{1-\alpha^2}{1-2\alpha\cos(2\pi F)+\alpha^2} \quad \text{or} \quad X(\Omega) = \frac{1-\alpha^2}{1-2\alpha\cos\Omega+\alpha^2}$$

**(e) Properties of the DTFT**

Find the DTFT of $x[n] = 4(0.5)^{n+3}u[n]$ and $y[n] = n(0.4)^{2n}u[n]$.

**1.** For $x[n]$, we rewrite it as $x[n] = 4(0.5)^3(0.5)^n u[n]$ to get

$$X(F) = \frac{0.5}{1-0.5e^{-j2\pi F}} \quad \text{or} \quad X(\Omega) = \frac{0.5}{1-0.5e^{-j\Omega}}$$

**2.** For $y[n]$, we rewrite it as $y[n] = n(0.16)^n u[n]$ to get

$$Y(F) = \frac{0.16e^{-j2\pi F}}{(1-0.16e^{-j2\pi F})^2} \quad \text{or} \quad Y(\Omega) = \frac{0.16e^{-j\Omega}}{(1-0.16e^{-j\Omega})^2}$$

**(f) Properties of the DTFT**

Let $x[n] \Leftarrow \boxed{\text{DTFT}} \Rightarrow \dfrac{4}{2-e^{-j2\pi F}} = X(F)$.

Find the DTFT of $y[n] = nx[n]$, $c[n] = x[-n]$, $g[n] = x[n] * x[n]$, and $h[n] = (-1)^n x[n]$.

**1.** By the times-$n$ property,

$$Y(F) = \frac{j}{2\pi}\frac{d}{dF}X(F) = \frac{4(-j/2\pi)(-j2\pi e^{-j2\pi F})}{(2-e^{-j2\pi F})^2} = \frac{-4e^{-j2\pi F}}{(2-e^{-j2\pi F})^2}$$

In the $\Omega$-form,

$$Y(\Omega) = j\frac{d}{d\Omega}X(\Omega) = \frac{4(-j/2\pi)(-j2\pi e^{-j\Omega})}{(2-e^{-j\Omega})^2} = \frac{-4e^{-j\Omega}}{(2-e^{-j\Omega})^2}$$

**2.** By the reversal property,

$$C(F) = X(-F) = \frac{4}{2-e^{-j2\pi F}} \quad \text{or} \quad C(\Omega) = X(-\Omega) = \frac{4}{2-e^{-j\Omega}}$$

**3.** By the convolution property,

$$G(F) = X^2(F) = \frac{16}{(2-e^{-j2\pi F})^2} \quad \text{or} \quad G(\Omega) = X^2(\Omega) = \frac{16}{(2-e^{-j\Omega})^2}$$

**4.** By the modulation property,

$$H(F) = X(F - 0.5) = \frac{4}{2 - e^{-j2\pi(F-0.5)}} = \frac{4}{2 + e^{-j2\pi F}}$$

In the $\Omega$-form,

$$H(\Omega) = X(\Omega - \pi) = \frac{4}{2 - e^{-j(\Omega-\pi)}} = \frac{4}{2 + e^{-j\Omega}}$$

**(g) Properties of the DTFT**
Let $x[n] = (0.5)^n u[n] \Leftarrow \boxed{\text{DTFT}} \Rightarrow X(F)$. Using properties, and without evaluating $X(F)$, find the time signals corresponding to

$$Y(F) = X(F) \circledast X(F), \quad H(F) = X(F + 0.4) + X(F - 0.4), \quad G(F) = X^2(F)$$

**1.** By the convolution property,

$$y[n] = x^2[n] = (0.25)^n u[n]$$

**2.** By the modulation property,

$$h[n] = 2\cos(2n\pi F_0)x[n] = 2(0.5)^n \cos(0.8n\pi)u[n] \quad (\text{where } F_0 = 0.4)$$

**3.** By the convolution property,

$$g[n] = x[n] * x[n] = (0.5)^n u[n] * (0.5)u[n] = (n + 1)(0.5)^n u[n]$$

## 5.3 The DTFT of Discrete-Time Periodic Signals

There is a unique relationship between the description of signals in the time domain and their spectra in the frequency domain. One useful result is that sampling in one domain results in a periodic extension in the other and vice versa. From analog theory (reviewed in Appendix A), we know that if a time-domain signal is made periodic by replication, the Fourier transform of the periodic signal is an impulse-sampled version of the original transform divided by the replication period (the time period). The frequency spacing of the impulses equals the reciprocal of the time period. For example, consider a periodic analog signal $x(t)$ with period $T$ whose one period is $x_1(t)$ with Fourier transform $X_1(f)$. When $x_1(t)$ is replicated every $T$ units to generate the periodic signal $x(t)$, the Fourier transform $X(f)$ of the periodic signal $x(t)$ becomes an impulse train of the form $\sum X_1(kf_0)\delta(f - kf_0)$. The frequency spacing $f_0$ of the impulses is given by $f_0 = \frac{1}{T}$, which is the reciprocal of the period. The impulse strengths $X_1(kf_0)$ are found by sampling $X_1(f)$ at $f = kf_0$ (integer multiples of $f_0 = 1/T$) and dividing the result by the period $T$ to give $\frac{1}{T}X_1(kf_0)$. These impulse strengths $\frac{1}{T}X_1(kf_0)$ also define the Fourier series coefficients of the periodic signal $x(t)$. Similarly, consider a discrete periodic signal $x[n]$ with period $N$ whose one period is $x_1[n]$, $0 \le n \le N - 1$ with DTFT $X_1(F)$. When $x_1[n]$ is replicated every $N$ samples to generate the discrete periodic signal $x[n]$, the DTFT $X(F)$ of the periodic signal $x[n]$

becomes an impulse train of the form $X_1(kF_0)\delta(F - kF_0)$. The frequency spacing $F_0$ of the impulses is given by $F_0 = \frac{1}{N}$, the reciprocal of the period. The impulse strengths $X_1(kF_0)$ are found by sampling $X_1(F)$ at $F = kF_0$ (integer multiples of $F_0 = \frac{1}{N}$) and dividing the result by the period $N$ to give $\frac{1}{N}X_1(kF_0)$. Note that $X_1(F)$ is periodic. As a result, $X(F)$ is also periodic. One period of $X(F)$ contains $N$ impulses and may be described by

$$X(F) = \frac{1}{N}\sum_{k=0}^{N-1} X_1(kF_0)\delta(F - kF_0) \quad \text{(over one period } 0 \le F < 1) \tag{5.16}$$

By convention, one period is chosen to cover the range $0 \le F < 1$ (and not the central period) in order to correspond to the summation index $n = 0 \le n \le N - 1$. Note that $X(F)$ exhibits conjugate symmetry about $k = 0$ (corresponding to $F = 0$ or $\Omega = 0$) and $k = \frac{N}{2}$ (corresponding to $F = 0.5$ or $\Omega = \pi$).

**The DTFT of $x[n]$ (Period $N$) Is a Periodic Impulse Train ($N$ Impulses per Period)**

If $x[n]$ is periodic with period $N$ and its one-period DTFT is $x_1[n] \Leftarrow \boxed{\text{DTFT}} \Rightarrow X_1(F)$, then

$$x[n] \Leftarrow \boxed{\text{DTFT}} \Rightarrow X(F) = \frac{1}{N}\sum_{k=0}^{N-1} X_1(kF_0)\delta(F - kF_0)$$

($N$ impulses per period $0 \le F < 1$)

**Example 5.4**

**DTFT of Periodic Signals**

Let one period of $x_p[n]$ be given by $x_1[n] = \{3, 2, 1, 2\}$, with $N = 4$. Then,

$$X_1(F) = 3 + 2e^{-j2\pi F} + e^{-j4\pi F} + 2e^{-j6\pi F}$$

The four samples of $X_1(kF_0)$ over $0 \le k \le 3$ are

$$X_1(kF_0) = 3 + 2e^{-j2\pi k/4} + e^{-j4\pi k/4} + 2e^{-j6\pi k/4} = \{8, 2, 0, 2\}$$

The DTFT of the periodic signal $x_p[n]$ for one period $0 \le F < 1$ is thus

$$X(F) = \frac{1}{4}\sum_{k=0}^{3} X_1(kF_0)\delta\left(F - \frac{k}{4}\right) = 2\delta(F) + 0.5\delta\left(F - \frac{1}{4}\right) + 0.5\delta\left(F - \frac{3}{4}\right)$$

(over one period $0 \le F < 1$)

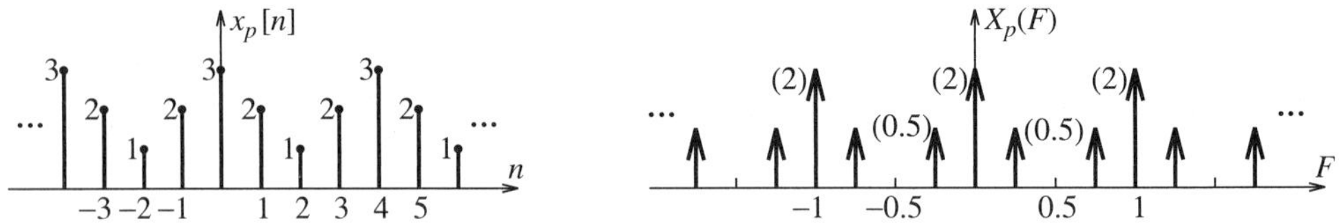


**FIGURE E.5.4**
Periodic signal for Example 5.4 and its DTFT

The signal $x_p[n]$ and its DTFT $X(F)$ are shown in Figure E5.4. Note that the DTFT is conjugate symmetric about $F = 0$ (or $k = 0$) and $F = 0.5$ (or $k = 0.5N = 2$).

# Drill Problem 5.12

Find the DTFT of a periodic signal $x[n]$ over $0 \le F < 1$ if its one period is given by

**(a)** $x_1[n] = \{\overset{\Downarrow}{4}, 0, 0, 0\}$ **(b)** $x_1[n] = \{\overset{\Downarrow}{4}, 4, 4, 4\}$ **(c)** $x_1[n] = \{\overset{\Downarrow}{4}, 0, 4, 0\}$

**Answers: (a)** $\sum_{k=0}^{3} \delta(F - 0.25k)$ **(b)** $4\delta(F)$ **(c)** $2\delta(F) + 2\delta(F - 0.5)$

## 5.3.1 The DFS and DFT

The **discrete Fourier transform** (DFT) of the signal $x_1[n]$ is defined as the sampled version of its DTFT

$$X_{\mathrm{DFT}}[k] = X_1(F)|_{F=kF_0=k/N} = \sum_{n=0}^{N-1} x_1[n] e^{-j2\pi nk/N}, \quad k = 0, 1, 2, \ldots, N-1 \quad (5.17)$$

The $N$-sample sequence that results when we divide the DFT by $N$ defines the **discrete Fourier series** (DFS) coefficients of the periodic signal $x[n]$ whose one period is $x_1[n]$.

$$X_{\mathrm{DFS}}[k] = \frac{1}{N} X_1(F)\bigg|_{F=kF_0=k/N} = \frac{1}{N} \sum_{n=0}^{N-1} x_1[n] e^{-j2\pi nk/N}, \quad k = 0, 1, 2, \ldots, N-1 \quad (5.18)$$

Note that the DFT and DFS differ only by a factor of $N$ with $X_{\mathrm{DFT}}[k] = N X_{\mathrm{DFS}}[k]$. The DFS result may be linked to the Fourier series coefficients of a periodic signal with period $T$

$$X[k] = \frac{1}{T} \int_0^T x_1(t) e^{-j2\pi kt/T}\, dt$$

Here, $x_1(t)$ describes one period of the periodic signal. The discrete version of this result using a sampling interval of $t_s$ allows us to set $t = nt_s$, $dt = t_s$, and $x_1(t) = x_1(nT_s) = x_1[n]$. Assuming $N$ samples per period ($T = Nt_s$), we replace the integral over one period (from $t = 0$ to $t = T$) by a summation over $N$ samples (from $n = 0$ to $n = N - 1$) to get the required expression for the DFS.

**The DFT is a Sampled Version of the DTFT**

If $x_1[n]$ is an $N$-sample sequence, its $N$-sample DFT is $X_{\text{DFT}}[k] = X_1(F)|_{F=k/N}$, $k = 0, 1, 2, \ldots, N-1$. The DFS coefficients of a periodic signal $x[n]$ whose one period is $x_1[n]$ are $X_{\text{DFS}}[k] = \frac{1}{N}X_{\text{DFT}}[k]$.

**Example 5.5**

**The DFT, DFS, and DTFT**

Let $x_1[n] = \{1, 0, 2, 0, 3\}$ describe one period of a periodic signal $x[n]$.

The DTFT of $x_1[n]$ is $X_1(F) = 1 + 2e^{-j4\pi F} + 3e^{-j8\pi F}$. The period of $x[n]$ is $N = 5$.

The discrete Fourier transform (DFT) of $x_1[n]$ is

$$X_{\text{DFT}}[k] = X_1(F)|_{F=k/N} = 1 + 2e^{-j4\pi F} + 3e^{-j8\pi F}|_{F=k/5}, \quad k = 0, 1, \ldots, 4$$

We find that

$$X_{\text{DFT}}[k] = \{6,\ 0.3090 + j1.6776,\ -0.8090 + j3.6655,\ -0.8090 - j3.6655,\ 0.3090 - j1.6776\}$$

The discrete Fourier series (DFS) coefficients of $x[n]$ are given by $X_{\text{DFS}}[k] = \frac{1}{N}X_{\text{DFT}}[k]$. We get

$$X_{\text{DFS}}[k] = \{1.2,\ 0.0618 + j0.3355,\ -0.1618 + j0.7331,\ -0.1618 - j0.7331,\ 0.0618 - j0.3355\}$$

The DTFT $X(F)$ of the periodic signal $x[n]$, for one period $0 \le F < 1$, is then

$$X(F) = \frac{1}{5}\sum_{k=0}^{4} X_1\left(\frac{k}{5}\right)\delta\left(F - \frac{k}{5}\right) \quad \text{(over one period } 0 \le F < 1\text{)}$$

Note that each of the transforms $X_{\text{DFS}}[k]$, $X_{\text{DFT}}[k]$, and $X(F)$ is conjugate symmetric about both $k = 0$ and $k = 0.5N = 2.5$.

# Drill Problem 5.13

Find the DFT of the following signals.

**(a)** $x[n] = \{\overset{\Downarrow}{4}, 0, 0, 0\}$ **(b)** $x[n] = \{\overset{\Downarrow}{4}, 4, 4, 4\}$ **(c)** $x[n] = \{\overset{\Downarrow}{4}, 4, 0, 0\}$

**Answers:** (a) $\{\underset{\Uparrow}{4}, 4, 4, 4\}$ (b) $\{\underset{\Uparrow}{16}, 0, 0, 0\}$ (c) $\{\underset{\Uparrow}{8}, 4 - j4, 0, 4 + j4\}$

## 5.4 The Inverse DTFT

For a finite sequence $X(F)$ whose DTFT is a polynomial in $e^{j2\pi F}$ (or $e^{j\Omega}$), the inverse DTFT $x[n]$ corresponds to the sequence of the polynomial coefficients. In many other situations, $X(F)$ can be expressed as a ratio of polynomials in $e^{j2\pi F}$ (or $e^{j\Omega}$). This allows us to split $X(F)$ into a sum of simpler terms (using partial fraction expansion) and find the inverse transform of these simpler terms through a table look-up. Only in special cases or as a last resort do we need to resort to the brute force method of finding the inverse DTFT by using the defining relation. Some examples follow.

**Example 5.6**

**The Inverse DTFT**

**(a)** Let $X(F) = 1 + 3e^{-j4\pi F} - 2e^{-j6\pi F}$.

Its IDFT is simply $x[n] = \delta[n] + 3\delta[n-2] - 2\delta[n-3]$ or $x[n] = \{\overset{\Downarrow}{1}, 0, 3, -2\}$.

**(b)** Let $X(\Omega) = \dfrac{2e^{-j\Omega}}{1 - 0.25e^{-j2\Omega}}$. We factor the denominator and use partial fractions to get

$$X(\Omega) = \frac{2e^{-j\Omega}}{(1-0.5e^{-j\Omega})(1+0.5e^{-j\Omega})} = \frac{2}{1-0.5e^{-j\Omega}} - \frac{2}{1+0.5e^{-j\Omega}}$$

We then find $x[n] = 2(0.5)^n u[n] - 2(-0.5)^n u[n]$.

**(c)** An **ideal differentiator** is described by $H(F) = j2\pi F$, $|F| < 0.5$. Its magnitude and phase spectrum are shown in Figure E5.6c.

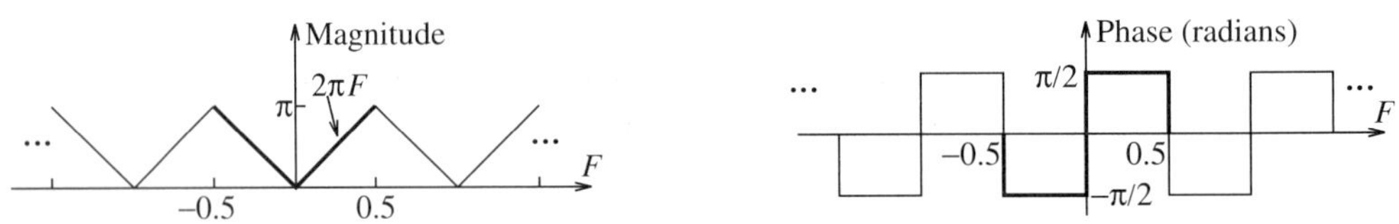


**FIGURE E 5.6.c** DTFT of the ideal differentiator for Example 5.6(c)

To find its inverse $h[n]$, we note that $h[0] = 0$ since $H(F)$ is odd. For $n \neq 0$, we also use the odd symmetry of $H(F)$ in the IDTFT to obtain

$$h[n] = \int_{-1/2}^{1/2} j2\pi F[\cos(2\pi nF) + j\sin(2\pi nF)]\,dF$$

$$= -4\pi \int_{0}^{1/2} F\sin(2\pi nF)\,dF$$

Using tables and simplifying the result, we get

$$h[n] = \frac{-4\pi[\sin(2\pi nF) - 2\pi nF\cos(2\pi nF)]}{(2\pi n)^2}\Bigg|_0^{1/2} = \frac{\cos(n\pi)}{n}$$

Since $H(F)$ is odd and imaginary, $h[n]$ is odd symmetric, as expected.

**(d)** A **Hilbert transformer** shifts the phase of a signal by $-90^\circ$. Its magnitude and phase spectrum are shown in Figure E5.6d.

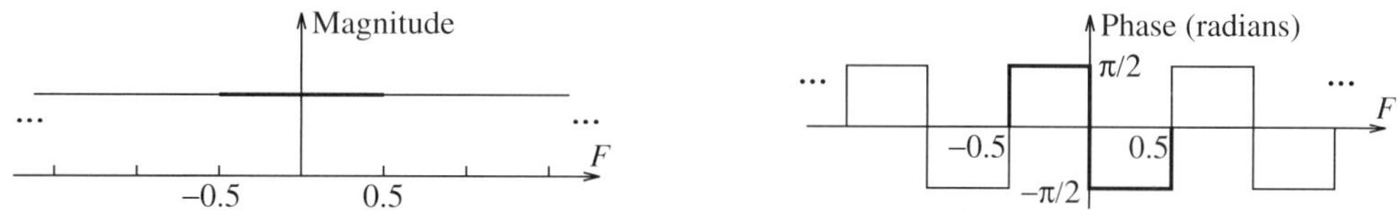


**FIGURE E 5.6.d** DTFT of the Hilbert transformer for Example 5.6(d)

Its DTFT given by $H(F) = -j\,\text{sgn}(F),\ |F| < 0.5$. This is imaginary and odd. To find its inverse $h[n]$, we note that $h[0] = 0$ and

$$h[n] = \int_{-1/2}^{1/2} -j\,\text{sgn}(F)[\cos(2\pi nF) + j\sin(2\pi nF)]\,dF$$

$$= 2\int_{0}^{1/2} \sin(2\pi nF)\,dF = \frac{1-\cos(n\pi)}{n\pi}$$

# Drill Problem 5.14

**(a)** Let $X(F) = 4 - 2e^{-j4\pi F}$. Find $x[n]$.
**(b)** Let $X(F) = (4 - 2e^{-j4\pi F})^2$. Find $x[n]$.
**(c)** Let $X(F) = 2,\ |F| < 0.2$. Find $x[n]$.

**Answers: (a)** $\{\overset{\Downarrow}{4}, 0, -2\}$ **(b)** $\{\overset{\Downarrow}{16}, 0, -16, 0, 4\}$ **(c)** $0.8\,\text{sinc}(0.4n)$

## 5.5 The Frequency Response

The time-domain response $y[n]$ of a relaxed discrete-time LTI system with impulse response $h[n]$ to the input $x[n]$ is given by the convolution

$$y[n] = x[n] * h[n] \tag{5.19}$$

Since convolution transforms to multiplication, transformation results in

$$Y(F) = X(F)H(F) \quad \text{or} \quad Y(\Omega) = X(\Omega)H(\Omega) \tag{5.20}$$

The **frequency response** or **steady-state transfer function** then equals

$$H(F) = \frac{Y(F)}{X(F)} \quad \text{or} \quad H(\Omega) = \frac{Y(\Omega)}{X(\Omega)} \tag{5.21}$$

We emphasize that the frequency response is defined only for a *relaxed* LTI system—either as the ratio $Y(F)/X(F)$ (or $Y(\Omega)/X(\Omega)$) of the DTFT of the output $y[n]$ and input $x[n]$ or as the DTFT of the impulse response $h[n]$. The equivalence between the time-domain and frequency-domain operations is illustrated in Figure 5.3.

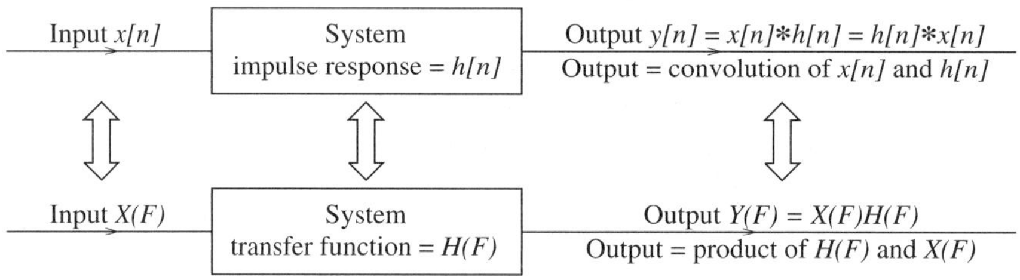


**FIGURE 5.3** The equivalence between the time domain and frequency domain. Note that convolution in the time domain transforms to multiplication in the frequency domain

A relaxed LTI system may also be described by the difference equation.

$$y[n]+A_1y[n-1]+\cdots+A_Ny[n-N] = B_0x[n]+B_1x[n-1]+\cdots+B_Mx[n-M] \tag{5.22}$$

The DTFT results in the transfer function, or **frequency response**, given by

$$H(F) = \frac{Y(F)}{X(F)} = \frac{B_0 + B_1e^{-j2\pi F} + \cdots + B_Me^{-j2\pi MF}}{1 + A_1e^{-j2\pi F} + \cdots + A_Ne^{-j2\pi NF}} \tag{5.23}$$

In the $\Omega$-form,

$$H(\Omega) = \frac{Y(\Omega)}{X(\Omega)} = \frac{B_0 + B_1e^{-j\Omega} + \cdots + B_Me^{-jM\Omega}}{1 + A_1e^{-j\Omega} + \cdots + A_Ne^{-jN\Omega}} \tag{5.24}$$

The transfer function is thus a ratio of polynomials in $e^{j2\pi F}$ (or $e^{j\Omega}$).

**Example 5.7**

### Frequency Response of a Recursive Filter

Let $y[n] = \alpha y[n-1] + x[n],\ 0 < \alpha < 1$.

To find its frequency response, we transform the difference equation and find $H(z)$ as

$$H(z) = \frac{Y(z)}{X(z)} = \frac{1}{1 - \alpha z^{-1}}$$

Next, we let $z = e^{j2\pi F}$ and evaluate $H(F)$ as

$$H(F) = \frac{1}{1 - \alpha e^{-j2\pi F}} = \frac{1}{1 - \alpha\cos(2\pi F) + j\alpha\sin(2\pi F)}$$

The magnitude and phase of $H(F)$ then equal

$$|H(F)| = \left[\frac{1}{1 - 2\alpha\cos(2\pi F) + \alpha^2}\right]^{1/2} \qquad \phi(F) = \tan^{-1}\left[\frac{1 - \alpha\cos(2\pi F)}{\alpha\sin(2\pi F)}\right]$$

Typical plots of the magnitude and phase are shown in Figure E5.7 over the principal period $(-0.5,\ 0.5)$. Note the conjugate symmetry (even symmetry of $|H(F)|$ and odd symmetry of $\phi(F)$).

**FIGURE E.5.7** Magnitude and phase of $H_p(F)$ for Example 5.7

---

# Drill Problem 5.15

**(a)** Find the frequency response $H(F)$ of the filter described by $h[n] = \{\overset{\Downarrow}{4},\ 0,\ -2\}$.
**(b)** Find the frequency response $H(F)$ of the filter described by $h[n] = 2(0.5)^n u[n] - \delta[n]$.
**(c)** Find the frequency response $H(F)$ of the filter described by $y[n] - y[n-2] = 2x[n] - 4x[n-1]$.

**Answers: (a)** $4 - 2e^{-j4\pi F}$ **(b)** $\dfrac{1 + 0.5e^{-j2\pi F}}{1 - 0.5e^{-j2\pi F}}$ **(c)** $\dfrac{2 - 4e^{-j2\pi F}}{1 - e^{-j4\pi F}}$

# 5.6 System Analysis using the DTFT

In concept, the DTFT may be used to find the zero-state response (ZSR) of relaxed LTI systems to arbitrary inputs. All it requires is the system transfer function $H(F)$ and the DTFT $X(F)$ of the input $x[n]$. We first find the response as $Y(F) = H(F)X(F)$ in the frequency domain and obtain the time-domain response $y[n]$ by using the inverse DTFT. We emphasize, once again, that the DTFT cannot handle the effect of initial conditions.

**Example 5.8** **The DTFT in System Analysis**

**(a)** Consider a system described by $y[n] = \alpha y[n-1] + x[n]$.
To find the response of this system to the input $\alpha^n u[n]$, we first set up the transfer function as $H(\Omega) = \dfrac{1}{1 - \alpha e^{-j\Omega}}$. Next, we find the DTFT of $x[n]$ as

$X(\Omega) = \dfrac{1}{1 - \alpha e^{-j\Omega}}$ and multiply the two to obtain

$$Y(\Omega) = H(\Omega)X(\Omega) = \frac{1}{(1 - \alpha e^{-j\Omega})^2}$$

Its inverse transform gives the response as $y[n] = (n+1)\alpha^n u[n]$. We could, of course, also use convolution to obtain $y[n] = h[n] * x[n]$ directly in the time domain.

**(b)** Consider the system described by $y[n] = 0.5y[n-1] + x[n]$.

Its response to the step $x[n] = 4u[n]$ is found using $Y(F) = H(F)X(F)$:

$$Y(F) = H(F)X(F) = \frac{1}{1 - 0.5e^{-j2\pi F}}\left[\frac{4}{1 - e^{-j2\pi F}} + 2\delta(F)\right]$$

We separate terms and use the product property of impulses to get

$$Y(F) = \frac{4}{(1 - 0.5e^{-j2\pi F})(1 - e^{-j2\pi F})} + 4\delta(F)$$

Splitting the first term into partial fractions, we obtain

$$Y(F) = \frac{-4}{1 - 0.5e^{-j2\pi F}} + \left[\frac{8}{1 - e^{-j2\pi F}} + 4\delta(F)\right]$$

The response $y[n]$ then equals $y[n] = -4(0.5)^n u[n] + 8u[n]$. The first term represents the *transient* response, and the second term describes the *steady-state* response, which can be found much more easily, as we now show.

# Drill Problem 5.16

**(a)** Let $H(F) = 4 - 2e^{-j4\pi F}$. If the input is $x[n] = 4\delta[n] + 2\delta[n-2]$, find the response $y[n]$.

**(b)** A filter is described by $h[n] = 2(0.5)^n u[n] - \delta[n]$. The input is $X(F) = 1 - 0.5e^{-j2\pi F}$. Find $y[n]$.

**(c)** A filter is described by $y[n] - 0.5y[n-1] = 2x[n]$. The input is $x[n] = 2(0.4)^n u[n]$. Find $y[n]$.

**Answers: (a)** $16\delta[n] - 4\delta[n-4]$ **(b)** $\delta[n] + 0.5\delta[n-1]$
**(c)** $[20(0.5)^n - 16(0.4)^n]u[n]$

## 5.6.1 The Steady-State Response to Discrete-Time Harmonics

The DTFT is much better suited to finding the *steady-state* response to discrete-time harmonics. Since everlasting harmonics are eigensignals of discrete-time linear systems, the response is simply a harmonic at the input frequency whose magnitude and phase is changed by the system function $H(F)$. We evaluate $H(F)$ at the input frequency, multiply its magnitude by the input magnitude to obtain the output magnitude, and add its phase to the input phase to obtain the output phase. The steady-state response is useful primarily for stable systems for which the natural response does indeed decay with time.

The response of an LTI system to a sinusoid (or harmonic) is called the **steady-state response** and is a sinusoid (or harmonic) at the input frequency. If the input is $x[n] = A\cos(2\pi F_0 n + \theta)$, the steady-state response is $y_{ss}[n] = AH_0\cos[2\pi Fn + \theta + \phi_0)]$, where $H_0$ and $\phi_0$ are the gain and phase of the frequency response $H(F)$ evaluated at the input frequency $F = F_0$. If the input consists of harmonics at different frequencies, we use superposition of the individual *time-domain* responses.

We can also find the steady-state component from the total response (using $z$-transforms), but this defeats the whole purpose if all we are after is the steady-state component.

### How to Find the Steady-State Response of an LTI System to a Sinusoidal Input

**Input:** $x[n] = A\cos(2\pi nF_0 + \theta)$
**Transfer Function:** Evaluate $H(F)$ at $F = F_0$ as $H_0\angle\phi_0$.
**Steady-State Output:** $y_{ss}[n] = AH_0\cos(2\pi nF_0 + \theta + \phi_0)$

**Example 5.9**

### The DTFT and Steady-State Response

**(a)** Consider a system described by $y[n] = 0.5y[n-1] + x[n]$.

We find its steady-state response to the sinusoidal input $x[n] = 10\cos(0.5n\pi + 60°)$. The transfer function $H(F)$ is given by

$$H(F) = \frac{1}{1 - 0.5e^{-j2\pi F}}$$

We evaluate $H(F)$ at the input frequency, $F = 0.25$:

$$H(F)|_{F=0.25} = \frac{1}{1 - 0.5e^{-j\pi/2}} = \frac{1}{1 + 0.5j} = 0.4\sqrt{5}\angle - 26.6° = 0.8944\angle - 26.6°$$

The steady-state response then equals

$$y_{ss}[n] = 10(0.4\sqrt{5})\cos(0.5n\pi + 60° - 26.6°) = 8.9443\cos(0.5n\pi + 33.4°)$$

Note that, if the input was $x[n] = 10\cos(0.5n\pi + 60°)u[n]$ (switched on at $n = 0$), the steady-state component would still be identical to what we calculated but the *total* response would differ and require a different method (such as $z$-transforms) to obtain.

**(b)** Consider a system described by $h[n] = (0.8)^n u[n]$.
We find its steady-state response to the step input $x[n] = 4u[n]$. The transfer function $H(F)$ is given by

$$H(F) = \frac{1}{1 - 0.8e^{-j2\pi F}}$$

We evaluate $H(F)$ at the input frequency $F = 0$ (corresponding to dc):

$$H(F)|_{F=0} = \frac{1}{1 - 0.8} = 5$$

The steady-state response is then $y_{ss}[n] = (5)(4) = 20$.

**(c)** Let $H(z) = \dfrac{2z-1}{z^2+0.5z+0.5}$.

We find its steady-state response to $x[n] = 6u[n]$. With $z = e^{j2\pi F}$, we obtain the frequency response $H(F)$ as

$$H(F) = \frac{2e^{j2\pi F} - 1}{e^{j4\pi F} + 0.5e^{j2\pi F} + 0.5}$$

Since the input is a constant for $n \ge 0$, the input frequency is $F = 0$.

At this frequency, $H(F)|_{F=0} = 0.5$. Then, $y_{ss}[n] = (6)(0.5) = 3$.

**(d)** Design a 3-point FIR filter with impulse response $h[n] = \{\alpha,\ \overset{\Downarrow}{\beta},\ \alpha\}$ that completely blocks the frequency $F = \frac{1}{3}$ and passes the frequency $F = 0.125$ with unit gain. What is the dc gain of this filter?

The filter transfer function is $H(F) = \alpha e^{j2\pi F} + \beta + \alpha e^{-j2\pi F} = \beta + 2\alpha\cos(2\pi F)$.

From the information given, we have

$$H\left(\frac{1}{3}\right) = 0 = \beta + 2\alpha\cos\left(\frac{2\pi}{3}\right) = \beta - \alpha \qquad H(0.125) = 1 = \beta + 2\alpha\cos\left(\frac{2\pi}{8}\right) = \beta - \alpha\sqrt{2}$$

This gives $\alpha = \beta = 0.4142$ and $h[n] = \{0.4142,\ \overset{\Downarrow}{0.4142},\ 0.4142\}$.

The dc gain of this filter is $H(0) = \sum h[n] = 3(0.4142) = 1.2426$.

# Drill Problem 5.17

**(a)** Let $H(F) = 4 - 2e^{-j4\pi F}$. If the input is $x[n] = 4\cos(0.4n\pi)$, what is the response $y[n]$.

**(b)** A filter is described by $h[n] = 2(0.5)^n u[n] - \delta[n]$. The input is $x[n] = 1 + \cos(0.5n\pi)$. Find $y[n]$.

**(c)** A filter is described by $y[n] = x[n] + \alpha x[n-1] + \beta x[n-2]$. Choose the values of $\alpha$ and $\beta$ such that the input $x[n] = 1 + 4\cos(n\pi)$ results in the output $y[n] = 4$.

**Answers:** **(a)** $22.96\cos(0.4n\pi + 12^\circ)$ **(b)** $3 + \cos(0.5n\pi - 53^\circ)$ **(c)** $\alpha = 2,\ \beta = 1$

## 5.7 Connections

A relaxed LTI system may be described by its difference equation, its impulse response, its transfer function, its frequency response, its pole-zero plot, or even its realization. Depending on what is required, one form may be better suited that others and, given one form, we should be able to obtain the others using time-domain methods and/or frequency-domain transformations. The connections are summarized here:

1. Given the transfer function $H(z)$, we can use it directly to generate a pole-zero plot. We can also use it to find the frequency response $H(F)$ by the substitution $z = e^{j2\pi F}$. The frequency response will allow us to sketch the gain and phase. The inverse

$z$-transform of $H(z)$ leads directly to the impulse response $h[n]$. Finally, if we express the transfer function $H(z) = Y(z)/X(z)$ where $X(z)$ and $Y(z)$ are polynomials in $z$, cross multiplication and inverse transformation can give us the system difference equation. The frequency response $H(F)$ may be used to find the system difference equation in a similar manner.

2. Given the system difference equation, we can use the $z$-transform to find the transfer function $H(z)$ and use $H(z)$ to find the remaining forms.
3. Given the impulse response $h[n]$, we can use the $z$-transform to find the transfer function $H(z)$ and use it to develop the remaining forms.
4. Given the pole-zero plot, obtain the transfer function $H(z)$ and use it to find the remaining forms. Note that we can find $H(z)$ directly from a pole-zero plot of the root locations but only to within the multiplicative factor $K$. If the value of $K$ is also shown on the plot, $H(z)$ is known in its entirety.

For quick computations of the dc gain and high-frequency gain (at $F = 0.5$), we need not even convert between forms. To find the dc gain,

1. If $H(F)$ is given, evaluate its absolute value at $F = 0$.
2. If $H(z)$ is given, evaluate its absolute value at $z = 1$ (because if $F = 0$, we have $z = e^{j2\pi F} = 1$).
3. If $h[n]$ is given, evaluate $|\sum h[n]|$ (by summing the samples).
4. If the difference equation is given in the form $\sum A_k y[n-k] = \sum B_m x[n-m]$, evaluate the absolute value of the ratio $\sum A_k / \sum B_m$ (after summing the coefficients on each side).

Similarly, to find the high-frequency gain (at $F = 0.5$),

1. If $H(F)$ is given, evaluate its absolute value at $F = 0.5$.
2. If $H(z)$ is given, evaluate its absolute value at $z = -1$ (because if $F = 0.5$, we have $z = e^{j2\pi F} = -1$).
3. If $h[n]$ is given, evaluate $|\sum(-1)^n h[n]|$ (by reversing the sign of alternate samples before summing them).
4. If the difference equation is given in the form $\sum A_k y[n-k] = \sum B_m x[n-m]$, evaluate the absolute value of the ratio $\sum(-1)^k A_k / \sum(-1)^k B_m$.

**Example 5.10**

### System Representation in Various Forms

**(a)** Let $y[n] = 0.8y[n-1] + 2x[n]$.

We obtain its transfer function and impulse response as follows:

$$Y(z) = 0.8z^{-1}Y(z) + 2X(z) \qquad H(z) = \frac{Y(z)}{X(z)} = \frac{2}{1-0.8z^{-1}} = \frac{2z}{z-0.8}$$

$$h[n] = 2(0.8)^n u[n]$$

**(b)** Let $y[n] - 0.6y[n-1] = x[n]$.

Its dc gain is $|\frac{1}{1-0.6}| = 2.5$.

Its high frequency gain is $|\frac{1}{1+0.6}| = 0.625$.

Its DTFT gives

$$Y(F) - 0.6e^{-j2\pi F}Y(F) = X(F) \quad \text{or} \quad Y(\Omega) - 0.6e^{-j\Omega}Y(\Omega) = X(\Omega)$$

We thus get the transfer function as

$$H(F) = \frac{Y(F)}{X(F)} = \frac{1}{1 - 0.6e^{-j2\pi F}} \quad \text{or} \quad H(\Omega) = \frac{Y(\Omega)}{X(\Omega)} = \frac{1}{1 - 0.6e^{-j\Omega}}$$

The system impulse response is thus $h[n] = (0.6)^n u[n]$.

**(c)** Let $h[n] = \delta[n] - (0.5)^n u[n]$.

Since $\sum_{n=0}^{\infty}(0.5)^n = \frac{1}{1-0.5} = -2$, the dc gain is $|1 - 2| = 1$.

Similarly, $\sum_{n=0}^{\infty}(-1)^n(0.5)^n = \sum_{n=0}^{\infty}(-0.5)^n = \frac{1}{1+0.5} = 2/3$ and the high-frequency gain is $|1 - 2/3| = 1/3$.

Of course, it is probably easier to find its dc gain and high frequency gain from $H(F)$. Its DTFT gives

$$H(F) = 1 - \frac{1}{1 - 0.5e^{-j2\pi F}} = \frac{-0.5e^{-j2\pi F}}{1 - 0.5e^{-j2\pi F}}$$

In the $\Omega$-form,

$$H(\Omega) = 1 - \frac{1}{1 - 0.5e^{-j\Omega}} = \frac{-0.5e^{-j\Omega}}{1 - 0.5e^{-j\Omega}}$$

From this we find

$$H(F) = \frac{Y(F)}{X(F)} \quad \text{or} \quad Y(F)(1 - 0.5e^{-j2\pi F}) = X(F)(-0.5e^{-j2\pi F})$$

In the $\Omega$-form,

$$H(\Omega) = \frac{Y(\Omega)}{X(\Omega)} \quad \text{or} \quad Y(\Omega)(1 - 0.5e^{-j\Omega}) = X(\Omega)(-0.5e^{-j\Omega})$$

Inverse transformation gives $y[n] - 0.5y[n-1] = -0.5x[n-1]$.

**(d)** Let $H(F) = 1 + 2e^{-j2\pi F} + 3e^{-j4\pi F}$.

Its dc gain is $|1 + 2 + 3| = 6$.

Its high-frequency gain is $|1 - 2 + 3| = 2$.

Its impulse response is $h[n] = \delta[n] + 2\delta[n-1] + 3\delta[n-2] = \{\overset{\Downarrow}{1}, 2, 3\}$.

Since $H(F) = \frac{Y(F)}{X(F)} = 1 + 2e^{-j2\pi F} + 3e^{-j4\pi F}$, we find

$$Y(F) = (1 + 2e^{-j2\pi F} + 3e^{-j4\pi F})X(F) \qquad y[n] = x[n] + 2x[n-1] + 3x[n-2]$$

---

# Drill Problem 5.18

**(a)** Let $h[n] = \delta[n] - 0.4(0.5)^n u[n]$. What is the dc gain and difference equation of this filter?

**(b)** Let $y[n] - 0.2y[n-1] = x[n] - 5x[n-1]$. What is the dc gain and high-frequency gain of this filter?

**(c)** Let $H(F) = 2 + 2\cos(2\pi F)$. What is the transfer function of this filter?

**Answers: (a)** 0.2, $y[n] - 0.5y[n-1] = 0.6x[n] - 0.5x[n-1]$ **(b)** 5, 5 **(c)** $H(z) = \frac{z}{(z+1)^2}$

# 5.8 Problems

**5.1.** **(DTFT of Sequences)** Find and simplify the DTFT of the following signals and evaluate at $F = 0$, $F = 0.5$, and $F = 1$.

**(a)** $x[n] = \{1, 2, \overset{\Downarrow}{3}, 2, 1\}$ **(b)** $y[n] = \{-1, 2, \overset{\Downarrow}{0}, -2, 1\}$

**(c)** $g[n] = \{\overset{\Downarrow}{1}, 2, 2, 1\}$ **(d)** $h[n] = \{\overset{\Downarrow}{-1}, -2, 2, 1\}$

[**Hints and Suggestions:** Use $e^{j\theta} + e^{-j\theta} = 2\cos\theta$ and $e^{j\theta} - e^{-j\theta} = j2\sin\theta$ to simplify the results. In (c) through (d), extract the factor $e^{-j3\pi F}$ before simplifying.]

**5.2.** **(DTFT from Definition)** Use the defining relation to find the DTFT $X(F)$ of the following signals.

**(a)** $x[n] = (0.5)^{n+2}u[n]$ **(b)** $x[n] = n(0.5)^{2n}u[n]$ **(c)** $x[n] = (0.5)^{n+2}u[n-1]$
**(d)** $x[n] = n(0.5)^{n+2}u[n-1]$ **(e)** $x[n] = (n+1)(0.5)^n u[n]$ **(f)** $x[n] = (0.5)^{-n}u[-n]$

[**Hints and Suggestions:** Pick appropriate limits in the defining summation and simplify using tables. For (a), (c) and (d), $(0.5)^{n+2} = (0.25)(0.5)^n$. For (b), $(0.5)^{2n} = (0.25)^n$. For (e), use superposition. For (f), use the change of variable $n \to -n$.]

**5.3.** **(Properties)** The DTFT of $x[n]$ is $X(F) = \dfrac{4}{2 - e^{-j2\pi F}}$. Find the DTFT of the following signals without first computing $x[n]$.

**(a)** $y[n] = x[n-2]$ **(b)** $d[n] = nx[n]$
**(c)** $p[n] = x[-n]$ **(d)** $g[n] = x[n] - x[n-1]$
**(e)** $h[n] = x[n] * x[n]$ **(f)** $r[n] = x[n]e^{jn\pi}$
**(g)** $s[n] = x[n]\cos(n\pi)$ **(h)** $v[n] = x[n-1] + x[n+1]$

[**Hints and Suggestions:** Use properties such as shifting for (a), times-$n$ for (b), reversal for (c), superposition for (d) and (h), convolution for (e), frequency shift for (f), and modulation for (g).]

**5.4.** **(Properties)** The DTFT of the signal $x[n] = (0.5)^n u[n]$ is $X(F)$. Find the time signal corresponding to the following transforms without first computing $X(F)$.

**(a)** $Y(F) = X(-F)$ **(b)** $G(F) = X(F - 0.25)$
**(c)** $H(F) = X(F + 0.5) + X(F - 0.5)$ **(d)** $P(F) = X'(F)$
**(e)** $R(F) = X^2(F)$ **(f)** $S(F) = X(F)\circledast X(F)$
**(g)** $D(F) = X(F)\cos(4\pi F)$ **(h)** $T(F) = X(F + 0.25) - X(F - 0.25)$

[**Hints and Suggestions:** Use properties such as reversal for (a), frequency shift for (b), superposition for (c) and (h), times-$n$ for (d), convolution for (e), multiplication for (f), and modulation for (g).]

**5.5.** **(DTFT)** Compute the DTFT of the following signals.

**(a)** $x[n] = 2^{-n}u[n]$ **(b)** $y[n] = 2^n u[-n]$
**(c)** $g[n] = 0.5^{|n|}$ **(d)** $h[n] = 0.5^{|n|}(u[n+1] - u[n-2])$
**(e)** $p[n] = (0.5)^n \cos(0.5n\pi)u[n]$ **(f)** $q[n] = \cos(0.5n\pi)(u[n+5] - u[n-6])$

[**Hints and Suggestions:** In (a), $2^{-n} = (0.5)^n$. In (b), use the reversal property on $x[n]$. In (c), $x[n] = (0.5)^n u[n] + (0.5)^{-n}u[-n] - \delta[n]$. In (d), find the 3-sample sequence first. In (e), use modulation. In (f), use modulation on $u[n+5] - u[n-6] = \text{rect}(0.1n)$.]

**5.6.** **(DTFT)** Compute the DTFT of the following signals.

**(a)** $x[n] = \text{sinc}(0.2n)$ **(b)** $h[n] = \sin(0.2n\pi)$ **(c)** $g[n] = \text{sinc}^2(0.2n)$

[**Hints and Suggestions:** In (a), $X(F)$ is a rect function. In (b), $X(F)$ is an impulse pair. In (c), $G(F) = X(F) \circledast X(F)$ is a triangular pulse (the periodic convolution requires no wraparound).]

**5.7.** **(DTFT)** Compute the DTFT of the following signals in the form $A(F)e^{\phi(F)}$ and plot the amplitude spectrum $A(F)$.

**(a)** $x[n] = \delta[n+1] + \delta[n-1]$
**(b)** $x[n] = \delta[n+1] - \delta[n-1]$
**(c)** $x[n] = \delta[n+1] + \delta[n] + \delta[n-1]$
**(d)** $x[n] = u[n+1] - u[n-1]$
**(e)** $x[n] = u[n+1] - u[n-2]$
**(f)** $x[n] = u[n] - u[n-4]$

[**Hints and Suggestions:** Use $e^{j\theta} + e^{-j\theta} = 2\cos\theta$ and $e^{j\theta} - e^{-j\theta} = j2\sin\theta$ to simplify the results. In (d), use $x[n] = \{1, \overset{\Downarrow}{1}\}$ and extract $e^{j\pi F}$ before simplifying. In (f), use $x[n] = \{\overset{\Downarrow}{1}, 1, 1\}$ and extract $e^{-j2\pi F}$ before simplifying.]

**5.8.** **(Properties)** The DTFT of a real signal $x[n]$ is $X(F)$. How is the DTFT of the following signals related to $X(F)$?

**(a)** $y[n] = x[-n]$
**(b)** $g[n] = x[n] * x[-n]$
**(c)** $r[n] = x[n/4]$ (zero interpolation)
**(d)** $s[n] = (-1)^n x[n]$
**(e)** $h[n] = (j)^n x[n]$
**(f)** $v[n] = \cos(2n\pi F_0)x[n]$
**(g)** $w[n] = \cos(n\pi)x[n]$
**(h)** $z[n] = [1 + \cos(n\pi)]x[n]$
**(i)** $b[n] = (-1)^{n/2}x[n]$
**(j)** $p[n] = e^{jn\pi}x[n-1]$

[**Hints and Suggestions:** In (c), $R(F)$ is a compressed version. In (d), use frequency shifting. In (e) and (i), use frequency shifting with $j^n = (-1)^{n/2} = e^{jn\pi/2}$. In (h), use superposition and modulation. In (j), use a time shift followed by a frequency shift.]

**5.9.** **(Properties)** Let $x[n] = \text{tri}(0.2n)$ and let $X(F)$ be its DTFT. Compute the following without evaluating $X(F)$.

**(a)** The DTFT of the odd part of $x[n]$
**(b)** The value of $X(F)$ at $F = 0$ and $F = 0.5$
**(c)** The phase of $X(F)$
**(d)** The phase of the DTFT of $x[-n]$
**(e)** The integral $\int_{-0.5}^{0.5} X(F)\,dF$
**(f)** The integral $\int_{-0.5}^{0.5} X(F-0.5)\,dF$
**(g)** The integral $\int_{-0.5}^{0.5} |X(F)|^2\,dF$
**(h)** The derivative $\frac{dX(F)}{dF}$
**(i)** The integral $\int_{-0.5}^{0.5} \left|\frac{dX(F)}{dF}\right|^2 dF$

[**Hints and Suggestions:** For (a), (c), and (d), note that $x[n]$ has even symmetry. In (e), find $x[0]$. In (f), find $(-1)^n x[n]$ at $n = 0$. In (g), use Parseval's theorem. In (h), use $-j2\pi n x[n] \Leftrightarrow X'(F)$.]

**5.10.** **(DTFT of Periodic Signals)** Find the DTFT of the following periodic signals with $N$ samples per period.

**(a)** $x[n] = \{\overset{\Downarrow}{1}, 1, 1, 1, 1\}$, $N = 5$
**(b)** $x[n] = \{\overset{\Downarrow}{1}, 1, 1, 1\}$, $N = 4$
**(c)** $x[n] = (-1)^n$
**(d)** $x[n] = 1$ ($n$ even) and $x[n] = 0$ ($n$ odd)

[**Hints and Suggestions:** Find the $N$-sample sequence $\frac{1}{N}X(F))|_{F=k/N}$, $k = 0, 1, \dots N-1$ to set up $X_p(F) = \sum X[k]\delta(F - k/N)$. In part (c), $x[n] = \{\overset{\Downarrow}{1}, -1\}$, $N = 2$. In part (d), $x[n] = \{\overset{\Downarrow}{1}, 0\}$, $N = 2$.]

**5.11.** **(IDTFT)** Compute the IDTFT $x[n]$ of the following $X(F)$ described over $|F| \leq 0.5$.

**(a)** $X(F) = \text{rect}(2F)$ **(b)** $X(F) = \cos(\pi F)$ **(c)** $X(F) = \text{tri}(2F)$

[**Hints and Suggestions:** In (a), use $\text{rect}(F/F_C) \Leftrightarrow 2F_C\text{sinc}(2nF_C)$. In (b), simplify the defining IDTFT relation. In (c), use $\text{rect}(2F)\circledast\text{rect}(2F) = 0.5\text{tri}(2F)$ and the results of (a).]

**5.12.** **(Properties)** Confirm that the DTFT of $x[n] = n\alpha^n u[n]$ (using each of the following methods) gives identical results.

**(a)** from the defining relation for the DTFT
**(b)** from the DTFT of $y[n] = \alpha^n u[n]$ and the times-$n$ property
**(c)** from the convolution result $\alpha^n u[n] * \alpha^n u[n] = (n+1)\alpha^n u[n+1]$ and the shifting property
**(d)** from the convolution result $\alpha^n u[n] * \alpha^n u[n] = (n+1)\alpha^n u[n+1]$ and superposition

**5.13.** **(Properties)** Find the DTFT of the following signals using the approach suggested.

**(a)** Starting with the DTFT of $u[n]$, show that the DTFT of $x[n] = \text{sgn}[n]$ is $X(F) = -j\cot(\pi F)$.
**(b)** Starting with $\text{rect}(n/2N) \Leftrightarrow (2N+1)\dfrac{\text{sinc}[(2N+1)F]}{\text{sinc}F}$, use the convolution property to find the DTFT of $x[n] = \text{tri}(n/N)$.
**(c)** Starting with $\text{rect}(n/2N) \Leftrightarrow (2N+1)\dfrac{\text{sinc}[(2N+1)F]}{\text{sinc}F}$, use the modulation property to find the DTFT of $x[n] = \cos(n\pi/2N)\text{rect}(n/2N)$.

[**Hints and Suggestions:** In part (a), note that $\text{sgn}[n] = u[n] - u[-n]$ and simplify the DTFT result. In part (b), start with $\text{rect}(\frac{n}{2N}) * \text{rect}(\frac{n}{2N}) = M\text{tri}(\frac{n}{M})$, where $M = 2N+1$.]

**5.14.** **(Spectrum of Discrete Periodic Signals)** Sketch the DTFT magnitude spectrum and phase spectrum of the following signals over $|F| \leq 0.5$.

**(a)** $x[n] = \cos(0.5n\pi)$
**(b)** $y[n] = \cos(0.5n\pi) + \sin(0.25n\pi)$
**(c)** $h[n] = \cos(0.5n\pi)\cos(0.25n\pi)$

[**Hints and Suggestions:** Over the principal period, the magnitude spectrum of each sinusoid is an impulse pair (with strengths that equal half the peak value) and the phase shows odd symmetry. For part (c), use $\cos\alpha\cos\beta = 0.5\cos(\alpha+\beta) + 0.5\cos(\alpha-\beta)$ to simplify $h[n]$.]

**5.15.** **(DTFT)** Compute the DTFT of the following signals and sketch the magnitude and phase spectrum over $-0.5 \leq F \leq 0.5$.

**(a)** $x[n] = \cos(0.4n\pi)$ **(b)** $x[n] = \cos(0.2n\pi + \frac{\pi}{4})$
**(c)** $x[n] = \cos(n)$ **(d)** $x[n] = \cos(1.2n\pi + \frac{\pi}{4})$
**(e)** $x[n] = \cos(2.4n\pi)$ **(f)** $x[n] = \cos^2(2.4n\pi)$

[**Hints and Suggestions:** Over the principal period, the magnitude spectrum of each sinusoid is an impulse pair (with strengths that equal half the peak value) and the phase shows odd symmetry. For part (f), use $\cos^2\alpha = 0.5 + 0.5\cos 2\alpha$ and use its frequency in the principal period.]

**5.16. (DTFT)** Compute the DTFT of the following signals and sketch the magnitude and phase spectrum over $-0.5 \le F \le 0.5$.

**(a)** $x[n] = \text{sinc}(0.2n)$ **(b)** $y[n] = \text{sinc}(0.2n)\cos(0.4n\pi)$
**(c)** $g[n] = \text{sinc}^2(0.2n)$ **(d)** $h[n] = \text{sinc}(0.2n)\cos(0.1n\pi)$
**(e)** $p[n] = \text{sinc}^2(0.2n)\cos(0.4n\pi)$ **(f)** $q[n] = \text{sinc}^2(0.2n)\cos(0.2n\pi)$

[**Hints and Suggestions:** In (a), $X(F)$ is a rect function. In (c), $G(F) = X(F)\circledast X(F)$ is a triangular pulse (the periodic convolution requires no wraparound). In (b), (d), (e), and (f) use modulation (and add the images if there is overlap).]

**5.17. (System Representation)** Find the transfer function $H(F)$ and the system difference equation for the following systems described by their impulse response $h[n]$.

**(a)** $h[n] = (\frac{1}{3})^n u[n]$ **(b)** $h[n] = [1 - (\frac{1}{3})^n]u[n]$
**(c)** $h[n] = n(\frac{1}{3})^n u[n]$ **(d)** $h[n] = 0.5\delta[n]$
**(e)** $h[n] = \delta[n] - (\frac{1}{3})^n u[n]$ **(f)** $h[n] = [(\frac{1}{3})^n + (\frac{1}{2})^n]u[n]$

[**Hints and Suggestions:** For the difference equation, set up $H(F)$ as the ratio of polynomials in $e^{-j2\pi F}$, equate with $\frac{Y(F)}{X(F)}$, cross multiply and use the shifting property to convert to the time domain.]

**5.18. (System Representation)** Find the transfer function and impulse response of the following systems described by their difference equation.

**(a)** $y[n] + 0.4y[n-1] = 3x[n]$ **(b)** $y[n] - \frac{1}{6}y[n-1] - \frac{1}{6}y[n-2] = 2x[n] + x[n-1]$
**(c)** $y[n] = 0.2x[n]$ **(d)** $y[n] = x[n] + x[n-1] + x[n-2]$

[**Hints and Suggestions:** Find $H(F) = \frac{Y(F)}{X(F)}$ after taking the DTFT and collecting terms in $X(F)$ and $Y(F)$. In (b), evaluate $h[n]$ after expanding $H(F)$ by partial fractions.]

**5.19. (System Representation)** Set up the system difference equation for the following systems described by their transfer function.

**(a)** $H(F) = \dfrac{6e^{j2\pi F}}{3e^{j2\pi F} + 1}$ **(b)** $H(F) = \dfrac{3}{e^{-j2\pi F} + 2} - \dfrac{e^{-j2\pi F}}{e^{-j2\pi F} + 3}$

**(c)** $H(F) = \dfrac{6}{1 - 0.3e^{-j2\pi F}}$ **(d)** $H(F) = \dfrac{6e^{j2\pi F} + 4e^{j4\pi F}}{(1 + 2e^{j2\pi F})(1 + 4e^{j2\pi F})}$

[**Hints and Suggestions:** Set up $H(F)$ as the ratio of polynomials in $e^{-j2\pi F}$, equate with $\frac{Y(F)}{X(F)}$, cross multiply and use the shifting property to convert to the time domain.]

**5.20. (Steady-State Response)** Consider the filter $y[n] + 0.25y[n-2] = 2x[n] + 2x[n-1]$. Find the filter transfer function $H(F)$ of this filter and use this to compute the steady-state response to the following inputs.

**(a)** $x[n] = 5$ **(b)** $x[n] = 3\cos(0.5n\pi + \frac{\pi}{4}) - 6\sin(0.5n\pi - \frac{\pi}{4})$
**(c)** $x[n] = 3\cos(0.5n\pi)u[n]$ **(d)** $x[n] = 2\cos(0.25n\pi) + 3\sin(0.5n\pi)$

[**Hints and Suggestions:** Set up $H(F)$. For each sinusoidal component, find the frequency $F_0$, evaluate the gain and phase of $H(F_0)$ and obtain the output. In (b) and (d), use superposition.]

**5.21.** **(Response of Digital Filters)** Consider the 3-point averaging filter described by the difference equation $y[n] = \frac{1}{3}(x[n] + x[n-1] + x[n-2])$.

**(a)** Find its impulse response $h[n]$.

**(b)** Find and sketch its frequency response $H(F)$.

**(c)** Find its response to $x[n] = \cos(\frac{n\pi}{3} + \frac{\pi}{4})$.

**(d)** Find its response to $x[n] = \cos(\frac{n\pi}{3} + \frac{\pi}{4}) + \sin(\frac{n\pi}{3} + \frac{\pi}{4})$.

**(e)** Find its response to $x[n] = \cos(\frac{n\pi}{3} + \frac{\pi}{4}) + \sin(\frac{2n\pi}{3} + \frac{\pi}{4})$.

[**Hints and Suggestions:** For (c) through (e), for each sinusoidal component, find the frequency $F_0$, evaluate the gain and phase of $H(F_0)$, and obtain the output. In (d) and (e), use superposition.]

**5.22.** **(Frequency Response)** Consider a system whose frequency response $H(F)$ in magnitude/phase form is $H(F) = A(F)e^{j\phi(F)}$. Find the response $y[n]$ of this system for the following inputs.

**(a)** $x[n] = \delta[n]$ **(b)** $x[n] = 1$ **(c)** $x[n] = \cos(2n\pi F_0)$ **(d)** $x[n] = (-1)^n$

[**Hints and Suggestions:** For (d), note that $(-1)^n = \cos(n\pi)$.]

**5.23.** **(Frequency Response)** Consider a system whose frequency response is given by $H(F) = 2\cos(\pi F)e^{-j\pi F}$. Find the response $y[n]$ of this system for the following inputs.

**(a)** $x[n] = \delta[n]$ **(b)** $x[n] = \cos(0.5n\pi)$
**(c)** $x[n] = \cos(n\pi)$ **(d)** $x[n] = 1$
**(e)** $x[n] = e^{j0.4n\pi}$ **(f)** $x[n] = (j)^n$

[**Hints and Suggestions:** For (a), use $2\cos\theta = e^{j\theta} + e^{-j\theta}$ and find $h[n]$. For the rest, evaluate $H(F)$ at the frequency of each input to find the output. For (f), note that $j = e^{j\pi/2}$.]

**5.24.** **(Frequency Response)** The signal $x[n] = \{\overset{\Downarrow}{1}, 0.5\}$ is applied to a digital filter and the resulting output is $y[n] = \delta[n] - 2\delta[n-1] - \delta[n-2]$.

**(a)** What is the frequency response $H(F)$ of this filter?

**(b)** Find the impulse response $h[n]$ of this filter.

**(c)** Find the filter difference equation.

[**Hints and Suggestions:** Set up $H(F) = \frac{Y(F)}{X(F)}$ and cross multiply to establish the system difference equation. Find $h[n]$ by first writing $H(F)$ as the sum of three terms.]

**5.25.** **(Frequency Response)** Consider the 2-point averager $y[n] = 0.5x[n] + 0.5x[n-1]$.

**(a)** Sketch its amplitude spectrum.

**(b)** Find its response to the input $x[n] = \cos(n\pi/2)$.

**(c)** Find its response to the input $x[n] = \delta[n]$.

**(d)** Find its response to the input $x[n] = 1$ and $x[n] = 3 + 2\delta[n] - 4\cos(n\pi/2)$.

**(e)** Show that its half-power frequency is given by $F_C = \dfrac{\cos^{-1}(\sqrt{0.5})}{\pi}$.

**(f)** What is the half-power frequency of a cascade of $N$ such 2-point averagers?

[**Hints and Suggestions:** For (a), extract the factor $e^{-j\pi F}$ in $H(F)$ to get $H(F) = A(F)e^{j\phi(F)}$ and sketch $A(F)$. For (b), find $h[n]$. For (c), evaluate $H(F)$ at the frequency of $x[n]$ to find the output. For (d), use superposition. For (e), use $|A(F)| = \sqrt{0.5}$. For (f), start with $|H_N(F)| = |H(F)|^N$.]

5.26. **(Frequency Response)** Consider the filter of Figure P5.26. Find the frequency response of the overall system, the difference equation relating $y[n]$ and $x[n]$, and examine the system stability if the impulse response $h_1[n]$ of the filter in the forward path is

**(a)** $h_1[n] = \delta[n] - \delta[n-1]$ **(b)** $h_1[n] = 0.5\delta[n] + 0.5\delta[n-1]$

**FIGURE P.5.26** Filter realization for Problem 5.26

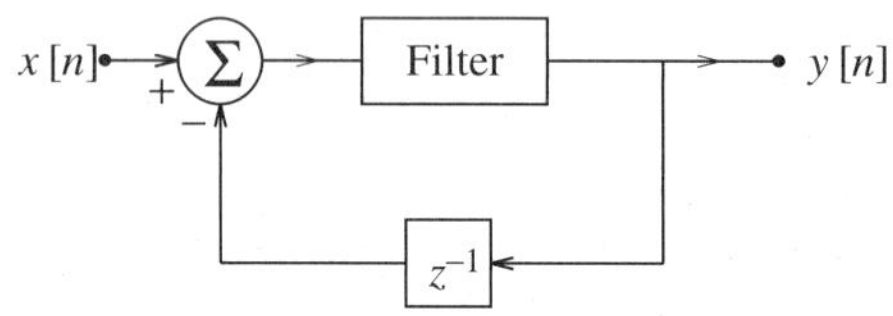


[**Hints and Suggestions:** Note that $y[n] = (x[n] - y[n-1]) * h_1[n]$. Use this to set up the difference equation, check stablity and find $H(F)$.]

5.27. **(Interconnected Systems)** Consider two systems with impulse response $h_1[n] = \delta[n] + \alpha\delta[n-1]$ and $h_2[n] = (0.5)^n u[n]$. Find the frequency response and impulse response of the combination if

**(a)** The two systems are connected in parallel with $\alpha = 0.5$.
**(b)** The two systems are connected in parallel with $\alpha = -0.5$.
**(c)** The two systems are connected in cascade with $\alpha = 0.5$.
**(d)** The two systems are connected in cascade with $\alpha = -0.5$.

[**Hints and Suggestions:** In each case, find $h[n]$ for the combination to find $H(F)$. You may also work directly in the frequency domain.]

5.28. **(Interconnected Systems)** Consider two systems with impulse response $h_1[n] = \delta[n] + \alpha\delta[n-1]$ and $h_2[n] = (0.5)^n u[n]$. Find the response of the overall system to the input $x[n] = \cos(0.5n\pi)$ if

**(a)** The two systems are connected in parallel with $\alpha = 0.5$.
**(b)** The two systems are connected in parallel with $\alpha = -0.5$.
**(c)** The two systems are connected in cascade with $\alpha = 0.5$.
**(d)** The two systems are connected in cascade with $\alpha = -0.5$.

[**Hints and Suggestions:** In each case, find $H(F)$ for the combination and evaluate at the frequency of the input to set up the output.]

5.29. **(DTFT of Periodic Signals)** Find the DTFT of the following periodic signals described over one period, with $N$ samples per period.

**(a)** $x[n] = \{\overset{\Downarrow}{1}, 0, 0, 0, 0\}, \; N = 5$
**(b)** $x[n] = \{\overset{\Downarrow}{1}, 0, -1, 0\}, \; N = 4$
**(c)** $x[n] = \{\overset{\Downarrow}{3}, 2, 1, 2\}, \; N = 4$
**(d)** $x[n] = \{\overset{\Downarrow}{1}, 2, 3\}, \; N = 3$

[**Hints and Suggestions:** Find the $N$-sample sequence $\frac{1}{N}X(F))|_{F=k/N}$, $k = 0, 1, \ldots, N-1$ to set up $X_p(F) = \sum X[k]\delta(F - k/N)$.]

**5.30.** **(System Response to Periodic Signals)** Consider the 3-point moving average filter described by $y[n] = x[n] + x[n-1] + x[n-2]$. Find one period of its periodic response to the following periodic inputs.

**(a)** $x[n] = \{\overset{\Downarrow}{1}, 0, 0, 0, 0\}, \ N = 5$

**(b)** $x[n] = \{\overset{\Downarrow}{1}, 0, -1, 0\}, \ N = 4$

**(c)** $x[n] = \{\overset{\Downarrow}{3}, 2, 1\}, \ N = 3$

**(d)** $x[n] = \{\overset{\Downarrow}{1}, 2\}, \ N = 2$

[**Hints and Suggestions:** Find $h[n]$, create its $N$-sample periodic extension, and find its periodic convolution with the input (you may use regular convolution and wraparound).]

**5.31.** **(DTFT of Periodic Signals)** One period of a periodic signal is given by

$$x[n] = \{\overset{\Downarrow}{1}, -2, 0, 1\}$$

**(a)** Find the DTFT of this periodic signal.

**(b)** The signal is passed through a filter whose impulse response is $h[n] = \text{sinc}(0.8n)$. What is the filter output?

[**Hints and Suggestions:** In (a), find the $N$-sample sequence $\frac{1}{N}X(F))|_{F=k/N}$, $k = 0, 1, \ldots, N-1$ to set up $X_p(F) = \sum X[k]\delta(F - k/N)$. In (b), $h[n]$ is an ideal filter that passes frequencies in the range $-0.4 \le F < 0.4$ with a gain of 1.25. So, extend $X_p(F)$ to the principal range $-0.5 \le F < 0.5$. Each impulse pair in $X_p(F)$ (in the principal range) passed by the filter is a sinusoid.]

**5.32.** **(System Response to Periodic Signals)** Consider the filter $y[n] - 0.5y[n-1] = 3x[n]$. Find its response to the following periodic inputs.

**(a)** $x[n] = \{\overset{\Downarrow}{4}, 0, 2, 1\}, \ N = 4$

**(b)** $x[n] = (-1)^n$

**(c)** $x[n] = 1$ ($n$ even) and $x[n] = 0$ ($n$ odd)

**(d)** $x[n] = 14(0.5)^n(u[n] - u[n-(N+1)]), \ N = 3$

[**Hints and Suggestions:** $h[n] = 3(0.5)^n u[n]$. Its periodic extension is given by $h_p[n] = 3(0.5)^n/(1-0.5^N)$ with period $N$. Find the periodic convolution of $h_p[n]$ and one period of $x[n]$ (by regular convolution and wraparound). For (b), $x[n] = \{\overset{\Downarrow}{1}, -1\}$ $(N = 2)$. For (c), $x[n] = \{\overset{\Downarrow}{1}, 0\}$ $(N = 2)$.]

**5.33.** **(System Response to Periodic Signals)** Find the response if a periodic signal whose one period is $x[n] = \{\overset{\Downarrow}{4}, 3, 2, 3\}$ (with period $N = 4$) is applied to the following filters.

**(a)** $h[n] = \text{sinc}(0.4n)$

**(b)** $H(F) = \text{tri}(2F)$

[**Hints and Suggestions:** Find the $N$-sample sequence $\frac{1}{N}X(F))|_{F=k/N}$, $k = 0, 1, \ldots, N-1$ to set up $X_p(F) = \sum X[k]\delta(F - k/N)$ and extend to the principal range $-0.5 \le F < 0.5$. Each impulse pair in $X_p(F)$ (in the principal range) passed by the filter defines a sinusoid. In (a), $H(F) = 2.5\text{rect}(F/0.4)$.]

**5.34.** **(Sampling, Filtering, and Aliasing)** The sinusoid $x(t) = \cos(2\pi f_0 t)$ is sampled at 1 kHz to yield the sampled signal $x[n]$. The signal $x[n]$ is passed through a 2-point averaging filter whose difference equation is $y[n] = 0.5(x[n] + x[n-1])$. The filtered output $y[n]$ is reconstructed using an ideal lowpass filter with a cutoff frequency of 0.5 kHz to generate the analog signal $y(t)$.

**(a)** Find an expression for $y(t)$ if $f_0 = 0.2$ kHz.
**(b)** Find an expression for $y(t)$ if $f_0 = 0.5$ kHz.
**(c)** Find an expression for $y(t)$ if $f_0 = 0.75$ kHz.

[**Hints and Suggestions:** In part (c), $y(t)$ has a lower frequency lower than 750 Hz due to aliasing.]

**5.35.** **(The Fourier Transform and DTFT)** The analog signal $x(t) = e^{-\alpha t}u(t)$ is ideally sampled at the sampling rate $S$ to yield the analog impulse train $x_i(t) = \sum_{k=-\infty}^{\infty} x(t)\delta(t - kt_s)$ where $t_s = 1/S$.

**(a)** Find the Fourier transform $X(f)$.
**(b)** Find the Fourier transform $X_i(f)$.
**(c)** If the sampled version of $x(t)$ is $x[n]$, find the DTFT $X(F)$.
**(d)** How is $X(F)$ related to $X_i(f)$?
**(e)** Under what conditions is $X(F)$ related to $X(f)$?

[**Hints and Suggestions:** In (b), $x_i(t) = \delta(t) + \delta(t - t_s) + \delta(t - 2t_s) + \cdots$. So, find $X_i(f)$ and simplify the geometric series. In (c), use $x[n] = e^{-\alpha n t_s}u[n] = (e^{-\alpha t_s})^n u[n]$. In (d), use $F = f t_s$.]

**5.36.** **(Echo Cancellation)** A microphone, whose frequency response is limited to 5 kHz, picks up not only a desired signal $x(t)$ but also its echoes. However, only the first echo, arriving after $t_d = 1.25$ ms, has a significant amplitude (of $\alpha = 0.5$) relative to the desired signal $x(t)$.

**(a)** Set up an equation relating the analog output and analog input of the microphone.
**(b)** The microphone signal is to be processed digitally in an effort to remove the echo. Set up a difference equation relating the sampled output and sampled input of the microphone using an arbitrary sampling rate $S$.
**(c)** Argue that if $S$ equals the Nyquist rate, the difference equation for the microphone will contain fractional delays and be difficult to implement.
**(d)** If the sampling rate $S$ can be varied only in steps of 1 kHz, choose the smallest $S$ that will ensure integer delays, and use it to find the difference equation of the microphone and describe the nature of this filter.
**(e)** Set up the difference equation of an echo-canceling system that can recover the original signal using the sampling rate of the previous part. Find its frequency response.
**(f)** If the microphone is cascaded with the echo-cancelling filter, what is the impulse response of the cascaded system?

[**Hints and Suggestions:** In (a), $y(t) = x(t) + \alpha x(t - t_d)$. In (b), $y[n] = x[n] + \alpha x[n - N]$ with $N = St_d$. In (d), choose the inverse system $y[n] + \alpha y[n - N] = x[n]$. In (f), note that $H_C(F) = 1$.]

**5.37.** **(Modulation)** A signal $x[n]$ is modulated by $\cos(0.5n\pi)$ to obtain the signal $x_1[n]$. The modulated signal $x_1[n]$ is filtered by a filter whose transfer function is $H(F)$ to obtain the signal $y_1[n]$.

**(a)** Sketch the spectra $X(F)$, $X_1(F)$, and $Y_1(F)$ if $X(F) = \mathrm{tri}(4F)$ and $H(F) = \mathrm{rect}(2F)$.

**(b)** The signal $y_1[n]$ is modulated again by $\cos(0.5n\pi)$ to obtain the signal $y_2[n]$ and filtered by $H(F)$ to obtain $y[n]$. Sketch $Y_2(F)$ and $Y(F)$.

**(c)** Are the signals $x[n]$ and $y[n]$ related in any way?

[**Hints and Suggestions:** In (a), the output is a version of $X(f)$ truncated in frequency. In (b), the output will show sinc distortion (due to the zero-order hold).]

## Computation and Design

5.38. **(Interconnected Systems)** The signal $x[n] = \cos(2\pi F_0 n)$ forms the input to a cascade of two systems whose impulse response is described by $h_1[n] = \{0.\overset{\Downarrow}{2}5,\ 0.5,\ 0.25\}$ and $h_2[n] = \delta[n] - \delta[n-1]$.

**(a)** Generate and plot $x[n]$ over $0 \le n \le 100$ if $F_0 = 0.1$. Can you identify its period from the plot?

**(b)** Let the output of the first system be fed to the second system. Plot the output of the each system. Are the two outputs periodic? If so, do they have the same period? Explain.

**(c)** Reverse the order of cascading and plot the output of the each system. Are the two outputs periodic? If so, do they have the same period? Does the order of cascading alter the overall response? Should it?

**(d)** Let the first system be described by $y[n] = x^2[n]$. Find and plot the output of each system in the cascade. Repeat after reversing the order of cascading. Does the order of cascading alter the intermediate response? Does the order of cascading alter the overall response? Should it?

5.39. **(Frequency Response)** This problem deals with the cascade and parallel connection of two FIR filters whose impulse response is given by

$$h_1[n] = \{\overset{\Downarrow}{1},\ 2,\ 1\} \qquad h_2[n] = \{\overset{\Downarrow}{2},\ 0,\ -2\}$$

**(a)** Plot the frequency response of each filter and identify the filter type.

**(b)** The frequency response of the parallel connection of $h_1[n]$ and $h_2[n]$ is $H_{P1}(F)$. If the second filter is delayed by one sample and then connected in parallel with the first, the frequency response changes to $H_{P2}(F)$. It is claimed that $H_{P1}(F)$ and $H_{P2}(F)$ have the same magnitude and differ only in phase. Use MATLAB to argue for or against this claim.

**(c)** Obtain the impulse response $h_{P1}[n]$ and $h_{P2}[n]$ and plot their frequency response. Use MATLAB to compare their magnitude and phase. Do the results justify your argument? What type of filters do $h_{P1}[n]$ and $h_{P2}[n]$ describe?

**(d)** The frequency response of the cascade of $h_1[n]$ and $h_2[n]$ is $H_{C1}(F)$. If the second filter is delayed by one sample and then cascaded with the first, the frequency response changes to $H_{C2}(F)$. It is claimed that $H_{C1}(F)$ and $H_{C2}(F)$ have the same magnitude and differ only in phase. Use MATLAB to argue for or against this claim.

**(e)** Obtain the impulse response $h_{C1}[n]$ and $h_{C2}[n]$ and plot their frequency response. Use MATLAB to compare their magnitude and phase. Do the results justify your argument? What type of filters do $h_{C1}[n]$ and $h_{C2}[n]$ represent?

**5.40.** **(Nonrecursive Forms of IIR Filters)** We can only approximately represent an IIR filter by an FIR filter by truncating its impulse response to $N$ terms. The larger the truncation index $N$, the better is the approximation. Consider the IIR filter described by $y[n]-0.8y[n-1]=x[n]$.

**(a)** Find its impulse response $h[n]$.

**(b)** Truncate $h[n]$ to three terms to obtain $h_N[n]$. Plot the frequency response $H(F)$ and $H_N(F)$. What differences do you observe?

**(c)** Truncate $h[n]$ to ten terms to obtain $h_N[n]$. Plot the frequency response $H(F)$ and $H_N(F)$. What differences do you observe?

**(d)** If the input to the original filter and truncated filter is $x[n]$, will the greatest mismatch in the response $y[n]$ of the two filters occur at earlier or later time instants $n$?

**5.41.** **(Frequency Response of Averaging Filters)** The averaging of data uses both FIR and IIR filters. Consider the following averaging filters:

**Filter 1:** $y[n] = \frac{1}{N}\sum_{k=0}^{N-1} x[n-k]$ ($N$-point moving average)

**Filter 2:** $y[n] = \frac{2}{N(N+1)}\sum_{k=0}^{N-1}(N-k)x[n-k]$ ($N$-point weighted moving average)

**Filter 3:** $y[n]-\alpha y[n-1]=(1-\alpha)x[n]$, $\alpha = \frac{N-1}{N+1}$ (first-order exponential average)

**(a)** Confirm that the dc gain of each filter is unity. Which of these are FIR filters?

**(b)** Sketch the frequency response magnitude of each filter with $N=4$ and $N=9$. How will the choice of $N$ affect the averaging?

**(c)** To test your filters, generate the signal $x[n]=1-(0.6)^n$, $0 \le n \le 300$, add some noise, and apply the noisy signal to each filter and compare the results. Which filter would you recommend?

CHAPTER 6

# Filter Concepts

## 6.0 Scope and Overview

This chapter introduces the terminology of filters and the time-domain and frequency-domain measures that characterize them. It deals with the connections between various techniques of digital filter analysis. It discusses the graphical interpretation of the frequency response and introduces the concept of filter design by pole-zero placement. It concludes by describing a variety of filters (such as minimum-phase, allpass, comb, and notch) that find use in digital signal processing.

### 6.0.1 Goals and Learning Objectives

The goals of this chapter are to provide an introduction to filter terminology and measures of filter performance and to show how the material of the previous chapters can be used for the analysis of digital filters in the time domain, $z$-domain, and frequency domain. After going through this chapter the reader should:

1. Understand the measures used to describe filter performance such as gain, phase delay, and group delay.
2. Understand the concept of minimum phase and know how to obtain a minimum phase filter from its magnitude spectrum.
3. Understand the significance of the unit circle in the $z$-plane and its connection to frequency response.
4. Know how to estimate filter gain graphically from its pole-zero plot.
5. Understand the concept of linear phase.
6. Know how to identify a linear-phase filter from its impulse response or zero locations.
7. Know how to identify the four types of linear-phase filters.
8. Understand the concept of signal averaging and know how to find and sketch the frequency response of FIR averaging filters.
9. Understand the concept of FIR comb filters and know how to find and sketch their frequency response.
10. Know how to identify a first-order lowpass and highpass filters from its impulse response, frequency response, difference equation, transfer function, or pole-zero plot.

11. Understand the concept of filter design by pole-zero placement.
12. Know how to design simple peaking filters or notch filters by pole-zero placement.
13. Understand the concept of allpass filters and know how to identify an allpass filter from its difference equation, transfer function, or pole-zero plot.
14. Know how to use various representations such as impulse response, frequency response, difference equations, transfer functions, or pole-zero plots to identify a filter or analyze filter performance.

## 6.1 Frequency Response and Filter Characteristics

In the frequency domain, a filter is described by its frequency response (spectrum) or the **gain** (magnitude) and phase of the filter transfer function against frequency. The frequency response is a very useful way of describing digital filters that tailor the frequency content of a signal in a desired fashion. The range of frequencies over which the filter gain is a significant fraction of its peak gain defines the **passband** of the filter. The range of frequencies for which the gain is insignificant defines the **stopband**. The band-edge frequencies are called the **cutoff frequencies**. An ideal filter shows perfect transmission in the passband and perfect rejection (zero gain) in the stopband. Perfect transmission implies a transfer function with constant gain and linear phase over the passband. The transfer function of such a filter may be described by

$$H(F) = Ke^{-j2\pi nF} \quad \text{(over the passband)}$$

Constant gain ensures that the output is an amplitude-scaled replica of the input. Linear phase ensures that the output is a time-shifted replica of the input. If the gain is not constant over the required frequency range, we have **amplitude distortion**. If the phase shift is not linear with frequency, we have **phase distortion** as the signal undergoes different delays for different frequencies. Traditional filters are often classified as lowpass (blocking high frequencies), highpass (blocking low frequencies), bandpass (transmitting a range of frequencies), or bandstop (blocking a range of frequencies). Of course, a filter may not correspond to any of these types. We may often get a fair idea of the filter behavior in the frequency domain even without detailed analysis. The dc gain (at $F = 0$ or $\Omega = 0$) and the high-frequency gain (at $F = 0.5$ or $\Omega = \pi$) serve as a useful starting point. For a lowpass filter, for example, whose gain typically decreases with frequency, we should expect the dc gain to exceed the high-frequency gain. Other clues are provided by the coefficients (and form) of the difference equation, the transfer function, and the impulse response. The more complex the filter (expression), the less likely it is for us to discern its nature by using only a few simple measures. In the frequency domain, performance measures of digital filters are based on the gain (magnitude), phase delay, and group delay.

### 6.1.1 Gain

The gain is given by the magnitude $|H(F)|$ of the transfer function. The dc gain (at $F = 0$ or $\Omega = 0$) and the high-frequency gain (at $F = 0.5$ or $\Omega = \pi$) are easily found from the filter impulse response, difference equation, or transfer function, as summarized in the following review panel.

---

**Finding the DC Gain and High-Frequency Gain**

**From $H(z)$:** Evaluate $H(z)$ at $z = 1$ (for dc gain) and $z = -1$ (for high-frequency gain).
**From $H(F)$ or $H(\Omega)$:** Evaluate $H(F)$ at $F = 0$ and $F = 0.5$, or evaluate $H(\Omega)$ at $\Omega = 0$ and $\Omega = \pi$.
**From impulse response:** Evaluate $\sum h[n] = H(0)$ and $\sum(-1)^n h[n] = H(F)|_{F=0.5} = H(\Omega)|_{\Omega=\pi}$.
**From difference equation:** For dc gain, take the ratio of the sum of the RHS and LHS coefficients.

For high-frequency gain, reverse the sign of alternate coefficients and then take the ratio of the sum.

---

### 6.1.2 Phase Delay and Group Delay

The **phase delay** and **group delay** of a digital filter are defined by

$$t_p(F_0) = -\frac{1}{2\pi}\frac{H(F_0)}{F_0} \quad \text{(phase delay)} \qquad t_g(F) = -\frac{1}{2\pi}\frac{dH(F)}{dF} \quad \text{(group delay)} \tag{6.1}$$

For a linear-phase filter (whose phase varies linearly with frequency), both the phase delay and the group delay are constant.

### 6.1.3 Minimum-Phase

Consider a system described in factored form by

$$H(z) = K\frac{(z - z_1)(z - z_2)(z - z_3)\cdots(z - z_m)}{(z - p_1)(z - p_2)(z - p_3)\cdots(z - p_n)} \tag{6.2}$$

If we replace $K$ by $-K$, or a factor $(z - \alpha)$ by $(\frac{1}{z} - \alpha)$ or by $(1 - \alpha z)$, the magnitude of $|H_p(F)|$ remains unchanged, and only the phase is affected. If $K > 0$ and all the poles and zeros of $H(z)$ lie inside the unit circle, $H(z)$ is stable and defines a **minimum-phase system**. It shows the *smallest group delay* (and the *smallest deviation from zero phase*) at every frequency among all systems with the same magnitude response. A stable system is called **mixed phase** if some of its zeros lie outside the unit circle and **maximum phase** if all its zeros lie outside the unit circle. *Of all stable systems with the same magnitude response, there is only one minimum-phase system.*

**Example 6.1**

**The Minimum-Phase Concept**

Consider the transfer function of the following systems:

$$H_1(z) = \frac{(z - \frac{1}{2})(z - \frac{1}{4})}{(z - \frac{1}{3})(z - \frac{1}{5})} \qquad H_2(z) = \frac{(1 - \frac{1}{2}z)(z - \frac{1}{4})}{(z - \frac{1}{3})(z - \frac{1}{5})} \qquad H_3(z) = \frac{(1 - \frac{1}{2}z)(1 - \frac{1}{4}z)}{(z - \frac{1}{3})(z - \frac{1}{5})}$$

Each system has poles inside the unit circle and is thus stable. All systems have the same magnitude. Their phase and delay are different, as shown Figure E6.1.

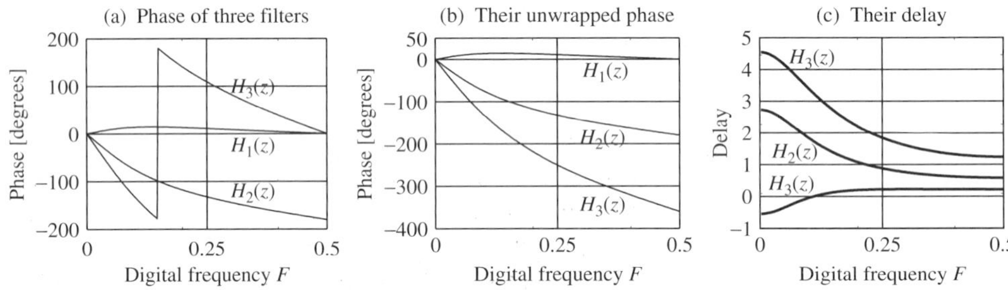


**FIGURE E.6.1** Response of the systems for Example 6.1

The phase response confirms that $H_1(z)$ is a minimum-phase system with no zeros outside the unit circle, $H_2(z)$ is a mixed-phase system with one zero outside the unit circle, and $H_3(z)$ is a maximum-phase system with all of its zeros outside the unit circle.

---

### 6.1.4 Minimum-Phase Filters from the Magnitude Spectrum

The design of many digital filters is often based on a specified magnitude response $|H_p(F)|$. The phase response is then selected to ensure a causal, stable system. The transfer function of such a system is unique and may be found by writing its magnitude squared function $|H_p(F)|^2$ as

$$|H_p(F)|^2 = H_p(F)H_p(-F) = H(z)H(1/z)|_{z\to\exp(j2\pi F)} \tag{6.3}$$

From $|H_p(F)|^2$, we can reconstruct $H_T(z) = H(z)H(1/z)$, which displays conjugate reciprocal symmetry. For every root $r_k$, there is a root at $1/r_k^*$. We thus select only the roots lying inside the unit circle to extract the minimum-phase transfer function $H(z)$. The following example illustrates the process.

**Example 6.2**

**Finding the Minimum-Phase Filter**

Find the minimum-phase transfer function $H(z)$ corresponding to $|H_p\Omega)|^2 = \dfrac{5+4\cos\Omega}{17+8\cos\Omega}$.

We use Euler's relation to give $|H_p(\Omega)|^2 = \dfrac{5+2e^{j\Omega}+2e^{-j\Omega}}{17+4e^{j\Omega}+4e^{-j\Omega}}$.

Upon substituting $e^{j\Omega}\to z$, we obtain $H_T(z)$ as

$$H_T(z) = H(z)H(1/z) = \frac{5+2z+2/z}{17+4z+4/z} = \frac{2z^2+5z+2}{4z^2+17z+4} = \frac{(2z+1)(z+2)}{(4z+1)(z+4)}$$

To extract $H(z)$, we pick the roots of $H_T(z)$ that correspond to $|z|<1$. This yields

$$H(z) = \frac{(2z+1)}{(4z+1)} = 0.5\frac{(z+0.5)}{(z+0.25)}$$

We find that $|H(z)|_{z=1} = |H_p(\Omega)|_{\Omega=0} = 0.6$, implying identical dc gains.

---

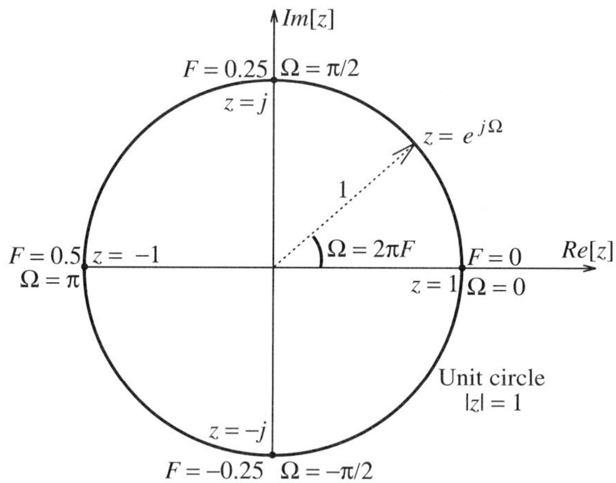


**FIGURE 6.1** Relating the variables $z$, $F$, and $\Omega$ through the unit circle in the $z$-plane. The angle in radians equals the digital frequency $\Omega$. On the unit circle, $F = 0$ corresponds to $z = 1$ while $F = 0.5$ corresponds to $z = -1$

### 6.1.5 The Frequency Response: A Graphical View

The frequency response of an LTI system may also be found by evaluating its transfer function $H(z)$ for values of $z = e^{j2\pi F} = e^{j\Omega}$ (on the unit circle). The quantity $\Omega = 2\pi F$ describes the angular orientation. Figure 6.1 shows how the values of $z$, $\Omega$, and $F$ are related on the unit circle.

**The Frequency Response is the *z*-Transform Evaluated on the Unit Circle**

It corresponds to the DTFT:

$$H(F) = H(z)|_{z=e^{j2\pi F}} \quad \text{or} \quad H(\Omega) = H(z)|_{z=e^{j\Omega}}$$

The factored form or pole-zero plot of $H(z)$ is quite useful if we want a qualitative picture of its frequency response $H(F)$ or $H(\Omega)$. Consider the stable transfer function $H(z)$ given in factored form by

$$H(z) = \frac{8(z-1)}{(z-0.6-j0.6)(z-0.6+j0.6)} \tag{6.4}$$

Its frequency response $H(\Omega)$ and magnitude $|H(\Omega)|$ at $\Omega = \Omega_0$ are given by

$$H(\Omega_0) = \frac{8(e^{j\Omega_0}-1)}{(e^{j\Omega_0}-0.6-j0.6)(e^{j\Omega_0}-0.6+j0.6]} = \frac{8\mathbf{N}_1}{\mathbf{D}_1\mathbf{D}_2} \tag{6.5}$$

$$|H(\Omega_0)| = 8\frac{|\mathbf{N}_1|}{|\mathbf{D}_1||\mathbf{D}_2|} = (\text{gain factor})\frac{\text{PRODUCT OF distances from zeros}}{\text{PRODUCT OF distances from poles}} \tag{6.6}$$

Analytically, the magnitude $|H(\Omega_0)|$ is the ratio of the magnitudes of each term. Graphically, the complex terms may be viewed in the $z$-plane as vectors $\mathbf{N}_1$, $\mathbf{D}_1$, and $\mathbf{D}_2$, directed from each pole or zero location to the location $\Omega = \Omega_0$ on the unit circle corresponding to $z = e^{j\Omega_0}$. The gain factor times the ratio of the vector magnitudes (the product of distances

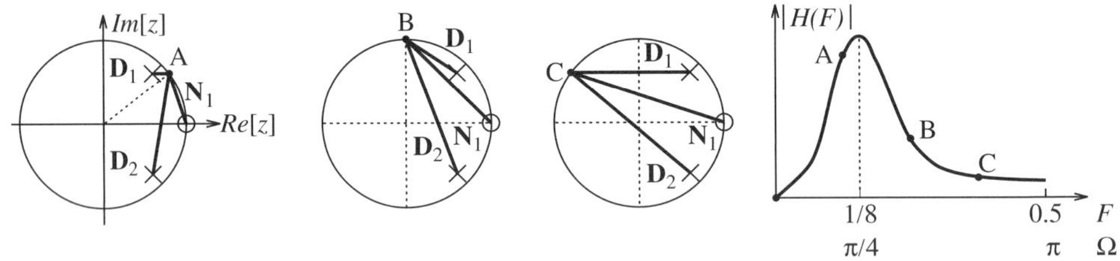


**FIGURE 6.2** Graphical interpretation of the frequency response. At the three frequencies indicated by the points A, B and C on the unit circle (in the first three panels), the gain is estimated by computing the ratio of the products of the distances to the zeros and products of distances to the poles. The corresponding values of the gain are labeled on the gain plot

from the zeros divided by the product of distances from the poles) yields $|H(\Omega_0)|$, which is the magnitude at $\Omega = \Omega_0$. The difference in the angles yields the phase at $\Omega = \Omega_0$. The vectors and the corresponding magnitude spectrum are sketched for several values of $\Omega$ in Figure 6.2.

A graphical evaluation can yield exact results but is much more suited to obtaining a qualitative estimate of the magnitude response. We observe how the vector ratio $\mathbf{N}_1/\mathbf{D}_1\mathbf{D}_2$ influences the magnitude as $\Omega$ is increased from $\Omega = 0$ to $\Omega = \pi$. For our example, at $\Omega = 0$, the vector $\mathbf{N}_1$ is zero, and the magnitude is zero. For $0 < \Omega < \pi/4$ (point A), both $|\mathbf{N}_1|$ and $|\mathbf{D}_2|$ increase, but $|\mathbf{D}_1|$ decreases. Overall, the response is small but increasing. At $\Omega = \pi/4$, the vector $\mathbf{D}_1$ attains its smallest length, and we obtain a peak in the response. For $\frac{\pi}{4} < \Omega < \pi$ (points B and C), $|\mathbf{N}_1|$ and $|\mathbf{D}_1|$ are of nearly equal length, while $|\mathbf{D}_2|$ is increasing. The magnitude is thus decreasing. The form of this response is typical of a bandpass filter.

### 6.1.6 The Rubber Sheet Analogy

If we imagine the $z$-plane as a rubber sheet, tacked down at the zeros of $H(z)$ and poked up to an infinite height at the pole locations—the curved surface of the rubber sheet approximates the magnitude of $H(z)$ for any value of $z$, as illustrated in Figure 6.3. The poles tend to poke

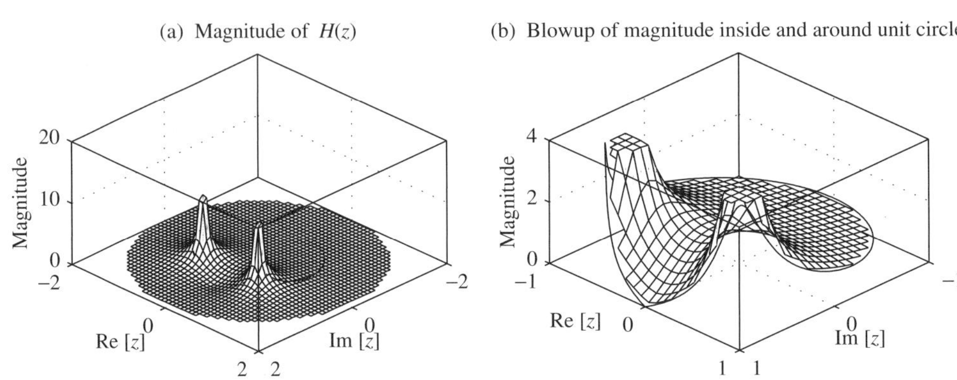


**FIGURE 6.3** A plot of the magnitude of $H(z) = 8(z-1)/(z^2 - 1.2z + 0.72)$ in the $z$-plane

up the surface, and the zeros try to pull it down. The slice around the unit circle ($|z| = 1$) approximates the frequency response $H(\Omega)$.

**Example 6.3**

### Filters and Pole-Zero Plots

Identify the filter types corresponding to the pole-zero plots of Figure E6.3.

FIGURE E.6.3
Pole-zero plots for Example 6.3

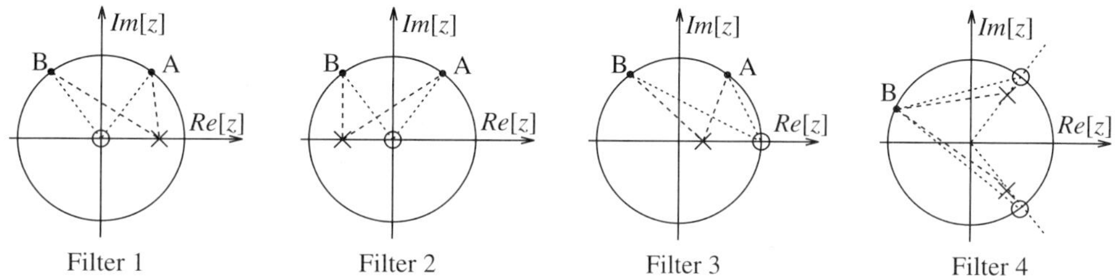


**(a)** For filter 1, the vector length of the numerator is always unity, but the vector length of the denominator keeps increasing as we increase frequency (the points A and B, for example). The magnitude (ratio of the numerator and denominator lengths) thus decreases with frequency and corresponds to a lowpass filter.

**(b)** For filter 2, the vector length of the numerator is always unity, but the vector length of the denominator keeps decreasing as we increase the frequency (the points A and B, for example). The magnitude increases with frequency and corresponds to a highpass filter.

**(c)** For filter 3, the magnitude is zero at $\Omega = 0$. As we increase the frequency (the points A and B, for example), the ratio of the vector lengths of the numerator and denominator increases, and this also corresponds to a highpass filter.

**(d)** For filter 4, the magnitude is zero at the zero location $\Omega = \Omega_0$. At any other frequency (the point B, for example), the ratio of the vector lengths of the numerator and denominator are almost equal and result in almost constant gain. This describes a bandstop filter.

## 6.2 FIR Filters and Linear Phase

The DTFT of a filter whose impulse response is symmetric about the origin is purely real or purely imaginary. The phase of such a filter is thus piecewise constant. If the symmetric sequence is shifted (to make it causal, for example), its phase is augmented by a linear-phase term and becomes piecewise linear. A filter whose impulse response is symmetric about its midpoint is termed a (generalized) *linear-phase filter*. Linear phase is important in filter design because it results in a pure time delay with no amplitude distortion. The transfer function of a causal linear-phase filter may be written as $H(F) = A(F)e^{j2\pi\alpha F}$ (for even symmetry) or $H(F) = jA(F)e^{j2\pi\alpha F} = A(F)e^{j(2\pi\alpha F+\frac{\pi}{2})}$ (for odd symmetry), where the real quantity $A(F)$ is the amplitude spectrum and $\alpha$ is the (integer or half-integer) index corresponding to the midpoint of its impulse response $h[n]$. The easiest way to obtain this form is to first set up an expression for $H(F)$, then extract the factor $e^{j2\pi\alpha F}$ from $H(F)$, and finally simplify using Euler's relation.

### 6.2.1 Pole-Zero Patterns of Linear-Phase Filters

Linear phase plays an important role in the design of digital filters because it results in a constant delay with no amplitude distortion. An FIR filter whose impulse response sequence is symmetric about the midpoint is endowed with linear phase and constant delay. The pole and zero locations of such a sequence cannot be arbitrary. The poles must lie at the origin if the sequence $h[n]$ is to be of finite length. Sequences that are symmetric about the origin also require $h[n] = \pm h[-n]$, and thus, $H(z) = \pm H(1/z)$. The zeros of a linear-phase sequence must occur in reciprocal pairs (and conjugate pairs if complex to ensure real coefficients) and exhibit what is called **conjugate reciprocal symmetry**. This is illustrated in Figure 6.4.

Each complex zero forms part of a quadruple, because each is paired with its conjugate and its reciprocal with the following exceptions. Zeros at $z = 1$ or at $z = -1$ can occur singly, because they form their own reciprocal and their own conjugate. Zeros on the real axis must occur in pairs with their reciprocals (with no conjugation required). Zeros on the unit circle (except at $z = \pm 1$) must occur in pairs with their conjugates (which also form their reciprocals). If there are no zeros at $z = 1$, a linear-phase sequence is always even symmetric about its midpoint. For odd symmetry about the midpoint, there must be *an odd number of zeros at* $z = 1$.

The frequency response of a linear-phase filter may be written as $H(F) = A(F)e^{j2\pi\alpha F}$ (for even symmetry) or $H(F) = jA(F)e^{j2\pi\alpha F} = A(F)e^{j(2\pi\alpha F+\frac{\pi}{2})}$ (for odd symmetry), where $A(F)$ is real.

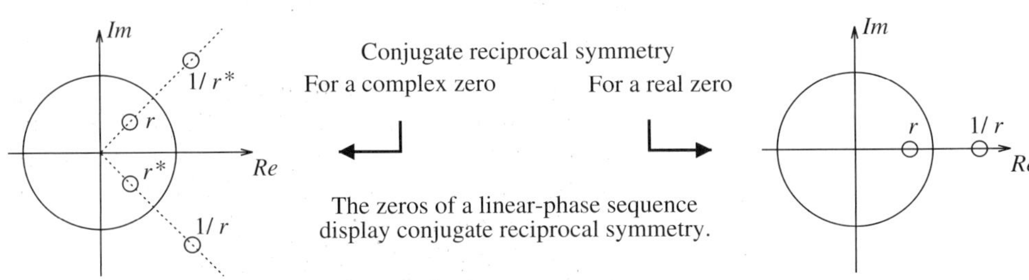


**FIGURE 6.4** Illustrating conjugate reciprocal symmetry. Each zero not on the real axis and not on the unit circle forms part of a quadruple, being paired with its reciprocal and its conjugate. A zero on the real axis but not on the unit circle need not be conjugated and is paired with just its reciprocal. A zero on the unit circle (but not at $z = 1$ or $z = -1$) forms its own reciprocal and is paired with just its conjugate. A zero at $z = 1$ or $z = -1$ can occur singly, because it forms its own reciprocal and its own conjugate

### Characteristics of a Linear-Phase Filter

1. The impulse response $h[n]$ is symmetric about its midpoint.
2. $H(F) = A(F)e^{j2\pi\alpha F}$ (for even symmetric $h[n]$) and $H(F) = jA(F)e^{j2\pi\alpha F}$ (for odd symmetric $h[n]$).
3. All poles are at $z = 0$ (for finite length).
4. Zeros occur in conjugate reciprocal quadruples, in general. Zeros on unit circle occur in conjugate pairs.
   Zeros on real axis occur in reciprocal pairs. Zeros at $z = 1$ or $z = -1$ can occur singly.
5. Odd symmetry about midpoint if an odd number of zeros at $z = 1$ (even symmetry otherwise).
6. If $h[n] = \pm h[-n]$, then $H(z) = \pm H(\frac{1}{z})$ (symmetry about the origin $n = 0$).

**Example 6.4**

### Linear-Phase Filters

**(a)** Does $H(z) = 1 + 2z^{-1} + 2z^{-2} + z^{-3}$ describe a linear-phase filter?
We express $H(z)$ as a ratio of polynomials in $z$ to get

$$H(z) = \frac{z^3 + 2z^2 + 2z + 1}{z^3}$$

All of its poles are at $z = 0$. Its zeros are at $z = -1$ and $z = -0.5 \pm j0.866$ and are consistent with a linear-phase filter, because the real zero at $z = -1$ can occur singly and the complex conjugate pair of zeros lie on the unit circle.

Since $H(z) \neq \pm H(1/z)$, the impulse response $h[n]$ cannot be symmetric about the origin $n = 0$ (even though it must be symmetric about its midpoint).

We could reach the same conclusions by recognizing that $h[n] = \{\overset{\Downarrow}{1}, 2, 2, 1\}$ describes a linear-phase sequence with even symmetry about its midpoint $n = 1.5$.

**(b)** Let $h[n] = \delta[n+2] + 4.25\delta[n] + \delta[n-2]$. Sketch the pole-zero plot of $H(z)$.
We find $H(z) = z^2 + 4.25 + z^{-2}$. Since $h[n]$ is even symmetric about $n = 0$ with $h[n] = h[-n]$, we must have $H(z) = H(1/z)$. This is, in fact, the case.

We express $H(z)$ as a ratio of polynomials in factored form to get

$$H(z) = \frac{z^4 + 4.25z^2 + 1}{z^2} = \frac{(z + j0.5)(z - j0.5)(z + j2)(z - j2)}{z^2}$$

The pole-zero plot is shown in Figure E6.4 (a). All of its poles are at $z = 0$. The four zeros at $z = j0.5$, $z = -j2$, $z = -j0.5$, and $z = j2$ display conjugate reciprocal symmetry.

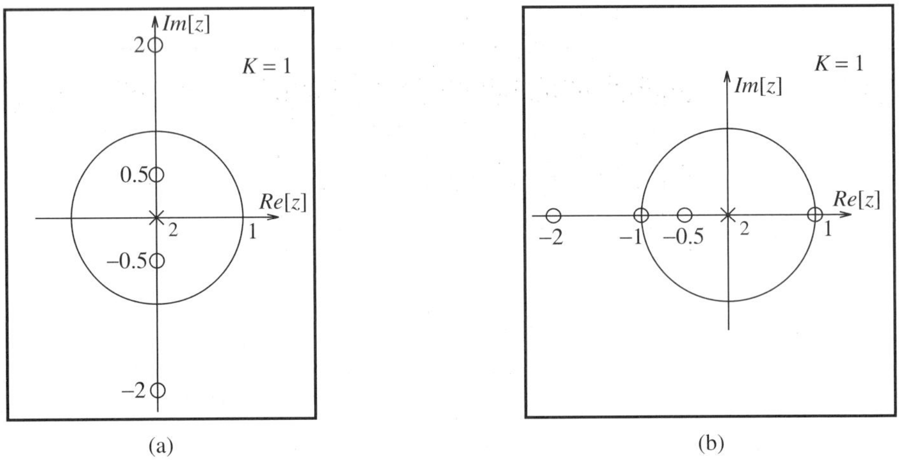


**FIGURE E.6.4** Pole-zero plots of the filters for Example 6.4

**(c)** Sketch the pole-zero plot for $H(z) = z^2 + 2.5z - 2.5z^{-1} - z^{-2}$. Is this a linear-phase filter?

We note that $H(z) = -H(1/z)$. This means that $h[n] = -h[-n]$. In fact, $h[n] = \{1, \ 2.5, \ \overset{\Downarrow}{0}, \ -2.5, \ -1\}$. This describes a linear-phase filter.

With $H(z)$ described as a ratio of polynomials in factored form, we get

$$H(z) = \frac{(z-1)(z+1)(z+0.5)(z+2)}{z^2}$$

The pole-zero plot is shown in Figure E6.4 (b). All its poles are at $z = 0$. There is a pair of reciprocal zeros at $z = -0.5$ and $z = -2$. The two zeros at $z = -1$ and $z = 1$ occur singly.

### 6.2.2 Types of Linear-Phase Sequences

Linear-phase sequences fall into four types. A type 1 sequence has even symmetry and odd length. A type 2 sequence has even symmetry and even length. A type 3 sequence has odd symmetry and odd length. A type 4 sequence has odd symmetry and even length.

To identify the type of sequence from its pole-zero plot, all we need to do is check for the presence of zeros at $z = \pm 1$ and count their *number*. A type 2 sequence must have an *odd number of zeros* at $z = -1$, a type 3 sequence must have an *odd number of zeros* at $z = -1$ and $z = 1$, and a type 4 sequence must have *an odd number of zeros* at $z = 1$. The number of other zeros, if present (at $z = 1$ for type 1 and type 2, or $z = -1$ for type 1 or type 4), must be even. The reason is simple. If we exclude $z = \pm 1$, the remaining zeros occur in pairs or quadruples and always yield a type 1 sequence with *odd length and even symmetry*. Including a zero at $z = -1$ increases the length by one without changing the symmetry. Including a zero at $z = -1$ increases the length by 1 and also changes the symmetry. Only an odd number of zeros at $z = 1$ results in an odd symmetric sequence. These ideas are illustrated in Figure 6.5.

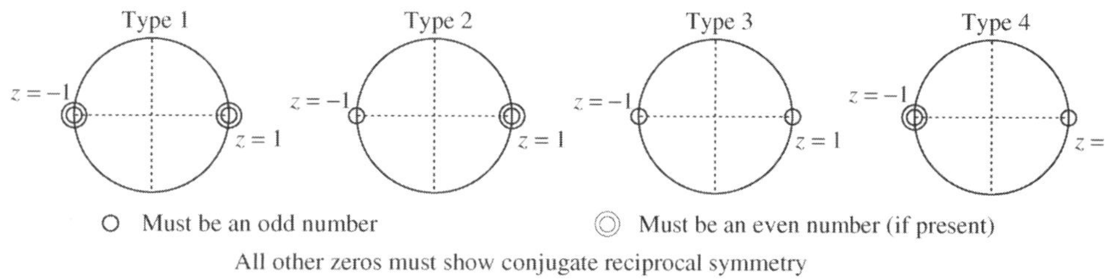


**FIGURE 6.5** Identifying the sequence type from its zeros at $z = \pm 1$. Type 2 sequences *require* an odd number of zeros at $z = -1$, type 4 sequences *require* an odd number of zeros at $z = 1$, and type 3 sequences *require* an odd number of zeros at both $z = 1$ and $z = -1$. The number of zeros (if present) at the locations $z = \pm 1$ (other than the ones required) must be even. Naturally, any zeros elsewhere must satisfy conjugate reciprocal symmetry

### How to Identify the Type of a Linear-Phase Sequence

**From the sequence** $x[n]$: Check the length and the symmetry to identify the type.
**From the pole-zero plot:** Count the number of zeros at $z = \pm 1$.

**Type 1:** Even number of zeros at $z = -1$ (if present) and at $z = 1$ (if present)
**Type 2:** Odd number of zeros at $z = -1$ (and even number of zeros at $z = 1$, if present)
**Type 3:** Odd number of zeros at $z = 1$ and odd number of zeros at $z = -1$
**Type 4:** Odd number of zeros at $z = 1$ (and even number of zeros at $z = -1$, if present)

**Example 6.5**

### Identifying Linear-Phase Sequences

**(a)** Find all of the zero locations of a type 1 sequence (assuming the smallest length) if it is known that there is a zero at $z = 0.5e^{j\pi/3}$ and a zero at $z = 1$.

Due to conjugate reciprocal symmetry, the zero at $z = 0.5e^{j\pi/3}$ implies the quadruple zeros

$$z = 0.5e^{j\pi/3} \qquad z = 0.5e^{-j\pi/3} \qquad z = 2e^{j\pi/3} \qquad z = 2e^{-j\pi/3}$$

For a type 1 sequence, the number of zeros at $z = 1$ must be even. So, there must be another zero at $z = 1$. Thus, we have a total of six zeros.

**(b)** Find all of the zero locations of a type 2 sequence (assuming the smallest length) if it is known that there is a zero at $z = 0.5e^{j\pi/3}$ and a zero at $z = 1$.

Due to conjugate reciprocal symmetry, the zero at $z = 0.5e^{j\pi/3}$ implies the quadruple zeros

$$z = 0.5e^{j\pi/3} \qquad z = 0.5e^{-j\pi/3} \qquad z = 2e^{j\pi/3} \qquad z = 2e^{-j\pi/3}$$

A type 2 sequence is even symmetric and requires an even number of zeros at $z = 1$. So, there must be a second zero at $z = -1$ to give six zeros. But this results in a sequence of odd length. For a type 2 sequence, the length is even. This requires a zero at $z = -1$. Thus, we have a total of seven zeros.

**(c)** Find the transfer function and impulse response of a causal type 3 linear-phase filter (assuming the smallest length and smallest delay) if it is known that there is a zero at $z = j$ and two zeros at $z = 1$.

The zero at $z = j$ must be paired with its conjugate (and reciprocal) $z = -j$. A type 3 sequence requires an odd number of zeros at $z = 1$ and $z = -1$. So, the minimum number of zeros required is one zero at $z = -1$ and three zeros at $z = 1$ (with two already present).

The transfer function has the form

$$H(z) = (z + j)(z - j)(z + 1)(z - 1)^3 = z^6 - 2z^5 + z^4 - z^2 + 2z - 1$$

The transfer function of the causal filter with the minimum delay is

$$H_C(z) = 1 - 2z^{-1} + z^{-2} - z^{-4} + 2z^{-5} - z^{-6} \qquad h_C[n] = \{\overset{\Downarrow}{1}, -2, 1, 0, -1, 2, -1\}$$

### 6.2.3 Averaging Filters

As an example of a linear-phase filter, we consider a causal $N$-point averaging filter whose system equation and impulse response are given by

$$y[n] = \frac{1}{N}(x[n] + x[n-1] + \cdots + x[n-(N-1)]) \qquad h[n] = \frac{1}{N}\{\underbrace{\overset{\Downarrow}{1}, 1, \ldots, 1, 1}_{N \text{ ones}}\} \tag{6.7}$$

Its transfer function then becomes

$$H(F) = \frac{1}{N}\sum_{k=0}^{N-1} h[k]e^{-j2\pi kF} = \frac{1}{N}\sum_{k=0}^{N-1}(e^{-j2\pi F})^k = \frac{1}{N}\left(\frac{1 - e^{-j2\pi NF}}{1 - e^{-j2\pi F}}\right)$$

We extract the factors $e^{-j\pi NF}$ and $e^{-j\pi F}$ from the numerator and denominator respectively to give

$$H(F) = \frac{1}{N}\left(\frac{e^{-j\pi NF}}{e^{-j\pi F}}\right)\left(\frac{e^{-j\pi NF} - e^{-j\pi NF}}{e^{-j\pi F} - e^{-j\pi F}}\right) = \frac{1}{N}e^{-j\pi(N-1)F}\left(\frac{j2\sin(\pi NF)}{j2\sin(\pi F)}\right)$$

This simplifies to

$$H(F) = e^{-j\pi(N-1)F}\frac{\sin(\pi NF)}{N\sin(\pi F)} = e^{-j\pi(N-1)F}\frac{\operatorname{sinc}(NF)}{\operatorname{sinc}(F)} = e^{-j\pi(N-1)F}A(F) \tag{6.8}$$

The term $A(F)$ represents the amplitude and the term $e^{-j\pi(N-1)F}$ describes a phase that varies linearly with frequency. The value of $H(F)$ equals unity at $F = 0$ for any length $N$. Also, $H(F) = 0$ when $F$ is a multiple of $\frac{1}{N}$ and there are a total of $N$ zeros in the principal range $0.5 < F \le 0.5$. At $F = 0.5$, the value of $H(F)$ equals zero if $N$ is even. For odd $N$, however, the value of $H(F)$ at $F = 0.5$ equals $(-1)^{(N-1)/2}$ (i.e., $+1$ if $\frac{N-1}{2}$ is even or $-1$ if $\frac{N-1}{2}$ is odd). As $|F|$ increases from 0 to 0.5, the value of $|\sin(\pi F)|$ increases from 0 to 1. Consequently, the magnitude spectrum shows a peak of unity at $F = 0$ and decreasing sidelobes of width $\frac{1}{N}$ on either side of the origin. If we look at the amplitude term $A(F)$ (not its absolute value), we find that it is even symmetric about $F = 0.5$ if $N$ is odd but odd symmetric about $F = 0.5$ if $N$ is even.

For an even symmetric sequence $w_d[n]$ of length $M = 2N + 1$, centered at the origin, whose coefficients are all unity, we have

$$w_d[n] = \{\underbrace{1,\ 1,\ \ldots,\ 1,\ 1}_{N \text{ ones}},\ \overset{\Downarrow}{1},\ \underbrace{1,\ 1,\ \ldots,\ 1,\ 1}_{N \text{ ones}}\} \tag{6.9}$$

Its frequency response may be written as

$$W_D(F) = \frac{\sin(\pi M F)}{\sin(\pi F)} = \frac{M\operatorname{sinc}(MF)}{\operatorname{sinc}(F)}, \qquad M = 2N + 1 \tag{6.10}$$

The quantity $W_D(F)$ is called the **Dirichlet kernel** or the **aliased sinc** function.

### 6.2.4 Zeros of Averaging Filters

Consider a causal $N$-point averaging filter whose system equation and impulse response are given by

$$y[n] = \frac{1}{N}(x[n]+x[n-1]+\ldots+x[n-(N-1)]) \qquad h[n] = \frac{1}{N}\{\underbrace{\overset{\Downarrow}{1},\ 1,\ \ldots,\ 1,\ 1}_{N \text{ ones}}\} \tag{6.11}$$

Its transfer function $H(z)$ then becomes

$$H(z) = \frac{1}{N}\sum_{k=0}^{N-1} h[k]z^{-k} = \frac{1}{N}\sum_{k=0}^{N-1}(z^{-1})^k = \frac{1}{N}\left(\frac{1 - z^{-N}}{1 - z^{-1}}\right) = \frac{1}{N}\left[\frac{z^N - 1}{z^{N-1}(z - 1)}\right]$$

The $N$ roots of $z^N - 1 = 0$ lie on the unit circle with an angular spacing of $\frac{2\pi}{N}$ radians starting at $\Omega = 0$ (corresponding to $z = 1$). However, the zero at $z = 1$ is cancelled by the pole at $z = 1$. So, we have $N - 1$ zeros on the unit circle and $N - 1$ poles at the origin ($z = 0$). The zeros occur in conjugate pairs and if the length $N$ is even, there is also a zero at $\Omega = \pi$ (corresponding to $z = -1$).

**Example 6.6** **Frequency Response and Filter Characteristics**

**(a)** Identify the pole and zero locations of a causal 10-point averaging filter. What is the dc gain and high-frequency gain of this filter? At what frequencies does the gain go to zero?

The impulse response and transfer function of a 10-point averaging filter is

$$h[n] = \{\overset{\Downarrow}{0.1},\ 0.1,\ 0.1,\ 0.1,\ 0.1,\ 0.1,\ 0.1,\ 0.1,\ 0.1,\ 0.1\} \qquad H(z) = \frac{0.1(z^{10} - 1)}{z^9(z - 1)}$$

There are nine poles at the origin $z = 0$. There are nine zeros with an angular spacing of $\frac{2\pi}{10}$ rad located on the unit circle at $z = e^{jk\pi/5},\ k = 1, 2, \ldots, 9$. The dc gain of the filter is $|H(0)| = |\sum h[n]| = 1$. The high-frequency gain is $|H(F)|_{F=0.5} = |\sum(-1)^n h[n]| = 0$. Over the principal range, the gain is zero at the frequencies $F_k = 0.1k,\ k = \pm 1, \pm 2, \ldots, 5$ and the magnitude response shows a central lobe with four successively decaying sidelobes on either side.

**(b)** Consider the 3-point averaging filter, $y[n] = \frac{1}{3}\{x[n-1] + x[n] + x[n+1]\}$. The filter replaces each input value $x[n]$ by an average of itself and its two neighbors. Its impulse response $h_1[n]$ is simply $h_1[n] = \{\frac{1}{3}, \overset{\Downarrow}{\frac{1}{3}}, \frac{1}{3}\}$. The frequency response is given by

$$H_1(F) = \sum_{n=-1}^{1} h[n]e^{-j2\pi Fn} = \frac{1}{3}\left[e^{j2\pi F} + 1 + e^{-j2\pi F}\right] = \frac{1}{3}[1 + 2\cos(2\pi F)]$$

The magnitude $|H_1(F)|$ decreases until $F = \frac{1}{3}$ (when $H_1(F) = 0$) and then increases to $\frac{1}{3}$ at $F = \frac{1}{2}$, as shown in Figure E6.6(a). This filter thus does a poor job of smoothing past $F = \frac{1}{3}$.

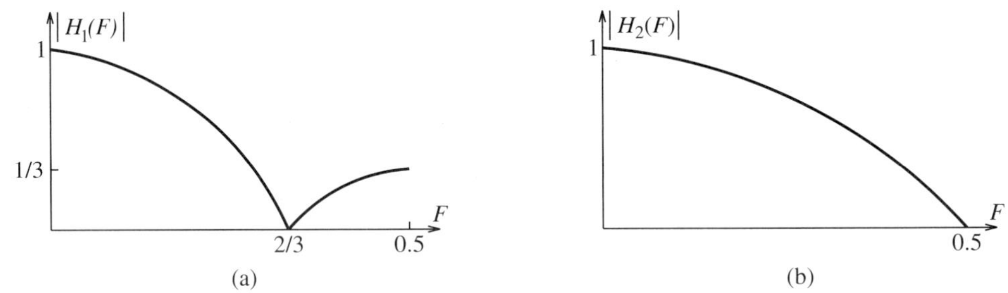


**FIGURE E.6.6** Frequency response of the filters for Example 6.6

**(c)** Consider the tapered 3-point moving-average filter $h_2[n] = \{\frac{1}{4}, \overset{\Downarrow}{\frac{1}{2}}, \frac{1}{4}\}$. Its frequency response is given by

$$H_2(F) = \sum_{n=-1}^{1} h[n]e^{-j2\pi Fn} = \frac{1}{4}e^{j2\pi F} + \frac{1}{2} + \frac{1}{4}e^{-j2\pi F} = \frac{1}{2} + \frac{1}{2}\cos(2\pi F)$$

Figure 6.6(b) shows that $|H_2(F)|$ decreases monotonically to zero at $F = 0.5$ and shows a much better smoothing performance. This lowpass filter actually describes a 3-point von Hann (or Hanning) smoothing window.

**(d)** Consider the differencing filter described by $y[n] = x[n] - x[n-1]$. Its impulse response may be written as $h[n] = \delta[n] - \delta[n-1] = \{1, -1\}$. This is actually odd symmetric about its midpoint ($n = 0.5$). Its frequency response (in the $\Omega$-form) is given by

$$H(\Omega) = 1 - e^{j\Omega} = e^{-j\Omega/2}(e^{j\Omega/2} - e^{-j\Omega/2}) = 2j\sin(0.5\Omega)e^{-j\Omega/2}$$

Its phase is $\phi(\Omega) = \frac{\pi}{2} - \frac{\Omega}{2}$ and shows a linear variation with $\Omega$. Its amplitude $A(\Omega) = 2\sin(\Omega/2)$ increases from zero at $\Omega = 0$ to two at $\Omega = \pi$. In other words, the difference operator enhances high frequencies and acts as a highpass filter.

### 6.2.5 FIR Comb Filters

A **comb filter** has a magnitude response that looks much like the rounded teeth of a comb. One such comb filter is described by the transfer function

$$H(z) = 1 - z^{-N} \qquad h[n] = \{\overset{\Downarrow}{1}, \underbrace{0,\ 0,\ \ldots,\ 0,\ 0}_{N-1 \text{ zeros}}, -1\} \tag{6.12}$$

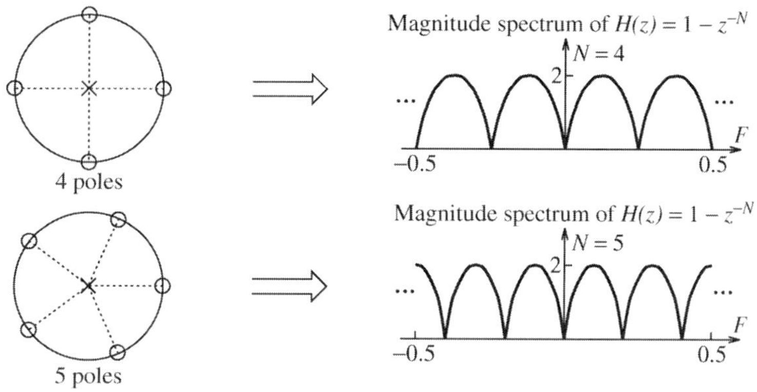


**FIGURE 6.6** Pole-zero plot and frequency response of the comb filter $H(z) = 1 - z^{-N}$. The number of lobes in one period corresponds to the number of zeros. The dc gain is always zero, but the high-frequency gain (at $F = 0.5$) is zero only if the filter length is odd (corresponding to an even number of zeros)

This corresponds to the system difference equation $y[n] = x[n] - x[n-N]$, and represents a linear-phase FIR filter whose impulse response $h[n]$ (which is *odd symmetric* about its midpoint) has $N+1$ samples with $h[0] = 1$, $h[N] = -1$. All other coefficients are zero. There is a pole of multiplicity $N$ at the origin, and the zeros lie on a unit circle with locations specified by

$$z^{-N} = e^{-jN\Omega} = 1 = e^{j2\pi k} \qquad \Omega_k = \frac{2k\pi}{N}, \quad k = 0, 1, \ldots, N-1 \tag{6.13}$$

The pole-zero pattern and magnitude spectrum of this filter are shown for two values of $N$ in Figure 6.6.

The zeros of $H(z)$ are uniformly spaced $2\pi/N$ radians apart around the unit circle, starting at $\Omega = 0$. For even $N$, there is also a zero at $\Omega = \pi$. Being an FIR filter, it is always stable for any $N$. Its frequency response is given by

$$H(F) = 1 - e^{-j2\pi FN} \tag{6.14}$$

Note that $H(0)$ always equals 0, but $H(0.5) = 0$ for even $N$ and $H(0.5) = 2$ for odd $N$. The frequency response $H(F)$ looks like a comb with $N$ rounded *teeth* over its principal period $-0.5 \le F \le 0.5$.

A more general form of this comb filter is described by

$$H(z) = 1 - \alpha z^{-N} \qquad h[n] = \{\overset{\Downarrow}{1}, \underbrace{0,\ 0,\ \ldots,\ 0,\ 0}_{N-1 \text{ zeros}}, -\alpha\} \tag{6.15}$$

It has $N$ poles at the origin, and its zeros are uniformly spaced $2\pi/N$ radians apart around a circle of radius $R = \alpha^{1/N}$, starting at $\Omega = 0$. Note that this filter is no longer a linear-phase filter, because its impulse response is not symmetric about its midpoint. Its frequency response $H(F) = 1 - \alpha e^{-j2\pi FN}$ suggests that $H(0) = 1 - \alpha$ for any $N$, and $H(0.5) = 1 - \alpha$ for even $N$ and $H(0.5) = 1 + \alpha$ for odd $N$. Thus, its magnitude varies between $1 - \alpha$ and $1 + \alpha$, as illustrated in Figure 6.7.

**FIGURE 6.7** Frequency response of the comb filter $H(z) = 1 - \alpha z^{-N}$

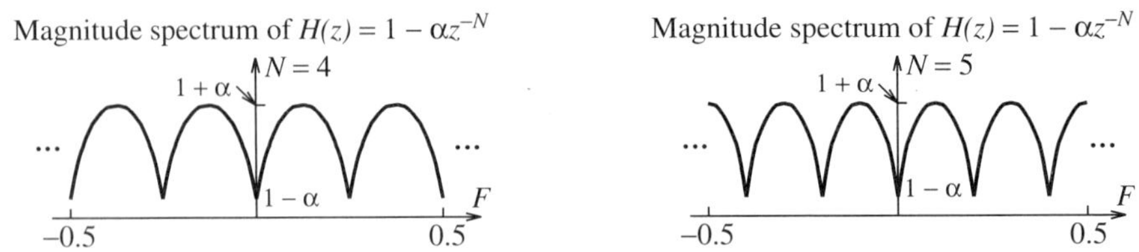


Another FIR comb filter is described by the transfer function

$$H(z) = 1 + z^{-N} \qquad h[n] = \{\overset{\Downarrow}{1},\ \underbrace{0,\ 0,\ \ldots,\ 0,\ 0}_{N-1 \text{ zeros}},\ 1\} \tag{6.16}$$

This corresponds to the system difference equation $y[n] = x[n] + x[n-N]$, and represents a linear-phase FIR filter whose impulse response $h[n]$ (which is *even symmetric* about its midpoint) has $N+1$ samples with $h[0] = h[N] = 1$. All other coefficients are zero. There is a pole of multiplicity $N$ at the origin, and the zero locations are specified by

$$z^{-N} = e^{-jN\Omega} = -1 = e^{j(2k+1)\pi} \qquad \Omega_k = \frac{(2k+1)\pi}{N}, \quad k = 0, 1, \ldots, N-1 \tag{6.17}$$

The pole-zero pattern and magnitude spectrum of this filter are shown for two values of $N$ in Figure 6.8.

The zeros of $H(z)$ are uniformly spaced $2\pi/N$ radians apart around the unit circle, starting at $\Omega = \pi/N$. For odd $N$, there is also a zero at $\Omega = \pi$. Its frequency response is given by

$$H(F) = 1 + e^{-j2\pi FN} \tag{6.18}$$

Note that $H(0) = 2$ for any $N$, $H(0.5) = 2$ for even $N$, and $H(0.5) = 0$ for odd $N$.

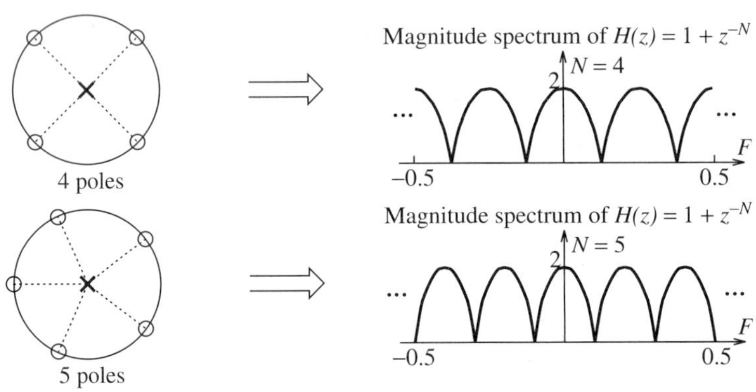


**FIGURE 6.8** Pole-zero plot and frequency response of the comb filter $H(z) = 1 + z^{-N}$. The number of lobes in one period corresponds to the number of zeros. The dc gain is always equal to the peak gain. The high-frequency gain (at $F = 0.5$) is zero if the filter length is even (corresponding to an odd number of zeros)

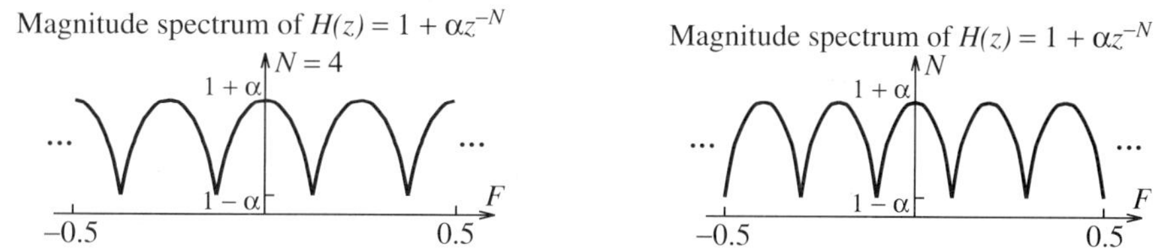


**FIGURE 6.9** Frequency response of the comb filter $H(z) = 1 + \alpha z^{-N}$

A more general form of this comb filter is described by

$$H(z) = 1 + \alpha z^{-N} \qquad h[n] = \{\overset{\Downarrow}{1}, \underbrace{0,\ 0,\ \ldots,\ 0,\ 0}_{N-1 \text{ zeros}}, \alpha\} \tag{6.19}$$

It has $N$ poles at the origin and its zeros are uniformly spaced $2\pi/N$ radians apart around a circle of radius $R = \alpha^{1/N}$, starting at $\Omega = \pi/N$. Note that this filter is no longer a linear-phase filter because its impulse response is not symmetric about its midpoint. Its frequency response $H(F) = 1 + \alpha e^{-j2\pi FN}$ suggests that $H(0) = 1 + \alpha$ for any $N$, $H(0.5) = 1 + \alpha$ for even $N$, and $H(0.5) = 1 - \alpha$ for odd $N$. Thus, its magnitude varies between $1 - \alpha$ and $1 + \alpha$, as illustrated in Figure 6.9.

## 6.3 IIR Filters

Consider the first-order filter described by

$$H_{\text{LP}}(F) = \frac{1}{1 - \alpha e^{-j2\pi F}}, \quad H_{\text{LP}}(0) = \frac{1}{1-\alpha}, \quad H_{\text{LP}}(F)|_{F=0.5} = \frac{1}{1+\alpha}, \quad 0 < \alpha < 1 \tag{6.20}$$

The filter $H_{\text{LP}}(F)$ describes a lowpass filter, because its dc gain (at $F = 0$) exceeds the high-frequency gain (at $F = 0.5$) and the gain shows a decrease with increasing frequency (from $F = 0$ to $F = 0.5$). The filter gain and impulse response are shown in Figure 6.10.

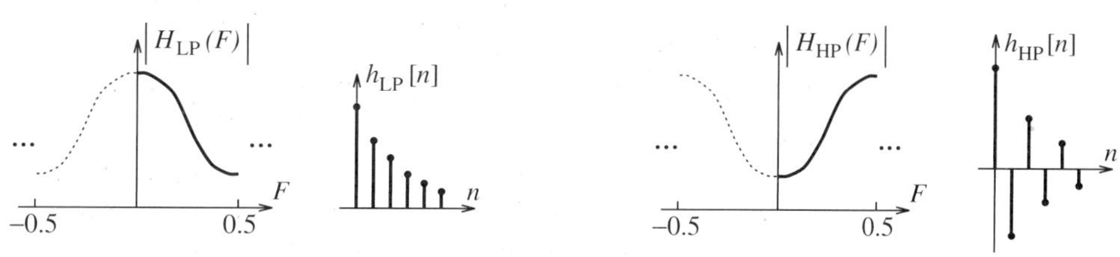


**FIGURE 6.10** Spectrum and impulse response of first-order lowpass and highpass filters. The impulse response of both filters shows an exponential decay with sample values that keep getting smaller. The difference is that, for the highpass form, the samples alternate in sign

We can find the half-power frequency of the lowpass filter from its magnitude squared function:

$$|H_{\text{LP}}(F)|^2 = \left|\frac{1}{1 - \alpha\cos(2\pi F) + j\alpha\sin(2\pi F)}\right|^2 = \frac{1}{1 - 2\alpha\cos(2\pi F) + \alpha^2} \tag{6.21}$$

At the half-power frequency, we have $|H_{\text{LP}}(F)|^2 = 0.5$, and this gives

$$\frac{1}{1 - 2\alpha\cos(2\pi F) + \alpha^2} = 0.5 \quad \text{or} \quad F = \frac{1}{2\pi}\cos^{-1}\left(\frac{\alpha^2 - 1}{2\alpha}\right), \quad |F| < 0.5 \tag{6.22}$$

The phase and group delay of the lowpass filter are given by

$$\phi(F) = -\tan^{-1}\left[\frac{\alpha\sin(2\pi F)}{1 - \alpha\cos(2\pi F)}\right], \quad t_g(F) = -\frac{1}{2\pi}\frac{d\phi(F)}{dF} = \frac{\alpha\cos(2\pi F) - \alpha^2}{1 - 2\alpha\cos(2\pi F) + \alpha^2} \tag{6.23}$$

For low frequencies, the phase is nearly linear, the group delay is nearly constant, and they can be approximated by

$$\phi(F) \approx \frac{-2\pi\alpha F}{1 - \alpha}, \quad |F| \ll 1 \qquad t_g(F) = -\frac{1}{2\pi}\frac{d\phi(F)}{dF} \approx \frac{\alpha}{1 - \alpha}, \quad |F| \ll 1 \tag{6.24}$$

For filters of higher order, we may obtain the group delay by expressing the filter transfer function as the sum of first-order sections.

### Time Constant and Reverberation Time

In the time domain, performance measures for a filter are typically based on features of the impulse response and step response. The impulse response of the first-order lowpass filter is given by

$$h_{\text{LP}}[n] = \alpha^n u[n] \tag{6.25}$$

The **time constant** of the lowpass filter describes how fast the impulse response decays to a specified fraction of its initial value. For a specified fraction of $\epsilon\%$, we obtain the effective time constant $\eta$ in samples as

$$\alpha^\eta = \epsilon \qquad \eta = \frac{\ln\epsilon}{\ln\alpha} \tag{6.26}$$

This value is rounded up to an integer, if necessary. We obtain the commonly measured 1% or 40-dB time constant if $\epsilon = 0.01$ (corresponding to an attenuation of 40 dB) and the 0.1% or 60-dB time constant if $\epsilon = 0.001$. If the sampling interval $t_s$ is known, the time constant $\tau$ in seconds can be computed from $\tau = \eta t_s$. The 60-dB time constant is also called the **reverberation time**. For higher-order filters whose impulse response contains several exponential terms, the effective time constant is dictated by the term with the slowest decay (that dominates the response).

## 6.3.1 First-Order Highpass Filters

Consider the first-order filter described by

$$H_{\text{HP}}(F) = \frac{1}{1 + \alpha e^{-j2\pi F}}, \quad H_{\text{HP}}(0) = \frac{1}{1 + \alpha}, \quad H_{\text{HP}}(F)\Big|_{F=0.5} = \frac{1}{1 - \alpha}, \qquad 0 < \alpha < 1 \tag{6.27}$$

The filter $H_{\mathrm{HP}}(F)$ describes a highpass filter because its high-frequency gain (at $F = 0.5$) exceeds the dc gain (at $F = 0$) and the gain shows an increase with increasing frequency (from $F = 0$ to $F = 0.5$). The filter gain and impulse response are shown in Figure 6.10. The impulse response of this filter is given by

$$h_{\mathrm{HP}}[n] = (-\alpha)^n u[n] = (-1)^n \alpha^n u[n] \tag{6.28}$$

It is the alternating sign changes in the samples of $h_{\mathrm{HP}}[n]$ that cause rapid time variations and lead to its highpass behavior. Its spectrum $H_{\mathrm{HP}}(F)$ is related to $H_{\mathrm{LP}}(F)$ by $H_{\mathrm{HP}}(F) = H_{\mathrm{LP}}(F - 0.5)$. This means that a lowpass cutoff frequency $F_0$ corresponds to the frequency $0.5 - F_0$ of the highpass filter.

### 6.3.2 Pole-Zero Placement and Filter Design

The qualitative effect of poles and zeros on the magnitude response can be used to advantage in understanding the pole-zero patterns of real filters. The basic strategy for pole and zero placement is based on mapping the passband and stopband frequencies on the unit circle and then positioning the poles and zeros based on the following reasoning:

1. **Conjugate symmetry:** All complex poles and zeros must be paired with their complex conjugates.
2. **Causality:** To ensure a causal system, the total number of zeros must be less than or equal to the total number of poles.
3. **Origin:** Poles or zeros at the origin do not affect the magnitude response.
4. **Stability:** For a stable system, the poles must be placed *inside* (not just on) the unit circle. The pole radius is proportional to the gain and inversely proportional to the bandwidth. Poles closer to the unit circle produce a large gain over a narrower bandwidth. Clearly, the passband should contain poles near the unit circle for large passband gains.
   **A rule of thumb:** For narrow-band filters with bandwidth $\Delta\Omega \le 0.2$ centered about $\Omega_0$, we place conjugate poles at $z = R\exp(\pm j\Omega_0)$, where $R \approx 1 - 0.5\Delta\Omega$ is the pole radius.
5. **Minimum phase:** Zeros can be placed anywhere in the $z$-plane. To ensure minimum phase, the zeros must lie within the unit circle. Zeros on the unit circle result in a null in the response. Thus, the stopband should contain zeros on or near the unit circle. A good starting choice is to place a zero (on the unit circle) in the middle of the stopband or two zeros at the edges of the stopband.
6. **Transition band:** To achieve a steep transition from the stopband to the passband, a good choice is to pair each stopband zero with a pole along (or near) the same radial line and close to the unit circle.
7. **Pole-zero interaction:** Poles and zeros interact to produce a composite response that may not match qualitative predictions. Their placement may have to be changed, or other poles and zeros may have to be added to tweak the response. Poles closer to the unit circle or farther away from each other produce small interaction. Poles closer to each other or farther from the unit circle produce more interaction. The closer we wish to approximate a given response, the more poles and zeros we require, and the higher is the filter order.

Bandstop and bandpass filters with real coefficients must have an order of at least two. Filter design using pole-zero placement involves trial and error. It is also possible to identify traditional filter types from their pole-zero patterns. The idea is to use the graphical approach to qualitatively observe how the magnitude changes from the lowest frequency $F=0$ (or $\Omega=0$) to the highest frequency $F=0.5$ (or $\Omega=\pi$) and use this to establish the filter type.

### 6.3.3 Second-Order IIR Filters

The general form for the transfer function of a second-order filter is

$$H(z) = K\frac{z^2+\alpha_1 z+\alpha 2}{z^2+\beta_1 z+\beta_2} = K\frac{(z-R_z e^{j\Omega_z})(z-R_z e^{-j\Omega_z})}{(z-R_p e^{j\Omega_p})(z-R_p e^{-j\Omega_p})}$$

This filter can describe a wide variety of responses. One interesting situation occurs when the poles and zeros lie along the same angular orientation with $\Omega_p=\Omega_z=\Omega_0$. The gain of such a filter shows a peak or a dip around $\Omega=\Omega_0$, depending on the values of $R_p$ and $R_z$. If we assume a causal, minimum-phase filter with $R_p<1$ and $R_z<1$, we observe the following:

- If $R_z>R_p$, the filter shows a dip in the gain around $\Omega=\Omega_0$.
- If $R_z<R_p$, the filter gain peaks around $\Omega=\Omega_0$.
- If $R_z=0$, the filter gain peaks around $\Omega=\Omega_0$.
- If $R_p=0$, we have an FIR filter with a dip or notch around $\Omega=\Omega_0$.

If we permit zeros to lie outside the unit circle ($R_z\ge 1$ but $R_p<1$), we observe the following:

- If $R_z=1$, we have a notch filter with a dc gain of zero at $\Omega=\Omega_0$.
- If $1<R_z<\frac{1}{R_p}$, there is a dip in the gain around $\Omega=\Omega_0$.
- If $R_z=\frac{1}{R_p}$, we have an allpass filter with constant gain.
- If $R_z>\frac{1}{R_p}$, the filter gain peaks around $\Omega=\Omega_0$.

**Example 6.7** **Filters and Pole-Zero Plots**

**(a)** Design a bandpass filter with center frequency = 100 Hz, passband 10 Hz, stopband edges at 50 Hz and 150 Hz, and sampling frequency 400 Hz.

We find $\Omega_0=\frac{\pi}{2}$, $\Delta\Omega=\frac{\pi}{20}$, $\Omega_s=[\frac{\pi}{4},\ \frac{3\pi}{4}]$, and $R=1-0.5\Delta\Omega=0.9215$.

***Passband:*** Place poles at $p_{1,2}=Re^{\pm j\Omega_0}=0.9215e^{\pm j\pi/2}=\pm j0.9215$.

***Stopband:*** Place conjugate zeros at $z_{1,2}=e^{\pm j\pi/4}$ and $z_{3,4}=e^{\pm j3\pi/4}$.

We then obtain the transfer function as

$$H(z)=\frac{(z-e^{j\pi/4})(z-e^{-j\pi/4})(z-e^{j3\pi/4})(z-e^{-j3\pi/4})}{(z-j0.9215)(z+j0.9215)} = \frac{z^4+1}{z^2+0.8941}$$

Note that this filter is noncausal. To obtain a causal filter $H_1(z)$, we could, for example, use double-poles at each pole location to get

$$H_1(z)=\frac{z^4+1}{(z^2+0.8941)^2}=\frac{z^4+1}{z^4+1.6982z^2+0.7210}$$

The pole-zero pattern and gain of the modified filter $H_1(z)$ are shown in Figure E6.7a.

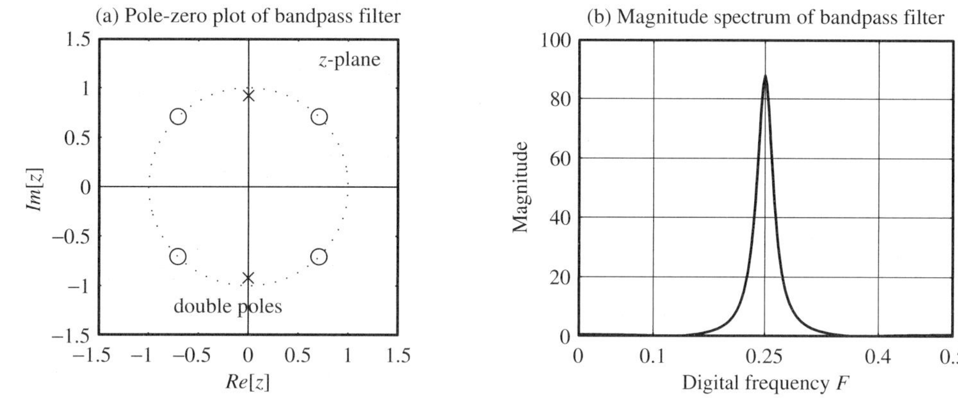


**FIGURE E.6.7A** Frequency response of the bandpass filter for Example 6.7(a)

**(b)** Design a notch filter with notch frequency 60 Hz, stopband 5 Hz, and sampling frequency 300 Hz.

We compute $\Omega_0 = \frac{2\pi}{5}$, $\Delta\Omega = \frac{\pi}{30}$, and $R = 1 - 0.5\Delta\Omega = 0.9476$.

***Stopband:*** We place zeros at the notch frequency to get $z_{1,2} = e^{\pm j\Omega_0} = e^{\pm j2\pi/5}$.

***Passband:*** We place poles along the orientation of the zeros at $p_{1,2} = Re^{\pm j\Omega_0} = 0.9476e^{\pm j2\pi/5}$.

We then obtain $H(z)$ as

$$H(z) = \frac{(z - e^{j2\pi/5})(z - e^{-j2\pi/5})}{(z - 0.9476e^{j2\pi/5})(z - 0.9476e^{-j2\pi/5})} = \frac{z^2 - 0.618z + 1}{z^2 - 0.5857 + 0.898}$$

The pole-zero pattern and magnitude spectrum of this filter are shown in Figure E6.7b.

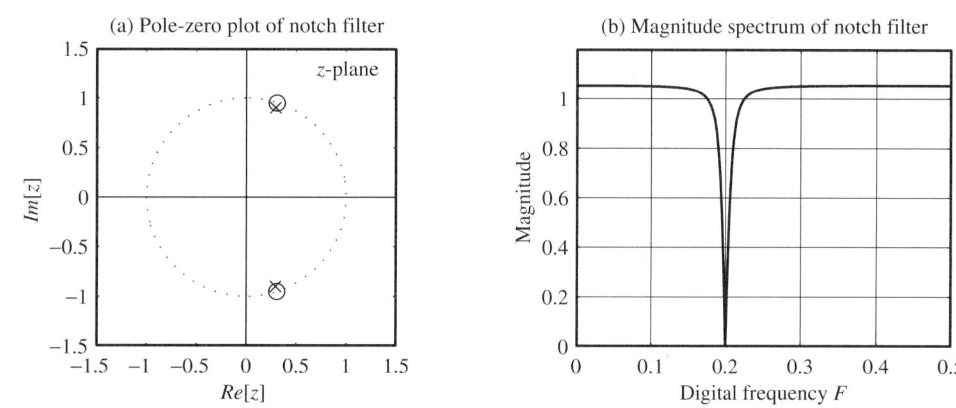


**FIGURE E.6.7B** Frequency response of the bandstop filter for Example 6.7(b)

## 6.3.4 Digital Resonators

A digital resonator is essentially a narrow-band bandpass filter. One way to realize a second-order resonator with a peak at $\Omega_0$ is to place a pair of poles with angular orientations of $\pm\Omega_0$ (at $z = Re^{j\Omega_0}$ and $z = Re^{-j\Omega_0}$) and a pair of zeros at $z = 0$, as shown in Figure 6.11. To ensure stability, we choose the pole radius $R$ to be less than (but close to) unity.

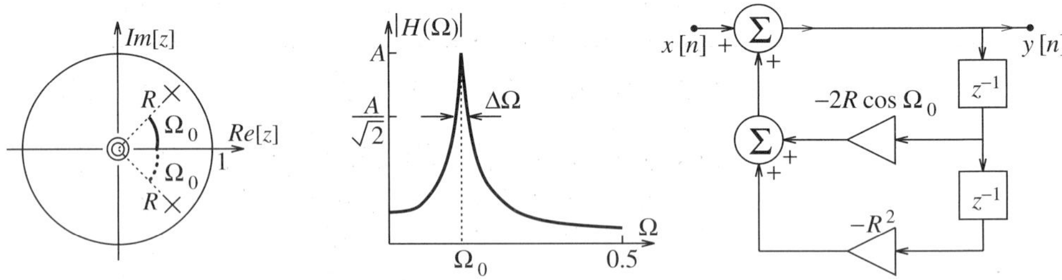


**FIGURE 6.11** A second-order digital resonator. The conjugate poles closer to the unit circle ensure a sharper peak. The zeros at the origin are not a requirement and may be removed or even moved elsewhere in order to change the shape of the response. A realization requires two delay elements.

The transfer function of such a digital resonator is

$$H(z) = \frac{z^2}{(z - Re^{j\Omega_0})(z - Re^{-j\Omega_0})} = \frac{z^2}{z^2 - 2zR\cos\Omega_0 + R^2} \tag{6.29}$$

Its magnitude spectrum $|H(\Omega)|$ and realization are also shown in Figure 6.11. The magnitude squared function $|H(\Omega)|^2$ is given by

$$|H(\Omega)|^2 = \left| \frac{e^{j2\Omega}}{(e^{j\Omega} - Re^{j\Omega_0})(e^{j\Omega} - Re^{-j\Omega_0})} \right|^2 \tag{6.30}$$

This simplifies to

$$|H(\Omega)|^2 = \frac{1}{[1 - 2R\cos(\Omega - \Omega_0) + R^2][1 - 2R\cos(\Omega + \Omega_0) + R^2]} \tag{6.31}$$

Its peak value $A$ occurs at (or very close to) the resonant frequency, and is given by

$$A^2 = |H(\Omega_0)|^2 = \frac{1}{(1 - R)^2(1 - 2R\cos 2\Omega_0 + R^2)} \tag{6.32}$$

The half-power bandwidth $\Delta\Omega$ is found by locating the frequencies at which $|H(\Omega)|^2 = 0.5|H(\Omega_0)|^2 = 0.5A^2$, to give

$$\frac{1}{[1 - 2R\cos(\Omega - \Omega_0) + R^2][1 - 2R\cos(\Omega + \Omega_0) + R^2]} = \frac{0.5}{(1 - R)^2(1 - 2R\cos 2\Omega_0 + R^2)} \tag{6.33}$$

Since the half-power frequencies are very close to $\Omega_0$, we have $\Omega + \Omega_0 \approx 2\Omega_0$ and $\Omega - \Omega_0 \approx 0.5\Delta\Omega$, and the above relation simplifies to

$$1 - 2R\cos(0.5\Delta\Omega) + R^2 = 2(1 - R)^2 \tag{6.34}$$

This relation between the pole radius $R$ and bandwidth $\Delta\Omega$ can be simplified when the poles are very close to the unit circle (with $R > 0.9$ or so). The values of $R$ and the peak magnitude $A$ are then reasonably well approximated by

$$R \approx 1 - 0.5\Delta\Omega, \qquad A \approx \frac{1}{(1 - R^2)\sin\Omega_0} \tag{6.35}$$

The peak gain can be normalized to unity by dividing the transfer function by $A$. The impulse response of this filter can be found by partial fraction expansion of $H(z)/z$, which gives

$$\frac{H(z)}{z} = \frac{z}{(z - Re^{j\Omega_0})(z - Re^{-j\Omega_0})} = \frac{K}{z - Re^{j\Omega_0}} + \frac{K^*}{z - Re^{-j\Omega_0}} \tag{6.36}$$

where

$$K = \left.\frac{z}{z - Re^{-j\Omega_0}}\right|_{z=Re^{j\Omega_0}} = \frac{Re^{j\Omega_0}}{Re^{j\Omega_0} - Re^{-j\Omega_0}} = \frac{Re^{j\Omega_0}}{2jR\sin\Omega_0} = \frac{e^{j(\Omega_0-\pi/2)}}{2\sin\Omega_0} \tag{6.37}$$

Then, from lookup tables, we obtain

$$h[n] = \frac{1}{\sin\Omega_0}R^n\cos\left(n\Omega_0 + \Omega_0 - \frac{\pi}{2}\right)u[n] = \frac{1}{\sin\Omega_0}R^n\sin[(n+1)\Omega_0]u[n] \tag{6.38}$$

To null the response at low and high frequencies ($\Omega = 0$ and $\Omega = \pi$), the two zeros in $H(z)$ may be relocated from the origin to $z = 1$ and $z = -1$, and this leads to the modified transfer function $H_1(z)$:

$$H_1(z) = \frac{z^2 - 1}{z^2 - 2zR\cos\Omega_0 + R^2} \tag{6.39}$$

**Example 6.8**

## Digital Resonator Design

We design a digital resonator with a peak gain of unity at 50 Hz and a 3-dB bandwidth of 6 Hz, assuming a sampling frequency of 300 Hz.

The digital-resonant frequency is $\Omega_0 = \frac{2\pi(50)}{300} = \frac{\pi}{3}$. The 3-dB bandwidth is $\Delta\Omega = \frac{2\pi(6)}{300} = 0.04\pi$. We compute the pole radius as $R = 1 - 0.5\Delta\Omega = 1 - 0.02\pi = 0.9372$. The transfer function of the digital resonator is thus

$$H(z) = \frac{Gz^2}{(z - Re^{j\Omega_0})(z - Re^{-j\Omega_0})} = \frac{Gz^2}{z^2 - 2R\cos\Omega_0 + R^2} = \frac{Gz^2}{z^2 - 0.9372z + 0.8783}$$

For a peak gain of unity, we choose $G = \frac{1}{A} = (1 - R^2)\sin\Omega_0 = 0.1054$. Thus,

$$H(z) = \frac{0.1054z^2}{z^2 - 0.9372z + 0.8783}$$

The magnitude spectrum and passband detail of this filter are shown in Figure E6.8.

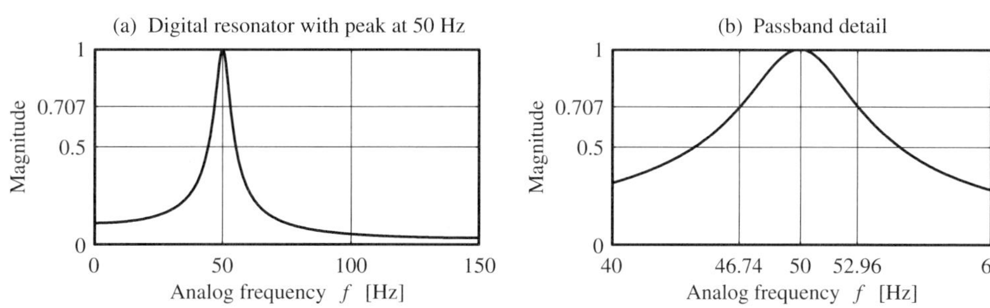


**FIGURE E.6.8** Frequency response of the digital resonator for Example 6.8

The passband detail reveals that the half-power frequencies are located at 46.74 Hz and 52.96 Hz, which is a close match to the bandwidth requirement (of 6 Hz).

### 6.3.5 Periodic Notch Filters

A **notch filter** with notches at periodic intervals may be constructed by placing zeros on the unit circle at the notch frequencies and poles in close proximity along the same angular orientations. One such form is

$$H(z) = \frac{N(z)}{N(z/R)} = \frac{1 - z^{-N}}{1 - (z/R)^{-N}} \tag{6.40}$$

Its zeros are uniformly spaced $2\pi/N$ radians apart around a unit circle, starting at $\Omega = 0$, and its poles are along the same angular orientations but on a circle of radius $R$. For stability, we require $R < 1$. Of course, the closer $R$ is to unity, the sharper are the notches and the more constant is the response at the other frequencies. Periodic notch filters are often used to remove unwanted components at the power line frequency and its harmonics.

**Example 6.9**

**Periodic Notch Filter Design**

We design a notch filter to filter out 60 Hz and its harmonics from a signal sampled at $S = 300$ Hz.

The digital frequency corresponding to 60 Hz is $F = 60/300 = 0.2$ or $\Omega = 0.4\pi$. There are thus $N = 2\pi/0.4\pi = 5$ notches in the principal range $\pi \le \Omega \le \pi$ (around the unit circle), and the transfer function of the notch filter is

$$H(z) = \frac{N(z)}{N(z/R)} = \frac{1 - z^{-5}}{1 - (z/R)^{-5}} = \frac{z^5 - 1}{z^5 - R^5}$$

Figure E6.9(a) shows the response of this filter for $R = 0.9$ and $R = 0.99$. The choice of $R$, the pole radius, is arbitrary (as long as $R < 1$) but should be close to unity for sharp notches.

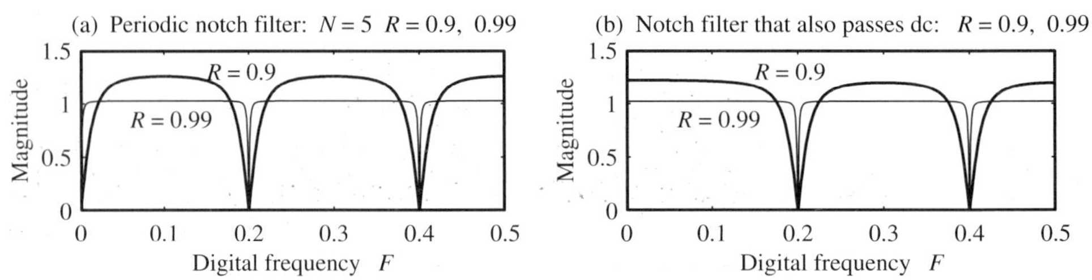


**FIGURE E.6.9** Frequency response of the notch filters for Example 6.9

*Comment:* This notch filter also removes the dc component. If we want to preserve the dc component, we must extract the zero at $z = 1$ (corresponding to $F = 0$) from $N(z)$ (by long division), to give

$$N_1(z) = \frac{1 - z^{-5}}{1 - z^{-1}} = 1 + z^{-1} + z^{-2} + z^{-3} + z^{-4}$$

and use $N_1(z)$ to compute the new transfer function $H_1(z) = N_1(z)/D(z)$ as

$$H_1(z) = \frac{N_1(z)}{N_1(z/R)} = \frac{1 + z^{-1} + z^{-2} + z^{-3} + z^{-4}}{1 + (z/R)^{-1} + (z/R)^{-2} + (z/R)^{-3} + (z/R)^{-4}}$$

$$= \frac{z^4 + z^3 + z^2 + z + 1}{z^4 + Rz^3 + R^2z^2 + R^3z + R^4}$$

Figure E6.9(b) compares the response of this filter for $R = 0.9$ and $R = 0.99$ and reveals that the dc component is indeed preserved by this filter.

# 6.4 Allpass Filters

Consider the first-order filter described by

$$H_A(F) = \frac{A + Be^{j2\pi F}}{B + Ae^{j2\pi F}} \qquad |H_A(0)| = \left|\frac{A + B}{B + A}\right| = 1 \qquad |H_A(0.5)| = \left|\frac{A - B}{B - A}\right| = 1 \tag{6.41}$$

Its magnitude at $F = 0$ and $F = 0.5$ is equal to unity. In fact, its magnitude is unity for all frequencies, and this describes an **allpass filter**. The fact that its magnitude is constant for all frequencies can be explained by a geometric argument, as illustrated in Figure 6.12.

The numerator and denominator may be regarded as the sum of two vectors of length $A$ and $B$ that subtend the same angle $\Omega = 2\pi F$. The length of their vector sum is thus equal for any $\Omega$. An interesting characteristic that allows us to readily identify an allpass filter is that its numerator and denominator coefficients (associated with powers of $e^{\pm j2\pi F}$ or $e^{j\Omega}$) appear in reversed order. Allpass filters are often used to shape the delay characteristics of digital filters.

## 6.4.1 Transfer Function of Allpass Filters

An **allpass filter** is characterized by a magnitude response that is constant for all frequencies. For an allpass filter with unit gain, $|H(F)| = 1$. Its transfer function $H(z)$ also satisfies the relationship

$$H(z)H(1/z) = 1 \tag{6.42}$$

This implies that each pole of an allpass filter is paired by a conjugate reciprocal zero. As a result, allpass filters cannot be minimum phase. Note that the cascade of allpass filters is

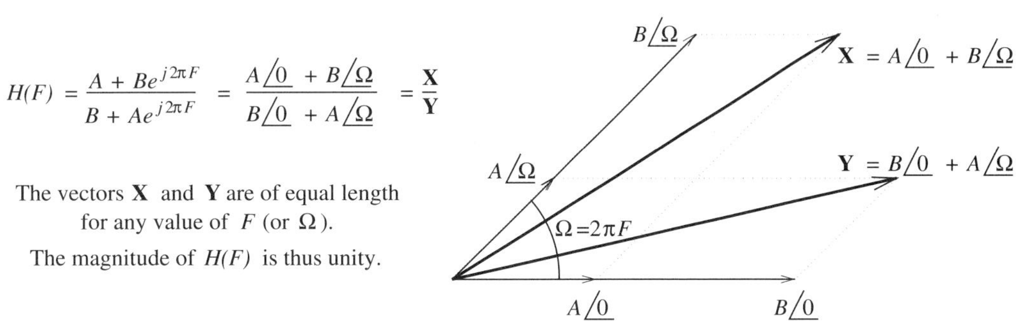


**FIGURE 6.12** The magnitude spectrum of an allpass filter is constant. The vectorial representation of the numerator and denominator of its transfer function reveals equal vector lengths. This implies unit gain

also an allpass filter. An allpass filter of order $N$ has a numerator and denominator of equal order $N$ with coefficients in reversed order:

$$H_{\text{AP}}(z) = \frac{N(z)}{D(z)} = \frac{C_N + C_{N-1}z^{-1} + \cdots + C_1 z^{N-1} + z^{-N}}{1 + C_1 z^{-1} + C_2 z^{-2} + \cdots + C_N z^{-N}} \tag{6.43}$$

Note that $D(z) = z^{-N}N(1/z)$. If the roots of $N(z)$ (the zeros) are at $r_k$, the roots of $D(z)$ (the poles) are at $1/r_k$, which are the reciprocal locations.

### How to Identify an Allpass Filter

**From $H(z) = N(z)/D(z)$:** The coefficients of $N(z)$ and $D(z)$ appear in reversed order.
**From pole-zero plot:** Each pole is paired with a conjugate reciprocal zero.
**From difference equation:** The coefficients of RHS and LHS appear in reversed order.

Consider a stable, first-order allpass filter whose transfer function $H(z)$ and frequency response $H(F)$ are described by

$$H(z) = \frac{1 + \alpha z}{z + \alpha} \qquad H_A(F) = \frac{1 + \alpha e^{j2\pi F}}{\alpha + e^{j2\pi F}}, \quad |\alpha| < 1 \tag{6.44}$$

If we factor out $e^{j\pi F}$ from the numerator and denominator of $H(F)$, we obtain the form

$$H(F) = \frac{1 + \alpha e^{j2\pi F}}{\alpha + e^{j2\pi F}} = \frac{e^{-j\pi F} + \alpha e^{j\pi F}}{\alpha e^{-j\pi F} + e^{j\pi F}}, \quad |\alpha| < 1 \tag{6.45}$$

The numerator and denominator are complex conjugates. This implies that their magnitudes are equal (an allpass characteristic) and that the phase of $H(F)$ equals twice the numerator phase. Now, the numerator may be simplified to

$$\begin{aligned} e^{-j\pi F} + \alpha e^{j\pi F} &= \cos(\pi F) - j\sin(\pi F) + \alpha\cos(\pi F) + j\alpha\sin(\pi F) \\ &= (1+\alpha)\cos(\pi F) - j(1-\alpha)\sin(\pi F) \end{aligned}$$

The phase $\phi(F)$ of the allpass filter $H(F)$ equals twice the numerator phase. The phase $\phi(F)$ and phase delay $t_p(F)$ may then be written as

$$\phi(F) = -2\tan^{-1}\left[\frac{1-\alpha}{1+\alpha}\tan(\pi F)\right], \qquad t_p(F) = -\frac{\phi(F)}{2\pi F} = \frac{1}{\pi F}\tan^{-1}\left[\frac{1-\alpha}{1+\alpha}\tan(\pi F)\right] \tag{6.46}$$

A low-frequency approximation for the phase delay is given by

$$t_p(F) = \frac{1-\alpha}{1+\alpha} \quad \text{(low-frequency approximation)} \tag{6.47}$$

The group delay $t_g(F)$ of this allpass filter equals

$$t_g(F) = -\frac{1}{2\pi}\frac{d\phi(F)}{dF} = \frac{1-\alpha^2}{1 + 2\alpha\cos(2\pi F) + \alpha^2} \tag{6.48}$$

At $F = 0$ and $F = 0.5$, the group delay is given by

$$t_g(0) = \frac{1-\alpha^2}{1+2\alpha+\alpha^2} = \frac{1-\alpha}{1+\alpha} \qquad t_g(0.5) = \frac{1-\alpha^2}{1-2\alpha+\alpha^2} = \frac{1+\alpha}{1-\alpha} \tag{6.49}$$

At low frequencies, $\cos(2\pi F) \approx 1$, and the group delay and phase delay are approximately equal. In particular, for $0 < \alpha < 1$, the delay is less than unity, and this allows us to use the allpass filter to generate fractional delays by appropriate choice of $\alpha$.

### 6.4.2 Minimum-Phase Filters using Allpass Filters

Even though allpass filters cannot be minimum phase, they can be used to convert stable nonminimum-phase filters to stable minimum-phase filters. We describe a nonminimum-phase transfer function by a cascade of a minimum-phase part $H_M(z)$ (with all poles and zeros inside the unit circle) and a portion with zeros outside the unit circle such that

$$H_{\rm NM}(z) = H_M(z)\prod_{m=1}^{P}(z+\alpha_M), \quad |\alpha_m| > 1 \tag{6.50}$$

We now seek an *unstable* allpass filter with

$$H_{\rm AP}(z) = \prod_{m=1}^{P}\frac{1+z\alpha_m^*}{z+\alpha_m}, \quad |\alpha_m| > 1 \tag{6.51}$$

The cascade of $H_{\rm NM}(z)$ and $H_{\rm AP}(z)$ yields a minimum-phase filter with

$$H(z) = H_{\rm NM}(z)H_{\rm AP}(z) = H_M(z)\prod_{m=1}^{P}(1+z\alpha_m^*) \tag{6.52}$$

Once again, $H(z)$ has the same order as the original filter.

### 6.4.3 Concluding Remarks

We have studied a variety of digital filters in previous sections. As mentioned before, the dc gain (at $F = 0$ or $\Omega = 0$) and the high-frequency gain (at $F = 0.5$ or $\Omega = \pi$) serve as a useful measures to identify filters. The coefficients (and form) of the difference equation, the transfer, function, and the impulse response reveal other interesting and useful characteristics, such as linear phase, allpass behaviour, and the like. The more complex the filter (expression), the less likely it is for us to discern its nature by using only a few simple measures.

**Example 6.10** **Traditional and Non-Traditional Filters**

**(a)** Let $h[n] = \{\overset{\Downarrow}{-1}, -2, 2, 1\}$. Is this a linear-phase sequence? What type of filter is this? The sequence $h[n]$ is linear-phase. It shows odd symmetry about the half-integer index $n = 1.5$.

We have $H(F) = -1 - 2e^{-j2\pi F} + 2e^{-j4\pi F} + e^{-j6\pi F}$. Extract the factor $(-e^{-j3\pi F})$ to get

$$H(F) = -e^{-j3\pi F}[e^{j3\pi F} + 2e^{j\pi F} - 2e^{-j\pi F} - e^{-j3\pi F}].$$

Using Euler's relation: $H(F) = -je^{-j3\pi F}[2\sin(3\pi F) + 4\sin(\pi F)]$
$H(0) = 0 \qquad H(0.5) = 2$

Amplitude: $A(F) = 2\sin(3\pi F) + 4\sin(\pi F)$
Phase $= -je^{-j3\pi F} = -3\pi F - \frac{\pi}{2}$
Since the magnitude at high frequencies increases, this appears to be a highpass filter.

*Comment:* We can also use the central ordinate relations to compute $H(0)$ and $H(0.5)$ directly from $h[n]$. Thus,

$$H(0) = \sum_{n=0}^{3} h[n] = 0 \text{ and } H(0.5) = \sum_{n=0}^{3} (-1)^n h[n] = -1+2+2-1 = 2$$

**(b)** Let $h[n] = (-0.8)^n u[n]$. Identify the filter type and establish whether the impulse response is a linear-phase sequence.

The sequence $h[n]$ is not linear-phase because it shows no symmetry. We have $H(F) = \frac{1}{1+0.8e^{-j2\pi F}}$. We find that

$$H(0) = \frac{1}{1+0.8} = 0.556 \text{ and } H(0.5) = \frac{1}{1+0.8e^{-j\pi}} = \frac{1}{1-0.8} = 5$$

Since the magnitude at high frequencies increases, this appears to be a highpass filter.

**(c)** Consider a system described by $y[n] = \alpha y[n-1] + x[n],\ 0 < \alpha < 1.$

**Solution:** This is an example of a *reverb filter* whose response equals the input plus a delayed version of the output. Its frequency response may be found by taking the DTFT of both sides to give $Y(F) = \alpha Y(F)e^{-j2\pi F} + X(F)$. Rearranging this equation, we obtain

$$H(F) = \frac{Y(F)}{X(F)} = \frac{1}{1-\alpha e^{-j2\pi F}}$$

Using Euler's relation, we rewrite this as

$$H(F) = \frac{1}{1-\alpha e^{-j2\pi F}} = \frac{1}{1-\alpha\cos(2\pi F) - j\alpha\sin(2\pi F)}$$

Its magnitude and phase are given by

$$|H(F)| = \frac{1}{[1-2\alpha\cos(2\pi F)+\alpha^2]^{1/2}} \qquad \phi(F) = -\tan^{-1}\left[\frac{\alpha\sin(2\pi F)}{1-\alpha\cos(2\pi F)}\right]$$

A typical magnitude and phase plot for this system (for $0 < \alpha < 1$) is shown in Figure E6.10c. The impulse response of this system equals $h[n] = \alpha^n u[n]$.

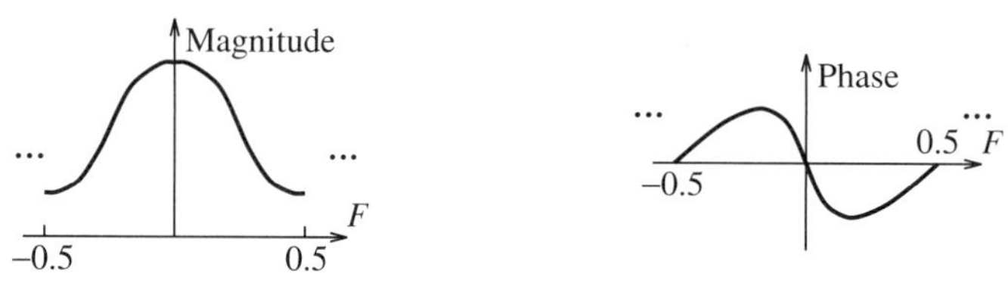


**FIGURE E.6.10c** Frequency response of the system for Example 6.10(c)

**(d)** Let $h[n] = \{\overset{\Downarrow}{5}, -4, 3, -2\}$. Identify the filter type and establish whether the impulse response is a linear-phase sequence.

The sequence $h[n]$ is not linear-phase because it shows no symmetry.

We have

$$H(F) = 5 - 4e^{-j2\pi F} + 3e^{-j4\pi F} - 2e^{-j6\pi F}$$

We find

$$H(0) = \sum h[n] = 2 \text{ and } H(0.5) = \sum (-1)^n h[n] = 5 + 4 + 3 + 2 = 14$$

Since the magnitude at high frequencies increases, this appears to be a highpass filter.

**(e)** Let $h[n] = (0.8)^n u[n]$. Identify the filter type, establish whether the impulse response is a linear-phase sequence, and find its 60-dB time constant.

We have

$$H(F) = \frac{1}{1 - 0.8e^{-j2\pi F}}, \qquad H(0) = \frac{1}{1 - 0.8} = 5, \quad \text{and}$$

$$H(0.5) = \frac{1}{1 - 0.8e^{-j\pi}} = \frac{1}{1 + 0.8} = 0.556$$

The sequence $h[n]$ is not a linear-phase phase because it shows no symmetry. Since the magnitude at high frequencies decreases, this appears to be a lowpass filter. The 60-dB time constant of this filter is found as

$$\eta = \frac{\ln \epsilon}{\ln \alpha} = \frac{\ln 0.001}{\ln 0.8} = 30.96 \quad \Rightarrow \quad \eta = 31 \text{ samples}$$

For a sampling frequency of $S = 100$ Hz, this corresponds to $\tau = \eta t_s = \eta/S = 0.31$ s.

**(f)** Let $h[n] = 0.8\delta[n] + 0.36(-0.8)^{n-1}u[n-1]$. Identify the filter type and establish whether the impulse response is a linear-phase sequence.

We find

$$H(F) = 0.8 + \frac{0.36e^{-j2\pi F}}{1 + 0.8e^{-j2\pi F}} = \frac{0.8 + e^{-j2\pi F}}{1 + 0.8e^{-j2\pi F}}$$

So,

$$H(0) = 1 \quad \text{and} \quad H(0.5) = \frac{0.8 - 1}{1 - 0.8} = -1$$

The sequence $h[n]$ is not a linear-phase sequence because it shows no symmetry. Since the magnitude is identical at low and high frequencies, this could be a bandstop or an allpass filter. Since the numerator and denominator coefficients in $H(F)$ appear in reversed order, it is an *allpass* filter.

**(g)** Let $h[n] = \delta[n] - \delta[n-8]$. Identify the filter type and establish whether the impulse response is a linear-phase sequence.

We find

$$H(F) = 1 - e^{-j16\pi F}$$

So,

$$H(0) = 0 \text{ and } H(0.5) = 0$$

This suggests a bandpass filter with a maximum between $F = 0$ and $F = 0.5$. On closer examination, we find that $H(F) = 0$ at $F = 0,\ 0.125,\ 0.25,\ 0.375,\ 0.5$ when $e^{-j16\pi F} = 1$ and $|H(F)| = 2$ at four frequencies halfway between these. This multi-humped response is sketched in Figure E10g and describes a **comb filter**.

We may write $h[n] = \{\overset{\Downarrow}{1}, 0, 0, 0, 0, 0, 0, 0, -1\}$. This is a linear-phase sequence because it shows odd symmetry about the index $n = 4$.

*Comment:* Note that the difference equation $y[n] = x[n] - x[n-8]$ is reminiscent of an echo filter.

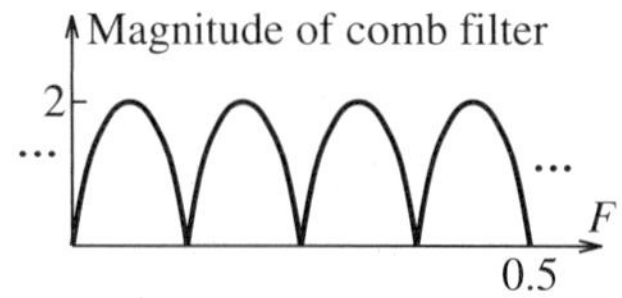


**FIGURE E.6.10G** Spectrum of the comb filter for Example 10(g)

## 6.5 Problems

**6.1. (Frequency Response)** For each filter, sketch the magnitude spectrum and identify the filter type.

**(a)** $h[n] = \delta[n] - \delta[n-2]$ **(b)** $y[n] - 0.25y[n-1] = x[n] - x[n-1]$

**(c)** $H(z) = \dfrac{z-2}{z-0.5}$ **(d)** $y[n] - y[n-1] + 0.25y[n-2] = x[n] + x[n-1]$

[**Hints and Suggestions:** To sketch the spectra, first find $H(F)$. To identify the filter type, start by computing the gain $|H(F)|$ at $F = 0$ and $F = 0.5$.]

**6.2. (Frequency Response)** Consider the 3-point averaging filter $h[n] = \frac{1}{3}\{1, \overset{\Downarrow}{1}, 1\}$.

**(a)** Sketch its amplitude spectrum.
**(b)** Find its phase delay and group delay. Is this a linear-phase filter?
**(c)** Find its response to the input $x[n] = \cos(n\pi/3)$.
**(d)** Find its response to the input $x[n] = \delta[n]$.
**(e)** Find its response to the input $x[n] = (-1)^n$.
**(f)** Find its response to the input $x[n] = 3 + 3\delta[n] - 6\cos(n\pi/3)$.

[**Hints and Suggestions:** For (a), use $e^{j\theta} + e^{-j\theta} = 2\cos\theta$ to set up $H(F) = A(F)e^{j\phi(F)}$ and sketch $A(F)$. For (b), use the symmetry in $h[n]$. For (c), evaluate $H(F)$ at the frequency of $x[n]$ to find the output. For (d), find $h[n]$. For (e), use $(-1)^n = \cos(n\pi)$. For (f), use superposition.]

**6.3. (Frequency Response)** Consider a filter described by $h[n] = \frac{1}{3}\{1, \overset{\Downarrow}{-1}, 1\}$.

**(a)** Sketch its amplitude spectrum.
**(b)** Find its phase delay and group delay. Is this a linear-phase filter?
**(c)** Find its response to the input $x[n] = \cos(n\pi/3)$.
**(d)** Find its response to the input $x[n] = \delta[n]$.
**(e)** Find its response to the input $x[n] = (-1)^n$.
**(f)** Find its response to the input $x[n] = 3 + 3\delta[n] - 3\cos(2n\pi/3)$.

[**Hints and Suggestions:** For (a), use $e^{j\theta} + e^{-j\theta} = 2\cos\theta$ to set up $H(F) = A(F)e^{j\phi(F)}$ and sketch $A(F)$. For (b), use the symmetry in $h[n]$. For (c), evaluate $H(F)$ at the frequency of $x[n]$ to find the output. For (d), find $h[n]$. For (e), use $(-1)^n = \cos(n\pi)$. For (f), use superposition.]

**6.4. (Frequency Response)** Consider the tapered 3-point averaging filter $h[n] = \{\overset{\Downarrow}{0.5}, 1, 0.5\}$.

**(a)** Sketch its amplitude spectrum.
**(b)** Find its phase delay and group delay. Is this a linear-phase filter?
**(c)** Find its response to the input $x[n] = \cos(n\pi/2)$.

**(d)** Find its response to the input $x[n] = \delta[n-1]$.
**(e)** Find its response to the input $x[n] = 1 + (-1)^n$.
**(f)** Find its response to the input $x[n] = 3 + 2\delta[n] - 4\cos(n\pi/2)$.

[**Hints and Suggestions:** For (a), extract $e^{-j2\pi F}$ in $H(F)$ to set up $H(F) = A(F)e^{j\phi(F)}$ and sketch $A(F)$. For (b), use $\phi(F)$. For (c), evaluate $H(F)$ at the frequency of $x[n]$ to find the output. For (d), find $h[n]$ and shift. For (e), use $(-1)^n = \cos(n\pi)$ and superposition. For (f), use superposition.]

**6.5.** **(Frequency Response)** Consider the 2-point differencing filter

$$h[n] = \delta[n] - \delta[n-1]$$

**(a)** Sketch its amplitude spectrum.
**(b)** Find its phase and group delay. Is this a linear-phase filter?
**(c)** Find its response to the input $x[n] = \cos(n\pi/2)$.
**(d)** Find its response to the input $x[n] = u[n]$.
**(e)** Find its response to the input $x[n] = (-1)^n$.
**(f)** Find its response to the input $x[n] = 3 + 2u[n] - 4\cos(n\pi/2)$.

[**Hints and Suggestions:** For (a), extract $e^{-j\pi F}$ in $H(F)$ to set up $H(F) = A(F)e^{j\phi(F)}$ and sketch $A(F)$. For (b), use $\phi(F)$. For (c), evaluate $H(F)$ at the frequency of $x[n]$ to find the output. For (d), find $y[n] = u[n] - u[n-1]$ and simplify. For (e), use $(-1)^n = \cos(n\pi)$. For (f), use superposition.]

**6.6.** **(Frequency Response)** Consider the 5-point tapered averaging filter $h[n] = \frac{1}{9}\{\overset{\Downarrow}{1}, 2, 3, 2, 1\}$.

**(a)** Sketch its amplitude spectrum.
**(b)** Find its phase delay and group delay. Is this a linear-phase filter?
**(c)** Find its response to the input $x[n] = \cos(n\pi/4)$.
**(d)** Find its response to the input $x[n] = \delta[n]$.
**(e)** Find its response to the input $x[n] = (-1)^n$.
**(f)** Find its response to the input $x[n] = 9 + 9\delta[n] - 9\cos(n\pi/4)$.

[**Hints and Suggestions:** For (a), extract $e^{-j4\pi F}$ in $H(F)$ to set up $H(F) = A(F)e^{j\phi(F)}$ and sketch $A(F)$. For (b), use $\phi(F)$. For (c), evaluate $H(F)$ at the frequency of $x[n]$ to find the output. For (d), find $h[n]$. For (e), use $(-1)^n = \cos(n\pi)$. For (f), use superposition.]

**6.7.** **(Frequency Response)** Sketch the gain $|H(F)|$ of the following digital filters and identify the filter type.

**(a)** $y[n] + 0.9y[n-1] = x[n]$ **(b)** $y[n] - 0.9y[n-1] = x[n]$
**(c)** $y[n] + 0.9y[n-1] = x[n-1]$ **(d)** $y[n] = x[n] - x[n-4]$

[**Hints and Suggestions:** To sketch, start with the gain $|H(F)|$ at $F = 0$ and $F = 0.5$ and add a few others such as $F = 0.25$.]

**6.8.** **(Frequency Response)** Consider a system whose frequency response $H(F)$ in magnitude/phase form is $H(F) = A(F)e^{j\phi(F)}$. Find the response $y[n]$ of this system for the following inputs.

**(a)** $x[n] = \delta[n]$ **(b)** $x[n] = 1$ **(c)** $x[n] = \cos(2n\pi F_0)$ **(d)** $x[n] = (-1)^n$

[**Hints and Suggestions:** For (d), note that $(-1)^n = \cos(n\pi)$.]

**6.9.** **(Poles and Zeros)** Find the transfer function corresponding to each pole-zero pattern shown in Figure P6.9 and identify the filter type.

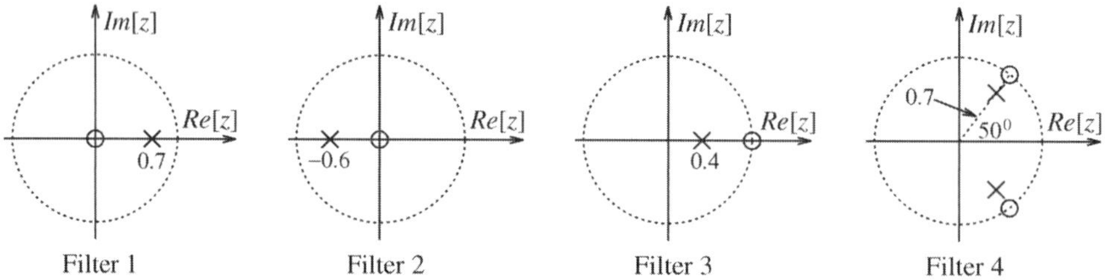


**FIGURE P.6.9** Figure for Problem 6.9

[**Hints and Suggestions:** To identify the filter type, evaluate $|H(z)|$ at $z = 1$ (dc gain) and at $z = -1$ (high-frequency gain).]

**6.10.** **(Inverse Systems)** Find the transfer function of the inverse systems for each of the following. Which inverse systems are causal? Which inverse systems are stable?

**(a)** $H(z) = \dfrac{z^2+0.1}{z^2-0.2}$ **(b)** $H(z) = \dfrac{z+2}{z^2+0.25}$

**(c)** $y[n] - 0.5y[n-1] = x[n] + 2x[n-1]$ **(d)** $h[n] = n(2)^n u[n]$

[**Hints and Suggestions:** For parts (c) and (d), set up $H(z)$ and find $H_I(z) = 1/H(z)$.]

**6.11.** **(Minimum-Phase Systems)** Classify each causal system as minimum phase, mixed phase, or maximum phase. Which of the systems are stable?

**(a)** $H(z) = \dfrac{z^2+0.16}{z^2-0.25}$ **(b)** $H(z) = \dfrac{z^2-4}{z^2+9}$

**(c)** $h[n] = n(2)^n u[n]$ **(d)** $y[n] + y[n-1] + 0.25y[n-2] = x[n] - 2x[n-1]$

**6.12.** **(Minimum-Phase Systems)** Find the minimum-phase transfer function corresponding to the systems described by the following:

**(a)** $H(z)H(1/z) = \dfrac{1}{3z^2 - 10z + 3}$

**(b)** $H(z)H(1/z) = \dfrac{3z^2 + 10z + 3}{5z^2 + 26z + 5}$

**(c)** $|H(F)|^2 = \dfrac{1.25 + \cos(2\pi F)}{8.5 + 4\cos(2\pi F)}$

[**Hints and Suggestions:** Pick $H(z)$ to have poles and zeros inside the unit circle. For part (c), set $\cos(2\pi F) = 0.5e^{j2\pi F} + 0.5e^{-j2\pi F} = 0.5z + 0.5z^{-1}$ and compare with $H(z)H(1/z)$.]

**6.13.** **(System Characteristics)** Consider the system $y[n] - \alpha y[n-1] = x[n] - \beta x[n-1]$.

**(a)** For what values of $\alpha$ and $\beta$ will the system be stable?
**(b)** For what values of $\alpha$ and $\beta$ will the system be minimum-phase?
**(c)** For what values of $\alpha$ and $\beta$ will the system be allpass?
**(d)** For what values of $\alpha$ and $\beta$ will the system be linear phase?

**6.14.** **(Frequency Response)** Consider the recursive filter $y[n] + 0.5y[n-1] = 0.5x[n] + x[n-1]$.

**(a)** Sketch the filter gain. What type of filter is this?
**(b)** Find its phase delay and group delay. Is this a linear-phase filter?
**(c)** Find its response to the input $x[n] = \cos(n\pi/2)$.
**(d)** Find its response to the input $x[n] = \delta[n]$.
**(e)** Find its response to the input $x[n] = (-1)^n$.
**(f)** Find its response to the input $x[n] = 4 + 2\delta[n] - 4\cos(n\pi/2)$.

[**Hints and Suggestions:** For (a), start with $|H(F)|$ at $F = 0$ and $F = 0.5$ to sketch the gain. For (b), use equations for the first-order allpass filter. For (c), evaluate $H(F)$ at the frequency of $x[n]$ to find the output. For (d), find $h[n]$. For (e), use $(-1)^n = \cos(n\pi)$. For (f), use superposition.]

**6.15. (Linear-Phase Filters)** Consider a filter whose impulse response is $h[n] = \{\overset{\Downarrow}{\alpha}, \beta, \alpha\}$.

**(a)** Find its frequency response in amplitude/phase form $H(F) = A(F)e^{j\phi(F)}$ and confirm that the phase $\phi(F)$ varies linearly with $F$.
**(b)** Determine the values of $\alpha$ and $\beta$ if this filter is to completely block a signal at the frequency $F = \frac{1}{3}$ while passing a signal at the frequency $F = \frac{1}{8}$ with unit gain.
**(c)** What will be the output of the filter in part(b) if the input is $x[n] = \cos(0.5n\pi)$?

[**Hints and Suggestions:** In (a), extract $e^{-j2\pi F}$ in $H(F)$ to set up the required form. In (b), set up $A(F)$ at the two frequencies and solve for $\alpha$ and $\beta$. In (c), evaluate $H(F)$ at the frequency of the input to set up the output.]

**6.16. (First-Order Filters)** For each filter, sketch the pole-zero plot, sketch the frequency response to establish the filter type, and evaluate the phase delay at low frequencies. Assume that $\alpha = 0.5$.

**(a)** $H(z) = \dfrac{z-\alpha}{z+\alpha}$ **(b)** $H(z) = \dfrac{z-1/\alpha}{z+\alpha}$ **(c)** $H(z) = \dfrac{z+1/\alpha}{z+\alpha}$

**6.17. (Filter Concepts)** Argue for or against the following. Use examples to justify your arguments.

**(a)** All the finite poles of an FIR filter must lie at $z = 0$.
**(b)** An FIR filter is always linear phase.
**(c)** An FIR filter is always stable.
**(d)** A causal IIR filter can never display linear phase.
**(e)** A linear-phase sequence is always symmetric about is midpoint.
**(f)** A minimum-phase filter can never display linear phase.
**(g)** An allpass filter can never display linear phase.
**(h)** An allpass filter can never be minimum phase.
**(i)** The inverse of a minimum-phase filter is also minimum phase.
**(j)** The inverse of a causal filter is also causal.
**(k)** For a causal minimum-phase filter to have a a causal inverse, the filter must have as many finite poles as finite zeros.

**6.18. (Filter Concepts)** Let $H(z) = \dfrac{0.2(z+1)}{z-0.6}$. What type of filter does $H(z)$ describe? Sketch its pole-zero plot.

**(a)** What filter does $H_1(z) = 1 - H(z)$ describe? Sketch its pole-zero plot.
**(b)** What filter does $H_2(z) = H(-z)$ describe? Sketch its pole-zero plot.

**(c)** What type of filter does $H_3(z) = 1 - H(z) - H(-z)$ describe? Sketch its pole-zero plot.
**(d)** Use a combination of the above filters to implement a bandstop filter.

**6.19. (Filter Design by Pole-Zero Placement)** Design the following filters by pole-zero placement.

**(a)** A bandpass filter with a center frequency of $f_0 = 200$ Hz, a 3-dB bandwidth of $\Delta f = 20$ Hz, zero gain at $f = 0$ and $f = 400$ Hz, and a sampling frequency of 800 Hz.
**(b)** A notch filter with a notch frequency of 1 kHz, a 3-dB stopband of 10 Hz, and sampling frequency 8 kHz.

[**Hints and Suggestions:** Convert to digital frequencies and put the poles at $z = Re^{\pm j2\pi F_0}$ where $R \approx 1 - \pi\Delta F$. For (a), put zeros at $z = \pm 1$. For (b), put zeros at $z = e^{\pm j2\pi F_0}$.]

**6.20. (Recursive and IIR Filters)** The terms *recursive* and *IIR* are not always synonymous. A recursive filter could in fact have a finite impulse response and even linear phase. For each of the following recursive filters, find the transfer function $H(z)$ and the impulse response $h[n]$. Which filters (if any) describe IIR filters? Which filters (if any) are linear phase?

**(a)** $y[n] - y[n-1] = x[n] - x[n-2]$
**(b)** $y[n] - y[n-1] = x[n] - x[n-1] - 2x[n-2] + 2x[n-3]$

[**Hints and Suggestions:** Set up $H(z)$ and cancel common factors to generate $h[n]$.]

**6.21. (Systems in Cascade and Parallel)** Consider the filter realization of Figure P6.21.

**(a)** Find its transfer function and impulse response if $\alpha \neq \beta$. Is the overall system FIR or IIR?
**(b)** Find its transfer function and impulse response if $\alpha = \beta$. Is the overall system FIR or IIR?
**(c)** Find its transfer function and impulse response if $\alpha = \beta = 1$. What does the overall system represent?

**FIGURE P.6.21** Filter realization for Problem 6.21

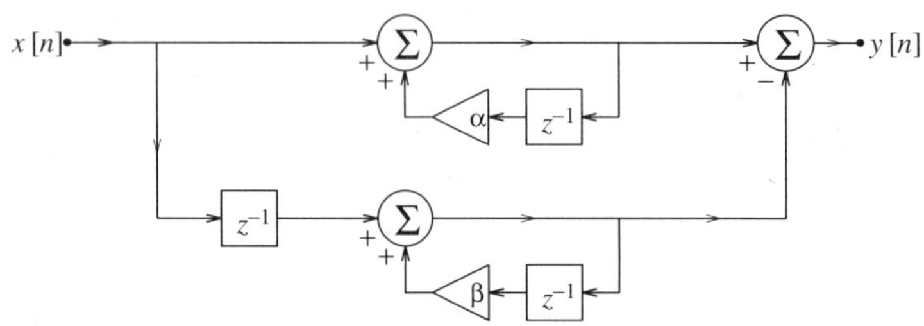


**6.22. (Poles and Zeros)** It is known that the transfer function $H(z)$ of a filter has two poles at $z = 0$, two zeros at $z = -1$, and a dc gain of 8.

**(a)** Find the filter transfer function $H(z)$ and impulse response $h[n]$.
**(b)** Is this an IIR or FIR filter?
**(c)** Is this a causal or noncausal filter?
**(d)** Is this a linear-phase filter? If so, what is the symmetry in $h[n]$?
**(e)** Repeat parts (a) through (d) if another zero is added at $z = -1$.

**6.23. (Frequency Response)** Sketch the pole-zero plot and frequency response of the following systems and describe the function of each system.

**(a)** $y[n] = 0.5x[n] + 0.5x[n-1]$

**(b)** $y[n] = 0.5x[n] - 0.5x[n-1]$

**(c)** $h[n] = \frac{1}{3}\{1, \overset{\Downarrow}{1}, 1\}$

**(d)** $h[n] = \frac{1}{3}\{1, \overset{\Downarrow}{-1}, 1\}$

**(e)** $h[n] = \{\overset{\Downarrow}{0.5}, 1, 0.5\}$

**(f)** $h[n] = \{\overset{\Downarrow}{0.5}, -1, 0.5\}$

**(g)** $H(z) = \dfrac{z - 0.5}{z + 0.5}$

**(h)** $H(z) = \dfrac{z - 2}{z + 0.5}$

**(i)** $H(z) = \dfrac{z + 2}{z + 0.5}$

**(j)** $H(z) = \dfrac{z^2 + 2z + 3}{3z^2 + 2z + 1}$

**6.24. (Inverse Systems)** Consider a system described by $h[n] = 0.5\delta[n] + 0.5\delta[n-1]$.

**(a)** Sketch the frequency response $H(F)$ of this filter.

**(b)** In an effort to recover the input $x[n]$, it is proposed to cascade this filter with another filter whose impulse response is $h_1[n] = 0.5\delta[n] - 0.5\delta[n-1]$, as shown:

$$x[n] \longrightarrow \boxed{h[n]} \longrightarrow \boxed{h_1[n]} \longrightarrow y[n]$$

What is the output of the cascaded filter to the input $x[n]$? Sketch the frequency response $H_1(F)$ and the frequency response of the cascaded filter.

**(c)** What must be the impulse response $h_2[n]$ of a filter connected in cascade with the original filter such that the output of the cascaded filter equals the input $x[n]$, as shown?

$$x[n] \longrightarrow \boxed{h[n]} \longrightarrow \boxed{h_2[n]} \longrightarrow x[n]$$

**(d)** Are $H_2(F)$ and $H_1(F)$ related in any way?

**6.25. (Linear Phase and Symmetry)** Assume a sequence $x[n]$ with real coefficients with all its poles at $z = 0$. Argue for or against the following statements. You may want to exploit two useful facts. First, each pair of terms with reciprocal roots such as $(z - \alpha)$ and $(z - 1/\alpha)$ yields an even symmetric impulse response sequence. Second, the convolution of symmetric sequences is also endowed with symmetry.

**(a)** If all the zeros lie on the unit circle, $x[n]$ must be linear phase.

**(b)** If $x[n]$ is linear phase, its zeros must *always* lie on the unit circle.

**(c)** If there are no zeros at $z = 1$ and $x[n]$ is linear phase, it is also even symmetric.

**(d)** If there is one zero at $z = 1$ and $x[n]$ is linear phase, it is also odd symmetric.

**(e)** If $x[n]$ is even symmetric, there can be no zeros at $z = 1$.

**(f)** If $x[n]$ is odd symmetric, there must be an odd number of zeros at $z = 1$.

**6.26. (Comb Filters)** For each comb filter, identify the pole and zero locations and determine whether it is a notch filter or a peaking filter.

**(a)** $H(z) = \dfrac{z^4 - 0.4096}{z^4 - 0.6561}$

**(b)** $H(z) = \dfrac{z^4 - 1}{z^4 - 0.6561}$

**6.27. (Linear-Phase Filters)** Argue for or against the following:

**(a)** The cascade connection of two linear-phase filters is also a linear-phase filter.

**(b)** The parallel connection of two linear-phase filters is also a linear-phase filter.

6.28. **(Minimum-Phase Filters)** Argue for or against the following:

**(a)** The cascade connection of two minimum-phase filters is also a minimum-phase filter.
**(b)** The parallel connection of two minimum-phase filters is also a minimum-phase filter.

6.29. **(System Analysis)** The impulse response of a system is $h[n] = \delta[n] - \alpha\delta[n-1]$. Determine $\alpha$ and make a sketch of the pole-zero plot for this system to act as

**(a)** a lowpass filter **(b)** a highpass filter **(c)** an allpass filter

6.30. **(System Analysis)** The impulse response of a system is $h[n] = \alpha^n u[n]$, $\alpha \neq 0$. Determine $\alpha$ and make a sketch of the pole-zero plot for this system to act as

**(a)** a stable lowpass filter **(b)** a stable highpass filter **(c)** an allpass filter

6.31. **(Allpass Filters)** Argue for or against the following:

**(a)** The cascade connection of two allpass filters is also an allpass filter.
**(b)** The parallel connection of two allpass filters is also an allpass filter.

6.32. **(Minimum-Phase Systems)** Consider the filter $y[n] = x[n] - 0.65x[n-1] + 0.1x[n-2]$.

**(a)** Find its transfer function $H(z)$ and verify that it is minimum phase.
**(b)** Find an allpass filter $A(z)$ with the same denominator as $H(z)$.
**(c)** Is the cascade $H(z)A(z)$ minimum phase? Is it causal? Is it stable?

6.33. **(Causality, Stability, and Minimum Phase)** Consider two causal, stable, minimum-phase digital filters whose transfer functions are given by

$$F(z) = \frac{z}{z-0.5} \qquad G(z) = \frac{z-0.5}{z+0.5}$$

Argue that the following filters are also causal, stable, and minimum phase.

**(a)** The inverse filter $M(z) = 1/F(z)$
**(b)** The inverse filter $P(z) = 1/G(z)$
**(c)** The cascade $H(z) = F(z)G(z)$
**(d)** The inverse of the cascade $R(z) = 1/H(z)$
**(e)** The parallel connection $N(z) = F(z) + G(z)$

6.34. **(Allpass Filters)** Consider the filter $H(z) = \frac{z+2}{z+0.5}$. The input to this filter is $x[n] = \cos(2n\pi F_0)$.

**(a)** Is $H(z)$ an allpass filter? If so, what is its gain?
**(b)** What is the response $y[n]$ and the phase delay if $F_0 = 0$?
**(c)** What is the response $y[n]$ and the phase delay if $F_0 = 0.25$?
**(d)** What is the response $y[n]$ and the phase delay if $F_0 = 0.5$?

6.35. **(Allpass Filters)** Consider two causal, stable, allpass digital filters whose transfer function is described by

$$F(z) = \frac{0.5z-1}{0.5-z} \qquad G(z) = \frac{0.5z+1}{0.5+z}$$

**(a)** Is the filter $L(z) = F^{-1}(z)$ causal? Stable? Allpass?
**(b)** Is the filter $H(z) = F(z)G(z)$ causal? Stable? Allpass?
**(c)** Is the filter $M(z) = H^{-1}(z)$ causal? Stable? Allpass?
**(d)** Is the filter $N(z) = F(z) + G(z)$ causal? Stable? Allpass?

**6.36. (Stabilization by Allpass Filters)** The transfer function of an unstable filter is

$$H(z) = \frac{z+3}{z-2}$$

A conceptual approach to stabilizing this filter is to cascade it with an allpass filter (so as to preserve the magnitude response).

**(a)** What is the transfer function $A_1(z)$ of a first-order allpass filter that could be used to stabilize this filter? What is the transfer function $H_S(z)$ of the stabilized filter?
**(b)** If $H_S(z)$ is not minimum phase, pick an allpass filter $A_2(z)$ that converts $H_S(z)$ to a minimum-phase filter $H_M(z)$.
**(c)** Verify that $|H(F)| = |H_S(F)| = |H_M(F)|$.
**(d)** Comment on the practical aspects of this approach for stabilizing an unstable filter

**6.37. (Allpass Filters)** Consider a lowpass filter with impulse response $h[n] = (0.5)^n u[n]$. If its input is $x[n] = \cos(0.5n\pi)$, the output will have the form $y[n] = A\cos(0.5n\pi + \theta)$.

**(a)** Find the values of $A$ and $\theta$.
**(b)** What should be the transfer function $H_1(z)$ of a first-order allpass filter that can be cascaded with the lowpass filter to correct for the phase distortion and produce the signal $z[n] = B\cos(0.5n\pi)$ at its output?
**(c)** What should be the gain of the allpass filter in order that $z[n] = x[n]$?

**6.38. (Allpass Filters)** Consider an unstable digital filter whose transfer function is

$$H(z) = \frac{(z+0.5)(2z+0.5)}{(z+5)(2z+5)}$$

This is to be stabilized in a way that does not affect its magnitude spectrum.

**(a)** What must be the transfer function $H_1(z)$ of a filter such that the cascaded filter described by $H_S(z) = H(z)H_1(z)$ is stable?
**(b)** What is the transfer function $H_S(z)$ of the stabilized filter?
**(c)** Is $H_S(z)$ causal? Minimum phase? Allpass?

**6.39. (Signal Delay)** The delay $D$ of a discrete-time energy signal $x[n]$ is defined by

$$D = \frac{\sum_{k=-\infty}^{\infty} kx^2[k]}{\sum_{k=-\infty}^{\infty} x^2[k]}$$

**(a)** Verify that the delay of the linear-phase sequence $x[n] = \{4, 3, 2, 1, \overset{\Downarrow}{0}, 1, 2, 3, 4\}$ is zero.
**(b)** Compute the delay of the signal $f[n] = x[n-1]$.
**(c)** Compute the delay of the signal $g[n] = x[n-2]$.
**(d)** What is the delay for the impulse response of the filter described by $H(z) = \dfrac{1-0.5z}{z-0.5}$?
**(e)** Compute the delay for the impulse response of the first-order allpass filter $H(z) = \dfrac{1+\alpha z}{z+\alpha}$.

[**Hints and Suggestions:** For part (c), $h[n] = -2\delta[n] + 1.5(0.5)^n u[n]$. So, $h[0] = -0.5$ and $h[n] = 1.5(0.5)^n,\ n > 0$. Use this with a table of summations to find the delay.]

6.40. **(Group Delay)** Show that the group delay $t_g$ of a filter described by its transfer function $H(F)$ may be expressed as

$$t_g = \frac{H'_R(F)H_I(F) - H'_I(F)H_R(F)}{2\pi|H(F)|^2}$$

Here, $H(F) = H_R(F) + jH_I(F)$ and the primed quantities describe derivatives with respect to $F$. This result may be used to find the group delay of both FIR and FIR filters.

**(a)** Find the group delay of the FIR filter described by $h[n] = \{\overset{\Downarrow}{\alpha},\ 1\}$.
**(b)** Find the group delay of the FIR filter described by $H(F) = 1 + \alpha e^{-j2\pi F}$.
**(c)** For an IIR filter described by $H(F) = \frac{N(F)}{D(F)}$, the overall group delay may be found as the difference of the group delays of $N(F)$ and $D(F)$. Use this concept to find the group delay of the filter given by $H(F) = \dfrac{\alpha + e^{-j2\pi F}}{1 + \alpha e^{-j2\pi F}}$.

[**Hints and Suggestions:** Start with $t_g = -\dfrac{1}{2\pi}\dfrac{d\phi(F)}{dF} = -\dfrac{1}{2\pi}\dfrac{d}{dF}[\tan^{-1}\dfrac{H_I(F)}{H_R(F)}]$. For (a) and (b), use the given expression to find $t_g$. For (c), use the results from (a) and (b).]

## Computation and Design

6.41. **(FIR Filter Design)** A 22.5-Hz signal is corrupted by 60-Hz hum. It is required to sample this signal at 180 Hz and filter out the interference from the the sampled signal.

**(a)** Design a minimum-length, linear-phase filter that passes the desired signal with unit gain and completely rejects the interference signal.
**(b)** Test your design by applying a sampled version of the desired signal, adding 60-Hz interference, filtering the noisy signal, and comparing the desired signal and the filtered signal.

6.42. **(Comb Filters)** Plot the frequency response of the following filters over $0 \le F \le 0.5$ and describe the action of each filter.

**(a)** $y[n] = x[n] + \alpha x[n-4], \quad \alpha = 0.5$
**(b)** $y[n] = x[n] + \alpha x[n-4] + \alpha^2 x[n-8], \quad \alpha = 0.5$
**(c)** $y[n] = x[n] + \alpha x[n-4] + \alpha^2 x[n-8] + \alpha^3 x[n-12], \quad \alpha = 0.5$
**(d)** $y[n] = \alpha y[n-4] + x[n], \quad \alpha = 0.5$

6.43. **(Filter Design)** An ECG signal sampled at 300 Hz is contaminated by interference due to 60-Hz hum. It is required to design a digital filter to remove the interference and provide a dc gain of unity.

**(a)** Design a 3-point FIR filter (using zero placement) that completely blocks the interfering signal. Plot its frequency response. Does the filter provide adequate gain at other frequencies in the passband? Is this a good design?
**(b)** Design an IIR filter (using pole-zero placement) that completely blocks the interfering signal. Plot its frequency response. Does the filter provide adequate gain at other frequencies in the passband? Is this a good design?
**(c)** The MATLAB-based routine **ecgsim** (on the author's website) simulates one period of an ECG signal. Use the command **yecg=ecgsim(3,9)** to generate one period (300 samples) of the ECG signal. Generate a noisy ECG signal by adding 300 samples of a 60-Hz sinusoid to **yecg**. Obtain filtered signals, using each filter, and compare plots of

the filtered signal with the original signal **yecg**. Do the results support your conclusions of parts (a) and (b)? Explain.

**6.44. (Nonrecursive Forms of IIR Filters)** If we truncate the impulse response of an IIR filter to $N$ terms, we obtain an FIR filter. The larger the truncation index $N$, the better the FIR filter approximates the underlying IIR filter. Consider the IIR filter described by $y[n] - 0.8y[n-1] = x[n]$.

**(a)** Find its impulse response $h[n]$ and truncate it to $N$ terms to obtain $h_N[n]$, the impulse response of the approximate FIR equivalent. Would you expect the greatest mismatch in the response of the two filters to identical inputs to occur for lower or higher values of $n$?

**(b)** Plot the frequency response $H(F)$ and $H_N(F)$ for $N = 3$. Plot the poles and zeros of the two filters. What differences do you observe?

**(c)** Plot the frequency response $H(F)$ and $H_N(F)$ for $N = 10$. Plot the poles and zeros of the two filters. Does the response of $H_N(F)$ show a better match to $H(F)$? How do the pole-zero plots compare? What would you expect to see in the pole-zero plot if you increase $N$ to 50? What would you expect to see in the pole-zero plot as $N \to \infty$?

**6.45. (Phase Delay and Group Delay of Allpass Filters)** Consider the filter $H(z) = \dfrac{z + 1/\alpha}{z + \alpha}$.

**(a)** Verify that this is an allpass filter.

**(b)** Pick values of $\alpha$ that correspond to a phase delay of $t_p = 0.1,\ 0.5,\ 0.9$. For each value of $\alpha$, plot the unwrapped phase, phase delay, and group delay of the filter.

**(c)** Over what range of digital frequencies is the phase delay a good match to the value of $t_p$ computed in part (b)?

**(d)** How does the group delay vary with frequency as $\alpha$ is changed?

**(e)** For each value of $\alpha$, compute the minimum and maximum values of the phase delay and the group delay and the frequencies at which they occur.

**6.46. (Allpass Filters)** Consider a lowpass filter with impulse response $h[n] = (0.5)^n u[n]$. The input to this filter is $x[n] = \cos(0.2n\pi)$. We expect the output to be of the form $y[n] = A\cos(0.2n\pi + \theta)$.

**(a)** Find the values of $A$ and $\theta$.

**(b)** What should be the transfer function $H_1(F)$ of a first-order allpass filter that can be cascaded with the lowpass filter to correct for the phase distortion and produce the signal $z[n] = B\cos(0.2n\pi)$ at its output?

**(c)** What should be the gain of the allpass filter in order that $z[n] = x[n]$?

**6.47. (Decoding a Mystery Message)** During transmission, a message signal gets contaminated by a low-frequency signal and high-frequency noise. The message can be decoded only by displaying it in the time domain. The contaminated signal is provided on the author's website as **mystery1**. Load this signal into MATLAB (using the command **load mystery1**). In an effort to decode the message, try the following steps and determine what the decoded message says.

**(a)** Display the contaminated signal. Can you "read" the message?

**(b)** Display the DFT of the signal to identify the range of the message spectrum.

**(c)** Design a peaking filter (with unit gain) centered about the message spectrum.

**(d)** Filter the contaminated signal and display the filtered signal to decode the message.

CHAPTER 7

# Digital Processing of Analog Signals

## 7.0 Scope and Overview

This chapter begins with a discussion of sampling and quantization that form the critical link between analog and digital signals. It explains how the sampling theorem forms the basis for sampling signals with little or no loss of information and describes schemes for the recovery of analog signals from their samples using idealized and practical methods. It introduces the concept of signal quantization and its effects and concludes with several useful applications.

### 7.0.1 Goals and Learning Objectives

The goals of this chapter are to provide an introduction to the methods and tools used for the digital processing of analog signals and to introduce some practical applications related to audio signal processing. After going through this chapter, the reader should:

1. Understand the concept of sampling and signal reconstruction.
2. Know how to obtain the spectrum of an ideally sampled or zero-order-hold (ZOH) sampled signal.
3. Know how to obtain the reconstructed signal from samples using various interpolating functions.
4. Understand the connection between zero interpolation and spectrum compression.
5. Know how to achieve an increase or reduction in the sampling rate without resampling the original signal.
6. Understand the concept of quantization and quantization noise.
7. Understand the building blocks of a system for the digital processing of analog signals.
8. Know how to design anti-aliasing and anti-imaging analog filters.
9. Understand the process of recording and playback for CD digital audio.
10. Understand the concept of graphic equalizers and parametric equalizers.
11. Understand various digital audio effects.
12. Understand the concept of quantization, oversampling, and sigma-delta DAC.

# 7.1 Ideal Sampling

Ideal sampling describes a sampled signal as a weighted sum of impulses, with the weights being equal to the values of the analog signal at the impulse locations. An ideally sampled signal $x_I(t)$ may be regarded as the product of an analog signal $x(t)$ and a periodic impulse train $i(t)$, as illustrated in Figure 7.1.

The ideally sampled signal may be described mathematically as

$$x_I(t) = x(t)i(t) = x(t)\sum_{n=-\infty}^{\infty}\delta(t - nt_s) = \sum_{n=-\infty}^{\infty}x(nt_s)\delta(t - nt_s) = \sum_{n=-\infty}^{\infty}x[n]\delta(t - nt_s) \tag{7.1}$$

Here, the discrete signal $x[n]$ simply represents the sequence of sample values $x(nt_s)$. Clearly, the sampling operation leads to a potential loss of information in the ideally sampled signal $x_I(t)$ when compared with its underlying analog counterpart $x(t)$. The smaller the sampling interval $t_s$, the less is this loss of information. Intuitively, there must always be some loss of information, no matter how small an interval we use. Fortunately, our intuition notwithstanding, it is indeed possible to sample signals without any loss of information. The catch is that the signal $x(t)$ must be band limited to some finite frequency $B$.

**The Ideally Sampled Signal is a Train of Impulses**

$$x_I(t) = x(t)\sum_{n=-\infty}^{\infty}\delta(t - nt_s) = \sum_{n=-\infty}^{\infty}x(nt_s)\delta(t - nt_s) = \sum_{n=-\infty}^{\infty}x[n]\delta(t - nt_s)$$

The discrete signal $x[n]$ corresponds to the sequence of the sample values $x(nt_s)$.

The spectra associated with the various signals in ideal sampling are illustrated in Figure 7.2. The impulse train $i(t)$ is a periodic signal with period $T = t_s = 1/S$. From analog theory (reviewed in Appendix A), its Fourier series coefficients are given by $I[k] = S$, and

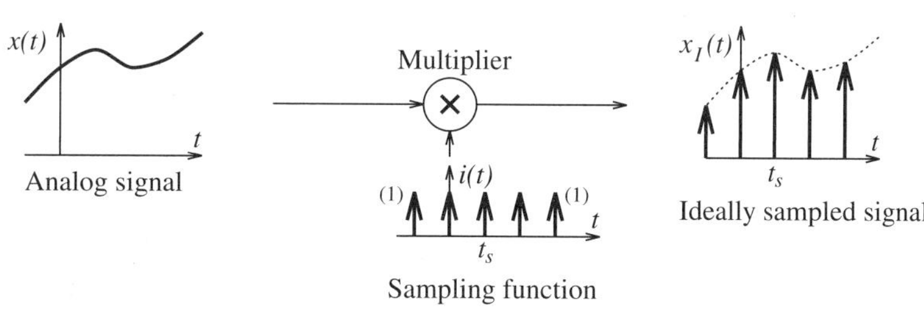


**FIGURE 7.1** The ideal sampling operation. An analog signal multiplied by a periodic impulse train (the ideal sampling function) results in a train of impulses whose strengths match the values of the analog signal at the sampling instants

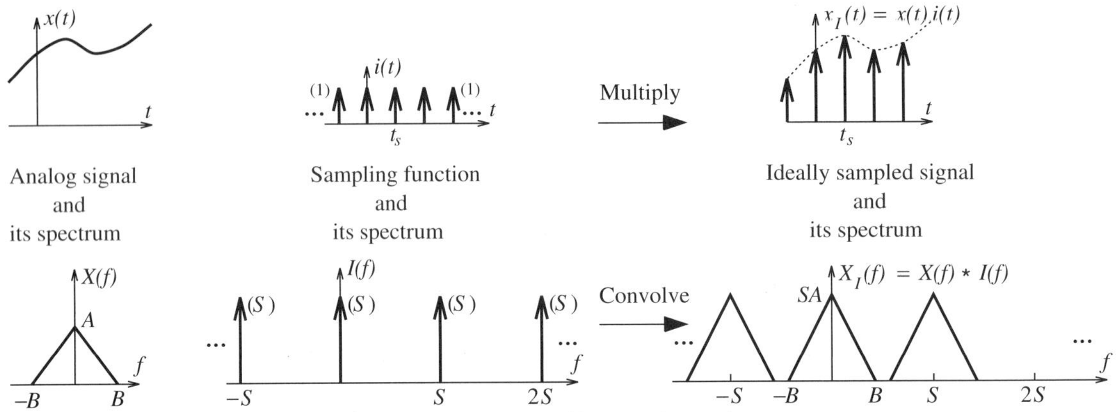


**FIGURE 7.2** Spectra of the signals for ideal sampling. The spectrum $X(f)$ of the analog signal is assumed to be bandlimited to $B$. The spectrum of the impulse sampling function is also a periodic impulse train with equal strengths. Multiplication of the analog signal and the ideal sampling function results in the convolution of their respective spectra. The spectrum of the sampled signal thus consists of replicas of $X(f)$ at multiples of the sampling rate $S$

its Fourier transform $I(f)$ is a train of impulses (at $f = kS$) whose strengths equal $I[k]$.

$$I(f) = \sum_{k=-\infty}^{\infty} I[k]\delta(f - kS) = S \sum_{k=-\infty}^{\infty} \delta(f - kS) \tag{7.2}$$

The ideally sampled signal $x_I(t)$ is the product of $x(t)$ and $i(t)$. Its spectrum $X_I(f)$ is thus described by the convolution

$$X_I(f) = X(f) * I(f) = X(f) * S \sum_{k=-\infty}^{\infty} \delta(f - kS) = S \sum_{k=-\infty}^{\infty} X(f - kS) \tag{7.3}$$

The spectrum $X_I(f)$ consists of $X(f)$ and its shifted replicas or **images**. It is periodic in frequency, with a period that equals the sampling rate $S$.

**The Spectrum of an Ideally Sampled Signal is Periodic with Period $S$**

$$X_I(f) = X(f) * S \sum_{k=-\infty}^{\infty} \delta(f - kS) = S \sum_{k=-\infty}^{\infty} X(f - kS)$$

The spectrum is the periodic extension of $SX(f)$ with period $S$.

Since the spectral image at the origin extends over $(-B, B)$, and the next image (centered at $S$) extends over $(S - B, S + B)$, the images will not overlap if

$$S - B > B \quad \text{or} \quad S > 2B \tag{7.4}$$

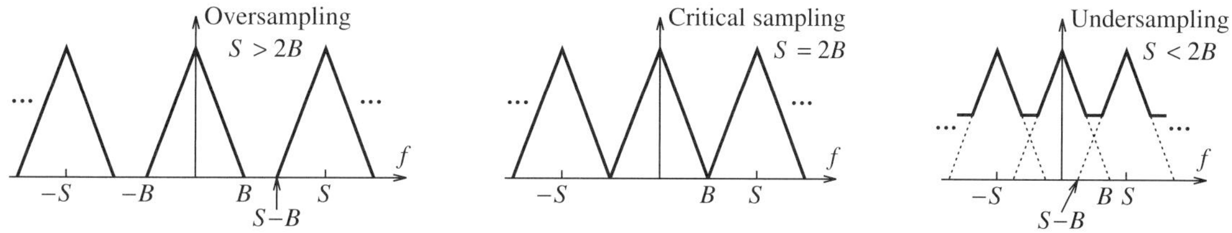


**FIGURE 7.3** Spectrum of an ideally sampled signal for three choices of the sampling frequency $S$. The spectrum $X(f)$ of the analog signal is bandlimited to $B$. If $S > 2B$ (oversampling), the periodic spectrum of the sampled signal shows non-overlapping replicas of $X(f)$. If $S < 2B$ (undersampling), the replicas overlap and one period of the periodic spectrum no longer matches $X(f)$. The critical case $S = 2B$ forms the transition between oversampling and undersampling

Figure 7.3 illustrates the spectra of an ideally sampled band limited signal for three choices of the sampling frequency $S$. As long as the images do not overlap, each period is a replica of the scaled analog-signal spectrum $SX(f)$. We can thus extract $X(f)$ (and hence $x(t)$) as the principal period of $X_I(f)$ (between $-0.5S$ and $0.5S$) by passing the ideally sampled signal through an ideal lowpass filter with a cutoff frequency of $0.5S$ and a gain of $1/S$ over the frequency range $-0.5S \le f \le 0.5S$.

This is the celebrated **sampling theorem**, which tells us that an analog signal band limited to a frequency $B$ can be sampled without loss of information if the sampling rate $S$ exceeds $2B$ (or the sampling interval $t_s$ is smaller than $\frac{1}{2B}$). The *critical* sampling rate $S_N = 2B$ is often called the **Nyquist rate** or **Nyquist frequency** and the critical sampling interval $t_N = 1/S_N = 1/2B$ is called the **Nyquist interval**.

---

**Band Limited Analog Signals can be Sampled without Loss of Information**

If $x(t)$ is band limited to $B$, it must be sampled at $S > 2B$ to prevent loss of information.
The images of $X(f)$ do not overlap in the periodic spectrum $X_I(f)$ of the ideally sampled signal.
We can recover $x(t)$ *exactly* from the principal period $(-0.5S,\ 0.5S)$, using an ideal lowpass filter.

---

If the sampling rate $S$ is less than $2B$, the spectral images overlap and the principal period $(-0.5S,\ 0.5S)$ of $X_I(f)$ is no longer an exact replica of $X(f)$. In this case, we cannot exactly recover $x(t)$, and there is loss of information due to **undersampling**. Undersampling results in spectral overlap. Components of $X(f)$ outside the principal range $(-0.5S,\ 0.5S)$ fold back into this range (due to the spectral overlap from adjacent images). Thus, frequencies higher than $0.5S$ appear as *lower frequencies* in the principal period. This is *aliasing*. The frequency $0.5S$ is also called the **folding frequency**.

Sampling is a band-limiting operation in the sense that (in practice) we typically extract only the principal period of the spectrum, which is band limited to the frequency range

$(-0.5S, 0.5S)$. Thus, the highest frequency we can recover or identify is $0.5S$ and depends only on the sampling rate $S$.

**Example 7.1**

### The Nyquist Rate

Let the signal $x_1(t)$ be band limited to 2 kHz and $x_2(t)$ be band limited to 3 kHz. Using properties of the Fourier transform, we find the Nyquist rate for the following signals.

**(a)** The spectrum of $x_1(2t)$ (time compression) stretches to 4 kHz. Thus, $S_N = 8$ kHz.
**(b)** The spectrum of $x_2(t-3)$ extends to 3 kHz (a time shift changes only the phase). Thus, $S_N = 6$ kHz.
**(c)** The spectrum of $x_1(t) + x_2(t)$ (sum of the spectra) extends to 3 kHz. Thus, $S_N = 6$ kHz.
**(d)** The spectrum of $x_1(t)x_2(t)$ (convolution in the frequency domain) extends to 5 kHz. Thus, $S_N = 10$ kHz.
**(e)** The spectrum of $x_1(t) * x_2(t)$ (product of the spectra) extends only to 2 kHz. Thus, $S_N = 4$ kHz.
**(f)** The spectrum of $x_1(t)\cos(1000\pi t)$ (modulation) is stretched by 500 Hz to 2.5 kHz. Thus, $S_N = 5$ kHz.

## 7.1.1 Sampling of Sinusoids and Periodic Signals

The Nyquist frequency for a sinusoid $x(t) = \cos(2\pi f_0 t + \theta)$ is $S_N = 2f_0$. The Nyquist interval is $t_N = 1/2f_0$ or $t_N = T/2$. This amounts to taking *more than* two samples per period. If, for example, we acquire just two samples per period, starting at a zero crossing, then all sample values will be zero and will yield no information.

If a signal $x(t) = \cos(2\pi f_0 t + \theta)$ is sampled at $S$, the sampled signal is $x[n] = \cos(2\pi n f_0/S + \theta)$. Its spectrum is periodic, with principal period $(-0.5S, 0.5S)$. If $f_0 < 0.5S$, there is no aliasing, and the principal period shows a pair of impulses at $\pm f_0$ (with strength 0.5). If $f_0 > 0.5S$, we have aliasing. The components at $\pm f_0$ are aliased to a lower frequency $\pm f_a$ in the principal range. To find the aliased frequency $|f_a|$, we subtract integer multiples of the sampling frequency from $f_0$ until the result $f_a = f_0 - NS$ lies in the principal range $(-0.5S, 0.5S)$. The spectrum then describes a sampled version of the *lower-frequency* aliased signal $x_a(t) = \cos(2\pi f_a t + \theta)$. The relation between the aliased frequency, original frequency, and sampling frequency is illustrated in Figure 7.4. The aliased frequency always lies in the principal range.

**Finding the Frequencies of Aliased Signals**

The signal $x(t) = \cos(2\pi f_0 t + \theta)$ is recovered as $x_a(t) = \cos(2\pi f_a t + \theta)$ if $S < 2f_0$. $f_a = f_0 - NS$, where $N$ is an integer that places $f_a$ in the principal period $(-0.5S, 0.5S)$.

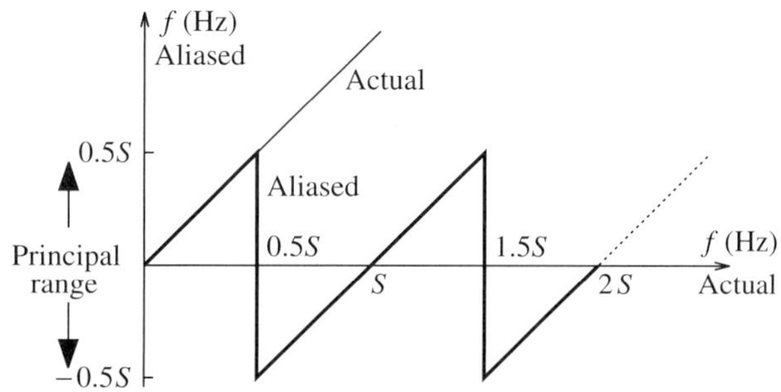


**FIGURE 7.4** Relation between the actual and aliased frequency. There a unique correspondence between the actual and aliased frequency only if $f < 0.5S$. If the actual frequency exceeds $0.5S$, it gets aliased to a lower frequency between $-0.5S$ and $0.5S$. The frequency $0.5S$ at which the aliased frequency flips (changes sign) is called the folding frequency

A periodic signal $x_p(t)$ with period $T$ can be described by a sum of sinusoids at the fundamental frequency $f_0 = 1/T$ and its harmonics $kf_0$. In general, such a signal may not be band limited and cannot be sampled without aliasing for any choice of sampling rate.

**Example 7.2**

### Aliasing and Sampled Sinusoids

**(a)** Consider the sinusoid $x(t) = A\cos(2\pi f_0 t + \theta)$ with $f_0 = 100$ Hz.

1. If $x(t)$ is sampled at $S = 300$ Hz, no aliasing occurs, since $S > 2f_0$.
2. If $x(t)$ is sampled at $S = 80$ Hz, we obtain $f_a = f_0 - S = 20$ Hz, The sampled signal describes a sampled version of the aliased signal $A\cos[2\pi(20)t + \theta]$.
3. If $S = 60$ Hz, we obtain $f_0 - 2S = -20$ Hz. The aliased sinusoid corresponds to $A\cos[2\pi(-20)t + \theta] = A\cos[2\pi(20)t - \theta$. Note the phase reversal.

**(b)** Let
$x_p(t) = 8\cos(2\pi t) + 6\cos(8\pi t) + 4\cos(22\pi t) + 6\sin(32\pi t) + \cos(58\pi t) + \sin(66\pi t)$.
If it is sampled at $S = 10$ Hz, the last four terms will be aliased. The reconstruction of the sampled signal will describe an analog signal whose first two terms are identical to $x_p(t)$, but whose other components are at the aliased frequencies. The following table shows what to expect.

| $f_0$ (Hz) | Aliasing? | Aliased Frequency $f_a$ | Analog Equivalent |
|---|---|---|---|
| 1 | No ($f_0 < 0.5S$) | No aliasing | $8\cos(2\pi t)$ |
| 4 | No ($f_0 < 0.5S$) | No aliasing | $6\cos(8\pi t)$ |
| 11 | Yes | $11 - S = 1$ | $4\cos(2\pi t)$ |
| 16 | Yes | $16 - 2S = -4$ | $6\sin(-8\pi t) = -6\sin(8\pi t)$ |
| 29 | Yes | $29 - 3S = -1$ | $\cos(-2\pi t) = \cos(2\pi t)$ |
| 33 | Yes | $33 - 3S = 3$ | $\sin(6\pi t)$ |

The reconstructed signal corresponds to $x_S(t) = 13\cos(2\pi t) + \sin(6\pi t) + 6\cos(8\pi t) - 6\sin(8\pi t)$, which cannot be distinguished from $x_p(t)$ at the sampling instants $t = nt_s$, where $t_s = 0.1$ s. To avoid aliasing and recover $x_p(t)$, we must choose $S > 2B = 66$ Hz.

**(c)** Suppose we sample a sinusoid $x(t)$ at 30 Hz and obtain the periodic spectrum of the sampled signal, as shown in Figure E7.2C. Is it possible to uniquely identify $x(t)$?

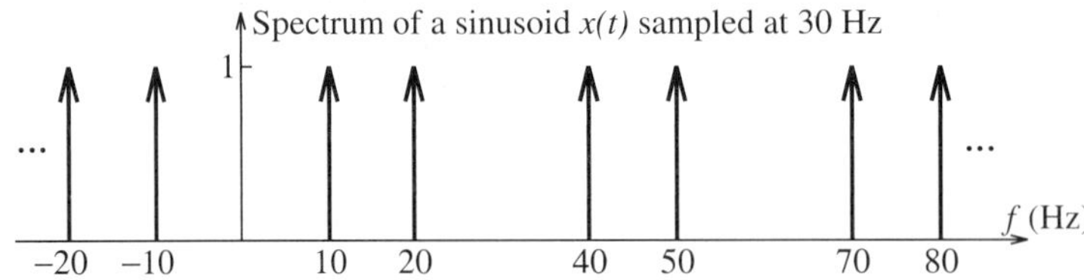


**FIGURE E 7.2.c** Spectrum of sampled sinusoid for Example 7.2(c)

We can certainly identify the period as 30 Hz, and thus $S = 30$ Hz. But we cannot *uniquely* identify $x(t)$, because it could be a sinusoid at 10 Hz (with no aliasing) or a sinusoid at 20 Hz, 50 Hz, 80 Hz, etc. (all aliased to 10 Hz). However, the analog signal $y(t)$ reconstructed from the samples will describe a 10-Hz sinusoid because reconstruction extracts only the principal period, $(-15, 15)$ Hz, of the periodic spectrum. In the absence of any *a priori* information, we almost invariably use the *principal period* as a means to uniquely identify the underlying signal from its periodic spectrum—for better or worse.

### 7.1.2 Application Example: The Sampling Oscilloscope

The sampling theorem tells us that, in order to sample an analog signal without aliasing and loss of information, we must sample above the Nyquist rate. However, some applications depend on the very act of aliasing for their success. One example is the sampling oscilloscope. A conventional oscilloscope cannot directly display a waveform whose bandwidth exceeds the bandwidth of the oscilloscope. However, if the signal is periodic (or can be repeated periodically) and band limited, a new, time-expanded waveform can be built up and displayed by sampling the periodic signal at successively later instants in successive periods. The idea is illustrated in Figure 7.5 for a periodic signal $x(t) = 1 + \cos(2\pi f_0 t)$ with fundamental frequency $f_0$.

If the sampling rate is chosen to be less than $f_0$ (i.e., $S < f_0$), the spectral component at $f_0$ will alias to the smaller frequency $f_a = f_0 - S$. To ensure no phase reversal, the

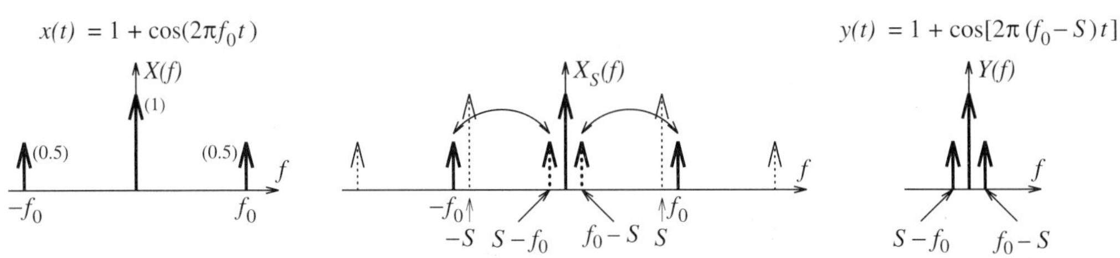


**FIGURE 7.5** The principle of the sampling oscilloscope. A high-frequency sinusoid is sampled at a rate far less than the Nyquist rate. The spectrum of the sampled signal corresponds to a much lower frequency and the signal can be displayed on a conventional oscilloscope

aliased frequency must be less than $0.5S$. Subsequent recovery by a lowpass filter with a cutoff frequency of $f_C = 0.5S$ will yield the signal $y(t) = 1 + \cos[2\pi(f_0 - S)t]$, which represents a time-stretched version of $x(t)$. With $y(t) = x(t/\alpha)$, the stretching factor $\alpha$ and highest sampling rate $S$ are related by

$$\alpha = \frac{f_0}{f_a} = \frac{f_0}{f_0 - S} \qquad S = \frac{(\alpha - 1)f_0}{\alpha} \tag{7.5}$$

The closer $S$ is to the fundamental frequency $f_0$, the larger is the stretching factor, and the more slowed down the signal $y(t)$. This reconstructed signal $y(t)$ is what is displayed on the oscilloscope (with appropriate scale factors to reflect the parameters of the original signal). For a given stretching factor $\alpha$, the sampling rate may also be reduced to $S_m = S/m$ (where $m$ is an integer), as long as it exceeds $2f_0/\alpha$ and ensures that the aliased component appears at $f_0/\alpha$ with no phase reversal. The only disadvantage of choosing smaller sampling rates (corresponding to fewer than one sample per period) is that we must wait much longer in order to acquire enough samples to build up a one-period display.

More generally, if a periodic signal $x(t)$ with a fundamental frequency of $f_0$ is band limited to the $N$th harmonic frequency $f_{\max} = Nf_0$, the stretching factor and sampling rate are still governed by the same equations. As before, the sampling rate may also be reduced to $S_m = S/m$ (where $m$ is an integer), as long as it exceeds $2f_{\max}/\alpha$ and ensures that each harmonic $kf_0$ is aliased to $kf_0/\alpha$ with no phase reversal.

**Example 7.3**

### Sampling Oscilloscope Concepts

**(a)** We wish to see a 100-MHz sinusoid slowed down by a factor of 50. The highest sampling rate we can choose is

$$S = \frac{(\alpha - 1)f_0}{\alpha} = 98 \text{ MHz}$$

This causes the slowed down (aliased) signal to appear at 2 MHz. We may even choose a lower sampling rate (as long as it exceeds 4 MHz) such as $S_2 = S/2 = 49$ MHz, $S_7 = S/7 = 14$ MHz, or $S_{20} = S/20 = 4.9$ MHz and the original signal will still alias to 2 MHz with no phase reversal. However, if we choose $S_{28} = S/28 = 3.5$ MHz (which is less than 4 MHz), the aliased signal will appear at $-1.5$ MHz and no longer reflect the correct stretching factor.

**(b)** We wish to slow down a signal made up of components at 100 MHz and 400 MHz by a factor of 50. The fundamental frequency is $f_0 = 100$ MHz, and the highest sampling rate we can choose is

$$S = \frac{(\alpha - 1)f_0}{\alpha} = 98 \text{ MHz}$$

The frequency of 100 MHz slows down to 2 MHz, and the highest frequency of 400 MHz slows down (aliases) to 8 MHz. We may also choose a lower sampling rate (as long as it exceeds 16 MHz), such as $S_2 = S/2 = 49$ MHz or $S_5 = S/5 = 19.6$ MHz and the frequencies will will still alias to 4 MHz and 8 Mhz with no phase reversal. However, if we choose $S_7 = S/7 = 14$ MHz (which is less than 16 MHz), the 400 MHz component will alias to $-6$ MHz and no longer reflect the correct stretching factor.

### 7.1.3 Sampling of Bandpass Signals

The spectrum of **baseband** signals includes the origin whereas the spectrum of **bandpass signals** occupies a range of frequencies between $f_L$ and $f_H$, where $f_L$ is greater than zero. The quantity $B = f_H - f_L$ is a measure of the bandwidth of the bandpass signal. Even though a Nyquist rate of $2f_H$ can be used to recover such signals, we can often get by with a lower sampling rate. This is especially useful for narrow-band signals (or modulated signals) centered about a very high frequency. To retain all the information in a bandpass signal, we actually need a sampling rate that aliases the entire spectrum to a lower frequency range without overlap. The smallest such frequency is $S = 2f_H/N$, where $N = \text{int}(f_H/B)$ is the integer part of $f_H/B$. Other choices are also possible and result in the following bounds on the sampling frequency $S$

$$\frac{2f_H}{k} \le S \le \frac{2f_L}{k-1}, \qquad k = 1, 2, \ldots, N \tag{7.6}$$

The integer $k$ can range from 1 to $N$. The value $k = N$ yields the smallest sampling rate $S = 2f_H/N$, and $k = 1$ yields the highest sampling rate corresponding to the Nyquist rate $S \ge 2f_H$. If $k$ is even, the spectrum of the sampled signal shows reversal in the baseband.

**Example 7.4**

**Sampling of Bandpass Signals**

Consider a bandpass signal $x_C(t)$ with $f_L = 4$ kHz and $f_H = 6$ kHz. Then $B = f_H - f_L = 2$ kHz, and we compute $N = \text{int}(f_H/B) = 3$. Possible choices for the sampling rate $S$ (in kHz) are given by

$$\frac{12}{k} \le S \le \frac{8}{k-1}, \qquad k = 1,\ 2,\ 3$$

For $k = 3$, we have $S = 4$ kHz. This represents the smallest sampling rate.
For $k = 2$, we have 6 kHz $\le S \le$ 8 kHz. Since $k$ is even, the spectrum shows reversal in the baseband.
For $k = 1$, we have $S \ge 12$ kHz and this corresponds to exceeding the Nyquist rate.

Figure E7.4 shows the spectra of the analog signal and its sampled versions for $S = 4$, 7, 14 kHz.

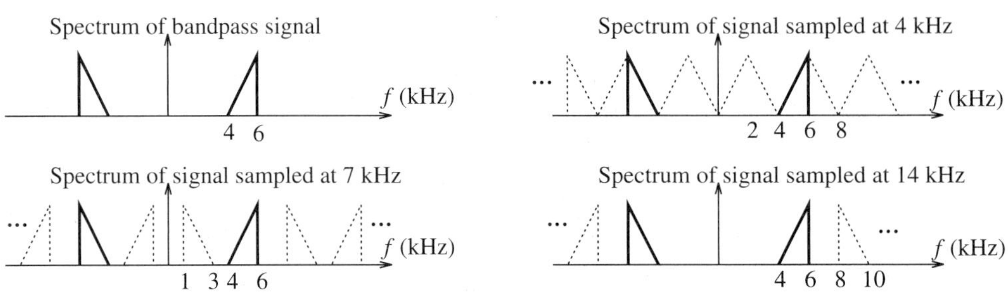


**FIGURE E.7.4** Spectra of bandpass signal and its sampled versions for Example 7.4

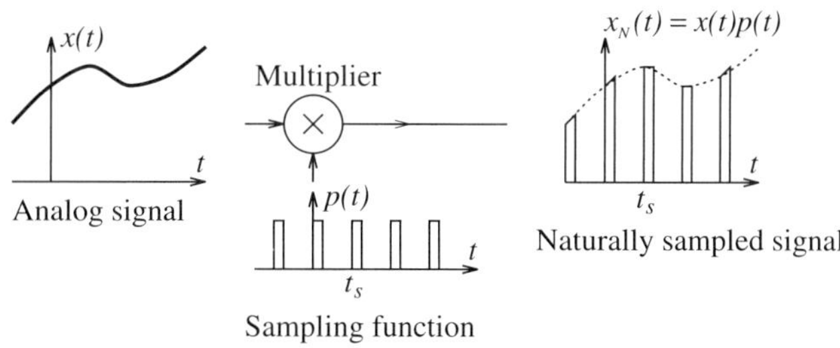


**FIGURE 7.6** Illustrating natural sampling or pulse-amplitude modulation (PAM). An analog signal multiplied by a periodic pulse train (the natural-sampling function) results in a signal that consists of strips of the the analog signal

### 7.1.4 Natural Sampling or Pulse-Amplitude Modulation

Natural sampling is also called **pulse-amplitude modulation** (PAM). Conceptually, natural sampling is equivalent to passing the signal $x(t)$ through a switch that opens and closes every $t_s$ seconds. The action of the switch can be modeled as a periodic pulse train $p(t)$ of unit height, with period $t_s$ and pulse width $t_d$. The sampled signal $x_N(t)$ equals $x(t)$ for the $t_d$ seconds that the switch remains closed and is zero when the switch is open, as illustrated in Figure 7.6.

The naturally sampled or pulse-amplitude modulated (PAM) signal is described by $x_N(t) = x(t)p(t)$. The Fourier transform $P(f)$ is a train of impulses whose strengths $P[k]$ equal the Fourier series coefficients $P[k]$ of $p(t)$:

$$P(f) = \sum_{k=-\infty}^{\infty} P[k]\delta(f - kS) \tag{7.7}$$

The spectrum $X_N(f)$ of the sampled signal $x_S(t) = x(t)p(t)$ is described by the convolution

$$X_N(f) = X(f) * P(f) = X(f) * \sum_{k=-\infty}^{\infty} P[k]\delta(f - kS) = \sum_{k=-\infty}^{\infty} P[k]X(f - kS) \tag{7.8}$$

The various spectra for natural sampling are illustrated in Figure 7.7. Again, $X_N(f)$ is a superposition of $X(f)$ and its amplitude-scaled (by $|P[k]|$), shifted replicas. We know that the coefficients $|P[k]|$ decay as $1/k$. As a result, the spectral images get smaller in height as we move away from the origin. In other words, $X_N(f)$ does not describe a periodic spectrum. However, if there is no spectral overlap, the image centered at the origin equals $P[0]X(f)$, and $X(f)$ can still be recovered by passing the sampled signal through an ideal lowpass filter with a cutoff frequency of $0.5S$ and a gain of $1/P[0]$, $-0.5S \le f \le 0.5S$. In theory, natural sampling can be performed by any periodic signal with a nonzero value of $P[0]$.

### 7.1.5 Zero-Order-Hold Sampling

Zero-order-hold sampling is also called flat-top pulse-amplitude modulation. In practice, analog signals are sampled using **zero-order-hold** (ZOH) devices that hold a sample value constant until the next sample is acquired. This operation is equivalent to ideal sampling followed by a system whose impulse response $h(t) = \text{rect}[(t - 0.5t_s)/t_s]$ is a pulse of unit height and duration $t_s$ (to stretch the incoming impulses). This is illustrated in Figure 7.8.

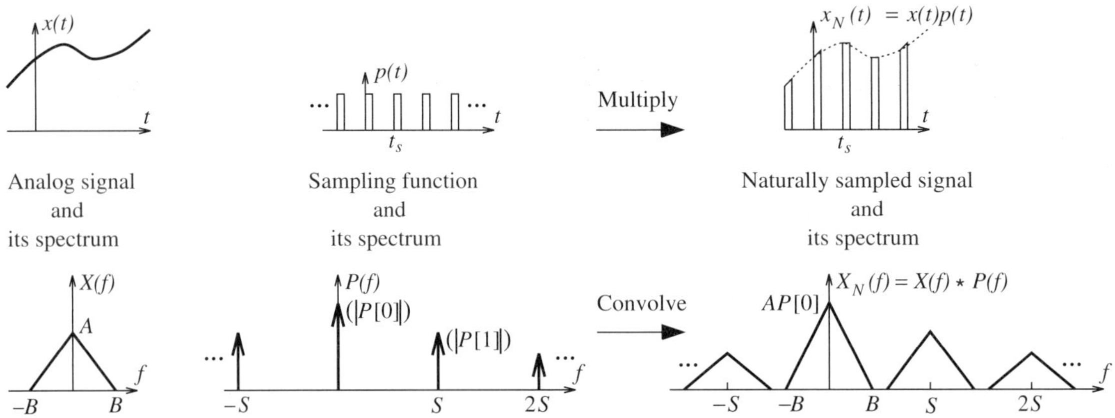


**FIGURE 7.7** Spectra of the signals for natural-sampling. The spectrum $X(f)$ of the analog signal is bandlimited to $B$. The spectrum of the natural-sampling function shows impulses whose strengths equal the Fourier series coefficients of the pulse train. Multiplication of the analog signal and the natural-sampling function results in the convolution of their respective spectra. The spectrum of the sampled signal thus consists of *scaled* replicas of $X(f)$ (scaled by the impulse strengths) at multiples of the sampling rate $S$

The sampled zero-order-hold signal $x_{\text{ZOH}}(t)$ can be regarded as the convolution of $h(t)$ and an ideally sampled signal:

$$x_{\text{ZOH}}(t) = h(t) * x_I(t) = h(t) * \left[\sum_{n=-\infty}^{\infty} x(nt_s)\delta(t - nt_s)\right] \tag{7.9}$$

The transfer function $H(f)$ of the zero-order-hold circuit is the sinc function

$$H(f) = t_s \operatorname{sinc}(ft_s)e^{-j\pi f t_s} = \frac{1}{S}\operatorname{sinc}\left(\frac{f}{S}\right)e^{-j\pi f/S} \tag{7.10}$$

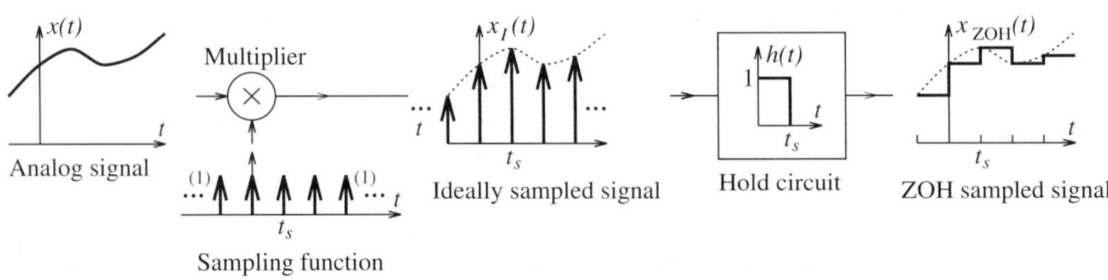


**FIGURE 7.8** Zero-order-hold sampling is equivalent to ideal sampling followed by a hold operation. Multiplying an analog signal by a periodic-impulse train (the ideal sampling function) results in a train of impulses whose strengths match the values of the analog signal at the sampling instants. The hold operation (convolution of the impulses with a rectangular pulse) produces a staircase approximation of the analog signal

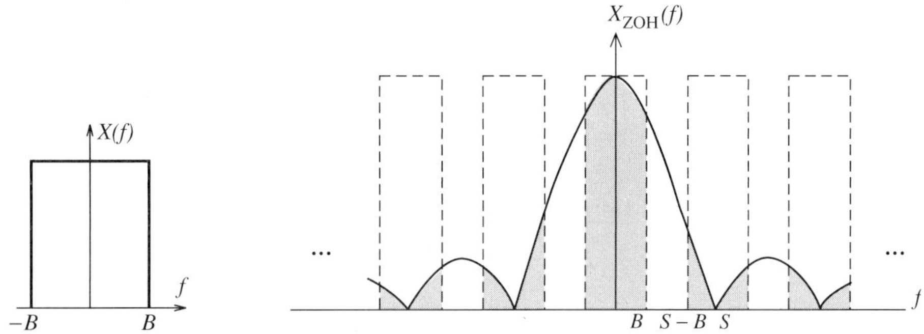


**FIGURE 7.9** Illustrating the spectrum of a zero-order-hold sampled signal. The spectrum of the analog signal is $X(f)$. The spectrum of the ideally sampled signal (images of $X(f)$) is multiplied by the spectrum of the hold system (a sinc with a main lobe and sidelobes) to produce the spectrum of the ZOH signal (shown shaded). Note how the sinc shape results in distortion of the images

Since the spectrum of the ideally sampled signal is $S\sum X(f-kS)$, the spectrum of the zero-order-hold sampled signal $x_{\text{ZOH}}(t)$ is given by the product

$$X_{\text{ZOH}}(f) = \operatorname{sinc}\left(\frac{f}{S}\right)e^{-j\pi f/S}\sum_{k=-\infty}^{\infty} X(f-kS) \tag{7.11}$$

This spectrum is illustrated in Figure 7.9. The term $\operatorname{sinc}(f/S)$ attenuates the spectral images $X(f-kS)$ and causes *sinc distortion*. The higher the sampling rate $S$, the less is the distortion in the spectral image $X(f)$ centered at the origin.

An ideal lowpass filter with unity gain over $-0.5S \le f \le 0.5S$ recovers the *distorted* signal

$$\tilde{X}(f) = X(f)\operatorname{sinc}\left(\frac{f}{S}\right)e^{-j\pi f/S}, \qquad -0.5S \le f \le 0.5S \tag{7.12}$$

To recover $X(f)$ with no amplitude distortion, we must use a *compensating* filter that negates the effects of the sinc distortion by providing a concave-shaped magnitude spectrum corresponding to the reciprocal of the sinc function over the principal period $|f| \le 0.5S$, as shown in Figure 7.10.

The magnitude spectrum of the compensating filter is given by

$$|H_r(f)| = \frac{1}{\operatorname{sinc}(f/S)}, \qquad |f| \le 0.5S \tag{7.13}$$

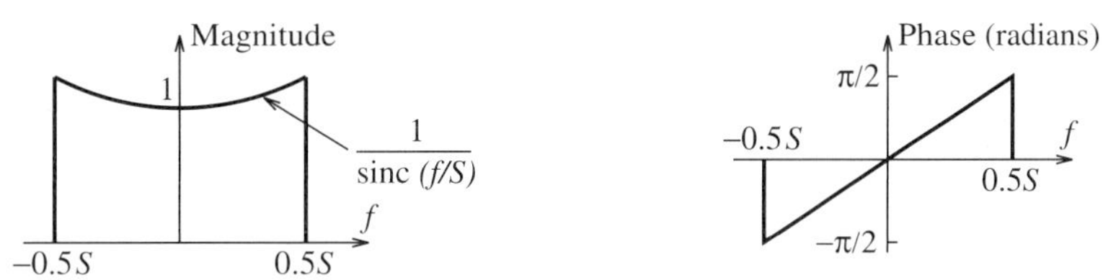


**FIGURE 7.10** Spectrum of a filter that compensates for sinc distortion. To undo its effects, the filter gain is chosen as the reciprocal of the sinc (in the principal range).The filter is chosen to have linear phase

**Example 7.5** **Sampling Operations**

**(a)** The signal $x(t) = \text{sinc}(4000t)$ is sampled at a sampling frequency $S = 5$ kHz. The Fourier transform of $x(t) = \text{sinc}(4000t)$ is $X(f) = (1/4000)\text{rect}(f/4000)$ (a rectangular pulse over $\pm 2$ kHz). The signal $x(t)$ is band limited to $f_B = 2$ kHz. The spectrum of the ideally sampled signal is periodic with replication every 5 kHz and may be written as

$$X_I(f) = \sum_{k=-\infty}^{\infty} SX(f - kS) = 1.25 \sum_{k=-\infty}^{\infty} \text{rect}\left(\frac{f - 5000k}{4000}\right)$$

**(b)** The magnitude spectrum of the zero-order-hold sampled signal is a version of the ideally sampled signal distorted (multiplied) by $\text{sinc}(f/S)$ and described by

$$X_{\text{ZOH}}(f) = \text{sinc}\left(\frac{f}{S}\right) \sum_{k=-\infty}^{\infty} X(f - kS) = 1.25\,\text{sinc}(0.0002f) \sum_{k=-\infty}^{\infty} \text{rect}\left(\frac{f - 5000k}{4000}\right)$$

**(c)** The spectrum of the naturally sampled signal, assuming a rectangular-sampling pulse train $p(t)$ with unit height and a duty ratio of 0.5 is given by

$$X_N(f) = \sum_{k=-\infty}^{\infty} P[k]X(f - kS) = \frac{1}{4000} \sum_{k=-\infty}^{\infty} P[k]\text{rect}\left(\frac{f - 5000k}{4000}\right)$$

Here, $P[k]$ are the Fourier series coefficients of $p(t)$, with $P[k] = 0.5\,\text{sinc}(0.5k)$.

## 7.2 Sampling, Interpolation, and Signal Recovery

For a sampled sequence obtained from an analog signal $x(t)$, an important aspect is the recovery of the original signal from its samples. This requires "filling in" the missing details or interpolating between the sampled values. The nature of the interpolation that recovers $x(t)$ may be discerned by considering the sampling operation in the time domain. Of the three sampling operations considered earlier, only ideal (or impulse) sampling leads to a truly discrete signal $x[n]$, whose samples equal the strengths of the impulses $x(nt_s)\delta(t - nt_s)$ at the sampling instants $nt_s$. This is the only case we pursue.

### 7.2.1 Ideal Recovery and the Sinc Interpolating Function

The ideally sampled signal $x_I(t)$ is the product of the impulse train $i(t) = \sum \delta(t - nt_s)$ and the analog signal $x(t)$ and may be written as

$$x_I(t) = x(t) \sum_{n=-\infty}^{\infty} \delta(t - nt_s) = \sum_{n=-\infty}^{\infty} x(nt_s)\delta(t - nt_s) = \sum_{n=-\infty}^{\infty} x[n]\delta(t - nt_s) \quad (7.14)$$

The discrete signal $x[n]$ is just the sequence of samples $x(nt_s)$. We can recover $x(t)$ by passing $x_I(t)$ through an ideal lowpass filter with a gain of $t_s$ and a cutoff frequency of $0.5S$. The frequency-domain and time-domain equivalents are illustrated in Figure 7.11.

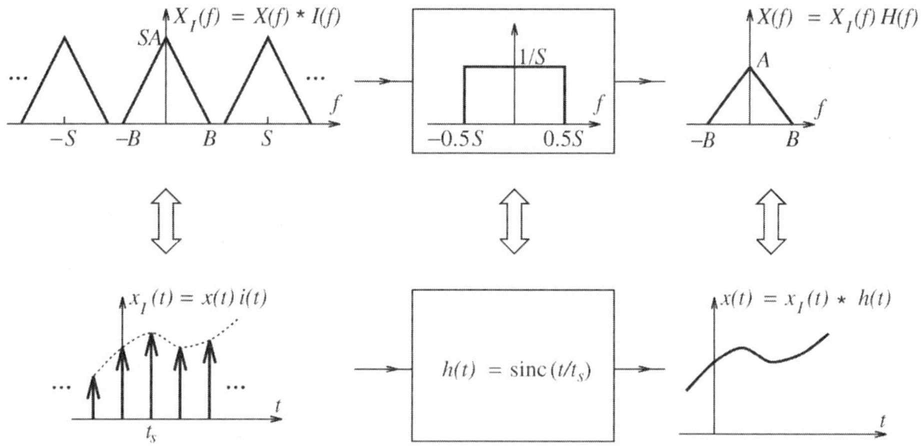


**FIGURE 7.11** Recovery of an analog signal from its sampled version. Ideal recovery uses an ideal lowpass filter that extracts the central period of the periodic spectrum of the sampled signal. The equivalent operation in the time domain corresponds to the convolution of the sampled signal with the impulse response of the ideal filter. If the sampling rate exceeds the Nyquist rate, the central period contains just the spectrum of the original signal, and we recover the original signal

The impulse response of the ideal lowpass filter is a sinc function given by $h(t) = \text{sinc}(t/t_s)$. The recovered signal $x(t)$ may therefore be described as the convolution

$$x(t) = x_I(t) * h(t) = \left[\sum_{n=-\infty}^{\infty} x(nt_s)\delta(t - nt_s)\right] * h(t) = \sum_{n=-\infty}^{\infty} x[n]h(t - nt_s) \qquad (7.15)$$

This describes the superposition of shifted versions of $h(t)$ weighted by the sample values $x[n]$. Substituting for $h(t)$, we obtain the following result that allows us to recover $x(t)$ exactly from its samples $x[n]$ as a sum of scaled shifted versions of sinc functions:

$$x(t) = \sum_{n=-\infty}^{\infty} x[n]\text{sinc}\left(\frac{t - nt_s}{t_s}\right) \qquad (7.16)$$

The signal $x(t)$ equals the superposition of shifted versions of $h(t)$ weighted by the sample values $x[n]$. At each sampling instant, we replace the sample value $x[n]$ by a sinc function whose peak value equals $x[n]$ and whose zero crossings occur at all the other sampling instants. The sum of these sinc functions yields the analog signal $x(t)$, as illustrated in Figure 7.12.

If we use a lowpass filter whose impulse response is $h_f(t) = 2t_s B\,\text{sinc}(2Bt)$ (i.e., whose cutoff frequency is $B$ instead of $0.5S$), the recovered signal $x(t)$ may be described by the convolution

$$x(t) = x_I(t) * h_f(t) = \sum_{n=-\infty}^{\infty} 2t_s B\, x[n]\text{sinc}[2B(t - nt_s)] \qquad (7.17)$$

This general result is valid for *any oversampled signal* with $t_s \le 0.5/B$ and reduces to the previously obtained result if the sampling rate $S$ equals the Nyquist rate (i.e., $S = 2B$).

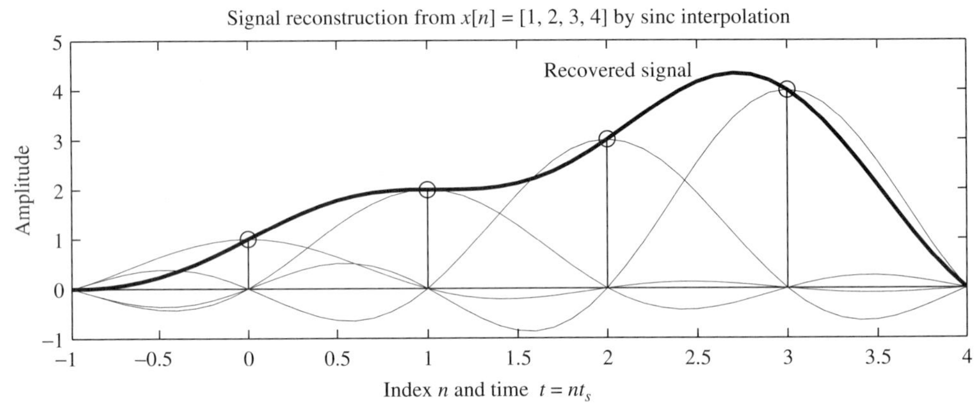


**FIGURE 7.12** Ideal recovery of an analog signal by sinc interpolation. The recovered signal is the sum of sinc functions centered at the sampling instants. The zeros of these sinc functions occur at multiples of the sampling interval

Sinc interpolation is unrealistic from a practical viewpoint. The infinite extent of the sinc means that it cannot be implemented on-line, and perfect reconstruction requires all its past and future values. We could truncate it on either side after its magnitude becomes small enough. But, unfortunately, it decays very slowly and must be preserved for a fairly large duration (covering many past and future sampling instants) in order to provide a reasonably accurate reconstruction. Since the sinc function varies smoothly, it cannot properly reconstruct a discontinuous signal at the discontinuities even with a large number of values. Sinc interpolation is also referred to as **band limited interpolation** and forms the yardstick by which all other schemes are measured in their ability to reconstruct band limited signals.

### 7.2.2 Interpolating Functions

Since the sinc interpolating function is a poor choice in practice, we must look to other interpolating signals. If an analog signal $x(t)$ is to be recovered from its sampled version $x[n]$ using an interpolating function $h_i(t)$, what must we require of $h_i(t)$ to obtain a good approximation to $x(t)$? At the very least, the interpolated approximation $\tilde{x}(t)$ should match $x(t)$ at the sampling instants $nt_s$. This suggests that $h_i(t)$ should equal zero at all sampling instants, except the origin where it must equal unity, such that

$$h_i(t) = \begin{cases} 1, & t = 0 \\ 0, & t = nt_s, \quad (n = \pm 1, \pm 2, \pm 3, \ldots) \end{cases} \tag{7.18}$$

In addition, we also require $h_i(t)$ to be absolutely integrable to ensure that it stays finite between the sampling instants. The interpolated signal $\tilde{x}(t)$ is simply the convolution of $h_i(t)$ with the ideally sampled signal $x_I(t)$ or a summation of shifted versions of the interpolating function

$$\tilde{x}(t) = h_i(t) * x_I(t) = \sum_{n=-\infty}^{\infty} x[n]h_i(t - nt_s) \tag{7.19}$$

At each instant $nt_s$, we erect the interpolating function $h_i(t - nt_s)$, scale it by $x[n]$, and sum to obtain $\tilde{x}(t)$. At a sampling instant $t = kt_s$, the interpolating function $h_i(kt_s - nt_s)$ equals zero, unless $n = k$ when it equals unity. As a result, $\tilde{x}(t)$ *exactly* equals $x(t)$ at each sampling instant. At all other times, the interpolated signal $\tilde{x}(t)$ is only an approximation to the actual signal $x(t)$.

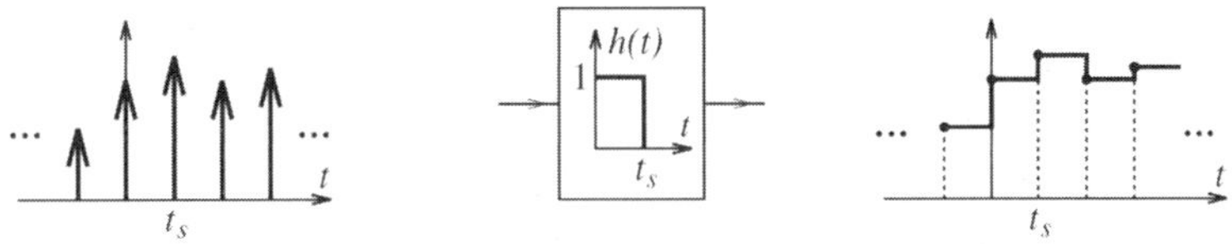


**FIGURE 7.13** Step interpolation is just a zero-order-hold operation on an ideally sampled signal. It produces a staircase approximation of the original analog signal

## 7.2.3 Interpolation in Practice

The nature of $h_i(t)$ dictates the nature of the interpolating system both in terms of its causality, stability, and physical realizability. It also determines how good the reconstructed approximation is. There is no "best" interpolating signal. Some are better in terms of their accuracy, others are better in terms of their cost effectiveness, still others are better in terms of their numerical implementation.

### Step Interpolation

Step interpolation is illustrated in Figure 7.13 and uses a rectangular interpolating function of width $t_s$, given by $h(t) = \text{rect}[(t - 0.5t_s)/t_s]$, to produce a stepwise or staircase approximation to $x(t)$. Even though it appears crude, it is quite useful in practice.

At any time between two sampling instants, the reconstructed signal equals the previously sampled value and does not depend on any future values. This is useful for *on-line* or *real-time* processing where the output is produced at the same rate as the incoming data. Step interpolation results in exact reconstruction of signals that are piecewise constant.

A system that performs step interpolation is just a *zero-order-hold*. A practical digital-to-analog converter (DAC) for sampled signals uses a zero-order-hold for a staircase approximation (step interpolation) followed by a lowpass (anti-imaging) filter (for smoothing the steps).

### Linear Interpolation

Linear interpolation is illustrated in Figure 7.14 and uses the interpolating function $h(t) = \text{tri}(t/t_s)$ to produce a linear approximation to $x(t)$ between the sample values.

At any instant $t$ between adjacent sampling instants $nt_s$ and $(n+1)t_s$, the reconstructed signal equals $x[n]$ plus an increment that depends on the slope of the line joining $x[n]$ and $x[n+1]$. We have

$$\tilde{x}(t) = x[n] + (t - nt_s)\frac{x[n+1] - x[n]}{t_s}, \quad nt_s \le t < (n+1)t_s \tag{7.20}$$

This operation requires one future value of the input and cannot actually be implemented on-line. It can, however, be realized with a delay of one sampling interval $t_s$, which is tolerable

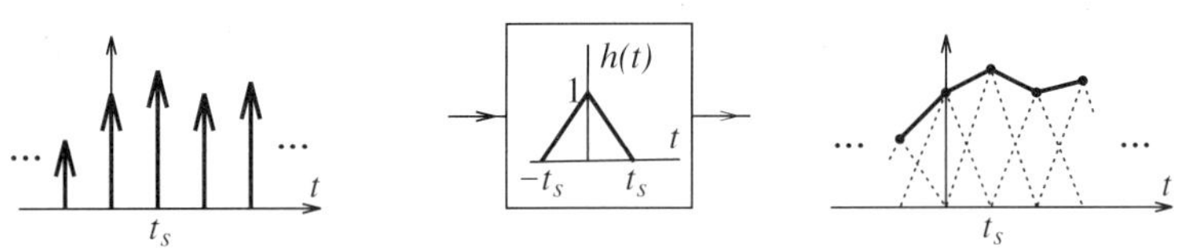


**FIGURE 7.14** Linear interpolation is equivalent to using tri functions centered at the sampling instants of an ideally sampled signal to produce a piecewise linear approximation of the original analog signal

in many situations. Systems performing linear interpolation are also called **first-order-hold systems**. They yield exact reconstructions of piecewise linear signals.

### Raised-Cosine Interpolating Function

The sinc interpolating function forms the basis for several others described by the generic relation

$$h(t) = g(t)\text{sinc}(t/t_s), \quad \text{where} \quad g(0) = 1 \tag{7.21}$$

One of the more commonly used of these is the **raised-cosine interpolating function** described by

$$h_{\text{rc}}(t) = \frac{\cos(\pi Rt/t_s)}{1 - (2Rt/t_s)^2}\,\text{sinc}(t/t_s), \quad 0 \le R \le 1 \tag{7.22}$$

Here, $R$ is called the **roll-off factor**. Like the sinc interpolating function, $h_{\text{rc}}(t)$ equals 1 at $t = 0$ and 0 at the other sampling instants. It exhibits faster decaying oscillations on either side of the origin for $R > 0$ as compared to the sinc function. This faster decay results in improved reconstruction if the samples are not acquired at exactly the sampling instants (in the presence of *jitter*, that is). It also allows fewer past and future values to be used in the reconstruction as compared with the sinc interpolating function. The terminology *raised cosine* is actually based on the shape of its spectrum. For $R = 0$, the raised-cosine interpolating function reduces to the sinc interpolating function.

**Example 7.6**

#### Signal Reconstruction from Samples

Let $x[n] = \{-1, 2, 3, 2\}$, $t_s = 1$. What is the value of the reconstructed signal $x(t)$ at 2.5 s that results from step, linear, sinc, and raised cosine (with $R = 0.5$) interpolation?

**(a)** For step interpolation, the signal value at $t = 2.5$ s is simply the value at $t = 2$. Thus, $\tilde{x}(2.5) = 3$.

**(b)** If we use linear interpolation, the signal value at $t = 2.5$ s is simply the average of the values at $t = 2$ and $t = 3$. Thus, $\tilde{x}(2.5) = 0.5(3 + 2) = 2.5$.

**(c)** If we use sinc interpolation, we obtain

$$\tilde{x}(t) = \sum_{k=0}^{3} x[k]\text{sinc}(t - kt_s) = -\text{sinc}(t) + 2\,\text{sinc}(t-1) + 3\,\text{sinc}(t-2) + 2\,\text{sinc}(t-3)$$

So, $\tilde{x}(2.5) = 0.1273 - 0.4244 + 1.9099 + 2.5465 = 4.1592$.

**(d)** If we use raised-cosine interpolation (with $R = 0.5$), we obtain

$$\tilde{x}(t) = \sum_{k=0}^{3} x[k]\text{sinc}(t-k)\frac{\cos[0.5\pi(t-k)]}{1-(t-k)^2}$$

We use this equation to compute

$$\tilde{x}(t) = \frac{\text{sinc}(t)\cos(0.5\pi t)}{1-t^2} + 2\frac{\text{sinc}(t-1)\cos[0.5\pi(t-1)]}{1-(t-1)^2} + 3\frac{\text{sinc}(t-2)\cos[0.5\pi(t-2)]}{1-(t-2)^2} + 4\frac{\text{sinc}(t-3)\cos[0.5\pi(t-3)]}{1-(t-3)^2}$$

Thus, $\tilde{x}(2.5) = 0.0171 - 0.2401 + 1.8006 + 2.4008 = 3.9785$.

# 7.3 Sampling Rate Conversion

In practice, different parts of a DSP system are often designed to operate at different sampling rates because of the advantages it offers. Sampling rate conversion can be performed directly on an already sampled signal (without having to re-sample the original analog signal) and requires concepts based on interpolation and decimation. Figure 7.15 shows the spectrum of sampled signals obtained by sampling a band limited analog signal $x(t)$ whose spectrum is $X(f)$ at a sampling rate $S$, a higher rate $NS$, and a lower rate $S/M$. All three rates exceed the Nyquist rate to prevent aliasing.

The spectrum of the oversampled signal shows a gain of $N$ but covers a smaller fraction of the principal period. The spectrum of a signal sampled at the lower rate $S/M$ is a stretched version with a gain of $1/M$. In terms of the digital frequency $F$, the period of all three sampled versions is unity. One period of the spectrum of the signal sampled at $S$ Hz extends to $B/S$, whereas the spectrum of the same signal sampled at $NS$ Hz extends only to $B/NS$ Hz, and the spectrum of the same signal sampled at $S/M$ Hz extends farther out to $BM/S$ Hz. After an analog signal is first sampled, all subsequent sampling rate changes are typically made by manipulating the signal samples (and not resampling the analog signal). The key to the process lies in interpolation and decimation.

## 7.3.1 Zero Interpolation and Spectrum Compression

A property of the DTFT that forms the basis for signal interpolation and sampling rate conversion is that $M$-fold zero interpolation of a discrete-time signal $x[n]$ leads to an $M$-fold spectrum compression and replication, as illustrated in Figure 7.16.

The zero-interpolated signal $y[n] = x^{\uparrow}[n/N]$ is nonzero only if $n = kN$, $k = 0, \pm 1, \pm 2, \ldots$ (i.e., if $n$ is an integer multiple of $N$). The DTFT $Y_p(F)$ of $y[n]$ (with $n = kN$) may be expressed as

$$Y_p(F) = \sum_{n=-\infty}^{\infty} y[n]e^{-j2\pi nF} = \sum_{k=-\infty}^{\infty} y[kN]e^{-j2\pi kNF} = \sum_{k=-\infty}^{\infty} x[k]e^{-j2\pi kNF} = X_p(NF) \tag{7.23}$$

This describes $Y_p(F)$ as a scaled (compressed) version of the periodic spectrum $X_p(F)$ and leads to $N$-fold spectrum replication. This is exactly analogous to the Fourier series

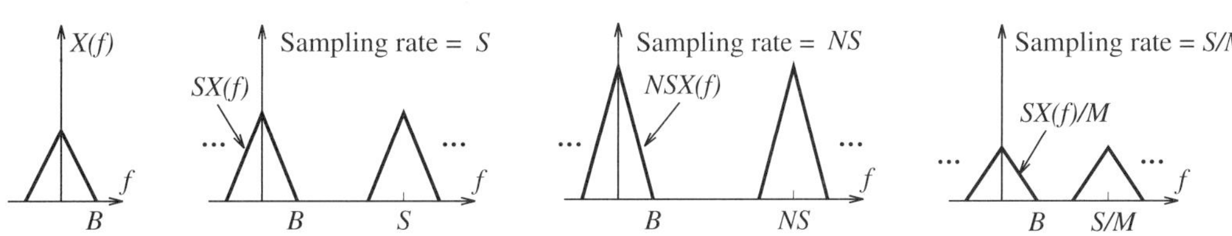


**FIGURE 7.15** Spectra of a signal sampled at three sampling rates. The original spectrum is $X(f)$. The next spectrum corresponds to the signal sampled at the sampling rate $S$ that exceeds the Nyquist rate. If the sampling rate is increased to $NS$, the images move further apart and are scaled $N$. If the sampling rate is decreased to $S/M$, the images move closer together and are scaled by $1/M$. As long as $S/M$ also exceeds the Nyquist rate, the images will not overlap

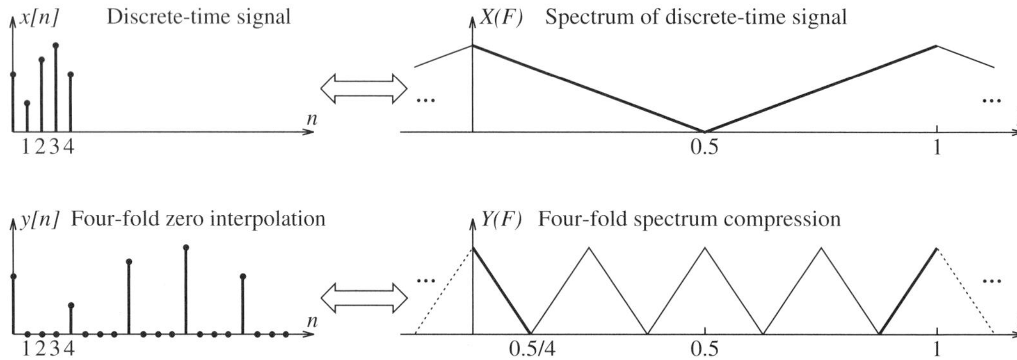


**FIGURE 7.16** Zero interpolation of a signal leads to spectrum compression. Zero interpolation of the signal $x[n]$ by a factor of 4 results in the interpolated signal $y[n]$. The spectrum $Y(F)$ shows four-fold compression compared to the spectrum $Y(F)$

result for analog periodic signals where spectrum zero interpolation produces replication (compression) of the periodic signal. The spectrum of the interpolated signal $y[n]$ shows $N$ compressed images per period, centered at $F = \frac{k}{N}$. The image centered at $F = 0$ occupies the frequency range $|F| \le 0.5/N$.

Similarly, the spectrum of a decimated signal $y[n] = x_{\downarrow}[nM]$ is a stretched version of the original spectrum and is described by

$$Y_p(F) = \frac{1}{M} X_p\left(\frac{F}{M}\right) \tag{7.24}$$

The factor $1/M$ ensures that we satisfy Parseval's relation. If the spectrum of the original signal $x[n]$ extends to $|F| \le 0.5/M$ in the central period, the spectrum of the decimated signal extends over $|F| \le 0.5$, and there is no aliasing or overlap between the spectral images. As a result, the central image represents a stretched version of the spectrum of the original signal $x[n]$. Otherwise, there is aliasing and the stretched images overlap. Because the images are added where overlap exists, the central image gets distorted and no longer represents a stretched version of the spectrum of the original signal. This is a situation that is best avoided.

### The Spectra of Zero-Interpolated and Decimated Signals

Zero interpolation by $N$ gives $N$ compressed images per period centered at $F = \frac{k}{N}$. In decimation by $M$, the gain is reduced by$M$ and images are stretched by $M$ (and added if they overlap).

**Example 7.7**

### Zero Interpolation and Spectrum Replication

The spectrum of a signal $x[n]$ is $X(F) = 2\,\text{tri}(5F)$. Sketch $X(F)$ and the spectra of the following signals and explain how they are related to $X(F)$.

1. The zero-interpolated signal $y[n] = x^{\uparrow}[n/2]$
2. The decimated signal $d[n] = x_{\downarrow}[2n]$
3. The signal $g[n]$ that equals $x[n]$ for even $n$, and 0 for odd $n$

Refer to Figure E7.7 for the spectra.

**(a)** The spectrum $Y(F)$ is a compressed version of $X(F)$ with $Y(F) = X(2F)$.

**(b)** The spectrum $D(F) = 0.5X(0.5F)$ is a stretched version of $X(F)$ with a gain factor of 0.5.

**(c)** The signal $g[n]$ may be expressed as $g[n] = 0.5(x[n] + (-1)^n x[n])$. Its spectrum (or DTFT) is described by $G(F) = 0.5[X(F) + X(F - 0.5)]$. We may also obtain $g[n]$ by first decimating $x[n]$ by 2 (to get $d[n]$), and then zero interpolating $d[n]$ by 2.

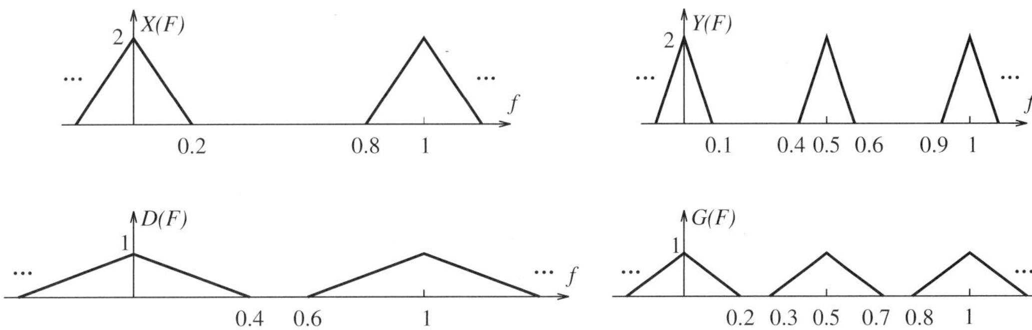


**FIGURE E.7.7** The spectra of the signals for Example 7.7

## 7.3.2 Sampling Rate Increase

A sampling rate increase by an integer factor $N$ involves zero interpolation and digital lowpass filtering, as shown in Figure 7.17. The signals and their spectra at various points in the system are illustrated in Figure 7.18 for $N = 2$.

An up-sampler inserts $N - 1$ zeros between signal samples and results in an $N$-fold zero-interpolated signal corresponding to the higher sampling rate $NS$. Zero interpolation by $N$ results in $N$-fold replication of the spectrum whose central image over $|F| \leq 0.5/N$ corresponds to the spectrum of the oversampled signal (except for a gain of $N$). The spurious images are removed by passing the zero-interpolated signal through a lowpass filter with a

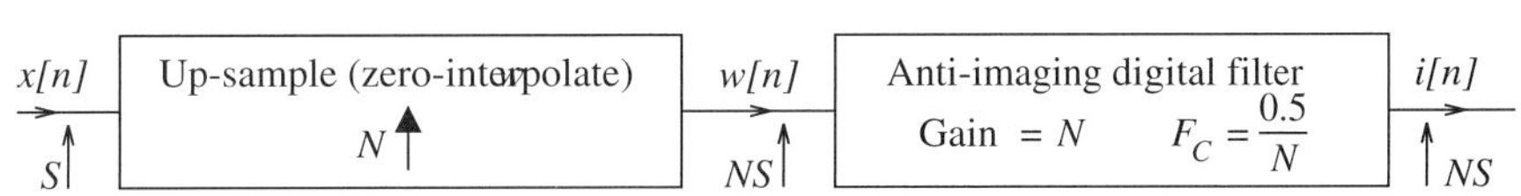


**FIGURE 7.17** Sampling rate increase by an integer factor $N$ requires zero interpolation followed by lowpass filtering. Zero interpolation results in $N$-fold compression of the spectrum. The central period contains $N$ compressed images. All but one is filtered out by the anti-imaging filter with cutoff frequency $F_C = 0.5/N$. This produces a signal sampled at $N$ times the original rate

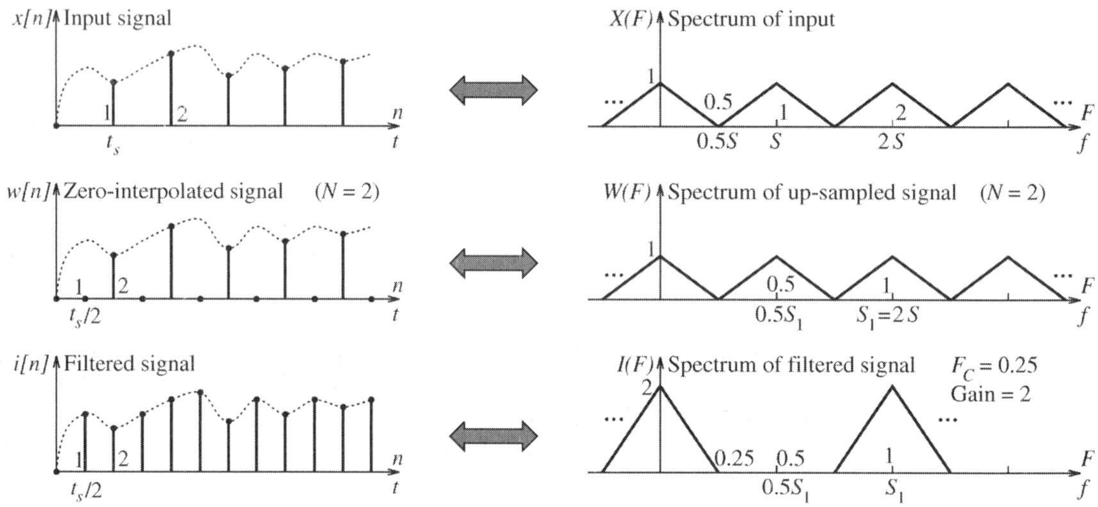


**FIGURE 7.18** Spectra of signals when increasing the sampling rate by two. The original spectrum has one image per period. Zero interpolation produces two compressed images per period. A lowpass filter with a cutoff frequency of $F_C = 0.25$ and a gain of 2 eliminates one image. Filtering of the spurious image in the spectrum is equivalent to replacing the zeros in the zero-interpolated signal by actual values of the original signal. The signal $i[n]$ is thus sampled at twice the original rate

gain of $N$ and a cutoff frequency of $F_C = 0.5/N$ to obtain the required oversampled signal. If the original samples were acquired at a rate that *exceeds* the Nyquist rate, the cutoff frequency of the digital lowpass filter can be made smaller than $0.5/N$. This process yields exact results as long as the underlying analog signal has been sampled above the Nyquist rate and the filtering operation is assumed ideal.

**Example 7.8**

### Up-Sampling and Filtering

In the system of Figure E7.8(1), the spectrum of the input signal is given by $X(F) = \text{tri}(2F)$, the up-sampling is by a factor of 2, and the impulse response of the digital filter is given by $h[n] = 0.25\,\text{sinc}(0.25n)$. Sketch $X(F)$, $W(F)$, $H(F)$, and $Y(F)$ over $-0.5 \le F \le 0.5$.

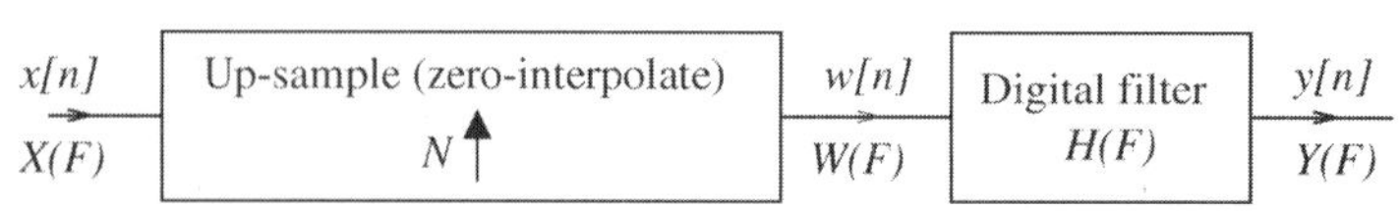


**FIGURE E.7.8(1)** The system for Example 7.8

The various spectra are shown in Figure E7.8(2).

The spectrum $X(F)$ is a triangular pulse of unit width. Zero interpolation by two results in the compressed spectrum $W(F)$. Now, $h[n] = 2F_C\,\text{sinc}(2nF_C)$ corresponds to an ideal filter with a cutoff frequency of $F_C$. Thus, $F_C = 0.125$, and the filter passes only the frequency range $|F| \le 0.125$.

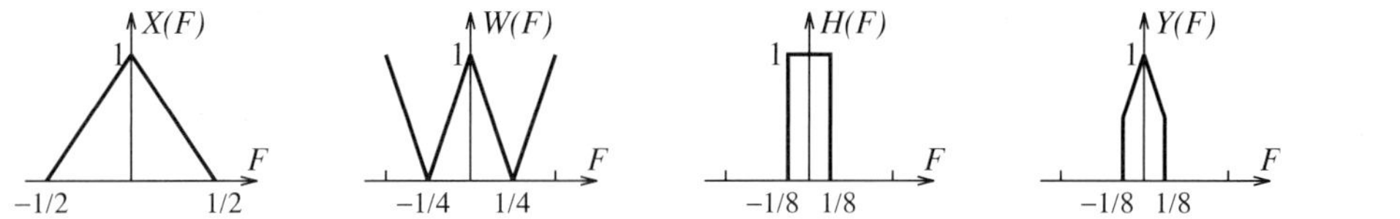


**FIGURE E.7.8(2)** The spectra at various points for the system of Example 7.8

### 7.3.3 Sampling Rate Reduction

A reduction in the sampling rate by an integer factor $M$ involves digital lowpass filtering and decimation, as shown in Figure 7.19. The signals and their spectra at various points in the system for $M = 2$ are shown in Figure 7.20.

First, the sampled signal is passed through a lowpass filter (with unit gain) to ensure that it is band limited to $|F| \le 0.5/M$ in the principal period and to prevent aliasing during the decimation stage. It is then decimated by a down-sampler that retains every $M$th sample. The spectrum of the decimated signal is a stretched version that extends to $|F| \le 0.5$ in the principal period. Note that the spectrum of the decimated signal has a gain of $1/M$ as required, and it is for this reason that we use a lowpass filter with *unit* gain. This process yields exact results, as long as the underlying analog signal has been sampled at a rate that is $M$ times the Nyquist rate (or higher) and the filtering operation is ideal.

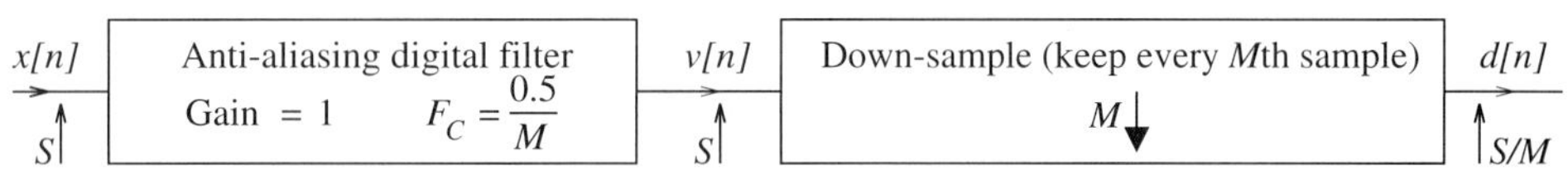


**FIGURE 7.19** Sampling rate reduction by an integer factor $M$ requires a lowpass filter followed by decimation (down-sampling). The lowpass filter bandlimits the signal to $F = 0.5/M$. Decimation by $M$ stretches the spectrum and produces a signal sampled at $1/M$ times the original rate

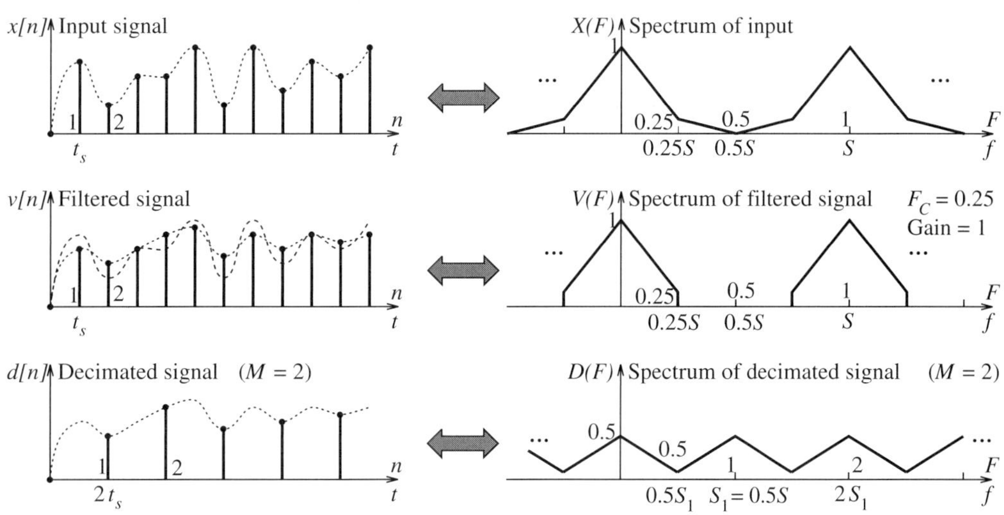


**FIGURE 7.20** The spectra of various signals during sampling rate reduction by two. A lowpass filter with a cutoff frequency of $F_C = 0.25$ bandlimits the spectrum. Decimation by 2 stretches the spectrum. The signal $d[n]$ is thus sampled at half the original rate

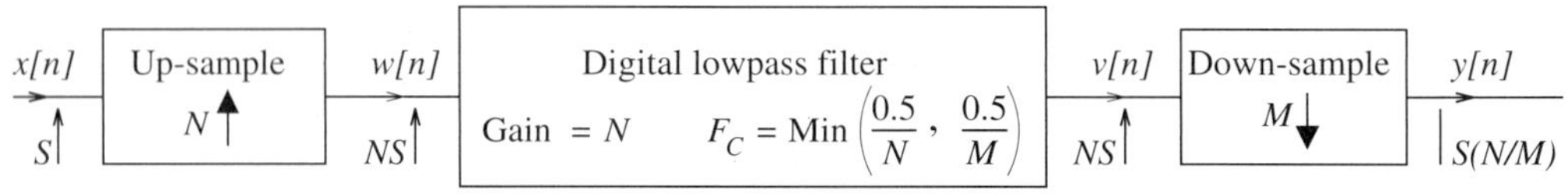


**FIGURE 7.21** Illustrating a sampling-rate change by $M/N$. The first step is up-sampling by $N$. The second step is lowpass filtering using a gain of $N$ and a cutoff frequency that is the smaller of $0.5/M$ and $0.5/N$. The final step is down-sampling by $M$

Fractional sampling-rate changes by a factor $M/N$ can be implemented by cascading a system that increases the sampling rate by $N$ (interpolation) and a system that reduces sampling rate by $M$ (decimation). In fact, we can replace the two lowpass filters that are required in the cascade (both of which operate at the sampling rate $NS$) by a single lowpass filter whose gain is $1/N$ and whose cutoff frequency is the smaller of $0.5/M$ and $0.5/N$, as shown in Figure 7.21.

# 7.4 Quantization

The importance of digital signals stems from the proliferation of high-speed digital computers for signal processing. Due to the finite memory limitations of such machines, we can process only finite data sequences. We must not only sample an analog signal in time but also *quantize* (round or truncate) the signal amplitudes to a finite set of values. Since quantization affects only the signal amplitude, both analog and discrete-time signals can be quantized. Quantized discrete-time signals are called *digital* signals.

Each quantized sample is represented as a group (*word*) of zeros and ones (*bits*) that can be processed digitally. The finer the quantization, the longer the word. Like sampling, improper quantization leads to loss of information. But unlike sampling, no matter how fine the quantization, its effects are irreversible, since word lengths must necessarily be finite. The systematic treatment of quantization theory is very difficult, because finite word lengths appear as nonlinear effects. Quantization always introduces some noise, whose effects can be described only in statistical terms, and is usually considered only in the final stages of any design, and many of its effects (such as overflow and limit cycles) are beyond the realm of this text.

## 7.4.1 Uniform Quantizers

Quantizers are devices that operate on a signal to produce a finite number of amplitude levels or quantization levels. It is common practice to use **uniform quantizers** with equal quantization levels.

The number of levels $L$ in most quantizers used in an analog-to-digital converter (ADC) is invariably a power of 2. If $L = 2^B$, each of the $L$ levels is coded to a binary number, and each signal value is represented in binary form as a $B$-bit word corresponding to its quantized value. A 4-bit quantizer is thus capable of $2^4$ (or 16) levels, and a 12-bit quantizer yields $2^{12}$ (or 4096) levels.

A signal may be quantized by **rounding** to the nearest quantization level, by **truncation** to a level smaller than the next higher one, or by **sign-magnitude truncation**, which is rather

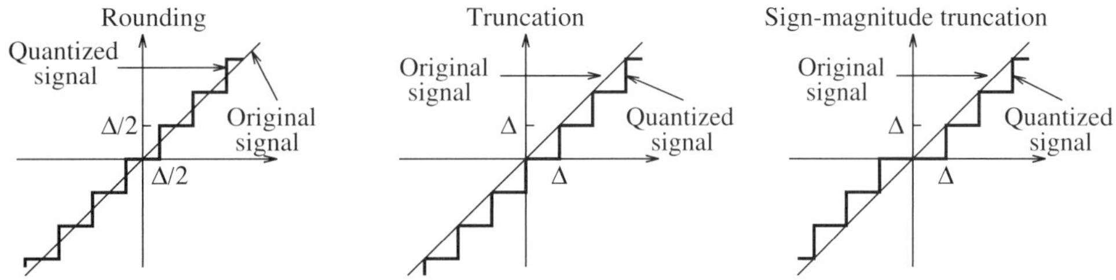


**FIGURE 7.22** Various ways of quantizing a signal. The quantized value is chosen as the nearest quantization level when rounding or the next lower level when truncating. In sign-magnitude truncation, the absolute value (magnitude) is truncated and the actual sign is restored afterwards

like truncating absolute values and then using the appropriate sign. These operations are illustrated in Figure 7.22.

For a quantizer with a full-scale amplitude of $\pm X$, input values outside the range $|X|$ will get clipped and result in overflow. The observed value may be set to the full-scale value (*saturation*) or zero (*zeroing*), leading to the overflow characteristics shown in Figure 7.23.

The quantized signal value is usually represented as a group (word) with a specified number of bits called the word length. Several number representations are in use and illustrated in Table 7.1 for $B = 3$. Some forms of number representation are better suited to some combinations of quantization and overflow characteristics and, in practice, certain combinations are preferred. In any case, the finite length of binary words leads to undesirable and often irreversible effects, collectively known as finite-word-length effects.

## 7.4.2 Quantization Error and Quantization Noise

The **quantization error**, naturally enough, depends on the number of levels. If the quantized signal corresponding to a discrete signal $x[n]$ is denoted by $x_Q[n]$, the quantization error $\epsilon[n]$ equals

$$\epsilon[n] = x[n] - x_Q[n] \tag{7.25}$$

It is customary to define the **quantization signal-to-noise ratio** ($\text{SNR}_Q$) as the ratio of the power $P_S$ in the signal and the power $P_N$ in the error $\epsilon[n]$ (or noise). This is usually

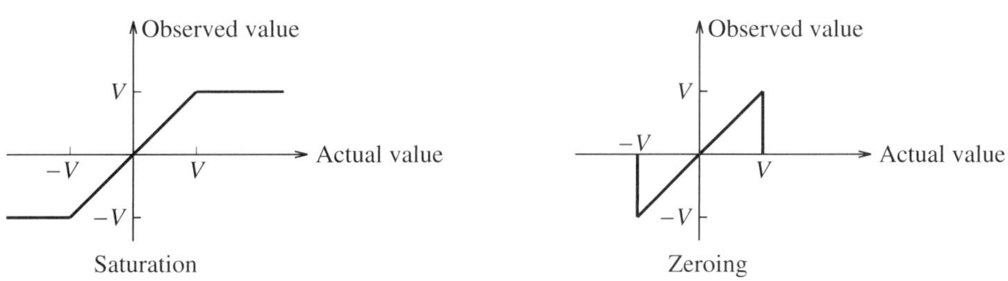


**FIGURE 7.23** Two overflow characteristics. In saturation, values outside the full-scale range are set to the full-scale value itself. In zeroing, values outside the full-scale range are set to zero

**TABLE 7.1 ➤** Various Number Representations for $B = 3$ Bits

| Decimal Value | Sign and Magnitude | One's Complement | Two's Complement | Offset Binary |
|---|---|---|---|---|
| +4 | — | — | — | 111 |
| +3 | 011 | 011 | 011 | 110 |
| +2 | 010 | 010 | 010 | 101 |
| +1 | 001 | 001 | 001 | 000 |
| +0 | 000 | 000 | 000 | — |
| −0 | 100 | 111 | — | 011 |
| −1 | 101 | 110 | 111 | 010 |
| −2 | 110 | 101 | 110 | 001 |
| −3 | 111 | 100 | 101 | 000 |
| −4 | — | — | 100 | — |

measured in decibels, and we get

$$P_S = \frac{1}{N}\sum_{n=0}^{N-1} x^2[n] \quad P_N = \frac{1}{N}\sum_{n=0}^{N-1} \epsilon^2[n] \quad \mathrm{SNR}_Q(\mathrm{dB}) = 10\log\frac{P_S}{P_N} = 10\log\frac{\sum x^2[n]}{\sum \epsilon^2[n]} \tag{7.26}$$

The effect of quantization errors due to rounding or truncation is quite difficult to quantify analytically unless statistical estimates are used. The **dynamic range** or full-scale range of a signal $x(t)$ is defined as its maximum variation $D = x_{\max} - x_{\min}$. If $x(t)$ is sampled and quantized to $L$ levels using a quantizer with a full-scale range of $D$, the **quantization step size**, or *resolution*, $\Delta$, is defined as

$$\Delta = D/L \tag{7.27}$$

This step size also corresponds to the least significant bit (LSB). The dynamic range of a quantizer is often expressed in decibels. For a 16-bit quantizer, the dynamic range is $20\log 2^{16} \approx 96$ dB.

For quantization by rounding, the maximum value of the quantization error must lie between $-\Delta/2$ and $\Delta/2$. If $L$ is large, the error is equally likely to take on any value between $-\Delta/2$ and $\Delta/2$ and is thus *uniformly distributed*. Its probability density function $f(\epsilon)$ has the form shown in Figure 7.24.

The noise power $P_N$ equals its variance $\sigma^2$ (the second central moment) and is given by

$$P_N = \sigma^2 = \int_{-\Delta/2}^{\Delta/2} \epsilon^2 f(\epsilon)\, d\epsilon = \frac{1}{\Delta}\int_{-\Delta/2}^{\Delta/2} \epsilon^2\, d\epsilon = \frac{\Delta^2}{12} \tag{7.28}$$

The quantity $\sigma = \Delta/\sqrt{12}$ defines the **rms quantization error**. With $\Delta = D/L$, we compute

$$10\log P_N = 10\log\frac{D^2}{12L^2} = 20\log D - 20\log L - 10.8 \tag{7.29}$$

A *statistical estimate* of the SNR in decibels, denoted by $\mathrm{SNR}_S$, is thus provided by

$$\mathrm{SNR}_S(\mathrm{dB}) = 10\log P_S - 10\log P_N = 10\log P_S + 10.8 + 20\log L - 20\log D \tag{7.30}$$

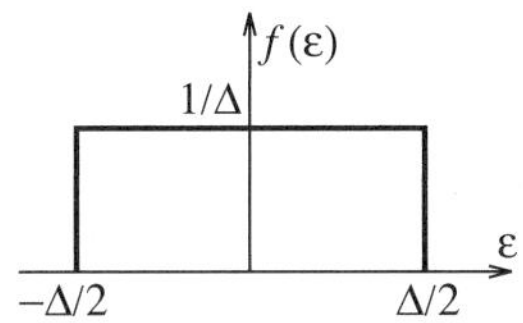


**FIGURE 7.24** The probability density function of a signal quantized by rounding is uniform because the error is equally likely to take on any value between $-\Delta/2$ and $\Delta/2$

For a $B$-bit quantizer with $L = 2^B$ levels (and $\Delta = D/2^B$), we obtain

$$\text{SNR}_S(\text{dB}) = 10\log P_S + 10.8 + 6B - 20\log D \qquad (7.31)$$

This result suggests a 6-dB improvement in the SNR for each additional bit. It also suggests a reduction in the SNR if the dynamic range $D$ is chosen to exceed the signal limits. In practice, signal levels do not often reach the extreme limits of their dynamic range. This allows us to increase the SNR by choosing a smaller dynamic range $D$ for the quantizer but at the expense of some distortion (at very high signal levels).

**Example 7.9**

### Quantization Effects

**(a)** A sampled signal that varies between −2 V and 2 V is quantized using $B$ bits. What value of $B$ will ensure an rms quantization error of less than 5 mV?
The full-scale range is $D = 4$ V. The rms error is given by $\sigma = \Delta/\sqrt{12}$. With $\Delta = D/2^B$, we obtain

$$2^B = \frac{D}{\sigma\sqrt{12}} = \frac{4}{0.005\sqrt{12}} = 230.94 \qquad B = \log_2 230.94 = 7.85$$

Rounding up this result, we get $B = 8$ bits.

**(b)** Consider the ramp $x(t) = 2t$ over (0, 1). For a sampling interval of 0.1 s and $L = 4$, we obtain the sampled signal, quantized (by rounding) signal, and error signal as

$$x[n] = \{0, 0.2, 0.4, 0.6, 0.8, 1.0, 1.2, 1.4, 1.6, 1.8, 2.0\}$$
$$x_Q[n] = \{0, 0.0, 0.5, 0.5, 1.0, 1.0, 1.0, 1.5, 1.5, 2.0, 2.0\}$$
$$e[n] = \{0, 0.2, -0.1, 0.1, -0.2, 0.0, 0.2, -0.1, 0.1, -0.2, 0.0\}$$

We can now compute the SNR in several ways.

1. $\text{SNR}_Q = 10\log\frac{P_S}{P_N} = 10\log\frac{\sum x^2[n]}{\sum e^2[n]} = 10\log\frac{15.4}{0.2} = 18.9$ dB
2. We also could use $\text{SNR}_Q = 10\log P_S + 10.8 + 20\log L - 20\log D$. With $D = 2$ and $N = 11$,

$$\text{SNR}_S = 10\log\left[\frac{1}{N}\sum_{n=0}^{N-1} x^2[n]\right] + 10.8 + 20\log 4 - 20\log 2 = 18.7 \text{ dB}$$

3. If $x(t)$ forms one period of a periodic signal with $T = 1$, we can also find $P_S$ and $\text{SNR}_S$ as

$$P_S = \frac{1}{T}\int x^2(t)\,dt = \frac{4}{3}$$

$$\text{SNR}_S = 10\log\left(\frac{4}{3}\right) + 10\log 12 + 20\log 4 - 20\log 2 = 18.062 \text{ dB}$$

Why the differences between the various results? Because $\text{SNR}_S$ is a *statistical estimate*. The larger the number of samples $N$, the less $\text{SNR}_Q$ and $\text{SNR}_S$ differ. For $N = 500$, for example, we find that $\text{SNR}_Q = 18.0751$ dB and $\text{SNR}_S = 18.0748$ dB are very close indeed.

**(c)** Consider the sinusoid $x(t) = A\cos(2\pi f t)$. The power in $x(t)$ is $P_S = 0.5A^2$. The dynamic range of $x(t)$ is $D = 2A$ (the peak-to-peak value). For a $B$-bit quantizer, we obtain the widely used result

$$\text{SNR}_S = 10\log P_S + 10.8 + 6B - 20\log D = 6B + 1.76 \text{ dB}$$

# 7.5 Digital Processing of Analog Signals

The crux of the sampling theorem is not just the choice of an appropriate sampling rate. More important, the processing of an analog signal is equivalent to the processing of its Nyquist sampled version, because it retains the same information content as the original. This is how the sampling theorem is used in practice. It forms the link between analog and digital signal processing and allows us to use digital techniques to manipulate analog signals. When we sample a signal $x(t)$ at the instants $nt_s$, we imply that the spectrum of the sampled signal is periodic with period $S = 1/t_s$ and band-limited to a highest frequency $B = 0.5S$. Figure 7.25 illustrates a typical system for analog-to-digital conversion.

An analog lowpass **pre-filter** or **anti-aliasing filter** (not shown) limits the highest analog-signal frequency to allow a suitable choice of the sampling rate and ensure freedom from aliasing. The **sampler** operates above the Nyquist sampling rate and is usually a zero-order-hold device. The **quantizer** limits the sampled signal values to a finite number of levels (16-bit quantizers allow a signal-to-noise ratio close to 100 dB). The **encoder** converts the quantized signal values to a string of *binary* **bits** or zeros and ones (**words**) whose length is determined by the number of quantization levels of the quantizer.

A digital signal-processing system, in hardware or software (consisting of digital filters), processes the encoded digital signal (or **bit stream**) in a desired fashion. Digital-to-analog conversion essentially reverses the process and is accomplished by the system shown in Figure 7.26.

A **decoder** converts the processed bit stream to a discrete signal with quantized signal values. The zero-order-hold device reconstructs a staircase approximation of the discrete

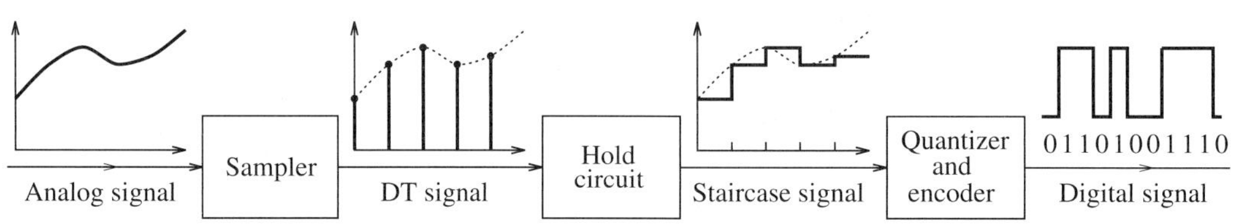


**FIGURE 7.25** Block diagram of a system for analog-to-digital conversion. The sampler produces a discrete signal. The hold circuit yields the staircase approximation. The quantizer and encoder yield the digital signal as a stream of zeros and ones

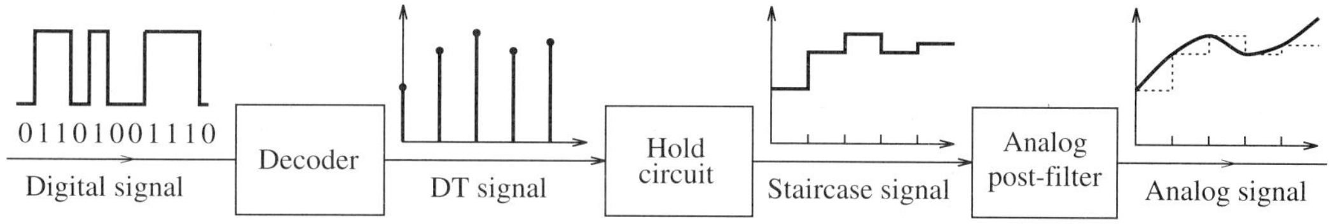


**FIGURE 7.26** Block diagram of a system for analog-to-digital conversion. The decoder converts the digital bit stream to a discrete signal. The hold circuit yields a staircase approximation. The analog post-filter helps round out the edges to yield the smoothed analog signal

signal. The lowpass analog **post-filter**, or **anti-imaging filter**, extracts the central period from the periodic spectrum, removes the unwanted replicas (images), and results in a smoothed reconstructed signal.

### 7.5.1 Practical ADC Considerations

An analog-to-digital converter (ADC) for converting analog signals to-digital signals consists of a sample-and-hold circuit followed by a quantizer and encoder. A block diagram of a typical sample-and-hold circuit and its practical realization is shown in Figure 7.27.

A clock signal controls a switch (an FET, for example) that allows the hold capacitor $C$ to charge very rapidly to the sampled value (when the switch is closed) and to discharge very slowly through the high input resistance of the buffer amplifier (when the switch is open). Ideally, the sampling should be as instantaneous as possible, and once acquired, the level should be held constant until it can be quantized and encoded. Practical circuits also include an operational amplifier at the input to isolate the source from the hold capacitor and provide better tracking of the input. In practice, the finite **aperture time** $T_A$ (during which the signal is being measured), the finite **acquisition time** $T_H$ (to switch from the hold mode to the sampling mode), the **droop** in the capacitor voltage (due to the leakage of the hold capacitor), and the finite **conversion time** $T_C$ (of the quantizer) are all responsible for less than perfect performance.

A finite aperture time limits both the accuracy with which a signal can be measured and the highest frequency that can be handled by the ADC. Consider the sinusoid $x(t) = A\sin(2\pi f_0 t)$. Its derivative $x'(t) = 2\pi A f_0 \cos(2\pi f_0 t)$ describes the rate at which the signal changes. The fastest rate of change equals $2\pi A f_0$ at the zero crossings of $x(t)$. If we assume

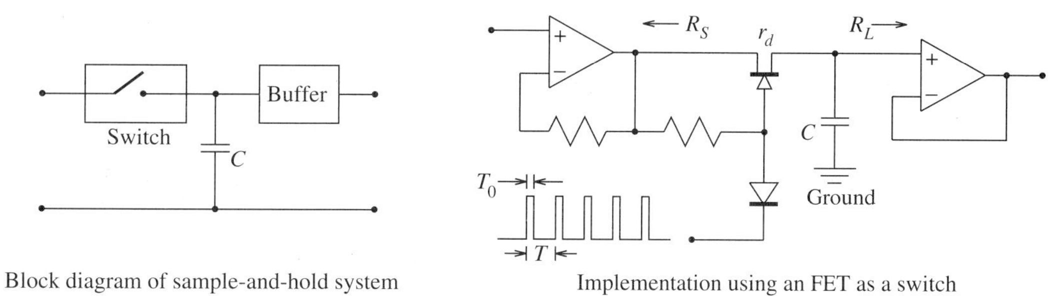


**FIGURE 7.27** Block diagram of a sample-and-hold system and a practical realization

that the signal level can change by no more than $\Delta X$ during the aperture time $T_A$, we must satisfy

$$2\pi A f_0 \leq \frac{\Delta X}{T_A} \quad \text{or} \quad f_0 \leq \frac{\Delta X}{\pi D T_A} \tag{7.32}$$

where $D = 2A$ corresponds to the full-scale range of the quantizer. Typically, $\Delta X$ may be chosen to equal the rms quantization error (which equals $\Delta/\sqrt{12}$) or 0.5 LSB (which equals $0.5\Delta$). In the absence of a sample-and-hold circuit (using only a quantizer), the aperture time must correspond to the conversion time of the quantizer. This time may be much too large to permit conversion of frequencies encountered in practice. An important reason for using a sample-and-hold circuit is that it holds the sampled value constant and allows much higher frequencies to be handled by the ADC. The maximum sampling rate $S$ that can be used for an ADC (with a sample-and-hold circuit) is governed by the aperture time $T_A$, and hold time $T_H$, of the sample-and-hold circuit, as well as the conversion time $T_C$ of the quantizer, and corresponds to

$$S \leq \frac{1}{T_A + T_H + T_C} \tag{7.33}$$

Naturally, this sampling rate must also exceed the Nyquist rate. During the hold phase, the capacitor discharges through the high input resistance $R$ of the buffer amplifier, and the capacitor voltage is given by $v_C(t) = V_0 e^{-t/\tau}$, where $V_0$ is the acquired value and $\tau = RC$. The voltage during the hold phase is not strictly constant but shows a droop. The maximum rate of change occurs at $t = 0$ and is given by $v'(t)|_{t=0} = -V_0/\tau$. A proper choice of the holding capacitor $C$ can minimize the droop. If the maximum droop is restricted to $\Delta V$ during the hold time $T_H$, we must satisfy

$$\frac{V_0}{RC} \leq \frac{\Delta V}{T_H} \quad \text{or} \quad C \geq \frac{V_0 T_H}{R\,\Delta V} \tag{7.34}$$

To impose a lower bound, $V_0$ is typically chosen to equal the full-scale range, and $\Delta V$ may be chosen to equal the rms quantization error (which equals $\Delta/\sqrt{12}$) or 0.5 LSB (which corresponds to $0.5\Delta$). Naturally, an upper bound on $C$ is also imposed by the fact that during the capture phase the capacitor must be able to charge very rapidly to the input level.

A digital-to-analog converter (DAC) allows the coded and quantized signal to be converted to an analog signal. In its most basic form, it consists of a summing operational amplifier, as shown in Figure 7.28.

**FIGURE 7.28** A system for digital-to-analog conversion

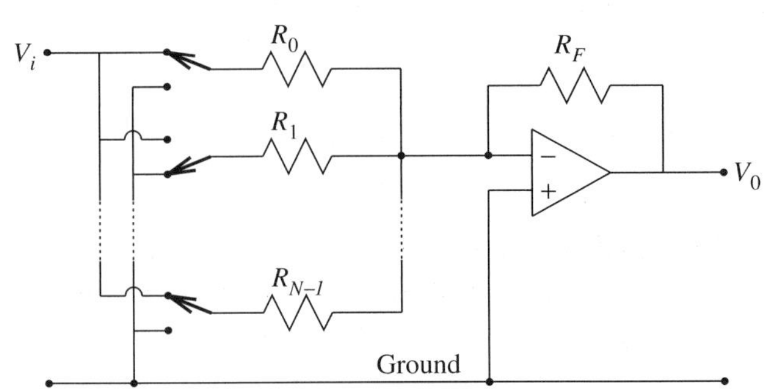

The switches connect either to ground or to the input. The output voltage $V_0$ is a weighted sum of only those inputs that are switched in. If the resistor values are selected as

$$R_0 = R \quad R_1 = \frac{R}{2} \quad R_2 = \frac{R}{2^2} \quad \cdots \quad R_{N-1} = \frac{R}{2^{N-1}} \tag{7.35}$$

the output is given by

$$V_0 = -V_i \frac{R_F}{R}(b_{N-1}2^{N-1} + b_{N-2}2^{N-2} + \cdots + b_1 2^1 + b_0 2^0) \tag{7.36}$$

where the coefficients $b_k$ correspond to the position of the switches and equal 1 (if connected to the input) or 0 (if connected to the ground). Practical circuits are based on modifications that use only a few different resistor values (such as $R$ and $2R$) to overcome the problem of choosing a wide range of resistor values (especially for high-bit converters). Of course, a 1-bit DAC simply corresponds to a constant gain.

**Example 7.10**

### ADC Considerations

**(a)** Suppose we wish to digitize a signal band limited to 15 kHz using only a 12-bit quantizer. If we assume that the signal level can change by no more than the rms quantization error during capture, the aperture time $T_A$ must satisfy

$$T_A \le \frac{\Delta X}{\pi D f_0} = \frac{D/2^B\sqrt{12}}{\pi D f_0} = \frac{1}{2^{12}\pi(15)(10^3)} \approx 5.18 \text{ ns}$$

The conversion time of most practical quantizers is much larger (in the microsecond range), and as a result, we can use such a quantizer only if it is preceded by a sample-and-hold circuit whose aperture time is less than 5 ns.

**(b)** If we digitize a signal band limited to 15 kHz, using a sample-and-hold circuit (with a capture time of 4 ns and a hold time of 10 $\mu$s) followed by a 12-bit quantizer, the conversion time of the quantizer can be computed from

$$S \le \frac{1}{T_A + T_H + T_C} \qquad T_A + T_H + T_C \le \frac{1}{S} = \frac{1}{(30)10^3} \qquad T_C = 23.3\ \mu\text{s}$$

The value of $T_C$ is well within the capability of practical quantizers.

**(c)** Suppose the sample-and-hold circuit is buffered by an amplifier with an input resistance of 1 M$\Omega$. To ensure a droop of no more that 0.5 LSB during the hold phase, we require

$$C \ge \frac{V_0 T_H}{R\Delta V}$$

Now, if $V_0$ corresponds to the full-scale value, $\Delta V = 0.5\,\text{LSB} = V_0/2^{B+1}$, and

$$C \ge \frac{V_0 T_H}{R(V_0/2^{B+1})} = \frac{(10)10^{-6}2^{13}}{10^6} \approx 81.9 \text{ nF}$$

## 7.5.2 Anti-Aliasing Filter Considerations

The purpose of the anti-aliasing filter is to band limit the input signal. In practice, however, we cannot design brick-wall filters, and some degree of aliasing is inevitable. The design of anti-aliasing filters must ensure that the effects of aliasing are kept small. One way to do this is to attenuate components above the Nyquist frequency to a level that cannot be detected by the ADC. The choice of sampling frequency $S$ is thus dictated not only by the highest frequency of interest but also the resolution of the ADC. If the aliasing level is $\Delta V$ and the maximum passband level is $V$, we require a filter with a stopband attenuation of $A_s > 20 \log \frac{V}{\Delta V}$ dB. If the peak signal level equals $A$, and the passband edge is defined as the half-power (or 3-dB) frequency, and $\Delta V$ is chosen as the rms quantization error for a $B$-bit quantizer, we have

$$A_s > 20 \log \frac{\text{maximum rms passband level}}{\text{minimum rms stopband level}} = \frac{A/\sqrt{2}}{\Delta/\sqrt{12}} = \frac{A/\sqrt{2}}{A/(2^B\sqrt{12})} = 20 \log(2^B\sqrt{6}) \text{ dB} \tag{7.37}$$

**Example 7.11**

### Anti-Aliasing Filter Considerations

Suppose we wish to process a noisy speech signal with a bandwidth of 4 kHz using an 8-bit ADC.

We require a minimum stopband attenuation of $A_s = 20 \log(2^B\sqrt{6}) \approx 56$ dB.

If we use a fourth-order Butterworth anti-aliasing filter with a 3-dB passband edge of $f_p$, the normalized frequency $\nu_s = f_s/f_p$ at which the stopband attenuation of 56 dB occurs is computed from

$$A_s = 10 \log\left(1 + \nu_s^{2n}\right) \qquad \nu_s = (10^{0.1A_s} - 1)^{1/2n} \approx 5$$

If the passband edge $f_p$ is chosen as 4 kHz, the actual stopband frequency is $f_s = \nu_s f_p = 20$ kHz.

If the stopband edge also corresponds to highest frequency of interest, then $S = 40$ kHz.

The frequency $f_a$ that gets aliased to the passband edge $f_p$ corresponds to $f_a = S - f_p = 36$ kHz.

The attenuation $A_a$ (in dB) at this frequency $f_a$ is

$$A_a = 10 \log\left(1 + \nu_a^{2n}\right) = 10 \log(1 + 9^8) \approx 76.3 \text{ dB}$$

This corresponds to a signal level (relative to unity) of $(1.5)10^{-4}$, well below the rms quantization error (which equals $1/(2^B\sqrt{12}) = 0.0011$).

## 7.5.3 Anti-Imaging Filter Considerations

The design of reconstruction filters requires that we extract only the central image and minimize the sinc distortion caused by practical zero-order-hold sampling. The design of such filters must meet stringent specifications unless oversampling is employed during the digital-to-analog conversion. In theory, **oversampling** (at a rate much higher than the Nyquist rate) may appear wasteful, because it can lead to the manipulation of a large number of samples. In practice, however, it offers several advantages. First, it provides adequate separation between spectral images and thus allows filters with less stringent cutoff requirements to be used for signal reconstruction. Second, it minimizes the effects of the sinc distortion caused by practical zero-order-hold sampling. Figure 7.29 shows staircase reconstructions at two sampling rates.

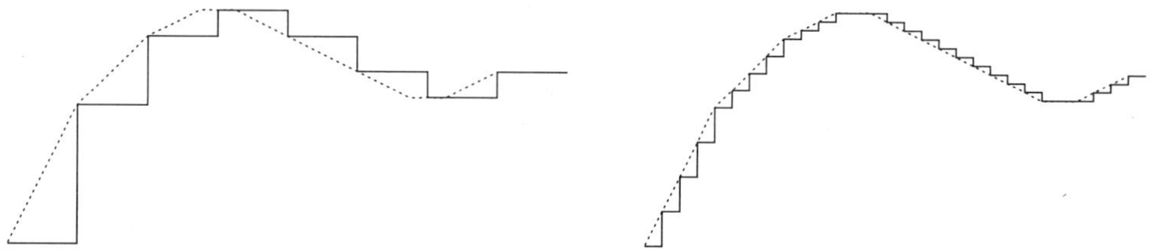

**FIGURE 7.29** Staircase reconstruction of a signal at low (left) and high (right) sampling rates. Higher sampling rates produce a much better approximation to the underlying signal and allow the use of a simpler post-filter to smooth out the edges

The smaller steps in the staircase reconstruction for higher sampling rates lead to a better approximation of the analog signal, and these smaller steps are much easier to smooth out using lower-order filters with less stringent specifications.

**Example 7.12**

### Anti-Imaging Filter Considerations

Consider the reconstruction of an audio signal band limited to 20 kHz, assuming ideal sampling. The reconstruction filter is required to have a maximum signal attenuation of 0.5 dB and to attenuate all images by at least 60 dB.

**(a)** The passband attenuation is $A_p = 0.5$ dB at the passband edge $f_p = 20$ kHz. The stopband attenuation is $A_s = 60$ dB at the stopband edge $f_s = S - f_p = S - 20$ kHz, where $S$ is the sampling rate in kilohertz. This is illustrated in Figure E7.12a.

**FIGURE E 7.12.a** Spectra for Example 7.12(a)

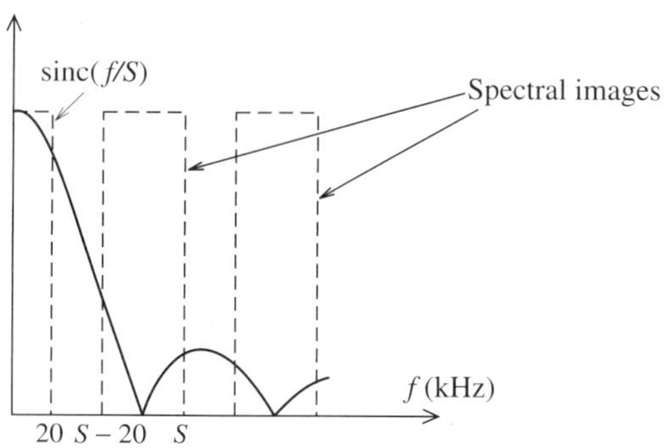


If we design a Butterworth filter, its order $n$ is given by

$$n = \frac{\log[(10^{0.1A_s} - 1)/\epsilon^2]^{1/2}}{\log(f_s/f_p)} \qquad \epsilon^2 = 10^{0.1A_p} - 1$$

1. If we choose $S = 44.1$ kHz, we compute $f_s = 44.1 - 20 = 22.1$ kHz and $n = 80$.
2. If we choose $S = 176.4$ kHz (four-times oversampling), we compute $f_s = 176.4 - 20 = 156.4$ kHz and $n = 4$. This is an astounding reduction in the filter order

**(b)** If we assume zero-order-hold sampling and $S = 176.4$ kHz, the signal spectrum is multiplied by $\text{sinc}(f/S)$. This already provides a signal attenuation of $-20\log[\text{sinc}(20/176.4)] = 0.184$ dB at the passband edge of 20 kHz and $-20\log[\text{sinc}(156.4/176.4)] = 18.05$ dB at the stopband edge of 156.4 kHz. The new filter attenuation specifications are thus $A_p = 0.5 - 0.184 = 0.316$ dB and $A_s = 60 - 18.05 = 41.95$ dB, and we require a Butterworth filter whose order is given by

$$n = \frac{\log[(10^{0.1A_s} - 1)/\epsilon^2]^{1/2}}{\log(f_s/f_p)} = 2.98 \quad \Rightarrow \quad n = 3$$

We see that oversampling allows us to use reconstruction filters of much lower order. In fact, the earliest commercial CD players used four-times oversampling (during the DSP stage) and second-order or third-order Bessel reconstruction filters (for linear phase).

The effects of sinc distortion may also be minimized by using digital filters with a 1/sinc response during the DSP phase itself (prior to reconstruction).

## 7.6 Compact Disc Digital Audio

One application of digital signal processing that has had a profound effect on the consumer market is in compact disc (CD) digital audio systems. The human ear is an incredibly sensitive listening device that depends on the pressure of the air molecules on the eardrum for the perception of sound. It covers a whopping dynamic range of 120 dB, from the Brownian motion of air molecules (the threshold of hearing) all the way up to the threshold of pain. The technology of recorded sound has come a long way since Edison's invention of the phonograph in 1877. The analog recording of audio signals on long-playing (LP) records suffers from poor signal-to-noise ratio (about 60 dB), inadequate separation between stereo channels (about 30 dB), wow and flutter, and wear due to mechanical tracking of the grooves. The CD overcomes the inherent limitations of LP records and cassette tapes and yields a signal-to-noise ratio, dynamic range, and stereo separation, all in excess of 90 dB. It makes full use of digital signal processing techniques during both recording and playback.

### 7.6.1 Recording

A typical CD recording system is illustrated in Figure 7.30. The analog signal recorded from each microphone is passed through an anti-aliasing filter, sampled at the industry standard of 44.1 kHz and quantized to 16 bits in each channel. The two signal channels are then multiplexed, the multiplexed signal is encoded, and *parity bits* are added for later

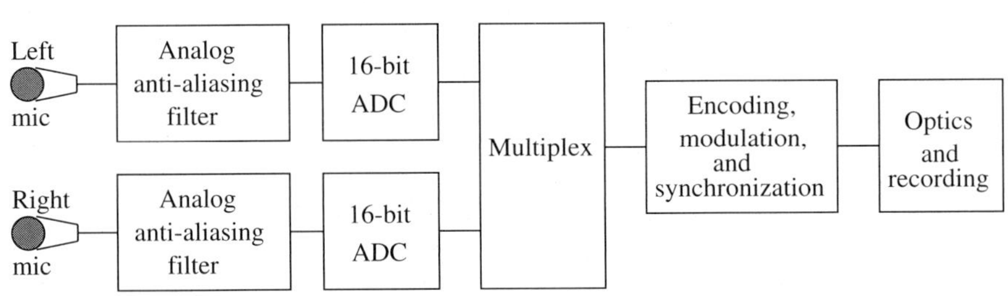


**FIGURE 7.30** Components of a compact-disc recording system

error correction and detection. Additional bits are also added to provide information for the listener (such as playing time and track number). The encoded data is then modulated for efficient storage, and more bits (synchronization bits) are added for subsequent recovery of the sampling frequency. The modulated signal is used to control a laser beam that illuminates the photosensitive layer of a rotating glass disc. As the laser turns on or off, the digital information is etched on the photosensitive layer as a pattern of *pits* and *lands* in a spiral track. This master disk forms the basis for mass production of the commercial CD from thermoplastic material.

How much information can a CD store? With a sampling rate of 44.1 kHz and 32 bits per sample (in stereo), the bit rate is $(44.1)(32)10^3 = (1.41)10^6$ audio bits per second. After encoding, modulation, and synchronization, the number of bits roughly triples to give a bit rate of $(4.23)10^6$ channel bits per second. For a recording time of an hour, this translates to about 600 megabytes (with 8 bits corresponding to a byte and 1024 bytes corresponding to a kilobyte).

## 7.6.2 Playback

The CD player reconstructs the audio signal from the information stored on the compact disc in a series of steps that essentially reverses the process at the recording end. A typical CD player is illustrated in Figure 7.31. During playback, the tracks on a CD are scanned optically by a laser to produce a digital signal. This digital signal is demodulated, and the parity bits are used for the detection of any errors (due to manufacturing defects or dust, for example) and to correct the errors by interpolation between samples (if possible) or to mute the signal (if correction is not possible). The demodulated signal is now ready for reconstruction using a DAC. However, the analog reconstruction filter following the DAC must meet tight specifications in order to remove the images that occur at multiples of 44.1 kHz. Even though the images are well above the audible range, they must be filtered out to prevent overloading of the amplifier and speakers. What is done in practice is to digitally oversample the signal (by a factor of 4) to a rate of 176.4 kHz and pass it through the DAC. A digital filter that compensates for the sinc distortion of the hold operation is also used prior to digital-to-analog conversion. Oversampling relaxes the requirements of the analog filter, which must now smooth out much smaller steps. The sinc compensating filter also provides an additional attenuation of 18 dB for the spectral images and further relaxes the stopband specifications of the analog reconstruction filter. The earliest systems used a third-order Bessel filter with a 3-dB passband of 30 kHz. Another advantage of oversampling is that it reduces the noise floor and spreads the quantization noise over a wider bandwidth. This allows us to round the oversampled signal to 14 bits and use a 14-bit DAC to provide the same level of performance as a 16-bit DAC.

**FIGURE 7.31** Components of a compact-disc playback system

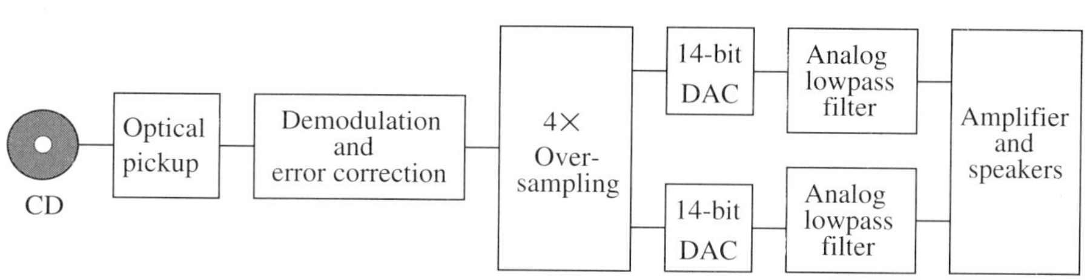

## 7.7 Dynamic-Range Processors

There are many instances in which an adjustment of the dynamic range of a signal is desirable. Sound levels above 100 dB appear uncomfortably loud, and the dynamic range should be matched to the listening environment for a pleasant listening experience. Typically, compression is desirable when listening to music with a large dynamic range in small enclosed spaces (unlike a concert hall) such as a living room or an automobile where the ambient (background) noise is not very low. For example, if the music has a dynamic range of 80 dB and the background noise is 40 dB above the threshold of hearing, it can lead to sound levels as high as 120 dB (close to the threshold of pain). Dynamic-range compression is also desirable for background music (in stores or elevators). It is also used to prevent distortion when recording on magnetic tape and in studios to adjust the dynamic range of the individual tracks in a piece of recorded music.

There are also situations where we may like to expand the dynamic range of a signal. For example, the dynamic range of LP records and cassette tapes is not very high (typically, between 50 dB and 70 dB) and can benefit from dynamic-range expansion, which in effect makes loud passages louder and soft passages softer and in so doing also reduces the record or tape hiss.

A compressor is a variable gain device whose gain is unity for low signal levels and decreases at higher signal levels. An expander is an amplifier with variable gain (which never exceeds unity). For high signal levels it provides unity gain, whereas for low signal levels it decreases the gain and makes the signal level even lower. Typical compression and expansion characteristics are shown in Figure 7.32 and are usually expressed as a ratio, such as a 2:1 compression or a 1:4 expansion. A 10:1 compression ratio (or higher) describes a **limiter** and represents an extreme form of compression. A 1:10 expansion ratio describes a **noise gate** whose output for low signal levels may be almost zero. Noise gating is one way to eliminate noise or hiss during moments of silence in a recording.

Control of the dynamic range requires variable gain devices whose gain is controlled by the signal level, as shown in Figure 7.33. Dynamic-range processing uses a control signal $c(t)$ to adjust the gain. If the gain decreases with increasing $c(t)$, we obtain compression; if the gain increases with increasing $c(t)$, we obtain expansion. Following the signal too closely is undesirable, because it would eliminate the dynamic variation altogether. Once the signal level exceeds the threshold, a typical compressor takes up to 0.1 s (called the

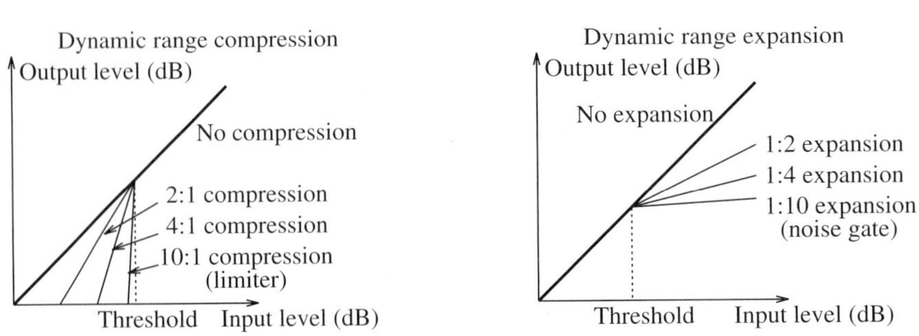


**FIGURE 7.32** Input-output characteristics of dynamic-range processors

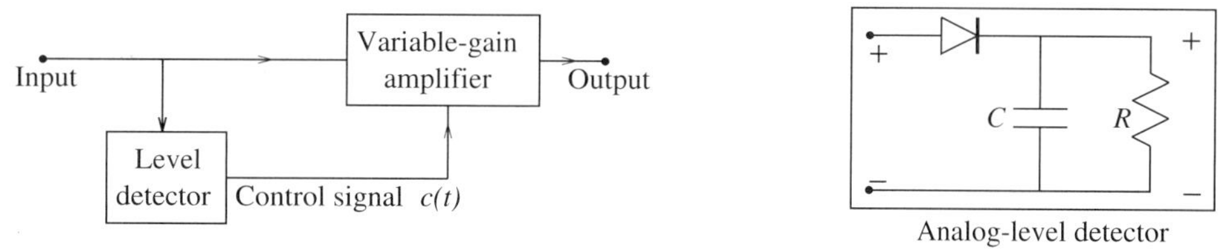


**FIGURE 7.33** Block digram of a dynamic-range processor

*attack time*) to respond, and once the level drops below threshold, it takes another second or two (called the *release time*) to restore the gain to unity. Analog circuits for dynamic range processing may use a peak detector (much like the one used for the detection of AM signals) that provides a control signal. Digital circuits replace the rectifier by simple binary operations and simulate the control signal and the attack characteristics and release time by using digital filters. In concept, the compression ratio (or gain), the delay, attack, and release characteristics may be adjusted independently.

### 7.7.1 Companders

Dynamic-range expanders and compressors are often used to combat the effects of noise during transmission of signals, especially if the dynamic range of the channel is limited. A compander is a combination of a compressor and expander. Compression allows us to increase the signal level relative to the noise level. An expander at the receiving end returns the signal to its original dynamic range. This is the principle behind noise reduction systems for both professional and consumer use. An example is the Dolby noise-reduction system.

In the professional Dolby A system, the input signal is split into four bands by a lowpass filter with a cutoff frequency of 80 Hz, a bandpass filter with band edges at [80, 3000] Hz, and two highpass filters with cutoff frequencies of 3 kHz and 8 kHz. Each band is compressed separately before being mixed and recorded. During playback, the process is reversed. The characteristics of the compression and expansion are shown in Figure 7.34.

During compression, signal levels below −40 dB are boosted by a constant 10 dB, signal levels between −40 dB and −20 dB are compressed by 2:1, and signal levels above −20 dB are not affected. During playback (expansion), signal levels above −20 dB are cut by 10 dB, signal levels between −30 dB and −20 dB face a 1:2 expansion, and signal levels below −30 dB are not affected. In the immensely popular Dolby B system found in consumer products (and also used by some FM stations), the input signal is not split but a pre-emphasis circuit is used to provide a high-frequency boost above 600 Hz. Another popular system is dbx, which uses pre-emphasis above 1.6 kHz with a maximum high-frequency boost of 20 dB.

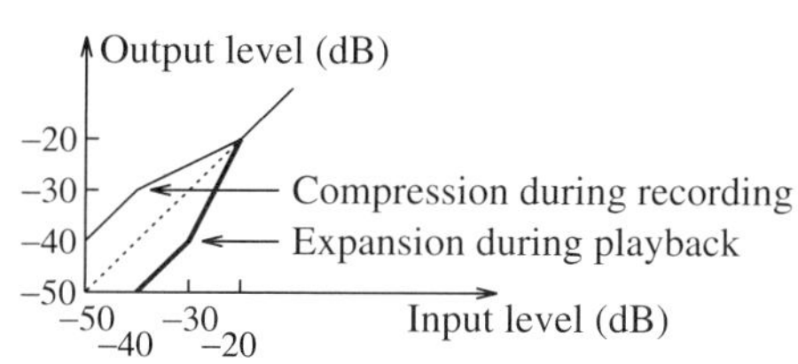


**FIGURE 7.34** Compression and expansion characteristics for Dolby A noise reduction

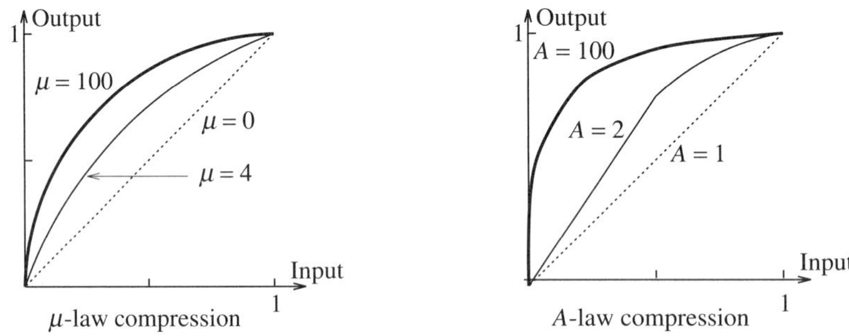


**FIGURE 7.35** Characteristics of $\mu$-law and $A$-law compressors

Voice-grade telephone communication systems also make use of dynamic-range compression, because the distortion caused is not significant enough to affect speech intelligibility. Two commonly used compressors are the *$\mu$-law compander* (used in North America and Japan) and the *$A$-law compander* (used in Europe). For a signal $x(t)$ whose peak level is normalized to unity, the two compression schemes are defined by

$$y_\mu(x) = \frac{\ln(1+\mu|x|)}{\ln(1+\mu)}\,\mathrm{sgn}(x) \qquad y_A(x) = \begin{cases} \dfrac{A|x|}{1+\ln A}\,\mathrm{sgn}(x), & 0 \le |x| \le \dfrac{1}{A} \\ \dfrac{1+\ln(A|x|)}{1+\ln A}\,\mathrm{sgn}(x), & \dfrac{1}{A} \le |x| \le 1 \end{cases} \tag{7.38}$$

The characteristics of these compressors are illustrated in Figure 7.35. The value $\mu = 255$ has become the standard in North America, and $A = 100$ is typically used in Europe. For $\mu = 0$ (and $A = 1$), there is no compression or expansion. The $\mu$-law compander is nearly linear for $\mu|x| \ll 1$. In practice, compression is based on a piecewise linear approximation of the theoretical $\mu$-law characteristic and allows us to use fewer bits to digitize the signal. At the receiving end, an expander (ideally, a true inverse of the compression law) restores the dynamic range of the original signal (except for the effects of quantization). The inverse for $\mu$-law compression is

$$|x| = \frac{(1+\mu)^{|y|} - 1}{\mu} \tag{7.39}$$

The quantization of the compressed signal using the same number of bits as the uncompressed signal results in a higher quantization SNR. For example, the value of $\mu = 255$ increases the SNR by about 24 dB. Since the SNR improves by 6 dB per bit, we can use a quantizer with fewer (only $B - 4$) bits to achieve the same performance as a $B$-bit quantizer with no compression.

## 7.8 Audio Equalizers

Audio equalizers are typically used to tailor the sound to suit the taste of the listener. The most common form of equalization is the tone controls (for bass and treble, for example) found on most low-cost audio systems. **Tone controls** employ *shelving filters* that boost or cut the response over a selected frequency band while leaving the rest of the spectrum unaffected (with unity gain). As a result, the filters for the various controls are typically connected *in cascade*. **Graphic equalizers** offer the next level in sophistication and employ a bank of (typically, second-order) bandpass filters covering a fixed number of frequency bands, and with a fixed bandwidth and center frequency for each range. Only the gain of each filter can be adjusted by the user. Each filter isolates a selected frequency range and provides almost zero gain elsewhere. As a result, the individual sections are connected *in parallel*. **Parametric equalizers** offer the ultimate in versatility and comprise filters that allow the user to vary not only the gain but also the filter parameters (such as the cutoff frequency, center frequency, and bandwidth). Each filter in a parametric equalizer affects only a selected portion of the spectrum (providing unity gain elsewhere), and as a result, the individual sections are connected *in cascade.*

### 7.8.1 Shelving Filters

To vary the bass and treble content, it is common to use first-order *shelving filters* with adjustable gain and cutoff frequency. The transfer function of a first-order lowpass filter whose dc gain is unity and whose response goes to zero at $F = 0.5$ (or $z = -1$) is described by

$$H_{\mathrm{LP}}(z) = \left(\frac{1-\alpha}{2}\right)\frac{z+1}{z-\alpha} \tag{7.40}$$

The half-power or 3-dB cutoff frequency $\Omega_C$ of this filter is given by

$$\Omega_C = \cos^{-1}\left(\frac{2\alpha}{1+\alpha^2}\right) \tag{7.41}$$

The transfer function of a first-order highpass filter with the same cutoff frequency $\Omega_C$ is simply $H_{\mathrm{HP}}(z) = 1 - H_{\mathrm{LP}}(z)$, and gives

$$H_{\mathrm{HP}}(z) = \left(\frac{1+\alpha}{2}\right)\frac{z-1}{z-\alpha} \tag{7.42}$$

Its gain equals zero at $F = 0$ and unity at $F = 0.5$.

A *lowpass shelving filter* consists of a first-order lowpass filter with adjustable gain $G$ in parallel with a highpass filter. With $H_{\mathrm{LP}}(z) + H_{\mathrm{HP}}(z) = 1$, its transfer function may be written as

$$H_{\mathrm{SL}} = GH_{\mathrm{LP}}(z) + H_{\mathrm{HP}}(z) = 1 + (G-1)H_{\mathrm{LP}}(z) \tag{7.43}$$

A *highpass shelving filter* consists of a first-order highpass filter with adjustable gain $G$ in parallel with a lowpass filter. With $H_{\mathrm{LP}}(z) + H_{\mathrm{HP}}(z) = 1$, its transfer function may be written as

$$H_{\mathrm{SH}} = GH_{\mathrm{HP}}(z) + H_{\mathrm{LP}}(z) = 1 + (G-1)H_{\mathrm{HP}}(z) \tag{7.44}$$

The realizations of these lowpass and highpass shelving filters are shown in Figure 7.36.

The response of a lowpass and highpass shelving filter is shown for various values of gain (and a fixed $\alpha$) in Figure 7.37. For $G > 1$, the lowpass shelving filter provides a low-frequency boost, and for $0 < G < 1$, it provides a low-frequency cut. For $G = 1$, we have $H_{\mathrm{SL}} = 1$, and the gain is unity for all frequencies. Similarly, for $> 1$, the highpass shelving filter provides a high-frequency boost, and for $0 < G < 1$, it provides a high-frequency cut.

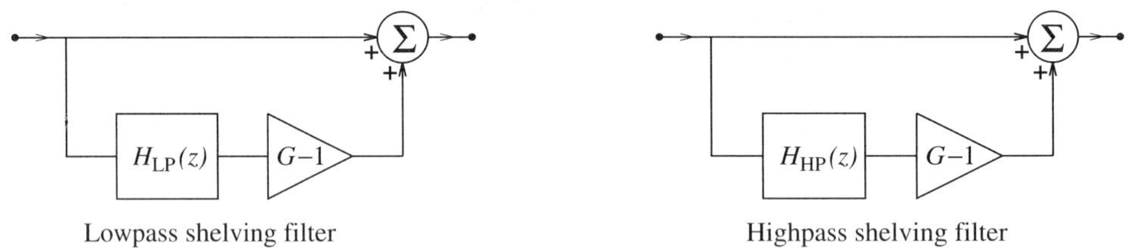


**FIGURE 7.36** Realizations of lowpass and highpass shelving filters

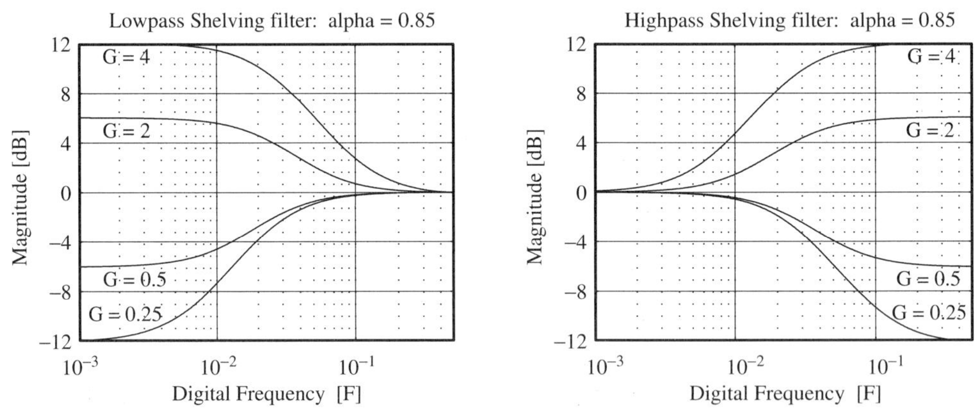


**FIGURE 7.37** Frequency response of lowpass and highpass shelving filters

In either case, the parameter $\alpha$ allows us to adjust the cutoff frequency. Practical realizations of shelving filters and parametric equalizers typically employ allpass structures.

## 7.8.2 Graphic Equalizers

Graphic equalizers permit adjustment of the tonal quality of the sound (in each channel of a stereo system) to suit the personal preference of a listener and represent the next step in sophistication, after tone controls. They employ a bank of (typically, second-order) bandpass filters covering the audio frequency spectrum, and with a fixed bandwidth and center frequency for each range. Only the gain of each filter can be adjusted by the user. Each filter isolates a selected frequency range and provides almost zero gain elsewhere. As a result, the individual sections are connected in *parallel*, as shown in Figure 7.38.

The input signal is split into as many channels as there are frequency bands, and the weighted sum of the outputs of each filter yields the equalized signal. A control panel, usually

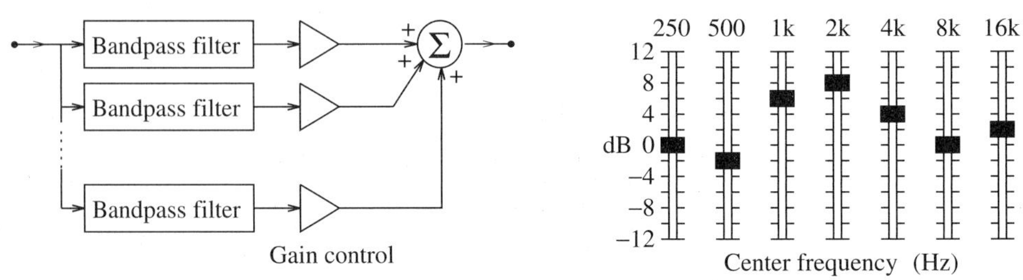


**FIGURE 7.38** (a) A graphic equalizer and (b) the display panel

calibrated in decibels, allows for gain adjustment by sliders, as illustrated in Figure 7.38. The slider settings provide a rough visual indication of the equalized response, hence the name *graphic equalizer*. The design of each second-order bandpass section is based on its center frequency and its bandwidth or quality factor $Q$. A typical set of center frequencies for a ten-band equalizer is [31.5, 63, 125, 250, 500, 1000, 2000, 4000, 8000, 16000] Hz. A typical range for the gain of each filter is $\pm 12$ dB (or 0.25 times to 4 times the nominal gain).

**Example 7.13**

### A Digital Graphic Equalizer

We design a five-band audio equalizer operating at a sampling frequency of 8192 Hz, with center frequencies of [120, 240, 480, 960, 1920] Hz. We select the 3-dB bandwidth of each section as 0.75 times its center frequency to cover the whole frequency range without overlap. This is equivalent to choosing each section with $Q = 4/3$. Each section is designed as a second-order bandpass IIR filter. Figure E7.13 shows the spectrum of each section and the overall response. The constant $Q$ design becomes much more apparent when the gain is plotted in dB against $\log(f)$, as shown in Figure E7.13(b). A change in the gain of any section results in a change in the equalized response.

**FIGURE E.7.13** Frequency response of the audio equalizer for Example 7.13

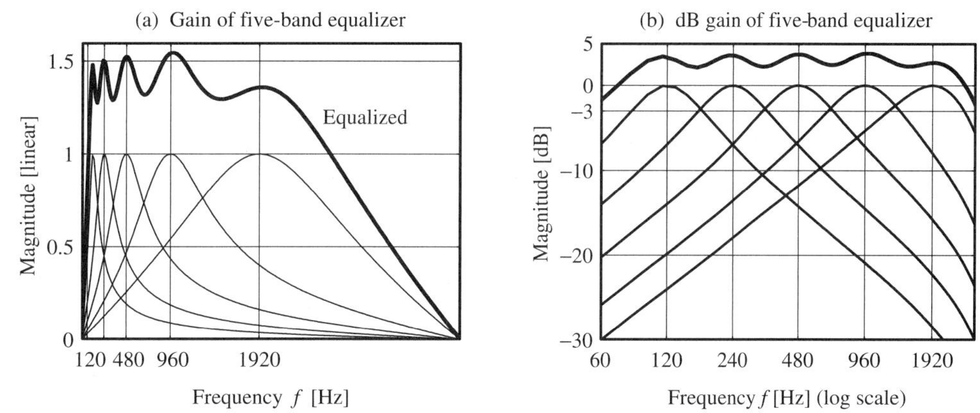


## 7.8.3 Parametric Equalizers

Parametric equalizers offer the ultimate in versatility and allow the user to vary not only the gain but also the filter parameters (such as the cutoff frequency, center frequency, and bandwidth). Each filter in a parametric equalizer affects only a selected portion of the spectrum (while providing unity gain elsewhere), and as a result, the individual sections are connected in *cascade*. Most parametric equalizers are based on second-order IIR peaking or notch filters whose transfer functions we reproduce here:

$$H_{\text{BP}}(z) = \frac{C}{1+C}\left(\frac{z^2-1}{z^2 - \frac{2\beta}{1+C}z + \frac{1-C}{1+C}}\right) \qquad H_{\text{BS}}(z) = \frac{1}{1+C}\left(\frac{z^2-2\beta z+1}{z^2 - \frac{2\beta}{1+C}z + \frac{1-C}{1+C}}\right) \tag{7.45}$$

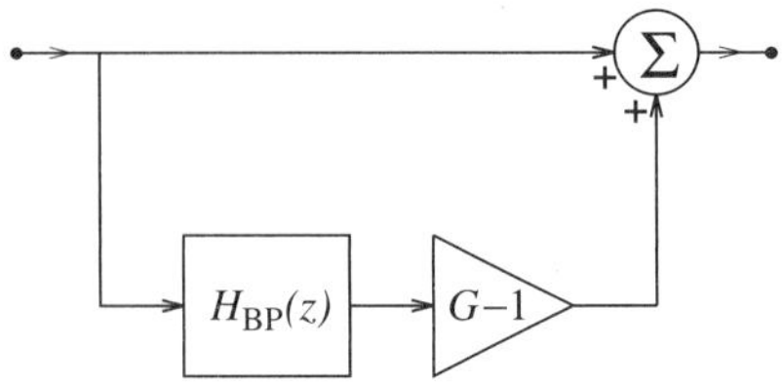


**FIGURE 7.39** A second-order parametric equalizer

The center frequency and bandwidth for the filters are given by

$$\Omega_0 = \cos^{-1}\beta \qquad \Delta\Omega = \cos^{-1}\left(\frac{2\alpha}{1+\alpha^2}\right), \qquad \text{where } \alpha = \frac{1-C}{1+C} \tag{7.46}$$

It is interesting to note that $H_{BS}(z) = 1 - H_{BP}(z)$. A tunable second-order equalizer stage $H_{PAR}(z)$ consists of the bandpass filter $H_{BP}(z)$, with a peak gain of $G$, in parallel with the bandstop filter $H_{BS}(z)$:

$$H_{PAR}(z) = GH_{BP}(z) + H_{BS}(z) = 1 + (G-1)H_{BP}(z) \tag{7.47}$$

A realization of this filter is shown in Figure 7.39.

For $G = 1$, we have $H_{PAR} = 1$, and the gain is unity for all frequencies. The parameters $\alpha$ and $\beta$ allow us to adjust the bandwidth and the center frequency, respectively. The response of this filter is shown for various values of these parameters in Figure 7.40.

## 7.9 Digital Audio Effects

Signal-processing techniques allow us not only to compensate for the acoustics of a listening environment but also to add special musical effects such as delay, echo, and reverb, if desired. The most common example is the audio equalizer that allows us to enhance the bass or treble content of an audio signal. What a listener hears in a room depends not only on the acoustic characteristics of that room, but also on the location of the listener. A piece of music recorded in a concert hall does not sound quite the same elsewhere. The ambiance of a concert hall can, however, be duplicated in the living room environment by first compensating for the room acoustics, and then adding the reverberation characteristics of the concert hall. The

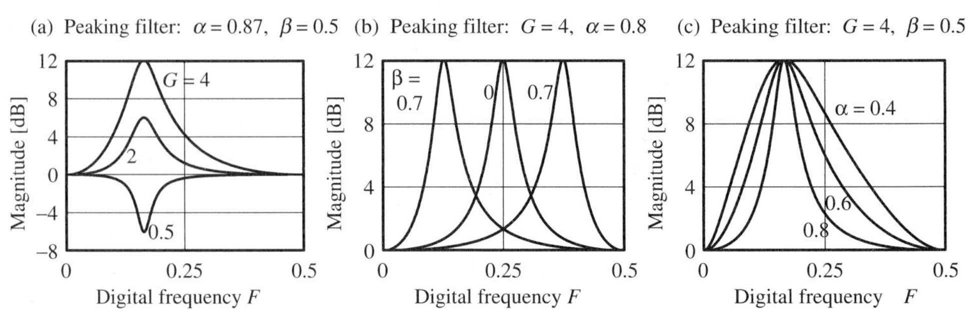


**FIGURE 7.40** Typical response of a parametric equalizer

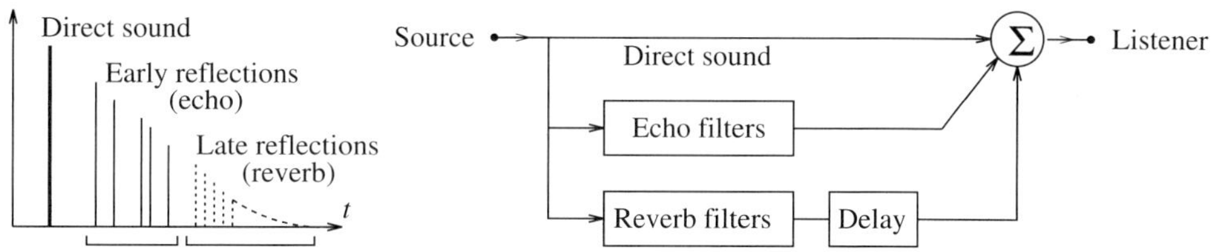


**FIGURE 7.41** Illustrating echo and reverb

compensation is achieved by passing the audio signal through an inverse filter corresponding to the impulse response of the room (which can be measured). The concert hall ambiance is then added by a filter that simulates the concert hall acoustics. High-end audio systems allow us not only to measure the room acoustics but also to simulate the acoustics of a variety of well-known concert-hall environments.

In any listening space, what we hear from an audio source consists not only of the direct sound, but also early echoes reflected directly from the walls and other structures, and a late sound or reverberation that describes multiple reflections (that get added to other echoes). This is illustrated in Figure 7.41.

The direct sound provides clues to the location of the source, the early echoes provide an indication of the physical size of the listening space, and the reverberation characterizes the warmth and liveliness that we usually associate with sounds. The amplitude of the echoes and reverberation decays exponentially with time. Together, these characteristics determine the psycho-acoustic qualities we associate with any perceived sound. Typical 60-dB reverberation times (for the impulse response to decay to 0.001 of its peak value) for concert halls are fairly long, up to two seconds. A conceptual model of a listening environment, also shown in Figure 7.41, consists of echo filters and reverb filters.

A single echo can be modeled by a feed-forward system of the form

$$y[n] = x[n] + \alpha x[n-D] \qquad H(z) = 1 + \alpha z^{D} \qquad h[n] = \delta[n] - \alpha\delta[n-D] \quad (7.48)$$

This is just a comb filter in disguise. The zeros of this filter lie on a circle of radius $R = \alpha^{1/D}$, with angular orientations of $\Omega = (2k+1)\pi/D$. Its comb-like magnitude spectrum $H(F)$ shows minima of $1-\alpha$ at the frequencies $F = (2k+1)/D$, and peaks of $1+\alpha$ midway between the dips. To perceive an echo, the index $D$ must correspond to a delay of at least about 50 ms.

A reverb filter that describes multiple reflections has a feedback structure of the form

$$y[n] = \alpha y[n-D] + x[n] \qquad H(z) = \frac{1}{1-\alpha z^{-D}} \quad (7.49)$$

This filter has an *inverse-comb* structure, and its poles lie on a circle of radius $R = \alpha^{1/D}$, with an angular separation of $\Omega = 2\pi/D$. The magnitude spectrum shows peaks of $1/(1-\alpha)$ at the pole frequencies $F = k/D$, and minima of $1/(1+\alpha)$ midway between the peaks.

Conceptually, the two systems just described can form the building blocks for simulating the acoustics of a listening space. Many reverb filters actually use a combination of reverb filters and allpass filters. A typical structure is shown in Figure 7.42. In practice, however, it is more of an art than a science to create realistic effects, and many of the commercial designs are propriety information.

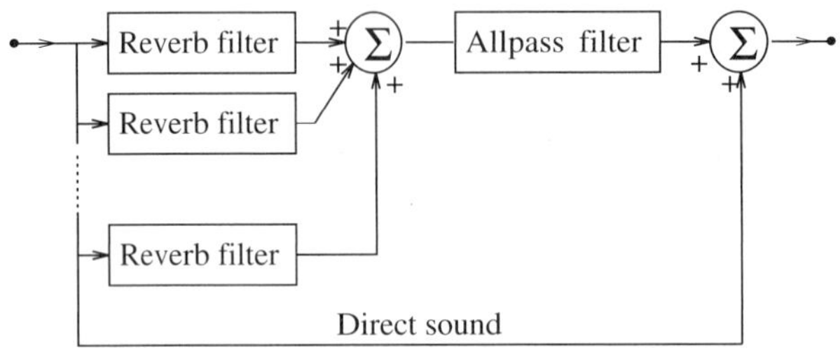


**FIGURE 7.42** Echo and reverb filters for simulating acoustic effects

The reverb filters in Figure 7.42 typically incorporate irregularly spaced delays to allow the blending of echoes, and the allpass filter serves to create the effect of early echoes. Some structures for the reverb filter and allpass filter are shown in Figure 7.43. The first structure is the *plain reverb*. In the second structure, the feedback path incorporates a first-order lowpass filter that accounts for the dependence (increase) of sound absorption with frequency. The allpass filter has the form

$$H(z) = \frac{-\alpha + z^{-L}}{1 - \alpha z^{L}} = -\frac{1}{\alpha} + \frac{\frac{1}{\alpha} - \alpha}{1 - \alpha z^{-L}} \tag{7.50}$$

The second form of this expression (obtained by long division) shows that, except for the constant term, the allpass filter has the same form as a reverb filter.

### 7.9.1 Gated Reverb and Reverse Reverb

Two other types of audio effects can also be generated by reverb filters. A **gated reverb** results from the truncation (abrupt or gradual) of the impulse response of a reverb filter, resulting in an FIR filter. A **reverse reverb** is essentially a flipped version of the impulse response of a gated reverb filter. The impulse response increases with time, leading to a sound that first gets louder, and then dies out abruptly.

### 7.9.2 Chorusing, Flanging, and Phasing

The echo filter also serves as the basis for several other audio special effects. Two of these effects, chorusing and phasing, are illustrated in Figure 7.44.

**Chorusing** mimics a chorus (or group) singing (or playing) in unison. In practice, of course, the voices (or instruments) are not in perfect synchronization nor identical in pitch. The chorusing effect can be implemented by a weighted combination of echo filters, each with a time-varying delay $d_n$ of the form

$$y[n] = x[n] + \alpha x[n - d_n] \tag{7.51}$$

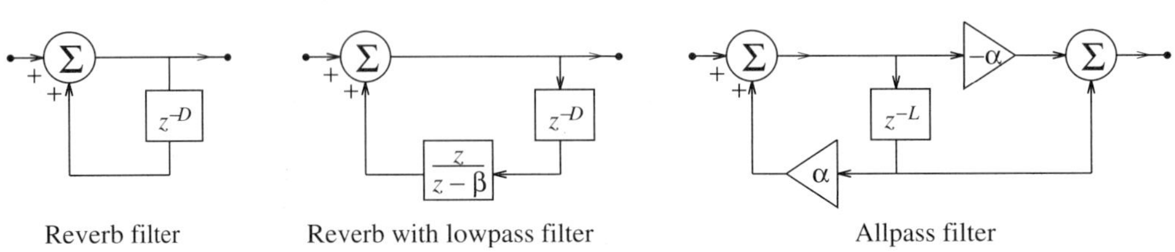


**FIGURE 7.43** Realization of typical reverb filters

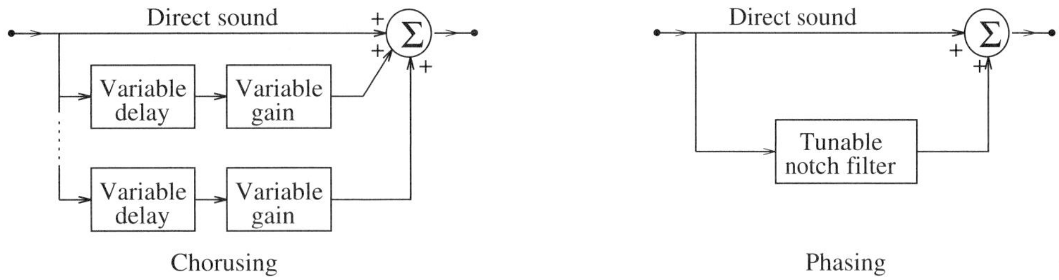


**FIGURE 7.44** Illustrating special audio effects

Typical delay times used in chorusing are between 20 ms and 30 ms. If the delays are less than 10 ms (but still variable), the resulting "whooshing" sound is known as **flanging**.

Phase shifting or **phasing** also creates many interesting effects, and may be achieved by passing the signal through a notch filter whose frequency can be tuned by the user. It is the sudden phase jumps at the notch frequency that are responsible for the phasing effect. The effects may also be enhanced by the addition of feedback.

### 7.9.3 Plucked-String Filters

Comb filters, reverb filters, and allpass filters have also been used to synthesize the sounds of plucked instruments, such as the guitar. What we require is a filter that has a comb-like response, with resonances at multiples of the fundamental frequency of the note. We also require that the harmonics decay in time, with higher frequencies decaying at a faster rate. A typical structure, first described by Karplus and Strong, is illustrated in Figure 7.45.

In the Karplus-Strong structure, the lowpass filter has the transfer function $G_{LP}(z) = 0.5(1 + z^{-1})$ and contributes a 0.5-sample phase delay. The delay line contributes an additional $D$-sample delay, and the *loop delay* is thus $D + 0.5$ samples. The overall transfer function of the first structure is

$$H(z) = \frac{1}{1 - Az^{-D}G_{LP}(z)} \qquad G_{LP}(z) = 0.5(1 + z^{-1}) \tag{7.52}$$

The frequency response of this Karplus-Strong filter is shown in Figure 7.46 for $D = 8$ and $D = 16$ and clearly reveals a resonant structure with sharp peaks. The lowpass filter $G_{LP}(z)$ is responsible for the decrease in the amplitude of the peaks and for making them broader as the frequency increases.

If $A$ is close to unity, the peaks occur at (or very close to) multiples of the fundamental digital frequency $F_0 = 1/(D + 0.5)$. For a sampling rate $S$, the note frequency thus corresponds to $f_0 = S/(D + 0.5)$ Hz. However, since $D$ is an integer, we cannot achieve precise tuning (control of the frequency). For example, to generate a 1600-Hz note at a

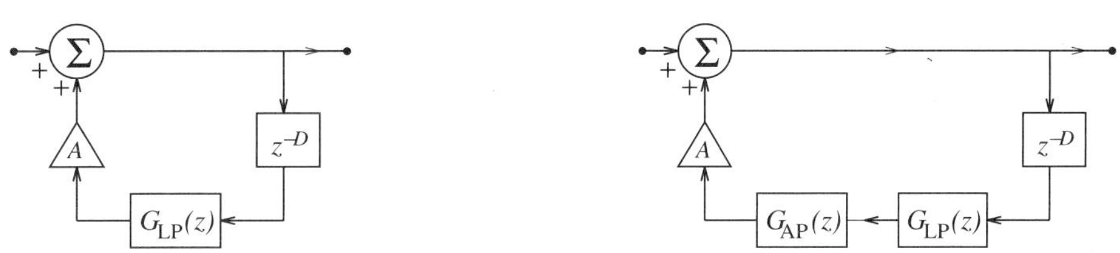


**FIGURE 7.45** The Karplus-Strong plucked-string filter

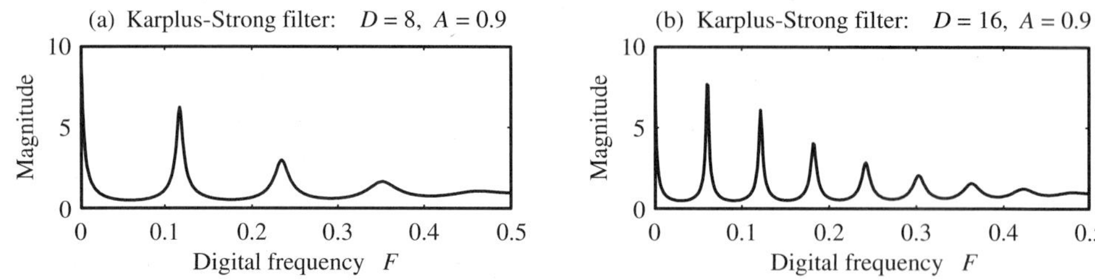


**FIGURE 7.46** Frequency response of the Karplus-Strong plucked-string filter

sampling frequency of 14 kHz requires that $D+0.5 = 8.75$. The closest we can come is by picking $D = 8$ (to give a frequency of $f_0 = \frac{14}{8.5} \approx 1647$ Hz). To generate the exact frequency requires an additional 0.25-sample phase delay. The second structure of Figure 7.46 includes an allpass filter of the form $G_{\text{AP}}(z) = (1 + \alpha z)/(z + \alpha)$ in the feedback path that allows us to implement such fractional delays by appropriate choice of the allpass parameter $\alpha$. Recall that the delay of this allpass filter, at low frequencies at least, is approximated by

$$D_{\text{AP}} \approx \frac{1-\alpha}{1+\alpha} \qquad \alpha \approx \frac{1-D_{\text{AP}}}{1+D_{\text{AP}}} \tag{7.53}$$

In order to implement the 0.25-sample delay, we require

$$\alpha = \frac{1-0.25}{1+0.25} = 0.6$$

In order to synthesize the sound of a plucked instrument, we require a reverb (or comb) filter (with its integer delay) to provide the multiple resonances, a lowpass filter to ensure that the resonances decay with frequency, and an allpass filter (with its fractional delay and constant gain) to fine-tune the note frequency.

# 7.10 Digital Oscillators and DTMF Receivers

The impulse response and transfer function of a causal oscillator that generates a pure cosine are

$$h[n] = \cos(n\Omega)u[n] \qquad H(z) = \frac{z^2 - z\cos\Omega}{z^2 - 2z\cos\Omega + 1} \tag{7.54}$$

Similarly, a system whose impulse response is a pure sine is given by

$$h[n] = \sin(n\Omega)u[n] \qquad H(z) = \frac{z\sin\Omega}{z^2 - 2z\cos\Omega + 1} \tag{7.55}$$

The realizations of these two systems, called *digital oscillators*, are shown in Figure 7.47.

## 7.10.1 DTMF Receivers

A touch-tone phone or *dual-tone multi-frequency* (DTMF) transmitter/receiver makes use of digital oscillators to generate audible tones by pressing buttons on a keypad, as shown

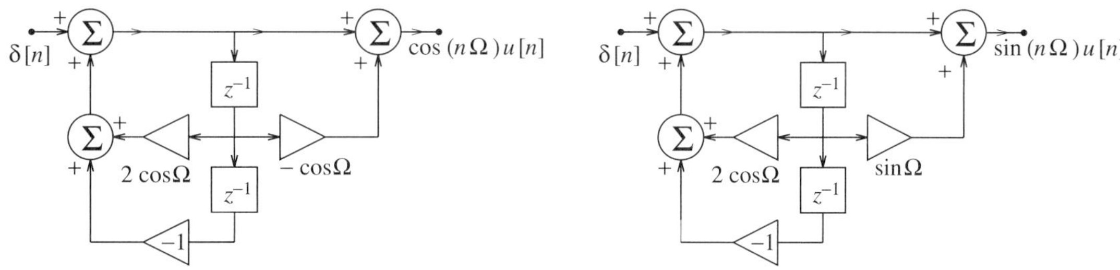


**FIGURE 7.47** Realization of cosine and sine digital oscillators

in Figure 7.48. Pressing a button produces a two-tone signal containing a high- and a low-frequency tone. Each button is associated with a unique pair of low-frequency and high-frequency tones. For example, pressing the button marked 5 would generate a combination of 770-Hz and 1336-Hz tones. There are four low frequencies and four high frequencies. The low- and high-frequency groups have been chosen to ensure that the paired combinations do not interfere with speech. The highest frequency (1633 Hz) is not currently in commercial use. The tones can be generated by using a parallel combination of two programmable digital oscillators, as shown in Figure 7.49.

The sampling rate typically used is $S = 8$ kHz. The digital frequency corresponding to a typical high-frequency tone $f_H$ is $\Omega_H = 2\pi f_H/S$. The code for each button selects the appropriate filter coefficients.

The keys pressed are identified at the receiver by first separating the low- and high-frequency groups using a lowpass filter (with a cutoff frequency of around 1000 Hz) and a highpass filter (with a cutoff frequency of around 1200 Hz) in parallel, and then isolating each tone, using a parallel bank of narrow-band bandpass filters tuned to the (eight) individual frequencies. The (eight) outputs are fed to a level detector and decision logic that establishes the presence or absence of a tone. The keys may also be identified by computing the FFT of the tone signal, followed by threshold detection.

**FIGURE 7.48** Layout of a DTMF touch-tone keypad

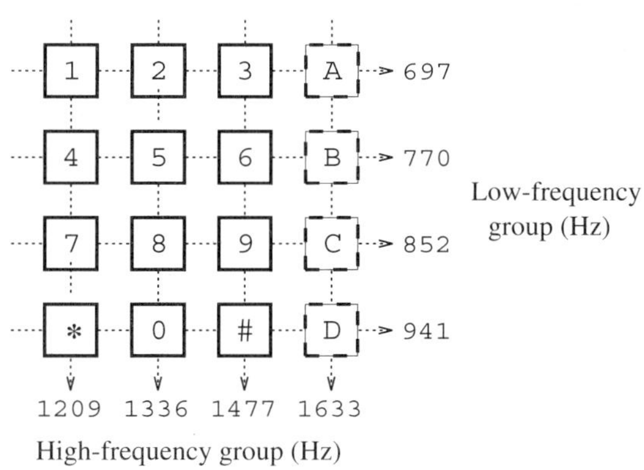

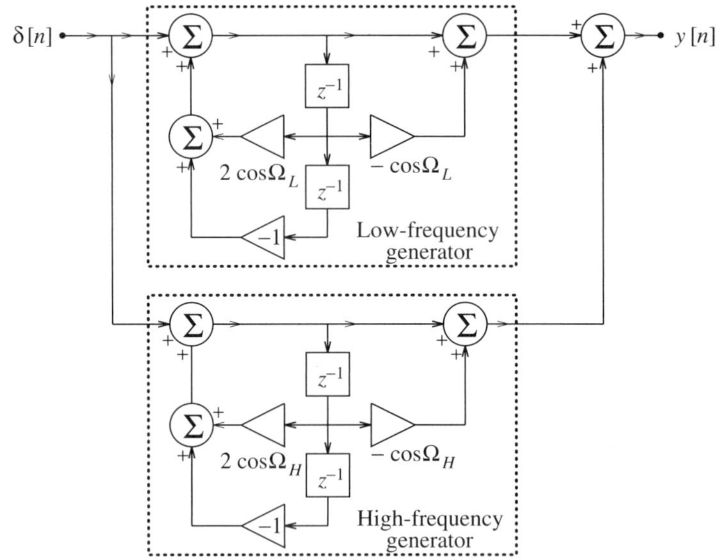


**FIGURE 7.49** Digital oscillators for DTMF tone generation

# 7.11 Multirate Signal Processing

In practice, different parts of a DSP system are often designed to operate at different sampling rates because of the advantages it offers. Since real-time digital filters must complete all algorithmic operations in one sampling interval, a smaller sampling interval (i.e., a higher sampling rate) can impose an added computational burden during the digital-processing stage. It is for this reason that the sampling rate is often reduced (by decimation) before performing DSP operations and increased (by interpolation) before reconstruction. This leads to the concept of multirate signal processing where different subsystems operate at different sampling rates best suited for the given task.

## 7.11.1 Quantization and Oversampling

Oversampling offers several advantages. At the input end, oversampling an analog signal prior to quantization allows the use of simple anti-aliasing filters. It also allows the use of quantizers with lower resolution (fewer bits) to achieve a given SNR because oversampling reduces the quantization noise level by spreading the quantization noise over a wider bandwidth. At the output end, oversampling of the processed digital signal (by sampling rate conversion) prior to reconstruction can reduce the errors (sinc distortion) caused by zero-order-hold devices. It also allows us to use a DAC with lower resolution and a lower-order anti-imaging filter for final analog reconstruction.

Even though sampling and quantization are independent operations, it turns out that oversampling allows us to use quantizers with fewer bits per sample. The idea is that the loss of accuracy in the sample values (by using fewer bits) is offset by the larger number of

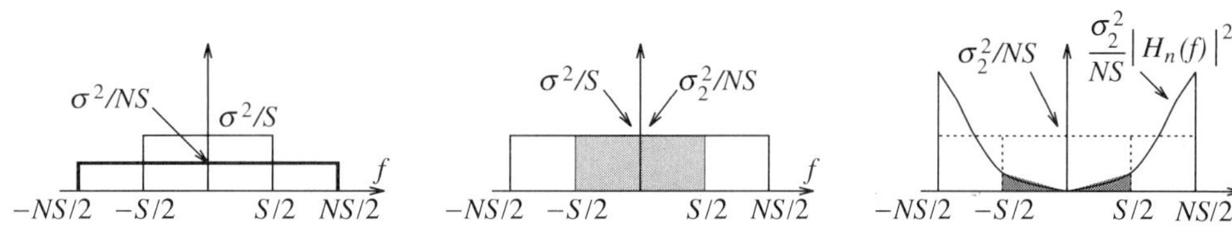


**FIGURE 7.50** Quantization noise spectrum of an oversampled signal

samples (by using a higher sampling rate). This is a consequence of how the quantization noise power is distributed.

The quantized signal is $x_Q[n] = x[n] + \epsilon[n]$. The quantization error $\epsilon[n]$ may be assumed to be a white-noise sequence that is uncorrelated with the signal $x[n]$, with a uniform probability density over the range $(-0.5\Delta,\ 0.5\Delta)$. The variance (average power) of $\epsilon[n]$ is $\sigma^2 = \Delta^2/12$. The spectrum of $\epsilon[n]$ is flat over the principal range $(0.5S,\ 0.5S)$, and the average power $\sigma^2$ is equally distributed over this range. The power spectral density of $\epsilon[n]$ is thus $P_{ee}(f) = \sigma^2/S$. If a signal is oversampled at the rate $NS$, the noise spectrum is spread over a wider frequency range $(-0.5NS,\ 0.5NS)$. The power spectral density is thus $N$ times less for the same total quantization noise power. In fact, the in-band quantization noise in the region of interest, $(-0.5S,\ 0.5S)$, is also decreased by a factor of $N$. This is illustrated in Figure 7.50.

If we were to keep the in-band noise power the same as before, we can recover the signal using fewer bits. For such a quantizer with $B_2 = (B - \Delta B)$ bits per sample, an average power of $\sigma_2^2$, and a sampling rate of $S_2 = NS$ Hz, we obtain the same in-band quantization noise power when its power spectral density is also $\sigma^2/S$, as illustrated in Figure 7.50. Thus,

$$\frac{\sigma^2}{S} = \frac{\sigma_2^2}{NS} \qquad \sigma^2 = \frac{\sigma_2^2}{N} \tag{7.56}$$

Assuming the same full-scale range $D$ for each quantizer, we obtain

$$\frac{D^2}{(12)2^{2B}} = \frac{D^2}{(12N)2^{2(B-\Delta B)}} \qquad N = 2^{2\Delta B} \qquad \Delta B = 0.5\log_2 N \tag{7.57}$$

This result suggests that we gain 0.5 bits for every doubling of the sampling rate. For example, $N = 4$ (four-times oversampling) leads to a gain of 1 bit. In practice, a better trade-off between bits and samples is provided by quantizers that not only use oversampling but also *shape* the noise spectrum (using filters) to further reduce the in-band noise, as shown in Figure 7.50. A typical noise shape is the sine function, and a $p$th-order noise-shaping filter has the form $H_{\text{NS}}(f) = |2\sin(\pi f/NS)|^p,\ -0.5NS \le f \le 0.5NS$, where $N$ is the oversampling factor. Equating the filtered in-band noise power to $\sigma^2$, we obtain

$$\sigma^2 = \left(\frac{\sigma_2^2}{N}\right)\frac{1}{S}\int_{-S/2}^{S/2} |H_{\text{NS}}(f)|^2\,df \tag{7.58}$$

If $N$ is large, $H_{\text{NS}}(f) \approx |(2\pi f/NS)|^p$ over the much smaller in-band range $(-0.5S,\ 0.5S)$, and we get

$$\sigma^2 = \left(\frac{\sigma_2^2}{N}\right)\frac{1}{S}\int_{-S/2}^{S/2} \left(\frac{2\pi f}{NS}\right)^{2p} df = \frac{\sigma_2^2\pi^{2p}}{(2p+1)N^{2p+1}} \tag{7.59}$$

Simplification, with $\sigma_2^2/\sigma^2 = 2^{2\Delta B}$, gives

$$\Delta B = (p + 0.5)\log_2 N - 0.5\log_2\left(\frac{\pi^{2p}}{2p+1}\right) \tag{7.60}$$

This shows that noise shaping (or *error-spectrum shaping*) results in a savings of $p\log_2 N$ additional bits. With $p = 1$ and $N = 4$, for example, $\Delta B \approx 2$ bits. This means that we can make do with a $(B - 2)$-bit DAC during reconstruction if we use noise shaping with four-times oversampling. In practical implementation, noise shaping is achieved by using an oversampling *sigma-delta* ADC. State-of-the-art CD players make use of this technology. What does it take to achieve 16-bit quality using a 1-bit quantizer (which is just a sign detector)? Since $\Delta B = 15$, we could, for example, use oversampling by $N = 64$ (to 2.8 MHz for audio signals sampled at 44.1 kHz) and $p = 3$ (third-order noise shaping).

### 7.11.2 Single-Bit Oversampling Sigma-Delta ADC

Oversampling yields a signal with a large number of closely spaced samples. As a result, the adjacent sample values differ only by small amounts and may be digitized using very few bits. Even a 1-bit quantizer (a comparator with just two levels) can be used if the oversampling ratio is high enough. Oversampling also spreads the quantization noise floor over a wider frequency range. The use of a noise-shaping filter allows use to further reduce the in-band noise. The out-of-band quantization noise can be removed if the signal is decimated. A 1-bit delta-sigma oversampling ADC uses these ideas and can serve as the front end of multirate DSP systems. To achieve a given resolution using fewer bits typically requires very large oversampling ratios. As a result, practical applications of oversampling sigma-delta ADC are limited primarily to low frequency signals such as voice-grade or hi-fi audio.

#### Delta Modulation

In delta modulation, the difference between an analog input signal and the delayed output at each sampling instant (after being fed to a 1-bit DAC) is quantized by a 1-bit ADC (with just two output levels). The quantizer output is one if the input is larger than the delayed output or zero otherwise. The delayed output is decreased (or increased) by a fixed step if it is smaller (or larger) than the input. For a rapidly rising input, the comparator output is a stream of ones and the delayed output is increased by a fixed step at each sampling instant until it matches the input. This scheme is called *delta* modulation because the output changes at a constant rate and the relative number of ones and zeros is indicative of the signal slope.

#### Sigma-Delta Modulation

Multirate DSP uses a modification of delta modulation called sigma-delta modulation. The output at each sampling instant is first accumulated by a summer (hence the name *sigma*) before being quantized by a 1-bit ADC, delayed, and fed back for comparison. As a result, the relative number of ones and zeros now indicates the input signal level rather than its slope (and can even be converted directly to an analog signal by passing through an analog filter if desired). The feedback loop, in fact, acts as a highpass filter that reduces the quantization noise at lower frequencies. Hence, it also serves as a noise-shaping filter.

### Decimation

The output of the sigma-delta modulator is decimated to the original sampling rate by using a lowpass filter followed by a down-sampler. The lowpass filter removes the images (and out-of-band quantization noise) and the down-sampler output has a b-bit resolution (determined by its word length). Since the decimated output corresponds to a much smaller sampling rate, it can be processed digitally without taxing the hardware.

### Single-Bit Oversampling Sigma-Delta DAC

An oversampling sigma-delta DAC can be used prior to signal reconstruction to ease the design requirements of the anti-imaging analog reconstruction filter. A b-bit signal at the original sampling rate is interpolated (using an up-sampler and digital lowpass filter). The output of the interpolator is then fed to a sigma-delta modulator, whose output is a digital stream with just two levels (single bit). A 1-bit DAC followed by an analog (anti-imaging) lowpass filter reconstructs the analog signal.

## 7.12 Problems

**7.1.** **(Sampling Operations)** The signal $x(t) = \cos(4000\pi t)$ is sampled at intervals of $t_s$. Sketch the sampled signal over $0 \le t \le 2$ ms and also sketch the spectrum of the sampled signal over a frequency range of at least $-5$ kHz $\le f \le 5$ kHz for the following choices of the sampling function and $t_s$.

**(a)** impulse train (ideal sampling function) with $t_s = 0.2$ ms
**(b)** impulse train (ideal sampling function) with $t_s = 0.8$ ms
**(c)** rectangular pulse train with pulse width $t_d = 0.1$ ms and $t_s = 0.2$ ms

[**Hints and Suggestions:** The spectrum $X(f) = 0.5\delta(f - B) + 0.5\delta(f + B)$ is an impulse pair. The spectrum of the sampled signals is $\sum X[k]X(f - k/t_s)$. In parts (a)–(b), $X[k] = 1/t_s$ and the impulses are replicated every $S = 1/t_s$ Hz. For (c), $X[k] = 0.5\text{sinc}(0.5k)$ and the replicated impulses decay in strength.]

**7.2.** **(Sampling Operations)** The signal $x(t) = \text{sinc}(4000t)$ is sampled at intervals of $t_s$. Sketch the sampled signal over $0 \le t \le 2$ ms and also sketch the spectrum of the sampled signal for the following choices of the sampling function and $t_s$

**(a)** impulse train (ideal sampling function) with $t_s = 0.2$ ms
**(b)** impulse train (ideal sampling function) with $t_s = 0.25$ ms
**(c)** impulse train (ideal sampling function) with $t_s = 0.4$ ms
**(d)** rectangular pulse train with pulse width $t_d = 0.1$ ms and $t_s = 0.2$ ms
**(e)** flat-top (zero-order-hold) sampling with $t_s = 0.2$ ms

[**Hints and Suggestions:** $X(f) = \dfrac{1}{4000}\text{rect}(\dfrac{f}{4000})$ is a rectangular pulse. The spectrum of the sampled signals is $\sum X[k]X(f - k/t_s)$. In parts (a)–(c), $X[k] = 1/t_s$ and the pulses are replicated every $S = 1/t_s$ Hz and added (where they overlap). For (d), $X[k] = 0.5\text{sinc}(0.5k)$ and the replicated pulses decrease in height. For (e), the spectrum of (d) is multiplied by $\text{sinc}(f/S)$.]

**7.3. (Digital Frequency)** Express the following signals using a digital frequency $|F| < 0.5$.

**(a)** $x[n] = \cos\left(\dfrac{4n\pi}{3}\right)$

**(b)** $x[n] = \cos\left(\dfrac{4n\pi}{7}\right) + \sin\left(\dfrac{8n\pi}{7}\right)$

**7.4. (Sampling Theorem)** Establish the Nyquist sampling rate for the following signals.

**(a)** $x(t) = 5\sin(300\pi t + \pi/3)$
**(b)** $x(t) = \cos(300\pi t) - \sin(300\pi t + 51°)$
**(c)** $x(t) = 3\cos(300\pi t) + 5\sin(500\pi t)$
**(d)** $x(t) = 3\cos(300\pi t)\sin(500\pi t)$
**(e)** $x(t) = 4\cos^2(100\pi t)$
**(f)** $x(t) = 6\,\mathrm{sinc}(100t)$
**(g)** $x(t) = 10\,\mathrm{sinc}^2(100t)$
**(h)** $x(t) = 6\,\mathrm{sinc}(100t)\cos(200\pi t)$

[**Hints and Suggestions:** For the product signals, time-domain multiplication means convolution of their spectra. The width of this convolution gives the highest frequency. For example, in (d), the spectra extend to $\pm 150$ Hz and $\pm 250$ Hz, their convolution covers $\pm 400$ Hz, and $f_{max} = 400$ Hz.]

**7.5. (Sampling Theorem)** A sinusoid $x(t) = A\cos(2\pi f_0 t)$ is sampled at three times the Nyquist rate for six periods. How many samples are acquired?

[**Hints and Suggestions:** For sinusoids, the Nyquist rate means taking two samples per period.]

**7.6. (Sampling Theorem)** The sinusoid $x(t) = A\cos(2\pi f_0 t)$ is sampled at twice the Nyquist rate for 1 s. A total of 100 samples is acquired. What is $f_0$ and the digital frequency of the sampled signal?

[**Hints and Suggestions:** For sinusoids, the Nyquist rate means taking two samples per period.]

**7.7. (Sampling Theorem)** A sinusoid $x(t) = \sin(150\pi t)$ is sampled at a rate of five samples per three periods. What fraction of the Nyquist sampling rate does this correspond to? What is the digital frequency of the sampled signal?

[**Hints and Suggestions:** For sinusoids, the Nyquist rate means taking two samples per period.]

**7.8. (Sampling Theorem)** A periodic square wave with period $T = 1$ ms whose value alternates between $+1$ and $-1$ for each half-period is passed through an ideal lowpass filter with a cutoff frequency of 4 kHz. The filter output is to be sampled. What is the smallest sampling rate we can choose? Consider both the symmetry of the signal as well as the Nyquist criterion.

[**Hints and Suggestions:** Even-indexed harmonics are absent if the signal is half-wave symmetric.]

**7.9. (Spectrum of Sampled Signals)** Given the spectrum $X(f)$ of an analog signal $x(t)$, sketch the spectrum of its ideally sampled version $x[n]$, assuming a sampling rate of 50, 40, and 30 Hz.

**(a)** $X(f) = \mathrm{rect}(f/40)$
**(b)** $X(f) = \mathrm{tri}(f/20)$

[**Hints and Suggestions:** The spectrum of the sampled signals is $\sum SX(f - kS)$. The images of $SX(f)$ are replicated every $S$ Hz and added (where they overlap).]

**7.10.** **(Spectrum of Sampled Signals)** Sketch the spectrum of the following signals against the digital frequency $F$.

**(a)** $x(t) = \cos(200\pi t)$ (ideally sampled at 450 Hz)
**(b)** $x(t) = \sin(400\pi t - \frac{\pi}{4})$ (ideally sampled at 300 Hz)
**(c)** $x(t) = \cos(200\pi t) + \sin(350\pi t)$ (ideally sampled at 300 Hz)
**(d)** $x(t) = \cos(200\pi t + \frac{\pi}{4}) + \sin(250\pi t - \frac{\pi}{4})$ (ideally sampled at 120 Hz)

[**Hints and Suggestions:** The spectrum $X(f)$ contains impulse pairs. The spectrum of the ideally sampled signals is $\sum SX(f - kS)$. The images of $SX(f)$ are replicated every $S$ Hz.]

**7.11.** **(Sampling and Aliasing)** A signal $x(t)$ is made up of the sum of pure sines with unit peak value at the frequencies 10, 40, 200, 220, 240, 260, 300, 320, 340, 360, 380, and 400 Hz.

**(a)** Sketch the magnitude and phase spectra of $x(t)$.
**(b)** If $x(t)$ is sampled at $S = 140$ Hz, which components, if any, will show aliasing?
**(c)** The sampled signal is passed through an ideal reconstruction filter whose cutoff frequency is $f_C = 0.5S$. Write an expression for the reconstructed signal $y(t)$ and sketch its magnitude spectrum and phase spectrum. Is $y(t)$ identical to $x(t)$? Should it be? Explain.
**(d)** What minimum sampling rate $S$ will allow ideal reconstruction of $x(t)$ from its samples?

**7.12.** **(Sampling and Aliasing)** The signal $x(t) = \cos(100\pi t)$ is applied to the following systems. Is it possible to find a minimum sampling rate required to sample the system output $y(t)$? If so, find the Nyquist sampling rate.

**(a)** $y(t) = x^2(t)$
**(b)** $y(t) = x^3(t)$
**(c)** $y(t) = |x(t)|$
**(d)** $h(t) = \text{sinc}(200t)$
**(e)** $h(t) = \text{sinc}(500t)$
**(f)** $h(t) = \delta(t - 1)$
**(g)** $y(t) = x(t)\cos(400\pi t)$
**(h)** $y(t) = u[x(t)]$

[**Hints and Suggestions:** The Nyquist rate applies only if $y(t)$ is bandlimited. In (b), $y(t)$ is a periodic full-rectified cosine. In (h), $y(t)$ is a periodic square wave.]

**7.13.** **(Sampling and Aliasing)** The sinusoid $x(t) = \cos(2\pi f_0 t + \theta)$ is sampled at 400 Hz and shows up as a 150-Hz sinusoid upon reconstruction. When the signal $x(t)$ is sampled at 500 Hz, it again shows up as a 150-Hz sinusoid upon reconstruction. It is known that $f_0 < 2.5$ kHz.

**(a)** If each sampling rate exceeds the Nyquist rate, what is $f_0$?
**(b)** By sampling $x(t)$ again at a different sampling rate, explain how you might establish whether aliasing has occurred?
**(c)** If aliasing occurs and the reconstructed phase is $\theta$, find all possible values of $f_0$.
**(d)** If aliasing occurs but the reconstructed phase is $-\theta$, find all possible values of $f_0$.

[**Hints and Suggestions:** In (b), $S > 2f_0$ for no aliasing. In (c), the aliased frequency is positive and so, $150 = f_0 - 500k = f_0 - 400m$ giving $\frac{k}{m} = \frac{4}{5}$ where $k$ and $m$ are chosen as integers to ensure $f_0 < 2.5$ kHz. In (d), start with $-150 = f_0 - 500k = f_0 - 400m$.]

**7.14.** **(Sampling and Aliasing)** One period of a periodic signal with period $T = 4$ is given by $x(t) = 2\text{tri}(0.5t) - 1$. The periodic signal is ideally sampled by the impulse train $i(t) = \sum_{k=-\infty}^{\infty} \delta(t - k)$.

**(a)** Sketch the signals $x(t)$ and $x_s(t)$ and confirm that $x_S(t)$ is a periodic signal with $T = 4$ seconds whose one period is given by $x_s(t) = \delta(t) - \delta(t-2),\ 0 \le t < 4$.
**(b)** If $x_s(t)$ is passed through an ideal lowpass filter with a cutoff frequency of 0.6 Hz, what is the filter output $y(t)$.
**(c)** How do $x_s(t)$ and $y(t)$ change if the sampling function is $i(t) = \sum_{k=-\infty}^{\infty}(-1)^k\delta(t-k)$?

[**Hints and Suggestions:** For (b), $x_S(t)$ is half-wave symmetric with $T = 4$ and Fourier series coefficients $X[k] = 0.5(1 - e^{-jk\pi})$. For (c), $x_S(t)$ is periodic with $T = 2$ seconds and $x_S(t) = \delta(t),\ 0 \le t < 2$. Its Fourier series coefficients are $X[k] = 0.5$.]

**7.15.** **(Bandpass Sampling)** The signal $x(t)$ is band limited to 500 Hz. The smallest frequency present in $x(t)$ is $f_0$. Find the minimum rate $S$ at which we can sample $x(t)$ without aliasing if

**(a)** $f_0 = 0$
**(b)** $f_0 = 300$ Hz
**(c)** $f_0 = 400$ Hz

**7.16.** **(Bandpass Sampling)** A signal $x(t)$ is band limited to 40 Hz and modulated by a 320-Hz carrier to generate the modulated signal $y(t)$. The modulated signal is processed by a square law device that produces $g(t) = y^2(t)$.

**(a)** What is the minimum sampling rate for $x(t)$ to prevent aliasing?
**(b)** What is the minimum sampling rate for $y(t)$ to prevent aliasing?
**(c)** What is the minimum sampling rate for $g(t)$ to prevent aliasing?

**7.17.** **(Sampling Oscilloscopes)** It is required to ideally sample a signal $x(t)$ at $S$ Hz and pass the sampled signal $s(t)$ through an ideal lowpass filter with a cutoff frequency of $0.5S$ Hz such that its output $y(t) = x(t/\alpha)$ is a stretched-by-$\alpha$ version of $x(t)$.

**(a)** Suppose $x(t) = 1 + \cos(20\pi t)$. What values of $S$ will ensure that the output of the lowpass filter is $y(t) = x(0.1t)$. Sketch the spectra of $x(t),\ s(t)$, and $y(t)$ for the chosen value of $S$.
**(b)** Suppose $x(t) = 2\cos(80\pi t) + \cos(160\pi t)$ and the sampling rate is chosen as $S = 48$ Hz. Sketch the spectra of $x(t),\ s(t)$, and $y(t)$. Will the output $y(t)$ be a stretched version of $x(t)$ with $y(t) = x(t/\alpha)$? If so, what will be the value of $\alpha$?
**(c)** Suppose $x(t) = 2\cos(80\pi t) + \cos(100\pi t)$. What values of $S$ will ensure that the output of the lowpass filter is $y(t) = x(t/20)$? Sketch the spectra of $x(t),\ s(t)$, and $y(t)$ for the chosen value of $S$ to confirm your results.

[**Hints and Suggestions:** We require $\alpha = \dfrac{f_0}{f_0 - S}$ where $f_0$ is the fundamental frequency. In part (a), try $S_m = S/m$ (with integer $m$) for other choices (as long as $S_m > 2f_{\max}/\alpha$).]

**7.18.** **(Sampling and Reconstruction)** A periodic signal whose one full period is $x(t) = \mathrm{tri}(20t)$ is band limited by an ideal analog lowpass filter whose cutoff frequency is $f_C$. It is then ideally sampled at 80 Hz. The sampled signal is reconstructed using an ideal lowpass filter whose cutoff frequency is 40 Hz to obtain the signal $y(t)$. Find $y(t)$ if $f_C = 20$, 40, and 60 Hz.

**7.19.** **(Sampling and Reconstruction)** Sketch the spectra at the intermediate points and at the output of the following cascaded systems. Assume that the input is $x(t) = 5\mathrm{sinc}(5t)$, the sampler operates at $S = 10$ Hz and performs ideal sampling, and the cutoff frequency of the ideal lowpass filter is 5 Hz.

**(a)** $x(t) \longrightarrow$ [sampler] $\longrightarrow$ [ideal LPF] $\longrightarrow y(t)$

**(b)** $x(t) \longrightarrow$ [sampler] $\longrightarrow$ [$h(t) = u(t) - u(t - 0.1)$] $\longrightarrow$ [ideal LPF] $\longrightarrow y(t)$

**(c)** $x(t) \longrightarrow$ [sampler] $\longrightarrow$ [$h(t) = u(t) - u(t - 0.1)$] $\longrightarrow$ [ideal LPF] $\longrightarrow$ [$|H(f)| = \frac{1}{|\text{sinc}(0.1f)|}$] $\longrightarrow y(t)$

[**Hints and Suggestions:** $X(f)$ is a rectangular pulse. In part (a), replicate $X(f)$. In part (b), the lowpass filter preserves only the central image distorted (multiplied) by sinc$(f/S)$. In (c), the compensating filter removes the sinc distortion.]

**7.20. (Sampling and Aliasing)** The signal $x(t) = e^{-t}u(t)$ is sampled at a rate $S$ such that the maximum aliased magnitude is less than 5% of the peak magnitude of the un-aliased image. Estimate the sampling rate $S$.

[**Hints and Suggestions:** Sketch the replicated spectra and observe that the maximum aliasing occurs at $f = 0.5S$. So, we require $|X(0.5S)| \le 0.05|X(0)|$.]

**7.21. (Signal Recovery)** A sinusoid $x(t) = \sin(150\pi t)$ is ideally sampled at 80 Hz. Describe the signal $y(t)$ that is recovered if the sampled signal is passed through the following filters.

**(a)** an ideal lowpass filter with cutoff frequency $f_C = 10$ Hz
**(b)** an ideal lowpass filter with cutoff frequency $f_C = 100$ Hz
**(c)** an ideal bandpass filter with a passband between 60 Hz and 80 Hz
**(d)** an ideal bandpass filter with a passband between 60 Hz and 100 Hz

[**Hints and Suggestions:** The spectrum of the sampled signal contains impulse pairs at $\pm 75$ Hz replicated every 80 Hz. Sketch this with the two-sided spectrum of each filter to obtain each output.]

**7.22. (Sampling and Aliasing)** A speech signal $x(t)$ band limited to 4 kHz is sampled at 10 kHz to obtain $x[n]$. The sampled signal $x[n]$ is filtered by an ideal bandpass filter whose passband extends over $0.03 \le F \le 0.3$ to obtain $y[n]$. Will the sampled output $y[n]$ be contaminated if $x(t)$ also includes an undesired signal at the following frequencies?

**(a)** 60 Hz **(b)** 360 Hz **(c)** 8.8 kHz **(d)** 9.8 kHz

[**Hints and Suggestions:** Use the digital frequency of $x[n]$ in the principal range.]

**7.23. (DTFT and Sampling)** The transfer function of a digital filter is $H(F) = \text{rect}(2F)e^{-j0.5\pi F}$.

**(a)** What is the impulse response $h[n]$ of this filter?
**(b)** The analog signal $x(t) = \cos(0.25\pi t)$ applied to the system shown.

$x(t) \longrightarrow$ [sampler] $\longrightarrow$ [$H(F)$] $\longrightarrow$ [ideal LPF] $\longrightarrow y(t)$

The sampler is ideal and operates at $S = 1$ Hz. The cutoff frequency of the ideal lowpass filter is $f_C = 0.5$ Hz. How are $X(f)$ and $Y(f)$ related? What is $y(t)$?

[**Hints and Suggestions:** In (b), find the digital frequency of $x[n]$ in the principal range and use it to find $y[n]$ and $y(t)$. Then, find the Fourier transform of $y(t)$.]

**7.24. (Sampling and Filtering)** The analog signal $x(t) = \cos(2\pi f_0 t)$ is applied to the following system:

$x(t) \longrightarrow$ [sampler] $\longrightarrow$ [$H(F)$] $\longrightarrow$ [ideal LPF] $\longrightarrow y(t)$

The sampler is ideal and operates at $S = 60$ Hz. The filter $H(F)$ describes the frequency response of a three-point averaging filter whose impulse response is $h[n] = \frac{1}{3}\{1, \overset{\Downarrow}{1}, 1\}$. The cutoff frequency of the ideal lowpass filter is $f_C = 40$ Hz. Find $y(t)$ if

**(a)** $f_0 = 20$ Hz **(b)** $f_0 = 50$ Hz **(c)** $f_0 = 75$ Hz

[**Hints and Suggestions:** Find the digital frequency of $x[n]$ in the principal range and use it to find $y[n]$ and the reconstructed signal $y(t)$.]

7.25. **(Sampling and Filtering)** A periodic signal $x(t)$ with period $T = 0.1$ s, whose Fourier series coefficients are given by $X[k] = 2.5\text{sinc}^2(0.5k)$, is applied to the following system:

$$x(t) \longrightarrow \boxed{\text{prefilter}} \longrightarrow \boxed{\text{sampler}} \longrightarrow \boxed{H(F)} \longrightarrow \boxed{\text{ideal LPF}} \longrightarrow y(t)$$

The prefilter band limits the signal $x(t)$ to $B$ Hz. The sampler is ideal and operates at $S = 80$ Hz. The transfer function of the digital filter is $H(F) = \text{tri}(2F)$. The cutoff frequency of the ideal lowpass filter is $f_C = 40$ Hz. Find $y(t)$ if

**(a)** $B = 20$ Hz **(b)** $B = 40$ Hz **(c)** $B = 80$ Hz

[**Hints and Suggestions:** Find the digital frequency of each nonzero harmonic in $x(t)$ up to $B$ Hz in the principal range and use it to find the output of the digital filter and the reconstructed signal $y(t)$ (by superposition).]

7.26. **(Sampling)** A speech signal band limited to 4 kHz is ideally sampled at a sampling rate $S$, and the sampled signal $x[n]$, is processed by the squaring filter $y[n] = x^2[n]$, whose output is ideally reconstructed to obtain $y(t)$ as follows:

$$x(t) \longrightarrow \boxed{\text{sampler}} \longrightarrow \boxed{y[n] = x^2[n]} \longrightarrow \boxed{\text{ideal reconstruction}} \longrightarrow y(t)$$

What minimum sampling rate $S$ will ensure that $y(t) = x^2(t)$?

[**Hints and Suggestions:** The spectrum of $y[n]$ extends to twice the frequency of $x[n]$.]

7.27. **(Sampling)** Consider the signal $x(t) = \sin(200\pi t)\cos(120\pi t)$. This signal is sampled at a sampling rate $S_1$, and the sampled signal $x[n]$ is ideally reconstructed at a sampling rate $S_2$ to obtain $y(t)$. What is the reconstructed signal if

**(a)** $S_1 = 400$ Hz and $S_2 = 200$ Hz
**(b)** $S_1 = 200$ Hz and $S_2 = 100$ Hz
**(c)** $S_1 = 120$ Hz and $S_2 = 120$ Hz

[**Hints and Suggestions:** Use $2\sin\alpha\cos\beta = \sin(\alpha + \beta) + \sin(\alpha - \beta)$, find the digital frequency of each component in the principal range, and set up the reconstructed signal (by superposition).]

7.28. **(Sampling and Filtering)** A speech signal $x(t)$ whose spectrum extends to 5 kHz is filtered by an ideal analog lowpass filter with a cutoff frequency of 4 kHz to obtain the filtered signal $x_f(t)$.

**(a)** If $x_f(t)$ is to be sampled to generate $y[n]$, what is the minimum sampling rate that will permit recovery of $x_f(t)$ from $y[n]$?
**(b)** If $x(t)$ is first sampled to obtain $x[n]$ and then digitally filtered to obtain $z[n]$, what must be the minimum sampling rate and the impulse response $h[n]$ of the digital filter such that $z[n]$ is identical to $y[n]$?

**7.29. (Sampling and Filtering)** The analog signal $x(t) = \mathrm{sinc}^2(10t)$ is applied to the following system:

$$x(t) \longrightarrow \boxed{\text{sampler}} \longrightarrow \boxed{H(F)} \longrightarrow \boxed{\text{reconstruction}} \longrightarrow y(t)$$

The sampler is ideal and operates at a sampling rate $S$ that corresponds to the Nyquist rate. The transfer function of the digital filter is $H(F) = \mathrm{rect}(2F)$. Sketch the spectra at the output of each subsystem if the reconstruction is based on

**(a)** an ideal analog filter whose cutoff frequency is $f_C = 0.5S$
**(b)** a zero-order-hold followed by an ideal analog filter with $f_C = 0.5S$

[**Hints and Suggestions:** In (a), the output is a version of $X(f)$ truncated in frequency. In (b), the output will show sinc distortion (due to the zero-order-hold).]

**7.30. (Interpolation)** The signal $x[n] = \{\overset{\Downarrow}{-1}, 2, 3, 2\}$ (with $t_s = 1$ s) is passed through an interpolating filter.

**(a)** Sketch the output if the filter performs step interpolation.
**(b)** Sketch the output if the filter performs linear interpolation.
**(c)** What is the interpolated value at $t = 2.5$ s if the filter performs sinc interpolation?
**(d)** What is the interpolated value at $t = 2.5$ s if the filter performs raised cosine interpolation (assume that $R = 0.5$)?

[**Hints and Suggestions:** For (a), sketch the staircase signal. In (b), simply 'connect the dots'. For (c), evaluate $x(t) = \sum x[k]\mathrm{sinc}(t-k)$ at $t = 2.5$. For (d), use the raised-cosine formula.]

**7.31. (DTFT and Sampling)** A signal is reconstructed using a filter that performs step interpolation between samples. The reconstruction sampling interval is $t_s$.

**(a)** What is the impulse response $h(t)$ and transfer function $H(f)$ of the interpolating filter?
**(b)** What is the transfer function $H_C(f)$ of a filter that can compensate for the non-ideal reconstruction?

[**Hints and Suggestions:** For the step interpolation filter, $h(t) = \frac{1}{t_s}[u(t) - u(t - t_s)]$.]

**7.32. (DTFT and Sampling)** A signal is reconstructed using a filter that performs linear interpolation between samples. The reconstruction sampling interval is $t_s$.

**(a)** What is the impulse response $h(t)$ and transfer function $H(f)$ of the interpolating filter?
**(b)** What is the transfer function $H_C(f)$ of a filter that can compensate for the non-ideal reconstruction?

[**Hints and Suggestions:** For the linear interpolation filter, $h(t) = \frac{1}{t_s}\mathrm{tri}(\frac{t}{t_s})$.]

**7.33. (Interpolation)** We wish to sample a speech signal band limited to $B = 4$ kHz using zero-order-hold sampling.

**(a)** Select the sampling frequency $S$ if the spectral magnitude of the sampled signal at 4 kHz is to be within 90% of its peak magnitude.
**(b)** On recovery, the signal is filtered using a Butterworth filter with an attenuation of less than 1 dB in the passband and more than 30 dB for all image frequencies. Compute the total attenuation in decibels due to both the sampling and filtering operations at 4 kHz and 12 kHz.
**(c)** What is the order of the Butterworth filter?

[**Hints and Suggestions:** The zero-order-hold causes sinc distortion of the form $\text{sinc}(f/S)$. For (a), we require $\text{sinc}(B/S) \geq 0.9$. For (b), it gives an additional attenuation of $-20\log|\text{sinc}(f/S)|$.]

**7.34. (Up-Sampling)** The analog signal $x(t) = 4000\text{sinc}^2(4000t)$ is sampled at 12 kHz to obtain the signal $x[n]$. The sampled signal is up-sampled (zero interpolated) by $N$ to obtain the signal $y[n]$ as follows:

$$x(t) \longrightarrow \boxed{\text{sampler}} \rightarrow x[n] \rightarrow \boxed{\text{up-sample} \uparrow N} \longrightarrow y[n]$$

Sketch the spectra of $X(F)$ and $Y(F)$ over $-1 \leq F \leq 1$ for $N = 2$ and $N = 3$.

[**Hints and Suggestions:** After up-sampling by $N$, there are $N$ compressed images per period (centered at $F = \frac{k}{N}$).]

**7.35. (Up-Sampling)** The signal $x[n]$ is up-sampled (zero interpolated) by $N$ to obtain the signal $y[n]$. Sketch $X(F)$ and $Y(F)$ over $-1 \leq F \leq 1$ for the following cases.

**(a)** $x[n] = \text{sinc}(0.4n),\ N = 2$
**(b)** $X(F) = \text{tri}(4F),\ N = 2$
**(c)** $X(F) = \text{tri}(6F),\ N = 3$

[**Hints and Suggestions:** After up-sampling by $N$, there are $N$ compressed images per period (centered at $F = \frac{k}{N}$).]

**7.36. (Linear Interpolation)** Consider a system that performs linear interpolation by a factor of $N$. One way to construct such a system is to perform up-sampling by $N$ (zero interpolation between signal samples) and pass the up-sampled signal through an interpolating filter with impulse response $h[n]$ whose output is the linearly interpolated signal $y[n]$, as shown.

$$x[n] \longrightarrow \boxed{\text{up-sample} \uparrow N} \longrightarrow \boxed{\text{filter } h[n]} \longrightarrow y[n]$$

**(a)** What impulse response $h[n]$ will result in linear interpolation by a factor of $N = 2$?
**(b)** Sketch the frequency response $H(F)$ of the interpolating filter for $N = 2$.
**(c)** Let the input to the system be $X(F) = \text{rect}(2F)$. Sketch the spectrum at the output of the up-sampler and the interpolating filter.

[**Hints and Suggestions:** In (a), linear interpolation requires $h[n] = \text{tri}(n/N)$. In (c), there are $N$ compressed images per period after up-sampling (centered at $F = \frac{k}{N}$). The digital filter removes any images outside $|F_C|$.]

**7.37. (Interpolation)** The input $x[n]$ is applied to a system that up-samples by $N$ followed by an ideal lowpass filter with a cutoff frequency of $F_C$ to generate the output $y[n]$. Draw a block diagram of the system. Sketch the spectra at various points of this system and find $y[n]$ and $Y(F)$ for the following cases.

**(a)** $x[n] = \text{sinc}(0.4n),\ N = 2,\ F_C = 0.4$
**(b)** $X(F) = \text{tri}(4F),\ N = 2,\ F_C = 0.375$

[**Hints and Suggestions:** After up-sampling by $N$, there are $N$ compressed images per period (centered at $F = \frac{k}{N}$). The digital filter removes any images outside $|F_C|$.]

**7.38. (Decimation)** The signal $x(t) = 2\cos(100\pi t)$ is ideally sampled at 400 Hz to obtain the signal $x[n]$, and the sampled signal is decimated by $N$ to obtain the signal $y[n]$. Sketch the spectra $X(F)$ and $Y(F)$ over $-1 \leq F \leq 1$ for the cases $N = 2$ and $N = 3$.

[**Hints and Suggestions:** In decimation by $N$, the images are stretched by $N$, amplitude scaled by $\frac{1}{N}$, and added (where overlap exists).]

**7.39.** **(Decimation)** The signal $x[n]$ is decimated by $N$ to obtain the decimated signal $y[n]$. Sketch $X(F)$ and $Y(F)$ over $-1 \le F \le 1$ for the following cases.

**(a)** $x[n] = \text{sinc}(0.4n), \quad N = 2$
**(b)** $X(F) = \text{tri}(4F), \quad N = 2$
**(c)** $X(F) = \text{tri}(3F), \quad N = 2$

[**Hints and Suggestions:** In decimation by $N$, the images are stretched by $N$, amplitude scaled by $\frac{1}{N}$, and added (where overlap exists).]

**7.40.** **(Interpolation and Decimation)** Consider the following system:

$$x[n] \longrightarrow \boxed{\text{up-sample} \uparrow N} \longrightarrow \boxed{\text{digital LPF}} \longrightarrow \boxed{\text{down-sample} \downarrow M} \longrightarrow y[n]$$

The signal $x[n]$ is up-sampled (zero-interpolated) by $N = 2$, the digital lowpass filter is ideal and has a cutoff frequency of $F_C$, and down-sampling is by $M = 3$. Sketch $X(F)$ and $Y(F)$ and explain how they are related for the following cases.

**(a)** $X(F) = \text{tri}(4F)$ and $F_C = 0.125$ **(b)** $X(F) = \text{tri}(2F)$ and $F_C = 0.25$

[**Hints and Suggestions:** After up-sampling by $N$, there are $N$ compressed images per period (centered at $F = \frac{k}{N}$). The digital filter removes any images outside $|F_C|$. In down-sampling by $M$, the images are stretched by $M$, amplitude scaled by $\frac{1}{M}$, and added (where overlap exists).]

**7.41.** **(Interpolation and Decimation)** For each of the following systems, $X(F) = \text{tri}(4F)$. The digital lowpass filter is ideal, has a cutoff frequency of $F_C = 0.25$, and a gain of 2. Sketch the spectra at the various points over $-1 \le F \le 1$ and determine whether any systems produce identical outputs.

**(a)** $x[n] \rightarrow \boxed{\text{up-sample} \uparrow N = 2} \rightarrow \boxed{\text{digital LPF}} \rightarrow \boxed{\text{down-sample} \downarrow M = 2} \rightarrow y[n]$

**(b)** $x[n] \rightarrow \boxed{\text{down-sample} \downarrow M = 2} \rightarrow \boxed{\text{digital LPF}} \rightarrow \boxed{\text{up-sample} \uparrow N = 2} \rightarrow y[n]$

**(c)** $x[n] \rightarrow \boxed{\text{down-sample} \downarrow M = 2} \rightarrow \boxed{\text{up-sample} \uparrow N = 2} \rightarrow \boxed{\text{digital LPF}} \rightarrow y[n]$

[**Hints and Suggestions:** After up-sampling by $N$, there are $N$ compressed images per period (centered at $F = \frac{k}{N}$). The digital filter removes any images outside $|F_C|$. In down-sampling by $M$, the images are stretched by $M$, amplitude scaled by $\frac{1}{M}$, and added (where overlap exists).]

**7.42.** **(Interpolation and Decimation)** For each of the following systems, $X(F) = \text{tri}(3F)$. The digital lowpass filter is ideal, has a cutoff frequency of $F_C = \frac{1}{3}$, and a gain of 2. Sketch the spectra at the various points over $-1 \le F \le 1$ and determine whether any systems produce identical outputs.

**(a)** $x[n] \rightarrow \boxed{\text{up-sample} \uparrow N = 2} \rightarrow \boxed{\text{digital LPF}} \rightarrow \boxed{\text{down-sample} \downarrow M = 2} \rightarrow y[n]$

**(b)** $x[n] \rightarrow \boxed{\text{down-sample} \downarrow M = 2} \rightarrow \boxed{\text{digital LPF}} \rightarrow \boxed{\text{up-sample} \uparrow N = 2} \rightarrow y[n]$

**(c)** $x[n] \rightarrow \boxed{\text{down-sample} \downarrow M = 2} \rightarrow \boxed{\text{up-sample} \uparrow N = 2} \rightarrow \boxed{\text{digital LPF}} \rightarrow y[n]$

[**Hints and Suggestions:** After up-sampling by $N$, there are $N$ compressed images per period (centered at $F = \frac{k}{N}$). The digital filter removes any images outside $|F_C|$. In down-sampling by $M$, the images are stretched by $M$, amplitude scaled by $\frac{1}{M}$, and added (where overlap exists).]

**7.43. (Interpolation and Decimation)** You are asked to investigate the claim that interpolation by $N$ and decimation by $N$ performed in any order, as shown, will recover the original signal.

**Method 1:** $x[n] \rightarrow$ [up-sample $\uparrow N$] $\rightarrow$ [digital LPF] $\rightarrow$ [down-sample $\downarrow N$] $\rightarrow y[n]$

**Method 2:** $x[n] \rightarrow$ [down-sample $\downarrow N$] $\rightarrow$ [up-sample $\uparrow N$] $\rightarrow$ [digital LPF] $\rightarrow y[n]$

**(a)** Let $X(F) = \text{tri}(4F)$ and $N = 2$. Let the lowpass filter have a cutoff frequency of $F_C = 0.25$ and a gain of 2. Sketch the spectra over $-1 \le F \le 1$ at the various points. For which method does $y[n]$ equal $x[n]$? Do the results justify the claim?

**(b)** Let $X(F) = \text{tri}(3F)$ and $N = 2$. Let the lowpass filter have a cutoff frequency of $F_C = \frac{1}{3}$ and a gain of 2. Sketch the spectra over $-1 \le F \le 1$ at the various points. For which method does $y[n]$ equal $x[n]$? Do the results justify the claim?

**(c)** Are any restrictions necessary on the input for $x[n]$ to equal $y[n]$ in each method? Explain.

[**Hints and Suggestions:** After up-sampling by $N$, there are $N$ compressed images per period (centered at $F = \frac{k}{N}$). The digital filter removes any images outside $|F_C|$. In down-sampling by $M$, the images are stretched by $M$, amplitude scaled by $\frac{1}{M}$, and added (where overlap exists).]

**7.44. (Fractional Delay)** The following system is claimed to implement a half-sample delay:

$x(t) \longrightarrow$ [sampler] $\longrightarrow$ [$H(F)$] $\longrightarrow$ [ideal LPF] $\longrightarrow y(t)$

The signal $x(t)$ is band-limited to $f_C$, and the sampler is ideal and operates at the Nyquist rate. The digital filter is described by $H_1(F) = e^{-j\pi F}$, $|F| \le F_C$, and the cutoff frequency of the ideal lowpass filter is $f_C$.

**(a)** Sketch the magnitude and phase spectra at the various points in this system.

**(b)** Show that $y(t) = x(t - 0.5t_s)$ (corresponding to a half-sample delay).

**7.45. (Fractional Delay)** In practice, the signal $y[n] = x[n - 0.5]$ may be generated from $x[n]$ using interpolation by 2 (to give $x[0.5n]$) followed by a one-sample delay (to give $x[0.5(n-1)]$) and decimation by 2 (to give $x[n-0.5]$). This is implemented as follows:

$x[n] \longrightarrow$ [up-sample $\uparrow 2$] $\longrightarrow$ [ideal LPF] $\longrightarrow$

$\longrightarrow$ [1-sample delay] $\longrightarrow$ [down-sample $\downarrow 2$] $\longrightarrow y[n]$

**(a)** Let $X(F) = \text{tri}(4F)$. Sketch the magnitude and phase spectra at the various points.

**(b)** What should be the gain and the cutoff frequency of the ideal lowpass filter if $y[n] = x[n - 0.5]$ or $Y(F) = X(F)e^{-j\pi F}$ (implying a half-sample delay).

[**Hints and Suggestions:** Up-sampling by $N$ gives $N$ compressed images per period (centered at $F = \frac{k}{N}$). The digital filter removes images outside $|F_C|$. The delay adds the linear phase $-2\pi F$, $|F| < F_C$. In down-sampling by $M$, the images are stretched by $M$ and amplitude scaled by $\frac{1}{M}$.]

**7.46.** **(Quantization SNR)** Consider the signal $x(t) = t^2$, $0 \le t < 2$. Choose $t_s = 0.1$ s, four quantization levels, and rounding to find the following:

**(a)** the sampled signal $x[n]$
**(b)** the quantized signal $x_Q[n]$
**(c)** the actual quantization signal-to-noise ratio $\text{SNR}_Q$
**(d)** the statistical estimate of the quantization $\text{SNR}_S$
**(e)** an estimate of the SNR, assuming $x(t)$ to be periodic, with period $T = 2$ s

[**Hints and Suggestions:** For (e), use $P_S = \frac{1}{T}\int_0^2 t^4\,dt$ to estimate the SNR.]

**7.47.** **(Quantization SNR)** A sinusoid with a peak value of 4 V is sampled and then quantized by a 12-bit quantizer whose full-scale range is $\pm 5$ V. What is the quantization SNR of the quantized signal?

[**Hints and Suggestions:** The dynamic range is $D = 10$ V and the signal power is $P_S = 0.5(4)^2$.]

**7.48.** **(Quantization Noise Power)** The quantization noise power based on quantization by rounding is $\sigma^2 = \Delta^2/12$, where $\Delta$ is the quantization step size. Find an expression for the quantization noise power based on

**(a)** quantization by truncation
**(b)** quantization by sign-magnitude truncation

[**Hints and Suggestions:** For part (a), the error is equally distributed between $-\Delta$ and 0, and so $f(\epsilon) = \frac{1}{\Delta}$, $-\Delta < \epsilon < 0$ with mean $m = -0.5\Delta$. For part (b), the error is equally distributed between $-\Delta$ and $\Delta$, and so $f(\epsilon) = \frac{1}{2\Delta}$, $-\Delta < \epsilon < \Delta$ with mean $m = 0$.]

**7.49.** **(Sampling and Quantization)** A sinusoid with amplitude levels of $\pm 1$ V is quantized by rounding, using a 12-bit quantizer. What is the rms quantization error and the quantization SNR?

[**Hints and Suggestions:** With $\Delta = \dfrac{V_{\text{FS}}}{2^B}$, the quantization error is $\sigma = \dfrac{\Delta}{\sqrt{12}}$.]

**7.50.** **(Anti-Aliasing Filters)** A speech signal is to be band limited using an anti-aliasing third-order Butterworth filter with a half-power frequency of 4 kHz and then sampled and quantized by an 8-bit quantizer. Determine the minimum stopband attenuation and the corresponding stopband frequency to ensure that the maximum stopband aliasing level (relative to the passband edge) is less than the rms quantization error.

[**Hints and Suggestions:** The attenuation is $A_s = 20\log(2^B\sqrt{6}) = 10\log(1 + \nu_s^{2n})$ with $\nu_s = f_s/f_p$.]

**7.51.** **(Anti-Aliasing Filters)** A noise-like signal with a flat magnitude spectrum is filtered using a third-order Butterworth filter with a half-power frequency of 3 kHz, and the filtered signal is sampled at 10 kHz. What is the aliasing level (relative to the signal level) at the half-power frequency in the sampled signal?

[**Hints and Suggestions:** The signal level at the passband edge (half-power frequency) is 0.707. The aliasing level at the passband edge is $(1+\nu^{2n})^{-1/2}$ with $\nu = (S - f_p)/f_p$.]

**7.52.** **(Anti-Aliasing Filters)** A speech signal with amplitude levels of $\pm 1$ V is to be band limited using an anti-aliasing second-order Butterworth filter with a half-power frequency of 4 kHz, and then sampled and quantized by an 8-bit quantizer. What minimum sampling rate $S$ will ensure that the maximum aliasing error at the passband edge is less than the rms quantization level?

[**Hints and Suggestions:** With $V_{\text{FS}} = 2$, find $\Delta = V_{\text{FS}}/2^B$ and the quantization noise level $\Delta/\sqrt{12}$. The aliasing level at the passband edge is $(1+\nu^{2n})^{-1/2}$ with $\nu = (S - f_p)/f_p$.]

**7.53.** **(Sampling and Quantization)** The signal $x(t) = 2\cos(2000\pi t) - 4\sin(4000\pi t)$ is quantized by rounding, using a 12-bit quantizer. What is the rms quantization error and the quantization SNR?

[**Hints and Suggestions:** The signal power is $P_S = 0.5(2^2 + 4^2)$. The peak value cannot exceed 6, so choose $D = 12$ (or twice the actual peak value).]

**7.54.** **(Sampling and Quantization)** A speech signal, band limited to 4 kHz, is to be sampled and quantized by rounding, using an 8-bit quantizer. What is the conversion time of the quantizer if it is preceded by a sample-and-hold circuit with an aperture time of 20 ns and an acquisition time of 2 $\mu$s? Assume sampling at the Nyquist rate.

[**Hints and Suggestions:** Use $S \le \frac{1}{T_A + T_H + T_C}$.]

**7.55.** **(Sampling and Quantization)** A 10-kHz sinusoid with amplitude levels of $\pm 1$ V is to be sampled and quantized by rounding. How many bits are required to ensure a quantization SNR of 45 dB? What is the bit rate (number of bits per second) of the digitized signal if the sampling rate is chosen as twice the Nyquist rate?

[**Hints and Suggestions:** The signal power is $P_S = 0.5$. The noise power is $\Delta^2/12$ where $\Delta = 2/2^B$. The SNR equals $10\log(P_S/P_N)$. The bit rate equals $SB$ bits/second.]

**7.56.** **(Anti-Imaging Filters)** A digitized speech signal, band limited to 4 kHz, is to be reconstructed using a zero-order-hold. What minimum reconstruction sampling rate will ensure that the signal level in the passband is attenuated by less than 1.2 dB due to the sinc distortion of the zero-order-hold? What will be the image rejection at the stopband edge?

[**Hints and Suggestions:** The sinc distortion is $\text{sinc}(f/S)$. So, $20\log|\text{sinc}(f_p/S)| = -1.2$. To find $S$, you will need to find a way to compute the inverse sinc!]

**7.57.** **(Anti-Imaging Filters)** A digitized speech signal, band limited to 4 kHz, is to be reconstructed using a zero-order-hold followed by an analog Butterworth lowpass filter. The signal level in the passband should be attenuated less than 1.5 dB, and an image rejection of better than 45 dB in the stopband is required. What are the specifications for the Butterworth filter if the reconstruction sampling rate is 16 kHz? What is the order of the Butterworth filter?

[**Hints and Suggestions:** Assume $f_s = S - f_p$ and subtract the attenuations $-20\log|\text{sinc}(f_p/S)|$ and $-20\log|\text{sinc}(f_s/S)|$ due to sinc distortion from the given values to find the filter specifications.]

**7.58.** **(Anti-Aliasing and Anti-Imaging Filters)** A speech signal is to be band limited using an anti-aliasing Butterworth filter with a half-power frequency of 4 kHz. The sampling frequency is 20 kHz. What filter order will ensure that the in-band aliasing level is be less 1% of the signal level? The processed signal is to be reconstructed using a zero-order-hold. What is the stopband attenuation required of an anti-imaging filter to ensure image rejection of better than 50 dB?

[**Hints and Suggestions:** The signal level at the half-power frequency is 0.707. The aliasing level at the passband edge is $(1 + \nu^{2n})^{-1/2}$ with $\nu = (S - f_p)/f_p$. Find the filter order $n$ and subtract the attenuation due to sinc distortion from 50 dB to set the filter stopband attenuation.]

## Computation And Design

**7.59.** **(Interpolating Functions)** Consider the signal $x(t) = \cos(0.5\pi t)$ sampled at $S = 1$ Hz to generate the sampled signal $x[n]$. We wish to compute the value of $x(t)$ at $t = 0.5$ by interpolating between its samples.

**(a)** Superimpose a plot of $x(t)$ and its samples $x[n]$ over one period. What is the value of $x(t)$ predicted by step interpolation and linear interpolation of $x[n]$? How good are these estimates? Can these estimates be improved by taking more signal samples (using the same interpolation schemes)?

**(b)** Use the sinc interpolation formula

$$x(t) = \sum_{n=-\infty}^{\infty} x[n]\text{sinc}[(t - nt_s)/t_s]$$

to obtain an estimate of $x(0.5)$. With $t = 0.5$ and $t_s = 1/S = 1$, compute the summation for $|n| \le 10,\ 20,\ 50$ to generate three estimates of $x(0.5)$. How good are these estimates? Would you expect the estimate to converge to the actual value as more terms are included in the summation (i.e., as more signal samples are included)? Compare the advantages and disadvantages of sinc interpolation with the schemes in part (a).

**7.60.** **(Interpolating Functions)** To interpolate a signal $x[n]$ by $N$, we use an up-sampler (that places $N - 1$ zeros after each sample) followed by a filter that performs the appropriate interpolation as shown:

$$x[n] \longrightarrow \boxed{\text{up-sample} \uparrow N} \longrightarrow \boxed{\text{interpolating filter}} \longrightarrow y[n]$$

The filter impulse response for step interpolation, linear interpolation, and ideal (sinc) interpolation is

$$h_S[n] = u[n] - u[n - (N - 1)] \qquad h_L[n] = \text{tri}(n/N) \qquad h_I[n] = \text{sinc}(n/N),\ |n| \le M$$

Note that the ideal interpolating function is actually of infinite length but must be truncated in practice. Generate the test signal $x[n] = \cos(0.5n\pi),\ 0 \le n \le 3$. Up-sample this by $N = 8$ (seven zeros after each sample) to obtain the signal $x_U[n]$. Use the MATLAB routine **filter** to filter $x_U[n]$, using:

**(a)** The step-interpolation filter to obtain the filtered signal $x_S[n]$. Plot $x_U[n]$ and $x_S[n]$ on the same plot. Does the system perform the required interpolation? Does the result look like a sine wave?

**(b)** The linear-interpolation filter to obtain the filtered signal $x_L[n]$. Plot $x_U[n]$ and a delayed (by 8) version of $x_L[n]$ (to account for the noncausal nature of $h_L[n]$) on the same plot. Does the system perform the required interpolation? Does the result look like a sine wave?

**(c)** The ideal-interpolation filter (with $M = 4,\ 8,\ 16$) to obtain the filtered signal $x_I[n]$. Plot $x_U[n]$ and a delayed (by $M$) version of $x_I[n]$ (to account for the noncausal nature of $h_I[n]$) on the same plot. Does the system perform the required interpolation? Does the result look like a sine wave? What is the effect of increasing $M$ on the interpolated signal? What is the effect of increasing both $M$ and the signal length? Explain.

**7.61. (Compensating Filters)** Digital filters are often used to compensate for the sinc distortion of a zero-order-hold DAC by providing a $1/\text{sinc}(F)$ boost. Two such filters are described by

$$\textbf{Compensating Filter 1: } y[n] = \frac{1}{16}(x[n] - 18x[n-1] + x[n-2])$$

$$\textbf{Compensating Filter 2: } y[n] + \frac{1}{8}y[n-1] = \frac{9}{8}x[n]$$

**(a)** For each filter, state whether it is FIR (and if so, linear phase) or IIR.

**(b)** Plot the frequency response of each filter and compare with $|1/\text{sinc}(F)|$.

**(c)** Over what digital-frequency range does each filter provide the required sinc boost? Which of these filters provides better compensation?

**7.62. (Up-Sampling and Decimation)** Let $x[n] = \cos(0.2n\pi) + 0.5\cos(0.4n\pi),\ 0 \le n \le 100$.

**(a)** Plot the spectrum of this signal.

**(b)** Generate the zero-interpolated signal $y[n] = x[n/2]$ and plot its spectrum. Can you observe the spectrum replication? Is there a correspondence between the frequencies in $y[n]$ and $x[n]$? Should there be? Explain.

**(c)** Generate the decimated signal $d[n] = x[2n]$ and plot its spectrum. Can you observe the stretching effect in the spectrum? Is there a correspondence between the frequencies in $d[n]$ and $x[n]$? Should there be? Explain.

**(d)** Generate the decimated signal $g[n] = x[3n]$ and plot its spectrum. Can you observe the stretching effect in the spectrum? Is there a correspondence between the frequencies in $g[n]$ and $x[n]$? Should there be? Explain.

**7.63. (Frequency Response of Interpolating Functions)** The impulse response of filters for step interpolation, linear interpolation, and ideal (sinc) interpolation by $N$ are given by

$$h_S[n] = u[n] - u[n-(N-1)] \qquad h_L[n] = \text{tri}(n/N) \qquad h_I[n] = \text{sinc}(n/N)$$

Note that the ideal-interpolating function is of infinite length.

**(a)** Plot the frequency response of each interpolating function for $N = 4$ and $N = 8$.

**(b)** How does the response of the step-interpolation and linear-interpolation schemes compare with ideal interpolation?

**7.64. (Interpolating Functions)** To interpolate a signal $x[n]$ by $N$, we use an up-sampler (that places $N-1$ zeros after each sample) followed by a filter that performs the appropriate interpolation. The filter impulse response for step interpolation, linear interpolation, and ideal (sinc) interpolation is chosen as

$$h_S[n] = u[n] - u[n-(N-1)] \qquad h_L[n] = \text{tri}(n/N) \qquad h_I[n] = \text{sinc}(n/N), \ |n| \le M$$

Note that the ideal-interpolating function is actually of infinite length and must be truncated in practice. Generate the test signal $x[n] = \cos(0.5n\pi),\ 0 \le n \le 3$. Up-sample this by $N = 8$ (seven zeros after each sample) to obtain the signal $x_U[n]$. Use the MATLAB routine **filter** to filter $x_U[n]$ as follows:

**(a)** Use the step-interpolation filter to obtain the filtered signal $x_S[n]$. Plot $x_U[n]$ and $x_S[n]$ on the same plot. Does the system perform the required interpolation? Does the result look like a sine wave?

**(b)** Use the step-interpolation filter followed by the compensating filter $y[n] = \{x[n] - 18x[n-1] + x[n-2]\}/16$ to obtain the filtered signal $x_C[n]$. Plot $x_U[n]$ and $x_C[n]$ on the same plot. Does the system perform the required interpolation? Does the result look like a sine wave? Is there an improvement compared to part (a)?

**(c)** Use the linear-interpolation filter to obtain the filtered signal $x_L[n]$. Plot $x_U[n]$ and a delayed (by 8) version of $x_L[n]$ (to account for the noncausal nature of $h_L[n]$) on the same plot. Does the system perform the required interpolation? Does the result look like a sine wave?

**(d)** Use the ideal-interpolation filter (with $M = 4,\ 8,\ 16$) to obtain the filtered signal $x_I[n]$. Plot $x_U[n]$ and a delayed (by $M$) version of $x_I[n]$ (to account for the noncausal nature of $h_I[n]$) on the same plot. Does the system perform the required interpolation? Does the result look like a sine wave? What is the effect of increasing $M$ on the interpolated signal? Explain.

**7.65.** **(FIR Filter Design)** A 22.5-Hz sinusoid is contaminated by 60-Hz interference. We wish to sample this signal and design a causal 3-point linear-phase FIR digital filter operating at a sampling frequency of $S = 180$ Hz to eliminate the interference and pass the desired signal with unit gain.

**(a)** Argue that an impulse response of the form $h[n] = \{\overset{\Downarrow}{\alpha},\ \beta,\ \alpha\}$ can be used. Choose $\alpha$ and $\beta$ to satisfy the design requirements.

**(b)** To test your filter, generate two signals $x[n]$ and $s[n]$, $0 \le n \le 50$, by sampling $x(t) = \cos(45\pi t)$ and $s(t) = 3\cos(120\pi t)$ at 180 Hz. Generate the noisy signal $g[n] = x[n] + s[n]$ and pass it through your filter to obtain the filtered signal $y[n]$. Compare $y[n]$ with the noisy signal $g[n]$ and the desired signal $x[n]$ to confirm that the filter meets design requirements. What is the phase of $y[n]$ at the desired frequency? Can you find an exact expression for $y[n]$?

**7.66.** **(Sampling and Quantization)** The signal $x(t) = \cos(2\pi t) + \cos(6\pi t)$ is sampled at $S = 20$ Hz.

**(a)** Generate 200 samples of $x[n]$ and superimpose plots of $x(t)$ versus $t$ and $x[n]$ versus $n/S$. Use a 4-bit quantizer to quantize $x[n]$ by rounding and obtain the signal $x_R[n]$. Superimpose plots of $x(t)$ versus $t$ and $x_R[n]$ versus $n/S$. Obtain the quantization error signal $e[n] = x[n] - x_R[n]$ and plot $e[n]$ and its 10-bin histogram. Does it show a uniform distribution? Compute the quantization SNR in decibels. Repeat for 800 samples. Does an increase in the signal length improve the SNR?

**(b)** Generate 200 samples of $x[n]$ and superimpose plots of $x(t)$ versus $t$ and $x[n]$ versus $n/S$. Use an 8-bit quantizer to quantize $x[n]$ by rounding and obtain the signal $x_R[n]$. Superimpose plots of $x(t)$ versus $t$ and $x_R[n]$ versus $n/S$. Obtain the quantization error signal $e[n] = x[n] - x_R[n]$ and plot $e[n]$ and its 10-bin histogram. Compute the quantization SNR in decibels. Compare the quantization error, the histogram, and the quantization SNR for the 8-bit and 4-bit quantizer and comment on any differences. Repeat for 800 samples. Does an increase in the signal length improve the SNR?

**(c)** Based on your knowledge of the signal $x(t)$, compute the theoretical SNR for a 4-bit and 8-bit quantizer. How does the theoretical value compare with the quantization SNR obtained in parts (a) and (b)?

**(d)** Repeat parts (a) and (b), using quantization by truncation of $x[n]$. How does the quantization SNR for truncation compare with the the quantization SNR for rounding?

**7.67.** **(LORAN)** A LORAN (long-range radio and navigation) system for establishing positions of marine craft uses three transmitters that send out short bursts (10 cycles) of 100-kHz signals in a precise phase relationship. Using phase comparison, a receiver (on the craft) can establish the position (latitude and longitude) of the craft to within a few hundred meters. Suppose the LORAN signal is to be digitally processed by first sampling it at 500 kHz and filtering the sampled signal using a second-order peaking filter with a half-power bandwidth of 1 kHz. Design the peaking filter and use MATLAB to plot its frequency response.

**7.68.** **(Plucked-String Filters)** Figure P7.68 shows three variants of the Karplus-Strong filter for synthesizing plucked-string instruments. Assume that $G_{LP}(z) = 0.5(1 + z^{-1})$.

**FIGURE P.7.68** Plucked-string filters for Problem 7.68

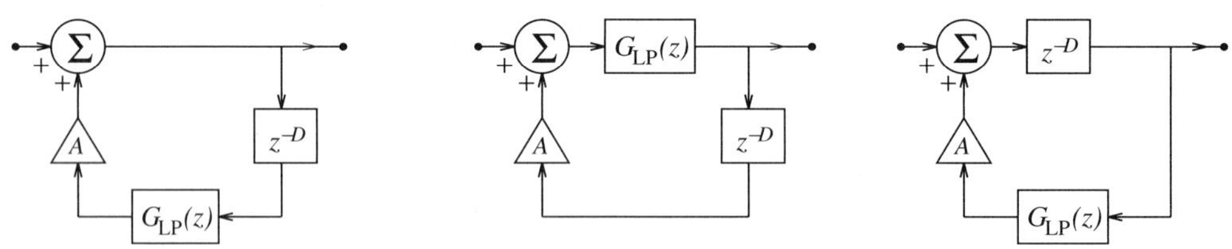


**(a)** Find the transfer function of each circuit.

**(b)** Plot the frequency response of each filter for $D = 12$ (choose an appropriate value for $A$).

**(c)** Describe the differences in the frequency response and impulse response of each filter.

**7.69.** **(Plucked-String Filters)** We wish to use the Karplus-Strong filter to synthesize a guitar note played at exactly 880 Hz, using a sampling frequency of 10 kHz, and by including a first-order allpass filter $G_{AP}(z) = (1+\alpha z)/(z+\alpha)$ in the feedback loop. Choose the second variant from the previous problem and assume that $G_{LP}(z) = 0.5(1 + z^{-1})$.

**(a)** What is the value of $D$ and the value of the allpass parameter $\alpha$?

**(b)** Plot the frequency response of the designed filter, using an appropriate value for $A$.

**(c)** How far off is the fundamental frequency of the designed filter from 880 Hz?

**(d)** Show that the exact relation for finding the parameter $\alpha$ from the phase delay $t_p$ at any frequency is given by

$$\alpha = \frac{\sin[(1 - t_p)\pi F]}{\sin[(1 + t_p)\pi F]}$$

**(e)** Use the result of part (d) to compute the exact value of $\alpha$ at the digital frequency corresponding to 880 Hz and plot the frequency response of the designed filter. Is the fundamental frequency of the designed filter any closer to 880 Hz? Will the exact value of $\alpha$ be more useful for lower or higher sampling rates?

**7.70. (Generating DTMF Tones)** In dual-tone multi-frequency (DTMF) or touch-tone telephone dialing, each number is represented by a dual-frequency tone (as described in the text). It is required to generate the signal at each frequency as a pure cosine, using a digital oscillator operating at a sampling rate of $S = 8192$ Hz. By varying the parameters of the digital filter, it should be possible to generate a signal at any of the required frequencies. A DTMF tone can then be generated by adding the appropriate low-frequency and high-frequency signals.

**(a)** Design the digital oscillator and use it to generate tones corresponding to all the digits. Each tone should last for 0.1 s.

**(b)** Use the FFT to obtain the spectrum of each tone and confirm that its frequencies correspond to the appropriate digit.

**7.71. (Decoding DTMF Tones)** To decode a DTMF signal, we must be able to isolate the tones and then identify the digits corresponding to each tone. This may be accomplished in two ways: using the FFT or by direct filtering.

**(a)** Generate DTMF tones (by any method) at a sampling rate of 8192 Hz.

**(b)** For each tone, use the FFT to obtain the spectrum and devise a test that would allow you to identify the digit from the frequencies present in its FFT. How would you automate the decoding process for an arbitrary DTMF signal?

**(c)** Apply each tone to a parallel bank of bandpass (peaking) filters, compute the output signal energy, compare with an appropriate threshold to establish the presence or absence of a frequency, and relate this information to the corresponding digit. You will need to design eight peaking filters (centered at the appropriate frequencies) to identify the frequencies present and the corresponding digit. How would you automate the decoding process for an arbitrary DTMF signal?

CHAPTER 8

# The Discrete Fourier Transform and Its Applications

## 8.0 Scope and Overview

This chapter describes the *discrete Fourier transform* (DFT) as a means of examining sampled signals in the frequency-domain. It starts with the definition of the DFT, its properties and its relationship to other frequency-domain transforms. It then discusses efficient algorithms for implementing the DFT that go by the generic name of *fast Fourier transforms* (FFTs). It concludes with applications of the DFT and FFT to digital signal processing.

### 8.0.1 Goals and Learning Objectives

The goals of this chapter are to provide an introduction to the DFT and its applications. After going through this chapter, the reader should:

1. Understand the DFT and IDFT and know how to compute the DFT and IDFT of finite sequences.
2. Understand the properties of the DFT and the implied periodicity in the DFT results.
3. Understand how to relate the DFT and DTFT of discrete signals.
4. Understand how to relate the Fourier series coefficients of a periodic signal to the DFT of its sampled version.
5. Understand the effects of aliasing and leakage.
6. Understand how to relate the Fourier transform of an analog signal to the DFT of its sampled version.
7. Understand the concept of zero padding and how to improve the DFT results.
8. Understand the concept of windowing and the spectrum of windowed signals.
9. Understand the concept of frequency resolution and dynamic-range resolution.
10. Understand how to use the DFT for convolution and bandlimited signal interpolation.
11. Understand the concept of spectrum estimation.
12. Understand the concept of the cepstrum and homomorphic signal processing.
13. Understand the matrix formulation of the DFT and IDFT.
14. Understand the FFT algorithms for computing the DFT and how they reduce the computation effort.

# 8.1 Introduction

The processing of analog signals using digital methods continues to gain widespread popularity. The Fourier series of analog periodic signals and the DTFT of discrete-time signals are duals of each other and are similar in many respects. In theory, both offer great insight into the spectral description of signals. In practice, both suffer from (similar) problems in their implementation. The finite memory limitations and finite precision of digital computers constrain us to work with a finite set of quantized numbers for describing signals in both time and frequency. This brings out two major problems inherent in the Fourier series and the DTFT as tools for digital signal processing. Both typically require an infinite number of samples (the Fourier series for its spectrum and the DTFT for its time signal). Both deal with one continuous variable (time $t$ or digital frequency $F$). A numerical approximation that can be implemented using digital computers requires that we replace the continuous variable with a discrete one and limit the number of samples to a finite value in both domains.

## 8.1.1 Connections between Frequency-Domain Transforms

Sampling and duality provide the basis for the connection between the various frequency-domain transforms and the concepts are worth repeating. Sampling in one domain induces a periodic extension in the other. The sample spacing in one domain is the reciprocal of the period in the other. Periodic analog signals have discrete spectra, and discrete-time signals have continuous periodic spectra. A consequence of these concepts is that a sequence that is both discrete and periodic in one domain is also discrete and periodic in the other. This leads to the development of the **discrete Fourier transform** (DFT) and **discrete Fourier series** (DFS), allowing us a practical means of arriving at the sampled spectrum of sampled signals using digital computers. The connections are illustrated in Figure 8.1 and summarized in Table 8.1.

Since sampling in one domain leads to a periodic extension in the other, a sampled representation in both domains also forces periodicity in both domains. This leads to two slightly different but functionally equivalent sets of relations, depending on the order in which we sample time and frequency, as listed in Table 8.2. If we first sample an analog signal $x(t)$, the sampled signal has a periodic spectrum $X(F)$ (the DTFT), and sampling of $X(F)$ leads to the DFT representation. If we first sample the Fourier transform $X(f)$

**TABLE 8.1** ➤ Connections between Various Transforms

| Operation in the Time Domain | Result in the Frequency Domain | Transform |
|---|---|---|
| Aperiodic, continuous $x(t)$ | Aperiodic, continuous $X(f)$ | FT |
| Periodic extension of $x(t) \Rightarrow x_p(t)$<br>Period $= T$ | Sampling of $X(f) \Rightarrow X[k]$<br>Sampling interval $= 1/T = f_0$ | FS |
| Sampling of $x_p(t) \Rightarrow x_p[n]$<br>Sampling interval $= t_s$ | Periodic extension of $X[k] \Rightarrow X_{\mathrm{DFS}}[k]$<br>Period $= S = 1/t_s$ | DFS |
| Sampling of $x(t) \Rightarrow x[n]$<br>Sampling interval $= 1$ | Periodic extension of $X(f) \Rightarrow X(F)$<br>Period $= 1$ | DTFT |
| Periodic extension of $x[n] \Rightarrow x_p[n]$<br>Period $= N$ | Sampling of $X(F) \Rightarrow X_{\mathrm{DFT}}[k]$<br>Sampling interval $= 1/N$ | DFT |

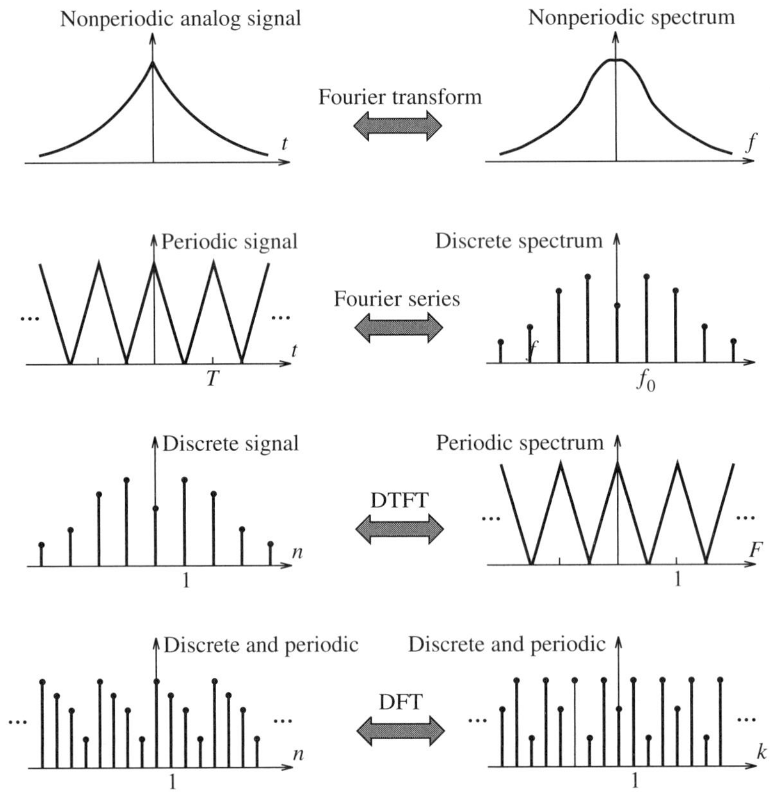


**FIGURE 8.1** Features of the various transforms. A nonperiodic analog signal has a nonperiodic analog spectrum described by its Fourier transform. Sampling in one domain induces periodicity in the other. Thus, a periodic analog signal has a discrete spectrum (described by its Fourier series) and a discrete nonperiodic signal has a periodic spectrum (described by the DTFT). If a signal is periodic and discrete, its spectrum is also periodic and discrete (and described by the DFT)

**TABLE 8.2** ➤ Relating Frequency-Domain Transforms

| Fourier Transform | |
|---|---|
| Aperiodic and Continuous Signal: $x(t) = \int_{-\infty}^{\infty} X(f)e^{j2\pi ft}\,df$ | Aperiodic and Continuous Spectrum: $X(f) = \int_{-\infty}^{\infty} x(t)e^{-j2\pi ft}\,dt$ |
| ⇓ | ⇓ |
| **Sampling $x(t)$ (DTFT)**<br>Sampled signal:<br>$x[n] = \int_1 X(F)e^{j2\pi nF}\,dF$<br>Periodic spectrum (period = 1):<br>$X(F) = \sum_{n=-\infty}^{\infty} x[n]e^{-j2\pi nF}$ | **Sampling $X(f)$ (Fourier series)**<br>Sampled spectrum:<br>$X[k] = \dfrac{1}{T}\displaystyle\int_T x_p(t)e^{-j2\pi kf_0 t}\,dt$<br>Periodic signal (period = $T$):<br>$x_p(t) = \sum_{k=-\infty}^{\infty} X[k]e^{j2\pi kf_0 t}$ |
| ⇓ | ⇓ |
| **Sampling $X(F)$ (DFT)**<br>Sampled and periodic spectrum:<br>$X_{\mathrm{DFT}}[k] = \sum_{n=0}^{N-1} x[n]e^{-j2\pi nk/N}$<br>Sampled and periodic signal:<br>$x[n] = \dfrac{1}{N}\sum_{k=0}^{N-1} X_{\mathrm{DFT}}[k]e^{j2\pi nk/N}$ | **Sampling $x_p(t)$ (DFS)**<br>Sampled and periodic signal:<br>$x[n] = \sum_{k=0}^{N-1} X_{\mathrm{DFS}}[k]e^{j2\pi nk/N}$<br>Sampled and periodic spectrum:<br>$X_{\mathrm{DFS}}[k] = \dfrac{1}{N}\sum_{n=0}^{N-1} x[n]e^{-j2\pi nk/N}$ |

in the frequency domain, the samples represent the Fourier series coefficients of a periodic time signal $x_p(t)$, and sampling of $x_p(t)$ leads to the **discrete Fourier series** (DFS) as the periodic extension of the frequency-domain samples. The DFS differs from the DFT only by a constant scale factor.

## 8.2 The DFT

The $N$-point **discrete Fourier transform** (DFT) $X_{\text{DFT}}[k]$ of an $N$-sample signal $x[n]$ and the **inverse discrete Fourier transform** (IDFT), which transforms $X_{\text{DFT}}[k]$ to $x[n]$, are defined by

$$X_{\text{DFT}}[k] = \sum_{n=0}^{N-1} x[n]e^{-j2\pi nk/N}, \quad k = 0, 1, 2, \ldots, N-1 \tag{8.1}$$

$$x[n] = \frac{1}{N}\sum_{k=0}^{N-1} X_{\text{DFT}}[k]e^{j2\pi nk/N}, \quad n = 0, 1, 2, \ldots, N-1 \tag{8.2}$$

Each relation is a set of $N$ equations. Each DFT sample is found as a weighted sum of all the samples in $x[n]$. One of the most important properties of the DFT and its inverse is implied periodicity. The exponential $\exp(\pm j2\pi nk/N)$ in the defining relations is periodic in both $n$ and $k$ with period $N$:

$$e^{j2\pi nk/N} = e^{j2\pi(n+N)k/N} = e^{j2\pi n(k+N)/N} \tag{8.3}$$

As a result, the DFT and its inverse are also periodic with period $N$, and it is sufficient to compute the results for only one period (0 to $N-1$). *Both $x[n]$ and $X_{\text{DFT}}[k]$ have a starting index of zero.*

---

**The $N$-point DFT and $N$-Point IDFT are Periodic with Period $N$**

$$X_{\text{DFT}}[k] = \sum_{n=0}^{N-1} x[n]e^{-j2\pi nk/N}, k=0, 1, 2, \ldots, N-1$$

$$x[n] = \frac{1}{N}\sum_{k=0}^{N-1} X_{\text{DFT}}[k]e^{j2\pi nk/N}, n=0, 1, 2, \ldots, N-1$$

---

**Example 8.1** **DFT from the Defining Relation**

Let $x[n] = \{\overset{\Downarrow}{1}, 2, 1, 0\}$. With $N = 4$, and $e^{-j2\pi nk/N} = e^{-jnk\pi/2}$, we successively compute

$$k = 0: \quad X_{\text{DFT}}[0] = \sum_{n=0}^{3} x[n]e^{0} = 1 + 2 + 1 + 0 = 4$$

$$k = 1: \quad X_{\text{DFT}}[1] = \sum_{n=0}^{3} x[n]e^{-jn\pi/2} = 1 + 2e^{-j\pi/2} + e^{-j\pi} + 0 = -j2$$

$$k = 2: \quad X_{\mathrm{DFT}}[2] = \sum_{n=0}^{3} x[n]e^{-jn\pi} = 1 + 2e^{-j\pi} + e^{-j2\pi} + 0 = 0$$

$$k = 3: \quad X_{\mathrm{DFT}}[3] = \sum_{n=0}^{3} x[n]e^{-j3n\pi/2} = 1 + 2e^{-j3\pi/2} + e^{-j3\pi} + 0 = j2$$

The DFT is thus $X_{\mathrm{DFT}}[k] = \{\overset{\Downarrow}{4}, -j2, 0, j2\}$.

## Drill Problem 8.1

**(a)** Find the DFT of $x[n] = \{\overset{\Downarrow}{2}, -1, 4, 3\}$.
**(b)** Find the DFT of $y[n] = \{\overset{\Downarrow}{1}, 0, 1, 0, 1\}$.

**Answers:** **(a)** $\{\overset{\Uparrow}{8}, -2 + j4, 4, -2 - j4\}$
**(b)** $\{\overset{\Uparrow}{3}, 0.5 + j0.36, 0.5 + j1.54, 0.5 - j1.54, 0.5 - j0.36\}$

# 8.3 Properties of the DFT

The properties of the DFT are summarized in Table 8.3. They are strikingly similar to other frequency-domain transforms, but must always be used in keeping with implied periodicity (of the DFT and IDFT) in both domains.

**TABLE 8.3** ➤ Properties of the $N$-Sample DFT

| Property | Signal | DFT | Remarks |
|---|---|---|---|
| Shift | $x[n - n_0]$ | $X_{\mathrm{DFT}}[k]e^{-j2\pi k n_0/N}$ | No change in magnitude |
| Shift | $x[n - 0.5N]$ | $(-1)^k X_{\mathrm{DFT}}[k]$ | Half-period shift for even $N$ |
| Modulation | $x[n]e^{j2\pi n k_0/N}$ | $X_{\mathrm{DFT}}[k - k_0]$ | |
| Modulation | $(-1)^n x[n]$ | $X_{\mathrm{DFT}}[k - 0.5N]$ | Half-period shift for even $N$ |
| Reversal | $x[-n]$ | $X_{\mathrm{DFT}}[-k]$ | This is *circular* flipping |
| Product | $x[n]y[n]$ | $\frac{1}{N} X_{\mathrm{DFT}}[k] \circledast Y_{\mathrm{DFT}}[k]$ | The convolution is *periodic* |
| Convolution | $x[n] \circledast y[n]$ | $X_{\mathrm{DFT}}[k]Y_{\mathrm{DFT}}[k]$ | The convolution is *periodic* |
| Correlation | $x[n] \circledast\circledast y[n]$ | $X_{\mathrm{DFT}}[k]Y^*_{\mathrm{DFT}}[k]$ | The correlation is *periodic* |
| Central ordinates | $x[0] = \frac{1}{N}\sum_{k=0}^{N-1} X_{\mathrm{DFT}}[k]$ | $X_{\mathrm{DFT}}[0] = \sum_{n=0}^{N-1} x[n]$ | |
| Central ordinates | $x[\frac{N}{2}] = \frac{1}{N}\sum_{k=0}^{N-1} (-1)^k X_{\mathrm{DFT}}[k]$ ($N$ even) | $X_{\mathrm{DFT}}[\frac{N}{2}] = \sum_{n=0}^{N-1} (-1)^n x[n]$ ($N$ even) | |
| Parseval's relation | $\sum_{n=0}^{N-1} \lvert x[n]\rvert^2 = \frac{1}{N}\sum_{k=0}^{N-1} \lvert X_{\mathrm{DFT}}[k]\rvert^2$ | | |

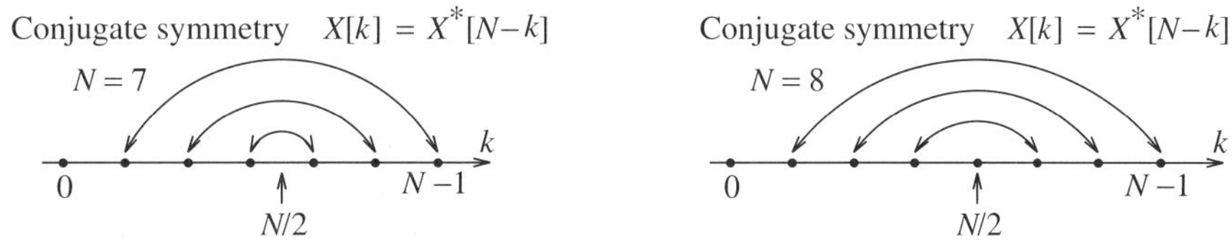


**FIGURE 8.2** Symmetry of the DFT for real signals. The DFT shows conjugate symmetry about the origin. The DFT also shows conjugate symmetry about the folding index $k = N/2$, which may or may not fall on a DFT sample

### 8.3.1 Symmetry

In analogy with all other frequency-domain transforms, the DFT of a real sequence possesses conjugate symmetry about the origin with $X_{\rm DFT}[-k] = X^*_{\rm DFT}[k]$. Since the DFT is periodic, $X_{\rm DFT}[-k] = X_{\rm DFT}[N-k]$. This also implies conjugate symmetry about the index $k = 0.5N$, and thus

$$X_{\rm DFT}[-k] = X^*_{\rm DFT}[k] = X_{\rm DFT}[N-k] \tag{8.4}$$

If $N$ is odd, the conjugate symmetry is about the half-integer value $0.5N$. The index $k = 0.5N$ is called the **folding index.** This is illustrated in Figure 8.2.

Conjugate symmetry suggests that we need compute only half the DFT values to find the entire DFT sequence—another labor-saving concept! A similar result applies to the IDFT.

**The DFT of a Real Signal Shows Conjugate Symmetry**

$$X_{\rm DFT}[-k] = X^*_{\rm DFT}[k]$$

$$X_{\rm DFT}[-k] = X_{\rm DFT}[N-k]$$

**Example 8.2**

**The DFT and Conjugate Symmetry**

Let us find the 8-point DFT of $x[n] = \{\overset{\Downarrow}{1}, 1, 0, 0, 0, 0, 0, 0\}$.

**Solution:** Since only $x[0]$ and $x[1]$ are nonzero, the upper index in the DFT summation will be $n = 1$ and the DFT reduces to

$$X_{\rm DFT}[k] = \sum_{n=0}^{1} x[n]e^{-j2\pi nk/8} = 1 + e^{-j\pi k/4}, \qquad k = 0, 1, 2, \ldots, 7$$

Since $N = 8$, we need compute $X_{\rm DFT}[k]$ only for $k \le 0.5N = 4$. We compute

$$X_{\rm DFT}[0] = 1 + 1 = 2, \quad X_{\rm DFT}[1] = 1 + e^{-j\pi/4} = 1.707 - j0.707,$$

$$X_{\rm DFT}[2] = 1 + e^{-j\pi/2} = 1 - j,$$

$$X_{\rm DFT}[3] = 1 + e^{-j3\pi/4} = 0.293 - j0.707, \quad X_{\rm DFT}[4] = 1 - 1 = 0$$

With $N = 8$, conjugate symmetry says $X_{\text{DFT}}[k] = X^*_{\text{DFT}}[N - k] = X^*_{\text{DFT}}[8 - k]$ and we find

$$\begin{aligned}X_{\text{DFT}}[5] &= X^*_{\text{DFT}}[3] = 0.293 + j0.707,\\ X_{\text{DFT}}[6] &= X^*_{\text{DFT}}[2] = 1 + j,\\ X_{\text{DFT}}[7] &= X^*_{\text{DFT}}[1] = 1.707 + j0.707\end{aligned}$$

Thus,

$$\begin{aligned}X_{\text{DFT}}[k] = \{\overset{\Downarrow}{2}, 1.707 - j0.707, 0.293 - j0.707, 1 - j, 0,\\ 1 + j, 0.293 + j0.707, 1.707 + j0.707\}\end{aligned}$$

# Drill Problem 8.2

**(a)** Find the DFT of $x[n] = \{\overset{\Downarrow}{1}, 0, 0, 2, 0, 3\}$.

**(b)** The DFT of a real signal is $\{\overset{\Downarrow}{1}, \boxed{\text{A}}, -1, \boxed{\text{B}}, 0, -j2, \boxed{\text{C}}, -1 + j\}$. Find $A$, $B$ and $C$.

**Answers: (a)** $\{\underset{\Uparrow}{6}, 0.5 + j2.6, 1.5 + j2.6, -4, 1.5 - j2.6, 0.5 - j2.6\}$ **(b)** $-1 - j, j2, -1$

## 8.3.2 Central Ordinates and Parseval's Theorem

The computation of the DFT at the indices $k = 0$ and (for even $N$) at $k = \frac{N}{2}$ can be simplified using the central ordinate theorems that arise as a direct consequence of the defining relations. In particular, we find that $X_{\text{DFT}}[0]$ equals the sum of the $N$ signal samples $x[n]$, and $X_{\text{DFT}}[\frac{N}{2}]$ equals the sum of $(-1)^n x[n]$ (with alternating sign changes). This also implies that, if $x[n]$ is real valued, so are $X_{\text{DFT}}[0]$ and $X_{\text{DFT}}[\frac{N}{2}]$. Similar results hold for the IDFT.

Parseval's theorem says that the DFT is an energy-conserving transformation and allows us to find the signal energy either from the signal or its spectrum. This implies that the sum of squares of the signal samples is related to the sum of squares of the magnitude of the DFT samples.

# Drill Problem 8.3

**(a)** Let $x[n] = \{\overset{\Downarrow}{A}, 2, 3, 4, 5, 6, 7, B\}$. If $X_{\text{DFT}}[0] = 20$ and $X_{\text{DFT}}[4] = 0$, find $A$ and $B$.

**(b)** The DFT of a real signal is $\{\overset{\Downarrow}{1}, \boxed{\text{A}}, -1, \boxed{\text{B}}, -7, -j2, \boxed{\text{C}}, -1 + j\}$. What is its signal energy?

**Answers: (a)** $-5, -2$ **(b)** 8

### 8.3.3 Circular Shift and Circular Symmetry

The defining relation for the DFT requires signal values for $0 \le n \le N-1$. By implied periodicity, these values correspond to one period of a periodic signal. If we wish to find the DFT of a time-shifted signal $x[n-n_0]$, its values must also be selected over $(0, N-1)$ from its periodic extension. This concept is called **circular shifting**. To generate $x[n-n_0]$, we delay $x[n]$ by $n_0$, create the periodic extension of the shifted signal, and pick $N$ samples over $(0, N-1)$. This is equivalent to moving the last $n_0$ samples of $x[n]$ to the beginning of the sequence. Similarly, to generate $x[n+n_0]$, we move the first $n_0$ samples to the end of the sequence. **Circular flipping** generates the signal $x[-n]$ from $x[n]$. We flip $x[n]$, create the periodic extension of the flipped signal, and pick $N$ samples of the periodic extension over $(0, N-1)$.

Even symmetry of $x[n]$ requires that $x[n] = x[-n]$. Its implied periodicity also means $x[n] = x[N-n]$, and the periodic signal $x[n]$ is said to possess **circular even symmetry**. Similarly, for **circular odd symmetry**, we have $x[n] = -x[N-n]$.

***Example 8.3*** **Circular Shift and Flipping**

**(a)** Let $y[n] = \{\overset{\Downarrow}{1}, 2, 3, 4, 5, 0, 0, 6\}$. Find one period of the circularly shifted signal $f[n] = y[n-2]$.

To create $f[n] = y[n-2]$, we move the last two samples to the beginning. So,

$$f[n] = y[n-2] = \{\overset{\Downarrow}{0}, 6, 1, 2, 3, 4, 5, 0\}$$

**(b)** Let $y[n] = \{\overset{\Downarrow}{1}, 2, 3, 4, 5, 0, 0, 6\}$. Find one period of the circularly shifted signal $g[n] = y[n+2]$.

To create $g[n] = y[n+2]$, we move the first two samples to the end. So,

$$g[n] = y[n+2] = \{\overset{\Downarrow}{3}, 4, 5, 0, 0, 6, 1, 2\}$$

**(c)** Let $y[n] = \{\overset{\Downarrow}{1}, 2, 3, 4, 5, 0, 0, 6\}$. Find one period of the circularly flipped signal $h[n] = y[-n]$.

To create $h[n] = y[-n]$, we flip $y[n]$ to $\{\overset{\Downarrow}{0}, 0, 6, 5, 4, 3, 2, 1\}$ and create its periodic extension to get

$$h[n] = y[-n] = \{\overset{\Downarrow}{1}, 0, 0, 6, 5, 4, 3, 2\}$$

## Drill Problem 8.4

**(a)** Let $x[n] = \{\overset{\Downarrow}{1}, 2, 0, 3\}$. Find the circularly shifted signal $x[n-3]$.

**(b)** Let $x[n] = \{\overset{\Downarrow}{1}, 2, 0, 3\}$. Find the circularly shifted signal $x[n+2]$.

**(c)** Let $x[n] = \{\overset{\Downarrow}{1}, 2, 0, 3\}$. Find the circularly folded signal $x[-n]$.

**(d)** If $x[n] = \{\overset{\Downarrow}{6}, 2, A, 0, B\}$ is circularly even symmetric, find $A$ and $B$.

**Answers:** **(a)** $\{\overset{\Downarrow}{2}, 0, 3, 1\}$ **(b)** $\{\overset{\Downarrow}{0}, 3, 1, 2\}$ **(c)** $\{\overset{\Downarrow}{1}, 3, 0, 2\}$ **(d)** 0, 2

### 8.3.4 Shifting, Reversal, and Modulation Properties of the DFT

A time shift of the signal $x[n]$ to $y[n] = x[n-m]$ multiplies the DFT $X_{\text{DFT}}[k]$ by $e^{-j2\pi mk/N}$ to give $Y_{\text{DFT}}[k] = X_{\text{DFT}}[k]e^{-j2\pi mk/N}$. The shift provides a linear phase contribution. If $x[n]$ is circularly flipped to give $f[n] = x[-n]$, then $F_{\text{DFT}}[k] = X_{\text{DFT}}[-k]$ and the result implies a conjugation of the DFT of $x[n]$.

If the DFT $X_{\text{DFT}}[k]$ is shifted to $G_{\text{DFT}}[k] = X_{\text{DFT}}[k-m]$, the IDFT corresponds to the signal $g[n] = x[n]e^{j2\pi mn/N}$. This is the basis for the modulation property. If the signal length $N$ is even and $m = \frac{N}{2}$, we get $g[n]$ by changing the sign of the even-indexed samples in $x[n]$.

**Example 8.4**

**Properties of the DFT**

Consider the DFT pair $x[n] \Longleftarrow\boxed{\text{DFT}}\Longrightarrow X_{\text{DFT}}[k] = \{\overset{\Downarrow}{4}, -j2, 0, j2\}$ with $N = 4$.

**(a) Time Shift**

To find the DFT of $y[n] = x[n-2]$, the time-shift property gives

$$Y_{\text{DFT}}[k] = X_{\text{DFT}}[k]e^{-j2\pi k n_0/4} = X_{\text{DFT}}[k]e^{-jk\pi} = \{\overset{\Downarrow}{4}, j2, 0, -j2\}$$

**(b) Flipping**

To find the DFT of $g[n] = x[-n]$, the reversal property gives

$$G_{\text{DFT}}[k] = X_{\text{DFT}}[-k] = X^*_{\text{DFT}}[k] = \{\overset{\Downarrow}{4}, j2, 0, -j2\}$$

**(c) Conjugation**

To find the DFT of $p[n] = x^*[n]$, the conjugation property gives

$$P_{\text{DFT}}[k] = X^*_{\text{DFT}}[-k] = \{\overset{\Downarrow}{4}, j2, 0, -j2\}^* = \{\overset{\Downarrow}{4}, -j2, 0, j2\}$$

Since $P_{\text{DFT}} = X_{\text{DFT}}$, this means $x[n] = x^*[n]$ and implies that $x[n]$ is real valued.

# Drill Problem 8.5

**(a)** Let $x[n] \Longleftarrow\boxed{\text{DFT}}\Longrightarrow \{\overset{\Downarrow}{1}, j2, 4, -j2\}$. Find the DFT of $y[n] = x[n-1]$.

**(b)** Let $x[n] \Longleftarrow\boxed{\text{DFT}}\Longrightarrow \{\overset{\Downarrow}{1}, j2, 4, 0, 4, -j2\}$. Find the DFT of $g[n] = x[n+3]$.

**(c)** Let $x[n] \Longleftarrow\boxed{\text{DFT}}\Longrightarrow \{\overset{\Downarrow}{1}, j2, j4, 0, -1, 0, -j4, -j2\}$. Find the DFT of $h[n] = x[-n]$.

**Answers:** (a) $\{\overset{\Downarrow}{1}, 2, -4, 2\}$, (b) $\{\overset{\Downarrow}{1}, -j2, 4, 0, 4, j2\}$, (c) $\{\overset{\Downarrow}{1}, -j2, -j4, 0, -1, 0, j4, j2\}$

## 8.3.5 Product and Convolution Properties of the DFT

Convolution in one domain transforms to multiplication in the other. Due to the implied periodicity in both domains, the convolution operation describes *periodic* (not regular) convolution. This also applies to the correlation operation.

### Periodic Convolution

The DFT offers an indirect means of finding the periodic convolution $y[n] = x[n] \circledast h[n]$ of two sequences $x[n]$ and $h[n]$ of equal length $N$. We compute their $N$-sample DFTs $X_{\text{DFT}}[k]$ and $H_{\text{DFT}}[k]$, multiply them to obtain $Y_{\text{DFT}}[k] = X_{\text{DFT}}[k]H_{\text{DFT}}[k]$, and find the inverse of $Y_{\text{DFT}}$ to obtain the periodic convolution $y[n]$:

$$x[n] \circledast h[n] \Longleftarrow\boxed{\text{DFT}}\Longrightarrow X_{\text{DFT}}[k]H_{\text{DFT}}[k] \tag{8.5}$$

### Periodic Correlation

Periodic correlation can be implemented using the DFT in almost exactly the same way as periodic convolution, except for an extra conjugation step prior to taking the inverse DFT. The periodic correlation of two sequences $x[n]$ and $h[n]$ of equal length $N$ gives

$$r_{xh}[n] = x[n] \circledast\circledast\, h[n] \Longleftarrow\boxed{\text{DFT}}\Longrightarrow X_{\text{DFT}}[k]H^*_{\text{DFT}}[k] \tag{8.6}$$

If $x[n]$ and $h[n]$ are real, the final result $r_{xh}[n]$ must also be real (to within machine roundoff).

### Regular Convolution and Correlation

We can also find regular convolution (or correlation) using the DFT. For two sequences of length $M$ and $N$, the regular convolution (or correlation) contains $M + N - 1$ samples. We must thus pad each sequence with enough zeros (to make each sequence of length $M + N - 1$) before finding the DFT.

## 8.3.6 The FFT

The DFT describes a set of $N$ equations, each with $N$ product terms and thus requires a total of $N^2$ multiplications for its computation. Computationally efficient algorithms to obtain the DFT go by the generic name FFT (fast Fourier transform) and need far fewer multiplications. In particular, **radix-2** FFT algorithms require the number of samples $N$ to be a power of 2 and compute the DFT using only $N \log_2 N$ multiplications. We discuss such algorithms in a later section.

**Example 8.5**

### Properties of the DFT

**(a)** Consider the DFT pair $x[n] \Longleftarrow\boxed{\text{DFT}}\Longrightarrow X_{\text{DFT}}[k] = \{\overset{\Downarrow}{4}, -j2, 0, j2\}$ with $N = 4$.

1. **Product**
   To find the DFT of $h[n] = x^2[n] = x[n]x[n]$, the product property gives

$$H_{\text{DFT}}[k] = \frac{1}{4}X_{\text{DFT}}[k] \circledast X_{\text{DFT}}[k]$$

$$= \frac{1}{4}\{\overset{\Downarrow}{4}, -j2, 0, j2\} \circledast \{\overset{\Downarrow}{4}, -j2, 0, j2\} = \{\overset{\Downarrow}{6}, -j4, 0, j4\}$$

   Note that the result is based on periodic convolution.

2. **Periodic Convolution**
To find the DFT of $v[n] = x[n] \circledast x[n]$, the convolution property gives

$$V_{\text{DFT}}[k] = X_{\text{DFT}}[k]X_{\text{DFT}}[k] = \{\overset{\Downarrow}{16}, -4, 0, -4\}$$

This result is just the pointwise product of the DFTs.

3. **Central Ordinates and Parseval's Theorem**
We find $x[0] = \frac{1}{4}\sum X_{\text{DFT}}[k] = 0$.
By Parseval's theorem, the signal energy is given by

$$\text{Signal energy } = \frac{1}{4}\sum |X_{\text{DFT}}[k]|^2 = \frac{1}{4}(16+4+4) = 6$$

**(b)** Consider the DFT pair $x[n] = \{\overset{\Downarrow}{1}, 2, 1, 0\} \Leftarrow\boxed{\text{DFT}}\Rightarrow X_{\text{DFT}}[k]$ with $N = 4$.

1. **Modulation**
To find the IDFT of $Z_{\text{DFT}}[k] = X_{\text{DFT}}[k-1]$, the modulation property gives

$$z[n] = x[n]e^{j2\pi n/4} = x[n]e^{j\pi n/2} = \{\overset{\Downarrow}{1}, j2, -1, 0\}$$

2. **Periodic Convolution**
To find the IDFT of $H_{\text{DFT}}[k] = X_{\text{DFT}}[k] \circledast X_{\text{DFT}}[k]$, we use the periodic convolution property to get

$$h[n] = Nx^2[n] = \{\overset{\Downarrow}{4}, 16, 4, 0\}$$

3. **Product**
The IDFT of $W_{\text{DFT}}[k] = X_{\text{DFT}}[k]X_{\text{DFT}}[k]$ is found from the periodic convolution of $x[n]$ with itself to give

$$w[n] = \{\overset{\Downarrow}{1}, 2, 1, 0\} \circledast \{\overset{\Downarrow}{1}, 2, 1, 0\} = \{\overset{\Downarrow}{2}, 4, 6, 4\}$$

4. **Central Ordinates and Signal Energy**
We find $X_{\text{DFT}}[0] = \sum x[n] = 4$
The signal energy is given by

$$\text{Signal energy } = \sum |x[n]|^2 = 1+4+1+0 = 6$$

**(c) Regular Convolution**
Let $x[n] = \{\overset{\Downarrow}{1}, 2, 1, 0\}$
The regular convolution $s[n] = x[n] * x[n]$ contains seven samples. To find its DFT $S_{\text{DFT}}[k]$, we zero-pad $x[n]$ to length 7 to get $y[n] = \{\overset{\Downarrow}{1}, 2, 1, 0, 0, 0, 0\}$. Its DFT $Y_{\text{DFT}}[k]$ is found as

$$Y_{\text{DFT}}[k] = \{\overset{\Downarrow}{4}, 2.02 - j2.54, -0.35 - j1.52, -0.17 - j0.09, \\ -0.17 + j0.09, -0.35 + j1.52, 2.02 + j2.54\}$$

This leads to $S_{\text{DFT}}[k] = Y^2_{\text{DFT}}[k]$ and we get (upon simplification)

$$S_{\text{DFT}}[k] = \{\overset{\Downarrow}{16}, -2.35 - j10.28, -2.18 + j1.05, 0.02 + j0.03, \\ 0.02 - j0.03, -2.18 - j1.05, -2.35 + j10.28\}$$

Its IDFT gives $s[n]$ as

$$s[n] = \{\overset{\Downarrow}{1}, 4, 6, 4, 1, 0, 0\}$$

This result can be confirmed by using conventional (time-domain) convolution.

## 8.3.7 Signal Replication and Spectrum Zero Interpolation

In analogy with the DTFT and Fourier series, two useful DFT results are that replication in one domain leads to zero interpolation in the other. Formally, if $x[n] \Leftarrow\boxed{\text{DFT}}\Rightarrow X_{\text{DFT}}[k]$ form a DFT pair, $M$-fold replication of $x[n]$ to $y[n] = \{x[n], x[n], \ldots, x[n]\}$ leads to zero interpolation of the DFT to $Y_{\text{DFT}}[k] = MX^{\uparrow}_{\text{DFT}}[k/M]$. The multiplying factor $M$ ensures that we satisfy Parseval's theorem and the central ordinate relations. Its dual is the result that zero interpolation of $x[n]$ to $z[n] = x^{\uparrow}[n/M]$ leads to $M$-fold replication of the DFT to $Z_{\text{DFT}}[k] = \{X_{\text{DFT}}[k], X_{\text{DFT}}[k], \ldots, X_{\text{DFT}}[k]\}$.

### Replication in One Domain Corresponds to Zero Interpolation in the Other

**If a signal is replicated by $M$, its DFT is zero interpolated and scaled by $M$.**

If $x[n] \Leftarrow\boxed{\text{DFT}}\Rightarrow X_{\text{DFT}}[k]$, then

$$\underbrace{\{x[n], x[n], \ldots, x[n]\}}_{M\text{-fold replication}} \Leftarrow\boxed{\text{DFT}}\Rightarrow MX^{\uparrow}_{\text{DFT}}[k/M]$$

**If a signal is zero-interpolated by $M$, its DFT shows $M$-fold replication.**

If $x[n] \Leftarrow\boxed{\text{DFT}}\Rightarrow X_{\text{DFT}}[k]$, then

$$x^{\uparrow}[n/M] \Leftarrow\boxed{\text{DFT}}\Rightarrow \underbrace{\{X_{\text{DFT}}[k], X_{\text{DFT}}[k], \ldots, X_{\text{DFT}}[k]\}}_{M\text{-fold replication}}$$

**Example 8.6**

### Signal and Spectrum Replication

Let $x[n] = \{\overset{\Downarrow}{2}, 3, 2, 1\}$ and $X_{\text{DFT}}[k] = \{\overset{\Downarrow}{8}, -j2, 0, j2\}$. Find the DFT of the 12-point signal described by $y[n] = \{x[n], x[n], x[n]\}$ and the 12-point zero-interpolated signal $h[n] = x^{\uparrow}[n/3]$.

**(a)** Signal replication by 3 leads to spectrum zero interpolation and multiplication by 3. Thus,

$$Y_{\text{DFT}}[k] = 3X^{\uparrow}_{\text{DFT}}[k/3] = \{\overset{\Downarrow}{24},\ 0,\ 0,\ -j6,\ 0,\ 0,\ 0,\ 0,\ 0,\ j6,\ 0,\ 0\}$$

**(b)** Signal zero interpolation by 3 leads to spectrum replication by 3. Thus,

$$H_{\text{DFT}}[k] = \{X_{\text{DFT}}[k], X_{\text{DFT}}[k], X_{\text{DFT}}[k]\}$$
$$= \{\overset{\Downarrow}{8}, -j2, 0, j2, 8, -j2, 0, j2, 8, -j2, 0, j2\}$$

# Drill Problem 8.6

**(a)** Let $x[n] \Leftarrow \boxed{\text{DFT}} \Rightarrow \{\overset{\Downarrow}{1}, j2, 4, -j2\}$. Find the DFT of the zero-interpolated signal $y[n] = x[n/2]$.

**(b)** Let $\{\overset{\Downarrow}{4}, 2, 0, 2\} \Leftarrow \boxed{\text{DFT}} \Rightarrow X_{\text{DFT}}[k]$. If $G_{\text{DFT}}[k] = X_{\text{DFT}}[k/2]$, find $g[n]$.

**Answers: (a)** $\{\overset{\Downarrow}{1}, j2, 4, -j2, 1, j2, 4, -j2\}$ **(b)** $\{\overset{\Downarrow}{2}, 1, 0, 1, 2, 1, 0, 1\}$

## 8.3.8 Some Useful DFT Pairs

The DFT of finite sequences defined mathematically often results in very unwieldy expressions and explains the lack of many "standard" DFT pairs. However, the following DFT pairs are quite useful and easy to obtain from the defining relation and properties:

$$\{\overset{\Downarrow}{1}, 0, 0, \ldots, 0\}(\text{impulse}) \Leftarrow \boxed{\text{DFT}} \Rightarrow \{\overset{\Downarrow}{1}, 1, 1, \ldots, 1\} \quad (\text{constant}) \tag{8.7}$$

$$\{\overset{\Downarrow}{1}, 1, 1, \ldots, 1\}(\text{constant}) \Leftarrow \boxed{\text{DFT}} \Rightarrow \{\overset{\Downarrow}{N}, 0, 0, \ldots, 0\} \quad (\text{impulse}) \tag{8.8}$$

$$\alpha^n(\text{exponential}) \Leftarrow \boxed{\text{DFT}} \Rightarrow \frac{1-\alpha^N}{1-\alpha e^{-j2\pi k/N}} \tag{8.9}$$

$$\cos\left(2\pi n \frac{k_0}{N}\right)(\text{sinusoid}) \Leftarrow \boxed{\text{DFT}} \Rightarrow 0.5N\delta[k-k_0] + 0.5N\delta[k-(N-k_0)] \quad (\text{impulse pair}) \tag{8.10}$$

The first result is a direct consequence of the defining relation. For the second result, the DFT is $\sum e^{j2\pi nk/N}$, the sum of $N$ equally spaced vectors of unit length and equals zero (except when $k = 0$). For the third result, we use the defining relation $\sum \alpha^n e^{-j2\pi nk/N}$ and the fact that $e^{j2\pi k} = 1$ to obtain

$$X_{\text{DFT}}[k] = \sum_{k=0}^{N-1} (\alpha e^{-j2\pi k/N})^n = \frac{1-(\alpha e^{-j2\pi k/N})^N}{1-\alpha e^{-j2\pi k/N}} = \frac{1-\alpha^N}{1-\alpha e^{-j2\pi k/N}} \tag{8.11}$$

Finally, the transform pair for the sinusoid says that, for a periodic sinusoid $x[n] = \cos(2\pi nF)$ whose digital frequency is $F = k_0/N$, the DFT is a pair of impulses at $k = k_0$ and $k = N - k_0$. By Euler's relation, $x[n] = 0.05e^{j2\pi nk_0/N} + 0.5e^{-j2\pi nk_0/N}$ and by periodicity, $0.5e^{-j2\pi nk_0/N} = 0.5e^{j2\pi n(N-k_0)/N}$. Then, with the DFT pair $1 \Leftarrow \boxed{\text{DFT}} \Rightarrow N\delta[k]$, and the modulation property, we get the required result.

**The $N$-Point DFT of a Sinusoid with Period $N$ and $F = k_0/N$ Has Two Nonzero Samples**

$$\underbrace{\cos(2\pi n \frac{k_0}{N} + \theta)}_{N\text{-sample sinusoid with } F_0 = k_0/N} \Leftarrow \boxed{\text{DFT}} \Rightarrow \{\overset{\Downarrow}{0}, \ldots, 0, \underbrace{0.5Ne^{j\theta}}_{k=k_0}, 0, \ldots, 0, \underbrace{0.5Ne^{-j\theta}}_{k=N-k_0}, 0, \ldots, 0\}$$

# Drill Problem 8.7

**(a)** Let $x[n] = \{\overset{\Downarrow}{2}, 0, 0, 0, 0\}$. Find $X_{\text{DFT}}[k]$.
**(b)** Let $y[n] = \{\overset{\Downarrow}{3}, 1, 1, 1, 1\}$. Find $Y_{\text{DFT}}[k]$.
**(c)** Let $g[n] = \cos(\frac{2n\pi}{3})$. Let $N = 6$. Express $F_0$ as $\frac{k_0}{N}$ and find the 6-point DFT $G_{\text{DFT}}[k]$.
**(d)** Let $h[n] = 8(0.5)^n, 0 \le n \le 3$. Obtain its 4-point DFT from the defining relation or otherwise.

**Answers: (a)** $\{\overset{\Downarrow}{2}, 2, 2, 2, 2\}$, **(b)** $\{\overset{\Downarrow}{7}, 2, 2, 2, 2\}$, **(c)** $\{\overset{\Downarrow}{0}, 0, 3, 0, 3, 0\}$, **(d)** $\{\overset{\Downarrow}{15}, 6 - j3, 5, 6 + j3\}$

## 8.3.9 The Inverse DFT

The **inverse discrete Fourier transform** (IDFT) is defined by

$$x[n] = \frac{1}{N}\sum_{k=0}^{N-1} X_{\text{DFT}}[k]e^{j2\pi nk/N}, \quad n = 0, 1, \ldots, N-1 \tag{8.12}$$

This recovers the sequence $x[n]$ from $N$ samples of its DFT $X_{\text{DFT}}[k]$. Note that the form of the DFT and IDFT relations is identical except for the positive exponent in the exponential factor and the scale factor of $\frac{1}{N}$ in the IDFT. The IDFT is periodic with period $N$ and implies that $x[n]$ actually corresponds to one period of a periodic signal. If the DFT is conjugate symmetric, $x[n]$ may also be expressed as a sum of sinusoids because each pair of conjugate symmetric DFT samples, $Ae^{j\theta}$ at $k = k_0$ and $Ae^{-j\theta}$ at $k = N - k_0$ corresponds to the sinusoid $\frac{2A}{N}\cos(2n\pi\frac{k_0}{N} + \theta)$. The DFT sample $X_{\text{DFT}}[0]$ corresponds to the dc component $X_{\text{DFT}}[0]/N$ while the DFT sample $X_{\text{DFT}}\left[\frac{N}{2}\right]$ (if $N$ is even) corresponds to $F = 0.5$ and gives the signal $K\cos(n\pi)$ where $K = \frac{1}{N}X_{\text{DFT}}[\frac{N}{2}]$.

**Example 8.7**

### Relating the DFT and the DTFT

**(a)** Let $X_{\text{DFT}}[k] = \{\overset{\Downarrow}{4}, -j2, 0, j2\}$. Find its IDFT.

**Solution:** With $N = 4$ and $e^{j2\pi nk/N} = e^{jnk\pi/2}$, we compute the IDFT as

$$n = 0: \quad x[0] = 0.25\sum_{k=0}^{3} X_{\text{DFT}}[k]e^{0} = 0.25(4 - j2 + 0 + j2) = 1$$

$$n = 1: \quad x[1] = 0.25\sum_{k=0}^{3} X_{\text{DFT}}[k]e^{jk\pi/2} = 0.25(4 - j2e^{j\pi/2} + 0 + j2e^{j3\pi/2}) = 2$$

$$n = 2: \quad x[2] = 0.25\sum_{k=0}^{3} X_{\text{DFT}}[k]e^{jk\pi} = 0.25(4 - j2e^{j\pi} + 0 + j2e^{j3\pi}) = 1$$

$$n = 3: \quad x[3] = 0.25 \sum_{k=0}^{3} X_{\text{DFT}}[k] e^{j3k\pi/2}$$
$$= 0.25(4 - j2e^{j3\pi/2} + 0 + j2e^{j9\pi/2}) = 0$$

The IDFT is thus $x[n] = \{\overset{\Downarrow}{1}, 2, 1, 0\}$.

**(b)** Let $X_{\text{DFT}}[k] = \{\overset{\Downarrow}{12}, -j24, 0, 4e^{j\pi/4}, 0, 4e^{-j\pi/4}, 0, j24\}$. Find its IDFT.
We have $N = 8$. The DFT shows a dc component at $k = 0$. There is a conjugate symmetric pair of samples ($24e^{-j\pi/2}$ and $24e^{j\pi/2}$) at $k = 1$ and $k = 7$ and another conjugate pair of samples ($4e^{j\pi/4}$ and $4e^{-j\pi/4}$) at $k = 3$ and $k = 5$. The IDFT may thus be expressed as a sum of a dc component and two sinusoids at $F = \frac{1}{8}$ and $F = \frac{3}{8}$. The dc value is found by dividing $X_{\text{DFT}}[0]$ by $N$ and the peak value of each sinusoid is found by dividing the DFT magnitude by $0.5N$. This gives

$$x[n] = 1.5 + 6\cos\left(\frac{n\pi}{4} - \frac{\pi}{2}\right) + \cos\left(\frac{3n\pi}{4} + \frac{\pi}{4}\right)$$

If we evaluate this expression for $n = 0, 1, \ldots, 7$, we obtain the sequence

$$x[n] = \{\overset{\Downarrow}{2.2071}, 4.7426, 8.2071, 5.7426, -0.7929, -1.7426, -5.2071, -2.7426\}$$

The same result could also have been obtained from the defining IDFT relation.

# Drill Problem 8.8

**(a)** Let $X_{\text{DFT}}[k] = \{\overset{\Downarrow}{1}, j4, 3, -j4\}$. Find $x[n]$ as a sequence and as a sum of sinusoids.

**(b)** Let $Y_{\text{DFT}}[k] = \{\overset{\Downarrow}{20}, 10, j5, -j5, 10\}$. Express $y[n]$ as a sum of sinusoids.

**Answers: (a)** $\{\overset{\Downarrow}{1}, -2.5, 1, 1.5\}$, $0.25 - 2\sin(0.5n\pi) + 0.75\cos(n\pi)$, **(b)** $4 + 4\cos(0.4n\pi) - 2\sin(0.8n\pi)$

## 8.4 Some Practical Guidelines

From a purely mathematical or computational standpoint, the DFT simply tells us how to transform a set of $N$ numbers into another set of $N$ numbers. Its physical significance (what the numbers mean), however, stems from its ties to the spectra of both analog and discrete signals. In general, the DFT is only an approximation to the actual (Fourier series or transform) spectrum of the underlying analog signal. The DFT spectral spacing and DFT magnitude are affected by the choice of sampling rate and how the sample values are chosen. The DFT phase is affected by the location of sampling instants. The DFT spectral spacing is affected by the sampling duration. Here are some practical guidelines on how to obtain samples of an analog signal $x(t)$ for spectrum analysis and interpret the DFT results.

**Choice of Sampling Instants:** The defining relation for the DFT (or DFS) mandates that samples of $x[n]$ be chosen over the range $0 \le n \le N-1$ (through periodic extension, if necessary). Otherwise, the DFT phase will not match the expected phase.

**Choice of Samples:** If a sampling instant corresponds to a jump discontinuity, the sample value should be chosen as the midpoint of the discontinuity. The reason is that the Fourier series (or transform) converges to the midpoint of any discontinuity.

**Choice of Frequency Axis:** The computation of the DFT is independent of the sampling frequency $S$ or sampling interval $t_s = 1/S$. However, if an analog signal is sampled at a sampling rate $S$, its spectrum is periodic with period $S$. The DFT spectrum describes one period ($N$ samples) of this spectrum *starting at the origin*. For sampled signals, it is useful to plot the DFT magnitude and phase against the analog frequency $f = kS/N$ Hz, $k = 0, 1, \ldots, N-1$ (with spacing $S/N$). For discrete-time signals, we can plot the DFT against the digital frequency $F = k/N, k = 0, 1, \ldots, N-1$ (with spacing $1/N$). These choices are illustrated in Figure 8.3.

**Choice of Frequency Range:** To compare the DFT results with conventional two-sided spectra, just remember that, by periodicity, a negative frequency $-f_0$ (at the index $-k_0$) in the two-sided spectrum corresponds to the frequency $S - f_0$ (at the index $N - k_0$) in the (one-sided) DFT spectrum.

**Identifying the Highest Frequency:** *The highest frequency in the* DFT *spectrum corresponds to the folding index* $k = 0.5N$ *and equals* $F = 0.5$ *for discrete signals or* $f = 0.5S$ Hz *for sampled analog signals.* This highest frequency is also called the **folding frequency**. For purposes of comparison, its is sufficient to plot the DFT spectra only over $0 \le k < 0.5N$ (or $0 \le F < 0.5$ for discrete-time signals or $0 \le f < 0.5S$ Hz for sampled analog signals).

**Plotting Reordered Spectra:** The DFT (or DFS) may also be plotted as two-sided spectra to reveal conjugate symmetry about the origin by creating its periodic extension. This is equivalent to creating a *reordered spectrum* by relocating the DFT samples at indices past the folding index $k = 0.5N$ to the left of the origin (because $X[-k] = X[N-k]$). This process is illustrated in Figure 8.4.

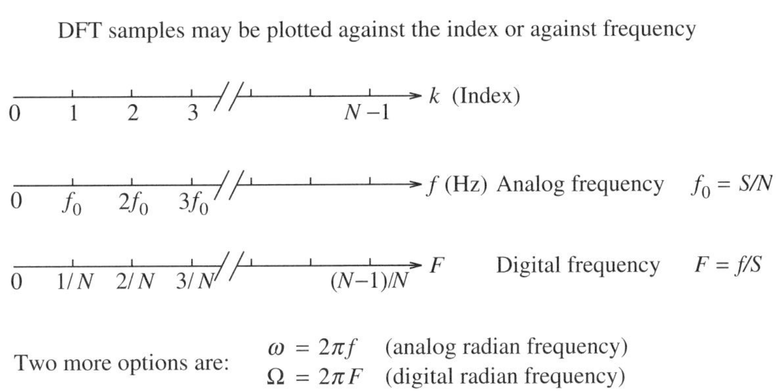


**FIGURE 8.3** Various ways of plotting the DFT. The DFT may be plotted against its discrete index or the analog frequency or digital frequency corresponding to each index

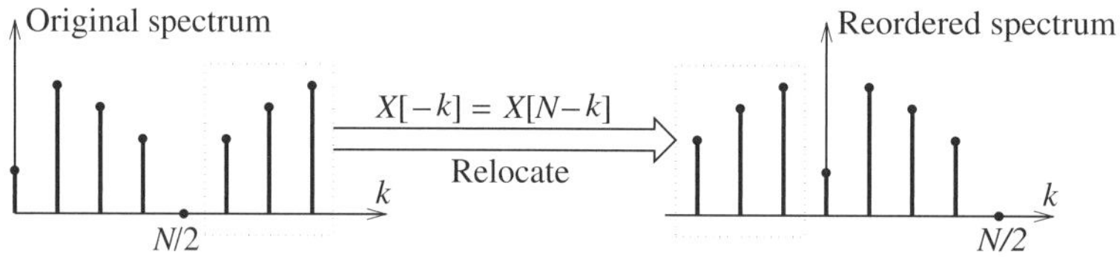


**FIGURE 8.4** Plotting the DFT or its reordered samples. The $N$-sample DFT is usually calculated and plotted over the range $0 \le k \le N-1$. However, the DFT results past the folding index may be shifted left by one period ($N$ samples) to create the reordered spectrum that now displays conjugate symmetry about the origin

# Drill Problem 8.9

Let $X_{\text{DFT}}[k] = \{\overset{\Downarrow}{1}, j5, 4-j, -j2, 0, j2, 4+j, -j5\}$. Express this in reordered form.

**Answer:** $X_{\text{DFT}}[k] = \{j2, 4+j, -j5, \overset{\Downarrow}{1}, j5, 4-j, -j2, 0\}$

---

**Practical Guidelines for Sampling a Signal and Interpreting the DFT Results**

**Sampling:** Start at $t = 0$. Choose the midpoint value at jumps. Sample above the Nyquist rate.
**Plotting:** Plot DFT against index $k = 0, 1, \ldots, N-1$ or $F = \frac{k}{N}$ or $f = k\frac{S}{N}$ Hz.
**Frequency Spacing of DFT Samples:** $\Delta f = S/N$ Hz (analog) or $\Delta F = 1/N$ (digital frequency).
**Highest Frequency:** This equals $F = 0.5$ or $f = 0.5S$ corresponding to the index $k = \frac{N}{2}$.
**For Long Sequences:** The DFT magnitude/phase are usually plotted as *continuous functions*.

---

## 8.5 The DTFT and the DFT

The DTFT relation and its inverse are

$$X(F) = \sum_{n=-\infty}^{\infty} x[n]e^{-j2\pi nF} \qquad x[n] = \int_1 X(F)e^{j2\pi nF}\,dF$$

where $X(F)$ is periodic with unit period. If $x[n]$ is a finite $N$-point sequence with $n = 0, 1, \ldots, N-1$, we obtain $N$ samples of the DTFT over one period at intervals of $1/N$ as

$$X_{\mathrm{DFT}}[k] = \sum_{n=0}^{N-1} x[n]e^{-j2\pi nk/N}, \qquad k = 0, 1, \ldots, N-1$$

This describes the DFT of $x[n]$ as a sampled version of its DTFT evaluated at the frequencies $F = k/N, k = 0, 1, \ldots, N-1$. The DFT spectrum thus corresponds to the frequency range $0 \le F < 1$ and is plotted at the frequencies $F = k/N, k = 0, 1, \ldots, N-1$.

To recover the finite sequence $x[n]$ from $N$ samples of $X_{\mathrm{DFT}}[k]$, we use $dF \approx 1/N$ and $F \to k/N$ to approximate the integral expression in the inversion relation by

$$x[n] = \frac{1}{N}\sum_{k=0}^{N-1} X_{\mathrm{DFT}}[k]e^{j2\pi nk/N}, \qquad n = 0, 1, \ldots, N-1$$

This, of course, is the defining relation for the IDFT.

### 8.5.1 Approximating the DTFT by the DFT

If $x[n]$ is a finite $N$-point signal with $n = 0, 1, \ldots, N-1$, the DFT is an exact match to its DTFT $X_p(F)$ at $F = k/N, k = 0, 1, \ldots, N-1$, and the IDFT results in perfect recovery of $x[n]$. This is evident from the defining relation for the DTFT and DFT.

If $x[n]$ is of infinite length, its $N$-point DFT is only an approximation to its DTFT $X(F)$ evaluated at $F = k/N, k = 0, 1, \ldots, N-1$. Due to implied periodicity, the DFT, in fact, exactly matches the DTFT of the *periodic extension* of $x[n]$ with period $N$ at these frequencies.

If $x[n]$ is a discrete periodic signal with period $N$, its scaled DFT ($\frac{1}{N}X_{\mathrm{DFT}}[k]$) is an exact match to the impulse strengths in its DTFT $X(F)$ at $F = k/N, k = 0, 1, \ldots, N-1$. In this case also, the IDFT results in perfect recovery of one period of $x[n]$ over $0 \le n \le N-1$.

**Example 8.8** **Relating the DFT and the DTFT**

**(a)** Let $x[n] = \{\overset{\Downarrow}{1}, 2, 1, 0\}$. If we use the DTFT, we first find

$$X(F) = 1 + 2e^{-j2\pi F} + e^{-j4\pi F} + 0 = [2 + 2\cos(2\pi F)]e^{-j2\pi F}$$

With $N = 4$, we have $F = k/4,\ k = 0, 1, 2, 3$. We then obtain the DFT as

$$X_{\mathrm{DFT}}[k] = [2 + 2\cos(2\pi k/4)]e^{-j2\pi k/4}, \qquad k = 0, 1, 2, 3, \qquad \text{or}$$

$$X_{\mathrm{DFT}}[k] = \{\overset{\Downarrow}{4}, -j2, 0, j2\}$$

Since $x[n]$ is a finite sequence, the DFT and DTFT show an exact match at $F = k/N$, $k = 0, 1, 2, 3$.

**(b)** Let $x[n] = \alpha^n u[n]$. Its DTFT is $X(F) = \frac{1}{1-\alpha e^{-j2\pi F}}$. Sampling $X(F)$ at intervals $F = k/N$ gives

$$X(F)\Big|_{F=k/N} = \frac{1}{1-\alpha e^{-j2\pi k/N}}$$

The $N$-point DFT of $x[n]$ is

$$\alpha^n,\ (n = 0, 1, \ldots, N-1) \Leftarrow\boxed{\mathrm{DFT}}\Rightarrow \frac{1-\alpha^N}{1-\alpha e^{-j2\pi k/N}}$$

Clearly, the $N$-sample DFT of $x[n]$ does not match the DTFT of $x[n]$ at $F = k/N$ (unless $N \to \infty$).

*Comment:* However, what matches is the DFT of the $N$-sample periodic extension $x_{\text{pe}}[n]$ and the DTFT of $x[n]$. We obtain one period of the periodic extension by wrapping around $N$-sample sections of $x[n] = \alpha^n u[n]$ and adding them to give

$$x_{\text{pe}}[n] = \alpha^n + \alpha^{n+N} + \alpha^{n+2N} + \cdots = \alpha^n(1 + \alpha^N + \alpha^{2N} + \cdots) = \frac{\alpha^n}{1 - \alpha^N}$$

From the DFT of $x[n]$ found earlier, we obtain the $N$-point DFT of $x_{\text{pe}}[n]$ as

$$\frac{\alpha^n}{1 - \alpha^N}, \ (n = 0, 1, \ldots, N-1) \Leftarrow\!\boxed{\text{DFT}}\!\Rightarrow \frac{1}{1 - \alpha e^{-j2\pi k/N}}$$

This is an exact match to the DTFT of $x[n]$ at $F = \frac{k}{N}$, $k = 0, 1, \ldots, N-1$.

## 8.6 The DFT of Periodic Signals and the DFS

The Fourier series relations for a periodic signal $x(t)$ are

$$x(t) = \sum_{k=-\infty}^{\infty} X[k] e^{j2\pi k f_0 t} \qquad X[k] = \frac{1}{T}\int_T x(t) e^{-j2\pi k f_0 t}\, dt \tag{8.13}$$

If we acquire $x[n]$, $n = 0, 1, \ldots, N-1$ as $N$ samples of $x(t)$ over one period using a sampling rate of $S$ Hz (corresponding to a sampling interval of $t_s$) and approximate the integral expression for $X[k]$ by a summation using $dt \to t_s$, $t \to nt_s$, $T = Nt_s$, and $f_0 = \frac{1}{T} = \frac{1}{Nt_s}$, we obtain

$$X_{\text{DFS}}[k] = \frac{1}{Nt_s}\sum_{n=0}^{N-1} x[n] e^{-j2\pi k f_0 n t_s} t_s = \frac{1}{N}\sum_{n=0}^{N-1} x[n] e^{-j2\pi nk/N}, \quad k = 0, 1, \ldots, N-1 \tag{8.14}$$

The quantity $X_{\text{DFS}}[k]$ defines the **discrete Fourier series** (DFS) as an approximation to the Fourier series coefficients of a periodic signal and equals $N$ times the DFT.

To recover $x[n]$ from one period of $X_{\text{DFS}}[k]$, we use the Fourier series reconstruction relation whose summation index covers one period (from $k = 0$ to $k = N-1$) to obtain

$$x[n] = \sum_{k=0}^{N-1} X_{\text{DFS}}[n] e^{j2\pi k f_0 n t_s} = \sum_{k=0}^{N-1} X_{\text{DFS}}[k] e^{j2\pi nk/N}, \quad n = 0, 1, 2, \ldots, N-1 \tag{8.15}$$

This relation describes the **inverse discrete Fourier series** (IDFS). The sampling interval $t_s$ does not enter into the computation of the DFS or its inverse. Except for a scale factor, we see that the DFS and DFT relations are identical.

### 8.6.1 Understanding the DFS Results

The DFS describes the spectrum of a sampled periodic signal. It is periodic with period $N$. Its $N$ samples are located at the indices $k = 0, 1, 2, \ldots, N-1$ corresponding to the digital frequencies $F = \frac{k}{N}$ or the analog frequencies $f = k\frac{S}{N}$. The highest frequency in the

DFS spectrum is $f = 0.5S$, corresponding to the digital frequency $F = 0.5$ or the index $k = 0.5N$. The Fourier series coefficients $X[k]$ of a periodic signal $x(t)$ exactly match its $N$-sample DFS $X_{\text{DFS}}[k] = \frac{1}{N}X_{\text{DFT}}[k]$ only if all of the following conditions are met:

1. The signal samples are acquired from $x(t)$ starting at $t = 0$ (using the periodic extension of the signal, if necessary). If not, the phase of the DFS coefficients will not match the phase of the corresponding Fourier series coefficients.
2. The periodic signal contains a finite number of sinusoids (to ensure a band-limited signal with a finite highest frequency) and is sampled above the Nyquist rate. If not, there will be *aliasing*, whose effects become more pronounced near the folding frequency $0.5S$. For a pure sinusoid, the Nyquist rate corresponds to two samples per period. Periodic signals that are not band limited (such as a rectangular, triangular or sawtooth pulse train) contain an infinite number of harmonics and no sampling rate is high enough to prevent aliasing.
3. The signal $x(t)$ is sampled for an *integer* number of periods (to ensure a match between the periodic extension of $x(t)$ and the implied periodic extension of the sampled signal). If not, the periodic extension of its samples will not match that of $x(t)$, and the DFS samples will describe the Fourier series coefficients of a different periodic signal whose harmonic frequencies do not match those of $x(t)$. This phenomenon is called **leakage** and results in nonzero spectral components at frequencies other than the harmonic frequencies in the original signal $x(t)$.

If we sample a periodic signal for an integer number of periods, the DFS (or DFT) also preserves the effects of symmetry. In other words, the DFS of an even symmetric signal will be real, the DFS of an odd symmetric signal will be imaginary, and the DFS of a half-wave symmetric signal will be zero at even values of the index $k$ (corresponding to an absence of even-indexed harmonics).

**The DFT of a Sampled Periodic Signal *x*(*t*) Is Related to its Fourier Series Coefficients *X*[*k*]**

If $x(t)$ is band limited and sampled for an integer number of periods, the DFT is an exact match to the Fourier series coefficients $X[k]$, with $X_{\text{DFT}}[k] = NX[k]$.
If $x(t)$ is not band limited, there is *aliasing*.
If $x(t)$ is not sampled for an integer number of periods, there is *leakage*.

## 8.6.2 The DFT and DFS of Sinusoids

Consider the sinusoid $x(t) = \cos(2\pi f_0 t + \theta)$ whose Fourier series coefficients we know to be $0.5\angle\theta = 0.5e^{j\theta}$ at $f = f_0$ and $0.5\angle-\theta = 0.5e^{-j\theta}$ at $f = -f_0$. If $x(t)$ is sampled at the rate $S$, starting at $t = 0$, the sampled signal is $x[n] = \cos(2\pi nF + \theta)$, where $F = f_0/S$ is the digital frequency. As long as $F$ is a rational fraction of the form $F = k_0/N$, we obtain $N$ samples of $x[n]$ from $k_0$ full periods of $x(t)$. In this case, there is no leakage, and the $N$-point DFS will match the expected results and show only two nonzero DFS values, $X_{\text{DFS}}[k_0] = 0.5e^{j\theta}$ and $X_{\text{DFS}}[N - k_0] = 0.5e^{-j\theta}$. The DFT is obtained by multiplying the DFS by $N$.

$$\underbrace{\cos(2\pi n \tfrac{k_0}{N} + \theta)}_{N\text{-sample sinusoid with } F_0 = k_0/N} \Longleftarrow \boxed{\text{DFS}} \Longrightarrow \{\overset{\Downarrow}{0}, \ldots, 0, \underbrace{0.5e^{j\theta}}_{k=k_0}, 0, \ldots, 0, \underbrace{0.5e^{-j\theta}}_{k=N-k_0}, 0, \ldots, 0\} \tag{8.16}$$

Here, $k_0$ is assumed to be a positive index in the range $0 \le k_0 \le N-1$. The analog frequency corresponding to $k_0$ will then be $k_0 S/N$ and will equal $f_0$ (if $S > 2f_0$) or its alias (if $S < 2f_0$). The nonzero DFT values will equal $X_{\text{DFT}}[k_0] = 0.5Ne^{j\theta}$ and $X_{\text{DFT}}[N - k_0] = 0.5Ne^{-j\theta}$. These results are straightforward to obtain without extensive computation and can be easily extended (by superposition) to the DFT of a combination of sinusoids, as sampled over an integer number of periods (starting at $t = 0$).

**The $N$-Point DFT or DFS of a Sinusoid $x[n] = \cos(2\pi n\frac{k_0}{N} + \theta)$ has Only Two Nonzero Samples**

$$\underbrace{\cos(2\pi n \tfrac{k_0}{N} + \theta)}_{N\text{-sample sinusoid with } F_0 = k_0/N} \Longleftarrow \boxed{\text{DFT}} \Longrightarrow \{\overset{\Downarrow}{0}, \ldots, 0, \underbrace{0.5Ne^{j\theta}}_{k=k_0}, 0, \ldots, 0, \underbrace{0.5Ne^{-j\theta}}_{k=N-k_0}, 0, \ldots, 0\}$$

$$\underbrace{\cos(2\pi n \tfrac{k_0}{N} + \theta)}_{N\text{-sample sinusoid with } F_0 = k_0/N} \Longleftarrow \boxed{\text{DFS}} \Longrightarrow \{\overset{\Downarrow}{0}, \ldots, 0, \underbrace{0.5e^{j\theta}}_{k=k_0}, 0, \ldots, 0, \underbrace{0.5e^{-j\theta}}_{k=N-k_0}, 0, \ldots, 0\}$$

**Example 8.9**

**The DFT and DFS of Sinusoids**

**(a)** The signal $x(t) = 4\cos(100\pi t)$ is sampled at twice the Nyquist rate for three full periods. Find and sketch its DFT.

The frequency of $x(t)$ is 50 Hz, the Nyquist rate is 100 Hz, and the sampling frequency is $S = 200$ Hz. The digital frequency is $F = 50/200 = 1/4 = 3/12 = k/N$. This means $N = 12$ for three full periods. The two nonzero DFT values will appear at $k = 3$ and $k = N - 3 = 9$. The nonzero DFT values will be $X[3] = X[9] = (0.5)(4)(N) = 24$. The signal and its DFT are sketched in Figure E8.9A.

**FIGURE E 8.9.a** The signal and DFT for Example 8.9(a)

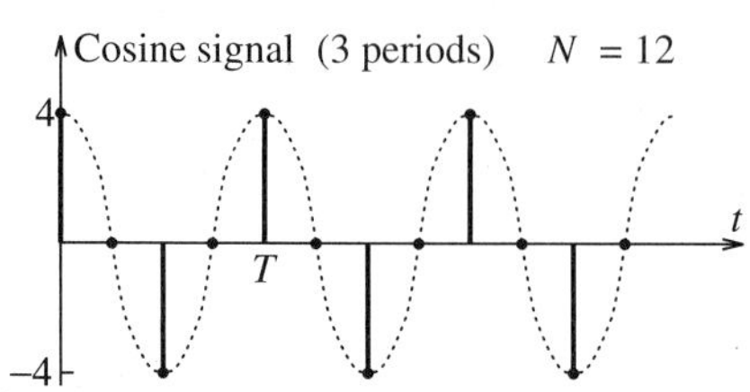


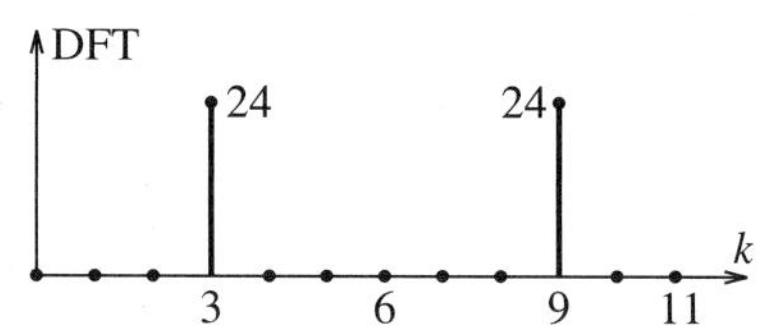

**(b)** Let $x(t) = 4\sin(72\pi t)$ be sampled at $S = 128$ Hz. Choose the minimum number of samples necessary to prevent leakage and find the DFS and DFT of the sampled signal.
The frequency of $x(t)$ is 36 Hz, so $F = 36/128 = 9/32 = k_0/N$. Thus, $N = 32$, $k_0 = 9$, and the frequency spacing is $S/N = 4$ Hz. The DFS components will appear at $k_0 = 9$ (36 Hz) and $N - k_0 = 23$ (92 Hz). The Fourier series coefficients of $x(t)$ are $-j2$ (at 36 Hz) and $j2$ (at $-36$ Hz).

The DFS samples will be $X_{\text{DFS}}[9] = -j2$ and $X_{\text{DFS}}[23] = j2$. Since $X_{\text{DFT}}[k] = NX_{\text{DFS}}[k]$, we get $X_{\text{DFT}}[9] = -j64$, $X_{\text{DFT}}[23] = j64$, and thus

$$X_{\text{DFT}}[k] = \{\overset{\Downarrow}{0}, \ldots, 0, \underbrace{-j64}_{k=9}, 0, \ldots, 0, \underbrace{j64}_{k=23}, 0, \ldots, 0\}$$

**(c)** Let $x(t) = 4\sin(72\pi t) - 6\cos(12\pi t)$ be sampled at $S = 21$ Hz. Choose the minimum number of samples necessary to prevent leakage and find the DFS and DFT of the sampled signal.

Clearly, the 36-Hz term will be aliased. The digital frequencies (between 0 and 1) of the two terms are $F_1 = 36/21 = 12/7 \Rightarrow 5/7 = k_0/N$ and $F_2 = 6/21 = 2/7$.

Thus, $N = 7$, and the frequency spacing is $S/N = 3$ Hz. The DFS components of the first term will be $-j2$ at $k = 5$ (15 Hz) and $j2$ at $N - k = 2$ (6 Hz). The DFS components of the second term will be $-3$ at $k = 2$ and $-3$ at $k = 5$. The DFS values will add up at the appropriate indices to give
$X_{\text{DFS}}[5] = -3 - j2$, $X_{\text{DFS}}[2] = -3 + j2$, and

$$X_{\text{DFS}}[k] = \{\overset{\Downarrow}{0}, 0, \underbrace{-3+j2}_{k=2}, 0, 0, \underbrace{-3-j2}_{k=5}, 0\} \qquad X_{\text{DFT}}[k] = NX_{\text{DFS}}[k] = 7X_{\text{DFS}}[k]$$

Note how the 36-Hz component was aliased to 6 Hz (the frequency of the second component).

**(d)** The signal $x(t) = 1 + 8\sin(80\pi t)\cos(40\pi t)$ is sampled at twice the Nyquist rate for two full periods. Is leakage present? If not, find the DFS of the sampled signal.

First, note that $x(t) = 1 + 4\sin(120\pi t) + 4\sin(40\pi t)$. The frequencies are $f_1 = 60$ Hz and $f_2 = 20$ Hz. The Nyquist rate is thus 120 Hz, and hence, $S = 240$ Hz. The digital frequencies are $F_1 = 60/240 = 1/4$ and $F_2 = 20/240 = 1/12$. The fundamental frequency is $f_0 = \text{GCD}(f_1, f_2) = 20$ Hz. Thus, two full periods correspond to 0.1 s or $N = 24$ samples. There is no leakage because we acquire the samples over two full periods. The index $k = 0$ corresponds to the constant (dc value). To find the indices of the other nonzero DFS samples, we compute the digital frequencies (in the form $k/N$) as $F_1 = 60/240 = 1/4 = 6/24$ and $F_2 = 20/240 = 1/12 = 2/24$. The nonzero DFS samples are thus $X_{\text{DFS}}[0] = 1$, $X_{\text{DFS}}[6] = -j2$, and $X_{\text{DFS}}[2] = -j2$, and the conjugates are $X_{\text{DFS}}[18] = j2$ and $X_{\text{DFS}}[22] = j2$. Thus,

$$X_{\text{DFS}}[k] = \{\overset{\Downarrow}{0}, 0, \underbrace{-j2}_{k=2}, 0, 0, 0, \underbrace{-j2}_{k=6}, 0, \ldots, 0, \underbrace{j2}_{k=18}, 0, 0, 0, \underbrace{j2}_{k=22}, 0, 0\}$$

# Drill Problem 8.10

**(a)** Let $x[n] = 4\cos(0.4n\pi) + 2\sin(0.8n\pi)$. Find its 5-point DFT $X_{\mathrm{DFT}}[k]$.
**(b)** We wish to compute the $N$-point DFT of a 100-Hz sinusoid $x(t)$ sampled at 75 Hz. What is the smallest value of $N$ that prevents leakage. Over how many full periods of $x(t)$ are these $N$ samples obtained?
**(c)** The analog signal $x(t) = \cos(10\pi t) + \cos(20\pi t)$ is sampled at 1.25 times the Nyquist rate for 0.4 s. What is the number of samples $N$ in the sampled signal? Will the $N$-point DFT show the effects of leakage? If not, find the $N$-point DFT.

**Answers:** **(a)** $\{\underset{\Uparrow}{0}, 10, -j5, j5, 10\}$ **(b)** $N = 3, 4$ **(c)** $N = 10$, $\{\underset{\Uparrow}{0}, 0, 5, 0, 5, 0, 5, 0, 5, 0\}$

## 8.6.3 The DFT and DFS of Sampled Periodic Signals

For periodic signals that are not band limited, leakage can be prevented by sampling for an integer number of periods, but aliasing is unavoidable for any sampling rate. The aliasing error increases at higher frequencies. The effects of aliasing can be minimized (but not entirely eliminated) by choosing a high enough sampling rate. A useful rule of thumb is that for the DFT to produce an error of about 5% or less up to the $M$th harmonic frequency $f_M = Mf_0$, we must choose $S \ge 8f_M$ (corresponding to $N \ge 8M$ samples per period).

**Example 8.10**

### DFS of Sampled Periodic Signals

Consider a square wave $x(t)$ that equals 1 for the first half-period and $-1$ for the next half-period, as illustrated in Figure E8.10.

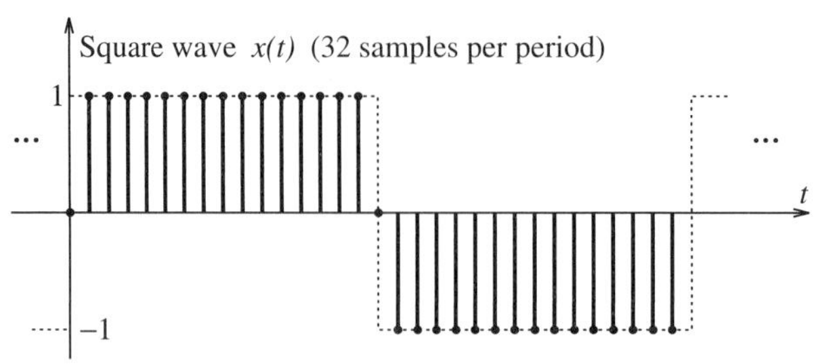


**FIGURE E.8.10** One period of the square-wave periodic signal for Example 8.10

Its non zero Fourier series coefficients are given by $X[k] = -\frac{j2}{k\pi}$ ($k$ odd). If we wish to find the DFT of its sampled version and ensure that the first four harmonics are in error by no more than about 5%, we choose a sampling rate $S = 32f_0$, where $f_0$ is the fundamental frequency. This means that we acquire $N = 32$ samples for one period. The 32 signal samples (where the value at the discontinuities is chosen as zero) and the first nine samples of the 32-point DFS (up to $k = 8$) are listed below, along with the error in the DFS values compared with the Fourier series coefficients.

$$x[n] = \{\overset{\Downarrow}{0}, \underbrace{1,\ 1,\ \ldots,\ 1,\ 1}_{\text{15 samples}}, 0, \underbrace{-1,\ -1,\ \ldots,\ -1,\ -1}_{\text{15 samples}}\}$$

$$X_{\text{DFS}}[k] = \{\overset{\Downarrow}{0},\ \underbrace{-j0.6346}_{\text{off by 0.3\%}},\ 0,\ \underbrace{-j0.206}_{\text{off by 2.9\%}},\ 0,\ \underbrace{-j0.1169}_{\text{off by 8.2\%}},\ 0,\ \underbrace{-j0.0762}_{\text{off by 16.3\%}},\ 0,\ \ldots\}$$

We see that the DFS coefficients are zero for odd $k$ (due to the half-wave symmetry in $x(t)$) and purely imaginary (due to the odd symmetry in $x(t)$). As expected, the error in the nonzero harmonics up to $k = 4$ is less than 5%.

### 8.6.4 The Effects of Leakage

Leakage is present in the DFT results if a periodic signal $x(t)$ is not sampled for an integer number of periods. The DFT shows nonzero components at frequencies other than $\pm f_0$ because the periodic extension of the sampled portion of the periodic signal (sampled for non-integer periods) does not match the original signal but describes a different signal altogether, as illustrated in Figure 8.5. The DFT results will also show aliasing because the periodic extension of the signal with non-integer periods will not, in general, be band limited. As a result, we must resort to the full force of the defining relation to compute the DFT.

Suppose we sample the signal $x(t) = \sin(2\pi t)$ at $S = 16$ Hz. Then, $f_0 = 1$ Hz and $F_0 = 1/16$. If we choose $N = 8$, the DFT spectral spacing equals $S/N = 2$ Hz. In other words, there is no DFT component at 1 Hz, the frequency of the sine wave! Where should we expect to see the DFT components? If we express the digital frequency $F_0 = 1/16$ as $F_0 = k_F/N$, we obtain $k_F = NF_0 = 0.5$. Thus, $F_0$ corresponds to the fractional index $k_F = 0.5$, and the largest DFT components should appear at the integer indices *closest* to

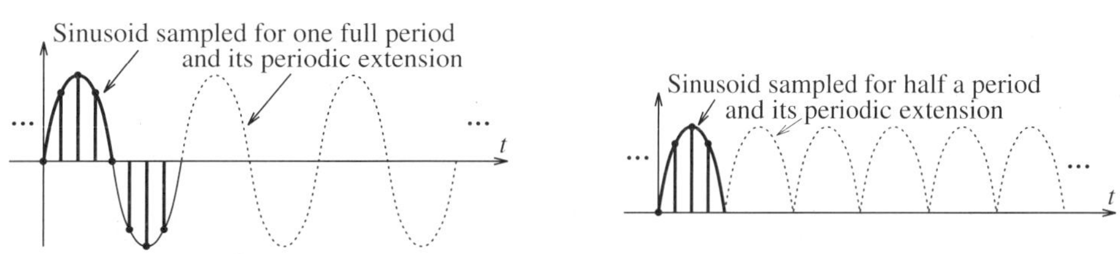


**FIGURE 8.5** Illustrating the concept of leakage. If we obtain samples of a pure sine over one full period, their periodic extension will match the underlying sine and the spectrum will reveal just the frequency of the sine. However, if we take samples of the pure sine only over half a cycle, their periodic extension will not match the pure sine but a full-rectified sine. Its spectrum will not contain just one frequency component but components at multiples of its fundamental frequency. The single-frequency component of the pure sine has leaked out to other frequencies that are not part of the original signal. This is leakage

$k_F$ at $k = 0$ (dc) and $k = 1$ (2 Hz). The signal values and their computed 16-point DFT are given by

$$x[n] = \{\overset{\Downarrow}{0},\ 0.3827,\ 0.7071,\ 0.9329,\ 1,\ 0.9329,\ 0.7071,\ 0.3827\}$$

$$X_{\text{DFS}}[k] = \{\ \underbrace{\overset{\Downarrow}{0.6284},}_{\text{off by 1.3\%}}\ \underbrace{-0.2207,}_{\text{off by 4\%}}\ \underbrace{-0.0518,}_{\text{off by 23\%}}\ \underbrace{-0.0293,}_{\text{off by 61\%}}\ \underbrace{-0.0249,}_{k=N/2}\ -0.0293,\ -0.0518,\ -0.2207\}$$

As expected, the largest components do appear at $k = 0$ and $k = 1$. Since $X_{\text{DFS}}[k]$ is real, the DFS results describe an even symmetric signal with nonzero average value (unlike the original single-frequency sine with zero average value). We can, however, make sense of the DFS results. The periodic extension of the sampled signal over half a period actually describes a full-rectified sine with even symmetry and a fundamental frequency of 2 Hz (see Figure 8.5). Now, the Fourier series coefficients of this full-rectified sine wave (with unit peak value) are given by

$$X[k] = \frac{2}{\pi(1 - 4k^2)}$$

We see that $X_{\text{DFS}}[0]$ and $X_{\text{DFS}}[1]$ match the coefficients $X[0]$ and $X[1]$ to within 5%. But the other DFS components $X_{\text{DFS}}[2]$, $X_{\text{DFS}}[3]$ and $X_{\text{DFS}}[4]$ deviate significantly from $X[2]$, $X[3]$, and $X[4]$. Why? Because the new periodic signal is no longer band limited, the sampling rate is not high enough, and we have aliasing. As a result, a DFT value $X_{\text{DFS}}[k]$ is actually the sum of the Fourier series coefficient $X[k]$ and all the other coefficients $X[k+8m]$ that correspond to frequencies that also alias to that index. The value $X_{\text{DFS}}[3]$, for example, equals the sum of the Fourier series coefficient $X[3]$ and all other Fourier series coefficients $X[11]$, $X[19]$, ... that alias to $k = 3$. In other words,

$$X_{\text{DFS}}[3] = \sum_{m=\infty}^{\infty} X[3 + 8m] = \frac{2}{\pi} \sum_{m=\infty}^{\infty} \frac{1}{1 - 4(3 + 8m)^2}$$

Although this sum is not easily amenable to a closed-form solution, it can be computed numerically and does in fact approach $X_{\text{DFS}}[3] = -0.0293$ (for a large but finite number of coefficients). The purpose of this exercise is to stress that the DFS results are related to the Fourier series coefficients in a predictable manner. If the sampling rate exceeds the Nyquist rate and samples are acquired for an integer number of periods (starting at $t = 0$), the DFS matches the Fourier series coefficients. If not, there is leakage, and the DFS results are related to the Fourier series coefficients of the periodic extension (and also incorporate the effects of aliasing).

## Minimizing Leakage

Ideally, we should sample periodic signals over an integer number of periods to prevent leakage. In practice, it may not be easy to identify the period of a signal in advance. In such cases, it is best to sample over as long a signal duration as possible (to reduce the mismatch between the periodic extension of the analog and sampled signal). Sampling for a larger time duration not only reduces the effects of leakage, but also yields a more closely spaced spectrum (by providing more signal samples and thus reducing the spectral spacing $S/N$), and gives a more accurate estimate of the spectrum of the original signal.

**The DFT of a Sinusoid at $f_0$ Hz Sampled for Non-Integer Periods Shows Leakage**

The largest DFT component appears at the integer index closest to $k_F = NF_0 = Nf_0/S$. **To Minimize Leakage:** Sample for the longest duration possible or (better still) for integer periods.

**Example 8.11**

### The Effects of Leakage

The signal $x(t) = 2\cos(20\pi t) + 5\cos(100\pi t)$ is sampled at intervals of $t_s = 0.005$ s for three different durations, 0.1 s, 0.125 s, and 1.125 s. Explain the DFT spectrum for each duration.

The sampling frequency is $S = 1/t_s = 200$ Hz. The frequencies in $x(t)$ are $f_1 = 10$ Hz and $f_2 = 50$ Hz. The two-sided spectrum of $x(t)$ will show a magnitude of 1 at 10 Hz and 2.5 at 50 Hz. The fundamental frequency is 10 Hz, and the common period of $x(t)$ is 0.1 s. We have the following results with reference to Figure E8.11, which shows the DFS magnitude ($|X_{\text{DFT}}|/N$) up to the folding index (or 100 Hz).

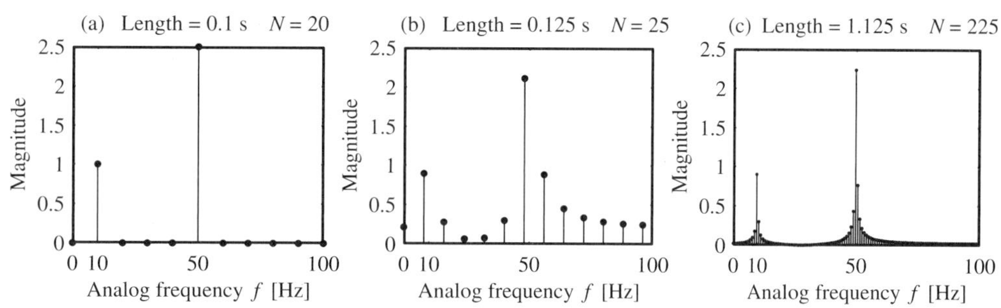


**FIGURE E.8.11** DFT results for Example 8.11

**(a)** The duration of 0.1 s corresponds to one full period, and $N = 20$. No leakage is present, and the DFS results reveal an *exact match* to the spectrum of $x(t)$. The nonzero components appear at the integer indices $k_1 = NF = Nf_1/S = 1$ and $k_2 = NF_2 = Nf_2/S = 5$ (corresponding to 10 Hz and 50 Hz, respectively).

**(b)** The duration of 0.125 s corresponds to 1.25 periods. So, leakage is present. The number of samples over 0.125 s is $N = 25$. The largest components in the 25-point DFS appear at the integer indices closest to $k_1 = NF = Nf_1/S = 1.25$ (i.e., $k = 1$ or 8 Hz) and $k_2 = Nf_2/S = 6.25$ (i.e., $k = 6$ or 48 Hz).

**(c)** The duration of 1.125 s corresponds to 11.25 periods. So, leakage is present. The number of samples over 1.125 s is $N = 225$. The largest components in the 225-point DFS appear at the integer indices closest to $k_1 = NF = Nf_1/S = 11.25$ (i.e., $k = 11$ or 9.78 Hz) and $k_2 = Nf_2/S = 56.25$ (i.e., $k = 56$ or 49.78 Hz).

*Comment:* The last two spectra show leakage. The spectrum of the signal sampled for the longer duration (1.125 s) produces the smaller leakage.

# 8.7 The DFT of Nonperiodic Signals

The Fourier transform $X(f)$ of a nonperiodic signal $x(t)$ is continuous. To find the DFT, $x(t)$ must be sampled over a finite duration ($N$ samples). The spectrum of a sampled signal over its principal period $(-0.5S, 0.5S)$ corresponds to the spectrum of the analog signal $SX(f)$, provided $x(t)$ is band limited to $B < 0.5S$. In practice, no analog signal is truly band limited. As a result, if $x[n]$ corresponds to $N$ samples of the analog signal $x(t)$, obtained at the sampling rate $S$, the DFT $X_{\text{DFT}}[k]$ of $x[n]$ yields essentially the Fourier series of its periodic extension, and is only approximately related to $X(f)$ by

$$X_{\text{DFT}}[k] \approx SX(f)|_{f=kS/N}, \qquad 0 \le k < 0.5N (0 \le f < 0.5S) \tag{8.17}$$

To find the DFT of an arbitrary signal with some confidence, we must decide on the number of samples $N$ and the sampling rate $S$, based on both theoretical considerations and practical compromises. For example, one way to choose a sampling rate is based on energy considerations. We pick the sampling rate as $2B$ Hz, where the frequency range up to $B$ Hz contains a significant fraction $P$ of the signal energy. The number of samples should cover a large enough duration to include significant signal values.

## 8.7.1 Spectral Spacing and Zero Padding

Often, the spectral spacing $S/N$ is not small enough to make appropriate visual comparisons with the analog spectrum $X(f)$, and we need a denser or interpolated version of the DFT spectrum. To decrease the spectral spacing, we must choose a larger number of samples $N$. This increase in $N$ cannot come about by increasing the sampling rate $S$ (which would leave the spectral spacing $S/N$ unchanged) but by increasing the duration over which we sample the signal. In other words, to reduce the frequency spacing, we must sample the signal for a longer duration at the given sampling rate. However, if the original signal is of finite duration, we can still increase $N$ by appending zeros (zero padding). Appending zeros does not improve accuracy because it adds no new signal information. It only decreases the spectral spacing and thus interpolates the DFT at a denser set of frequencies. To improve the accuracy of the DFT results, we must increase the number of signal samples by sampling the signal for a longer time (and not just zero padding).

**Example 8.12** **DFT of Finite-Duration Signals**

**(a) Spectral Spacing**

A 3-s signal is sampled at $S = 100$ Hz. The maximum spectral spacing is to be $\Delta f = 0.25$ Hz. How many samples are needed for the DFT and FFT?

If we use the DFT, the number of samples is $N = S/\Delta f = 400$. Since the 3-s signal gives only 300 signal samples, we must add 100 padded zeros. The spectral spacing is $S/N = 0.25$ Hz, as required. If we use the FFT, we need $N_{\text{FFT}} = 512$ samples (the next higher power of 2). There are now 212 padded zeros, and the spectral spacing is $S/N_{\text{FFT}} = 0.1953$ Hz, better (i.e., less) than required.

**(b)** Let $x(t) = \text{tri}(t)$
Its Fourier transform is $X(f) = \text{sinc}^2(f)$. Let us choose $S = 4$ Hz and $N = 8$. To obtain samples of $x(t)$ starting at $t = 0$, we sample the periodic extension of $x(t)$ as illustrated in Figure E8.12b and obtain $t_s X_{\text{DFT}}[k]$ to give

$$x[n] = \{\overset{\Downarrow}{1},\ 0.75,\ 0.5,\ 0.25,\ 0,\ 0.25,\ 0.5,\ 0.75\}$$

$$X(f) \approx t_s X_{\text{DFT}}[k] = \{\overset{\Downarrow}{1},\ \underbrace{0.4268,}_{\text{off by } 5.3\%}\ 0,\ \underbrace{0.0732,}_{\text{off by } 62.6\%}\ \underbrace{0,}_{k=N/2}\ 0.0732, 0, 0.4268\}$$

Since highest frequency present in the DFT spectrum is $0.5S = 2$ Hz, the DFT results are listed only up to $k = 4$. Since the frequency spacing is $S/N = 0.5$ Hz, we compare $t_s X_{\text{DFT}}[k]$ with $X(kS/N) = \text{sinc}^2(0.5k)$. At $k = 0$ (dc) and $k = 2$ (1 Hz), we see a perfect match. At $k = 1$ (0.5 Hz), $t_s X_{\text{DFT}}[k]$ is in error by about 5.3%, but at $k = 3$ (1.5 Hz), the error is a whopping 62.6%.

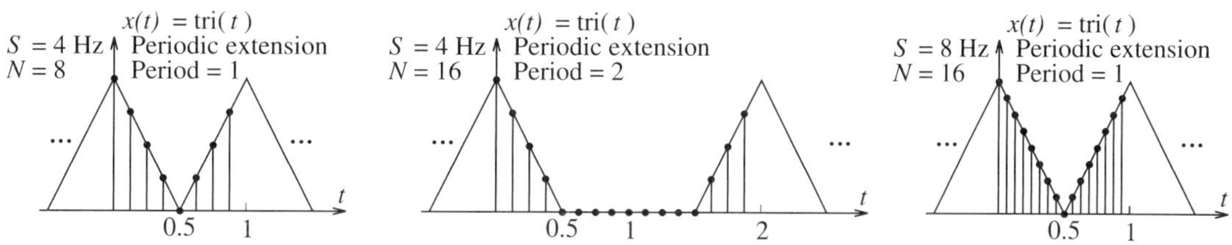


**FIGURE E 8.12.b** The triangular pulse signal for Example 8.12(b)

- **Reducing Spectral Spacing**
  Let us decrease the spectral spacing by zero padding to increase the number of samples to $N = 16$. We must sample the periodic extension of the zero-padded signal, as shown in Figure E8.12B, to give

$$x[n] = \{\overset{\Downarrow}{1}, 0.75, 0.5, 0.25, 0, \underbrace{0, 0, 0, 0, 0, 0, 0, 0,}_{8 \text{ zeros}} 0.25, 0.5, 0.75\}$$

Note how the padded zeros appear *in the middle*. Since highest frequency present in the DFT spectrum is 2 Hz, the DFT results are listed only up to the folding index $k = 8$.

$$t_s X_{\text{DFT}}[k] = \{\overset{\Downarrow}{1}, 0.8211,\ \underbrace{0.4268,}_{\text{off by } 5.3\%}\ 0.1012, 0, 0.0452,$$

$$\underbrace{0.0732,}_{\text{off by } 62.6\%}\ 0.0325,\ \underbrace{0,}_{k=N/2}\ \ldots\}$$

The frequency separation is reduced to $S/N = 0.25$ Hz. Compared with $X(kf_0) = \text{sinc}^2(0.25k)$, the DFT results for $k = 2$ (0.5 Hz) and $k = 6$ (1.5 Hz) are still off by 5.3% and 62.6%, respectively. In other words, zero padding reduces the spectral spacing, but the DFT results are no more accurate. To improve the accuracy, we must pick more signal samples.

- **Improving Accuracy**
  If we choose $S = 8$ Hz and $N = 16$, we obtain 16 samples shown in Figure E8.12b. We list the 16-sample DFT up to $k = 8$ (corresponding to the highest frequency of 4 Hz present in the DFT):

$$t_s X_{\mathrm{DFT}}[k] = \{\overset{\Downarrow}{1},\ \underbrace{0.4105,}_{\text{off by } 1.3\%}\ 0,\ \underbrace{0.0506,}_{\text{off by } 12.4\%}\ 0,\ \underbrace{0.0226,}_{\text{off by } 39.4\%}\ 0,\ \underbrace{0.0162,}_{\text{off by } 96.4\%}\ \underbrace{0,}_{k=N/2}\ \ldots\}$$

  The DFT spectral spacing is still $S/N = 0.5$ Hz. In comparison with $X(k/2) = \mathrm{sinc}^2(0.5k)$, the error in the DFT results for the 0.5-Hz component ($k = 1$) and the 1.5-Hz component ($k = 3$) is now only about 1.3% and 12.4%, respectively. In other words, increasing the number of signal samples improves the accuracy of the DFT results. However, the error at 2.5 Hz ($k = 5$) and 3.5 Hz ($k = 7$) is 39.4% and 96.4%, respectively, and implies that the effects of aliasing are more predominant at frequencies closer to the folding frequency.

**(c)** Consider the signal $x(t) = e^{-t}u(t)$ whose Fourier transform is $X(f) = 1/(1 + j2\pi f)$. Since the energy $E$ in $x(t)$ equals 1, we use Parseval's relation to estimate the bandwidth $B$ that contains the fraction $P$ of this energy as

$$P = \int_{-B}^{B} \frac{1}{1 + 4\pi^2 f^2}\, df \quad \text{or} \quad B = \frac{\tan(0.5\pi P)}{2\pi}$$

1. If we choose $B$ to contain 95% of the signal energy ($P = 0.95$), we find $B = 12.71/2\pi = 2.02$ Hz. Then, $S > 4.04$ Hz. Let us choose $S = 5$ Hz. For a spectral spacing of 1 Hz, we have $S/N = 1$ Hz and $N = 5$. So, we sample $x(t)$ at intervals of $t_s = 1/S = 0.2$ s, starting at $t = 0$, to obtain $x[n]$. Since $x(t)$ is discontinuous at $t = 0$, we pick $x[0] = 0.5$, not 1. The DFT results based on this set of choices will not be very good because with $N = 5$ we sample only a 1-s segment of $x(t)$.

2. A better choice is $N = 15$, a 3-s duration over which $x(t)$ decays to 0.05. A more practical choice is $N = 16$ (a power of 2, which allows efficient computation of the DFT, using the FFT algorithm). This gives a spectral spacing of $S/N = 5/16$ Hz. Our rule of thumb ($N > 8M$) suggests that with $N = 16$, the DFT values $X_{\mathrm{DFT}}[1]$ and $X_{\mathrm{DFT}}[2]$ should show an error of only about 5%. We see that $t_s X_{\mathrm{DFT}}[k]$ does compare well with $X(f)$ (see Figure E8.12c, even though the effects of aliasing are still evident.

**FIGURE E 8.12.c** DFT results for Example 8.12(c)

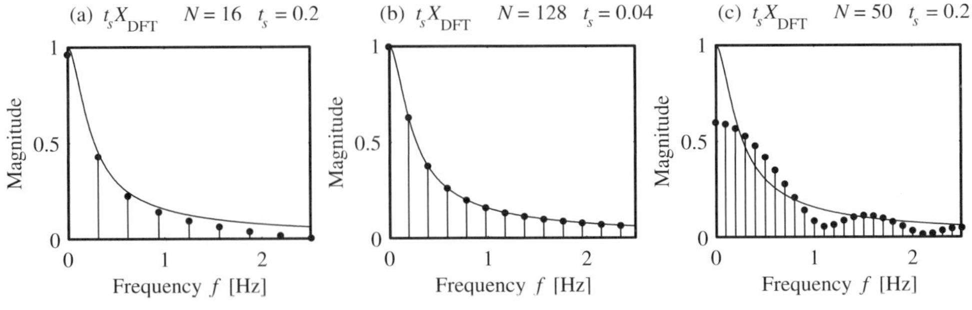

3. To improve the results (and minimize aliasing), we must increase $S$. For example, if we require the highest frequency based on 99% of the signal energy, we obtain $B = 63.6567/2\pi = 10.13$ Hz. Based on this, let us choose $S = 25$ Hz. If we sample $x(t)$ over 5 s (by which time it decays to less that 0.01), we compute $N = (25)(5) = 125$. Choosing $N = 128$ (the next higher power of 2), we find that the 128-point DFT result $t_s X_{\text{DFT}}[k]$ is almost identical to the true spectrum $X(f)$.
4. What would happen if we choose $S = 5$ Hz and five signal samples, but reduce the spectral spacing by zero padding to give $N = 50$? The 50-point DFT clearly shows the effects of truncation (as wiggles) and is a poor match to the true spectrum. This confirms that improved accuracy does not come by zero padding but by including more signal samples.

# 8.8 Spectral Smoothing by Time Windows

Sampling an analog signal for a finite number of samples $N$ is equivalent to multiplying the samples by a rectangular $N$-point window. Due to abrupt truncation of this rectangular window, the spectrum of the windowed signal shows a main lobe and sidelobes that do not decay rapidly enough. This phenomenon is similar to the Gibbs effect, which arises during the reconstruction of periodic signals from a finite number of harmonics (an abrupt truncation of its spectrum). Just as Fourier series reconstruction uses tapered *spectral windows* to smooth the time signal, the DFT uses tapered *time-domain windows* to smooth the spectrum but at the expense of making it broader. This is another manifestation of leakage in that the spectral energy is distributed (leaked) over a wider frequency range.

Unlike windows for Fourier series smoothing, we are not constrained by odd-length windows. As shown in Figure 8.6, an $N$-point DFT window is actually generated from a symmetric $(N+1)$-point window (sampled over $N$ intervals and symmetric about its midpoint) with its last sample discarded (in keeping with the implied periodicity of the signal and the window itself). To apply a window, we position it over the signal samples and create their pointwise product.

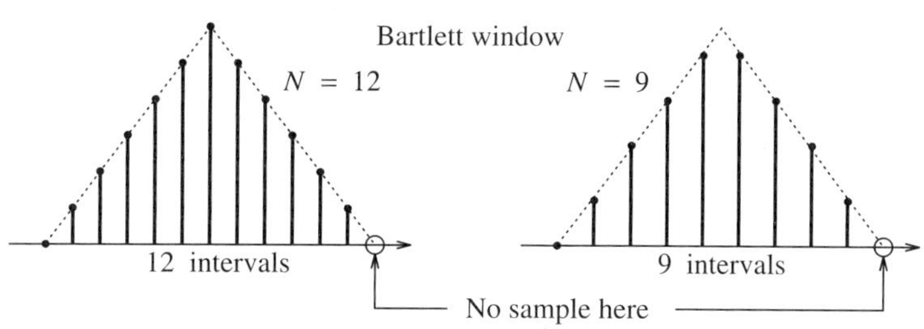


**FIGURE 8.6** Features of a DFT window. An $N$-point DFT window is based on a result that corresponds to generating a symmetric $(N+1)$-point window whose last sample is discarded

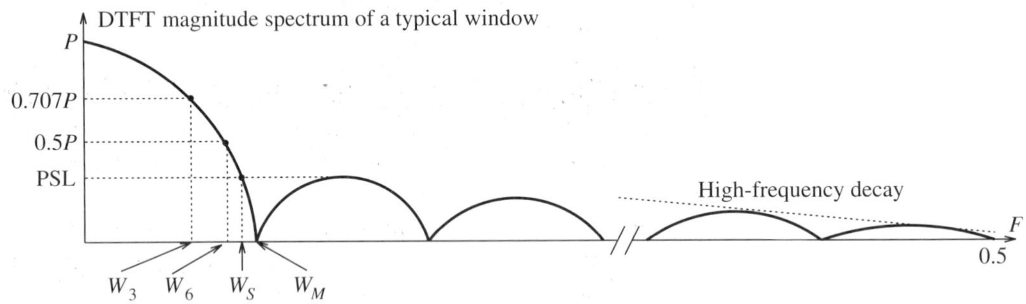


**FIGURE 8.7** Spectrum of a typical window. Measures include the peak gain P, the 3-dB and 6-dB gain $0.707P$ and $0.5P$ corresponding to the widths $W_3$ and $W_6$, the peak sidelobe level PSL corresponding to the width $W_S$, and the main-lobe half-width $W_M$. The high-frequency decay $D_S$ is typically measured in dB/decade or dB/octave

### 8.8.1 Performance Characteristics of Windows

The spectrum of all windows shows a main lobe and decaying sidelobes, as illustrated in Figure 8.7. Measures of magnitude are often normalized by the peak magnitude $P$ and expressed in decibels (dB) (as a gain or attenuation). These measures include the peak value $P$, the peak sidelobe level, the 3-dB level and the 6-dB level, and the high-frequency decay rate (in dB/decade or dB/octave). Measures of spectral width include the 3-dB width $W_3$, 6-dB width $W_6$, the width $W_S$ to reach the peak sidelobe level (PSL), and the main-lobe half-width $W_M$. These measures are illustrated in Figure 8.7.

Other measures of window performance include the **coherent gain** (CG), **equivalent noise bandwidth** (ENBW), and the **scallop loss** (SL). For an $N$-point window $w[n]$, these measures are defined by

$$\text{CG} = \frac{1}{N}\sum_{k=0}^{N-1}|w[k]| \qquad \text{ENBW} = \frac{N\sum_{k=0}^{N-1}|w[k]|^2}{\left|\sum_{k=0}^{N-1}w[k]\right|^2} \qquad \text{SL} = 20\log\frac{\left|\sum_{k=0}^{N-1}w[k]e^{-j\pi k/N}\right|}{\sum_{k=0}^{N-1}w[k]}\,\text{dB} \tag{8.18}$$

The reciprocal of the equivalent noise bandwidth is also called the **processing gain**. The larger the processing gain, the easier it is to reliably detect a signal in the presence of noise.

As we increase the window length, the main lobe width of all windows decreases, but the peak sidelobe level remains more or less constant. Ideally, for a given window length, the spectrum of a window should approach an impulse with as narrow (and tall) a main lobe as possible, and as small a peak sidelobe level as possible. The aim is to pack as much energy in a narrow main lobe as possible and make the sidelobe level as small as possible. These are conflicting requirements in that a narrow main-lobe width also translates to a higher peak sidelobe level. Some DFT windows and their spectral characteristics are illustrated in Figure 8.8, and summarized in Table 8.4.

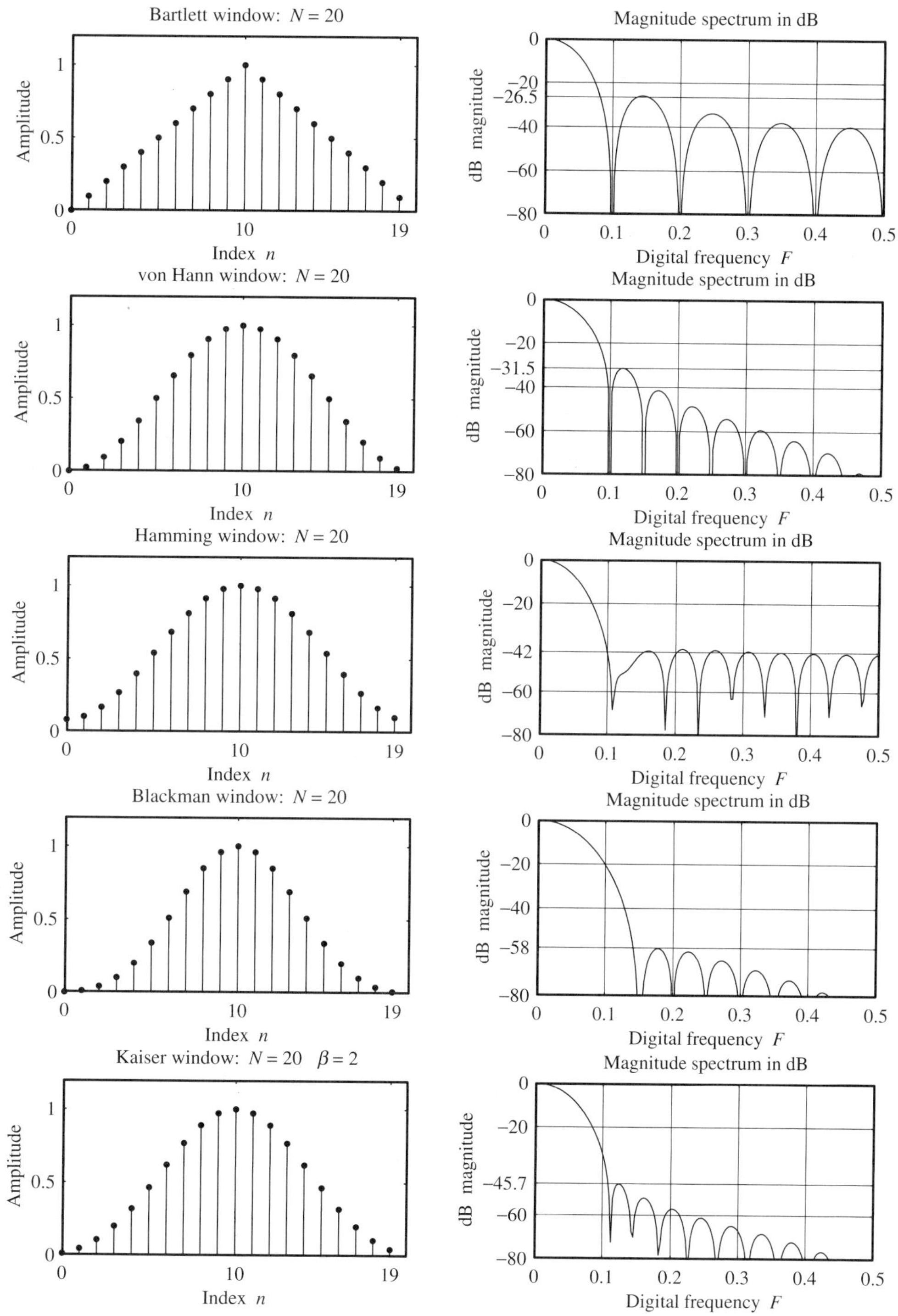


**FIGURE 8.8** Commonly used DFT windows and their spectral characteristics

**TABLE 8.4 ➤ Some Commonly Used $N$-Point DFT Windows**

| Entry | Window | Expression for $w[n]$ | $W_M = \frac{K}{N}$ | Normalized Peak Sidelobe |
|---|---|---|---|---|
| 1 | Boxcar | 1 | $2/N$ | $0.2172 \approx -13.3$ dB |
| 2 | Bartlett | $1 - \dfrac{2\lvert k\rvert}{N}$ | $4/N$ | $0.0472 \approx -26.5$ dB |
| 3 | von Hann | $0.5 + 0.5\cos\left(\dfrac{2\pi k}{N}\right)$ | $4/N$ | $0.0267 \approx -31.5$ dB |
| 4 | Hamming | $0.54 + 0.46\cos\left(\dfrac{2\pi k}{N}\right)$ | $4/N$ | $0.0073 \approx -42.7$ dB |
| 5 | Blackman | $0.42 + 0.5\cos\left(\dfrac{2\pi k}{N}\right) + 0.08\cos\left(\dfrac{4\pi k}{N}\right)$ | $6/N$ | $0.0012 \approx -58.1$ dB |
| 6 | Kaiser | $\dfrac{I_0(\pi\beta\sqrt{1-(2k/N)^2})}{I_0(\pi\beta)}$ | $\dfrac{2\sqrt{1+\beta^2}}{N}$ | $\dfrac{0.22\pi\beta}{\sinh(\pi\beta)} \approx -45.7$ dB (for $\beta = 2$) |

*NOTES:* $k = 0.5N - n$, where $n = 0, 1, \ldots, N-1$. $W_M$ is the main-lobe width.
For the Kaiser window, $I_0(.)$ is the modified Bessel function of order zero.
For the Kaiser window, the parameter $\beta$ controls the peak sidelobe level.
The von Hann window is also known as the *Hanning* window.

### 8.8.2 The Spectrum of Windowed Sinusoids

Consider the signal $x[n] = \cos(2\pi n F_0)$, $F_0 = k_0/M$. Its DTFT is $X_p(F) = 0.5\delta(F - F_0) + 0.5\delta(F + F_0)$. If this sinusoid is windowed by an $N$-point window $w[n]$ whose spectrum is $W(F)$, the DTFT of the windowed signal is given by the periodic convolution

$$X_w(F) = X_p(F) \circledast W(F) = 0.5W(F - F_0) + 0.5W(F + F_0) \tag{8.19}$$

The window thus smears out the true spectrum. To obtain a windowed signal that is a replica of the spectrum of the sinusoid, we require $W(F) = \delta(F)$, which corresponds to the impractical infinite-length time window. The more the spectrum $W(F)$ of an $N$-point window resembles an impulse, the better the windowed spectrum matches the original.

The $N$-point DFT of the signal $x[n]$ may be regarded as the DTFT of the product of the infinite-length $x[n]$ and an $N$-point rectangular window, evaluated at the frequency $F = k/N$, $k = 0, 1, \ldots, N-1$. Since the spectrum of the $N$-point rectangular window is $W(F) = N\frac{\operatorname{sinc}(NF)}{\operatorname{sinc} F}$, the DTFT spectrum of the windowed signal is

$$X_w(F) = 0.5N\frac{\operatorname{sinc}[N(F - F_0)]}{\operatorname{sinc}(F - F_0)} + 0.5N\frac{\operatorname{sinc}[N(F + F_0)]}{\operatorname{sinc}(F + F_0)} \tag{8.20}$$

The $N$-point DFT of the windowed sinusoid is given by $X_{\text{DFT}}[k] = X_w(F)|_{F=k/N}$. If the DFT length $N$ equals $M$ (the number of samples over $k_0$ full periods of $x[n]$) we see that $\operatorname{sinc}[N(F - F_0)] = \operatorname{sinc}(k - k_0)$, and this equals zero, unless $k = k_0$. Similarly, $\operatorname{sinc}[N(F + F_0)] = \operatorname{sinc}(k + k_0)$ is nonzero only if $k = -k_0$. The DFT thus contains only two nonzero terms and equals

$$X_{\text{DFT}}[k] = X_w(F)|_{F=k/N} = 0.5N\delta[k - k_0] + 0.5N\delta[k - k_0] \quad (\text{if } N = M) \tag{8.21}$$

In other words, using an $N$-point rectangular window that covers an integer number of periods ($M$ samples) of a sinusoid (i.e., with $M = N$) gives us exact results. The reason of course is that the DTFT sampling instants fall on the nulls of the sinc spectrum. If the window length $N$ does not equal $M$ (an integer number of periods), the sampling instants will fall between the nulls, and since the sidelobes of the sinc function are large, the DFT results will show considerable leakage. To reduce the effects of leakage, we must use windows whose spectrum shows small sidelobe levels.

### 8.8.3 Resolution

Windows are often used to reduce the effects of leakage and improve resolution. **Frequency resolution** refers to our ability to clearly distinguish between two closely spaced sinusoids of similar amplitudes. **Dynamic-range resolution** refers to our ability to resolve large differences in signal amplitudes. The spectrum of all windows reveals a main-lobe and smaller sidelobes. It smears out the true spectrum and makes components separated by less than the main-lobe width indistinguishable. The rectangular window yields the best frequency resolution for a given length $N$, since it has the smallest main lobe. However, it also has the largest peak sidelobe level of any window. This leads to significant leakage and the worst dynamic-range resolution, because small amplitude signals can get masked by the sidelobes of the window.

Tapered windows with less abrupt truncation show reduced sidelobe levels and lead to reduced leakage and improved dynamic-range resolution. They also show increased main-lobe widths $W_M$, leading to poorer frequency resolution. The choice of a window is based on a compromise between the two conflicting requirements of minimizing the main-lobe width (improving frequency resolution) and minimizing the sidelobe magnitude (improving dynamic-range resolution).

The main-lobe width of all windows decreases as we increase the window length. However, the peak sidelobe level remains more or less constant. To achieve a frequency resolution of $\Delta f$, the digital frequency $\Delta F = \frac{\Delta f}{S}$ must equal or exceed the main-lobe width $W_M$ of the window. This yields the window length $N$. For a given window to achieve the same frequency resolution as the rectangular window, we require a larger window length (a smaller main-lobe width) and hence a larger signal length. The increase in signal length must come by choosing more signal samples (and not by zero padding). To achieve a given dynamic-range resolution, however, we must select a window with small sidelobes, regardless of the window length.

**The Smallest Frequency We can Resolve Depends on the Main-lobe Width of the Window**

To resolve frequencies separated by $\Delta f$, we require $\frac{\Delta f}{S} = W_M = \frac{K}{N}$ (window main-lobe width).
$K$ depends on the window type. To decrease $\Delta f$, increase $N$ (more signal samples, not zero-padding).

**Example 8.13** **Frequency Resolution**

The signal $x(t) = A_1\cos(2\pi f_0 t) + A_2\cos[2\pi(f_0 + \Delta f)t]$, where $A_1 = A_2 = 1$, $f_0 = 30$ Hz is sampled at the rate $S = 128$ Hz. We acquire $N$ samples, zero-pad them to length $N_{\text{FFT}}$, and obtain the $N_{\text{FFT}}$-point FFT.

**(a)** What is the smallest $\Delta f$ that can be resolved for

$N = 256$, $N_{\text{FFT}} = 2048$ using a rectangular and von Hann (Hanning) window
$N = 512$, $N_{\text{FFT}} = 2048$ using a rectangular and von Hann window
$N = 256$, $N_{\text{FFT}} = 4096$ using a rectangular and von Hann window

**(b)** How do the results change if $A_2 = 0.05$?

**(a) Frequency Resolution**
Since $\Delta F = \frac{\Delta f}{S} = W_M$, we have $\Delta f = SW_M$. We compute:

$$\text{Rectangular window: } \Delta f = SW_M = \frac{2S}{N} = 1 \text{ Hz for } N = 256 \text{ and } 0.5 \text{ Hz for } N = 512$$

$$\text{von Hann window: } \Delta f = SW_M = \frac{4S}{N} = 2 \text{ Hz for } N = 256 \text{ and } 1 \text{ Hz for } N = 512$$

Note that $N_{\text{FFT}}$ governs only the FFT spacing $\frac{S}{N_{\text{FFT}}}$, whereas $N$ governs only the frequency resolution $S/N$ (which does not depend on the zero-padded length). Figure E8.13a shows the FFT spectra, plotted as continuous curves, over a selected frequency range. We make the following remarks:

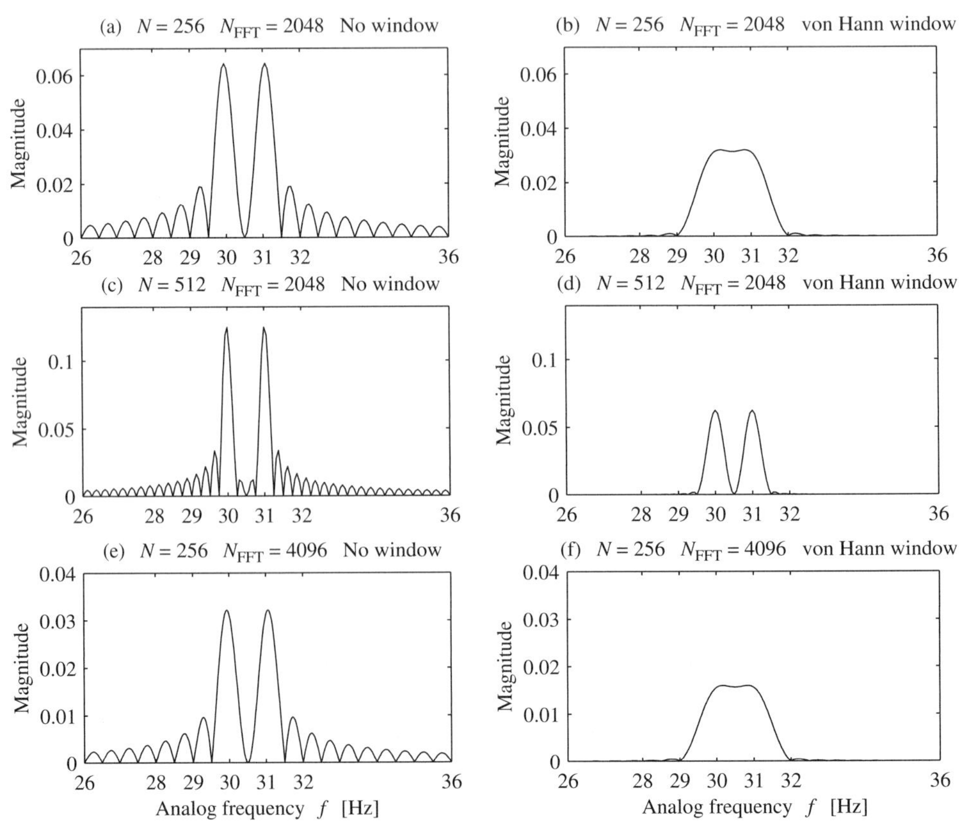


**FIGURE E.8.13a** DFT spectra for Example 8.13(a)

1. For a given signal length $N$, the rectangular window resolves a smaller $\Delta f$ but also has the largest sidelobes (Figure E8.13a(a) and (b)). This means that

the effects of leakage are more severe for a rectangular window than for any other.

2. We can resolve a smaller $\Delta f$ by increasing the signal length $N$ *alone* (Figure E8.13a(c)). To resolve the same $\Delta f$ with a von Hann window, we must *double* the signal length $N$ (Figure E8.13a(d)). This means that we can improve resolution only by increasing the number of signal samples (adding more signal information). How many more signal samples we require will depend on the desired resolution and the type of window used.
3. We cannot resolve a smaller $\Delta f$ by simply increasing the zero padded length $N_{\text{FFT}}$ *alone* (Figure E8.13a(e) and (f)). In other words, zero-padding cannot improve resolution. Zero padding simply interpolates the DFT at a denser set of frequencies. It cannot improve the accuracy of the DFT results, because adding more zeros does not add more signal information.

**(b) Dynamic-Range Resolution**

If $A_2 = 0.05$ (26 dB below $A_1$), the large sidelobes of the rectangular window (13 dB below the peak) will mask the second peak at 31 Hz, even if we increase $N$ and $N_{\text{FFT}}$. This is illustrated in Figure E8.13b(a) (where the peak magnitude is normalized to unity, or 0 dB) for $N = 512$ and $N_{\text{FFT}} = 4096$. For the same values of $N$ and $N_{\text{FFT}}$, however, the smaller sidelobes of the von Hann window (31.5 dB below the peak) do allow us to resolve two distinct peaks in the windowed spectrum, as shown in Figure E8.13b(b).

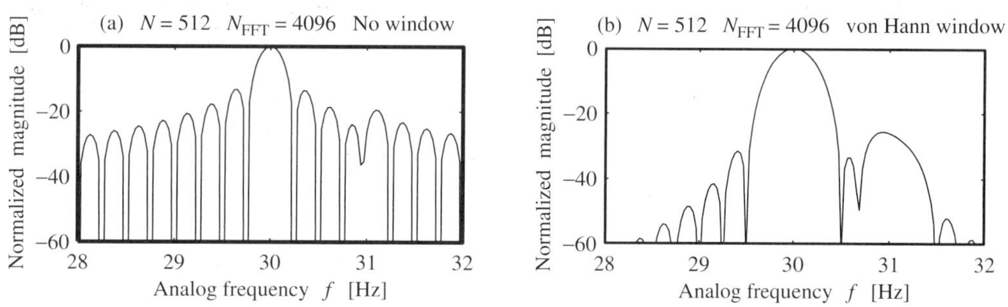


**FIGURE E.8.13b** DFT spectra for Example 8.13(b)

## 8.8.4 Detecting Hidden Periodicity using the DFT

Given an analog signal $x(t)$ known to contain periodic components, how do we estimate their frequencies and magnitude? There are several ways. Most rely on statistical estimates, especially if the signal $x(t)$ is also corrupted by noise. Here is a simpler, perhaps more intuitive—but by no means the best—approach based on the effects of aliasing. The location and magnitude of the components in the DFT spectrum can change with the sampling rate if this rate is below the Nyquist rate. Due to aliasing, the spectrum may not drop to zero at $0.5S$ and may even show increased magnitudes as we move toward the folding frequency. If we try to minimize the effects of noise by using a lowpass filter, we must ensure that its cutoff frequency exceeds the frequency of all the components of interest present in $x(t)$. We have no *a priori* way of doing this. A better way, if the data can be acquired repeatedly, is

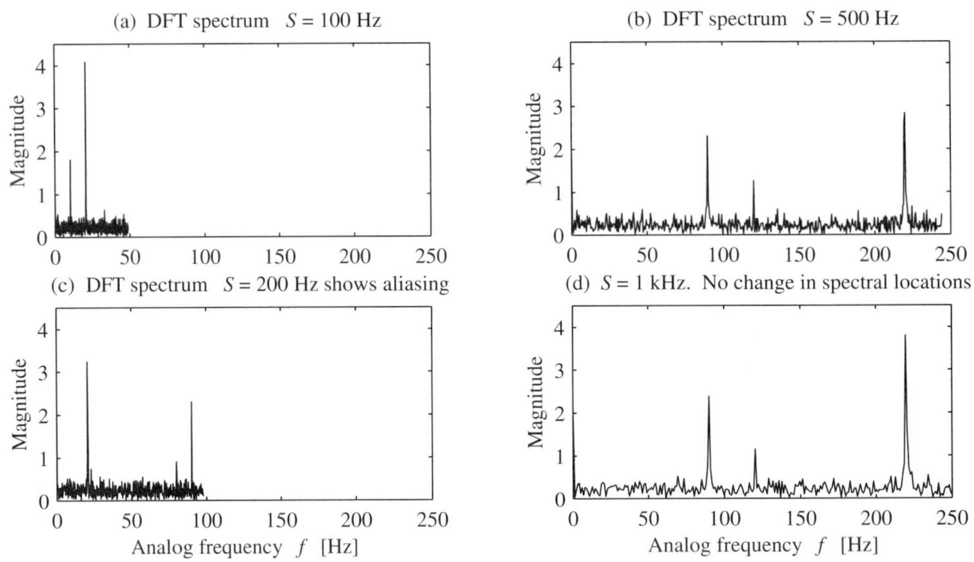


**FIGURE 8.9** Trial-and-error method for obtaining the DFT spectrum. If the sampling frequency is not high enough (100 Hz or 200 Hz in the plots shown), the spectral peaks can change location due to aliasing. The idea is to keep observing the spectrum as the sampling frequency is increased. Once it is high enough (500 Hz or 1 kHz in the plots shown), the locations of the spectral peaks are unaltered, and we have the correct spectrum

to use the average of many runs. Averaging minimizes noise while preserving the integrity of the signal.

A crude estimate of the sampling frequency may be obtained by observing the most rapidly varying portions of the signal. Failing this, we choose an arbitrary but small sampling frequency, sample $x(t)$, and observe the DFT spectrum. We repeat the process with increasing sampling rates and observe how the DFT spectrum changes, and when the spectrum shows little change in the *location* of its spectral components, we have the right spectrum and the right sampling frequency. This trial-and-error method is illustrated in Figure 8.9 and actually depends on aliasing for its success.

If $x(t)$ is a sinusoid, its magnitude $A$ is computed from $X_{\text{DFT}}[k_0] = 0.5NA$, where the index $k_0$ corresponds to the peak in the $N$-point DFT. However, if other nonperiodic signals are also present in $x(t)$, this may not yield a correct result. A better estimate of the magnitude of the sinusoid may be obtained by comparing two DFT results of different lengths—say, $N_1 = N$ and $N_2 = 2N$. The $N_1$-point DFT at the index $k_1$ of the peak will equal $0.5N_1A$ plus a contribution due to the nonperiodic signals. Similarly, the $N_2$-point DFT will show a peak at $k_2$ (where $k_2 = 2k_1$ if $N_2 = 2N_1$), and its value will equal $0.5N_2A$ plus a contribution due to the nonperiodic signals. If the nonperiodic components do not affect the spectrum significantly, the difference in these two values will cancel out the contribution due to the nonperiodic components and yield an estimate for the magnitude of the sinusoid from

$$X_{\text{DFT2}}[k_2] - X_{\text{DFT1}}[k_1] = 0.5N_2A - 0.5N_1A \tag{8.22}$$

**Example 8.14**

### Detecting Hidden Periodicity

A signal $x(t)$ is known to contain a sinusoidal component. The 80-point DFT result and a comparison of the 80-point and 160-point DFT are shown in Figure E8.14. Estimate the frequency and magnitude of the sinusoid and its DFT index. The sampling rate is $S = 10$ Hz.

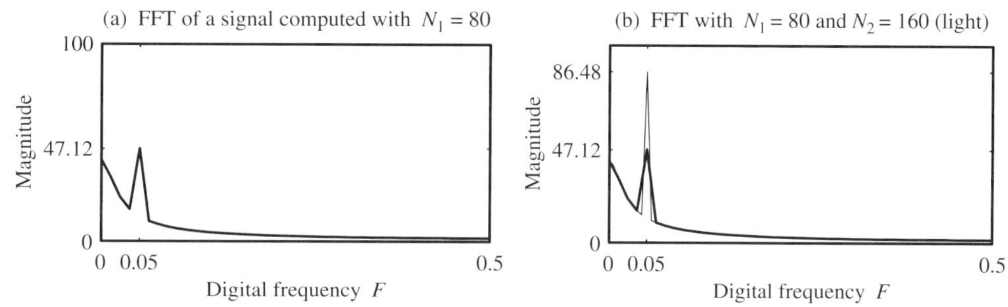


**FIGURE E.8.14** DFT spectra for Example 8.14

The comparison of the two DFT results suggests a peak at $F = 0.05$ and the presence of a sinusoid. Since the sampling rate is $S = 10$ Hz, the frequency of the sinusoid is $f = FS = 0.5$ Hz. Let $N_1 = 80$ and $N_2 = 160$. The peak in the $N_1$-point DFT occurs at the index $k_1 = 4$ because $F = 0.05 = k_1/N_1 = 4/80$. Similarly, the peak in the $N_2$-point DFT occurs at the index $k_2 = 8$ because $F = 0.05 = k_2/N_2 = 8/160$. Since the two spectra do not differ much except near the peak, the difference in the peak values allows us to compute the peak value $A$ of the sinusoid from

$$X_{\text{DFT2}}[k_2] - X_{\text{DFT1}}[k_1] = 86.48 - 47.12 = 0.5N_2A - 0.5N_1A = 40A$$

Thus, $A = 0.984$ and implies the presence of the 0.5-Hz sinusoidal component $0.984\cos(\pi t + \theta)$.

*Comment:* The DFT results shown in Figure E8.14 are actually for the signal $x(t) = e^{-t} + \cos(\pi t)$ sampled at $S = 10$ Hz. The sinusoidal component has unit peak value, and the DFT estimate ($A = 0.984$) differs from this value by less than 2%. Choosing larger DFT lengths would improve the accuracy of the estimate. However, the 80-point DFT alone yields the estimate $A = \dfrac{47.12}{40} = 1.178$ (an 18% difference), whereas the 160-point DFT alone yields $A = \dfrac{86.48}{80} = 1.081$ (an 8% difference).

## 8.9 Applications in Signal Processing

The applications of the DFT and FFT span a wide variety of disciplines. Here, we briefly describe some applications directly related to digital signal processing.

### 8.9.1 Convolution of Long Sequences

A situation that often arises in practice is the processing of a long stream of incoming data by a filter whose impulse response is much shorter than that of the incoming data. The convolution of a short sequence $h[n]$ of length $N$ (such as an averaging filter) with a very long sequence $x[n]$ of length $L \gg N$ (such as an incoming stream of data) can involve large amounts of computation and memory. There are two preferred alternatives, both of which are based on sectioning the long sequence $x[n]$ into shorter ones. The DFT offers a useful means of finding such a convolution. It even allows on-line implementation if we can tolerate a small processing delay.

#### The Overlap-Add Method

Suppose $h[n]$ is of length $N$, and the length of $x[n]$ is $L = mN$ (if not, we can always zero pad it to this length). We partition $x[n]$ into $m$ segments $x_0[n], x_1[n], \ldots, x_{m-1}[n]$, each of length $N$. We find the regular convolution of each section with $h[n]$ to give the partial results $y_0[n], y_1[n], \ldots, y_{m-1}[n]$. Using superposition, the total convolution is the sum of their shifted (by multiples of $N$) versions

$$y[n] = y_0[n] + y_1[n-N] + y_2[n-2N] + \cdots + y_{m-1}[n-(m-1)N] \qquad (8.23)$$

Since each regular convolution contains $2N-1$ samples, we zero pad $h[n]$ and each section $x_k[n]$ with $N-1$ zeros before finding $y_k[n]$ using the FFT. Splitting $x[n]$ into equal-length segments is not a strict requirement. We may use sections of different lengths—provided we keep track of how much each partial convolution must be shifted before adding the results.

#### The Overlap-Save Method

The regular convolution of sequences of length $L$ and $N$ has $L+N-1$ samples. If $L > N$ and we zero pad the second sequence to length $L$, their periodic convolution has $2L-1$ samples. Its first $N-1$ samples are contaminated by wraparound, and the rest correspond to the regular convolution. To understand this, let $L = 16$ and $N = 7$. If we pad $N$ by nine zeros, their regular convolution has 31 (or $2L-1$) samples with nine trailing zeros ($L-N = 9$). For periodic convolution, 15 samples ($L-1 = 15$) are wrapped around. Since the last nine (or $L-N$) are zeros, only the first six samples ($L-N-(N-1) = N-1 = 6$) of the periodic convolution are contaminated by wraparound. This idea is the basis for the overlap-save method. First, we add $N-1$ *leading* zeros to the longer sequence $x[n]$ and section it into $k$ *overlapping* (by $N-1$) segments of length $M$. Typically, we choose $M \approx 2N$. Next, we zero pad $h[n]$ (with *trailing* zeros) to length $M$, and find the *periodic* convolution of $h[n]$ with each section of $x[n]$. Finally, we discard the first $N-1$ (contaminated) samples from each convolution and glue (concatenate) the results to give the required convolution.

In either method, the FFT of the shorter sequence need be found only once, stored, and reused for all subsequent partial convolutions. Both methods allow on-line implementation if we can tolerate a small processing delay that equals the time required for each section of the long sequence to arrive at the processor (assuming the time taken for finding the partial convolutions is less than this processing delay). The correlation of two sequences also may be found in exactly the same manner, using either method, and provided we use a flipped version of one sequence.

**Example 8.15**

### Overlap-Add and Overlap-Save Methods of Convolution

Let $x[n] = \{\overset{\Downarrow}{1}, 2, 3, 3, 4, 5\}$ and $h[n] = \{\overset{\Downarrow}{1}, 1, 1\}$. Here $L = 6$ and $N = 3$.

**(a)** To find their convolution using the overlap-add method, we section $x[n]$ into two sequences given by $x_0[n] = \{1, 2, 3\}$ and $x_1[n] = \{3, 4, 5\}$, and obtain the two convolution results:

$$y_0[n] = x_0[n] * h[n] = \{1, 3, 6, 5, 3\} \qquad y_1[n] = x_1[n] * h[n] = \{3, 7, 12, 9, 5\}$$

Shifting and superposition results in the required convolution $y[n]$ as

$$y[n] = y_0[n] + y_1[n-3] = \begin{Bmatrix} 1, 3, 6, 5, 3 & \\ & 3, 7, 12, 9, 5 \end{Bmatrix} = \{1, 3, 6, 8, 10, 12, 9, 5\}$$

This result can be confirmed using any of the convolution algorithms described in the text.

**(b)** To find their convolution using the overlap-add method, we start by creating the zero padded sequence $x[n] = \{0, 0, 1, 2, 3, 3, 4, 5\}$. If we choose $M = 5$, we get three overlapping sections of $x[n]$ (we need to zero pad the last one) described by

$$x_0[n] = \{0, 0, 1, 2, 3\} \qquad x_1[n] = \{2, 3, 3, 4, 5\} \qquad x_2[n] = \{4, 5, 0, 0, 0\}$$

The zero-padded $h[n]$ becomes $h[n] = \{1, 1, 1, 0, 0\}$. Periodic convolution gives

$$x_0[n] \circledast h[n] = \{5, 3, 1, 3, 6\}$$

$$x_1[n] \circledast h[n] = \{11, 10, 8, 10, 12\}$$

$$x_2[n] \circledast h[n] = \{4, 9, 9, 5, 0\}$$

We discard the first two samples from each convolution and glue the results to obtain

$$y[n] = x[n] * h[n] = \{1, 3, 6, 8, 10, 12, 9, 5, 0\}$$

Note that the last sample (due to the zero padding) is redundant and may be discarded.

## 8.9.2 Deconvolution

Given a signal $y[n]$ that represents the output of some system with impulse response $h[n]$, how do we recover the input $x[n]$ where $y[n] = x[n] * h[n]$? One method is to undo the effects of convolution using deconvolution. The time-domain approach to deconvolution was studied earlier. Here, we examine a frequency-domain alternative based on the DFT (or FFT).

The idea is to transform the convolution relation using the FFT to obtain $Y_{FFT}[k] = X_{FFT}[k]H_{FFT}[k]$, compute $X_{FFT}[k] = Y_{FFT}[k]/H_{FFT}[k]$ by pointwise division, and then find $x[n]$ as the IFFT of $X_{FFT}[k]$. This process does work in many cases, but it has two disadvantages. First, it fails if $H_{FFT}[k]$ equals zero at some index because we get division by zero. Second, it is quite sensitive to noise in the input $x[n]$ and to the accuracy with which $y[n]$ is known.

### 8.9.3 Band-Limited Signal Interpolation

Interpolation of $x[n]$ by $M$ to a new signal $x_I[n]$ is equivalent to a sampling rate increase by $M$. If the signal has been sampled above the Nyquist rate, signal interpolation should add no new information to the spectrum. The idea of zero padding forms the basis for an interpolation method using the DFT in the sense that the $(MN)$-sample DFT of the interpolated signal should contain $N$ samples corresponding to the DFT of $x[n]$, while the rest should be zero. Thus, if we find the DFT of $x[n]$, zero pad it (by inserting zeros about the folding index) to increase its length to $MN$ and find the inverse DFT of the zero-padded sequence, we should obtain the interpolated signal $x_I[n]$. This approach works well for band-limited signals (such as pure sinusoids sampled above the Nyquist rate over an integer number of periods). To implement this process, we split the $N$-point DFT $X_{DFT}[k]$ of $x[n]$ about the folding index $N/2$. If $N$ is even, the folding index falls on the sample value $X[N/2]$, and it must also be split in half. We then insert enough zeros in the middle to create a padded sequence $X_{zp}[k]$ with $MN$ samples. It has the form

$$X_{zp}[k] = \begin{cases} \{X[0], \ldots, X[N/2-1], \underbrace{0, \ldots, 0,}_{(M-1)N \text{ zeros}} X[N/2+1], \ldots, X[N-1]\} & (\text{odd } N) \\ \{X[0], \ldots, 0.5X[N/2], \underbrace{0, \ldots, 0,}_{N(M-1)-1 \text{ zeros}} 0.5X[N/2], \ldots, X[N-1]\} & (\text{even } N) \end{cases} \tag{8.24}$$

The inverse DFT of $X_{zp}[k]$ will include the factor $1/MN$, and its machine computation may show (small) imaginary parts. We therefore retain only its real part and divide by $M$ to obtain the interpolated signal $x_I[n]$, which contains $M-1$ interpolated values between each sample of $x[n]$:

$$x_I[n] = \frac{1}{M}\text{Re}[\text{IDFT}\{X_{zp}[k]\}] \tag{8.25}$$

This method is entirely equivalent to creating a zero-interpolated signal (which produces spectrum replication) and filtering the replicated spectrum (by zeroing out the spurious images). For periodic band-limited signals sampled above the Nyquist rate for an integer number of periods, the interpolation is exact. For all others, imperfections show up as a poor match, especially near the ends, since we are actually interpolating to zero outside the signal duration.

**Example 8.16** **Signal Interpolation Using the FFT**

**(a)** For a sinusoid sampled over one period with four samples, we obtain the signal $x[n] = \{\overset{\Downarrow}{0}, 1, 0, -1\}$. Its DFT is $X_{DFT}[k] = \{\overset{\Downarrow}{0}, -j2, 0, j2\}$. To interpolate this by $M = 8$, we generate the 32-sample zero-padded sequence

$$Z_T = \{\overset{\Downarrow}{0}, -j2, 0, (27 \text{ zeros}), 0, j2\}$$

The interpolated sequence (the IDFT of $Z_T$) shows an exact match to the sinusoid, as illustrated in Figure E8.16(a).

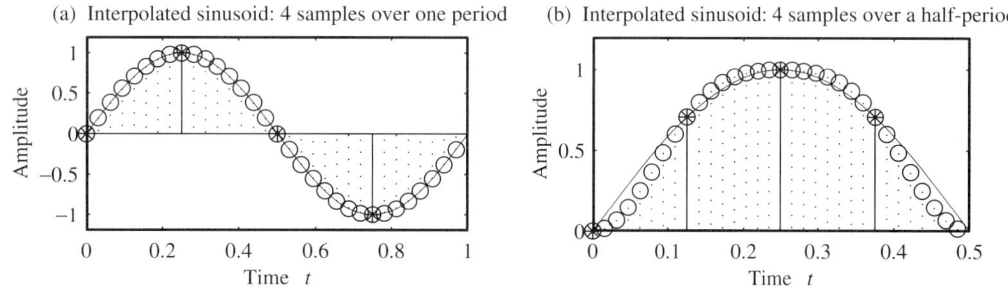


**FIGURE E 8.16.a** Interpolated sinusoids for Example 8.16

**(b)** For a sinusoid sampled over a half period with four samples, interpolation does not yield exact results, as shown in Figure E8.16(b). Since we are actually sampling one period of a full-rectified sine (the periodic extension), the signal is not band limited, and the chosen sampling frequency is too low. This shows up as a poor match, especially near the ends of the sequence.

### 8.9.4 The Discrete Hilbert Transform

The **Hilbert transform** describes an operation that shifts the phase of a signal $x(t)$ by $-\frac{\pi}{2}$. In the analog domain, the phase shift can be achieved by passing $x(t)$ through a filter whose transfer function $H(f)$ is

$$H(f) = -j\,\mathrm{sgn}(f) = \begin{cases} -j, & f > 0 \\ j, & f < 0 \end{cases} \tag{8.26}$$

In the time domain, the phase-shifted signal $\hat{x}(t)$ is given by the convolution

$$\hat{x}(t) = \frac{1}{\pi t} * x(t) \tag{8.27}$$

The phase-shifted signal $\hat{x}(t)$ defines the Hilbert transform of $x(t)$. The spectrum $\hat{X}(f)$ of the Hilbert-transformed signal equals the product of $X(f)$ with the transform of $\frac{1}{\pi t}$. In other words,

$$\hat{X}(f) = -j\,\mathrm{sgn}(f)X(f) \tag{8.28}$$

A system that shifts the phase of a signal by $\pm\frac{\pi}{2}$ is called a **Hilbert transformer** or a **quadrature filter**. Such a system can be used to generate a single-sideband amplitude modulated (SSB AM) signal.

Unlike most other transforms, the Hilbert transform belongs to the same domain as the signal transformed. Due to the singularity in $\frac{1}{\pi t}$ at $t = 0$, a formal evaluation of this relation requires complex variable theory.

The **discrete Hilbert transform** of a sequence may be obtained by using the FFT. The Hilbert transform of $x[n]$ involves the convolution of $x[n]$ with the impulse response $h[n]$ of the Hilbert transformer whose spectrum is $H(F) = -j\,\mathrm{sgn}(F)$. The easiest way to perform this convolution is by FFT methods. We find the $N$-point FFT of $x[n]$ and multiply by the *periodic extension* (from $k = 0$ to $k = N - 1$) of $N$ samples of $\mathrm{sgn}(F)$, which may be written as

$$\text{sgn}[k] = \{\overset{\Downarrow}{0}, \quad \underbrace{1, 1, \ldots, 1}_{\frac{N}{2}-1\text{samples}}, 0, \quad \underbrace{-1, -1, \ldots, -1}_{\frac{N}{2}-1\text{samples}}\} \tag{8.29}$$

The inverse FFT of the product multiplied by the omitted factor $-j$ yields the Hilbert transform.

## 8.10 Spectrum Estimation

The **power spectral density** (PSD) $R_{xx}(f)$ of an analog power signal or random signal $x(t)$ is the Fourier transform of its autocorrelation function $r_{xx}(t)$ and is a real, non-negative, even function with $R_{xx}(0)$ equal to the average power in $x(t)$. If a signal $x(t)$ is sampled and available only over a finite duration, the best we can do is *estimate* the PSD of the underlying signal $x(t)$ from the given finite record. This is because the spectrum of finite sequences suffers from leakage and poor resolution. The PSD estimate of a noisy analog signal $x(t)$ from a finite number of its samples is based on two fundamentally different approaches. The first, **non-parametric spectrum estimation**, makes no assumptions about the data. The second, **parametric spectrum estimation**, models the data as the output of a digital filter excited by a noise input with a constant power spectral density, estimates the filter coefficients, and in so doing, arrives at an estimate of the true PSD. Both rely on statistical measures to establish the quality of the estimate.

### 8.10.1 The Periodogram Estimate

The simplest non-parametric estimate is the **periodogram** $P[k]$ that is based on the DFT of an $N$-sample sequence $x[n]$. It is defined as

$$P[k] = \frac{1}{N}|X_{\text{DFT}}[k]|^2 \tag{8.30}$$

Although $P[k]$ provides a good estimate for deterministic, band-limited, power signals sampled above the Nyquist rate, it yields poor estimates for noisy signals because the quality of the estimate does not improve with increasing record length $N$ (even though the spectral spacing decreases).

**Example 8.17**

**The Concept of the Periodogram**

The sequence $x[n] = \{\overset{\Downarrow}{0}, \ 1, \ 0, \ -1\}$ is obtained by sampling a sinusoid for one period at twice the Nyquist rate. Find its periodogram estimate.

The DFT of $x[n]$ is $X_{\text{DFT}}[k] = \{\overset{\Downarrow}{0}, -j2, 0, j2\}$. Thus, $P[k] = \frac{1}{N}|X_{\text{DFT}}[k]|^2 = \{\overset{\Downarrow}{0}, 1, 0, 1\}$. This is usually plotted as a *bar graph* with each sample occupying a bin width $\Delta F = 1/N = 0.25$.

The total power thus equals $\Delta F \sum P[k] = 0.5$ and matches the true power in the sinusoid.

## 8.10.2 PSD Estimation by the Welch Method

The **Welch method** is based on the concept of averaging the results of overlapping periodogram estimates and finds widespread use in spectrum estimation. The $N$-sample signal $x[n]$ is sectioned into $K$ overlapping $M$-sample sections $x_M[n]$—each with an overlap of $D$ samples. A 50 to 75% overlap is common practice. Each section is windowed by an $M$-point window $w[n]$ to control leakage. The PSD of each windowed section is found using the periodogram $\frac{1}{M}|X_M[k]|^2$, and the $K$ periodograms are averaged to obtain an $M$-point **averaged periodogram**. It is customary to normalize (divide) this result by the power in the window signal ($\sum |W_{\text{DFT}}[k]|^2$ or $\frac{1}{M}\sum |w[n]|^2$). Sections of larger length result in a smaller spectral spacing, whereas more sections result in a smoother estimate (at the risk of masking sharp details). The number of sections chosen is a trade-off between decreasing the spectral spacing and smoothing the estimate. Averaging the results of the (presumably uncorrelated) segments reduces the statistical fluctuations (but at the expense of a larger spectral spacing). In the related **Bartlett method**, the segments are neither overlapped nor windowed. Figure 8.10(a) shows the Welch PSD of a 400-point chirp signal, whose digital frequency varies from $F = 0.2$ to $F = 0.4$, using a 45% overlap with a 64-point von Hann window (shown dark) and a rectangular window (no window). Note how the windowing results in significant smoothing.

## 8.10.3 PSD Estimation by the Blackman-Tukey Method

The **Blackman-Tukey method** relies on finding the PSD from the windowed autocorrelation, which is assumed to be zero past the window length. The $N$-sample signal $x[n]$ is zero padded to get the $2N$-sample signal $y[n]$. The *linear* autocorrelation of $x[n]$ equals the *periodic* autocorrelation of $y[n]$ and is evaluated by finding the FFT $Y_{\text{FFT}}[k]$ and taking the IFFT of $Y_{\text{FFT}}[k]Y^*_{\text{FFT}}[k]$ to obtain the $2N$-sample autocorrelation estimate $r_{xx}[n]$. This autocorrelation is then windowed by an $M$-point window (to smooth the spectrum and reduce the effects of poor autocorrelation estimates due to the finite data length). The FFT of the $M$-sample windowed autocorrelation yields the **smoothed periodogram**. Using a smaller $M$ (narrower windows) for the autocorrelation provides greater smoothing but may

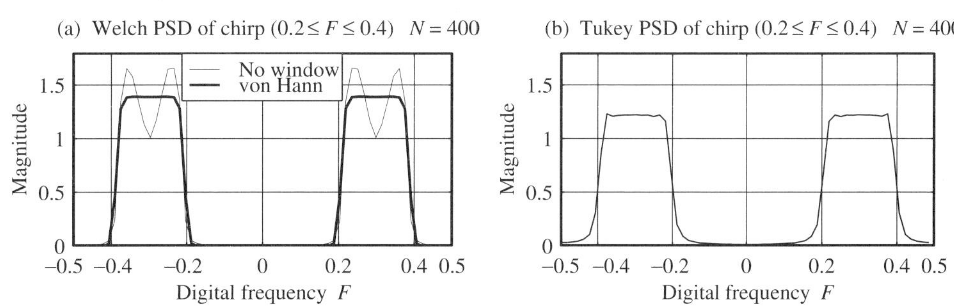


**FIGURE 8.10** Welch and Tukey PSD of a chirp signal. A chirp signal is characterized by a spectrum that is constant over the frequency range present in the chirp. The unwindowed spectrum of a chirp signal is shown (in light) in the first plot. The improvement using either the Welch method (using a von Hann window, shown dark in the first plot) or the Tukey method (shown in the second plot) is clearly visible

also mask any peaks or obscure the sharp details. Typical values of the window length $M$ range from $M = 0.1N$ to $M = 0.5N$. Only windows whose transform is entirely positive should be used. Of the few that meet this constraint, the most commonly used is the Bartlett (triangular) window. Figure 8.10(b) shows the Tukey PSD of a 1000-point chirp signal, whose digital frequency varies from $F = 0.2$ to $F = 0.4$, using a 64-point Bartlett window.

The Welch method is more useful for detecting closely spaced peaks of similar magnitudes (*frequency resolution*). The Blackman-Tukey method is better for detecting well-separated peaks with different magnitudes (*dynamic-range resolution*). Neither is very effective for short data lengths.

### 8.10.4 Non-Parametric System Identification

The Fourier transform $R_{yx}(f)$ of the cross correlation $r_{yx}(t) = y(t) \star\star\, x(t) = y(t) * x(-t)$ of two random signals $x(t)$ and $y(t)$ is called the **cross-spectral density**. For a system with impulse response $h(t)$, the input $x(t)$ yields the output $y(t) = x(t) * h(t)$, and we have

$$r_{yx}(t) = y(t) * x(-t) = h(t) * x(t) * x(-t) = h(t) * r_{xx}(t) \tag{8.31}$$

The Fourier transform of both sides yields

$$R_{yx}(f) = H(f)R_{xx}(f) \qquad H(f) = \frac{R_{yx}(f)}{R_{xx}(f)} \tag{8.32}$$

This relation allows us to identify an unknown transfer function $H(f)$ as the ratio of the cross-spectral density $R_{yx}(f)$ and the PSD $R_{xx}(f)$ of the input $x(t)$. If $x(t)$ is a noise signal with a constant power spectral density, its PSD is a constant of the form $R_{xx}(f) = K$. Then, $H(f) = R_{yx}(f)/K$ is directly proportional to the cross-spectral density.

This approach is termed **non-parametric** because it presupposes no model for the system. In practice, we use the FFT to approximate $R_{xx}(f)$ and $R_{yx}(f)$ (using the Welch method, for example) by the finite $N$-sample sequences $R_{xx}[k]$ and $R_{yx}[k]$. As a consequence, the transfer function $H[k] = R_{yx}[k]/R_{xx}[k]$ also has $N$ samples and describes an FIR filter. Its inverse FFT yields the $N$-sample impulse response $h[n]$. Figure 8.11 shows the spectrum of an IIR filter defined by $y[n] - 0.5y[n-1] = x[n]$ (or $h[n] = (0.5)^n u[n]$). Its 20-point FIR estimate is obtained by using a 400-sample noise sequence and the Welch method.

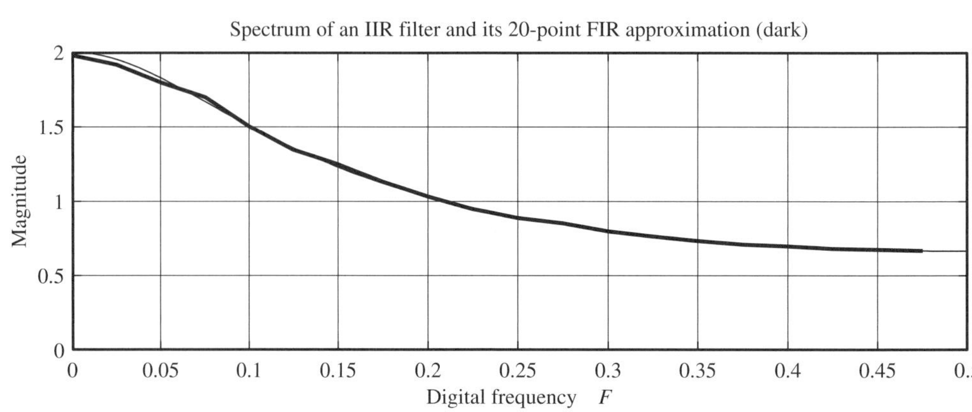


**FIGURE 8.11** The spectrum of an IIR filter and its non-parametric FIR filter estimate. Even though there is a close match in the magnitude spectra, the phase spectra of the FIR and IIR will show little correspondence

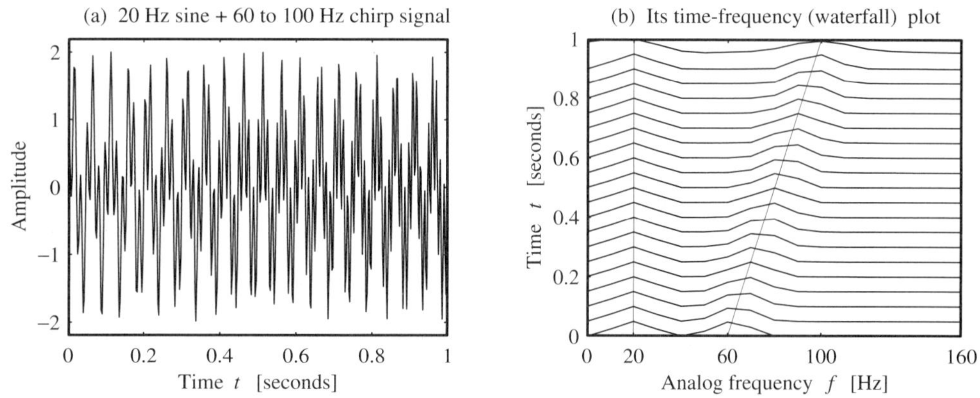


**FIGURE 8.12** Time-frequency plot for the sum of a sinusoid and a chirp signal. The spectra are stacked above each other at successive times. The thin vertical line is shown to indicate that the spectral peak of the sinusoid occurs at the same frequency and does not change with time. However, the chirp signal has a frequency that increases with time and this is clearly visible as a shift in the spectral peak (from 60 Hz to 100 Hz) as time increases

### 8.10.5 Time-Frequency Plots

In practical situations, we are often faced with the task of finding how the spectrum of signals varies with time. A simple approach is to section the signal into overlapping segments, window each section to reduce leakage, and find the PSD (using the Welch method, for example). The PSD for each section is then staggered and stacked to generate what is called a **waterfall plot** or **time-frequency plot**. Figure 8.12 shows such a waterfall plot for the sum of a single frequency sinusoid and a chirp signal whose frequency varies linearly with time. It provides a much better visual indication of how the spectrum evolves with time.

The fundamental restriction in obtaining the time-frequency information, especially for short time records, is that *we cannot localize both time and frequency to arbitrary precision*. Recent approaches are based on expressing signals in terms of **wavelets** (much like the Fourier series). Wavelets are functions that show the "best" possible localization characteristics in both the time-domain and the frequency-domain and are an area of intense ongoing research.

## 8.11 The Cepstrum and Homomorphic Filtering

Consider a signal $x[n]$. If we take the complex logarithm of its DTFT $X(F)$ to get $X_K(F) = \ln X(F)$, the inverse transform $x_K[n]$ is called the **cepstrum** (pronounced kepstrum) of $x[n]$. We have

$$X_K(F) = \ln X(F) \qquad x_K[n] = \int_{-1/2}^{1/2} [\ln X(F)]e^{j2n\pi F}\, dF \tag{8.33}$$

We can recover $X(F)$ from $X_K(F)$ by taking complex exponentials to give

$$X(F) = |X(F)|e^{j\phi(F)} = e^{X_K(F)}$$

This allows us to obtain $X_K(F)$ from the magnitude and phase of $X(F)$ as follows

$$X_K(F) = \ln X(F) = \ln|X(F)| + j\phi(F) \tag{8.34}$$

The cepstrum $x_K[n]$ may then be expressed as

$$x_K[n] = \int_{-1/2}^{1/2} (\ln|X(F)|)e^{j2n\pi F}\, dF + j\int_{-1/2}^{1/2} \phi(F)e^{j2n\pi F}\, dF \tag{8.35}$$

In other words, the cepstrum is a complex quantity. Its real part corresponds to the IDTFT of $\ln|X(F)|$ and is called the **real cepstrum**. The existence of the cepstrum is restricted by the nature of $X(F)$. The cepstrum does not exist if $X(F) = 0$ over a range of frequencies, because $\ln|X(F)|$ is undefined. The cepstrum does not exist if $X(F)$ is bipolar (zero at isolated frequencies where it changes sign) and exhibits phase jumps of $\pm\pi$. In fact, the cepstrum does not exist unless the phase $\phi(F)$ is a continuous single-valued function of frequency (with no phase jumps of any kind) and equals zero at $F = 0$ and $F = 0.5$. We can get around these problems by working with the unwrapped phase (to eliminate any phase jumps of $\pm 2\pi$) and, when necessary, adding a large enough positive constant to $X(F)$ to ensure $X(F) > 0$ for all frequencies (and eliminate any phase jumps of $\pm\pi$).

### 8.11.1 Homomorphic Filters and Deconvolution

Deconvolution is a process of extracting the input signal from a filter output and its transfer function. Since the convolution $y[n] = x[n] * h[n]$ transforms to the product $Y(F) = X(F)H(F)$, we can extract $x[n]$ from the inverse transform of the ratio $X(F) = Y(F)/H(F)$. We may also perform deconvolution using cepstral transformations. The idea is to first find the cepstrum of $y[n]$ by starting with start with $Y(F) = X(F)H(F)$ to obtain

$$Y_K(F) = \ln Y(F) = \ln X(F) + \ln H(F) = X_K(F) + H_K(F) \tag{8.36}$$

Inverse transformation gives

$$y_K[n] = x_K[n] + h_K[n] \tag{8.37}$$

Notice how the convolution operation transforms to a summation in the cepstral domain.

$$y[n] = x[n] * h[n] \qquad y_K[n] = x_K[n] + h_K[n] \tag{8.38}$$

Systems capable of this transformation are called **homomorphic systems**.

If $x_K[n]$ and $h_K[n]$ lie over different ranges, we may extract $x_K[n]$ by using a time window $w[n]$ whose samples equal 1 over the extent of $x_K[n]$ and zero over the extent of $h_K[n]$. This operation is analogous to filtering (albeit in the time domain) and is often referred to as **homomorphic filtering**. Finally, we can use $x_K[n]$ to recover $x[n]$ by first finding $X_K(F)$, then taking its exponential to give $X(F)$, and finally evaluating its inverse transform. A homomorphic filter performs deconvolution in several steps as outlined in the following review panel.

**Overview of Homomorphic Filtering and Cepstral Analysis**

**Transform of Cepstrum:**
$$Y_K(F) = \ln Y(F) = \ln[X(F)H(F)] = \ln X(F) + \ln H(F) = X_K(F) + H_K(F)$$
**Cepstrum:** $y_K[n] = x_K[n] + h_K[n]$
**Windowed Cepstrum:** $y_K[n]w[n] = x_K[n]$
**Signal Recovery:** $x_K[n] \Leftarrow\boxed{\text{DTFT}}\Rightarrow X_K(F) \qquad e^{X_K(F)} = X(F)$

Cepstral methods and homomorphic filtering have found practical applications in various fields including deconvolution, image processing (restoring degraded images), communications (echo cancellation), speech processing (pitch detection, dynamic-range expansion, digital restoration of old audio recordings), and seismic signal processing.

### 8.11.2 Echo Detection and Cancellation

Cepstral methods may be used to detect and remove unwanted echos from a signal. We illustrate echo detection and cancellation by considering a simple first-order echo system described by

$$y[n] = x[n] + \alpha x[n-D] \tag{8.39}$$

Here, $\alpha x[n-D]$ describes the echo of strength $\alpha$ delayed by $D$ samples. It is reasonable to assume that $\alpha < 1$, since the echo magnitude will be less than that of the original signal $x[n]$. In the frequency domain, we get

$$Y(F) = X(F) + \alpha X(F)e^{-j2\pi FD}$$

The transfer function of the system that produces the echo may be written as

$$H(F) = \frac{Y(F)}{X(F)} = 1 + \alpha e^{-j2\pi FD} \tag{8.40}$$

Now, we find $H_K(F) = \ln H(F)$ to yield the transform of the cepstrum $h_K[n]$ as

$$H_K(F) = \ln H(F) = \ln[1 + \alpha e^{-j2\pi FD}] \tag{8.41}$$

Note that $H_K(F)$ is periodic in $F$ with period $\frac{1}{D}$, and as a consequence, the cepstrum $h_K[n]$ describes a signal with sample spacing $D$. To formalize this result, we invoke the series expansion

$$\ln(1+r) = \sum_{k=1}^{\infty}(-1)^{k+1}\frac{r^k}{k}, \qquad |r| < 1$$

With $r = \alpha e^{-j2\pi FD}$ in the above result, we can express $H_K(F)$ as

$$H_K(F) = \ln[1 + \alpha e^{-j2\pi FD}] = \sum_{k=1}^{\infty}(-1)^{k+1}\frac{\alpha^k}{k}e^{-j2\pi kFD} \tag{8.42}$$

Its inverse transform leads to the cepstrum $h_K[n]$ of the echo-producing system as

$$h_K[n] = \sum_{k=1}^{\infty}(-1)^{k+1}\frac{\alpha^k}{k}\delta(n-kD) \tag{8.43}$$

Note that this cepstrum $h_K[n]$ is an impulse train with impulses located at integer multiples of the delay $D$. The impulses alternate in sign and their amplitude decreases with the index $k$ that equals integer multiples of the delay.

Since $y[n] = x[n] * h[n]$, its cepstral transformation leads to

$$y_K[n] = x_K[n] + h_K[n] \qquad Y_K(F) = X_K(F) + H_K(F) \tag{8.44}$$

We see that the cepstrum $y_K[n]$ of the contaminated signal $y[n]$ equals the sum of the cepstrum $x_K[n]$ of the original signal $x[n]$, and an impulse train $h_K[n]$, whose decaying impulses alternate in sign and are located at multiples of the delay $D$. If the spectra $X_K(F)$ and $H_K(F)$ occupy different frequency bands, $H_K(F)$ may be eliminated by ordinary frequency-domain filtering and lead to the recovery of $x[n]$. Since $h_K[n]$ contains impulses at multiples of $D$, we may also eliminate the unwanted echoes in the cepstral domain itself by using a **comb filter**, whose weights are zero at integer multiples of $D$ and unity elsewhere. Naturally, the comb filter can be designed only if the delay $D$ is known *a priori*. Even if $D$ is not known, its value may still be estimated from the location of the impulses in the cepstrum.

As an example of echo cancellation by homomorphic signal processing, let us start with a clean decaying exponential signal $x[n] = (0.9)^n u[n]$. If this signal is contaminated by its echo delayed by 20 samples and with a strength of 0.8, the contaminated signal $y[n]$ may be written as

$$y[n] = x[n] + 0.8x[n-20]$$

Our objective is to recover the clean signal $x[n]$ using cepstral techniques. A 64-sample portion of $x[n]$ and $y[n]$ is shown in Figure 8.13(b). The complex cepstrum of the contaminated signal $y[n]$ is displayed in Figure 8.13(a) and is a practical approximation based on the 64-point FFT (to approximate the DTFT).

The appearance of the cepstral spikes (deltas) at $n = 20, 40, 60$ indicates a delay of $D = 20$ as expected. Once these spikes are removed, we get the cepstrum shown in Figure 8.14(a). The signal corresponding to this cepstrum is shown in Figure 8.14(b) and clearly reveals the absence of the echo signal.

However, the correspondence with the clean signal is not perfect. The reason is that the complex cepstrum still contains spikes at the indices $n = 80, 100, 120, \ldots$ that get aliased

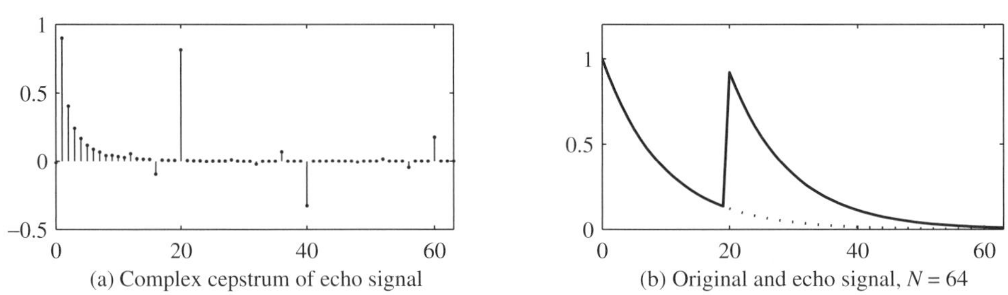


**FIGURE 8.13** Complex cepstrum (a) of the echo signal in (b). The spikes in the cepstrum are 20 samples apart and correspond to the delay in the echo signal

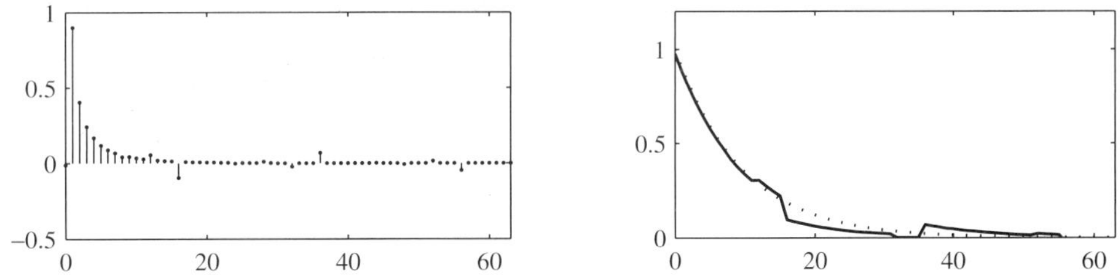


(a) Cepstrum with deltas removed (b) Original and echo-cancelled signal

**FIGURE 8.14** Complex cepstrum (a) with spikes removed and its corresponding signal (b). Note that the mismatch between the original and recovered signal is, in part, due to other spikes that still remain in the cepstrum at locations that correspond to the aliased values

to $n = 16, 36, 56, \ldots$. These aliases are visible in Figure 8.14(a) and are the consequence of using the FFT to approximate the true cepstrum. Once these aliased spikes are also removed, we get the cepstrum of Figure 8.15(a). The signal corresponding to this cepstrum, shown in Figure 8.15(b), is now an almost exact replica of the original clean signal.

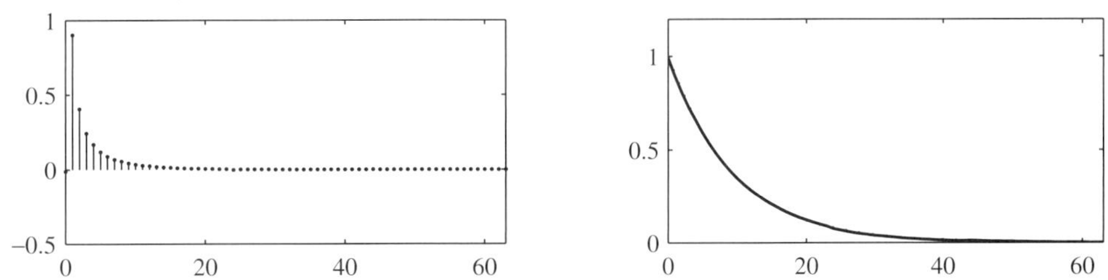


(a) Cepstrum with aliased deltas removed (b) Original and echo-cancelled signal

**FIGURE 8.15** Complex cepstrum (a) with aliased spikes removed and its corresponding signal (b). The recovered signal is impossible to distinguish from the original in the figure

# 8.12 Optimal Filtering

For a filter with impulse response $h[n]$, the output $y[n]$ is described by the convolution $y[n] = x[n] * h[n]$. Ideally, it is possible to recover $x[n]$ by deconvolution. However, if the output is contaminated by noise or interference, an optimal filter provides a means for recovering the input signal. Such a filter is optimized in the sense that its output, when deconvolved by $h[n]$, results in a signal $\hat{x}[n]$ that is the *best* (in some sense) approximation to $x[n]$. Suppose the contaminated response $w[n]$ equals the sum of the ideal response $y[n]$ and a noise component $s[n]$ such that

$$w[n] = \underbrace{y[n]}_{\text{ideal output}} + \underbrace{s[n]}_{\text{noise}} \tag{8.45}$$

If we pass this signal through an optimal filter with a transfer function $\Phi(F)$, the filter output in the frequency domain equals $W(F)\Phi(F)$. When we deconvolve this by $h[n]$, we obtain $\hat{x}[n]$. In the frequency domain, this is equivalent to finding $\hat{X}(F)$ as the ratio

$$\hat{X}(F) = \frac{W(F)\Phi(F)}{H(F)} = \frac{\Phi(F)[Y(F) + S(F)]}{H(F)}$$

If $\hat{x}[n]$ is to be the best approximation to $x[n]$ in the **least squares sense**, we must minimize the mean squared (or integral squared) error. Since the error is $x[n] - \hat{x}[n]$, the mean square error $\epsilon$ is given by

$$\epsilon = \sum_{-\infty}^{\infty} |x[n] - \hat{x}[n]|^2 \, dt = \int_{-1/2}^{1/2} |X(F) - \hat{X}(F)|^2 \, dF$$

Substituting for $X(F)$ and $\hat{X}(F)$, we get

$$\epsilon = \int_{-1/2}^{1/2} \frac{|Y(F) - \Phi(F)[Y(F) + S(F)]|^2}{|H(F)|^2} \, dF$$

$$= \int_{-1/2}^{1/2} \frac{|Y(F)[1 - \Phi(F)] - S(F)\Phi(F)|^2}{|H(F)|^2} \, dF$$

If the noise $s[n]$ and signal $y[n]$ are uncorrelated (as they usually are), the integral of the product $Y(F)S(F)$ equals zero, and we obtain

$$\epsilon = \int_{-1/2}^{1/2} \frac{|Y(F)|^2 |1 - \Phi(F)|^2 + |S(F)|^2 |\Phi(F)|^2}{|H(F)|^2} \, dF = \int_{-1/2}^{1/2} K(F) \, dF$$

To ensure that $\epsilon$ is a minimum, the kernel $K(F)$ must be minimized with respect to $\Phi(F)$. So, we set $dK(F)/d\Phi(F) = 0$, and (assuming that $\Phi(F)$ is real) this leads to the result

$$\Phi(F) = \frac{|Y(F)|^2}{|Y(F)|^2 + |S(F)|^2}$$

This result describes a **Wiener** (or **Wiener-Hopf**) optimal filter. Note that $\Phi(F) = 1$ in the absence of noise and $\Phi(F) \approx 0$ when noise dominates. In other words, it combats the effects of noise only when required. Interestingly, the result for $\Phi(F)$ does not depend on $X(F)$ even though it requires that we obtain estimates of $|Y(F)|$ and $|S(F)|^2$ separately. In theory, this may indeed be difficult without additional information. In practice however, the power spectral density (PSD) $|W(F)|^2$ may be approximated by

$$|W(F)|^2 \approx |Y(F)|^2 + |S(F)|^2$$

In addition, the noise spectral density $|S(F)|^2$ often has a form that can be deduced from its high-frequency behaviour. Once this form is known, we estimate $|Y(F)|^2 = |W(F)|^2 - |S(F)|^2$ (by graphical extrapolation, for example). Once known, we generate the optimal-filter transfer function as

$$\Phi(F) = \frac{|Y(F)|^2}{|Y(F)|^2 + |S(F)|^2} \approx \frac{|Y(F)|^2}{|W(F)|^2}$$

The spectrum of the estimated input is then

$$\hat{X}(F) = \frac{W(F)\Phi(F)}{H(F)}$$

The inverse transform of $\hat{X}(F)$ leads to the desired time-domain signal $\hat{x}[n]$. We remark that the transfer function $\Phi(F)$ is even symmetric. In turn, the impulse response is also even symmetric about $n = 0$. This means that the optimal filter is noncausal. The design of causal optimal filters is much more involved and has led to important developments such as the Kalman filter.

In practice, the DFT (or its FFT implementation) is often used as a tool in the optimal filtering of discrete-time signals. The idea is to implement the optimal-filter transfer function and signal estimate by the approximate relations

$$\Phi[k] = \frac{|Y[k]|^2}{|Y[k]|^2 + |S[k]|^2} \approx \frac{|Y[k]|^2}{|W[k]|^2} \qquad \hat{X}[k] = \frac{W[k]\Phi[k]}{H[k]}$$

Here, $W[k]$ is the DFT of $w[n]$ which allows us to generate $|W[k]|^2$, estimate $|Y[k]|^2$, and then compute the transfer function $\Phi[k]$ of the optimal filter. We also compute $H[k]$, the DFT of $h[n]$ which leads to the spectrum of the estimated input as $\hat{X}[k] = W[k]\Phi[k]/H[k]$. The IDFT of this result recovers the signal $\hat{x}[n]$. This approach has its pitfalls however and may fail, for example, if some samples of $H[k]$ are zero.

## 8.13 Matrix Formulation of the DFT and IDFT

If we let $W_N = e^{-j2\pi/N}$, the defining relations for the DFT and IDFT may be written as

$$X_{\text{DFT}}[k] = \sum_{n=0}^{N-1} x[n]W_N^{nk}, \quad k = 0, 1, \ldots, N-1 \tag{8.46}$$

$$x[n] = \frac{1}{N}\sum_{k=0}^{N-1} X_{\text{DFT}}[k][W_N^{nk}]^*, \quad n = 0, 1, \ldots, N-1 \tag{8.47}$$

The first set of $N$ DFT equations in $N$ unknowns may be expressed in matrix form as

$$\mathbf{X} = \mathbf{W}_N\mathbf{x} \tag{8.48}$$

Here, $\mathbf{X}$ and $\mathbf{x}$ are $(N \times 1)$ matrices, and $\mathbf{W}_N$ is an $(N \times N)$ square matrix called the **DFT matrix**. The full-matrix form is described by

$$\begin{bmatrix} X[0] \\ X[1] \\ X[2] \\ \vdots \\ X[N-1] \end{bmatrix} = \begin{bmatrix} W_N^0 & W_N^0 & W_N^0 & \ldots & W^0 \\ W_N^0 & W_N^1 & W_N^2 & \ldots & W_N^{N-1} \\ W_N^0 & W_N^2 & W_N^4 & \ldots & W_N^{2(N-1)} \\ \vdots & \vdots & \vdots & \ddots & \vdots \\ W_N^0 & W_N^{N-1} & W_N^{2(N-1)} & \ldots & W_N^{(N-1)(N-1)} \end{bmatrix} \begin{bmatrix} x[0] \\ x[1] \\ x[2] \\ \vdots \\ x[N-1] \end{bmatrix} \tag{8.49}$$

The exponents $t$ in the elements $W_N^t$ of $\mathbf{W}_N$ are called **twiddle factors**.

**Example 8.18**

**The DFT from the Matrix Formulation**

Let $x[n] = \{\overset{\Downarrow}{1}, 2, 1, 0\}$. Then, with $N = 4$ and $W_N = \exp(-j2\pi/4) = -j$, the DFT may be obtained by solving the matrix product:

$$\begin{bmatrix} X[0] \\ X[1] \\ X[2] \\ X[3] \end{bmatrix} = \begin{bmatrix} W_N^0 & W_N^0 & W_N^0 & W_N^0 \\ W_N^0 & W_N^1 & W_N^2 & W_N^3 \\ W_N^0 & W_N^2 & W_N^4 & W_N^6 \\ W_N^0 & W_N^3 & W_N^6 & W_N^9 \end{bmatrix} \begin{bmatrix} x[0] \\ x[1] \\ x[2] \\ x[3] \end{bmatrix} = \begin{bmatrix} 1 & 1 & 1 & 1 \\ 1 & -j & -1 & j \\ 1 & -1 & 1 & -1 \\ 1 & j & -1 & -j \end{bmatrix} \begin{bmatrix} 1 \\ 2 \\ 1 \\ 0 \end{bmatrix} = \begin{bmatrix} 4 \\ -j2 \\ 0 \\ j2 \end{bmatrix}$$

The result is $X_{\text{DFT}}[k] = \{\overset{\Downarrow}{4}, \ -j2, \ 0, \ j2\}$.

### 8.13.1 The IDFT from the Matrix Form

The matrix $\mathbf{x}$ may be expressed in terms of the inverse of $\mathbf{W}_N$ as

$$\mathbf{x} = \mathbf{W}_N^{-1}\mathbf{X} \tag{8.50}$$

The matrix $\mathbf{W}_N^{-1}$ is called the **IDFT matrix**. We may also obtain $\mathbf{x}$ directly from the IDFT relation in matrix form, where the change of index from $n$ to $k$ and the change in the sign of the exponent in $\exp(j2\pi nk/N)$ lead to a **conjugate transpose** of $\mathbf{W}_N$. We then have

$$\mathbf{x} = \frac{1}{N}[\mathbf{W}_N^*]^T\mathbf{X} \tag{8.51}$$

Comparison of the two forms suggests that

$$\mathbf{W}_N^{-1} = \frac{1}{N}[\mathbf{W}_N^*]^T \tag{8.52}$$

This very important result shows that $\mathbf{W}_N^{-1}$ requires only conjugation and transposition of $\mathbf{W}_N$—an obvious computational advantage.

The elements of the DFT and IDFT matrices satisfy $A_{ij} = A^{(i-1)(j-1)}$. Such matrices are known as **Vandermonde matrices**. They are notoriously ill conditioned insofar as their numerical inversion. This is not the case for $\mathbf{W}_N$, however. The product of the DFT matrix $\mathbf{W}_N$ with its conjugate transpose matrix equals the identity matrix $\mathbf{I}$. Matrices that satisfy such a property are called **unitary**. For this reason, the DFT and IDFT, which are based on unitary operators, are also called **unitary transforms**.

### 8.13.2 Using the DFT to Find the IDFT

Both the DFT and IDFT are matrix operations, and there is an inherent symmetry in the DFT and IDFT relations. In fact, we can obtain the IDFT by finding the DFT of the conjugate sequence and then conjugating the results and dividing by $N$. Mathematically,

$$x[n] = \text{IDFT}\{X_{\text{DFT}}[k]\} = \frac{1}{N}(\text{DFT}\{X_{\text{DFT}}^*[k]\})^* \tag{8.53}$$

This result invokes the conjugate symmetry and duality of the DFT and IDFT, and suggests that the DFT algorithm itself can also be used to find the IDFT. In practice, this is indeed what is done.

**Example 8.19**

### Using the DFT to Find the IDFT

Let us find the IDFT of $X_{\text{DFT}}[k] = \{\overset{\Downarrow}{4}, -j2, 0, j2\}$ using the DFT. First, we conjugate the sequence to get $X^*_{\text{DFT}}[k] = \{\overset{\Downarrow}{4}, j2, 0, -j2\}$. Next, we find the DFT of $X^*_{\text{DFT}}[k]$ using the $4 \times 4$ DFT matrix of the previous example, to give

$$\text{DFT}\{X^*_{\text{DFT}}[k]\} = \begin{bmatrix} 1 & 1 & 1 & 1 \\ 1 & -j & -1 & j \\ 1 & -1 & 1 & -1 \\ 1 & j & -1 & -j \end{bmatrix} \begin{bmatrix} 4 \\ -j2 \\ 0 \\ j2 \end{bmatrix} = \begin{bmatrix} 4 \\ 8 \\ 4 \\ 0 \end{bmatrix}$$

Finally, we conjugate this result (if complex) and divide by $N = 4$ to get the IDFT of $X_{\text{DFT}}[k]$ as

$$\text{IDFT}\,\{X_{\text{DFT}}[k]\} = \tfrac{1}{4}\{\overset{\Downarrow}{4}, 8, 4, 0\} = \{\overset{\Downarrow}{1}, 2, 1, 0\}$$

# 8.14 The FFT

The importance of the DFT stems from the fact that it is amenable to fast and efficient computation using algorithms called **fast Fourier transform** (FFT) algorithms. Fast algorithms reduce the problem of calculating an $N$-point DFT to that of calculating many smaller-sized DFTs. The key ideas in optimizing the computation are based on the following ideas.

## Symmetry and Periodicity

All FFT algorithms take advantage of the symmetry and periodicity of the exponential $W_N = e^{-j2\pi n/N}$, as listed in Table 8.5. The last entry in this table, for example, suggests that $W_{N/2} = e^{j4\pi n/N}$ is periodic with period $\frac{N}{2}$.

## Choice of Signal Length

We choose the signal length $N$ as a number that is the product of many smaller numbers $r_k$ such that $N = r_1 r_2 \ldots r_m$. A more useful choice results when the factors are equal, such that $N = r^m$. The factor $r$ is called the **radix**. By far the most practically implemented choice for $r$ is 2, such that $N = 2^m$ and leads to the **radix-2** FFT algorithms.

**TABLE 8.5** Symmetry and Periodicity of $W_N = \exp(-j2\pi/N)$

| Entry | Exponential Form | Symbolic Form |
|---|---|---|
| 1 | $e^{-j2\pi n/N} = e^{-j2\pi(n+N)/N}$ | $W_N^{n+N} = W_N^n$ |
| 2 | $e^{-j2\pi(n+N/2)/N} = -e^{-j2\pi n/N}$ | $W_N^{n+N/2} = -W_N^n$ |
| 3 | $e^{-j2\pi K} = e^{-j2\pi NK/N} = 1$ | $W_N^{NK} = 1$ |
| 4 | $e^{-j2(2\pi/N)} = e^{-j2\pi/(N/2)}$ | $W_N^2 = W_{N/2}$ |

### Index Separation and Storage

The computation is carried out separately on even-indexed and odd-indexed samples to reduce the computational effort. All algorithms allocate storage for computed results. The less the storage required, the more efficient is the algorithm. Many FFT algorithms reduce storage requirements by performing computations *in place* by storing results in the same memory locations that previously held the data.

## 8.14.1 Some Fundamental Results

We begin by considering two trivial, but extremely important, results.

**1-Point Transform:** The DFT of a single number $A$ is the number $A$ itself.

**2-Point Transform:** The DFT of a 2-point sequence is easily found to be

$$X_{\text{DFT}}[0] = x[0] + x[1] \quad \text{and} \quad X_{\text{DFT}}[1] = x[0] - x[1] \tag{8.54}$$

The single most important result in the development of a radix-2 FFT algorithm is that an $N$-sample DFT can be written as the sum of two $\frac{N}{2}$-sample DFTs formed from the even- and odd-indexed samples of the original sequence. Here is the development:

$$X_{\text{DFT}}[k] = \sum_{n=0}^{N-1} x[n]W_N^{nk} = \sum_{n=0}^{N/2-1} x[2n]W_N^{2nk} + \sum_{n=0}^{N/2-1} x[2n+1]W_N^{(2n+1)k}$$

$$X_{\text{DFT}}[k] = \sum_{n=0}^{N/2-1} x[2n]W_N^{2nk} + W_N^k \sum_{n=0}^{N/2-1} x[2n+1]W_N^{2nk}$$

$$= \sum_{n=0}^{N/2-1} x[2n]W_{N/2}^{nk} + W_N \sum_{n=0}^{N/2-1} x[2n+1]W_{N/2}^{nk}$$

If $X^e[k]$ and $X^o[k]$ denote the DFT of the even- and odd-indexed sequences of length $N/2$, we can rewrite this result as

$$X_{\text{DFT}}[k] = X^e[k] + W_N^k X^o[k], \quad k = 0, 1, 2, \ldots, N-1 \tag{8.55}$$

Note carefully that the index $k$ in this expression varies from 0 to $N-1$ and that $X^e[k]$ and $X^o[k]$ are both periodic in $k$ with period $N/2$; we thus have two periods of each to yield $X_{\text{DFT}}[k]$. Due to periodicity, we can split $X_{\text{DFT}}[k]$ and compute the first half and next half of the values as

$$X_{\text{DFT}}[k] = X^e[k] + W_N^k X^o[k], \quad k = 0, 1, 2, \ldots, \frac{1}{2}N - 1$$

$$X_{\text{DFT}}\left[k + \frac{N}{2}\right] = X^e\left[k + \frac{1}{2}N\right] + W_N^{k+\frac{N}{2}} X^o\left[k + \frac{N}{2}\right]$$

$$= X^e[k] - W_N^k X^o[k], \quad k = 0, 1, 2, \ldots, \frac{N}{2} - 1$$

This result is known as the **Danielson-Lanczos lemma**. Its *signal-flow graph* is shown in Figure 8.16 and is called a **butterfly** due to its characteristic shape.

The inputs $X^e$ and $X^o$ are transformed into $X^e + W_N^k X^o$ and $X^e - W_N^k X^o$. A butterfly operates on one pair of samples and involves *two complex additions and one complex multiplication*. For $N$ samples, there are $N/2$ butterflies in all. Starting with $N$ samples, this lemma reduces the computational complexity by evaluating the DFT of two $\frac{N}{2}$-point

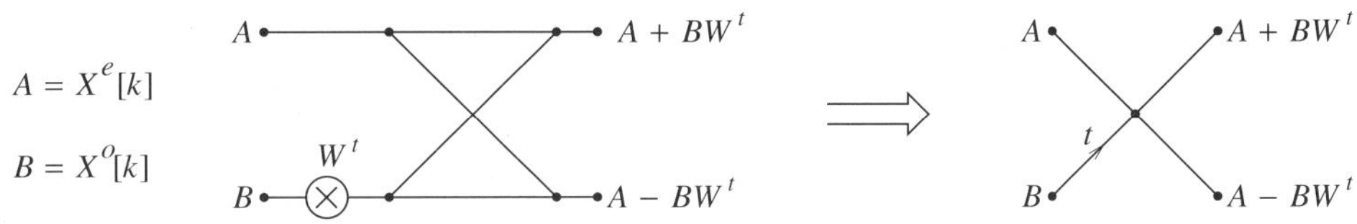


**FIGURE 8.16** A typical butterfly

sequences. The DFT of each of these can once again be reduced to the computation of sequences of length $N/4$ to yield

$$X^e[k] = X^{ee}[k] + W^k_{N/2}X^{eo}[k] \qquad X^o[k] = X^{oe}[k] + W^k_{N/2}X^{oo}[k] \tag{8.56}$$

Since $W^k_{N/2} = W^{2k}_N$, we can rewrite this expression as

$$X^e[k] = X^{ee}[k] + W^{2k}_N X^{eo}[k] \qquad X^o[k] = X^{oe}[k] + W^{2k}_N X^{oo}[k] \tag{8.57}$$

Carrying this process to its logical extreme, if we choose the number $N$ of samples as $N = 2^m$, we can reduce the computation of an $N$-point DFT to the computation of 1-point DFTs in $m$ stages. And the 1-point DFT is just the sample value itself (repeated with period 1). This process is called *decimation*. The FFT results so obtained are actually in **bit-reversed order**. If we let $e = 0$ and $o = 1$ and then reverse the order, we have the sample number in binary representation. The reason is that splitting the sequence into even and odd indices is equivalent to testing each index for the least significant bit (0 for even, 1 for odd). We describe two common in-place FFT algorithms based on decimation. A summary appears in Table 8.6.

### 8.14.2 The Decimation-in-Frequency FFT Algorithm

The decimation-in-frequency (DIF) FFT algorithm keeps reducing an $N$-point transform at each successive stage to two $\frac{N}{2}$-point transforms, then four $\frac{N}{4}$-point transforms, and so on, until we arrive at $N$ 1-point transforms that correspond to the actual DFT. With the

**TABLE 8.6** ➤ FFT Algorithms for Computing the DFT

| Entry | Characteristic | Decimation in Frequency | Decimation in Time |
|---|---|---|---|
| 1 | Number of samples | $N = 2^m$ | $N = 2^m$ |
| 2 | Input sequence | Natural order | Bit-reversed order |
| 3 | DFT result | Bit-reversed order | Natural order |
| 4 | Computations | In place | In place |
| 5 | Number of stages | $m = \log_2 N$ | $m = \log_2 N$ |
| 6 | Multiplications | $\frac{N}{2}\log_2 N$ (complex) | $\frac{N}{2}\log_2 N$(complex) |
| 7 | Additions | $N\log_2 N$ (complex) | $N\log_2 N$ (complex) |
| | | **Structure of the $i$th Stage** | |
| 8 | Number of butterflies | $\frac{N}{2}$ | $\frac{N}{2}$ |
| 9 | Butterfly input | A (top) and B (bottom) | A (top) and B (bottom) |
| 10 | Butterfly output | $(A + B)$ and $(A - B)W^t_N$ | $(A + BW^t_N)$ and $(A - BW^t_N)$ |
| 11 | Twiddle factors $t$ | $2^{i-1}Q,\ Q = 0, 1, \ldots, P-1$ | $2^{m-i}Q,\ Q = 0, 1, \ldots, P-1$ |
| 12 | Values of $P$ | $P = 2^{m-i}$ | $P = 2^{i-1}$ |

input sequence in natural order, computations can be done in place, but the DFT result is in bit-reversed order and must be reordered.

The algorithm slices the input sequence $x[n]$ into two halves and leads to

$$X_{\mathrm{DFT}}[k] = \sum_{n=0}^{N-1} x[n]W_N^{nk} = \sum_{n=0}^{N/2-1} x[n]W_N^{nk} + \sum_{n=N/2}^{N/2-1} x[n]W_N^{nk}$$

$$= \sum_{n=0}^{N/2-1} x[n]W_N^{nk} + \sum_{n=0}^{N/2-1} x[n+N/2]W_N^{(n+\frac{N}{2})k}$$

This may be rewritten as

$$X_{\mathrm{DFT}}[k] = \sum_{n=0}^{N/2-1} x[n]W_N^{nk} + W_N^{Nk/2}\sum_{n=0}^{N/2-1} x[n+\frac{N}{2}]W_N^{nk}$$

$$= \sum_{n=0}^{N/2-1} x[n]W_N^{nk} + (-1)^k\sum_{n=0}^{N/2-1} x[n+\frac{N}{2}]W_N^{nk}$$

Separating even and odd indices, and letting $x[n] = x^a$ and $x[n+N/2] = x^b$,

$$X_{\mathrm{DFT}}[2k] = \sum_{n=0}^{N/2-1} [x^a + x^b]W_N^{2nk}, \quad k = 0, 1, 2, \ldots, \frac{N}{2}-1 \tag{8.58}$$

$$X_{\mathrm{DFT}}[2k+1] = \sum_{n=0}^{N/2-1} [x^a - x^b]W_N^{(2k+1)n}$$

$$= \sum_{n=0}^{N/2-1} [x^a - x^b]W_N^{n}W_N^{2nk}, \quad k = 0, 1, \ldots, \frac{N}{2}-1 \tag{8.59}$$

Since $W_N^{2nk} = W_{N/2}^{nk}$, the even-indexed and odd-indexed terms describe a $\frac{N}{2}$-point DFT. The computations result in a butterfly structure with inputs $x[n]$ and $x[n+\frac{N}{2}]$ and outputs $X_{\mathrm{DFT}}[2k] = \{x[n] + x[n+\frac{N}{2}]\}$ and $X_{\mathrm{DFT}}[2k+1] = \{x[n] - x[n+\frac{N}{2}]\}W_N^n$. Its butterfly structure is shown in Figure 8.17.

The factors $W^t$, called **twiddle factors**, appear only in the lower corners of the butterfly wings at each stage. Their exponents $t$ have a definite order, described as follows for an $(N = 2^m)$-point FFT algorithm with $m$ stages:

1. The number $P$ of distinct twiddle factors $W^t$ at the $i$th stage is $P = 2^{m-i}$.
2. The values of $t$ in the twiddle factors $W^t$ are $t = 2^{i-1}Q$, $Q = 0, 1, 2, \ldots, P-1$.

The DIF algorithm is illustrated in Figure 8.18 for $N = 2$, $N = 4$, and $N = 8$.

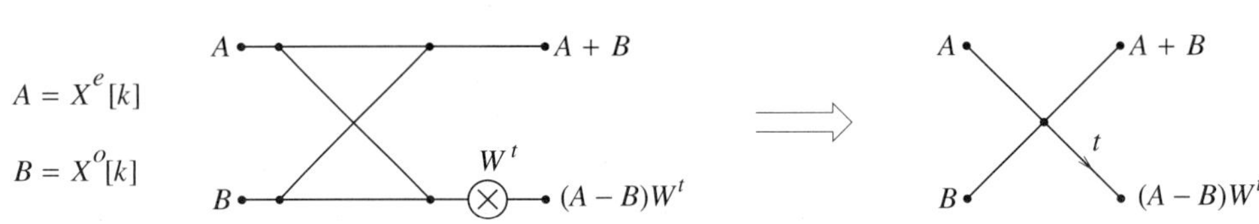


**FIGURE 8.17** A typical butterfly for the decimation-in-frequency FFT algorithm

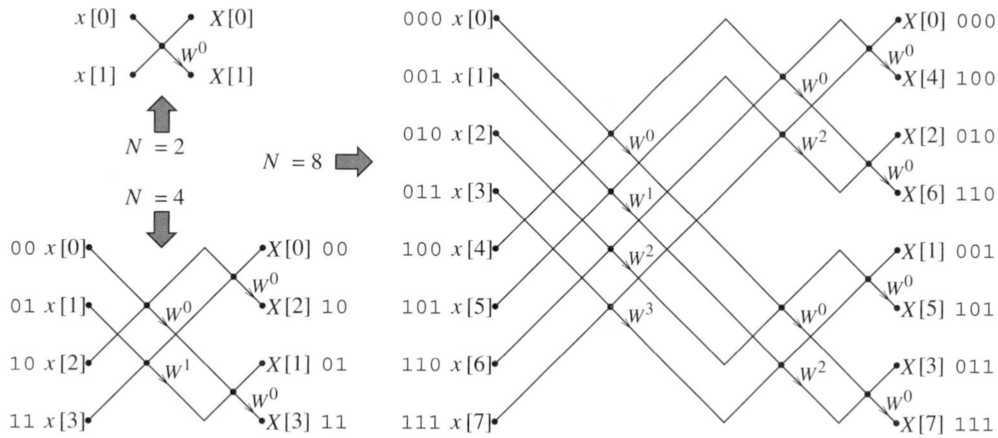


**FIGURE 8.18** The decimation-in-frequency FFT algorithm for $N = 2, 4, 8$

**Example 8.20** **A 4-Point Decimation-in-Frequency FFT Algorithm**

For a 4-point DFT, we use the above equations to obtain

$$X_{\text{DFT}}[2k] = \sum_{n=0}^{1}\{x[n] + x[n+2]\}W_4^{2nk}$$

$$X_{\text{DFT}}[2k+1] = \sum_{n=0}^{1}\{x[n] - x[n+2]\}W_4^{n}W_4^{2nk}, \qquad k = 0, 1$$

Since $W_4^0 = 1$ and $W_4^2 = -1$, we arrive at the following result:

$$X_{\text{DFT}}[0] = x[0] + x[2] + x[1] + x[3] \qquad X_{\text{DFT}}[2] = x[0] + x[2] - \{x[1] + x[3]\}$$
$$X_{\text{DFT}}[1] = x[0] - x[2] + W_4\{x[1] - x[3]\} \quad X_{\text{DFT}}[3] = x[0] - x[2] - W_4\{x[1] - x[3]\}$$

We do not reorder the input sequence before using it.

### 8.14.3 The Decimation-in-Time FFT Algorithm

In the decimation-in-time (DIT) FFT algorithm, we start with $N$ 1-point transforms, combine adjacent pairs at each successive stage into 2-point transforms, then 4-point transforms, and so on, until we get a single $N$-point DFT result. With the input sequence in bit-reversed order, the computations can be done in place, and the DFT is obtained in natural order. Thus, for a 4-point input, the binary indices {00, 01, 10, 11} reverse to {00, 10, 01, 11}, and we use the bit-reversed order $\{x[0], x[2], x[1], x[3]\}$.

For an 8-point input sequence, {000, 001, 010, 011, 100, 101, 110, 111}, the reversed sequence corresponds to {000, 100, 010, 110, 001, 101, 011, 111} or $\{x[0], x[4], x[6], x[2], x[1], x[5], x[3], x[7]\}$, and we use this sequence to perform the computations.

At a typical stage, we obtain

$$X_{\text{DFT}}[k] = X^e[k] + W_N^k X^o[k] \qquad X_{\text{DFT}}\left[k + \frac{N}{2}\right] = X^e[k] - W_N^k X^o[k] \tag{8.60}$$

Its butterfly structure is shown in Figure 8.19.

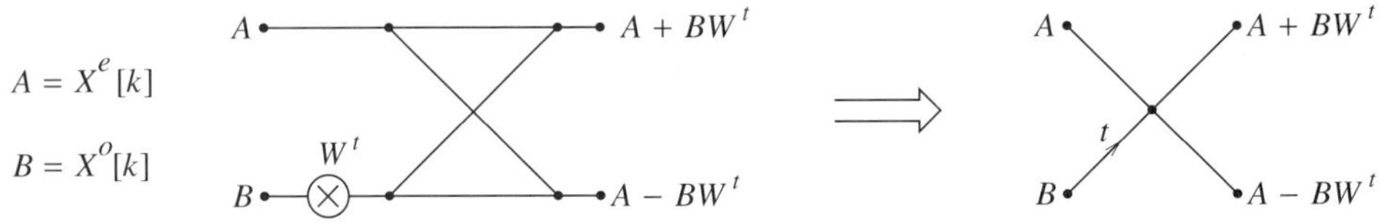


**FIGURE 8.19** A typical butterfly for the decimation-in-time FFT algorithm

As with the decimation-in-frequency algorithm, the twiddle factors $W^t$ at each stage appear only in the bottom wing of each butterfly. The exponents $t$ also have a definite (and almost similar) order described by

1. The number $P$ of distinct twiddle factors $W^t$ at the $i$th stage is $P = 2^{i-1}$.
2. The values of $t$ in the twiddle factors $W^t$ are $t = 2^{m-i}Q$, $Q = 0, 1, 2, \ldots, P-1$.

The DIT algorithm is illustrated in Figure 8.20 for $N = 2$, $N = 4$, and $N = 8$.

In both the DIF algorithm and DIT algorithm, it is possible to use a sequence in natural order and get DFT results in natural order. This, however, requires more storage, since the computations cannot now be done in place.

**Example 8.21**

### A 4-Point Decimation-in-Time FFT Algorithm

For a 4-point DFT, with $W_4 = e^{-j\pi/2} = -j$, we have

$$X_{\text{DFT}}[k] = \sum_{n=0}^{3} x[n]W_4^{nk}, \qquad k = 0, 1, 2, 3$$

We group by even-indexed and odd-indexed samples of $x[n]$ to obtain

$$X_{\text{DFT}}[k] = X^e[k] + W_4^k X^o[k] \qquad \begin{cases} X^e[k] = x[0] + x[2]W_4^{2k}, \\ X^o[k] = x[1] + x[3]W_4^{2k}, \end{cases} \qquad k = 0, 1, 2, 3$$

Using periodicity, we simplify this result to

$$\begin{aligned} X_{\text{DFT}}[k] &= X^e[k] + W_4^k X^o[k] \\ X_{\text{DFT}}[k + \tfrac{1}{2}N] &= X^e[k] - W_4^k X^o[k] \end{aligned} \qquad \begin{cases} X^e[k] = x[0] + x[2]W_4^{2k}, \\ X^o[k] = x[1] + x[3]W_4^{2k}, \end{cases} \qquad k = 0, 1$$

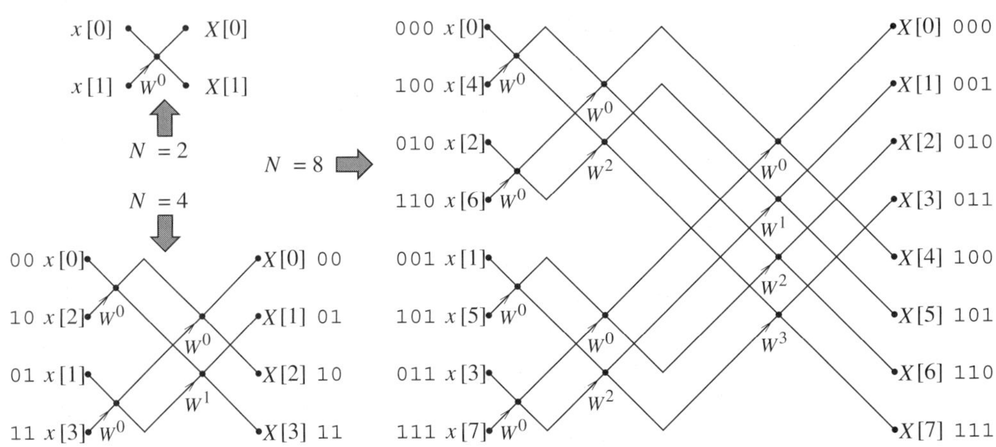


**FIGURE 8.20** The decimation-in-time FFT algorithm for $N = 2, 4, 8$

These equations yield $X_{\mathrm{DFT}}[0]$ through $X_{\mathrm{DFT}}[3]$ as

$$X_{\mathrm{DFT}}[0] = X^e[0] + W_4^0 X^o[0] = x[0] + x[2]W_4^0 + W_4^0\{x[1] + x[3]W_4^0\}$$

$$X_{\mathrm{DFT}}[1] = X^e[1] + W_4^1 X^o[1] = x[0] + x[2]W_4^2 + W_4^1\{x[1] + x[3]W_4^2\}$$

$$X_{\mathrm{DFT}}[2] = X^e[0] - W_4^0 X^o[0] = x[0] + x[2]W_4^0 - W_4^0\{x[1] + x[3]W_4^0\}$$

$$X_{\mathrm{DFT}}[3] = X^e[1] - W_4^1 X^o[1] = x[0] + x[2]W_4^2 - W_4^1\{x[1] + x[3]W_4^2\}$$

Upon simplification, using $W_4^0 = 1$, $W_4^1 = W_4 = 1$, and $W_4^2 = -1$, we obtain

$$X_{\mathrm{DFT}}[0] = x[0] + x[2] + x[1] + x[3] \qquad X_{\mathrm{DFT}}[1] = x[0] - x[2] + W_4\{x[1] - x[3]\}$$

$$X_{\mathrm{DFT}}[2] = x[0] + x[2] - \{x[1] + x[3]\} \qquad X_{\mathrm{DFT}}[3] = x[0] - x[2] - W_4\{x[1] - x[3]\}$$

### 8.14.4 Computational Cost

Both the DIF and DIT $N$-point FFT algorithms involve $m = \log_2 N$ stages and $0.5N$ butterflies per stage. The FFT computation thus requires $0.5mN = 0.5N \log_2 N$ complex multiplications and $mN = N \log_2 N$ complex additions. Table 8.7 shows how the FFT stacks up against direct DFT evaluation for $N = 2^m$, counting all operations. Note, however, that multiplications take the bulk of computing time.

The difference between $0.5 \log_2 N$ and $N^2$ may not seem like much for small $N$. For example, with $N = 16$, $0.5N \log_2 N = 64$ and $N^2 = 256$. For large $N$, however, the difference is phenomenal. For $N = 1024 = 2^{10}$, $0.5N \log_2 N \approx 5000$ and $N^2 \approx 10^6$. This is like waiting 1 minute for the FFT result and (more than) 3 hours for the identical direct DFT result. Note that $N \log_2 N$ is nearly linear with $N$ for large $N$, whereas $N^2$ shows a much faster, quadratic growth.

In the DFT relations, since there are $N$ factors that equal $W^0$ or 1, we require only $N^2 - N$ complex multiplications. In the FFT algorithms, the number of factors that equal 1 doubles (or halves) at each stage and is given by $1+2+2^2+\cdots+2^{m-1} = 2^m - 1 = N - 1$. We thus actually require only $0.5N \log_2 N - (N - 1)$ complex multiplications for the FFT. The DFT requires $N^2$ values of $W^k$, but the FFT requires at most $N$ such values at each stage. Due to the periodicity of $W^k$, only about $\frac{3}{4}N$ of these values are distinct. Once computed, they can be stored and used again. However, this hardly affects the comparison for large $N$. Since computers use real arithmetic, the number of real operations may be found by noting that one complex addition involves two real additions, and one complex multiplication involves four real multiplications and three real additions, because $(A + jB)(C + jD) = AC - BD + jBC + jAD$.

**TABLE 8.7 ➤** Computational Cost of the DFT and FFT

| Feature | $N$-Point DFT | $N$-Point FFT |
|---|---|---|
| Algorithm | Solution of $N$ equations in $N$ unknowns | $0.5N$ butterflies/stage, $m$ stages Total butterflies $= 0.5mN$ |
| Multiplications | $N$ per equation | 1 per butterfly |
| Additions | $N - 1$ per equation | 2 per butterfly |
| Total multiplications | $N^2$ | $0.5mN = 0.5N \log_2 N$ |
| Total additions | $N(N - 1)$ | $mN = N \log_2 N$ |

### Speed of Fast Convolution

A direct computation of the convolution of two $N$-sample signals requires $N^2$ complex multiplications. The FFT method works with sequences of length $2N$. The number of complex multiplications involved is $2N \log_2 2N$ to find the FFT of the two sequences, $2N$ to form the product sequence, and $N \log_2 N$ to find the IFFT sequence that gives the convolution. It thus requires $3N \log_2 2N + 2N$ complex multiplications. If $N = 2^m$, the FFT approach becomes computationally superior only for $m > 5$ $(N > 32)$ or so.

## 8.15 Why Equal Lengths for the DFT and IDFT?

The DTFT of an $N$-sample discrete time signal $x[n]$ is given by

$$X_p(F) = \sum_{n=0}^{N-1} x[n]e^{-j2\pi nF} \tag{8.61}$$

If we sample the digital frequency $F$ at $M$ intervals over one period, the frequency interval $F_0$ equals $1/M$ and $F \to kF_0 = \frac{k}{M}$, $k = 0, 1, \ldots, M-1$, and we get

$$X_{\text{DFT}}[k] = \sum_{n=0}^{N-1} x[n]e^{-j2\pi nk/M}, \quad k = 0, 1, \ldots, M-1 \tag{8.62}$$

With $W_M = e^{-j2\pi/M}$, this equation can be written as

$$X_{\text{DFT}}[k] = \sum_{n=0}^{N-1} x[n]W_M^{nk}, \quad k = 0, 1, \ldots, M-1 \tag{8.63}$$

This is a set of $M$ *equations in* $N$ *unknowns* and describes *the M-point DFT of the N-sample sequence* $x[n]$. It may be written in matrix form as

$$\mathbf{X} = \mathbf{W}_M\mathbf{x} \tag{8.64}$$

Here, $\mathbf{X}$ is an $M \times 1$ matrix, $\mathbf{x}$ is an $N \times 1$ matrix, and $\mathbf{W}_M$ is an $(M \times N)$ matrix. In full form,

$$\begin{bmatrix} X[0] \\ X[1] \\ X[2] \\ \vdots \\ X[M-1] \end{bmatrix} = \begin{bmatrix} W_M^0 & W_M^0 & W_M^0 & \cdots & W_M^0 \\ W_M^0 & W_M^1 & W_M^2 & \cdots & W_M^{N-1} \\ W_M^0 & W_M^2 & W_M^4 & \cdots & W_M^{2(N-1)} \\ \vdots & \vdots & \vdots & \ddots & \vdots \\ W_M^0 & W_M^{M-1} & W_M^{2(M-1)} & \cdots & W_M^{(N-1)(M-1)} \end{bmatrix} \begin{bmatrix} x[0] \\ x[1] \\ x[2] \\ \vdots \\ x[N-1] \end{bmatrix} \tag{8.65}$$

**Example 8.22** **A 4-Point DFT from a 3-Point Sequence**

Let $x[n] = \{\overset{\Downarrow}{1}, 2, 1\}$. We have $N = 3$. The DTFT of this signal is

$$X_p(F) = 1 + 2e^{-j2\pi F} + e^{-j4\pi F} = [2 + 2\cos(2\pi F)]e^{-j2\pi F}$$

If we pick $M = 4$, we have $F = k/4$, $k = 0, 1, 2, 3$, and obtain the DFT as

$$X_{\rm DFT}[k] = [2+2\cos(2\pi k/4)]e^{-j2\pi k/4}, \quad k = 0, 1, 2, 3 \quad \text{or} \quad X_{\rm DFT}[k] = \{\overset{\Downarrow}{4}, -j2, 0, j2\}$$

Using matrix notation with $W_M = e^{-j2\pi/4} = -j$, we can also find $X_{\rm DFT}[k]$ as

$$\begin{bmatrix} X[0] \\ X[1] \\ X[2] \\ X[3] \end{bmatrix} = \begin{bmatrix} W_M^0 & W_M^0 & W_M^0 \\ W_M^0 & W_M^1 & W_M^2 \\ W_M^0 & W_M^2 & W_M^4 \\ W_M^0 & W_M^3 & W_M^6 \end{bmatrix} \begin{bmatrix} x[0] \\ x[1] \\ x[2] \end{bmatrix} = \begin{bmatrix} 1 & 1 & 1 \\ 1 & -j & -1 \\ 1 & -1 & 1 \\ 1 & j & -1 \end{bmatrix} \begin{bmatrix} 1 \\ 2 \\ 3 \end{bmatrix} = \begin{bmatrix} 4 \\ -j2 \\ 0 \\ j2 \end{bmatrix}$$

Thus, $X_{\rm DFT}[k] = \{\overset{\Downarrow}{4}, -j2, 0, j2\}$, as before.

### 8.15.1 The Inverse DFT

How do we obtain the $N$-sample sequence $x[n]$ from the $M$-sample DFT? It would seem that we require the product of $\mathbf{X}$, an $M \times 1$ matrix with an $N \times M$ matrix to give $\mathbf{x}$ as an $N \times 1$ matrix. But what is this $(M \times N)$ matrix, and how is it related to the $M \times N$ matrix $\mathbf{W}_M$? To find out, recall that

$$x[n] = \int_0^1 X(F)e^{j2\pi nF}\,dF \tag{8.66}$$

Converting this to discrete form with $F = kF_0 = \frac{k}{M}$ results in periodicity of $x[n]$ with period 1, and we obtain $N$ samples of $x[n]$ using

$$x[n] = \frac{1}{M}\sum_{k=0}^{M-1} X_{\rm DFT}[k]e^{j2\pi nk/M}, \quad n = 0, 1, \ldots, N-1 \tag{8.67}$$

For $N < M$, one period of $x[n]$ is a zero-padded $M$-sample sequence. For $N > M$, however, one period of $x[n]$ is the *periodic extension* of the $N$-sample sequence with period $M$.

The sign of the exponent and the interchange of the indices $n$ and $k$ allow us to set up the matrix formulation for obtaining $x[n]$ using an $N \times M$ inversion matrix $\mathbf{W}_I$ that just equals $\frac{1}{M}$ times $[\mathbf{W}_M^*]^T$, which is the **conjugate transpose** of the $M \times N$ DFT matrix $\mathbf{W}_M$. Its product with the $N \times 1$ matrix corresponding to $X[k]$ yields the $M \times 1$ matrix for $x[n]$. We thus have the forward and inverse matrix relations:

$$\mathbf{X} = \mathbf{W}_M\mathbf{x} \quad \text{(DFT)} \qquad \mathbf{x} = \mathbf{W}_I\mathbf{X} = \frac{1}{M}[\mathbf{W}_M^*]^T\mathbf{X} \quad \text{(IDFT)} \tag{8.68}$$

These results are valid for any choice of $M$ and $N$. An interesting result is that the product of $\mathbf{W}_M$ with $\mathbf{W}_I$ is the $N \times N$ identity matrix.

**Example 8.23** **A 3-Point IDFT from a 4-Point DFT**

Let $X_{\rm DFT}[k] = \{\overset{\Downarrow}{4}, -j2, 0, j2\}$ and $M = 4$.

The IDFT matrix equals

$$\mathbf{W}_I = \frac{1}{4}\begin{bmatrix} 1 & 1 & 1 \\ 1 & j & -1 \\ 1 & -1 & 1 \\ 1 & -j & -1 \end{bmatrix}^{*T} = \frac{1}{4}\begin{bmatrix} 1 & 1 & 1 & 1 \\ 1 & -j & -1 & j \\ 1 & -1 & 1 & -1 \end{bmatrix}$$

We then get the IDFT as

$$\mathbf{x} = \mathbf{W}_I\mathbf{X} = \frac{1}{4}\begin{bmatrix} 1 & 1 & 1 & 1 \\ 1 & -j & -1 & j \\ 1 & -1 & 1 & -1 \end{bmatrix}\begin{bmatrix} 4 \\ -j2 \\ 0 \\ j2 \end{bmatrix} = \begin{bmatrix} 1 \\ 2 \\ 1 \end{bmatrix}$$

The important thing to realize is that $x[n]$ is actually periodic with $M = 4$, and one period of $x[n]$ is the zero-padded sequence $\{\overset{\Downarrow}{1}, 2, 1, 0\}$.

### 8.15.2 How Unequal Lengths Affect the DFT Results

Even though the $M$-point IDFT of an $N$-point sequence is valid for any $M$, the choice of $M$ affects the nature of $x[n]$ through the IDFT and its inherent periodic extension.

1. If $M = N$, the IDFT is periodic with period $M$, and its one period equals the $N$-sample $x[n]$. Both the DFT matrix and IDFT matrix are square ($M \times M$) and allow a simple inversion relation to go back and forth between the two.
2. If $M > N$, the IDFT is periodic with period $M$. Its one period is the original $N$-sample $x[n]$ with $M - N$ padded zeros. The choice $M > N$ is equivalent to using a zero-padded version of $x[n]$ with a total of $M$ samples and $M \times M$ square matrices for both the DFT matrix and IDFT matrix.
3. If $M < N$, the IDFT is periodic with period $M < N$. Its one period is the periodic extension of the $N$-sample $x[n]$ with period $M$. It thus yields a signal that corresponds to $x[n]$ *wrapped around* after $M$ samples and does not recover the original $x[n]$.

**Example 8.24** **The Importance of Periodic Extension**

Let $x[n] = \{\overset{\Downarrow}{1}, 2, 1\}$. We have $N = 3$. The DTFT of this signal is

$$X_p(F) = 1 + 2e^{-j2\pi F} + e^{-j4\pi F} = [2 + 2\cos(2\pi F)]e^{-j2\pi F}$$

We sample $X_p(F)$ at $M$ intervals and find the IDFT as $y[n]$. What do we get?

**(a)** For $M = 3$, we should get $y[n] = \{\overset{\Downarrow}{1}, 2, 1\} = x[n]$. Let us find out. With $M = 3$, we have $F = k/3$ for $k = 0, 1, 2$, and $X_{\text{DFT}}[k]$ becomes

$$X_{\text{DFT}}[k] = [2 + 2\cos(2k\pi/3)]e^{-j2k\pi/3} = \left\{\overset{\Downarrow}{4}, -\frac{1}{2} - j\sqrt{\frac{3}{4}}, -\frac{1}{2} + j\sqrt{\frac{3}{4}}\right\}$$

We also find $\mathbf{W}_I$ and the IDFT $\mathbf{x} = \mathbf{W}_I\mathbf{X}$ as

$$\mathbf{W}_I = \frac{1}{3}\begin{bmatrix} W_M^0 & W_M^0 & W_M^0 \\ W_M^0 & W_M^1 & W_M^2 \\ W_M^0 & W_M^2 & W_M^4 \end{bmatrix}^{*T} = \frac{1}{3}\begin{bmatrix} 1 & 1 & 1 \\ 1 & -\frac{1}{2}+j\sqrt{\frac{3}{4}} & -\frac{1}{2}-j\sqrt{\frac{3}{4}} \\ 1 & -\frac{1}{2}-j\sqrt{\frac{3}{4}} & -\frac{1}{2}+j\sqrt{\frac{3}{4}} \end{bmatrix}$$

$$\mathbf{x} = \mathbf{W}_I\mathbf{X} = \frac{1}{3}\begin{bmatrix} 1 & 1 & 1 \\ 1 & -\frac{1}{2}+j\sqrt{\frac{3}{4}} & -\frac{1}{2}-j\sqrt{\frac{3}{4}} \\ 1 & -\frac{1}{2}-j\sqrt{\frac{3}{4}} & -\frac{1}{2}+j\sqrt{\frac{3}{4}} \end{bmatrix}\begin{bmatrix} 4 \\ -\frac{1}{2}-j\sqrt{\frac{3}{4}} \\ -\frac{1}{2}+j\sqrt{\frac{3}{4}} \end{bmatrix} = \begin{bmatrix} 1 \\ 2 \\ 1 \end{bmatrix}$$

This result is periodic with $M = 3$, and one period of this equals $x[n]$.

**(b)** For $M = 4$, we should get a new sequence $y[n] = \{\overset{\Downarrow}{1},\ 2,\ 1,\ 0\}$ that corresponds to a zero-padded version of $x[n]$.

**(c)** For $M = 2$, we should get a new sequence $z[n] = \{\overset{\Downarrow}{2},\ 2\}$ that corresponds to the periodic extension of $x[n]$ with period 2.

With $M = 2$ and $k = 0, 1$, we have $Z_{\text{DFT}}[k] = [2 + 2\cos(\pi k)]e^{-jk\pi} = \{\overset{\Downarrow}{4}, 0\}$. Since $e^{-j2\pi/M} = e^{-j\pi} = -1$, we can find the IDFT $z[n]$ directly from the definition as

$$z[0] = 0.5\{X_{\text{DFT}}[0] + X_{\text{DFT}}[1]\} = 2 \qquad z[1] = 0.5\{X_{\text{DFT}}[0] - X_{\text{DFT}}[1]\} = 2$$

The sequence $z[n] = \{\overset{\Downarrow}{2}, 2\}$ is periodic with $M = 2$. As expected, this equals one period of the periodic extension of $x[n] = \{\overset{\Downarrow}{1}, 2, 1\}$ (with wraparound past two samples).

# 8.16 Problems

**8.1.** **(DFT from Definition)** Compute the DFT and DFS of the following signals.

**(a)** $x[n] = \{1, 2, 1, 2\}$ **(b)** $x[n] = \{2, 1, 3, 0, 4\}$
**(c)** $x[n] = \{2, 2, 2, 2\}$ **(d)** $x[n] = \{1, 0, 0, 0, 0, 0, 0, 0\}$

[**Hints and Suggestions:** Compute the DFT only for the indices $k \le \frac{N}{2}$ and use conjugate symmetry about $k = \frac{N}{2}$ and $X_{\text{DFT}}[N-k] = X^*_{\text{DFT}}[k]$ to find the rest.]

**8.2.** **(DFT from Definition)** Use the defining relation to compute the $N$-point DFT of the following:

**(a)** $x[n] = \delta[n], \quad 0 \le n \le N-1$
**(b)** $x[n] = \alpha^n, \quad 0 \le n \le N-1$
**(c)** $x[n] = e^{j\pi n/N}, \quad 0 \le n \le N-1$

**8.3. (IDFT from Definition)** Compute the IDFT of the following.

**(a)** $X_{\rm DFT}[k] = \{2, -j, 0, j\}$ **(b)** $X_{\rm DFT}[k] = \{4, -1, 1, 1, -1\}$
**(c)** $X_{\rm DFT}[k] = \{1, 2, 1, 2\}$ **(d)** $X_{\rm DFT}[k] = \{1, 0, 0, j, 0, -j, 0, 0\}$

[**Hints and Suggestions:** Each DFT has conjugate symmetry about $k = \frac{N}{2}$, so $x[n]$ should be real. For (b) through (c), the DFT is also real, so $x[n]$ has conjugate symmetry about $k = \frac{N}{2}$.]

**8.4. (Symmetry)** For the DFT of each real sequence, compute the boxed quantities.

**(a)** $X_{\rm DFT}[k] = \left\{0, \boxed{X_1}, 2+j, -1, \boxed{X_4}, j\right\}$
**(b)** $X_{\rm DFT}[k] = \left\{1, 2, \boxed{X_2}, \boxed{X_3}, 0, 1-j, -2, \boxed{X_7}\right\}$

[**Hints and Suggestions:** The DFT of a real signal shows conjugate symmetry about $k = \frac{N}{2}$.]

**8.5. (Properties)** The DFT of $x[n]$ is $X_{\rm DFT}[k] = \{1, 2, 3, 4\}$. Find the DFT of each of the following sequences, using properties of the DFT.

**(a)** $y[n] = x[n-2]$ **(b)** $f[n] = x[n+6]$ **(c)** $g[n] = x[n+1]$
**(d)** $h[n] = e^{jn\pi/2}x[n]$ **(e)** $p[n] = x[n] \circledast x[n]$ **(f)** $q[n] = x^2[n]$
**(g)** $r[n] = x[-n]$ **(h)** $s[n] = x^*[n]$ **(i)** $v[n] = x^2[-n]$

[**Hints and Suggestions:** In (a) through (c) use the shifting property. In (d), use circular shift of the DFT. In (e), square the DFT values. In (f), divide the periodic convolution by $N$. In (g), flip the DFT and create is periodic extension to get samples starting at $k = 0$. In (h), conjugate the results of (g).]

**8.6. (Properties)** Let $X_{\rm DFT}[k]$ be the $N$-point DFT of a real signal. How many DFT samples will always be real, and what will be their index $k$? [*Hint*: Use the concept of conjugate symmetry and consider the cases for odd $N$ and even $N$ separately.]
[**Hints and Suggestions:** Check the DFT at $k = 0$ and $k = \frac{N}{2}$ for even $N$ and odd $N$.]

**8.7. (Properties)** Let $X_{\rm DFT}[k]$ be the $N$-point DFT of a (possibly complex) signal $x[n]$. What can you say about the symmetry of $x[n]$ for the following $X_{\rm DFT}[k]$?

**(a)** conjugate symmetric
**(b)** real but with no symmetry
**(c)** real and even symmetric
**(d)** real and odd symmetric
**(e)** imaginary and odd symmetric

**8.8. (Properties)** For each DFT pair shown, compute the values of the boxed quantities.

**(a)** $\{\boxed{x_0}, 3, -4, 0, 2\} \Longleftrightarrow \{5, \boxed{X_1}, -1.28 - j4.39, \boxed{X_3}, 8.78 - j1.4\}$
**(b)** $\{\boxed{x_0}, 3, -4, 2, 0, 1\} \Longleftrightarrow \{4, \boxed{X_1}, 4 - j5.2, \boxed{X_3}, \boxed{X_4}, 4 - j1.73\}$

[**Hints and Suggestions:** Use conjugate symmetry. In (a) also use $x[0] = \sum X_{\rm DFT}[k]$ to find $x_0$. In (b), also use $X_{\rm DFT}[0] = \sum x[n]$ (or Parseval's relation) to find $X_3$.]

**8.9. (Properties)** Let $x[n] = \{\overset{\Downarrow}{1}, -2, 3, -4, 5, -6\}$. Without evaluating its DFT $X_{\rm DFT}[k]$, compute the following quantities:

**(a)** $X[0]$ **(b)** $\sum_{k=0}^{5} X[k]$ **(c)** $X[3]$ **(d)** $\sum_{k=0}^{5} |X[k]|^2$ **(e)** $\sum_{k=0}^{5} (-1)^k X[k]$

[**Hints and Suggestions:** In (a), also use $X_{\text{DFT}}[0] = \sum x[n]$. In (b), use $\sum X_{\text{DFT}}[k] = Nx[0]$. In (c), use $X_{\text{DFT}}[\frac{N}{2}] = \sum(-1)^n x[n]$. In (d), use Parseval's relation. In (e), use $\sum(-1)^k X_{\text{DFT}}[k] = Nx[\frac{N}{2}]$.]

**8.10.** **(DFT Computation)** Find the $N$-point DFT of each of the following signals.

**(a)** $x[n] = \delta[n]$
**(b)** $x[n] = \delta[n-K],\ K < N$
**(c)** $x[n] = \delta[n-0.5N]$ ($N$ even)
**(d)** $x[n] = \delta[n-0.5(N-1)]$ ($N$ odd)
**(e)** $x[n] = 1$
**(f)** $x[n] = \delta[n-0.5(N-1)] + \delta[n-0.5(N+1)]$ ($N$ odd)
**(g)** $x[n] = (-1)^n$ ($N$ even)
**(h)** $x[n] = e^{j4n\pi/N}$
**(i)** $x[n] = \cos(\frac{4n\pi}{N})$
**(j)** $x[n] = \cos(\frac{4n\pi}{N} + 0.25\pi)$

[**Hints and Suggestions:** In (a), the DFT has $N$ unit samples. Use this result with the shifting property in (b), (c), (d), and (f). In (e), only $X_{\text{DFT}}[0]$ is nonzero. In (g), $(-1)^n = \cos(n\pi)$. In (h), use frequency shift on the result of (e). In (i) and (j), $F_0 = \frac{2}{N}$ and the DFT is a pair of impulses with amplitude $0.5Ne^{j\pm\theta}$ at $k = 2,\ N-2$.]

**8.11.** **(Properties)** The DFT of a signal $x[n]$ is $X_{\text{DFT}}[k]$. If we use its conjugate $Y_{\text{DFT}}[k] = X^*_{\text{DFT}}[k]$ and obtain its DFT as $y[n]$, how is $y[n]$ related to $x[n]$?
[**Hints and Suggestions:** A typical DFT term is $Ae^{j\theta}$. Its IDFT gives terms of the form $\frac{1}{N}Ae^{j\theta}e^{j\phi}$ where $\phi = \frac{2nk\pi}{N}$. Similarly, examine the DFT of the conjugated DFT sequence and compare.]

**8.12.** **(Properties)** Let $X[k] = \{\overset{\Downarrow}{1},\ -2,\ 1-j,\ j2,\ 0,\ \ldots\}$ be the 8-point DFT of a real signal $x[n]$.

**(a)** Determine $X[k]$ in its entirety.
**(b)** What is the DFT $Y[k]$ of the signal $y[n] = (-1)^n x[n]$?
**(c)** What is the DFT $G[k]$ of the zero-interpolated signal $g[n] = x[n/2]$?
**(d)** What is the DFT $H[k]$ of $h[n] = \{x[n], x[n], x[n]\}$ obtained by threefold replication of $x[n]$?

[**Hints and Suggestions:** In (b), $(-1)^n x[n] \Leftrightarrow X[k - \frac{N}{2}]$, so use circular shift. In (c), replicate the DFT. In (d), zero interpolate $X[k]$ to $3X[k/3]$ (with two zeros after each DFT sample).]

**8.13.** **(Replication and Zero Interpolation)** The DFT of $x[n]$ is $X_{\text{DFT}}[k] = \{1, 2, 3, 4, 5\}$.

**(a)** What is the DFT of the replicated signal $y[n] = \{x[n], x[n]\}$?
**(b)** What is the DFT of the replicated signal $f[n] = \{x[n], x[n], x[n]\}$?
**(c)** What is the DFT of the zero-interpolated signal $g[n] = x[n/2]$?
**(d)** What is the DFT of the zero-interpolated signal $h[n] = x[n/3]$?

[**Hints and Suggestions:** In (a), insert a zero after each DFT sample. In (b), insert two zeros after each DFT sample. In (b) and (c), double or triple the DFT length by replicating it.]

**8.14.** **(DFT of Pure Sinusoids)** Determine the DFT of $x(t) = \sin(2\pi f_0 t + \frac{\pi}{3})$ (without doing any DFT computations) if we sample this signal, starting at $t = 0$, and acquire the following:

**(a)** 4 samples over 1 period
**(b)** 8 samples over 2 periods
**(c)** 8 samples over 1 period
**(d)** 18 samples over 3 periods
**(e)** 8 samples over 5 periods
**(f)** 16 samples over 10 periods

[**Hints and Suggestions:** The Nyquist rate corresponds to taking two samples per period. Set up the digital frequency as $F_0 = \frac{k_0}{N}$. The $N$-point DFT of $x[n] = A\cos(2\pi F_0 n + \theta)$, is a pair of nonzero samples, $0.5Ae^{j\theta}$ at $k = k_0$ and $0.5Ae^{-j\theta}$ at $k = N - k_0$ (all other samples are zero).]

8.15. **(DFT of Sinusoids)** The following signals are sampled starting at $t = 0$. Identify the indices of the nonzero DFT components and write out the DFT as a sequence.

**(a)** $x(t) = \cos(4\pi t)$ sampled at 25 Hz, using the minimum number of samples to prevent leakage
**(b)** $x(t) = \cos(20\pi t) + 2\sin(40\pi t)$ sampled at 25 Hz with $N = 15$
**(c)** $x(t) = \sin(10\pi t) + 2\sin(40\pi t)$ sampled at 25 Hz for 1 s
**(d)** $x(t) = \sin(40\pi t) + 2\sin(60\pi t)$ sampled at intervals of 0.004 s for four periods

[**Hints and Suggestions:** Let $F_0 = \frac{k_0}{N}$. The $N$-point DFT of $x[n] = A\cos(2\pi F_0 n + \theta)$ is a pair of nonzero samples, with $0.5Ae^{j\theta}$ at $k = k_0$ and $0.5Ae^{-j\theta}$ at $k = N - k_0$ (all other samples are zero).]

8.16. **(Aliasing and Leakage)** The following signals are sampled and the DFT of the sampled signal obtained. Which cases show the effects of leakage?

**(a)** a 100-Hz sinusoid for 2 periods at 400 Hz
**(b)** q 100-Hz sinusoid for 4 periods at 70 Hz
**(c)** q 100-Hz sinusoid at 400 Hz, using 10 samples
**(d)** q 100-Hz sinusoid for 2.5 periods at 70 Hz
**(e)** q sum of sinusoids at 100 Hz and 150 Hz for 100 ms at 450 Hz
**(f)** q sum of sinusoids at 100 Hz and 150 Hz for 130 ms period at 450 Hz

[**Hints and Suggestions:** Leakage occurs if we do not sample for integer periods. If $F_0 = \frac{k}{N}$, where $k$ is an integer, leakage is avoided if we sample for $k$ full periods (to obtain $N$ samples).]

8.17. **(DFT Concepts)** The signal $x(t) = \cos(150\pi t) + \cos(180\pi t)$ is to be sampled and analyzed using the DFT.

**(a)** What is the minimum sampling $S_{\min}$ rate to prevent aliasing?
**(b)** With the sampling rate chosen as $S = 2S_{\min}$, what is the minimum number of samples $N_{\min}$ required to prevent leakage?
**(c)** With the sampling rate chosen as $S = 2S_{\min}$ and $N = 3N_{\min}$, what is the DFT of the sampled signal?
**(d)** With $S = 160$ Hz and $N = 256$, at what DFT indices would you expect to see spectral peaks? Will leakage be present?

[**Hints and Suggestions:** In (b), $N_{\min}$ =LCM$(N_1, N_2)$ where $N_1$ and $N_2$ are the individual periods. In (c), the nonzero DFT samples occur at $k_1, k_2, N_{\min} - k_1, N_{\min} - k_2$ where $F_1 = \frac{k_1}{N_{\min}}$ and $F_2 = \frac{k_2}{N_{\min}}$. In (d), leakage occurs if any index of the form $k = Nf_k/S$ is not an integer.]

8.18. **(DFT Concepts)** The signal $x(t) = \cos(50\pi t) + \cos(80\pi t)$ is sampled at $S = 200$ Hz.

**(a)** What is the minimum number of samples required to prevent leakage?
**(b)** Find the DFT of the sampled signal if $x(t)$ is sampled for 1 s.
**(c)** What are the DFT indices of the spectral peaks if $N = 128$?

[**Hints and Suggestions:** In (a), $N_{\min}$ =LCM$(N_1, N_2)$ where $N_1$ and $N_2$ are the individual periods. In (b), the nonzero DFT samples occur at $k_1, k_2, N_{\min} - k_1, N_{\min} - k_2$ where $F_1 = \frac{k_1}{N_{\min}}$ and $F_2 = \frac{k_2}{N_{\min}}$. In (c), the peaks occur at the integer indices closest to $k = Nf_k/S$.]

8.19. **(DFT Concepts)** Let $x[n] = \cos(2n\pi F_0)$.

**(a)** Use the defining relation to compute its $N$-point DFT.

**(b)** For what values of $F_0$ would you expect aliasing and/or leakage?

**(c)** Compute the DFT if $N = 8$ and $F_0 = 0.25$. Has aliasing occurred? Has leakage occurred?

**(d)** Compute the DFT if $N = 8$ and $F_0 = 1.25$. Has aliasing occurred? Has leakage occurred?

**(e)** Compute the DFT if $N = 9$ and $F_0 = 0.2$. Has aliasing occurred? Has leakage occurred?

[**Hints and Suggestions:** There is no aliasing if $F_0 < 0.5$ and no leakage if $k = NF_0$ is an integer.]

8.20. **(Spectral Spacing)** What is the spectral spacing in the 500-point DFT of a sampled signal obtained by sampling an analog signal at 1 kHz?
[**Hints and Suggestions:** The spectral spacing equals $\frac{S}{N}$.]

8.21. **(Spectral Spacing)** We wish to sample a signal of 1-s duration, band limited to 50 Hz, and compute the DFT of the sampled signal.

**(a)** Using the minimum sampling rate that avoids aliasing, what is the spectral spacing $\Delta f$, and how many samples are acquired?

**(b)** How many padding zeros are needed to reduce the spacing to $0.5\Delta f$, using the minimum sampling rate to avoid aliasing if we use the DFT?

**(c)** How many padding zeros are needed to reduce the spacing to $0.5\Delta f$, using the minimum sampling rate to avoid aliasing if we use a radix-2 FFT?

[**Hints and Suggestions:** The spectral spacing is $\frac{S}{N}$. To reduce it, increase $N$ by zero padding.]

8.22. **(Spectral Spacing)** We wish to sample the signal $x(t) = \cos(50\pi t) + \sin(200\pi t)$ at 800 Hz and compute the $N$-point DFT of the sampled signal $x[n]$.

**(a)** Let $N = 100$. At what indices would you expect to see the spectral peaks? Will the peaks occur at the frequencies of $x(t)$?

**(b)** Let $N = 128$. At what indices would you expect to see the spectral peaks? Will the peaks occur at the frequencies of $x(t)$?

[**Hints and Suggestions:** Use $F_0 = \frac{k}{N}$ to compute $k = NF_0 = Nf_0/S$. A peak will occur at the exact frequency only for integer $k$.]

8.23. **(Spectral Spacing)** We wish to sample the signal $x(t) = \cos(50\pi t) + \sin(80\pi t)$ at 100 Hz and compute the $N$-point DFT of the sampled signal $x[n]$.

**(a)** Let $N = 100$. At what indices would you expect to see the spectral peaks? Will the peaks occur at the frequencies of $x(t)$?

**(b)** Let $N = 128$. At what indices would you expect to see the spectral peaks? Will the peaks occur at the frequencies of $x(t)$?

[**Hints and Suggestions:** Use $F_0 = \frac{k}{N}$ to compute $k = NF_0 = Nf_0/S$. A peak will occur at the exact frequency only for integer $k$.]

**8.24.** **(Spectral Spacing)** We wish to identify the 21-Hz component from the $N$-sample DFT of a signal. The sampling rate is 100 Hz, and only 128 signal samples are available.

**(a)** If $N = 128$, will there be a DFT component at 21 Hz? If not, what is the frequency closest to 21 Hz that can be identified? What DFT index does this correspond to?

**(b)** Assuming that all signal samples must be used and zero padding is allowed, what is the smallest value of $N$ that will result in a DFT component at 21 Hz? How many padding zeros will be required? At what DFT index will the 21-Hz component appear?

[**Hints and Suggestions:** Use $F_0 = \frac{k}{N}$ to compute $k = NF_0 = Nf_0/S$. In (a), the peak will occur at the exact frequency only if $k$ is an integer. In (b), pick $N$ such that $k$ is an integer.]

**8.25.** **(Sampling Frequency)** For each of the following signals, estimate the sampling frequency and sampling duration by arbitrarily choosing the bandwidth $B$ as the frequency where $|H(f)|$ is 5% of its maximum and the signal duration $D$ as the time at which $x(t)$ is 1% of its maximum.

**(a)** $x(t) = e^{-t}u(t)$ **(b)** $x(t) = te^{-t}u(t)$ **(c)** $x(t) = \text{tri}(t)$

[**Hints and Suggestions:** First, set $|X(f)| = 0.05|X(f)|_{\text{max}}$ to estimate $B$ and $S > 2B$. Next, estimate the time $D$ at which $x(t)$ is 1% of its peak and find $N = SD$ (as an integer). Numerical estimates give $D = 7.64$ s for (b) and $B \approx 0.81$ Hz for (c).]

**8.26.** **(Sampling Frequency and Spectral Spacing)** It is required to find the DFT of the signal $x(t) = e^{-t}u(t)$ after sampling it for a duration $D$ that contains 95% of the total signal energy $E$. How many samples are acquired if the sampling rate $S$ is chosen to ensure that

**(a)** the aliasing level at $f = 0.5S$ due to the first replica is less than 1% of the peak level?

**(b)** the energy in the aliased signal past $f = 0.5S$ is less than 1% of the total signal energy?

[**Hints and Suggestions:** Use Parseval's relation to set $0.95E = \int_0^D e^{-2t}\,dt$ and estimate $D$. In (b), set $|X(f)|_{f=S/2} = 0.01|X(f)|_{\text{max}}$ to estimate $B$. In (c), set $\int_{-S/2}^{S/2} |X(f)|^2\,df = 0.99E$ to estimate $B$. Use $S > 2b$ to compute $N = SD$ (as an integer). Note that $E = 0.5$ and $|X(f)|^2 = \frac{1}{1+4\pi^2 f^2}$.]

**8.27.** **(Sampling Rate and the DFT)** The pulse $u(t) - u(t-1)$ is replicated every two seconds to get a periodic signal $x(t)$ with period $T = 2$. The periodic signal is sampled for one full period to obtain $N$ samples. The $N$-point DFT of the samples corresponds to the signal $y(t) = A + B\sin(\pi t)$.

**(a)** What are the possible values of $N$ for which you could obtain such a result? For each such choice, compute the values of $A$ and $B$.

**(b)** What are the possible values of $N$ for which you could obtain such a result using the radix-2 FFT? For each such choice, compute the values of $A$ and $B$.

**(c)** Is it possible for $y(t)$ to be identical to $x(t)$ for any choice of sampling rate?

[**Hints and Suggestions:** The DFT contains a nonzero sample at $k = 0$ and a pair of nonzero imaginary samples $\pm j0.5$ at the indices $\pm k$ corresponding to $f_0 = 0.5$ Hz.]

**8.28.** **(Sampling Rate and the DFT)** A periodic signal $x(t)$ with period $T = 2$ is sampled for one full period to obtain $N$ samples. The signal reconstructed from the $N$-point DFT of the samples is $y(t)$.

**(a)** Will the DFT show the effects of leakage?

**(b)** Let $N = 8$. How many harmonics of $x(t)$ can be identified in the DFT? What constraints on $x(t)$ will ensure that $y(t) = x(t)$?

**(c)** Let $N = 12$. How many harmonics of $x(t)$ can be identified in the DFT? What constraints on $x(t)$ will ensure that $y(t) = x(t)$?

[**Hints and Suggestions:** The number of useful harmonics is $\frac{N-1}{2} - 1$. If $y(t) = x(t)$, the periodic signal $x(t)$ must be bandlimited to $\frac{N}{2} f_0$.]

**8.29. (Frequency Resolution)** A signal is sampled at 5 kHz to acquire $N$ samples of $x[n]$.

**(a)** Let $N = 1000$. What is the frequency resolution in the DFT of the sampled signal $x[n]$?

**(b)** Let $N = 1000$. What is the frequency resolution in the DFT of the von Hann windowed signal?

**(c)** Find the smallest value of $N$ for a frequency resolution of 2 Hz with no window.

**(d)** Find the smallest value of $N$ for a frequency resolution of 2 Hz with a von Hann window.

[**Hints and Suggestions:** In (a), the frequency resolution is $\Delta f = \frac{S}{N}$. In (b), it is $\Delta f = \frac{S}{N}$ where $C = 2$. In (c), $N = \frac{S}{\Delta f}$. In (d), $N = \frac{CS}{\Delta f}$.]

**8.30. (Convolution and the DFT)** Consider two sequences $x[n]$ and $h[n]$ of length 12 samples and 20 samples, respectively.

**(a)** How many padding zeros are needed for $x[n]$ and $h[n]$ in order to find their regular convolution $y[n]$, using the DFT?

**(b)** If we pad $x[n]$ with eight zeros and find the periodic convolution $y_p[n]$ of the resulting 20-point sequence with $h[n]$, for what indices will the samples of $y[n]$ and $y_p[n]$ be identical?

[**Hints and Suggestions:** Periodic convolution of sequences of equal length $N$, requires wraparound of the last $N - 1$ samples of the regular convolution. The last ($N$th sample) in the periodic convolution is not contaminated. If the last $L$ samples of the regular convolution are zero, the last $L + 1$ samples of the periodic convolution are not contaminated.]

**8.31. (FFT)** Write out the DFT sequence that corresponds to the following bit-reversed sequences obtained using the DIF FFT algorithm.

**(a)** $\{1, 2, 3, 4\}$ **(b)** $\{0, -1, 2, -3, 4, -5, 6, -7\}$

**8.32. (FFT)** Set up a flowchart showing all twiddle factors and values at intermediate nodes to compute the DFT of $x[n] = \{1, 2, 2, 2, 1, 0, 0, 0\}$, using

**(a)** the 8-point DIF algorithm **(b)** the 8-point DIT algorithm.

**8.33. (Spectral Spacing and the FFT)** We wish to sample a signal of 1-s duration and band limited to 100 Hz in order to compute its spectrum. The spectral spacing should not exceed 0.5 Hz. Find the minimum number $N$ of samples needed and the actual spectral spacing $\Delta f$ if we use

**(a)** the DFT **(b)** the radix-2 FFT

[**Hints and Suggestions:** The spectral spacing is $\frac{S}{N}$. For the FFT, $N$ is a power of 2.]

**8.34.** **(Convolution)** Find the linear convolution of $x[n] = \{\overset{\Downarrow}{1}, 2, 1\}$ and $h[n] = \{\overset{\Downarrow}{1}, 2, 3\}$, using

**(a)** the time-domain convolution operation
**(b)** the DFT and zero padding
**(c)** the radix-2 FFT and zero padding

Which of these methods yield identical results and why?
[**Hints and Suggestions:** In (a), the convolution length is $N = 5$. In (b), zero pad each signal to $N = 5$, multiply the DFTs (at each index) and find the 5-point IDFT. In (c), zero pad to $N = 8$.]

**8.35.** **(Convolution)** Find the *periodic* convolution of $x[n] = \{\overset{\Downarrow}{1}, 2, 1\}$ and $h[n] = \{\overset{\Downarrow}{1}, 2, 3\}$, using

**(a)** the time-domain convolution operation.
**(b)** the DFT operation. Is this result identical to that of part (a)?
**(c)** the radix-2 FFT and zero-padding. Is this result identical to that of part (a)? Should it be?

Which of these methods yield identical results and why?
[**Hints and Suggestions:** In (a), use regular convolution and wraparound. In (b), multiply the DFTs (at each index) and find the 3-point IDFT. In (c), zero pad each signal to $N = 4$, multiply the DFTs (at each index) and find the 4-point IDFT.]

**8.36.** **(Correlation)** Find one period (starting at $n = 0$) of the *periodic* correlation $r_{xh}$ of $x[n] = \{\overset{\Downarrow}{1}, 2, 1\}$ and $h[n] = \{\overset{\Downarrow}{1}, 2, 3\}$, using

**(a)** the time-domain correlation operation
**(b)** the DFT

[**Hints and Suggestions:** In (a), use circular flipping and replication to get samples of $h[-n]$ starting at the origin, compute $x[n] * h[-n]$ and wraparound. In (b), multiply the DFTs (at each index) and find the 3-point IDFT. In (b), multiply $X_{\text{DFT}}[k]$ and $H^*_{\text{DFT}}[k]$ (at each index) and find the IDFT.]

**8.37.** **(Convolution of Long Signals)** Let $x[n] = \{\overset{\Downarrow}{1}, 2, 1\}$ and $h[n] = \{\overset{\Downarrow}{1}, 2, 1, 3, 2, 2, 3, 0, 1, 0, 2, 2\}$.

**(a)** Find their convolution using the overlap-add method.
**(b)** Find their convolution using the overlap-save method.
**(c)** Are the results identical to the time-domain convolution of $x[n]$ and $h[n]$?

[**Hints and Suggestions:** In (a), split $h[n]$ into three-sample segments, find their convolution with $x[n]$, shift (by 0, 3, 6 and 9 samples) and add. In (b), generate five-sample segments from $h[n]$ with the first as $\{0, 0, 1, 2, 1\}$, the second as $\{2, 1, 3, 2, 2\}$, etc, with a two-sample overlap in each segment (zero pad the last segment to five samples, if required). Find their periodic convolutions with $\{x[n], 0, 0\}$, discard the first two samples of each convolution and concatenate (string together).]

## Computation and Design

**8.38.** **(DFT Properties)** Consider the signal $x[n] = n+1$, $0 \le n \le 7$. Use MATLAB to compute its DFT. Confirm the following properties by computing the DFT of

(a) $y[n] = x[-n]$ to confirm the (circular) flipping property
(b) $f[n] = x[n-2]$ to confirm the (circular) shift property
(c) $g[n] = x[n/2]$ to confirm the zero-interpolation property
(d) $h[n] = \{x[n], x[n]\}$ to confirm the signal-replication property
(e) $p[n] = x[n]\cos(0.5n\pi)$ to confirm the modulation property
(f) $r[n] = x^2[n]$ to confirm the multiplication property
(g) $s[n] = x[n] \circledast x[n]$ to confirm the periodic convolution property

8.39. **(IDFT from DFT)** Consider the signal $x[n] = (1 + j)n, 0 \le n \le 9$.

(a) Find its DFT $X[k]$. Find the DFT of the sequence $0.1X[k]$. Does this appear to be related to the signal $x[n]$?
(b) Find the DFT of the sequence $0.1X^*[k]$. Does this appear to be related to $x[n]$?
(c) Use your results to explain how you might find the IDFT of a sequence by using the forward DFT algorithm alone.

8.40. **(Resolution)** We wish to compute the radix-2 FFT of the signal samples acquired from the signal $x(t) = A\cos(2\pi f_0 t) + B\cos[2\pi(f_0 + \Delta f)t]$, where $f_0 = 100$ Hz. The sampling frequency is $S = 480$ Hz.

(a) Let $A = B = 1$. What is the smallest number of signal samples $N_{\text{min}}$ required for a frequency resolution of $\Delta f = 2$ Hz if no window is used? How does this change if we wish to use a von Hann window? What about a Blackman window? Plot the FFT magnitude for each case to confirm your expectations.
(b) Let $A = 1$ and $B = 0.02$. Argue that we cannot obtain a frequency resolution of $\Delta f = 2$ Hz if no window is used. Plot the FFT magnitude for various lengths to justify your argument. Of the Bartlett, Hamming, von Hann, and Blackman windows, which ones can we use to obtain a resolution of $\Delta f = 2$ Hz? Which one will require the minimum number of samples, and why? Plot the FFT magnitude for each applicable window to confirm your expectations.

8.41. **(Convolution)** Consider the sequences $x[n] = \{1, 2, 1, 2, 1\}$ and $h[n] = \{1, 2, 3, 3, 5\}$.

(a) Find their regular convolution using three methods: the convolution operation; zero padding and the DFT; and zero padding to length 16 and the DFT. How are the results of each operation related? What is the effect of zero padding?
(b) Find their periodic convolution using three methods: regular convolution and wraparound; the DFT; and zero padding to length 16 and the DFT. How are the results of each operation related? What is the effect of zero padding?

8.42. **(Convolution)** Consider the signals $x[n] = 4(0.5)^n, 0 \le n \le 4$, and $h[n] = n, 0 \le n \le 10$.

(a) Find their regular convolution $y[n] = x[n] * h[n]$, using the MATLAB routine **conv**.
(b) Use the FFT and IFFT to obtain the regular convolution, assuming the *minimum* length $N$ (that each sequence must be zero padded to) for correct results.
(c) How do the results change if each sequence is zero padded to length $N + 2$?
(d) How do the results change if each sequence is zero padded to length $N - 2$?

8.43. **(FFT of Noisy Data)** Sample the sinusoid $x(t) = \cos(2\pi f_0 t)$ with $f_0 = 8$ Hz at $S = 64$ Hz for 4 s to obtain a 256-point sampled signal $x[n]$. Also generate 256 samples of a uniformly distributed noise sequence $s[n]$ with zero mean.

**(a)** Display the first 32 samples of $x[n]$. Can you identify the period from the plot? Compute and plot the DFT of $x[n]$. Does the spectrum match your expectations?
**(b)** Generate the noisy signal $y[n] = x[n] + s[n]$ and display the first 32 samples of $y[n]$. Do you detect any periodicity in the data? Compute and plot the DFT of the noisy signal $y[n]$. Can you identify the frequency and magnitude of the periodic component from the spectrum? Do they match your expectations?
**(c)** Generate the noisy signal $z[n] = x[n]s[n]$ (by element-wise multiplication) and display the first 32 samples of $z[n]$. Do you detect any periodicity in the data? Compute and plot the DFT of the noisy signal $z[n]$. Can you identify the frequency the periodic component from the spectrum?

8.44. **(Filtering a Noisy ECG Signal: I)** During recording, an electrocardiogram (ECG) signal, sampled at 300 Hz, gets contaminated by a 60-Hz hum. Two beats of the original and contaminated signal (600 samples) are provided on the author's website as **ecgo** and **ecg**. Load these signals into MATLAB (for example, use the command **load ecgo**). In an effort to remove the 60-Hz hum, use the DFT as a filter to implement the following steps.

**(a)** Compute (but do not plot) the 600-point DFT of the contaminated ECG signal.
**(b)** By hand, compute the DFT indices that correspond to the 60-Hz signal.
**(c)** Zero out the DFT components corresponding the 60-Hz signal.
**(d)** Take the IDFT to obtain the filtered ECG and display the original and filtered signal.
**(e)** Display the DFT of the original and filtered ECG signal and comment on the differences.
**(f)** Is the DFT effective in removing the 60-Hz interference?

8.45. **(Filtering a Noisy ECG Signal: II)** Continuing with the previous problem, load the original and contaminated ECG signal sampled at 300 Hz with 600 samples provided on the author's website as **ecgo** and **ecg** (for example, use the command **load ecgo**). Truncate each signal to 512 samples. In an effort to remove the 60-Hz hum, use the DFT as a filter to implement the following steps.

**(a)** Compute (but do not plot) the 512-point DFT of the contaminated ECG signal.
**(b)** Compute the DFT indices *closest to* 60 Hz and zero out the DFT at these indices.
**(c)** Take the IDFT to obtain the filtered ECG and display the original and filtered signal.
**(d)** Display the DFT of the original and filtered ECG signal and comment on the differences.
**(e)** The DFT is not effective in removing the 60-Hz interference. Why?
**(f)** From the DFT plots, suggest and implement a method for improving the results (by zeroing out a larger portion of the DFT around 60 Hz, for example).

8.46. **(Decoding a Mystery Message)** During transmission, a message signal gets contaminated by a low-frequency signal and high-frequency noise. The message can be decoded only by displaying it in the time domain. The contaminated signal is provided on the author's website as **mystery1**. Load this signal into MATLAB (use the command **load mystery1**). In an effort to decode the message, use the DFT as a filter to implement the following steps and determine what the decoded message says.

**(a)** Display the contaminated signal. Can you "read" the message?
**(b)** Take the DFT of the signal to identify the range of the message spectrum.
**(c)** Zero out the DFT component corresponding to the low-frequency signal.
**(d)** Zero out the DFT components corresponding to the high-frequency noise.
**(e)** Take the IDFT to obtain the filtered signal and display it to decode the message.

**8.47. (Spectrum Estimation)** The FFT is extensively used in estimating the spectrum of various signals, detecting periodic components buried in noise, or detecting long-term trends. The monthly rainfall data, for example, tends to show periodicity (an annual cycle). However, long-term trends may also be present due to factors such as deforestation and soil erosion that tend to reduce rainfall amounts over time. Such long-term trends are often masked by the periodicity in the data and can be observed only if the periodic components are first removed (filtered).

**(a)** Generate a signal $x[n] = 0.01n + \sin(n\pi/6), 0 \le n \le 500$, and add some random noise to simulate monthly rainfall data. Can you observe any periodicity or long-term trend from a plot of the data?

**(b)** Find the FFT of the rainfall data. Can you identify the periodic component from the FFT magnitude spectrum?

**(c)** Design a notch filter to remove the periodic component from the rainfall data. You may identify the frequency to be removed from $x[n]$ (if you have not been able to identify it from the FFT). Filter the rainfall data through your filter and plot the filtered data. Do you observe any periodicity in the filtered data? Can you detect any long-term trends from the plot?

**(d)** To detect the long-term trend, pass the filtered data through a moving average filter. Experiment with different lengths. Does averaging of the filtered data reveal the long-term trend? Explain how you might go about quantifying the trend.

**8.48. (The FFT as a Filter)** We wish to filter out the 60-Hz interference from the signal

$$x(t) = \cos(100\pi t) + \cos(120\pi t)$$

by sampling $x(t)$ at $S = 500$ Hz and passing the sampled signal $x[n]$ through a lowpass filter with a cutoff frequency of $f_C = 55$ Hz. The $N$-point FFT of the filtered signal may be obtained by simply *zeroing out* the FFT $X[k]$ of the sampled signal between the indices $M = \text{int}(Nf_C/S)$ and $N - M$ (corresponding to the frequencies between $f_C$ and $S - f_C$). This is entirely equivalent to multiplying $X[k]$ by a *filter function* $H[k]$ of the form

$$H[k] = \{\ 1, (M \text{ ones}), (N - 2M - 1 \text{ zeros}), (M \text{ ones})\ \}$$

The FFT of the filtered signal equals $Y[k] = H[k]X[k]$, and the filtered signal $y[n]$ is obtained by computing its IFFT.

**(a)** Start with the smallest value of $N$ you need to resolve the two frequencies and successively generate the sampled signal $x[n]$, its FFT $X[k]$, the filter function $H[k]$, the FFT $Y[k]$ of the filtered signal as $Y[k] = H[k]X[k]$, and its IFFT $y[n]$. Plot $X[k]$ and $Y[k]$ on the same plot and $x[n]$ and $y[n]$ on the same plot. Is the 60-Hz signal completely blocked? Is the filtering effective?

**(b)** Double the value of $N$ several times and, for each case, repeat the computations and plot requested in part (a). Is there a noticeable improvement in the filtering?

**(c)** The filter described by $h[n] = \text{IFFT}\{H[k]\}$ is not very useful because its true frequency response $H(F)$ matches the $N$-point FFT $H[k]$ only at $N$ points and varies considerably in between. To see this, pick the value of $N$ from part (a), use the MATLAB routine, compute the DTFT $H(F)$ of $h[n]$ at $N$ points over $0 \le F \le 1$, and superimpose plots of $|H(F)|$ and $H[k]|$. How does $H(F)$ differ from $H[k]$? Compute $H(F)$ at $4N$ points over $0 \le F \le 1$ and superimpose plots of $|H(F)|$ and $H[k]|$. How does $H(F)$

differ from $H[k]$? What is the reason for this difference? Can a larger $N$ reduce the differences? If not, how can the differences be minimized? (This forms the subject of frequency sampling filters.)

8.49. **(Band-Limited Interpolation)** Consider the band-limited signal $x(t) = \sin(200\pi t)$.

**(a)** Sample $x(t)$ at $S = 400$ Hz to obtain the sampled signal $x[n]$ with $N = 4$ samples. Compute and plot the DFT $X[k]$ of $x[n]$. Can you identify the frequency and magnitude from the spectrum?

**(b)** Zero interpolate the signal $G[k] = 8X[k]$ by inserting 28 zeros about the middle (about index $k = 0.5N$) to obtain the interpolated 32-point spectrum $Y[k]$. Compute the IDFT of $Y[k]$ to obtain the signal $y[n]$. Sample $x(t)$ again, but at $8S$ Hz, to obtain the sampled signal $y_1[n]$. Plot $e[n] = y[n] - y_1[n]$. Are $y[n]$ and $y_1[n]$ identical (to within machine roundoff)? Should they be?

**(c)** Sample $x(t)$ at $S = 800$ Hz to obtain the sampled signal $x[n]$ with $N = 4$ samples. Compute its DFT $X[k]$ and zero interpolate the signal $G[k] = 8X[k]$ by inserting 28 zeros about the middle (about index $k = 0.5N$) to obtain the interpolated 32-point spectrum $Y[k]$. Compute the IDFT of $Y[k]$ to obtain the signal $y[n]$. Sample $x(t)$ again, but at $8S$ Hz, to obtain the sampled signal $y_1[n]$. Plot $e[n] = y[n] - y_1[n]$. Are $y[n]$ and $y_1[n]$ identical? Should they be?

**(d)** Explain the differences in the results of parts (a) and (b).

8.50. **(Decimation)** To decimate a signal $x[n]$ by $N$, we use a lowpass filter (to band limit the signal to $F = 0.5/N$), followed by a down-sampler (that retains only every $N$th sample). In this problem, ignore the lowpass filter.

**(a)** Generate the test signal $x[n] = \cos(0.2n\pi) + \cos(0.3n\pi)$, $0 \le n \le 59$. Plot its DFT. Can you identify the frequencies present?

**(b)** Decimate $x[n]$ by $N = 2$ to obtain the signal $x_2[n]$. Is the signal $x[n]$ sufficiently band limited in this case? Plot the DFT of $x_2[n]$. Can you identify the frequencies present? Do the results match your expectations? Would you be able to recover $x[n]$ from band-limited interpolation (by $N = 2$) of $x_2[n]$?

**(c)** Decimate $x[n]$ by $N = 4$ to obtain the signal $x_4[n]$. Is the signal $x[n]$ sufficiently band limited in this case? Plot the DFT of $x_4[n]$. Can you identify the frequencies present? If not, explain how the result differs from that of part (b). Would you be able to recover $x[n]$ from band-limited interpolation (by $N = 4$) of $x_4[n]$?

8.51. **(DFT of Large Data Sets)** The DFT of a large $N$-point data set may be obtained from the DFT of smaller subsets of the data. In particular, if $N = RC$, we arrange the data as an $R \times C$ matrix (by filling along columns), find the DFT of each row, multiply each result at the location $(r, c)$ by $W_{rc} = e^{-j2\pi rc/N}$, (where $r = 0, 1, \ldots, R-1$ and $c = 0, 1, \ldots, C-1$) find the DFT of each column, and reshape the result (by rows) into the required $N$-point DFT. Let $x[n] = n + 1$, $0 \le n \le 11$.

**(a)** Find the DFT of $x[n]$ by using this method with $R = 3$, $C = 4$.

**(b)** Find the DFT of $x[n]$ by using this method with $R = 4$, $C = 3$.

**(c)** Find the DFT of $x[n]$ using the MATLAB command **fft**.

**(d)** Do all methods yield identical results? Can you justify your answer?

**8.52.** **(Time-Frequency Plots)** This problem deals with time-frequency plots of a sum of sinusoids.

**(a)** Generate 600 samples of the signal $x[n] = \cos(0.1n\pi) + \cos(0.4n\pi) + \cos(0.7n\pi)$, the sum of three pure cosines at $F = 0.05, 0.2, 0.35$. Use the MATLAB command **fft** to plot its DFT magnitude. Use the MATLAB based routine **timefreq** (from the author's website) to display its time-frequency plot. What do the plots reveal?

**(b)** Generate 200 samples each of the three signals $y_1[n] = \cos(0.1n\pi)$, $y_2[n] = \cos(0.4n\pi)$, and $y_3[n] = \cos(0.7n\pi)$. Concatenate them to form the 600-sample signal $y[n] = \{y_1[n], y_2[n], y_3[n]\}$. Plot its DFT magnitude and display its time-frequency plot. What do the plots reveal?

**(c)** Compare the DFT magnitude and time-frequency plots of $x[n]$ and $y[n]$ How do they differ?

**8.53.** **(Deconvolution)** The FFT is a useful tool for deconvolution. Given an input signal $x[n]$ and the system response $y[n]$, the system impulse response $h[n]$ may be found from the IDFT of the ratio $H_{\text{DFT}}[k] = Y_{\text{DFT}}[k]/X_{\text{DFT}}[k]$. Let $x[n] = \{1, 2, 3, 4\}$ and $h[n] = \{1, 2, 3\}$.

**(a)** Obtain the convolution $y[n] = x[n] * h[n]$. Now zero pad $x[n]$ to the length of $y[n]$ and find the DFT of the two sequences and their ratio $H_{\text{DFT}}[k]$. Does the IDFT of $H_{\text{DFT}}[k]$ equal $h[n]$ (to within machine roundoff)? Should it?

**(b)** Repeat part (a) with $x[n] = \{1, 2, 3, 4\}$ and $h[n] = \{1, 2, -3\}$. Does the method work for this choice? Does the IDFT of $H_{\text{DFT}}[k]$ equal $h[n]$ (to within machine roundoff)?

**(c)** Repeat part (a) with $x[n] = \{1, 2, -3\}$ and $h[n] = \{1, 2, 3, 4\}$. Show that the method does not work because the division yields infinite or indeterminate results (such as $\frac{1}{0}$ or $\frac{0}{0}$). Does the method work if you replace zeros by very small quantities (for example, $10^{-10}$)? Should it?

**8.54.** **(FFT of Two Real Signals in One Step)** Show that it is possible to find the FFT of two real sequences $x[n]$ and $y[n]$ from a single FFT operation on the complex signal $g[n] = x[n] + jy[n]$ as

$$X_{\text{DFT}}[k] = 0.5(G^*_{\text{DFT}}[N-k] + G_{\text{DFT}}[k]) \qquad Y_{\text{DFT}}[k] = j0.5(G^*_{\text{DFT}}[N-k] - G_{\text{DFT}}[k])$$

Use this result to find the FFT of $x[n] = \{1, 2, 3, 4\}$ and $y[n] = \{5, 6, 7, 8\}$ and compare the results with their FFT computed individually.

**8.55.** **(Quantization Error)** Quantization leads to noisy spectra. Its effects can be studied only in statistical terms. Let $x(t) = \cos(20\pi t)$ be sampled at 50 Hz to obtain the 256-point sampled signal $x[n]$.

**(a)** Plot the linear and decibel magnitude of the DFT of $x[n]$.

**(b)** Quantize $x[n]$ by rounding to $B$ bits to generate the quantized signal $y[n]$. Plot the linear and decibel magnitude of the DFT of $y[n]$. Compare the DFT spectra of $x[n]$ and $y[n]$ for $B = 8, 4, 2$, and 1. What is the effect of decreasing the number of bits on the DFT spectrum of $y[n]$?

**(c)** Repeat parts (a) and (b), using quantization by truncation. How do the spectra differ in this case?

**(d)** Repeat parts (a) through (c) after windowing $x[n]$ by a von Hann window. What is the effect of windowing?

8.56. **(Sampling Jitter)** During the sampling operation, the phase noise on the sampling clock can result in **jitter**—or random variations in the time of occurrence of the true sampling instant. Jitter leads to a noisy spectral, and its effects can be studied only in statistical terms. Consider the analog signal $x(t) = \cos(2\pi f_0 t)$ sampled at a rate $S$ that equals three times the Nyquist rate.

**(a)** Generate a time array $t_n$ of 256 samples at intervals of $t_s = 1/S$. Generate the sampled signal $x[n]$ from values of $x(t)$ at the time instants in $t_n$. Plot the DFT magnitude of $x[n]$.

**(b)** Add some uniformly distributed random noise with a mean of zero and a noise amplitude of $At_s$ to $t_n$ to form the new time array $t_{nn}$. Generate the sampled signal $y[n]$ from values of $x(t)$ at the time instants in $t_{nn}$. Plot the DFT magnitude of $y[n]$ and compare with the DFT magnitude of $x[n]$ for $A = 0.01, 0.1, 1, and 10$. What is the effect of increasing the noise amplitude on the DFT spectrum of $y[n]$? What is the largest value of $A$ for which you can still identify the signal frequency from the DFT of $y[n]$?

**(c)** Repeat parts (a) and (b) after windowing $x[n]$ and $y[n]$ by a von Hann window. What is the effect of windowing?

CHAPTER 9

# Design of IIR Filters

## 9.0 Scope and Overview

Digital filters process discrete-time signals. They are essentially mathematical implementations of a filter equation in software or hardware. They suffer from few limitations. Among their many advantages are high noise immunity, high accuracy (limited only by the roundoff error in the computer arithmetic), easy modification of filter characteristics, freedom from component variations, and, of course, low and constantly decreasing cost. Digital filters are therefore rapidly replacing analog filters in many applications where they can be used effectively. The term *digital filtering* is to be understood in its broadest sense, not only as a smoothing or averaging operation, but also as any processing of the input signal.

This chapter begins with an introduction to IIR filters and the various mappings that are used to convert analog filters to digital filters. It then describes the design of IIR digital filters based on an analog *lowpass prototype* that meets the given specifications, followed by an appropriate *mapping* and *spectral transformation*. The bilinear transformation and its applications are discussed in detail.

### 9.0.1 Goals and Learning Objectives

The goals of this chapter are to provide an introduction to the techniques of IIR filter design. After going through this chapter, the reader should:

1. Understand the concept of filter specifications.
2. Understand the connection between the impulse response of an analog filter and its sampled version.
3. Understand the effects of aliasing in mapping the impulse response to the $z$-domain.
4. Know how to convert an analog filter to a digital filter by impulse invariance or response invariance.
5. Understand the concept of the matched $z$-transform and know how to use it to convert an analog filter to a digital filter.
6. Understand the concept of mapping the variable $s$ to the variable $z$ using numerical difference or numerical integration algorithms.
7. Know how to use the mappings to convert an analog filter to a digital filter.

8. Understand the concept of the bilinear transformation and warping.
9. Know how to use the bilinear transformation to convert an analog filter to a digital filter.
10. Understand the concept of D2D and A2D frequency transformations and know how to use them to convert a lowpass analog prototype to a digital filter.
11. Be able to design a digital filter from given specifications.

# 9.1 Introduction

A filter may be regarded as a *frequency-selective device* that allows us to shape the magnitude or phase response in a prescribed manner. We concentrate on the design of lowpass filters because it can easily be converted to other forms by suitable frequency transformations. The design process is essentially a three-step process that requires specification of the design parameters, design of the least-complex transfer function that meets or beats the design specifications, and realization of the transfer function in software or hardware. The design based on any set of performance specifications is (at best) a compromise at all three levels. At the first level, the actual filter may never meet performance specifications if they are too stringent; at the second, the same set of specifications may lead to several possible realizations; and at the third, quantization and roundoff errors may render the design useless if based on too critical a set of design values. The fewer or less stringent the specifications, the better is the possibility of achieving both design objectives and implementation.

## 9.1.1 Filter Specifications

Filters are typically described by frequency specifications and attenuation (or gain) specifications. Frequency specifications include the passband edge(s) and stopband edge(s). If analog frequencies are specified in the design or if the digital filter is to be designed for processing sampled analog signals, the sampling rate must also be given.

Typical magnitude (or gain) specifications for a lowpass digital filter are illustrated in Figure 9.1 and include the passband attenuation (or gain or ripple) and the stopband attenuation (or gain or ripple). The passband ripple (deviation from maximum gain) and stopband ripple (deviation from zero gain) are usually described by their attenuation (reciprocal of the linear gain) expressed in decibels (dB).

**How are the Gain and Attenuation Related on a Linear or Decibel Scale?**

**Linear:** Attenuation equals the *reciprocal* of gain:

$$G = |H(f)|, \quad A = 1/|H(f)|$$

**Decibel:** Attenuation equals the *negative* of dB gain:

$$G_{\text{dB}} = 20\log|H(f)|, \quad A_{\text{dB}} = -20\log|H(f)|$$

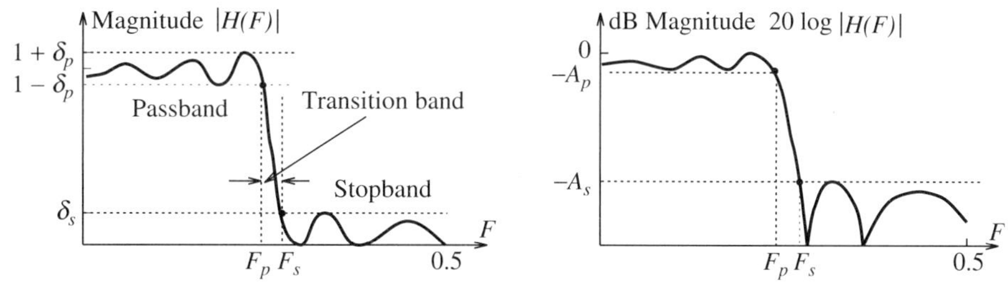


**FIGURE 9.1** The features of a typical lowpass digital filter. Up to the passband edge $F_p$, the gain is nearly unity, with small variations about unity measured by the passband ripple $\delta_p$. Past the stopband edge $F_s$, the gain is nearly zero, with a maximum variation measured by the stopband ripple $\delta_s$. On the decibel scale, unit gain corresponds to 0 dB. The *maximum* passband attenuation and *minimum* stopband attenuation are measured by $A_p$-dB and $A_s$-dB, respectively. These correspond to a *minimum* passband gain of $-A_p$-dB and a *maximum* stopband gain of $-A_s$-dB, respectively

### 9.1.2 Techniques of Digital Filter Design

Digital filter design revolves around two distinctly different approaches. If linear phase is not critical, IIR filters yield a much smaller filter order for a given application. Only FIR filters can be designed with linear phase (no phase distortion).

The design starts with an analog *lowpass prototype* based on the given specifications. It is then converted to the required digital filter, using an appropriate *mapping* and an appropriate *spectral transformation*. A causal, stable IIR filter can never display linear-phase for several reasons. The transfer function of a linear-phase filter must correspond to a symmetric sequence and ensure that $H(z) = \pm H(-1/z)$. For every pole inside the unit circle, there is a reciprocal pole outside the unit circle. This makes the system unstable (if causal) or noncausal (if stable). To make the infinitely long, symmetric impulse-response sequence of an IIR filter causal, we need an infinite delay, which is not practical, and symmetric truncation (to preserve linear phase) simply transforms the IIR filter into an FIR filter.

## 9.2 IIR Filter Design

There are two related approaches for the design of IIR digital filters. A popular method is based on using methods of well-established analog filter design, followed by a mapping that converts the analog filter to the digital filter. An alternative method is based on designing the digital filter directly, using digital equivalents of analog (or other) approximations. Any transformation of an analog filter to a digital filter should ideally preserve both the response and stability of the analog filter. In practice, this is seldom possible because of the effects of sampling.

### 9.2.1 Equivalence of Analog and Digital Systems

The impulse response $h(t)$ of an analog system may be approximated by

$$h(t) \approx \tilde{h}_a(t) = t_s \sum_{n=-\infty}^{\infty} h(t)\delta(t - nt_s) = t_s \sum_{n=-\infty}^{\infty} h(nt_s)\delta(t - nt_s) \tag{9.1}$$

Here, $t_s$ is the sampling interval corresponding to the sampling rate $S = 1/t_s$. The discrete-time impulse response $h_s[n]$ describes the samples $h(nt_s)$ of $h(t)$ and may be written as

$$h_s[n] = h(nt_s) = \sum_{k=-\infty}^{\infty} h_s[k]\delta[n-k] \tag{9.2}$$

The Laplace transform $H_a(s)$ of $\tilde{h}_a(t)$ and the $z$-transform $H_d(z)$ of $h_s[n]$ are

$$H(s) \approx H_a(s) = t_s \sum_{k=-\infty}^{\infty} h(kt_s)e^{-skt_s} \qquad H_d(z) = \sum_{k=-\infty}^{\infty} h_s[k]z^{-k} \tag{9.3}$$

Comparison suggests the equivalence $H_a(s) = t_s H_d(z)$ provided $z^{-k} = e^{-skt_s}$, or

$$z \Rightarrow e^{st_s} \qquad s \Rightarrow \ln(z)/t_s \tag{9.4}$$

These relations describe a **mapping** between the variables $z$ and $s$. Since $s = \sigma + j\omega$, where $\omega$ is the continuous frequency, we can express the complex variable $z$ as

$$z = e^{(\sigma + j\omega)t_s} = e^{\sigma t_s}e^{j\omega t_s} = e^{\sigma t_s}e^{j\Omega} \tag{9.5}$$

Here, $\Omega = \omega t_s = 2\pi f/S = 2\pi F$ is the digital frequency in radians/sample.

The sampled signal $h_s[n]$ has a periodic spectrum given by its DTFT:

$$H_p(f) = S \sum_{k=-\infty}^{\infty} H(f - kS) \tag{9.6}$$

If the analog signal $h(t)$ is band limited to $B$ and sampled above the Nyquist rate ($S > 2B$), the principal period ($-0.5 \le F \le 0.5$) of $H_p(f)$ equals $SH(f)$, which is a scaled version of the true spectrum $H(f)$. We may thus relate the analog and digital systems by

$$H(f) = t_s H_p(f) \quad \text{or} \quad H_a(s)|_{s=j2\pi f} \approx t_s H_d(z)|_{z=e^{j2\pi f/S}}, \qquad |f| < 0.5S \tag{9.7}$$

If $S < 2B$, we have aliasing, and this relationship no longer holds.

### 9.2.2 The Effects of Aliasing

The relations $z \Rightarrow e^{st_s}$ and $s \Rightarrow \ln(z)/t_s$ do not describe a one-to-one mapping between the $s$-plane and the $z$-plane. Since $e^{j\Omega}$ is periodic with period $2\pi$, all frequencies $\Omega_0 \pm 2k\pi$ (corresponding to $\omega_0 \pm k\omega_s$, where $\omega_s = 2\pi S$) are mapped to the same point in the $z$-plane. A one-to-one mapping is thus possible only if $\Omega$ lies in the principal range $-\pi \le \Omega \le \pi$ (or $0 \le \Omega \le 2\pi$), corresponding to the analog frequency range $-0.5\omega_s \le \omega \le 0.5\omega_s$ (or $0 \le \omega \le \omega_s$). Figure 9.2 illustrates how the mapping $z \Rightarrow e^{st_s}$ translates points in the $s$-domain to corresponding points in the $z$-domain.

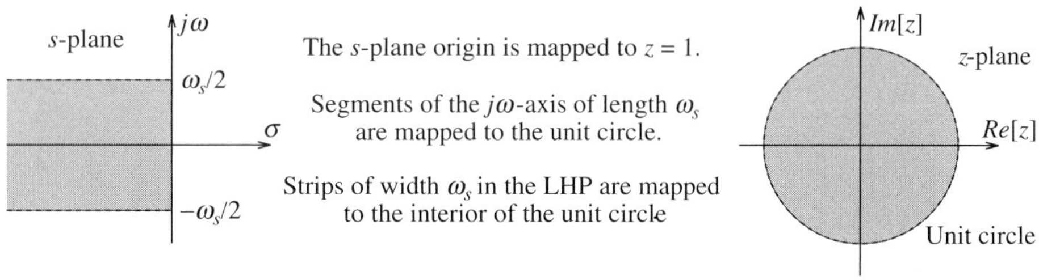


**FIGURE 9.2** Characteristics of the mapping $z \Rightarrow \exp(st_s)$. Each strip of width $\omega_s$ in the left half of the $s$-plane is mapped to the interior of the unit circle in the $z$-plane. Each segment of the $j\omega$-axis in the $s$-plane of length $\omega_s$ maps to the unit circle itself. Clearly, the mapping is not unique

**The Origin:** The origin $s = 0$ is mapped to $z = 1$, as are all other points corresponding to $s = 0 \pm jk\omega_s$ for which $z = e^{jk\omega_s t_s} = e^{jk2\pi} = 1$.

**The $j\omega$-Axis:** For points on the $j\omega$-axis, $\sigma = 0$, $z = e^{j\Omega}$, and $|z| = 1$. As $\omega$ increases from $\omega_0$ to $\omega_0 + \omega_s$, the frequency $\Omega$ increases from $\Omega_0$ to $\Omega_0 + 2\pi$, and segments of the $j\omega$-axis of length $\omega_s = 2\pi S$ thus map to the unit circle, *over and over.*

**The Left Half-Plane:** In the left half-plane, $\sigma < 0$. Thus, $z = e^{\sigma t_s} e^{j\Omega}$ or $|z| = e^{\sigma t_s} < 1$. This describes the interior of the unit circle in the $z$-plane. In other words, strips of width $\omega_s$ in the left half of the $s$-plane are mapped to the *interior* of the unit circle, *over and over.*

**The Right Half-Plane:** In the right half-plane, $\sigma > 0$, and we see that $|z| = e^{\sigma t_s} > 1$. Thus, strips of width $\omega_s$ in the right half of the $s$-plane are repeatedly mapped to the *exterior* of the unit circle.

## 9.2.3 Practical Mappings

The transcendental nature of the transformation $s \Rightarrow \ln(z)/t_s$ does not permit direct conversion of a rational transfer function $H(s)$ to a rational transfer function $H(z)$. Nor does it permit a one-to-one correspondence for frequencies higher than $0.5S$ Hz. A unique representation in the $z$-plane is possible only for band-limited signals sampled above the Nyquist rate. Practical mappings are based on one of the following methods:

1. Matching the time response (the response-invariant transformation)
2. Matching terms in a factored $H(s)$ (the matched $z$-transform)
3. Conversion of system differential equations to difference equations
4. Rational approximations for $z \Rightarrow e^{st_s}$ or $s \Rightarrow \ln(z)/t_s$

In general, each method results in different mapping rules and leads to different forms for the digital filter $H(z)$ from a given analog filter $H(s)$, and not all methods preserve stability. When comparing the frequency response, it is helpful to remember that the analog frequency range $0 \le f \le 0.5S$ for the frequency response of $H(s)$ corresponds to the digital frequency range $0 \le F \le 0.5$ for the frequency response of $H(z)$. The time-domain response can be compared only at the sampling instants $t = nt_s$.

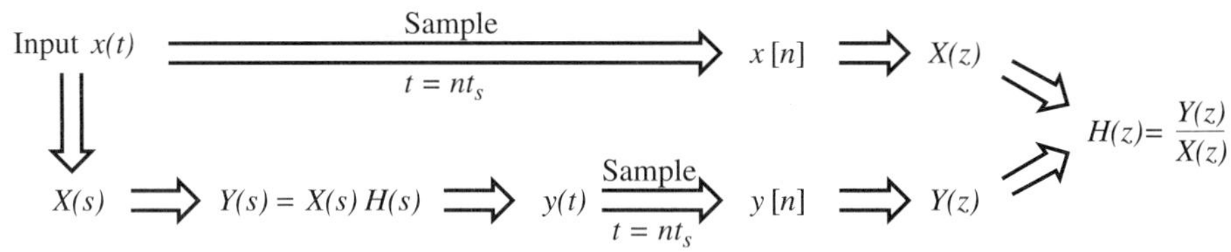


**FIGURE 9.3** The concept of response invariance. The signals $x(t)$ and $y(t)$ form the input and output of an analog filter. The ratio of the $z$-transform of their sampled versions $y[n]$ and $x[n]$ yields the transfer function $H(z)$ of the response-invariant digital filter. If the input to this digital filter is $x[n]$, its output $y[n]$ matches the the output $y(t)$ of the analog filter at the sampling instants

## 9.3 Response Matching

The idea behind response matching is to *match* the time-domain analog and digital response for a given input, typically the impulse response or step response. Given the analog filter $H(s)$ and the input $x(t)$ whose invariance we seek, we first find the analog response $y(t)$ as the inverse transform of $H(s)X(s)$. We then sample $x(t)$ and $y(t)$ at intervals $t_s$ to obtain their sampled versions $x[n]$ and $y[n]$. Finally, we compute $H(z) = Y(z)/X(z)$ to obtain the digital filter. The process is illustrated in Figure 9.3.

Response-invariant matching yields a transfer function that is a good match only for the time-domain response to the input for which it was designed. It may not provide a good match for the response to other inputs. The quality of the approximation depends on the choice of the sampling interval $t_s$, and a unique correspondence is possible only if the sampling rate $S = 1/t_s$ is above the Nyquist rate (to avoid aliasing). This mapping is thus useful only for analog systems, such as lowpass and bandpass filters, whose frequency response is essentially band limited. This also implies that the analog transfer function $H(s)$ must be *strictly proper* (with numerator degree *less than* the denominator degree).

**Example 9.1**

### Response-Invariant Mappings

**(a)** Convert $H(s) = \dfrac{1}{s+1}$ to a digital filter $H(z)$ using impulse invariance, with $t_s = 1$ s.

For impulse invariance, we select the input as $x(t) = \delta(t)$. We then find

$$X(s) = 1 \qquad Y(s) = H(s)X(s) = \frac{1}{s+1} \qquad y(t) = e^{-t}u(t)$$

The sampled versions of the input and output are

$$x[n] = \delta[n] \qquad y[n] = e^{-nt_s}u[n]$$

Taking the ratio of their $z$-transforms and using $t_s = 1$, we obtain

$$H(z) = \frac{Y(z)}{X(z)} = Y(z) = \frac{z}{z - e^{-t_s}} = \frac{z}{z - e^{-1}} = \frac{z}{z - 0.3679}$$

The frequency response of $H(s)$ and $H(z)$ is compared in Figure E9.1a(1)(a).

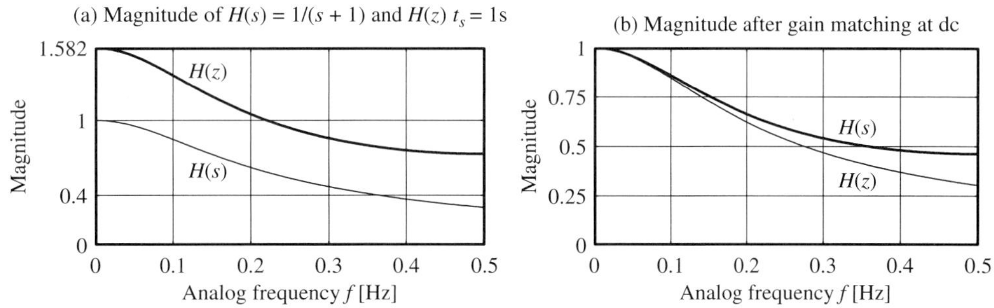


**FIGURE E.9.1.a(1)** Frequency response of $H(s)$ and $H(z)$ for Example 9.1(a)

The dc gain of $H(s)$ (at $s = 0$) is unity, but that of $H(z)$ (at $z = 1$) is 1.582. Even if we normalize the dc gain of $H(z)$ to unity, as in Figure E9.1a(1)(b), we see that the frequency response of the analog and digital filters is different. However, the analog impulse response $h(t) = e^{-t}$ matches $h[n] = e^{-n}u[n]$ at the sampling instants $t = nt_s = n$. A perfect match for the time-domain response, for which the filter was designed, lies at the heart of response-invariant mappings. The time-domain response to any other inputs will be different. For example, the step response of the analog filter is

$$S(s) = \frac{1}{s(s+1)} = \frac{1}{s} - \frac{1}{s+1} \qquad s(t) = (1 - e^{-t})u(t)$$

To find the step response $S(z)$ of the digital filter whose input is $u[n] \Leftrightarrow z/(z-1)$, we use partial fractions on $S(z)/z$ to obtain

$$S(z) = \frac{z^2}{(z-1)(z-e^{-1})} = \frac{ze/(e-1)}{z-1} + \frac{z/(1-e)}{z-e^{-1}} \qquad s[n] = \frac{e}{e-1}u[n] + \frac{1}{1-e}e^{-n}u[n]$$

The sampled version of $s(t)$ is quite different from $s[n]$. Figure E9.1a(2) reveals that, at the sampling instants $t = nt_s$, the impulse response of the two filters shows a perfect match, but the step response does not, and neither will the time-domain response to any other input.

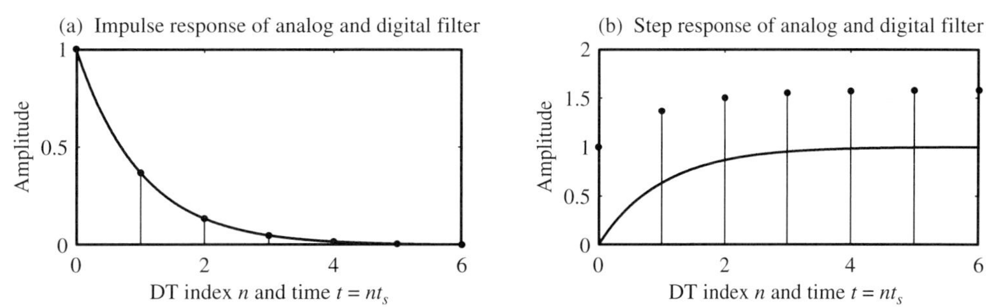


**FIGURE E.9.1.a(2)** Impulse response and step response of $H(s)$ and $H(z)$ for Example 9.1(a)

**(b)** Convert $H(s) = \dfrac{4}{(s+1)(s+2)}$ to $H(z)$, using various response-invariant transformations.

1. **Impulse invariance:** We choose $x(t) = \delta(t)$. Then, $X(s) = 1$, and

$$Y(s) = H(s)X(s) = \frac{4}{(s+1)(s+2)} = \frac{4}{s+1} - \frac{4}{s+2} \qquad y(t) = 4e^{-t}u(t) - 4e^{-2t}u(t)$$

The sampled input and output are then

$$x[n] = \delta[n] \qquad y[n] = 4e^{-nt_s}u[n] - 4e^{-2nt_s}u[n]$$

The ratio of their $z$-transforms yields the transfer function of the digital filter as

$$H_I(z) = \frac{Y(z)}{X(z)} = Y(z) = \frac{4z}{z - e^{-t_s}} - \frac{4z}{z - e^{-2t_s}}$$

2. **Step invariance:** We choose $x(t) = u(t)$. Then, $X(s) = 1/s$, and

$$Y(s) = H(s)X(s) = \frac{4}{s(s+1)(s+2)} = \frac{2}{s} - \frac{4}{s+1} + \frac{2}{s+2}$$

$$y(t) = (2 - 4e^{-t} + 2e^{-2t})u(t)$$

The sampled input and output are then

$$x[n] = u[n] \qquad y[n] = (2 - 4e^{-nt_s} + 2e^{-2nt_s})u[n]$$

Their $z$-transforms give

$$X(z) = \frac{z}{z-1} \qquad Y(z) = \frac{2z}{z-1} - \frac{4z}{z - e^{-t_s}} + \frac{2z}{z - e^{-2t_s}}$$

The ratio of their $z$-transforms yields the transfer function of the digital filter as

$$H_S(z) = \frac{Y(z)}{X(z)} = 2 - \frac{4(z-1)}{z - e^{-t_s}} + \frac{2(z-1)}{z - e^{-2t_s}}$$

3. **Ramp invariance:** We choose $x(t) = r(t) = tu(t)$. Then, $X(s) = 1/s^2$, and

$$Y(s) = \frac{4}{s^2(s+1)(s+2)} = \frac{-3}{s} + \frac{2}{s^2} + \frac{4}{s+1} - \frac{1}{s+2}$$

$$y(t) = (-3 + 2t + 4e^{-t} - e^{-2t})u(t)$$

The sampled input and output are then

$$x[n] = nt_s u[n] \qquad y[n] = (-3 + 2nt_s + 4e^{-nt_s} - e^{-2nt_s})u[n]$$

Their $z$-transforms give

$$X(z) = \frac{zt_s}{(z-1)^2} \qquad Y(z) = \frac{-3z}{z-1} + \frac{2zt_s}{(z-1)^2} + \frac{4z}{z - e^{-t_s}} - \frac{z}{z - e^{-2t_s}}$$

The ratio of their $z$-transforms yields the transfer function of the digital filter as

$$H_R(z) = \frac{-3(z-1)}{t_s} + 2 + \frac{4(z-1)^2}{t_s(z - e^{-t_s})} - \frac{(z-1)^2}{t_s(z - e^{-2t_s})}$$

### 9.3.1 The Impulse-Invariant Transformation

The impulse-invariant mapping yields some useful design relations. We start with a first-order analog filter described by $H(s) = 1/(s+p)$. The impulse response $h(t)$ and its sampled version $h[n]$ are

$$h(t) = e^{-pt}u(t) \qquad h[n] = e^{-pnt_s}u[n] = (e^{-pt_s})^n u[n] \tag{9.8}$$

**TABLE 9.1** ➤ Impulse-Invariant Transformations

| Term | Form of $H(s)$ | $H(z)$ (with $\alpha = e^{-pt_s}$) |
|---|---|---|
| Distinct | $\dfrac{A}{(s+p)}$ | $\dfrac{Az}{(z-\alpha)}$ |
| Complex conjugate | $\dfrac{Ae^{j\Omega}}{s+p+jq} + \dfrac{Ae^{-j\Omega}}{s+p-jq}$ | $\dfrac{2z^2 A\cos(\Omega) - 2A\alpha z\cos(\Omega + qt_s)}{z^2 - 2\alpha z\cos(qt_s) + \alpha^2}$ |
| Repeated twice | $\dfrac{A}{(s+p)^2}$ | $At_s\dfrac{\alpha z}{(z-\alpha)^2}$ |
| Repeated thrice | $\dfrac{A}{(s+p)^3}$ | $0.5At_s^2\dfrac{\alpha z(z+\alpha)}{(z-\alpha)^3}$ |

The $z$-transform of $h[n]$ (which has the form $\alpha^n u[n]$, where $\alpha = e^{-pt_s}$) yields the transfer function $H(z)$ of the digital filter as

$$H(z) = \frac{z}{z - e^{-pt_s}}, \qquad |z| > e^{-pt_s} \tag{9.9}$$

This relation suggests that we can go directly from $H(s)$ to $H(z)$ using the mapping

$$\frac{1}{s+p} \Longrightarrow \frac{z}{z - e^{-pt_s}} \tag{9.10}$$

We can now extend this result to filters of higher order. If $H(s)$ is in partial-fraction form, we can obtain simple expressions for impulse-invariant mapping. If $H(s)$ has no repeated roots, it can be described as a sum of first-order terms (using partial-fraction expansion), and each term can be converted by the impulse-invariant mapping to give

$$H(s) = \sum_{k=1}^{N} \frac{A_k}{s + p_k} \qquad H(z) = \sum_{k=1}^{N} \frac{zA_k}{z - e^{-p_k t_s}}, \qquad \text{ROC: } |z| > e^{-|p|_{\max} t_s} \tag{9.11}$$

Here, the region of convergence of $H(z)$ is in terms of the largest pole magnitude $|p|_{\max}$ of $H(s)$.

If the denominator of $H(s)$ also contains repeated roots, we start with a typical $k$th term $H_k(s)$ with a root of multiplicity $M$ and find

$$H_k(s) = \frac{A_k}{(s + p_k)^M} \qquad h_k(t) = \frac{A_k}{(M-1)!} t^{M-1} e^{-p_k t} u(t) \tag{9.12}$$

The sampled version $h_k[n]$ and its $z$-transform can then be found by the times-$n$ property of the $z$-transform. Similarly, quadratic terms corresponding to complex conjugate poles in $H(s)$ may also be simplified to obtain a real form. These results are listed in Table 9.1. Note that *impulse-invariant design requires $H(s)$ in partial-fraction form and yields a digital filter $H(z)$ in the same form.* It must be reassembled if we need a composite, rational-function form. The left half-plane poles of $H(s)$ (corresponding to $p_k > 0$) map into poles of $H(z)$ that lie inside the unit circle (corresponding to $z = e^{-p_k t_s} < 1$). Thus, a stable analog filter $H(s)$ is transformed into a stable digital filter $H(z)$.

**Example 9.2**

### Impulse-Invariant Mappings

**(a)** Convert $H(s) = \dfrac{4s+7}{s^2+5s+4}$ to $H(z)$ using impulse invariance at $S = 2$ Hz.

First, by partial fractions, we obtain

$$H(s) = \frac{4s+7}{s^2+5s+4} = \frac{4s+7}{(s+1)(s+4)} = \frac{3}{s+4} + \frac{1}{s+1}$$

The impulse-invariant transformation, with $t_s = 1/S = 0.5$ s, gives

$$H(z) = \frac{3z}{z - e^{-4t_s}} + \frac{z}{z - e^{-t_s}} = \frac{3z}{z - e^{-2}} + \frac{z}{z - e^{-0.5}} = \frac{4z^2 - 1.9549z}{z^2 - 0.7419z + 0.0821}$$

**(b)** Convert $H(s) = \dfrac{4}{(s+1)(s^2+4s+5)}$ to $H(z)$ using impulse invariance, with $t_s = 0.5$ s.

The partial-fraction form for $H(s)$ is

$$H(s) = \frac{2}{s+1} + \frac{-1-j}{s+2+j} + \frac{-1+j}{s+2-j}$$

For the second term, we write $K = (-1-j) = \sqrt{2}e^{-j3\pi/4} = Ae^{j\Omega}$. Thus, $A = \sqrt{2}$ and $\Omega = -3\pi/4$. We also have $p = 2$, $q = 1$, and $\alpha = e^{-pt_s} = 1/e$. With these values, Table 9.1 gives

$$H(z) = \frac{2z}{z - 1/\sqrt{e}} + \frac{2\sqrt{2}z^2\cos(-\frac{3\pi}{4}) - 2\sqrt{2}(z/e)\cos(0.5 - \frac{3\pi}{4})}{z^2 - 2(z/e)\cos(0.5) + e^{-2}}$$

This result simplifies to

$$H(z) = \frac{0.2146z^2 + 0.0930z}{z^3 - 1.2522z^2 + 0.5270z - 0.0821}$$

*Comment:* The first step involved partial fractions. Note that we cannot compute $H(z)$ as the cascade of the impulse-invariant digital filters for $H_1(s) = \dfrac{4}{s+1}$ and $H_2(s) = \dfrac{1}{s^2+4s+5}$.

## 9.3.2 Modifications to Impulse-Invariant Design

### Gain Matching

The mapping $H(s) = 1/(s + p_k) \Rightarrow H(z) = z/(z - e^{-p_k t_s})$ reveals that the dc gain of the analog term $H(s)$ equals $1/p_k$ (with $s = 0$) but the dc gain of the digital term $H(z)$ (with $z = 1$) is $1/(1 - e^{-p_k t_s})$. If the sampling interval $t_s$ is small enough such that $p_k t_s \ll 1$, we can use the approximation $e^{-p_k t_s} \approx 1 - p_k t_s$ to give the dc gain of $H(z)$ as $1/p_k t_s$. This suggests that the transfer function $H(z)$ must be multiplied by $t_s$ in order for its dc gain to closely match the dc gain of the analog filter $H(s)$. This scaling is not needed if we normalize (divide) the analog frequency specifications by the sampling frequency $S$ before designing the digital filter, because normalization is equivalent to choosing $t_s = 1$ during the mapping. In practice, regardless of normalization, it is customary to scale $H(z)$ to $KH(z)$, where the constant $K$ is chosen to match the gain of $H(s)$ and $KH(z)$ at a convenient frequency (typically, dc). Since the scale factor also changes the impulse response of the digital filter from $h[n]$ to $Kh[n]$, the design is now no longer strictly impulse invariant.

### Accounting for Sampling Errors

The impulse-invariant method suffers from errors in sampling $h(t)$ if it shows a jump at $t = 0$. If $h(0)$ is not zero, the sampled value at the origin should be chosen as $0.5h(0)$. As a result, the impulse response of the digital filter must be modified to $h_M[n] = h[n] - 0.5h(0)\delta[n]$. This leads to the modified transfer function $H_M(z) = H(z) - 0.5h(0)$. The simplest way to find $h(0)$ is to use the initial value theorem $h(0) = \lim_{s\to\infty} sH(s)$. Since $h(0)$ is nonzero only if the degree $N$ of the denominator of $H(s)$ exceeds the degree $M$ of its numerator by 1, we need this modification only if $N - M = 1$.

**Example 9.3**

## Modified Impulse-Invariant Design

**(a) Impulse-Invariant Design**

Convert the analog filter $H(s) = \dfrac{1}{s+1}$, with a cutoff frequency of 1 rad/s, to a digital filter with a cutoff frequency of $f_c = 10$ Hz and $S = 60$ Hz, using impulse- invariance and gain matching.

There are actually two ways to do this:

**1.** We normalize by the sampling frequency $S$, which allows us to use $t_s = 1$ in all subsequent computations. Normalization gives $\Omega_C = 2\pi f_C/S = \frac{\pi}{3}$. We denormalize $H(s)$ to $\Omega_C$ to get

$$H_1(s) = H\left(\frac{s}{\Omega_C}\right) = \frac{\frac{\pi}{3}}{s + \frac{\pi}{3}}$$

Finally, with $t_s = 1$, impulse invariance gives

$$H_1(z) = \frac{\frac{\pi}{3}z}{z - e^{-\pi/3}} = \frac{1.0472z}{z - 0.3509}$$

**2.** We first denormalize $H(s)$ to $f_C$ to get

$$H_2(s) = H\left(\frac{s}{2\pi f_C}\right) = \frac{20\pi}{s + 20\pi}$$

We use impulse invariance and multiply the resulting digital-filter transfer function by $t_s$ to get

$$H_2(z) = t_s\frac{20\pi z}{z - e^{-20\pi t_s}} = \frac{z\frac{\pi}{3}}{z - e^{-\pi/3}} = \frac{1.0472z}{z - 0.3509}$$

*Comment:* Both approaches yield identical results. The first method automatically accounts for the gain matching. For a perfect gain match at dc, we should multiply $H(z)$ by the gain factor

$$K = \frac{1}{H(1)} = \frac{1 - 0.3509}{1.0472} = 0.6198$$

**(b) Modified Impulse-Invariant Design**

Convert $H(s) = \dfrac{1}{s+1}$ to a digital filter with $t_s = 1$ s using modified impulse invariance to account for sampling errors and gain matching at dc.

Using impulse invariance, the transfer function of the digital filter is

$$H(z) = \frac{z}{z - e^{-1}} = \frac{z}{z - 0.3679}$$

Since $h(t) = e^{-t}u(t)$, $h(t)$ has a jump of 1 unit at $t = 0$, we get $h(0) = 1$. The modified impulse-invariant mapping thus gives

$$H_M(z) = H(z) - 0.5h(0) = \frac{z}{z - e^{-1}} - 0.5 = \frac{0.5(z + e^{-1})}{z - e^{-1}} = \frac{0.5(z + 0.3679)}{z - 0.3679}$$

The dc gain of $H(s)$ is unity. We compute the dc gain of $H(z)$ and $H_M(z)$ (with $z = 1$) as

$$H(z)|_{z=1} = \frac{1}{1-e^{-1}} = 1.582 \qquad H_M(z)|_{z=1} = 0.5\frac{1+e^{-1}}{1-e^{-1}} = 1.082$$

For unit dc gain, the transfer functions of the original and modified digital filter become

$$H_1(z) = \frac{z}{1.582(z-e^{-1})} = \frac{0.6321z}{z-0.3679}$$

$$H_{M1}(z) = \frac{0.5(z+e^{-1})}{1.082(z-e^{-1})} = \frac{0.4621(z+0.3679)}{z-0.3679}$$

Figure E9.3b compares the response of $H(s)$, $H(z)$, $H_1(z)$, $H_M(z)$, and $H_{M1}(z)$. It clearly reveals the improvement due to each modification.

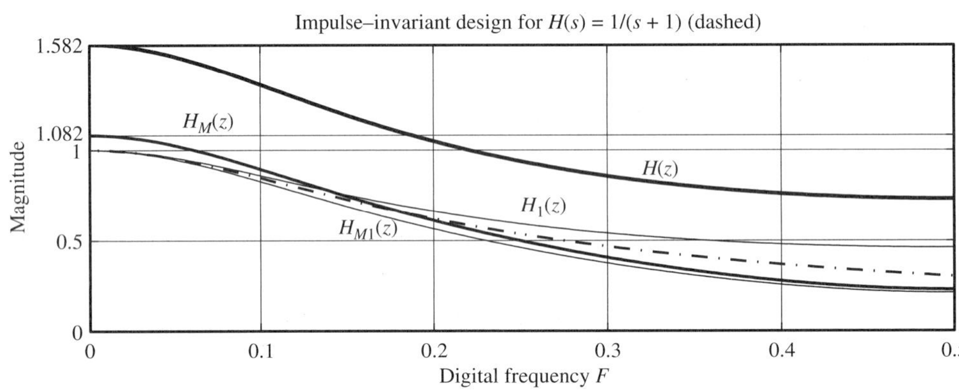


**FIGURE E 9.3.b** Response of the various filters for Example 9.3(b)

**(c) Modified Impulse-Invariant Design**

Convert the analog filter $H(s) = \dfrac{4s+7}{s^2+5s+4}$ to a digital filter with $t_s = 0.5$ s using modified impulse invariance to account for sampling errors.

Since the numerator degree is $M = 1$ and the denominator degree is $N = 2$, we have $N - M = 1$, and a modification is needed. The initial value theorem gives

$$h(0) = \lim_{s\to\infty} sH(s) = \lim_{s\to\infty} \frac{4s^2+7s}{s^2+5s+4} = \frac{4+7/s}{1+5/s+4/s^2} = 4$$

The transfer function of the digital filter using impulse-invariant mapping was found in part (a) as

$$H(z) = \frac{3z}{z-e^{-2}} + \frac{z}{z-e^{-0.5}}$$

The modified transfer function is thus

$$H_M(z) = H(z) - 0.5h(0) = \frac{3z}{z-e^{-2}} + \frac{z}{z-e^{-0.5}} - 2 = \frac{2z^2-0.4712z-0.1642}{z^2-0.7419z+0.0821}$$

**(d) Modified Impulse-Invariant Design**

Convert $H(s) = \dfrac{4}{(s+1)(s^2+4s+5)}$ to a digital filter with $t_s = 0.5$ s using modified impulse invariance (if required) to account for sampling errors.

Since the numerator degree is $M = 0$ and the denominator degree is $N = 3$, we have $N - M = 3$. Since this does not equal 1, the initial value is zero, and no modification is required. The transfer function of the digital filter using impulse-invariant mapping is thus

$$H(z) = \frac{0.2146z^2 + 0.0930z}{z^3 - 1.2522z^2 + 0.5270z - 0.0821}$$

## 9.4 The Matched $z$-Transform for Factored Forms

The impulse-invariant mapping may be expressed as

$$\frac{1}{s+\alpha} \Longrightarrow \frac{z}{z - e^{-\alpha t_s}} \qquad s + \alpha \Longrightarrow \frac{z - e^{-\alpha t_s}}{z} \tag{9.13}$$

The **matched $z$-transform** uses this mapping to convert each numerator and denominator term of a *factored* $H(s)$ to yield the digital filter $H(z)$ (also in factored form) as

$$H(s) = C\frac{\prod_{i=1}^{M}(s - z_i)}{\prod_{k=1}^{N}(s - p_k)} \qquad H(z) = (Kz^P)\frac{\prod_{i=1}^{M}(z - e^{z_i t_s})}{\prod_{k=1}^{N}(z - e^{p_k t_s})} \tag{9.14}$$

The power $P$ of $z^P$ in $H(z)$ is $P = N - M$, the difference in the degree of the denominator and numerator polynomials of $H(s)$. The constant $K$ is chosen to match the gains of $H(s)$ and $H(z)$ at some convenient frequency (typically, dc).

For complex roots, we can replace each conjugate pair using the mapping

$$(s + p - jq)(s + p + jq) \Longrightarrow \frac{z^2 - 2ze^{-pt_s}\cos(qt_s) + e^{-2pt_s}}{z^2} \tag{9.15}$$

Since poles in the left half of the $s$-plane are mapped inside the unit circle in the $z$-plane, the matched $z$-transform preserves stability. It converts an all-pole analog system to an all-pole digital system but may not preserve the frequency response of the analog system. As with the impulse-invariant mapping, the matched $z$-transform also suffers from aliasing errors.

### 9.4.1 Modifications to Matched $z$-Transform Design

The matched $z$-transform maps the zeros of $H(s)$ at $s = \infty$ (corresponding to the highest analog frequency $f = \infty$) to $z = 0$. This yields the term $z^P$ in the design relation. In the modified matched $z$-transform, some or all of these zeros are mapped to $z = -1$ (corresponding to the highest digital frequency $F = 0.5$ or the analog frequency $f = 0.5S$). Two modifications to $H(z)$ are in common use:

1. Move *all* zeros from $z = 0$ to $z = -1$ (or replace $z^P$ by $(z + 1)^P$) in $H(z)$.
2. Move *all but one* of the zeros at $z = 0$ to $z = -1$ (or replace $z^P$ by $z(z + 1)^{P-1}$ in $H(z)$).

These modifications allow us to use the matched $z$-transform even for highpass and bandstop filters.

**Example 9.4**

### The Matched z-Transform

Convert $H(s) = \frac{4}{(s+1)(s+2)}$ to $H(z)$ by the matched $z$-transform and its modifications, with $t_s = 0.5$ s.

**Solution:**

**(a) No modification:** The matched $z$-transform yields

$$H(z) = \frac{Kz^2}{(z - e^{-t_s})(z - e^{-2t_s})} = \frac{Kz^2}{(z - e^{-0.5})(z - e^{-1})} = \frac{Kz^2}{z^2 - 0.9744z + 0.2231}$$

**(b) First modification:** Replace both zeros in $H(z)$ (the term $z^2$) by $(z+1)^2$:

$$H_1(z) = \frac{K_1(z+1)^2}{(z - e^{-t_s})(z - e^{-2t_s})} = \frac{K_1(z+1)^2}{(z - e^{-0.5})(z - e^{-1})} = \frac{K_1(z+1)^2}{z^2 - 0.9744z + 0.2231}$$

**(c) Second modification:** Replace only one zero in $H(z)$ (the term $z^2$) by $z + 1$:

$$H_2(z) = \frac{K_2z(z+1)}{(z - e^{-t_s})(z - e^{-2t_s})} = \frac{K_2z(z+1)}{(z - e^{-0.5})(z - e^{-1})} = \frac{K_2z(z+1)}{z^2 - 0.9744z + 0.2231}$$

*Comment:* The constants $K$, $K_1$, and $K_2$ may be chosen for a desired gain.

## 9.5 Mappings from Discrete Algorithms

Discrete-time algorithms are often used to develop mappings to convert analog filters to digital filters. In the discrete domain, the operations of integration and differentiation are replaced by numerical integration and numerical differences, respectively. The mappings are derived by equating the transfer function $H(s)$ of the ideal analog operation with the transfer function $H(z)$ of the corresponding discrete operation.

### 9.5.1 Mappings from Difference Algorithms

Numerical differences are often used to convert differential equations to difference equations. Three such difference algorithms are listed in Table 9.2.

Numerical differences approximate the slope (derivative) at a point, as illustrated for the backward-difference and forward-difference algorithms in Figure 9.4.

**TABLE 9.2** ➤ Numerical Difference Algorithms

| Difference | Numerical Algorithm | Mapping for $s$ |
|---|---|---|
| Backward | $y[n] = \frac{x[n] - x[n-1]}{t_s}$ | $s = \frac{z-1}{zt_s}$ |
| Forward | $y[n] = \frac{x[n+1] - x[n]}{t_s}$ | $s = \frac{z-1}{t_s}$ |
| Central | $y[n] = \frac{x[n+1] - x[n-1]}{2t_s}$ | $s = \frac{z^2-1}{2zt_s}$ |

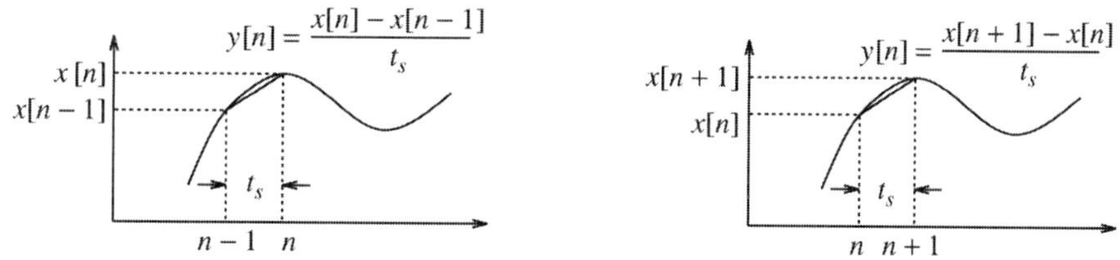


**FIGURE 9.4** Backward-difference and forward-difference algorithms approximate the current slope as the ratio of the rise and run. In the backward-difference algorithm, the rise is the difference between the current and previous sample values. The forward-difference measures the rise as the difference between the next and current sample values

The mappings that result from the operations (also listed in Table 9.2) are based on comparing the ideal derivative operator $H(s) = s$ with the transfer function $H(z)$ of each difference algorithm, as follows:

**Backward Difference:** $y[n] = \dfrac{x[n] - x[n-1]}{t_s}$

$$Y(z) = \frac{X(z) - z^{-1}X(z)}{t_s} \qquad H(z) = \frac{Y(z)}{X(z)} = \frac{z-1}{zt_s} \qquad s = \frac{z-1}{zt_s} \tag{9.16}$$

**Forward Difference:** $y[n] = \dfrac{x[n+1] - x[n]}{t_s}$

$$Y(z) = \frac{zX(z) - X(z)}{t_s} \qquad H(z) = \frac{Y(z)}{X(z)} = \frac{z-1}{t_s} \qquad s = \frac{z-1}{t_s} \tag{9.17}$$

**Central Difference:** $y[n] = \dfrac{x[n+1] - x[n-1]}{2t_s}$

$$Y(z) = \frac{zX(z) - z^{-1}X(z)}{2t_s} \qquad H(z) = \frac{Y(z)}{X(z)} = \frac{z^2-1}{2zt_s} \qquad s = \frac{z^2-1}{2zt_s} \tag{9.18}$$

### 9.5.2 Stability Properties of the Backward-Difference Algorithm

The backward-difference mapping $s \to (z-1)/zt_s$ results in

$$z = 0.5 + 0.5\frac{1+st_s}{1-st_s} \qquad z - 0.5 = 0.5\frac{1+st_s}{1-st_s} \tag{9.19}$$

To find where the $j\omega$-axis maps, we set $\sigma = 0$ to obtain

$$z - 0.5 = 0.5\frac{1+j\omega t_s}{1-j\omega t_s} \qquad |z - 0.5| = 0.5 \tag{9.20}$$

Thus, the $j\omega$-axis is mapped into a circle of radius 0.5 and centered at $z = 0.5$, as shown in Figure 9.5.

Since this region is within the unit circle, the mapping preserves stability. It does, however, restrict the pole locations of the digital filter. Since the frequencies are mapped into a smaller circle, this mapping is a good approximation only in the vicinity of $z = 1$ or $\Omega \approx 0$ (where it approximates the unit circle), which implies high sampling rates.

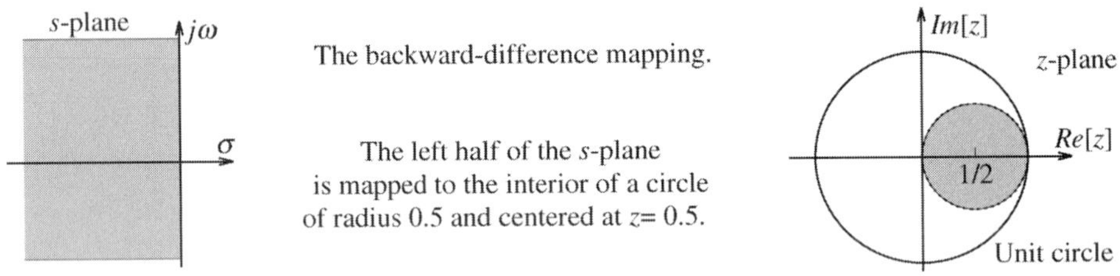


**FIGURE 9.5** Mapping region for the mapping based on the backward difference. The *entire* left half of the s-plane is mapped to the interior of a circle of radius 0.5 and centered at $z = 0.5$. This restricts the pole locations of a stable digital filter only to this circular region, instead of the interior of the entire unit circle

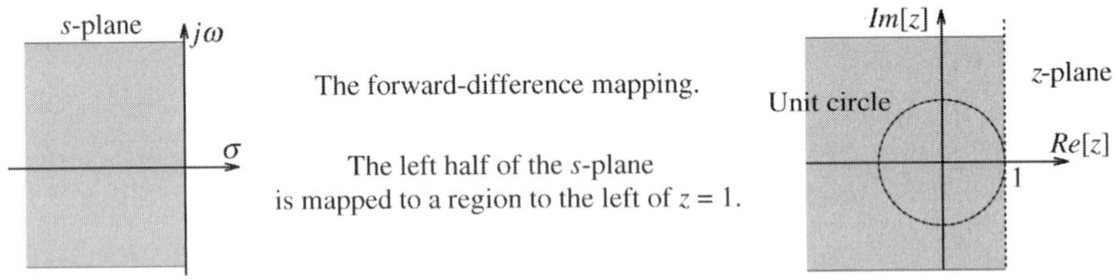


**FIGURE 9.6** Mapping region for the mapping based on the forward difference. The *entire* left half of the s-plane is mapped to the left of the line $z = 1$ in the z-plane that also includes the unit circle. This transformation is not always stable because some of the poles of a stable analog filter may be mapped to regions outside the unit circle

### 9.5.3 The Forward-Difference Algorithm

The mapping for the forward-difference is $s \rightarrow (z-1)/t_s$. With $z = u + jv$, we obtain

$$z = u + jv = 1 + t_s(\sigma + j\omega) \tag{9.21}$$

If $\sigma = 0$, then $u = 1$, and the $j\omega$-axis maps to $z = 1$. If $\sigma > 0$, then $u > 1$, and the right half of the $s$-plane maps to the right of $z = 1$. If $\sigma < 0$, then $u < m1$, and the left half of the $s$-plane maps to the left of $z = 1$, as shown in Figure 9.6. This region (in the $z$-plane) includes not only the unit circle but also a vast region outside it. Thus, a stable analog filter with poles in the left half of the $s$-plane may result in an unstable digital filter $H(z)$ with poles anywhere to the left of $z = 1$ (but not inside the unit circle) in the $z$-plane !

**Example 9.5** **Mappings from Difference Algorithms**

**(a)** We convert the stable analog filter $H(s) = \dfrac{1}{s+\alpha}$, $\alpha > 0$ to a digital filter using the backward-difference mapping $s = (z-1)/zt_s$ to obtain

$$H(z) = \frac{1}{\alpha + (z-1)/zt_s} = \frac{zt_s}{(1+\alpha t_s)z - 1}$$

The digital filter has a pole at $z = 1/(1+\alpha t_s)$. Since this is always less than unity if $\alpha > 0$ (for a stable $H(s)$) and $t_s > 0$, we have a stable $H(z)$.

**(b)** We convert the stable analog filter $H(s) = \frac{1}{s+\alpha}$, $\alpha > 0$ to a digital filter using the forward-difference mapping $s = (z-1)/t_s$ to obtain

$$H(z) = \frac{1}{\alpha + (z-1)/t_s} = \frac{t_s}{z - (1 - \alpha t_s)}$$

The digital filter has a pole at $z = 1 - \alpha t_s$ and is thus stable only if $0 < \alpha t_s < 2$ (to ensure $|z| < 1$). Since $\alpha > 0$ and $t_s > 0$, we are assured a stable system only if $\alpha < 2/t_s$. This implies that the sampling rate $S$ must be chosen to ensure that $S > 0.5\alpha$.

**(c)** We convert the stable analog filter $H(s) = \frac{1}{s+\alpha}$, $\alpha > 0$ to a digital filter using the central-difference mapping $s = (z^2-1)/2zt_s$ to obtain

$$H(z) = \frac{1}{\alpha + (z^2-1)/2zt_s} = \frac{2zt_s}{z^2 + 2\alpha t_s z - 1}$$

The digital filter has a pair of poles at $z = -\alpha t_s \pm \sqrt{(\alpha t_s)^2 + 1}$. The magnitude of one of the poles is always greater than unity, and the digital filter is thus unstable for any $\alpha > 0$.

*Comment:* Clearly, from a stability viewpoint, only the mapping based on the backward difference is useful for the filter $H(s) = 1/(s+\alpha)$. In fact, this mapping preserves stability for any stable analog filter.

## 9.5.4 Mappings from Integration Algorithms

Two commonly used algorithms for numerical integration are based on the rectangular rule and trapezoidal rule. These integration algorithms, listed in Table 9.3, estimate the "area" $y[n]$ from $y[n-1]$ by using step interpolation (for the rectangular rule) or linear interpolation (for the trapezoidal rule) between the samples of $x[n]$, as illustrated in Figure 9.7.

**TABLE 9.3 ➤** Numerical Integration Algorithms.

| Rule | Numerical Algorithm | Mapping for $s$ |
|---|---|---|
| Rectangular | $y[n] = y[n-1] + t_s x[n]$ | $s = \frac{1}{t_s}\left(\frac{z-1}{z}\right)$ |
| Trapezoidal | $y[n] = y[n-1] + 0.5t_s\,(x[n] + x[n-1])$ | $s = \frac{2}{t_s}\left(\frac{z-1}{z+1}\right)$ |

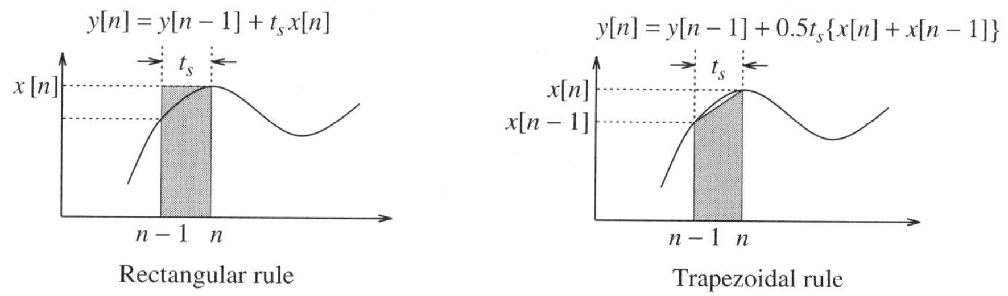


**FIGURE 9.7** Illustrating two numerical-integration algorithms. The rectangular rule is based on a staircase approximation of the signal and sums the area of rectangular strips. The trapezoidal rule is based on a piecewise-linear approximation of the signal and sums the area of trapezoidal strips

The mappings resulting from these operators are also listed in Table 9.3 and are based on comparing the transfer function $H(s) = 1/s$ of the ideal integrator with the transfer function $H(z)$ of each integration algorithm, as follows:

**Rectangular Rule:** $y[n] = y[n-1] + t_s x[n]$

$$Y(z) = z^{-1}Y(z) + t_s X(z) \qquad H(z) = \frac{Y(z)}{X(z)} = \frac{zt_s}{z-1} \qquad s = \frac{z-1}{zt_s} \tag{9.22}$$

We remark that the rectangular algorithm for integration and the backward difference for the derivative generate identical mappings.

**Trapezoidal Rule:** $y[n] = y[n-1] + 0.5t_s(x[n] + x[n-1])$

$$Y(z) = z^{-1}Y(z) + 0.5t_s[X(z) + z^{-1}X(z)$$

$$H(z) = \frac{Y(z)}{X(z)} = \frac{0.5t_s(z+1)}{z-1} \qquad s = \frac{2}{t_s}\left[\frac{z-1}{z+1}\right] \tag{9.23}$$

The mapping based on the trapezoidal rule is also called **Tustin's rule**.

### 9.5.5 Stability Properties of Integration-Algorithm Mappings

Mappings based on the rectangular and trapezoidal algorithms always yield a stable $H(z)$ for any stable $H(s)$, and any choice of $t_s$. The rectangular rule is equivalent to the backward-difference algorithm. It thus maps the left half of the $s$-plane to the interior of a circle of radius 0.5 in the $z$-plane, centered at $z = 0.5$, and is always stable. For the trapezoidal rule, we express $z$ in terms of $s = \sigma + j\omega$ to get

$$z = \frac{2 + st_s}{2 - st_s} = \frac{2 + \sigma t_s + j\omega t_s}{2 - \sigma t_s - j\omega t_s} \tag{9.24}$$

If $\sigma = 0$, we get $|z| = 1$, and for $\sigma < 0$, $|z| < 1$. Thus, the $j\omega$-axis is mapped to the unit circle, and the left half of the $s$-plane is mapped into the interior of the unit circle, as shown in Figure 9.8. This means that a stable analog system will always yield a stable digital system using this transformation. If $\omega = 0$, we have $z = 1$, and the dc gain of both the analog and digital filters is identical.

Discrete difference and integration algorithms are good approximations only for small digital frequencies ($F < 0.1$, say) or high sampling rates $S$ (small $t_s$) that may be well in excess of the Nyquist rate. This is why the sampling rate is a critical factor in the frequency-domain performance of these algorithms. Another factor is stability. For example,

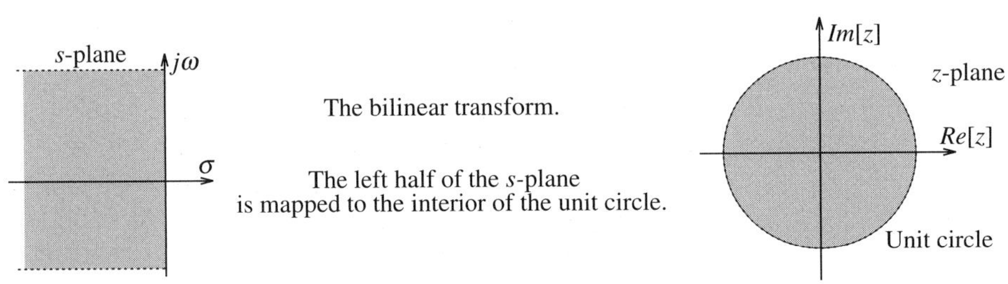


**FIGURE 9.8** Mapping region for the trapezoidal integration algorithm (or bilinear transform). The entire left-half of the $s$-plane is mapped just once to the interior of the unit circle in the $z$-plane while the entire $j\omega$-axis is mapped to the unit circle itself

the mapping based on the central-difference algorithm is not very useful because it always produces an unstable digital filter. Algorithms based on trapezoidal integration and the backward difference are popular, because they always produce stable digital filters.

**Example 9.6**

### Mappings from Integration Algorithms

Convert $H(s) = \frac{1}{s+\alpha}$, $\alpha > 0$ to a digital filter $H(z)$ using the trapezoidal numerical-integration algorithm and comment on the stability of $H(z)$.

**(a)** Using the the mapping based on the trapezoidal rule, we obtain

$$H(z) = H(s)\Big|_{s=\frac{2(z-1)}{t_s(z+1)}} = \frac{t_s(z+1)}{(2+\alpha t_s)z - (2-\alpha t_s)}$$

The pole location of $H(z)$ is $z = (2-\alpha t_s)/(2+\alpha t_s)$. Since this is always less than unity (if $\alpha > 0$ and $t_s > 0$), we have a stable $H(z)$.

**(b)** Simpson's algorithm for numerical integration finds $y[n]$ over two time steps from $y[n-2]$ and is given by

$$y[n] = y[n-2] + \frac{t_s}{3}(x[n] + 4x[n-1] + x[n-2])$$

Derive a mapping based on Simpson's rule, use it to convert $H(s) = \frac{1}{s+\alpha}$, $\alpha > 0$ to a digital filter $H(z)$ and comment on the stability of $H(z)$.
The transfer function $H_S(z)$ of this algorithm is found as follows:

$$Y(z) = z^{-2}Y(z) + \frac{t_s}{3}(1 + 4z^{-1} + z^{-2})X(z) \qquad H_S(z) = t_s\frac{z^2+4z+1}{3(z^2-1)}$$

Comparison with the transfer function of the ideal integrator $H(s) = 1/s$ gives

$$s = \frac{3}{t_s}\left[\frac{z^2-1}{z^2+4z+1}\right]$$

If we convert $H(s) = 1/(s+\alpha)$ using this mapping, we obtain

$$H(z) = \frac{t_s(z^2+4z+1)}{(3+\alpha t_s)z^2 + 4\alpha t_s z - (3-\alpha t_s)}$$

The poles of $H(z)$ are the roots of $(3+\alpha t_s)z^2 + 4\alpha t_s z - (3-\alpha t_s) = 0$. The magnitude of one of these roots is always greater than unity (if $\alpha > 0$ and $t_s > 0$), and $H(z)$ is thus an unstable filter.

## 9.5.6 Frequency Response of Discrete Algorithms

The frequency response of an ideal integrator is $H_I(F) = 1/j2\pi F$. For an ideal differentiator, $H_D(F) = j2\pi F$. In the discrete domain, the operations of integration and differentiation are replaced by numerical integration and numerical differences. Table 9.4 lists some of the algorithms and their frequency response $H(F)$.

Numerical differences approximate the slope (derivative) at a point and numerical-integration algorithms estimate the area $y[n]$ from $y[n-1]$ by using step, linear, or quadratic interpolation. Only Simpson's rule finds $y[n]$ over two time steps from $y[n-2]$. Discrete algorithms are good approximations only at low digital frequencies ($F < 0.1$, say). Even

**TABLE 9.4** ➤ Discrete Algorithms and their Frequency Response

| Algorithm | Difference Equation | Frequency Response |
|---|---|---|
| | **Numerical Differences** | |
| Backward | $y[n] = x[n] - x[n-1]$ | $H_B(F) = 1 - e^{-j2\pi F}$ |
| Central | $y[n] = \frac{1}{2}(x[n+1] - x[n-1])$ | $H_C(F) = j\sin(2\pi F)$ |
| Forward | $y[n] = x[n+1] - x[n]$ | $H_F(F) = e^{j2\pi F} - 1$ |
| | **Numerical Integration** | |
| Rectangular | $y[n-1] + x[n]$ | $H_R(F) = \dfrac{1}{1 - e^{-j2\pi F}}$ |
| Trapezoidal | $y[n-1] + \frac{1}{2}(x[n] + x[n-1])$ | $H_T(F) = \dfrac{1 + e^{-j2\pi F}}{2(1 - e^{-j2\pi F})}$ |
| Simpson's | $y[n-2] + \frac{1}{3}(x[n] + 4x[n-1] + x[n-2])$ | $H_S(F) = \dfrac{1 + 4e^{-j2\pi F} + e^{-j4\pi F}}{3(1 - e^{-j4\pi F})}$ |

when we satisfy the sampling theorem, low digital frequencies mean high sampling rates, well in excess of the Nyquist rate. This is why the sampling rate is a critical factor in the frequency-domain performance of these operators. Another factor is stability. For example, if Simpson's rule is used to convert analog systems to discrete systems, it results in an unstable system. The trapezoidal integration algorithm and the backward difference are two popular choices.

**Example 9.7**

### DTFT of Numerical Algorithms

**(a)** For the trapezoidal numerical-integration operator, the DTFT yields

$$Y_p(F) = Y_p(F)e^{-j2\pi F} + 0.5[X_p(F) + e^{-j2\pi F}X_p(F)]$$

A simple rearrangement leads to

$$H_T(F) = \frac{Y_p(F)}{X_p(F)} = 0.5\left[\frac{1 + e^{-j2\pi F}}{1 - e^{-j2\pi F}}\right] = \frac{1}{j2\tan(\pi F)}$$

Normalizing this by $H_I(F) = 1/j2\pi F$ and expanding the result, we obtain

$$\frac{H_T(F)}{H_I(F)} = \frac{j2\pi F}{j2\tan(\pi F)} = \frac{\pi F}{\tan(\pi F)} \approx 1 - \frac{(\pi F)^2}{3} - \frac{(\pi F)^4}{45} - \cdots$$

Figure E9.7 shows the magnitude and phase *error* by plotting the ratio $H_T(F)/H_I(F)$. For an ideal algorithm, this ratio should equal unity at all frequencies. At low frequencies ($F \ll 1$), we have $\tan(\pi F) \approx \pi F$ and $H_T(F)/H_I(F) \approx 1$, and the trapezoidal rule is a valid approximation to integration. The phase response matches the ideal phase at all frequencies.

**(b)** Simpson's numerical-integration algorithm yields the following normalized result:

$$\frac{H_S(F)}{H_I(F)} = 2\pi F\left(\frac{2 + \cos 2\pi F}{3\sin 2\pi F}\right)$$

It displays a perfect phase match for all frequencies, but has an overshoot in its magnitude response past $F = 0.25$, and thus amplifies high frequencies. Note that Simpson's algorithm yields a mapping rule that results in an unstable filter.

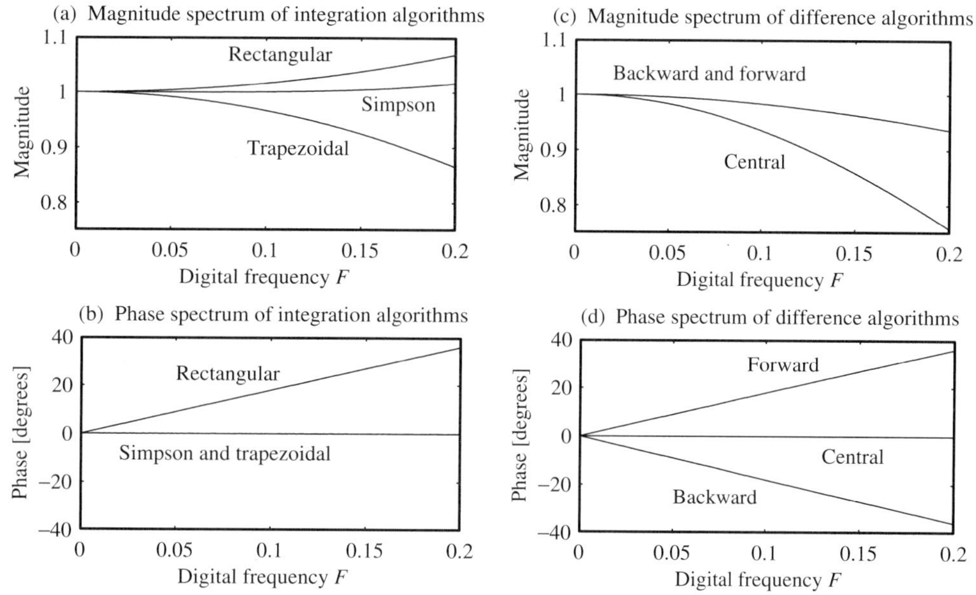


**FIGURE E 9.7.a** Frequency response of the numerical algorithms for Example 9.7

**(c)** For the forward difference operator, the DTFT yields

$$Y_p(F) = X_p(F)e^{j2\pi F} - X_p(F) = X_p(F)[e^{j2\pi F} - 1] \quad H_F(F) = \frac{Y_p(F)}{X_p(F)} = e^{j2\pi F} - 1$$

The ratio $H_F(F)/H_D(F)$ may be expanded as

$$\frac{H_F(F)}{H_D(F)} = 1 + j\frac{1}{2!}(2\pi F) - \frac{1}{3!}(2\pi F)^2 + \cdots$$

Again, we observe correspondence only at low digital frequencies (or high sampling rates). The high frequencies are amplified, making the algorithm susceptible to high-frequency noise. The phase response also deviates from the true phase, especially at high frequencies.

**(d)** For the central difference algorithm, we find $H_C(F)/H_D(F)$ as

$$\frac{H_C(F)}{H_D(F)} = \frac{\sin(2\pi F)}{2\pi F} = 1 - \frac{1}{3!}(2\pi F)^2 + \frac{1}{5!}(2\pi F)^4 + \cdots$$

We see a perfect match only for the phase at all frequencies.

## 9.5.7 Mappings from Rational Approximations

Some of the mappings that we have derived from numerical algorithms may also be viewed as rational approximations of the transformations $z \rightarrow e^{st_s}$ and $s \rightarrow \ln(z)/t_s$. The forward-difference mapping is based on a first-order approximation for $z = e^{st_s}$ and yields

$$z = e^{st_s} \approx 1 + st_s, \quad st_s \ll 1 \qquad s \approx \frac{z-1}{t_s} \tag{9.25}$$

The backward-difference mapping is based on a first-order approximation for $z^{-1} = e^{-st_s}$ and yields

$$\frac{1}{z} = e^{-st_s} \approx 1 - st_s, \quad st_s \ll 1 \qquad s \approx \frac{z-1}{zt_s} \tag{9.26}$$

The trapezoidal mapping is based on a first-order rational function approximation of $s = \ln(z)/t_s$ with $\ln(z)$ described by a power series and yields

$$s = \frac{\ln(z)}{t_s} = \frac{2}{t_s}\left[\frac{z-1}{z+1}\right] + \frac{2}{3t_s}\left[\frac{z-1}{z+1}\right]^2 + \frac{2}{5t_s}\left[\frac{z-1}{z+1}\right]^2 + \cdots \approx \frac{2}{t_s}\left[\frac{z-1}{z+1}\right] \tag{9.27}$$

## 9.6 The Bilinear Transformation

If we generalize the mapping based on the trapezoidal rule by letting $C = 2/t_s$, we obtain the **bilinear transformation**, which is defined by

$$s = C\frac{z-1}{z+1} \qquad z = \frac{C+s}{C-s} \tag{9.28}$$

If we let $\sigma = 0$, we obtain the complex variable $z$ in the form

$$z = \frac{C+j\omega}{C-j\omega} = e^{j2\tan^{-1}(\omega/C)} \tag{9.29}$$

Since $z = e^{j\Omega}$, where $\Omega = 2\pi F$ is the digital frequency, we find

$$\Omega = 2\tan^{-1}(\omega/C) \qquad \omega = C\tan(0.5\Omega) \tag{9.30}$$

This is a *nonlinear* relation between the analog frequency $\omega$ and the digital frequency $\Omega$. When $\omega = 0$, $\Omega = 0$, and as $\omega \to \infty$, $\Omega \to \pi$. It is thus a one-to-one mapping that *nonlinearly compresses* the analog frequency range $-\infty < f < \infty$ to the digital frequency range $-\pi < \Omega < \pi$. It avoids the effects of aliasing at the expense of distorting, compressing, or *warping* the analog frequencies, as shown in Figure 9.9.

The higher the frequency, the more severe is the warping. We can compensate for this warping (but not eliminate it) if we **prewarp** the frequency specifications before designing

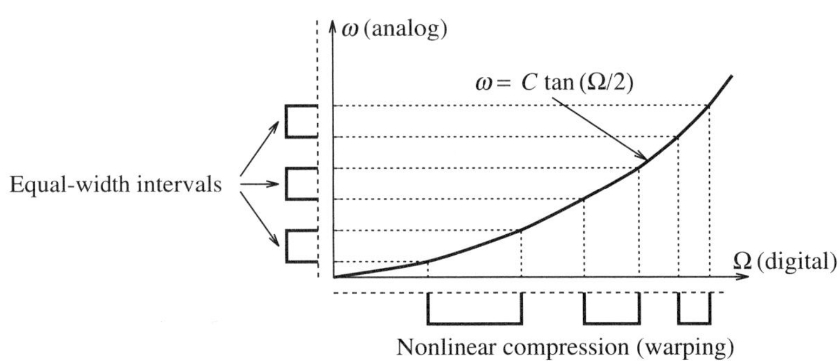


**FIGURE 9.9** The warping effect of the bilinear transformation is a nonlinear compression of the infinite analog frequency range to the finite digital frequency range (of $\Omega = 2\pi$). Thus, equal-width analog frequency regions are compressed more at higher frequencies, as shown

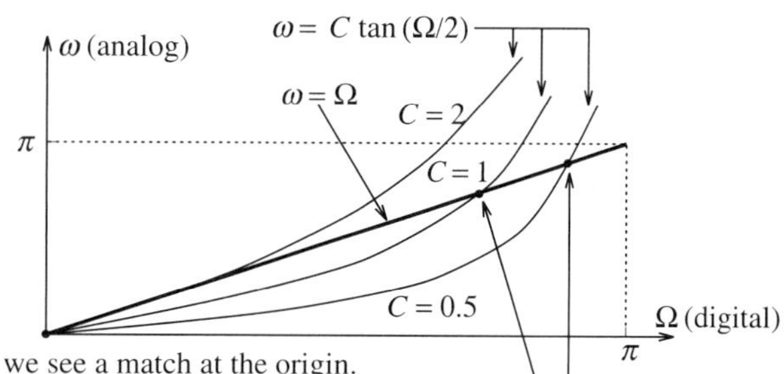


**FIGURE 9.10** The warping relation $\omega = C\tan(0.5\Omega)$ for various choices of $C$. The ideal relation is $\omega = \Omega$. At $\omega = 0$, both relations give $\Omega = 0$. For choices of $C < 2$ however, the graphs of the two plots intersect at one more point and the warping relation matches the ideal relation at one more frequency

the analog system $H(s)$ or applying the bilinear transformation. Prewarping of the frequencies prior to analog design is just a scaling (stretching) operation based on the inverse of the warping relation and is given by

$$\omega = C\tan(0.5\Omega) \tag{9.31}$$

Figure 9.10 shows a plot of $\omega$ versus $\Omega$ for various values of $C$, compared with the linear relation $\omega = \Omega$. The analog and digital frequencies always show a match at the origin ($\omega = \Omega = 0$) and at *one other value* dictated by the choice of $C$.

We point out that the nonlinear stretching effect of the prewarping often results in a filter of lower order, especially if the sampling frequency is not high enough. For high sampling rates, it turns out that the prewarping has little effect and may even be redundant.

The popularity of the bilinear transformation stems from its simple, stable, one-to-one mapping. It avoids problems caused by aliasing and can thus be used even for highpass and bandstop filters. Though it does suffer from warping effects, it can also compensate for these effects using a simple relation.

### 9.6.1 Using the Bilinear Transformation

Given an analog transfer function $H(s)$ whose response at the analog frequency $\omega_A$ is to be matched to $H(z)$ at the digital frequency $\Omega_D$, we may design $H(z)$ in one of two ways:

1. We pick $C$ by matching $\omega_A$ and the prewarped frequency $\Omega_D$ and obtain $H(z)$ from $H(s)$ using the transformation $s = C\frac{z-1}{z+1}$. This process may be summarized as follows:

$$\omega_A = C\tan(0.5\Omega_D) \qquad C = \frac{\omega_A}{\tan(0.5\Omega_D)} \qquad H(z) = H(s)|_{s=C(z-1)/(z+1)} \tag{9.32}$$

2. We pick a convenient value for $C$ (say, $C = 1$). This actually matches the response at an arbitrary prewarped frequency $\omega_x$ given by

$$\omega_x = \tan(0.5\Omega_D) \tag{9.33}$$

Next, we frequency scale $H(s)$ to $H_1(s) = H(s\omega_A/\omega_x)$, and obtain $H(z)$ from $H_1(s)$, using the transformation $s = \frac{z-1}{z+1}$ (with $C = 1$). This process may be summarized as follows (for $C = 1$):

$$\omega_x = \tan(0.5\Omega_D) \qquad H_1(s) = H(s)|_{s=s\omega_A/\omega_x} \qquad H(z) = H_1(s)|_{s=(z-1)/(z+1)} \tag{9.34}$$

These two methods yield an identical digital filter $H(z)$. The first method does away with the scaling of $H(s)$, and the second method allows a convenient choice for $C$.

**Example 9.8**

### Using the Bilinear Transformation

**(a)** Consider a Bessel filter described by $H(s) = \dfrac{3}{s^2 + 3s + 3}$. Design a digital filter whose magnitude at $f_0 = 3$ kHz equals the magnitude of $H(s)$ at $\omega_A = 4$ rad/s if the sampling rate $S = 12$ kHz.

The digital frequency is $\Omega = 2\pi f_0/S = 0.5\pi$. We can now proceed in one of two ways:

1. **Method 1:** We select $C$ by choosing the prewarped frequency to equal $\omega_A = 4$:

$$\omega_A = 4 = C\tan(0.5\Omega) \quad \text{or} \quad C = \frac{4}{\tan(0.5\Omega)} = 4$$

We transform $H(s)$ to $H(z)$ using $s = C\frac{z-1}{z+1} = \frac{4(z-1)}{z+1}$, to obtain

$$H(z) = H(s)|_{s=4(z-1)/(z+1)} = \frac{3(z+1)^2}{31z^2 - 26z + 7}$$

2. **Method 2:** We choose $C = 1$, say, and evaluate $\omega_x = \tan(0.5\Omega) = \tan(0.25\pi) = 1$.

Next, we frequency scale $H(s)$ to

$$H_1(s) = H(s\omega_A/\omega_x) = H(4s) = \frac{3}{16s^2 + 12s + 3}$$

Finally, we transform $H_1(s)$ to $H(z)$ using $s = \frac{z-1}{z+1}$ to obtain

$$H(z) = H(s)|_{s=(z-1)/(z+1)} = \frac{3(z+1)^2}{31z^2 - 26z + 7}$$

The magnitude $|H(s)|$ at $s = j\omega = j4$ matches the magnitude of $|H(z)|$ at $z = e^{j\Omega} = e^{j\pi/2} = j$.

We find that

$$|H(s)|_{s=j4} = \left|\frac{3}{-13 + j12}\right| = 0.1696 \qquad |H(z)|_{z=j} = \left|\frac{3(j+1)^2}{-24 - j26}\right| = 0.1696$$

Figure E9.8(a) compares the magnitude of $H(s)$ and $H(z)$. The linear phase of the Bessel filter is not preserved during the transformation (unless the sampling frequency is very high).

**(b)** The twin-$T$ notch filter $H(s) = \dfrac{s^2 + 1}{s^2 + 4s + 1}$ has a notch frequency $\omega_0 = 1$ rad/s. Design a digital notch filter with $S = 240$ Hz and a notch frequency $f = 60$ Hz.

The digital notch frequency is $\Omega = 2\pi f/S = 0.5\pi$. We pick $C$ by matching the analog notch frequency $\omega_0$ and the prewarped digital notch frequency $\Omega$ to get

$$\omega_0 = C\tan(0.5\Omega) \qquad 1 = C\tan(0.25\pi) \qquad C = 1$$

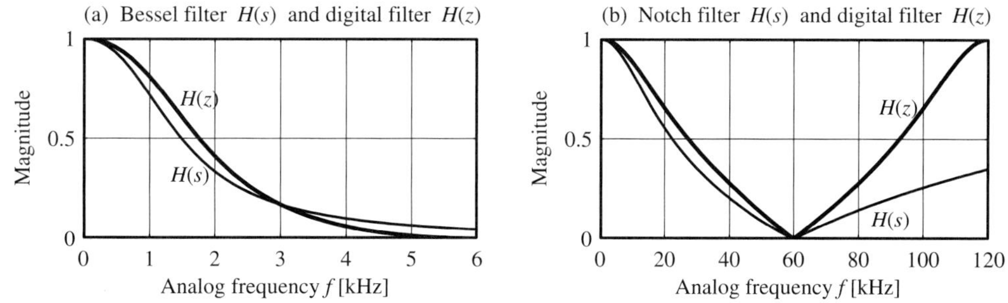


**FIGURE E 9.8.a** Magnitude of the analog and digital filters for Example 9.8(a and b)

Finally, we convert $H(s)$ to $H(z)$ using $s = C\frac{z-1}{z+1} = \frac{z-1}{z+1}$, to get

$$H(z) = H(s)|_{s=(z-1)/(z+1)} = \frac{(z-1)^2 + (z+1)^2}{(z-1)^2 + 4(z^2-1) + (z+1)^2} = \frac{z^2+1}{3z^2-1}$$

We confirm that $H(s) = 0$ at $s = j\omega_0 = j$ and $H(z) = 0$ at $z = e^{j\Omega} = e^{j\pi/2} = j$. Figure E9.8(b) shows the magnitude of the two filters and the perfect match at $f = 60$ Hz (or $F = 0.25$).

# 9.7 Spectral Transformations for IIR Filters

The design of IIR filters usually starts with an analog lowpass prototype, which is converted to a digital lowpass prototype by an appropriate mapping and transformed to the required filter type by an appropriate spectral transformation. For the bilinear mapping, we may even perform the mapping and the spectral transformation (in a single step) on the analog prototype itself.

## 9.7.1 Digital-to-Digital Transformations

If a digital lowpass prototype has been designed, the digital-to-digital (D2D) transformations of Table 9.5 can be used to convert it to the required filter type. These transformations preserve stability by mapping the unit circle (and all points within it) into itself.

**TABLE 9.5 ➤** Digital-to-Digital (D2D) Frequency Transformations

| Form | Band Edge(s) | Mapping $z \to$ | Mapping Parameters |
|---|---|---|---|
| LP2LP | $\Omega_C$ | $\dfrac{z-\alpha}{1-\alpha z}$ | $\alpha = \sin[0.5(\Omega_D - \Omega_C)]/\sin[0.5(\Omega_D + \Omega_C)]$ |
| LP2HP | $\Omega_C$ | $\dfrac{-(z+\alpha)}{1+\alpha z}$ | $\alpha = -\cos[0.5(\Omega_D + \Omega_C)]/\cos[0.5(\Omega_D - \Omega_C)]$ |
| LP2BP | $[\Omega_1, \Omega_2]$ | $\dfrac{-(z^2 + A_1 z + A_2)}{A_2 z^2 + A_1 z + 1}$ | $K = \tan(0.5\Omega_D)/\tan[0.5(\Omega_2 - \Omega_1)]$<br>$\alpha = -\cos[0.5(\Omega_2 + \Omega_1)]/\cos[0.5(\Omega_2 - \Omega_1)]$<br>$A_1 = 2\alpha K/(K+1) \quad A_2 = (K-1)/(K+1)$ |
| LP2BS | $[\Omega_1, \Omega_2]$ | $\dfrac{(z^2 + A_1 z + A_2)}{A_2 z^2 + A_1 z + 1}$ | $K = \tan(0.5\Omega_D)\tan[0.5(\Omega_2 - \Omega_1)]$<br>$\alpha = -\cos[0.5(\Omega_2 + \Omega_1)]/\cos[0.5(\Omega_2 - \Omega_1)]$<br>$A_1 = 2\alpha/(K+1) \quad A_2 = -(K-1)/(K+1)$ |

*NOTE:* The digital lowpass prototype cutoff frequency is $\Omega_D$.
All digital frequencies are normalized to $\Omega = 2\pi f/S$.

As with analog transformations, the lowpass-to-bandpass (LP2BP) and lowpass-to-bandstop (LP2BS) transformations yield a digital filter with twice the order of the lowpass prototype. The lowpass-to-lowpass (LP2LP) transformation is actually a special case of the more general allpass transformation:

$$z \Rightarrow \pm\frac{z-\alpha}{1-\alpha z}, \qquad |\alpha| < 1 \text{ (and } \alpha \text{ real)} \tag{9.35}$$

**Example 9.9**

### Using D2D Transformations

A lowpass filter $H(z) = \dfrac{3(z+1)^2}{31z^2 - 26z + 7}$ operates at $S = 8$ kHz, and its cutoff frequency is $f_C = 2$ kHz.

**(a)** Use $H(z)$ to design a highpass filter with a cutoff frequency of 1 kHz.

We find $\Omega_D = 2\pi f_C/S = 0.5\pi$ and $\Omega_C = 0.25\pi$. The LP2HP transformation (Table 9.5) requires

$$\alpha = -\frac{\cos[0.5(\Omega_D + \Omega_C)]}{\cos[0.5(\Omega_D - \Omega_C)]} = -\frac{\cos(3\pi/8)}{\cos(\pi/8)} = -0.4142$$

The LP2HP spectral transformation is thus $z \rightarrow \frac{-(z+\alpha)}{1+\alpha z} = \frac{-(z-0.4142)}{1-0.4142z}$ and yields

$$H_{\text{HP}}(z) = \frac{0.28(z-1)^2}{z^2 - 0.0476z + 0.0723}$$

**(b)** Use $H(z)$ to design a bandpass filter with band edges of 1 kHz and 3 kHz.

The various digital frequencies are $\Omega_1 = 0.25\pi$, $\Omega_2 = 0.75\pi$, $\Omega_2 - \Omega_1 = 0.5\pi$, and $\Omega_2 + \Omega_1 = \pi$.

From Table 9.5, the parameters needed for the LP2BP transformation are

$$K = \frac{\tan(\pi/4)}{\tan(\pi/4)} = 1 \qquad \alpha = -\frac{\cos(\pi/2)}{\cos(\pi/4)} = 0 \qquad A_1 = 0 \qquad A_2 = 0$$

The LP2BP transformation is thus $z \rightarrow -z^2$ and yields

$$H_{\text{BP}}(z) = \frac{3(z^2-1)^2}{31z^4 + 26z^2 + 7}$$

**(c)** Use $H(z)$ to design a bandstop filter with band edges of 1.5 kHz and 2.5 kHz.

Once again, we need $\Omega_1 = 3\pi/8$, $\Omega_2 = 5\pi/8$, $\Omega_2 - \Omega_1 = \pi/4$, and $\Omega_2 + \Omega_1 = \pi$. From Table 9.5, the LP2BS transformation requires the parameters

$$K = \frac{\tan(\pi/8)}{\tan(\pi/4)} = 0.4142 \qquad \alpha = -\frac{\cos(\pi/2)}{\cos(\pi/8)} = 0 \qquad A_1 = 0 \qquad A_2 = 0.4142$$

The LP2BS transformation is thus $z \rightarrow \frac{z^2+0.4142}{0.4142z^2+1}$ and yields

$$H_{\text{BS}}(z) = \frac{0.28(z^2+1)^2}{z^4 + 0.0476z^2 + 0.0723}$$

Figure E9.9 compares the magnitudes of each filter designed in this example.

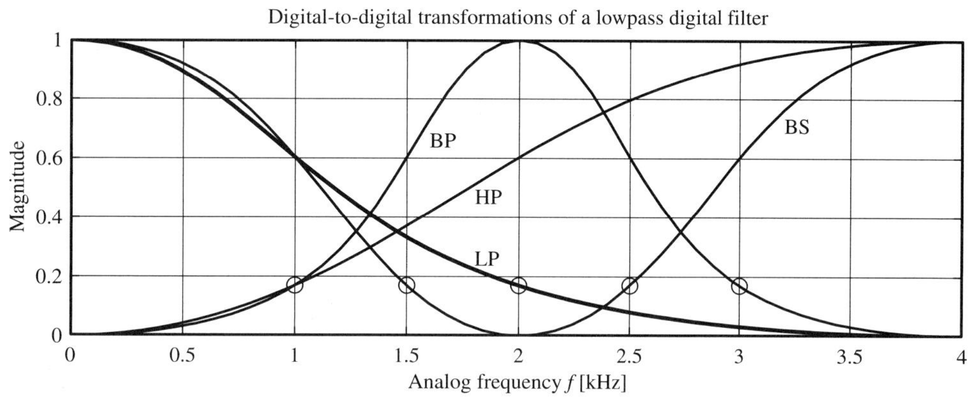


**FIGURE E 9.9.c** The digital filters for Example

## 9.7.2 Direct (A2D) Transformations for Bilinear Design

All stable transformations that are also free of aliasing introduce warping effects. Only the bilinear mapping offers a simple relation to compensate for the warping. Combining the bilinear mapping with the D2D transformations yields the analog-to-digital (A2D) transformations of Table 9.6 for bilinear design. These can be used to convert a prewarped analog lowpass prototype (*with a cutoff frequency of* 1 rad/s) directly to the required digital filter.

## 9.7.3 Bilinear Transformation for Peaking and Notch Filters

If we wish to use the bilinear transformation to design a second-order digital peaking (bandpass) filter with a 3-dB bandwidth of $\Delta\Omega$ and a center frequency of $\Omega_0$, we start with the lowpass analog prototype $H(s) = 1/(s+1)$ (whose cutoff frequency is 1 rad/s) and apply the A2D LP2BP transformation to obtain

$$H_{\rm BP}(z) = \frac{C}{1+C}\left(\frac{z^2-1}{z^2 - \frac{2\beta}{1+C}z + \frac{1-C}{1+C}}\right) \qquad \beta = \cos\Omega_0 \quad C = \tan(0.5\Delta\Omega) \qquad (9.36)$$

Similarly, if we wish to use the bilinear transformation to design a second-order digital notch (bandstop) filter with a 3-dB notch bandwidth of $\Delta\Omega$ and a notch frequency of $\Omega_0$,

**TABLE 9.6 ➤** Direct Analog-to-Digital (A2D) Transformations for Bilinear Design

| Form | Band Edge(s) | Mapping $s \rightarrow$ | Mapping Parameters |
|---|---|---|---|
| LP2LP | $\Omega_C$ | $\dfrac{z-1}{C(z+1)}$ | $C = \tan(0.5\Omega_C)$ |
| LP2HP | $\Omega_C$ | $\dfrac{C(z+1)}{z-1}$ | $C = \tan(0.5\Omega_C)$ |
| LP2BP | $\Omega_1 < \Omega_0 < \Omega_2$ | $\dfrac{z^2-2\beta z+1}{C(z^2-1)}$ | $C = \tan[0.5(\Omega_2-\Omega_1)]$, $\beta = \cos\Omega_0$ or $\beta = \cos[0.5(\Omega_2+\Omega_1)]/\cos[0.5(\Omega_2-\Omega_1)]$ |
| LP2BS | $\Omega_1 < \Omega_0 < \Omega_2$ | $\dfrac{C(z^2-1)}{z^2-2\beta z+1}$ | $C = \tan[0.5(\Omega_2-\Omega_1)]$, $\beta = \cos\Omega_0$ or $\beta = \cos[0.5(\Omega_2+\Omega_1)]/\cos[0.5(\Omega_2-\Omega_1)]$ |

*NOTE:* The analog lowpass prototype *prewarped* cutoff frequency is 1 rad/s.
The digital frequencies are normalized ($\Omega = 2\pi f/S$) but are *not prewarped.*

we once again start with the lowpass analog prototype $H(s) = 1/(s+1)$ and apply the A2D LP2BS transformation to obtain

$$H_{BS}(z) = \frac{1}{1+C}\left(\frac{z^2 - 2\beta z + 1}{z^2 - \frac{2\beta}{1+C}z + \frac{1-C}{1+C}}\right) \quad \beta = \cos\Omega_0 \quad C = \tan(0.5\Delta\Omega) \tag{9.37}$$

If either design calls for an $A$-dB bandwidth of $\Delta\Omega$, the constant $C$ is replaced by $KC$, where

$$K = \left(\frac{1}{10^{0.1A} - 1}\right)^{1/2} \quad (A \text{ in dB}) \tag{9.38}$$

This is equivalent to denormailzing the lowpass prototype such that its gain corresponds to an attenuation of $A$ decibels at unit radian frequency. For a 3-dB bandwidth, we obtain $K = 1$ as expected. These design relations prove quite helpful in the quick design of notch and peaking filters.

The center frequency $\Omega_0$ is used to determine the parameter $\beta = \cos\Omega_0$. If only the band edges $\Omega_1$ and $\Omega_2$ are specified (but the center frequency $\Omega_0$ is not) $\beta$ may also be found from the alternative relation of Table 9.6 in terms of $\Omega_1$ and $\Omega_2$. The center frequency of the designed filter will then be based on the geometric symmetry of the prewarped frequencies and can be computed from

$$\tan(0.5\Omega_0) = \sqrt{\tan(0.5\Omega_1)\tan(0.5\Omega_2)} \tag{9.39}$$

In fact, the digital band edges $\Omega_1$ and $\Omega_2$ do not show geometric symmetry with respect to the center frequency $\Omega_0$. We can find $\Omega_1$ and $\Omega_2$ in terms of $\Delta\Omega$ and $\Omega_0$ by equating the two expressions for finding $\beta$ (in Table 9.6) to obtain

$$\beta = \cos\Omega_0 = \frac{\cos[0.5(\Omega_2 + \Omega_1)]}{\cos[0.5(\Omega_2 - \Omega_1)]} \tag{9.40}$$

With $\Delta\Omega = \Omega_2 - \Omega_1$, we get

$$\Omega_2 = 0.5\Delta\Omega + \cos^{-1}[\cos\Omega_0 \cos(0.5\Delta\Omega)] \quad \Omega_1 = \Omega_2 - \Delta\Omega \tag{9.41}$$

**Example 9.10**

### Bilinear Design of Second-Order Filters

**(a)** Let us design a peaking filter with a 3-dB bandwidth of 5 kHz and a center frequency of 6 kHz. The sampling frequency is 25 kHz.

The digital frequencies are $\Delta\Omega = 2\pi(5/25) = 0.4\pi$ and $\Omega_0 = 2\pi(6/25) = 0.48\pi$. We compute $C = \tan(0.5\Delta\Omega) = 0.7265$ and $\beta = \cos\Omega_0 = 0.0628$. Substituting these into the form for the required filter, we obtain

$$H(z) = \frac{0.4208(z^2 - 1)}{z^2 - 0.0727z + 0.1584}$$

Figure E9.10(a) shows the magnitude spectrum. The center frequency is 6 kHz, as expected. The band edges $\Omega_1$ and $\Omega_2$ may be computed from

$$\Omega_2 = 0.5\Delta\Omega + \cos^{-1}[\cos\Omega_0 \cos(0.5\Delta\Omega)] = 2.1483 \qquad \Omega_1 = \Omega_2 - \Delta\Omega = 0.8917$$

These correspond to the frequencies $f_1 = \frac{S\Omega_1}{2\pi} = 3.55$ kHz and $f_2 = \frac{S\Omega_2}{2\pi} = 8.55$ kHz.

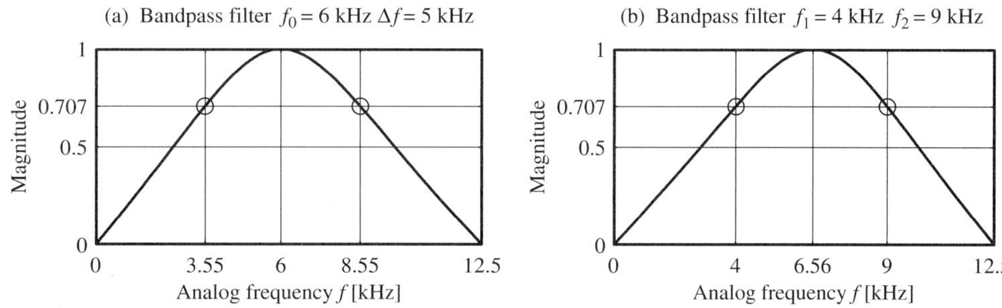


**FIGURE E 9.10.a** Response of bandpass filters for Example 9.10(a and b)

**(b)** Design a peaking (bandpass) filter with a 3-dB band edges of 4 kHz and 9 kHz. The sampling frequency is 25 kHz.

The digital frequencies are $\Omega_1 = 2\pi(4/25) = 0.32\pi$, $\Omega_2 = 2\pi(6/25) = 0.72\pi$, and $\Delta\Omega = 0.4\pi$.

We find $C = \tan(0.5\Delta\Omega) = 0.7265$ and $\beta = \cos[0.5(\Omega_2 + \Omega_1)]/\cos[0.5(\Omega_2 - \Omega_1)] = -0.0776$. Substituting these into the form for the required filter, we obtain

$$H(z) = \frac{0.4208(z^2 - 1)}{z^2 + 0.0899z + 0.1584}$$

Figure E9.10(b) shows the magnitude spectrum. The band edges are at 4 kHz and 9 kHz as expected. The center frequency, however, is at 6.56 kHz. This is because the digital center frequency $\Omega_0$ must be computed from $\beta = \cos\Omega_0 = -0.0776$. This gives $\Omega_0 = \cos^{-1}(-0.0776) = 1.6485$. This corresponds to $f_0 = S\Omega_0/(2\pi) = 6.5591$ kHz.

*Comment:* We could also have computed $\Omega_0$ from

$$\tan(0.5\Omega_0) = \sqrt{\tan(0.5\Omega_1)\tan(0.5\Omega_2)} = 1.0809$$

Then, $\Omega_0 = 2\tan^{-1}(1.0809) = 1.6485$, as before.

**(c)** Design a peaking filter with a center frequency of 40 Hz and a 6-dB bandwidth of 2 Hz operating at a sampling rate of 200 Hz.

We compute $\Delta\Omega = 2\pi(2/200) = 0.02\pi$, $\Omega_0 = 2\pi(40/200) = 0.4\pi$, and $\beta = \cos\Omega_0 = 0.3090$. Since we are given the 6-dB bandwidth, we compute $K$ and $C$ as follows:

$$K = \left(\frac{1}{10^{0.1A} - 1}\right)^{1/2} = \left(\frac{1}{10^{0.6} - 1}\right)^{1/2} = 0.577 \qquad C = K\tan(0.5\Delta\Omega) = 0.0182$$

Substituting these into the form for the required filter, we obtain

$$H(z) = \frac{0.0179(z^2 - 1)}{z^2 - 0.6070z + 0.9642}$$

Figure E9.10C shows the magnitude spectrum. The blowup reveals that the 6-dB bandwidth equals 2 Hz (where the gain is 0.5), as required.

**FIGURE E 9.10.c** Response of peaking filter for Example 9.10(c)

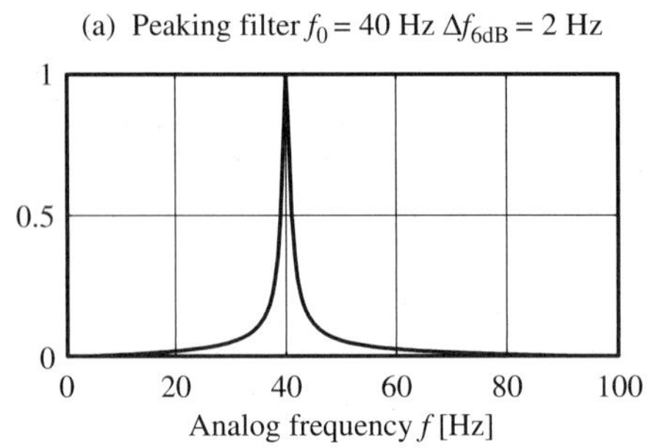


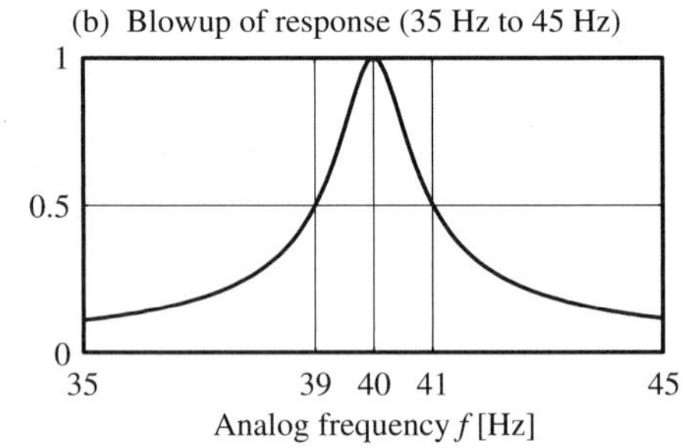


**Example 9.11**

## Interference Rejection

We wish to design a filter to remove 60-Hz interference in an ECG signal sampled at 300 Hz. A 2-s recording of the noisy signal is shown in Figure E9.11(1).

**FIGURE E.9.11(1)** Simulated ECG signal with 60-Hz interference for Example 9.11

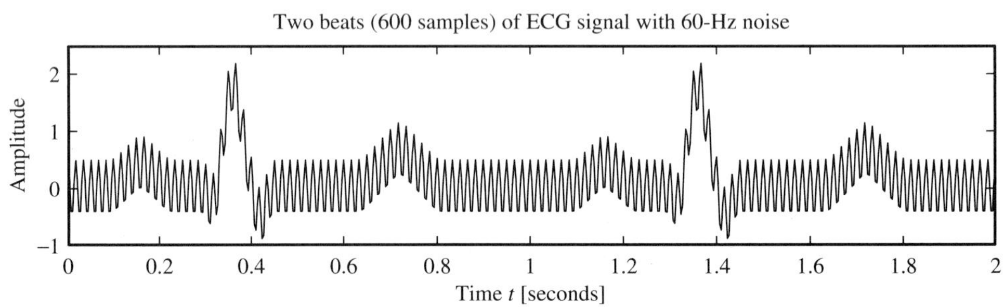


If we design a high-$Q$ notch filter with $Q = 50$ and a notch at $f_0 = 60$ Hz, we have a notch bandwidth of $\Delta f = f_0/Q = 1.2$ Hz. The digital notch frequency is $\Omega_0 = 2\pi f_0/S = 2\pi(60/300) = 0.4\pi$, and the digital bandwidth is $\Delta\Omega = 2\pi\Delta f/S = 2\pi(1.2/300) = 0.008\pi$.

We find $C = \tan(0.5\Delta\Omega) = 0.0126$ and $\beta = \cos\Omega_0 = 0.3090$. Substituting these into the form for the notch filter, we obtain

$$H_1(z) = \frac{0.9876(z^2 - 0.6180z + 1)}{z^2 - 0.6104z + 0.9752}$$

A low-$Q$ design with $Q = 5$ gives $\Delta f = f_0/Q = 12$ Hz, and $\Delta\Omega = 2\pi(12/300) = 0.08\pi$.

We find $C = \tan(0.5\Delta\Omega) = 0.1263$, and with $\beta = \cos\Omega_0 = 0.3090$ as before, we obtain

$$H_2(z) = \frac{0.8878(z^2 - 0.6180z + 1)}{z^2 - 0.5487z + 0.7757}$$

Figure E9.11(2) shows the magnitude spectrum of the two filters. Naturally, the filter $H_1(z)$ (with the higher $Q$) exhibits the sharper notch.

**FIGURE E.9.11(2)** Response of the notch filters for Example 9.11(3)

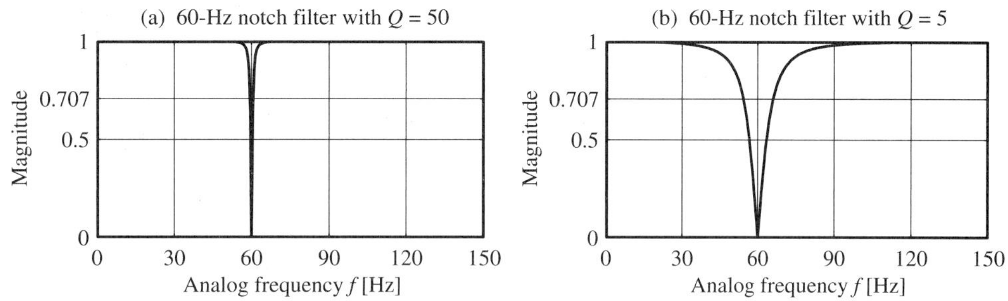


The filtered ECG signal corresponding to these two notch filters is shown in Figure E9.11(3). Although both filters are effective in removing the 60-Hz noise, the filter $H_2(z)$ (with the lower $Q$) shows a much shorter start-up transient (because the highly oscillatory transient response of the high-$Q$ filter $H_1(z)$ takes much longer to reach steady state).

**FIGURE E.9.11(3)** Output of the notch filters for Example 9.11

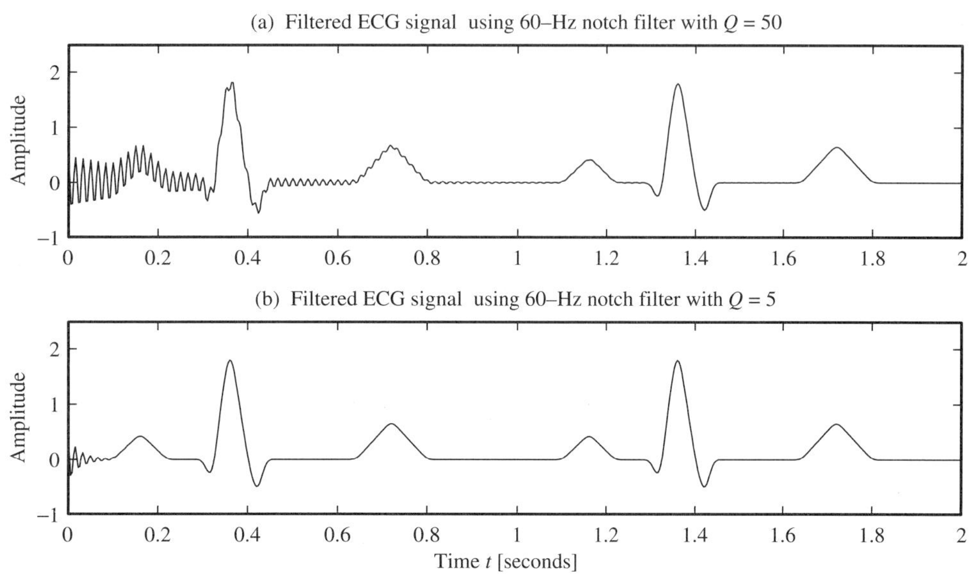


# 9.8 Design Recipe for IIR Filters

There are several approaches to the design of IIR filters, using the mappings and spectral transformations described in this chapter. The first approach, illustrated in Figure 9.11, is based on developing the analog filter $H(s)$ followed by the required mapping to convert $H(s)$ to $H(z)$.

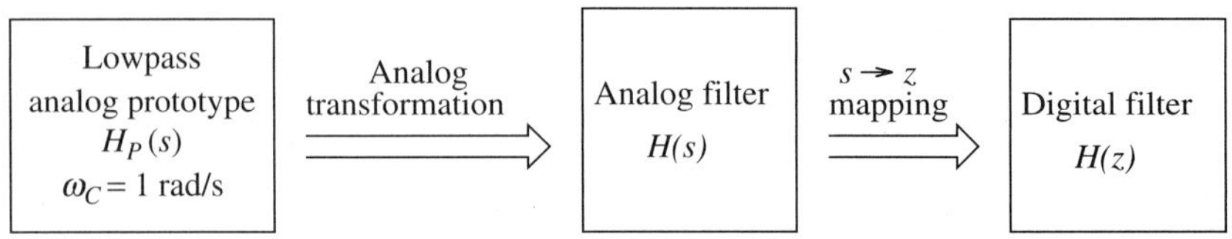


**FIGURE 9.11** Converting an analog filter to a digital filter. This process starts with a lowpass prototype with unit cutoff frequency. This is converted to the required analog filter meeting the given specifications. This analog filter is then converted to the required digital filter using appropriate mapping rules

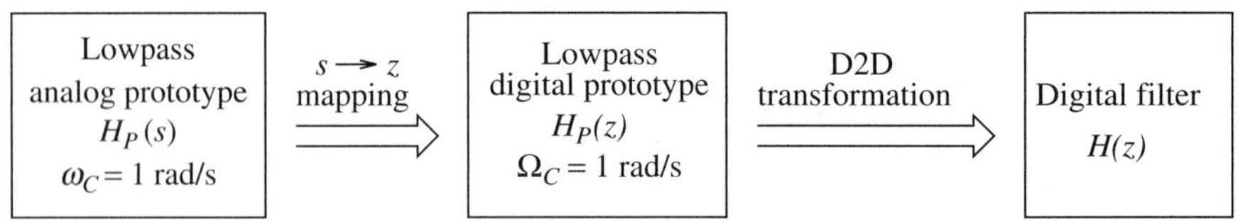


**FIGURE 9.12** Indirect conversion of an analog filter to a digital filter. This process also starts with a lowpass prototype with unit cutoff frequency. This is then converted to a digital lowpass prototype with unit cutoff frequency. This digital lowpass prototype is converted to the required digital filter using the D2D transformations described in the text

A major disadvantage of this approach is that it cannot be used for mappings that suffer from aliasing problems (such as the impulse-invariant mapping) to design highpass or bandpass filters.

The second, indirect approach, is illustrated in Figure 9.12 and tries to overcome this problem by designing only the lowpass prototype $H_P(s)$ in the analog domain. This is followed by the required mapping to obtain a digital lowpass prototype $H_P(z)$. The final step is the spectral (D2D) transformation of $H_P(z)$ to the required digital filter $H(z)$. This indirect approach allows us to use any mappings, including those (such as response invariance) that may otherwise lead to excessive aliasing for highpass and bandstop filters. Designing $H_P(z)$ also allows us to match its dc magnitude with $H_P(s)$ for subsequent comparison.

A third approach (that applies only to the bilinear transformation) is illustrated in Figure 9.13. We prewarp the frequencies, design an analog lowpass prototype (from *prewarped* specifications), and apply A2D transformations to obtain the required digital filter $H(z)$.

**FIGURE 9.13** Conversion of analog lowpass prototype directly to a digital filter. This process also starts with a lowpass prototype with unit cutoff frequency. This is then converted in a single step to the required digital filter using the A2D transformations described in the text

## A Step-by-Step Approach

Given the passband and stopband edges, the passband and stopband attenuation, and the sampling frequency $S$, a standard recipe for the design of IIR filters is as follows:

1. Normalize (divide) the design band edges by $S$. This allows us to use a sampling interval $t_s = 1$ in subsequent design. For bilinear design, we also *prewarp* the normalized band edges.
2. Use the normalized band edges and attenuation specifications to design an *analog lowpass prototype* $H_P(s)$ whose cutoff frequency is $\omega_C = 1$ rad/s.
3. Apply the chosen mapping (with $t_s = 1$) to convert $H_P(s)$ to a *digital lowpass prototype* filter $H_P(z)$ with $\Omega_D = 1$.
4. Use D2D transformations (with $\Omega_D = 1$) to convert $H_P(z)$ to $H(z)$.
5. For bilinear design, we can also convert $H_P(s)$ to $H(z)$ directly (using A2D transformations).

---

**Design Recipe for IIR Digital Filters**

1. Normalize (divide) band edges by $S$ (and prewarp if using bilinear design).
2. Use the normalized band edges to design *analog lowpass prototype* $H_P(s)$ with $\omega_C = 1$ rad/s.
3. Apply the chosen mapping (with $t_s = 1$) to convert $H_P(s)$ to $H_P(z)$ with $\Omega_D = 1$.
4. Use D2D transformations to convert $H_P(z)$ to $H(z)$.
5. For bilinear design, convert $H_P(s)$ to $H(z)$ (using A2D transformations).

---

**Example 9.12**

### IIR Filter Design

Design a Chebyshev IIR filter to meet the following specifications:

Passband edges at [1.8, 3.2] kHz, stopband edges at [1.6, 4.8] kHz, $A_p = 2$ dB, $A_s = 20$ dB, and sampling frequency $S = 12$ kHz.

**(a) Indirect Bilinear Design**

The normalized band edges $[\Omega_1,\ \Omega_2,\ \Omega_3,\ \Omega_4]$, in increasing order, are

$$[\Omega_1,\ \Omega_2,\ \Omega_3,\ \Omega_4] = 2\pi[1.6,\ 1.8,\ 3.2,\ 4.8]/12 = [0.84,\ 0.94,\ 1.68,\ 2.51]$$

The passband edges are $[\Omega_1, \Omega_2] = [0.94, 1.68]$. We choose $C = 2$ and prewarp each band-edge frequency using $\omega = 2\tan(0.5\Omega)$ to give the prewarped values $[0.89,\ 1.019,\ 2.221,\ 6.155]$.

The prewarped passband edges are $[\Omega_{p1},\ \Omega_{p2}] = [1.019,\ 2.221]$.

We design an analog filter meeting the prewarped specifications (as described in the appendix). This yields the lowpass prototype and actual transfer function as

$$H_P(s) = \frac{0.1634}{s^4 + 0.7162s^3 + 1.2565s^2 + 0.5168s + 0.2058}$$

$$H_{BP}(s) = \frac{0.34s^4}{s^8 + 0.86s^7 + 10.87s^6 + 6.75s^5 + 39.39s^4 + 15.27s^3 + 55.69s^2 + 9.99s + 26.25}$$

Finally, we transform the bandpass filter $H_{BP}(s)$ to $H(z)$ using $s \rightarrow 2(z-1)/(z+1)$, to obtain

$$H(z) = \frac{0.0026z^8 - 0.0095z^6 + 0.0142z^4 - 0.0095z^2 + 0.0026}{z^8 - 1.94z^7 + 4.44z^6 - 5.08z^5 + 6.24z^4 - 4.47z^3 + 3.44z^2 - 1.305z + 0.59}$$

Figure E9.12a compares the response of the digital filter $H(z)$ with the analog filter $H_{BP}(s)$, and with a digital filter designed from the *unwarped* frequencies.

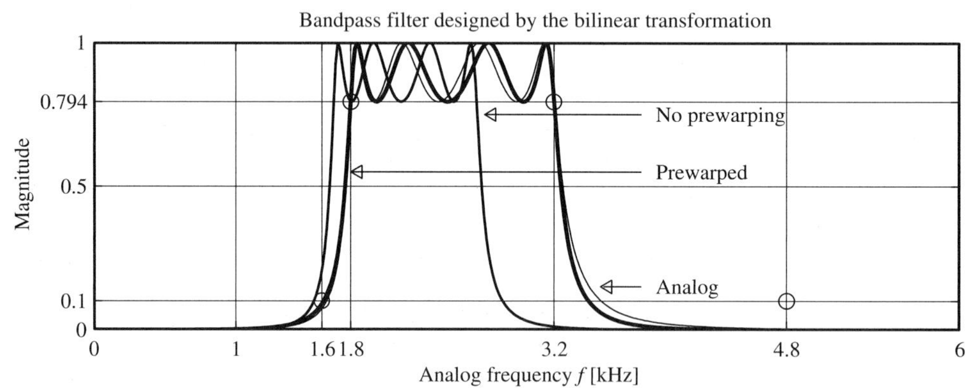


**FIGURE E 9.12.a** Bandpass filter for Example 9.12 designed by the bilinear transformation

**(b) Direct A2D Design**

For bilinear design, we can also convert $H_P(s)$ directly.

We use the A2D LP2BP transformation with the *unwarped* passband edges $[\Omega_1,\ \Omega_2] = [0.94,\ 1.68]$.

The constants $C$ and $\beta$ (from Table 9.6) are found as

$$C = \tan[0.5(\Omega_2 - \Omega_1)] = 0.3839 \qquad \beta = \frac{\cos[0.5(\Omega_2 + \Omega_1)]}{\cos[0.5(\Omega_2 - \Omega_1)]} = 0.277$$

We transform the prototype analog filter $H_P(s)$ to obtain

$$H(z) = \frac{0.0026z^8 - 0.0095z^6 + 0.0142z^4 - 0.0095z^2 + 0.0026}{z^8 - 1.94z^7 + 4.44z^6 - 5.08z^5 + 6.24z^4 - 4.47z^3 + 3.44z^2 - 1.305z + 0.59}$$

This expression is identical to the transfer function $H(z)$ of part (a).

**(c) Design using Other Mappings**

We can also design the digital filter based on other mappings by using the following steps:

1. Use the normalized (but *unwarped*) band edges [0.89, 0.94, 1.68, 2.51] to design an analog lowpass prototype $H_P(s)$ with $\omega_C = 1$ rad/s (fortunately, the unwarped and prewarped specifications yield the same $H_P(s)$ for the specifications of this problem).
2. Convert $H_P(s)$ to $H_P(z)$ with $\Omega_D = 1$, using the chosen mapping with $t_s = 1$. For the backward difference, for example, we would use $s = (z-1)/zt_s = (z-1)/z$. To use the impulse-invariant mapping, we would have to first convert $H_P(s)$ to partial-fraction form.

3. Convert $H_P(z)$ to $H(z)$ using the D2D LP2BP transformation with $\Omega_D = 1$ and the *unwarped* passband edges $[\Omega_1, \ \Omega_2] = [0.94, \ 1.68]$.

Figure E9.12c compares the response of two such designs, using the impulse-invariant mapping and the backward-difference mapping (both with gain matching at dc). The design based on the backward-difference mapping shows a poor match to the analog filter.

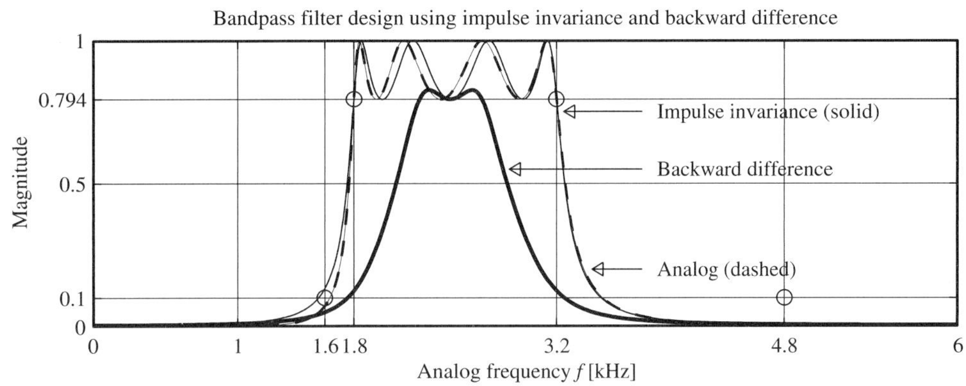


**FIGURE E 9.12.c** Bandpass filter for Example 9.12(c) designed by impulse invariance and backward difference

## 9.8.1 Finite-Word-Length Effects

The effects of quantization must be considered in the design and implementation of both IIR and FIR filters. Quantization implies that we choose a finite number of bits, and this less than ideal representation leads to problems collectively referred to as *finite-word-length effects*.

**Quantization Noise:** Quantization noise limits the signal-to-noise ratio. One way to improve the SNR is to increase the number of bits. Another is to use oversampling (as discussed earlier).

**Coefficient Quantization:** This refers to representation of the filter coefficients by a limited number of bits. Its effects can produce benign effects such as a slight change in the frequency response of the resulting filter (typically, a larger passband ripple and/or a smaller stopband attenuation) or disastrous effects (for IIR filters) that can lead to instability.

**Roundoff Errors:** When lower-order bits are discarded before storing results (of a multiplication, say), there is roundoff error. The amount of error depends on the type of arithmetic used and the filter structure. The effects of roundoff errors are similar to the effects of quantization noise and lead to a reduction in the signal-to-noise ratio.

**Overflow Errors:** Overflow errors occur when the filter output or the result of arithmetic operations (such as the sum of two large numbers with the same sign) exceeds the permissible wordlength. Such errors are avoided in practice by scaling the filter coefficients and/or the input in a manner that the output remains within the permissible word length.

### 9.8.2 Effects of Coefficient Quantization

For IIR filters, the effects of rounding or truncating the filter coefficients can range from minor changes in the frequency response to serious problems, including instability. Consider the stable analog filter:

$$H(s) = \frac{(s+0.5)(s+1.5)}{(s+1)(s+2)(s+4.5)(s+8)(s+12)} \tag{9.42}$$

Bilinear transformation of $H(s)$ with $t_s = 0.01$ s yields the digital transfer function $H(z) = B(z)/A(z)$ whose denominator coefficients to double precision ($A_k$) and truncated to seven significant digits ($A_k^t$) are given by

| Filter Coefficients $A_k$ | Truncated $A_k^t$ |
|---|---|
| 1.144168420199997e+0 | 1.144168e+0 |
| −5.418904483999996e+0 | −5.418904e+0 |
| 1.026166736200000e+1 | 1.026166e+1 |
| −9.712186808000000e+0 | −9.712186e+0 |
| 4.594164261000004e+0 | 4.594164e+0 |
| −8.689086648000011e−1 | −8.689086e−1 |

The poles of $H(z)$ all lie within the unit circle, and the designed filter is thus stable. However, if we use the truncated coefficients $A_k^t$ to compute the roots, the filter becomes unstable because one pole moves out of the unit circle! The bottom line is that stability is an important issue in the design of IIR digital filters.

### 9.8.3 Concluding Remarks

IIR filters are well suited to applications requiring frequency-selective filters with sharp cutoffs or where linear phase is relatively unimportant. Examples include graphic equalizers for digital audio, tone generators for digital touch-tone receivers, and filters for digital telephones. The main advantages are standardized easy design and low filter order. On the other hand, IIR filters cannot exhibit linear phase and are quite susceptible to the effects of coefficient quantization. If linear phase is important (as in biomedical signal processing) or stability is paramount as in many adaptive filtering schemes, it is best to use FIR filters.

## 9.9 Problems

9.1. **(Response Invariance)** Consider the analog filter $H(s) = \dfrac{1}{s+2}$.

**(a)** Convert $H(s)$ to a digital filter $H(z)$ using impulse invariance. Assume that the sampling frequency is $S = 2$ Hz.

**(b)** Will the impulse response $h[n]$ match the impulse response $h(t)$ of the analog filter at the sampling instants? Should it? Explain.

**(c)** Will the step response $s[n]$ match the step response $s(t)$ of the analog filter at the sampling instants? Should it? Explain.

[**Hints and Suggestions:** For part (a), find $h(t)$, sample it ($t \to nt_s$) to get $h[n]$, and find its $z$-transform to obtain $H(z)$.]

**9.2.** **(Response Invariance)** Consider the analog filter $H(s) = \dfrac{1}{s+2}$.

**(a)** Convert $H(s)$ to a digital filter $H(z)$ using step invariance at a sampling frequency of $S = 2\,\text{Hz}$.

**(b)** Will the impulse response $h[n]$ match the impulse response $h(t)$ of the analog filter at the sampling instants? Should it? Explain.

**(c)** Will the step response $s[n]$ match the step response $s(t)$ of the analog filter at the sampling instants? Should it? Explain.

[**Hints and Suggestions:** For part (a), find $y(t)$ from $Y(s) = H(s)X(s) = H(s)/s$, sample it ($t \to nt_s$) to get $y[n]$, and find the ratio $H(z) = Y(z)/X(z)$ where $x[n] = u[n]$.]

**9.3.** **(Response Invariance)** Consider the analog filter $H(s) = \dfrac{1}{s+2}$.

**(a)** Convert $H(s)$ to a digital filter $H(z)$ using ramp invariance at a sampling frequency of $S = 2\,\text{Hz}$.

**(b)** Will the impulse response $h[n]$ match the impulse response $h(t)$ of the analog filter at the sampling instants? Should it? Explain.

**(c)** Will the step response $s[n]$ match the step response $s(t)$ of the analog filter at the sampling instants? Should it? Explain.

**(d)** Will the response $v[n]$ to a unit-ramp match the unit-ramp response $v(t)$ of the analog filter at the sampling instants? Should it? Explain.

[**Hints and Suggestions:** For part (c), find $y(t)$ from $Y(s) = H(s)X(s) = H(s)/s^2$, sample it ($t \to nt_s$) to get $y[n]$, and find the ratio $H(z) = Y(z)/X(z)$ where $x[n] = nt_s u[n]$.]

**9.4.** **(Response Invariance)** Consider the analog filter $H(s) = \dfrac{s+1}{(s+1)^2 + \pi^2}$.

**(a)** Convert $H(s)$ to a digital filter $H(z)$ using impulse invariance. Assume that the sampling frequency is $S = 2$ Hz.

**(b)** Convert $H(s)$ to a digital filter $H(z)$ using invariance to the input $x(t) = e^{-t}u(t)$ at a sampling frequency of $S = 2$ Hz.

[**Hints and Suggestions:** For part (c), find $y(t)$ from $Y(s) = H(s)X(s) = H(s)/(s+1)$, sample it ($t \to nt_s$) to get $y[n]$, and find the ratio $H(z) = Y(z)/X(z)$ where $x[n] = e^{-nt_s}u[n] = (e^{-t_s})^n$.]

**9.5.** **(Impulse Invariance)** Use the impulse-invariant transformation with $t_s = 1$ s to transform the following analog filters to digital filters.

**(a)** $H(s) = \dfrac{1}{s+2}$ **(b)** $H(s) = \dfrac{2}{s+1} + \dfrac{2}{s+2}$ **(c)** $H(s) = \dfrac{1}{(s+1)(s+2)}$

[**Hints and Suggestions:** For (c), find the partial fractions for $H(s)$. For all, use the transformation $\frac{1}{s+\alpha} \Rightarrow \frac{z}{z-e^{-\alpha t_s}}$ for each partial-fraction term.]

**9.6.** **(Impulse-Invariant Design)** We are given the analog lowpass filter $H(s) = \dfrac{1}{s+1}$ whose cutoff frequency is known to be 1 rad/s. It is required to use this filter as the basis for designing a digital filter by the impulse-invariant transformation. The digital filter is to have a cutoff frequency of 50 Hz and operate at a sampling frequency of 200 Hz.

**(a)** What is the transfer function $H(z)$ of the digital filter if no gain matching is used?

**(b)** What is the transfer function $H(z)$ of the digital filter if the gain of the analog filter and digital filter are matched at dc? Does the gain of the two filters match at their respective cutoff frequencies?

**(c)** What is the transfer function $H(z)$ of the digital filter if the gain of the analog filter at its cutoff frequency (1 rad/s) is matched to the gain of the digital filter at its cutoff frequency (50 Hz)? Does the gain of the two filters match at dc?

[**Hints and Suggestions:** For (a), pick $\Omega_C = 2\pi f_C/S$, obtain $H_A(s) = H(s/\Omega_C)$, and convert to $H(z)$ using $\dfrac{1}{s+\alpha} \Rightarrow \dfrac{z}{z-e^{-\alpha t_s}}$. For (b), find $G_A = |H(s)|_{s=0}$, $G_D = |H(z)|_{z=1}$, and multiply $H(z)$ by $G_A/G_D$. For (c), find $G_A = |H(s)|_{s=j1}$, $G_D = |H(z)|$ at $z = e^{j2\pi(50)/S}$, and multiply $H(z)$ by $G_A/G_D$.]

**9.7. (Impulse Invariance)** The impulse-invariant method allows us to take a digital filter described by $H_1(z) = \dfrac{z}{z-\alpha t_1}$ at a sampling interval of $t_1$ and convert this to a new digital filter $H_2(z) = \dfrac{z}{z-\alpha t_2}$ at a different sampling interval $t_2$.

**(a)** Using the fact that $s = \dfrac{\ln(\alpha t_1)}{t_1} = \dfrac{\ln(\alpha t_2)}{t_2}$, show that $\alpha t_2 = (\alpha t_1)^M$ where $M = t_2/t_1$.

**(b)** Use the result of part (a) to convert the digital filter $H_1(z) = \frac{z}{z-0.5} + \frac{z}{z-0.25}$ with $t_s = 1$ s to a digital filter $H_2(z)$ with $t_s = 0.5$ s.

**9.8. (Matched $z$-Transform)** Use the matched $z$-transform $s+\alpha \Longrightarrow \frac{z-e^{-\alpha t_s}}{z}$ with $t_s = 0.5$ s and gain matching at dc to transform each analog filter $H(z)$ to a digital filter $H(z)$.

**(a)** $H(s) = \dfrac{1}{s+2}$ **(b)** $H(s) = \dfrac{1}{(s+1)(s+2)}$

**(c)** $H(s) = \dfrac{1}{s+1} + \frac{1}{s+2}$ **(d)** $H(s) = \dfrac{s+1}{(s+1)^2+\pi^2}$

[**Hints and Suggestions:** Set up $H(s)$ in factored form and use $(s+\alpha) \Rightarrow \frac{z-e^{-\alpha t_s}}{z}$ for each factor to obtain $H(z)$. Then, find $G_A = |H(s)|_{s=0}$, $G_D = |H(z)|_{z=1}$, and multiply $H(z)$ by $G_A/G_D$.]

**9.9. (Matched $z$-Transform)** The analog filter $H(s) = \dfrac{4s(s+1)}{(s+2)(s+3)}$ is to be converted to a digital filter $H(z)$ at a sampling rate of $S = 4$ Hz.

**(a)** Convert $H(s)$ to a digital filter using the matched $z$-transform $s+\alpha \Longrightarrow \frac{z-e^{-\alpha t_s}}{z}$.

**(b)** Convert $H(s)$ to a digital filter using the *modified* matched $z$-transform by moving all zeros at the origin ($z = 0$) to $z = -1$.

**(c)** Convert $H(s)$ to a digital filter using the *modified* matched $z$-transform by moving *all but one* zero at the origin ($z = 0$) to $z = -1$.

**9.10. (Backward Euler Algorithm)** The backward Euler algorithm for numerical integration is given by $y[n] = y[n-1] + t_s x[n]$.

**(a)** Derive a mapping rule for converting an analog filter to a digital filter based on this algorithm.

**(b)** Apply the mapping to convert the analog filter $H(s) = \dfrac{4}{s+4}$ to a digital filter $H(z)$ using a sampling interval of $t_s = 0.5$ s.

[**Hints and Suggestions:** For (a), find $H(z)$ and compare with $H(s) = 1/s$ (ideal integrator).]

**9.11.** **(Mapping from Difference Algorithms)** Consider the analog filter $H(s) = \dfrac{1}{s+\alpha}$.

**(a)** For what values of $\alpha$ is this filter stable?

**(b)** Convert $H(s)$ to a digital filter $H(z)$ using the mapping based on the forward difference at a sampling rate $S$. Is $H(z)$ always stable if $H(s)$ is stable?

**(c)** Convert $H(s)$ to a digital filter $H(z)$ using the mapping based on the backward difference at a sampling rate $S$. Is $H(z)$ always stable if $H(s)$ is stable?

[**Hints and Suggestions:** For (b) through (c), find $H(z)$ from the difference equations for the forward and backward difference, respectively, and compare with $H(s) = s$ (ideal differentiator).]

**9.12.** **(Simpson's Algorithm)** Simpson's numerical integration algorithm is described by

$$y[n] = y[n-2] + \tfrac{t_s}{3}(x[n] + 4x[n-1] + x[n-2])$$

**(a)** Derive a mapping rule to convert an analog filter $H(s)$ to a digital filter $H(z)$ based on this algorithm.

**(b)** Let $H(s) = \dfrac{1}{s+1}$. Convert $H(s)$ to $H(z)$ using the mapping derived in part (a).

**(c)** Is the filter $H(z)$ designed in part (b) stable for any choice of $t_s > 0$?

**9.13.** **(Response of Numerical Algorithms)** Simpson's and Tick's rules for numerical integration find $y[k]$ (the approximation to the area) over two time steps from $y[k-2]$ and are described by

**Simpson's Rule:** $y[n] = y[n-2] + \dfrac{x[n] + 4x[n-1] + x[n-2]}{3}$

**Tick's Rule:** $y[n] = y[n-2] + 0.3584x[n] + 1.2832x[n-1] + 0.3584x[n-2]$

**(a)** Find the transfer function $H(F)$ corresponding to each rule.

**(b)** For each rule, sketch $|H(F)|$ over $0 \le F \le 0.5$ and compare with the spectrum of an ideal integrator.

**(c)** It is claimed that the coefficients in Tick's rule optimize $H(F)$ in the range $0 < F < 0.25$. Does your comparison support this claim?

**9.14.** **(Bilinear Transformation)** Consider the lowpass analog Bessel filter $H(s) = \dfrac{3}{s^2 + 3s + 3}$.

**(a)** Use the bilinear transformation to convert this analog filter $H(s)$ to a digital filter $H(z)$ at a sampling rate of $S = 2$ Hz.

**(b)** Use $H(s)$ and the bilinear transformation to design a digital lowpass filter $H(z)$ whose gain at $f_0 = 20$ kHz matches the gain of $H(s)$ at $\omega_a = 3$ rad/s. The sampling frequency is $S = 80$ kHz.

[**Hints and Suggestions:** For (b), use $s \Rightarrow C\frac{z-1}{z+1}$ with $C\tan(2\pi f_0/S) = \omega_a$.]

**9.15.** **(Bilinear Transformation)** Consider the analog filter $H(s) = \frac{s}{s^2+s+1}$.

**(a)** What type of filter does $H(s)$ describe?

**(b)** Use $H(s)$ and the bilinear transformation to design a digital filter $H(z)$ operating at $S = 1$ kHz such that its gain at $f_0 = 250$ Hz matches the gain of $H(s)$ at $\omega_a = 1$ rad/s. What type of filter does $H(z)$ describe?

**(c)** Use $H(s)$ and the bilinear transformation to design a digital filter $H(z)$ operating at $S = 10$ Hz such that gains of $H(z)$ and $H(s)$ match at $f_m = 1$ Hz. What type of filter does $H(z)$ describe?

[**Hints and Suggestions:** For (a), use $s \Rightarrow C\frac{z-1}{z+1}$ with $C\tan(2\pi f_0/S) = \omega_a$. For (b), use $s \Rightarrow C\frac{z-1}{z+1}$ with $C\tan(2\pi f_m/S) = 2\pi\phi_m$.]

**9.16.** **(Bilinear Transformation)** A second-order Butterworth lowpass analog filter with a half-power frequency of 1 rad/s is converted to a digital filter $H(z)$ using the bilinear transformation at a sampling rate of $S = 1$ Hz.

**(a)** What is the transfer function $H(s)$ of the analog filter?
**(b)** What is the transfer function $H(z)$ of the digital filter?
**(c)** Are the dc gains of $H(z)$ and $H(s)$ identical? Should they be? Explain.
**(d)** Are the gains of $H(z)$ and $H(s)$ at their respective half-power frequencies identical? Explain.

**9.17.** **(IIR Filter Design)** Lead-lag systems are often used in control systems and have the generic form $H(s) = \dfrac{1+s\tau_1}{1+s\tau_2}$. Use a sampling frequency of $S = 10$ Hz and the bilinear transformation to design IIR filters from this lead-lag compensator if

**(a)** $\tau_1 = 1$ s, $\tau_2 = 10$ s  **(b)** $\tau_1 = 10$ s, $\tau_2 = 1$ s

**9.18.** **(Spectral Transformation of Digital Filters)** The digital lowpass filter described by $H(z) = \dfrac{z+1}{z^2 - z + 0.2}$ has a cutoff frequency $f = 0.5$ kHz and operates at a sampling frequency $S = 10$ kHz. Use this filter to design the following:

**(a)** a lowpass digital filter with a cutoff frequency of 2 kHz
**(b)** a highpass digital filter with a cutoff frequency of 1 kHz
**(c)** a bandpass digital filter with band edges of 1 kHz and 3 kHz
**(d)** a bandstop digital filter with band edges of 1.5 kHz and 3.5 kHz

[**Hints and Suggestions:** Use the tables for digital-to-digital (D2D) transformations.]

**9.19.** **(Spectral Transformation of Analog Prototypes)** The analog lowpass filter $H(s) = \dfrac{2}{s^2 + 2s + 2}$ has a cutoff frequency of 1 rad/s. Use this prototype to design the following digital filters.

**(a)** a lowpass filter with a passband edge of 100 Hz and $S = 1$ kHz
**(b)** a highpass filter with a cutoff frequency of 500 Hz and $S = 2$ kHz
**(c)** a bandpass filter with band edges at 400 Hz and 800 Hz and $S = 3$ kHz
**(d)** a bandstop filter with band edges at 1 kHz and 1200 Hz and $S = 4$ kHz

[**Hints and Suggestions:** Use the tables in this chapter for analog-to-digital (A2D) transformations.]

**9.20.** **(Notch Filters)** A notch filter is required to remove 50-Hz interference. Assuming a bandwidth of 4 Hz and a sampling rate of 300 Hz, design the simplest such filter using the bilinear transformation. Compute the filter gain at 40 Hz, 50 Hz, and 60 Hz.

[**Hints and Suggestions:** Use the standard form of the second-order notch filter.]

**9.21.** **(Peaking Filters)** A peaking filter is required to isolate a 100-Hz signal with unit gain. Assuming a bandwidth of 5 Hz and a sampling rate of 500 Hz, design the simplest such filter using the bilinear transformation. Compute the filter gain at 90 Hz, 100 Hz, and 110 Hz.

[**Hints and Suggestions:** Use the standard form of the second-order peaking (bandpass) filter.]

**9.22.** **(IIR Filter Design)** A fourth-order digital filter operating at a sampling frequency of 40 kHz is required to have a passband between 8 kHz and 12 kHz and a maximum passband ripple that equals 5% of the peak magnitude. Design the digital filter using the bilinear transformation.

[**Hints and Suggestions:** You need a Chebyshev bandpass filter of order 4. So, start with a Chebyshev analog lowpass prototype of order 2 (not 4) and apply A2D transformations.]

**9.23.** **(IIR Filter Design)** Design IIR filters that meet each of the following sets of specifications. Assume a passband attenuation of $A_p = 2$ dB and a stopband attenuation of $A_s = 30$ dB.

**(a)** A Butterworth lowpass filter with passband edge at 1 kHz, stopband edge at 3 kHz, and $S = 10$ kHz. Use the backward Euler transformation.

**(b)** A Butterworth highpass filter with passband edge at 400 Hz, stopband edge at 100 Hz, and $S = 2$ kHz, using the impulse-invariant transformation.

**(c)** A Chebyshev bandpass filter with passband edges at 800 Hz and 1600 Hz, stopband edges at 400 Hz and 2 kHz, and $S = 5$ kHz. Use the bilinear transformation.

**(d)** An inverse Chebyshev bandstop filter with passband edges at 200 Hz and 1.2 kHz, stopband edges at 500 Hz and 700 Hz, and $S = 4$ kHz. Use the bilinear transformation.

[**Hints and Suggestions:** Normalize all frequencies. For (a) through (b), design an analog lowpass prototype at $\omega_C = 1$ rad/s, convert to a digital prototype at $S = 1$ Hz, and use digital-to-digital (D2D) transformations with $\Omega_C = 1$. For (c) through (d), design the analog lowpass prototype using prewarped frequencies. Then, apply A2D transformations using unwarped digital band edges.]

**9.24.** **(Digital-to-Analog Mappings)** In addition to the bilinear transformation, the backward Euler method also allows a linear mapping to transform a digital filter $H(z)$ to an analog equivalent $H(s)$.

**(a)** Develop such a mapping based on the backward Euler algorithm.

**(b)** Use this mapping to convert a digital filter $H(z) = \dfrac{z}{z - 0.5}$ operating at $S = 2$ Hz to its analog equivalent $H(s)$.

**9.25.** **(Digital-to-Analog Mappings)** The forward Euler method also allows a linear mapping to transform a digital filter $H(z)$ to an analog equivalent $H(s)$.

**(a)** Develop such a mapping based on the forward Euler algorithm.

**(b)** Use this mapping to convert a digital filter $H(z) = \dfrac{z}{z - 0.5}$ operating at $S = 2$ Hz to its analog equivalent $H(s)$.

**9.26.** **(Digital-to-Analog Mappings)** Two other methods that allow us to convert a digital filter $H(z)$ to an analog equivalent $H(z)$ are impulse invariance and the matched $z$-transform $s + \alpha \Longrightarrow \frac{z - e^{-\alpha t_s}}{z}$. Let $H(z) = \dfrac{z(z+1)}{(z-0.25)(z-0.5)}$. Find the analog filter $H(s)$ from which $H(z)$ was developed, assuming a sampling frequency of $S = 2$ Hz and

**(a)** impulse invariance **(b)** matched $z$-transform **(c)** bilinear transformation

**9.27.** **(Digital-to-Analog Mappings)** The bilinear transformation allows us to use a linear mapping to transform a digital filter $H(z)$ to an analog equivalent $H(s)$.

**(a)** Develop such a mapping based on the bilinear transformation.

**(b)** Use this mapping to convert a digital filter $H(z) = \dfrac{z}{z - 0.5}$ operating at $S = 2$ Hz to its analog equivalent $H(s)$.

**9.28.** **(Group Delay)** A digital filter $H(z)$ is designed from an analog filter $H(s)$ using the bilinear transformation $\omega_A = C\tan(\Omega_D/2)$.

**(a)** Show that the group delays $T_g(\omega_A)$ and $T_g(\Omega_D)$ of $H(s)$ and $H(z)$, respectively, are related by

$$T_g(\Omega_D) = 0.5C(1 + \omega_A^2)T_g(\omega_A)$$

**(b)** Design a digital filter $H(z)$ from the analog filter $H(s) = \dfrac{5}{s + 5}$ at a sampling frequency of $S = 4$ Hz such that the gain of $H(s)$ at $\omega = 2$ rad/s matches the gain of $H(z)$ at 1 Hz.

**(c)** What is the group delay $T_g(\omega_A)$ of the analog filter $H(s)$?

**(d)** Use the results to find the group delay $T_g(\Omega_D)$ of the digital filter $H(z)$ designed in part (b).

**9.29.** **(Pade Approximations)** A delay of $t_s$ may be approximated by

$$e^{-st_s} \approx 1 - st_s + \frac{(st_s)^2}{2!} - \cdots$$

An $n$th-order Pade approximation is based on a rational function of order $n$ that minimizes the truncation error of this approximation. The first-order and second-order Pade approximations are

$$P_1(s) = \frac{1 - \frac{1}{2}st_s}{1 + \frac{1}{2}st_s} \qquad P_2(s) = \frac{1 - \frac{1}{2}st_s + \frac{1}{12}(st_s)^2}{1 + \frac{1}{2}st_s + \frac{1}{12}(st_s)^2}$$

Since $e^{-st_s}$ describes a delay of one sample (or $z^{-1}$), Pade approximations can be used to generate inverse mappings for converting a digital filter $H(z)$ to an analog filter $H(s)$.

**(a)** Generate mappings for converting a digital filter $H(z)$ to an analog filter $H(s)$ based on the first-order and second-order Pade approximations.

**(b)** Use each mapping to convert $H(z) = \dfrac{z}{z - 0.5}$ to $H(s)$, assuming $t_s = 0.5$ s.

**(c)** Show that the first-order mapping is bilinear. Is this mapping related in any way to the bilinear transformation?

## Computation and Design

**9.30.** **(IIR Filter Design)** It is required to design a lowpass digital filter $H(z)$ from the analog filter $H(s) = \dfrac{1}{s + 1}$. The sampling rate is $S = 1$ kHz. The half-power frequency of $H(z)$ is to be $\Omega_C = \pi/4$.

**(a)** Use impulse invariance to design $H(z)$ such that gain of the two filters matches at dc. Compare the frequency response of both filters (after appropriate frequency scaling). Which filter would you expect to yield better performance? To confirm your expectations, define the (in-band) signal to (out-of-band) noise ratio (SNR) in dB as

$$\text{SNR} = 20\log\left(\frac{\text{signal level}}{\text{noise level}}\right)\ \text{dB}$$

**(b)** What is the SNR at the input and output of each filter if the input is $x(t) = \cos(0.2\omega_C t) + \cos(1.2\omega_C t)$ for $H(s)$ and $x[n] = \cos(0.2n\Omega_C) + \cos(1.2n\Omega_C)$ for $H(z)$?

**(c)** What is the SNR at the input and output of each filter if the input is $x(t) = \cos(0.2\omega_C t) + \cos(3\omega_C t)$ for $H(s)$ and $x[n] = \cos(0.2n\Omega_C) + \cos(3n\Omega_C)$ for $H(z)$?

**(d)** Use the bilinear transformation to design another filter $H_1(z)$ such that gain of the two filters matches at dc. Repeat the computations of parts (a) and (b) for this filter. Of the two digital filters $H(z)$ and $H_1(z)$, which one would you recommend using, and why?

9.31. **(IIR Filter Design)** A digital filter is required to have a monotonic response in the passband and stopband. The half-power frequency is to be 4 kHz, and the attenuation past 5 kHz is to exceed 20 dB. Design the digital filter using impulse invariance and a sampling frequency of 15 kHz.

9.32. **(The Effect of Group Delay)** The nonlinear phase of IIR filters is responsible for signal distortion. Consider a lowpass filter with a 1-dB passband edge at $f = 1$ kHz, a 50-dB stopband edge at $f = 2$ kHz, and a sampling frequency of $S = 10$ kHz.

**(a)** Design a Butterworth filter $H_B(z)$ and an elliptic filter $H_E(z)$ to meet these specifications. Using the MATLAB routine **grpdelay** (or otherwise), compute and plot the group delay of each filter. Which filter has the lower order? Which filter has a more nearly constant group delay in the passband? Which filter would cause the least phase distortion in the passband? What are the group delays $N_B$ and $N_E$ (expressed as the number of samples) of the two filters?

**(b)** Generate the signal $x[n] = 3\sin(0.03n\pi) + \sin(0.09n\pi) + 0.6\sin(0.15n\pi)$ over $0 \le n \le 100$. Use the ADSP routine **filter** to compute the response $y_B[n]$ and $y_E[n]$ of each filter. Plot the filter outputs $y_B[n]$ and $y_E[n]$ (delayed by $N_B$ and $N_E$, respectively) and the input $x[n]$ on the same plot to compare results. Does the filter with the more nearly constant group delay also result in smaller signal distortion?

**(c)** Are all the frequency components of the input signal in the filter passband? If so, how can you justify that the distortion is caused by the nonconstant group delay and not by the filter attenuation in the passband?

9.33. **(LORAN)** A LORAN (long-range radio and navigation) system for establishing positions of marine craft uses three transmitters that send out short bursts (10 cycles) of 100-kHz signals in a precise phase relationship. Using phase comparison, a receiver (on the craft) can establish the position (latitude and longitude) of the craft to within a few hundred meters. Suppose the LORAN signal is to be digitally processed by first sampling it at 500 kHz and filtering the sampled signal using a second order peaking filter with a half-power bandwidth of 100 Hz. Use the bilinear transformation to design the filter from an analog filter with unit half-power frequency. Compare your design with the digital filter designed in Problem 7.67 (to meet the same specifications).

**9.34. (Decoding a Mystery Message)** During transmission, a message signal gets contaminated by a low-frequency signal and high-frequency noise. The message can be decoded only by displaying it in the time domain. The contaminated signal $x[n]$ is provided on the author's website as **mystery1**. Load this signal into MATLAB (use the command **load mystery1**). In an effort to decode the message, try the following methods and determine what the decoded message says.

**(a)** Display the contaminated signal. Can you "read" the message? Display the DFT of the signal to identify the range of the message spectrum.

**(b)** Use the bilinear transformation to design a second-order IIR bandpass filter capable of extracting the message spectrum. Filter the contaminated signal and display the filtered signal to decode the message. Use both the **filter** (filtering) and **filifilt** (zero-phase filtering) commands.

**(c)** As an alternative method, first zero out the DFT component corresponding to the low-frequency contamination and obtain the IDFT $y[n]$. Next, design a lowpass IIR filter (using impulse invariance) to reject the high-frequency noise. Filter the signal $y[n]$ and display the filtered signal to decode the message. Use both the **filter** and **filtfilt** commands.

**(d)** Which of the two methods allows better visual detection of the message? Which of the two filtering routines (in each method) allows better visual detection of the message?

**9.35. (Interpolation)** The signal $x[n] = \cos(2\pi F_0 n)$ is to be interpolated by 5 using up-sampling followed by lowpass filtering. Let $F_0 = 0.4$.

**(a)** Generate and plot 20 samples of $x[n]$ and up-sample by 5.

**(b)** What must be the cutoff frequency $F_C$ and gain $A$ of a lowpass filter that follows the up-sampler to produce the interpolated output?

**(c)** Design a fifth-order digital Butterworth filter (using the bilinear transformation) whose half-power frequency equals $F_C$ and whose peak gain equals $A$.

**(d)** Filter the up-sampled signal through this filter and plot the result. Is the result an interpolated version of the input signal? Do the peak amplitudes of the interpolated signal and original signal match? Should they? Explain.

**9.36. (Coefficient Quantization)** Consider the analog filter described by

$$H(s) = \frac{(s+0.5)(s+1.5)}{(s+1)(s+2)(s+4.5)(s+8)(s+12)}$$

**(a)** Is this filter stable? Why?

**(b)** Use the bilinear transformation with $S = 100$ Hz to convert this to a digital filter $H(z)$.

**(c)** Truncate the filter coefficients to seven significant digits to generate the filter $H_2(z)$.

**(d)** Compare the frequency response of $H(z)$ and $H_2(z)$. Are there any significant differences?

**(e)** Is the filter $H_2(z)$ stable? Should it be? Explain.

**(f)** Suppose the coefficients are to be quantized to $B$ bits by rounding. What is the smallest number of bits $B$ required in order to preserve the stability of the quantized filter?

**9.37. (Numerical-Integration Algorithms)** It is claimed that mapping rules to convert an analog filter to a digital filter, based on numerical-integration algorithms that approximate the area

$y[n]$ from $y[n-2]$ or $y[n-3]$ (two or more time steps away), do not usually preserve stability. Consider the following integration algorithms.

**(a)** $y[n] = y[n-1] + \frac{t_s}{12}(5x[n] + 8x[n-1] - x[n-2])$ (Adams-Moulton rule)
**(b)** $y[n] = y[n-2] + \frac{t_s}{3}(x[n] + 4x[n-1] + x[n-2])$ (Simpson's rule)
**(c)** $y[n] = y[n-3] + \frac{3t_s}{8}(x[n] + 3x[n-1] + 3x[n-2] + x[n-3])$ (Simpson's three-eighths rule)

Derive mapping rules for each algorithm, convert the analog filter $H(s) = \frac{1}{s+1}$ to a digital filter using each mapping with $S = 5$ Hz, and use MATLAB to compare their frequency response. Which of these mappings (if any) allow us to convert a stable analog filter to a stable digital filter? Is the claim justified?

9.38. **(RIAA Equalization)** Audio signals usually undergo a high-frequency boost (and low-frequency cut) before being used to make the master for commercial production of phonograph records. During playback, the signal from the phono cartridge is fed to a preamplifier (equalizer) that restores the original signal. The frequency response of the preamplifier is based on the RIAA (Recording Industry Association of America) equalization curve whose Bode plot is shown in Figure P9.38, with break frequencies at 50, 500, and 2122 Hz.

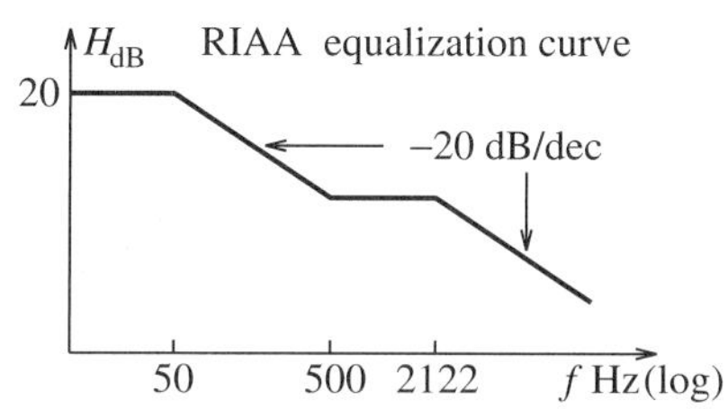


**FIGURE P.9.38** Figure for Problem 9.38

**(a)** What is the transfer function $H(s)$ of the RIAA equalizer?
**(b)** It is required to implement RIAA equalization using a digital filter. Assume that the signal from the cartridge is band limited to 15 kHz. Design an IIR filter $H_I(z)$ using impulse invariance that implements the equalization characteristic.
**(c)** Use the bilinear transformation to design an IIR filter $H_B(z)$ that implements the equalization characteristic. Assume that the gains of $H(s)$ and $H_B(z)$ are to match at 1 kHz.
**(d)** Compare the performance of $H(s)$ with $H_B(z)$ and $H_I(z)$ and comment on the results. Which IIR design method results in better implementation?

CHAPTER 10

# Design of FIR Filters

## 10.0 Scope and Overview

FIR filters are always stable and can be designed with linear phase and, hence, no phase distortion. Their design is typically based on selecting a symmetric impulse response sequence whose length is chosen to meet design specifications. For given specifications, FIR filters require many more elements in their realization than do IIR filters. This chapter starts with an introduction to filter design. It describes symmetric sequences and their application to FIR filters. It examines various methods of FIR filter design, including the window method and frequency sampling. It concludes with special-purpose FIR filters, such as differentiators and Hilbert transformers, and an introduction to adaptive signal processing.

### 10.0.1 Goals and Learning Objectives

The goals of this chapter are to provide an introduction to the techniques of FIR filter design. After going through this chapter, the reader should:

1. Understand the concept of ideal filters.
2. Understand the concept of truncation and windowing and its effect on the spectrum.
3. Know how to convert a lowpass prototype to a highpass, bandpass, or bandstop filter.
4. Understand the central issues in the design of FIR filters and linear-phase filters.
5. Understand the concept of window-based design.
6. Know how to design an FIR filter by windowing.
7. Understand the concept of half-band filters and know how to design a half-band filter.
8. Understand the concept of filter design by frequency sampling.
9. Understand the concept of optimal linear-phase FIR filters and their design.
10. Understand the concept of multistage interpolation and decimation using FIR filters.
11. Understand the concept of FIR differentiators and their design.
12. Understand the concept of Hilbert transformers and their design.

# 10.1 Ideal Filters

Ideal filters form the yardstick by which the performance of other filters is assessed. An ideal lowpass filter is described by constant (typically unity) gain in the passband and zero gain in the stopband. The frequency response of such a filter may be described over its principal range by

$$H_{\mathrm{LP}}(F) = 1, \quad |F| < F_C \tag{10.1}$$

The spectrum of an ideal lowpass filter is illustrated in Figure 10.1.

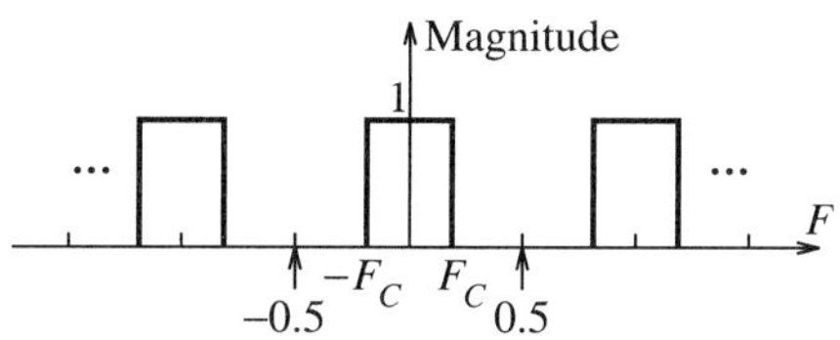


**FIGURE 10.1**
Spectrum of an ideal lowpass filter

Its impulse response $h_{\mathrm{LP}}[n]$ is found using the IDTFT to give

$$\begin{aligned} h_{\mathrm{LP}}[n] &= \int_{0.5}^{-0.5} H_{\mathrm{LP}}(F) e^{j2\pi nF}\, dF = \int_{-F_C}^{F_C} e^{j2\pi nF}\, dF \\ &= \int_{-F_C}^{F_C} \cos(2\pi nF)\, dF = \left.\frac{\sin(2\pi nF)}{2\pi n}\right|_{-F_C}^{F_C} \end{aligned} \tag{10.2}$$

Simplifying this result, we obtain

$$\begin{aligned} h_{\mathrm{LP}}[n] &= \frac{\sin(2\pi nF_C) - \sin(-2\pi nF_C)}{2\pi n} = \frac{2\sin(2\pi nF_C)}{2\pi n} \\ &= \frac{2F_C \sin(2\pi nF_C)}{2\pi nF_C} = 2F_C\, \mathrm{sinc}(2nF_C) \end{aligned} \tag{10.3}$$

### 10.1.1 Frequency Transformations

The impulse response and transfer function of highpass, bandpass, and bandstop filters may be related to those of a lowpass filter using frequency transformations based on the properties of the DTFT. The impulse response of an ideal lowpass filter with a cutoff frequency of $F_C$ and unit passband gain is given by

$$h[n] = 2F_C\, \mathrm{sinc}(2nF_C) \tag{10.4}$$

Figure 10.2 shows two ways to obtain a highpass transfer function from this lowpass filter.

The first way is to shift the spectrum $H(F)$ of the lowpass filter by 0.5 to obtain $H_{\mathrm{H1}}(F) = H(F - 0.5)$, which is a highpass filter whose cutoff frequency is given by $F_{\mathrm{H1}} = 0.5 - F_C$. This leads to

$$h_{\mathrm{H1}}[n] = (-1)^n h[n] = 2(-1)^n F_C\, \mathrm{sinc}(2nF_C) \quad \text{(with cutoff frequency } 0.5 - F_C) \tag{10.5}$$

Alternatively, we see that $H_{\mathrm{H2}}(F) = 1 - H(F)$ also describes a highpass filter but with a cutoff frequency given by $F_{\mathrm{H2}} = F_C$, and this leads to

$$h_{\mathrm{H2}}[n] = \delta[n] - 2F_C\, \mathrm{sinc}(2nF_C) \quad \text{(with cutoff frequency } F_C) \tag{10.6}$$

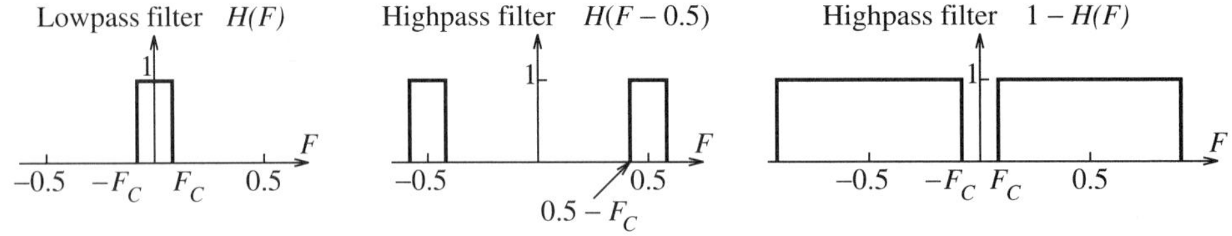


**FIGURE 10.2** Two lowpass-to-highpass transformations. The first transformation shifts the lowpass spectrum $H(F)$ to $H(F - 0.5)$. The second transformation subtracts the lowpass spectrum $H(F)$ from unity.

Figure 10.3 shows how to transform a lowpass filter either to a bandpass filter or to a bandstop filter. To obtain a bandpass filter with a center frequency of $F_0$ and band edges $[F_0 - F_C,\ F_0 + F_C]$, we simply shift $H(F)$ by $F_0$ to get $H_{\text{BP}}(F) = H(F + F_0) + H(F - F_0)$, and obtain

$$h_{\text{BP}}[n] = 2h[n]\cos(2\pi n F_0) = 4F_C \cos(2\pi n F_0)\operatorname{sinc}(2nF_C) \tag{10.7}$$

A bandstop filter with center frequency $F_0$ and band edges $[F_0 - F_C,\ F_0 + F_C]$ can be obtained from the bandpass filter using $H_{\text{BS}}(F) = 1 - H_{\text{BP}}(F)$ to give

$$h_{\text{BS}}[n] = \delta[n] - h_{\text{BP}}[n] = \delta[n] - 4F_C \cos(2\pi n F_0)\operatorname{sinc}(2nF_C) \tag{10.8}$$

### Transformation of an Ideal Lowpass Filter *h*[*n*] and *H*(*F*) to Other Forms

$$h[n] = 2F_C \operatorname{sinc}(2nF_C); \quad H(F) = \operatorname{rect}(F/2F_C)$$

$$h_{\text{HP}}[n] = \delta[n] - h[n]; \quad H_{\text{HP}}(F) = 1 - H(F)$$

$$h_{\text{BP}}[n] = 2h[n]\cos(2\pi n F_0); \quad H_{\text{BP}}(F) \text{ (by modulation)}$$

$$h_{\text{BS}}[n] = \delta[n] - h_{\text{BP}}[n]; \quad H_{\text{BS}}(F) = 1 - H_{\text{BP}}(F)$$

*Note:* $F_C$ = cutoff frequency of lowpass prototype; $F_0$ = center frequency (for BPF and BSF)

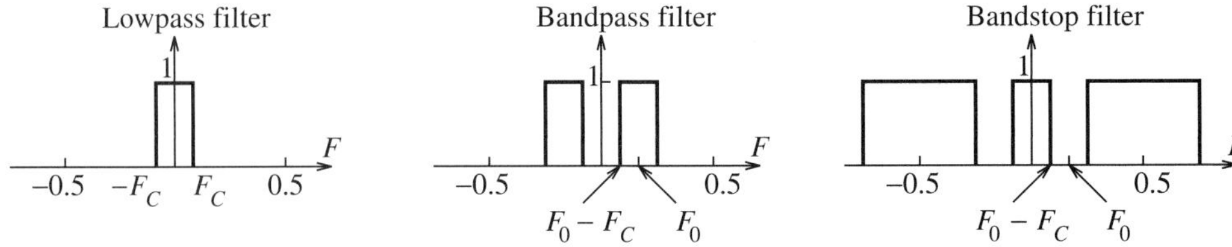


**FIGURE 10.3** Transforming a lowpass filter to a bandpass or bandstop filter. The bandstop transformation shifts the lowpass spectrum $H(F)$ by $F_0$ and $-F_0$. The bandstop filter is found by subtracting the bandpass spectrum from unity

**Example 10.1**

**Frequency Transformations**

Use frequency transformations to find the transfer function and impulse response of:

**(a)** An ideal *lowpass* filter with a digital cutoff frequency of 0.25.
**(b)** An ideal *highpass* filter with a digital cutoff frequency of 0.3.
**(c)** An ideal *bandpass* filter with a passband between $F = 0.1$ and $F = 0.3$.
**(d)** An ideal *bandstop* filter with a stopband between $F = 0.2$ and $F = 0.4$.

**Solution:**

**(a)** For the lowpass filter, pick the LPF cutoff $F_C = 0.25$. Then, $h[n] = 0.5\,\text{sinc}(0.5n)$.

**(b)** For a highpass filter whose cutoff is $F_{\text{HP}} = 0.3$:

Pick an LPF with $F_C = 0.5 - 0.3 = 0.2$. Then, $h_{\text{HP}}[n] = (-1)^n(0.4)\text{sinc}(0.4n)$.

Alternatively, pick $F_C = 0.3$. Then, $h_{\text{HP}}[n] = \delta[n] - (0.6)\text{sinc}(0.6n)$.

**(c)** For a bandpass filter with band edges $[F_1, F_2] = [0.1,\ 0.3]$:

Pick the LPF cutoff as $F_C = F_2 - F_0 = 0.1,\ F_0 = 0.2$. Then, $h_{\text{BP}}[n] = 0.4\cos(0.4n\pi)\text{sinc}(0.2n)$.

**(d)** For a bandstop filter with band edges $[F_1, F_2] = [0.2,\ 0.4]$:

Pick the LPF cutoff as $F_C = F_2 - F_0 = 0.1,\ F_0 = 0.3$. Then, $h_{\text{BS}}[n] = \delta[n] - 0.4\cos(0.6n\pi)\text{sinc}(0.2n)$.

## 10.1.2 Truncation and Windowing

An ideal filter possesses linear phase because its impulse response $h[n]$ is a symmetric sequence. It can never be realized in practice because $h[n]$ is a sinc function that goes on forever, making the filter *noncausal*. Moreover, the ideal filter is *unstable* because $h[n]$ (a sinc function) is *not absolutely summable*. One way to approximate an ideal lowpass filter is by symmetric truncation (about $n = 0$) of its impulse response $h[n]$ (which ensures linear phase). Truncation of $h[n]$ to $|n| \le N$ is equivalent to multiplying $h[n]$ by a rectangular **window** $w_D[n] = 1,\ |n| \le N$. To obtain a causal filter, we must also delay the impulse response by $N$ samples so that its first sample appears at the origin.

## 10.1.3 The Rectangular Window and its Spectrum

Consider the rectangular window $w_d[n] = \text{rect}(n/2N)$ with $M = 2N+1$ samples. We have

$$w_d[n] = \text{rect}\left(\frac{n}{2N}\right) = \{\underbrace{1,\ 1,\ \ldots,\ 1,\ 1}_{N\text{ ones}},\ \overset{\Downarrow}{1},\ \underbrace{1,\ 1,\ \ldots,\ 1,\ 1}_{N\text{ ones}}\} \tag{10.9}$$

Its DTFT has been computed earlier as

$$W_D(F) = \frac{\sin(\pi MF)}{\sin(\pi F)} = \frac{M\text{sinc}(MF)}{\text{sinc}(F)}, \qquad M = 2N+1 \tag{10.10}$$

The quantity $W_D(F)$ describes the **Dirichlet kernel** or the **aliased sinc** function. It also equals the *periodic extension* of sinc($MF$) with period $F = 1$. Figure 10.4 shows the Dirichlet kernel for $N = 3,\ 6$, and 9.

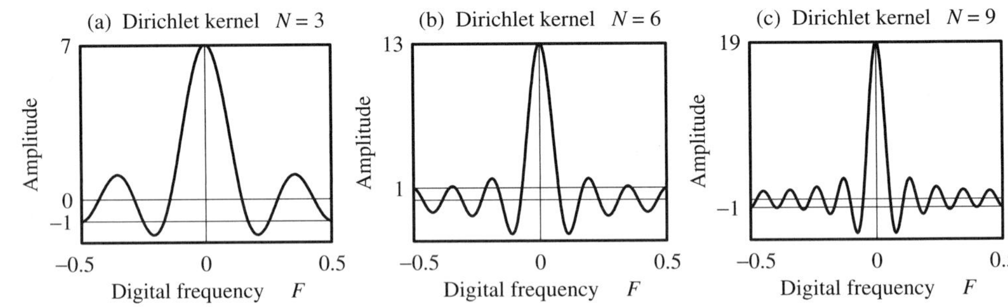


**FIGURE 10.4** The Dirichlet kernel is the spectrum of a rectangular window

The Dirichlet kernel has some very interesting properties. Over one period, we observe the following:

1. It has $N$ maxima. The width of the main lobe is $2/M$. The width of the decaying positive and negative sidelobes is $1/M$. There are $2N$ zeros located at the frequencies $F = \frac{1}{M}, \frac{2}{M}, \ldots, \frac{2N}{M}$.
2. Its area equals unity, and it attains a maximum peak value of $M$ at the origin. Its value at $|F| = 0.5$ is $+1$ (for even $N$) or $-1$ (for odd $N$). Increasing $M$ increases the main lobe height and compresses the sidelobes. The ratio $R$ of the main lobe magnitude and peak sidelobe magnitude, however, stays more or less constant, varying between 4 for small $M$ and $1.5\pi = 4.71$ (or 13.5 dB) for very large $M$. As $M \to \infty$, the spectrum approaches a unit impulse.

## 10.1.4 The Triangular Window and its Spectrum

A triangular window may be regarded as the convolution of two rectangular windows. For a rectangular window $w_d[n] = \text{rect}(n/2N)$, we find

$$w_d[n] * w_d[n] = \text{rect}\left(\frac{n}{2N}\right) * \text{rect}\left(\frac{n}{2N}\right) = (2N+1)\text{tri}\left(\frac{n}{2N+1}\right)$$

$$= M\text{tri}\left(\frac{n}{M}\right), \quad M = 2N+1$$

We may thus express a triangular window $w_f[n] = \text{tri}(n/M)$ in terms of the convolution

$$\text{tri}\left(\frac{n}{M}\right) = \frac{1}{M} w_d[n] * \omega_d[n], \quad M = 2N+1 \tag{10.11}$$

Since convolution in the time domain corresponds to multiplication in the frequency domain, the spectrum of the triangular window may be written as

$$W_F(F) = \frac{1}{M} W_D^2(F) = \frac{\sin^2(\pi M F)}{M \sin^2(\pi F)} = \frac{M \text{sinc}^2(MF)}{\text{sinc}^2(F)}$$

The quantity $W_F(F)$ is called the **Fejer kernel**. Figure 10.5 shows the Fejer kernel for $M = 4, 5,$ and 8. Note that the spectrum is always positive. Its area equals unity and it attains a peak value of $M$ at the origin. Its value at $|F| = 0.5$ is 0 (for even $M$) or $1/M$ (for odd $M$). Increasing $M$ increases the main lobe height and compresses the sidelobes. There are $M$ maxima over one period. As $M \to \infty$, the spectrum approaches a unit impulse. For a given finite length $M$, the main lobe of the Fejer kernel is twice as wide as that of the Dirichlet kernel, while the sidelobe magnitudes of the Fejer kernel show a faster decay and are much smaller.

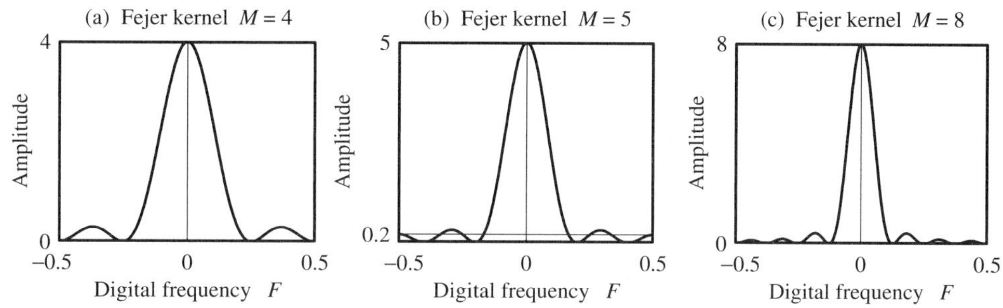


**FIGURE 10.5** The Fejer kernel is the spectrum of a triangular window

## 10.1.5 The Consequences of Windowing

Windowing (or multiplication) of the filter impulse response $h[n]$ by a window $w[n]$ in the time domain is equivalent to the *periodic convolution* of the filter spectrum $H(F)$ and the window spectrum $W(F)$ in the frequency domain. If the impulse response, an ideal filter given by $h[n] = 2F_C\text{sinc}(2F_Cn)$ is truncated (multiplied) by a rectangular window $w_d[n]$, the spectrum of the windowed filter exhibits overshoot and oscillations, as illustrated in Figure 10.6. This is reminiscent of the **Gibbs effect** that occurs in the Fourier series reconstruction of a periodic signal containing sudden jumps (discontinuities).

What causes the overshoot and oscillation is the *slowly* decaying sidelobes of $W_D(F)$ (the spectrum of the rectangular window). The overshoot and oscillation persists even if we increase the window length. The only way to reduce and/or eliminate the overshoot and oscillation is to use a tapered window, such as the triangular window, whose spectral sidelobes decay much faster. If the impulse response $h[n]$ an ideal filter is truncated (multiplied) by a triangular window $w_f[n]$, the spectrum of the windowed filter exhibits a monotonic response and a complete absence of overshoot, as illustrated in Figure 10.7. All useful windows are tapered in some fashion to minimize (or even eliminate) the overshoot in the spectrum of the windowed filter while maintaining as sharp a cutoff as possible. The spectrum of any window should have a narrow main lobe (to ensure a sharp cutoff in the windowed spectrum) and small sidelobe levels (to minimize the overshoot and oscillation in the windowed spectrum). For a given window length, it is not possible to minimize both

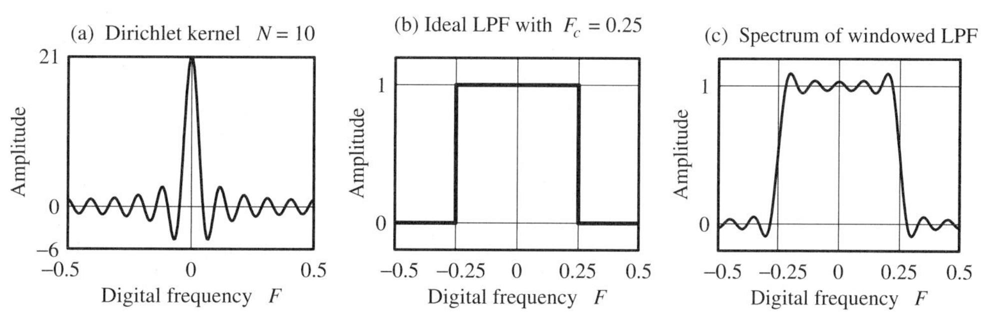


**FIGURE 10.6** The spectrum of an ideal filter windowed by a rectangular window shows overshoot near the edges and oscillations that persist even if we increase the filter length. The peak overshoot stays at about 0.9% of the jump

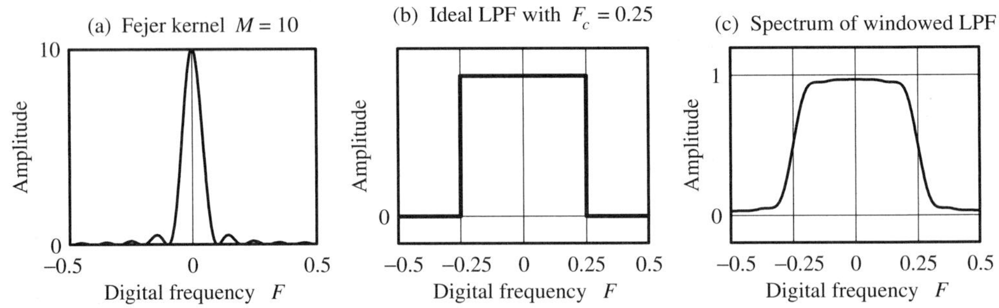


**FIGURE 10.7** The spectrum of an ideal filter windowed by a triangular window results in smoothing of the edges, an absence of overshoot and a broader transition at the edges. The transition width is inversely proportional to the filter length

the main lobe width and the sidelobe levels simultaneously. Consequently, the design of any window entails a compromise between these inherently conflicting requirements.

**Windowing of $h[n]$ in Time Domain Means Periodic Convolution in Frequency Domain**

Windowing makes the transition in the spectrum more gradual (less abrupt).
A rectangular window (truncation) leads to overshoot and oscillations in spectrum (Gibbs effect).
Any other window leads to reduced (or no) overshoot in spectrum.

### 10.1.6 Design Specifications for FIR Filters

FIR filters are typically described by frequency specifications and attenuation (or gain) specifications. Frequency specifications include the passband edge(s) and stopband edge(s). If analog frequencies are specified in the design or if the digital filter is to be designed for processing sampled analog signals, the sampling rate must also be given.

Typical magnitude (or gain) specifications for a lowpass digital filter are illustrated in Figure 10.8 and include the passband attenuation (or gain or ripple) and the stopband attenuation (or gain or ripple). The passband ripple (deviation from maximum gain) and stopband ripple (deviation from zero gain) are usually described by their attenuation (reciprocal of the linear gain) expressed in decibels (dB).

### 10.1.7 FIR Filters and Linear Phase

Only FIR filters can be designed with linear phase (no phase distortion). The design of such filters is typically based on a symmetric impulse-response sequence (a linear-phase sequence) whose length is chosen to meet design specifications. For example, the starting point may be the infinite-length impulse response of an ideal filter that is truncated by a symmetric window of finite length The *smallest* length that allows the filter to meet

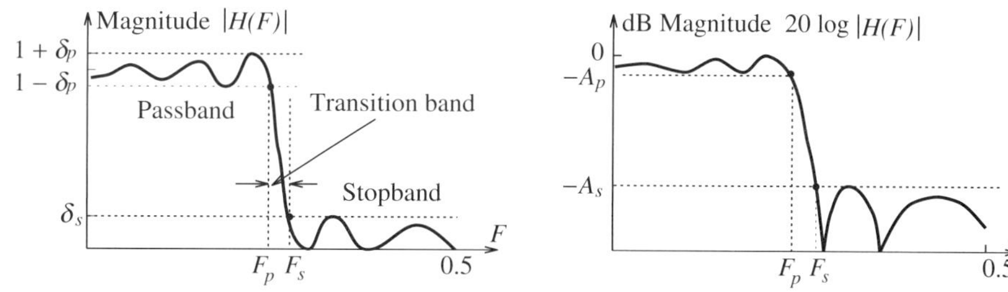


**FIGURE 10.8** The features of a typical lowpass digital filter. Up to the passband edge $F_p$, the gain is nearly unity, with small variations about unity measured by the passband ripple $\delta_p$. Past the stopband edge $F_s$, the gain is nearly zero, with a maximum variation measured by the stopband ripple $\delta_s$. On the decibel scale, unit gain corresponds to 0 dB. The *maximum* passband attenuation and *minimum* stopband attenuation are measured by $A_p$-dB and $A_s$-dB, respectively. These correspond to a *minimum* passband gain of $-A_p$-dB and a *maximum* stopband gain of $-A_s$-dB, respectively

specifications is often determined by iterative techniques (or even trial and error). For given specifications, FIR filters require many more elements in their realization than do IIR filters.

# 10.2 Symmetric Sequences and Linear Phase

Symmetric sequences possess linear phase and result in a *constant delay* with no distortion. This is an important consideration in filter design. The DTFT of a real, even symmetric sequence $x[n]$ is of the form $H(F) = A(F)$ and always real, and the DTFT of a real, odd symmetric sequence is of the form $H(F) = jA(F)$ and purely imaginary. Symmetric sequences also imply noncausal filters. However, such filters can be made causal if the impulse response is suitably delayed. A time shift of $x[n]$ to $x[n - M]$ introduces only a linear phase of $\phi(F) = e^{-j2M\pi F}$. The DTFT of sequences that are symmetric about their midpoint is said to possess **generalized linear phase**. Generalized linear phase is illustrated in Figure 10.9.

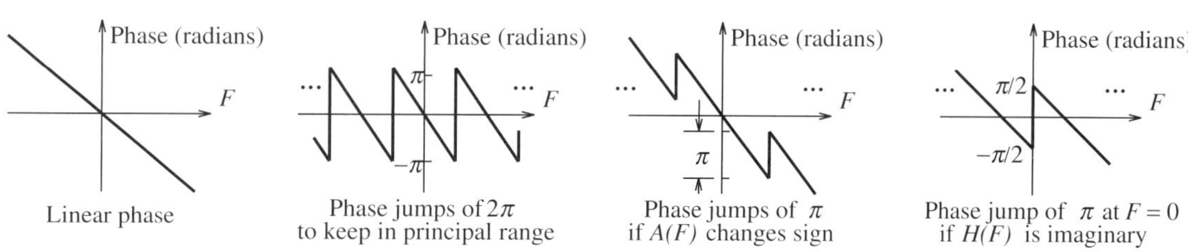


**FIGURE 10.9** Examples of linear phase and generalized linear phase. Generalized linear phase may include phase jumps of $\pi$ or $2\pi$ as long as the piecewise-linear portions have the same slope

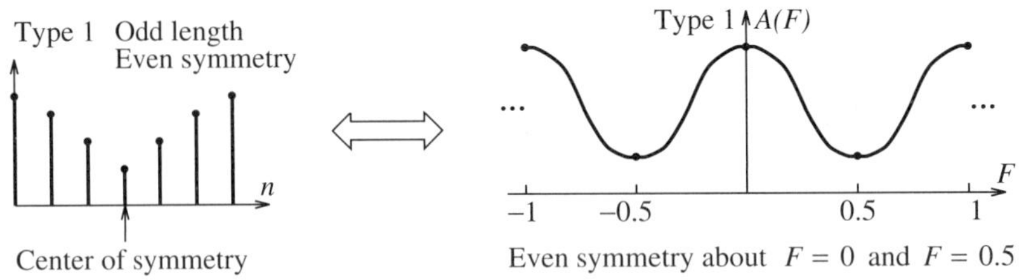


**FIGURE 10.10** Features of a type 1 symmetric sequence. The impulse response has odd length and even symmetry about its midpoint which falls on a sample value. The amplitude spectrum $A(F)$ is periodic with unit period in $F$ and shows even symmetry about both $F = 0$ and $F = 0.5$

The term *generalized* means that $\phi(F)$ may include a jump (of $\pi$ at $F = 0$, if $H(F)$ is imaginary). There may also be phase jumps of $2\pi$ (if the phase is restricted to the principal range $\pi \le \phi(F) \le \pi$). If we plot the magnitude $|H(F)|$, there will also be phase jumps of $\pi$ (where the amplitude $A(F)$ changes sign).

### 10.2.1 Types of Linear-Phase Sequences for FIR Filter Design

Symmetric sequences may possess even or odd length. If the length is odd, the center of symmetry lies on a sample point, but if the length is even, the center of symmetry lies midway between samples. Symmetric sequences may also possess even symmetry or odd symmetry. This gives rise to four possible types of symmetric sequences.

#### Type 1 Sequences

A type 1 sequence $h_1[n]$ and its amplitude spectrum $A_1(F)$ are illustrated in Figure 10.10.

This sequence is even symmetric with odd length $N$, and it has a center of symmetry at the integer value $M = (N - 1)/2$. Using Euler's relation, its frequency response $H_1(F)$ may be expressed as

$$H_1(F) = \left[ h[M] + 2\sum_{k=0}^{M-1} h[k]\cos[(M-k)2\pi F] \right] e^{-j2\pi MF} = A_1(F)e^{-j2\pi MF} \qquad (10.12)$$

Thus, $H_1(F)$ shows a linear phase of $-2\pi MF$ and a constant group delay of $M$. The amplitude spectrum $A_1(F)$ is even symmetric about both $F = 0$ and $F = 0.5$, and both $|H_1(0)|$ and $|H_1(0.5)|$ can be nonzero.

#### Type 2 Sequences

A type 2 sequence $h_2[n]$ and its amplitude spectrum $A_2(F)$ are illustrated in Figure 10.11.

This sequence is also even symmetric but of even length $N$, and it has a center of symmetry at the half-integer value $M = (N - 1)/2$. Using Euler's relation, its frequency response $H_2(F)$ may be expressed as

$$H_2(F) = \left[ 2\sum_{k=0}^{M-1/2} h[k]\cos[(M-k)2\pi F] \right] e^{-j2\pi MF} = A_2(F)e^{-j2\pi MF} \qquad (10.13)$$

Thus, $H_2(F)$ also shows a linear phase of $-2\pi MF$ and a constant group delay of $M$. The amplitude spectrum $A_2(F)$ is even symmetric about $F = 0$, and odd symmetric about $F = 0.5$ and as a result, $|H_2(0.5)|$ is always zero.

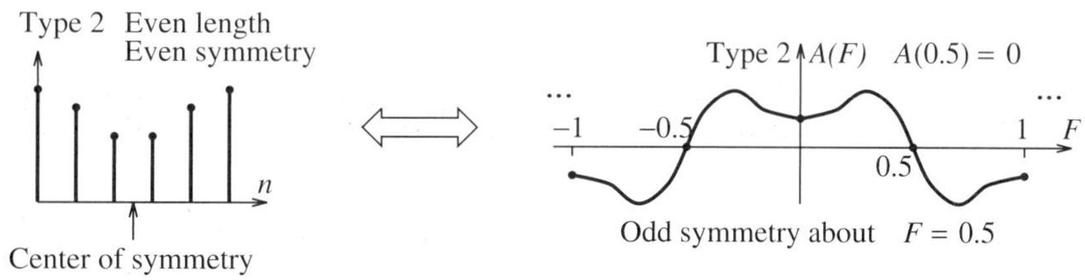


**FIGURE 10.11** Features of a type 2 symmetric sequence. The impulse response has even length and even symmetry about its midpoint, which falls between two samples. The amplitude spectrum $A(F)$ is periodic with a period of $F = 2$ and shows even symmetry about $F = 0$ and $F = 1$

## Type 3 Sequences

A type 3 sequence $h_3[n]$ and its amplitude spectrum $A_3(F)$ are illustrated in Figure 10.12.

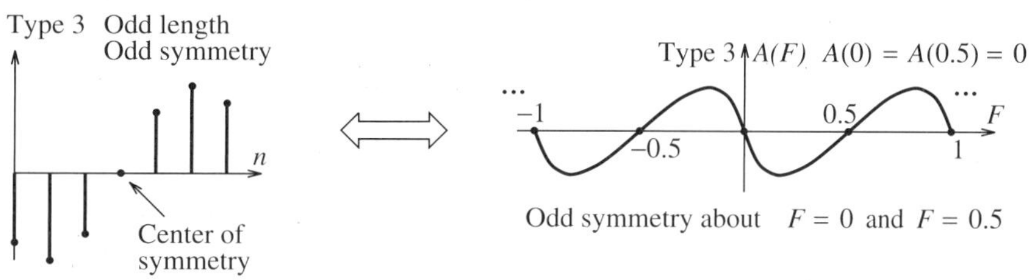


**FIGURE 10.12** Features of a type 3 symmetric sequence. The impulse response has odd length and odd symmetry about its midpoint, which falls on a zero-valued sample. The amplitude spectrum $A(F)$ is periodic with unit period in $F$ and shows odd symmetry about both $F = 0$ and $F = 0.5$

This sequence is odd symmetric with odd length $N$, and it has a center of symmetry at the integer value $M = (N - 1)/2$. Using Euler's relation, its frequency response $H_3(F)$ may be expressed as

$$H_3(F) = j\left[2\sum_{k=0}^{M-1} h[k]\sin[(M-k)2\pi F]\right]e^{-j2\pi MF} = A_3(F)e^{j(0.5\pi - 2\pi MF)} \qquad (10.14)$$

Thus, $H_3(F)$ shows a *generalized* linear phase of $\frac{\pi}{2} - 2\pi MF$ and a constant group delay of $M$. The amplitude spectrum $A_3(F)$ is odd symmetric about both $F = 0$ and $F = 0.5$, and as a result, $|H_3(0)|$ and $|H_3(0.5)|$ are always zero.

## Type 4 Sequences

A type 4 sequence $h_4[n]$ and its amplitude spectrum $A_4(F)$ are illustrated in Figure 10.13.

This sequence is odd symmetric with even length $N$, and it has a center of symmetry at the half-integer value $M = (N - 1)/2$. Using Euler's relation, its frequency response $H_4(F)$ may be expressed as

$$H_4(F) = j\left[2\sum_{k=0}^{M-1/2} h[k]\sin[2(M-k)\pi F]\right]e^{-j2\pi MF} = A_4(F)e^{j(0.5\pi - 2\pi MF)} \qquad (10.15)$$

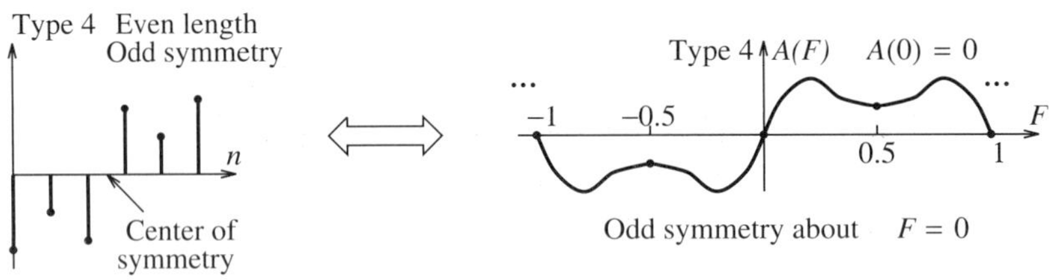


**FIGURE 10.13** Features of a type 4 symmetric sequence. The impulse response has even length and odd symmetry about its midpoint which falls between sample values. The amplitude spectrum $A(F)$ is periodic with period $F = 2$ and shows odd symmetry about $F = 0$ and $F = 1$

Thus, $H_4(F)$ also shows a generalized linear phase of $\frac{\pi}{2} - 2\pi MF$ and a constant group delay of $M$. The amplitude spectrum $A_4(F)$ is odd symmetric about $F = 0$, and even symmetric about $F = 0.5$ and as a result, $|H_4(0)|$ is always zero.

### 10.2.2 Applications of Linear-Phase Sequences

Table 10.1 summarizes the amplitude response characteristics of the four types of linear-phase sequences and their use in FIR digital filter design. Type 1 sequences are by far the most widely used, because they allow us to design any filter type by appropriate choice of filter coefficients. Type 2 sequences can be used for lowpass and bandpass filters but not for bandstop or highpass filters (whose response is not zero at $F = 0.5$). Type 3 sequences are useful primarily for bandpass filters, differentiators, and Hilbert transformers. Type 4 sequences are suitable for highpass or bandpass filters and for differentiators and Hilbert transformers.

Bandstop filters (whose response is nonzero at $F = 0$ and $F = 0.5$) can be designed only with type 1 sequences. Only antisymmetric sequences (whose transfer function is imaginary) or their causal versions (which correspond to type 3 and type 4 sequences) can be used to design digital differentiators and Hilbert transformers.

**TABLE 10.1 ➤** Applications of Symmetric Sequences

| Type | $H(F) = 0$ (or $H(\Omega) = 0$) at | Application |
|---|---|---|
| 1 | | All filter types. Only sequence for BSF |
| 2 | $F = 0.5$ $(\Omega = \pi)$ | Only LPF and BPF |
| 3 | $F = 0$ $(\Omega = 0)$, $F = 0.5$ $(\Omega = \pi)$ | BPF, differentiators, Hilbert transformers |
| 4 | $F = 0$ $(\Omega = 0)$ | HPF, BPF, differentiators, Hilbert transformers |

### 10.2.3 FIR Filter Design

The design of FIR filters involves the selection of a finite sequence that best represents the impulse response of an ideal filter. FIR filters are always stable. Even more important, FIR filters are capable of perfectly linear phase (a pure time delay), meaning total freedom from phase distortion. For given specifications, however, FIR filters typically require a much higher filter order or length than do IIR filters. And sometimes, we must go to great lengths to ensure linear phase! The three most commonly used methods for FIR filter design are window-based design using the impulse response of ideal filters, frequency sampling, and iterative design based on optimal constraints.

# 10.3 Window-Based Design

The window method starts by selecting the impulse response $h_N[n]$ as a symmetrically truncated version of the impulse response $h[n]$ of an ideal filter with frequency response $H(F)$. The impulse response of an ideal lowpass filter with a cutoff frequency $F_C$ is $h[n] = 2F_C \operatorname{sinc}(2nF_C)$. Its symmetric truncation yields

$$h_N[n] = 2F_C \operatorname{sinc}(2nF_C), \quad |n| \le 0.5(N-1) \tag{10.16}$$

Note that for an even length $N$, the index $n$ is not an integer. Even though the designed filter is an approximation to the ideal filter, it is the best approximation (in the mean-square sense) compared to any other filter of the same length. The problem is that it shows certain undesirable characteristics. Truncation of the ideal impulse response $h[n]$ is equivalent to multiplication of $h[n]$ by a rectangular window $w[n]$ of length $N$. The spectrum of the windowed impulse response $h_W[n] = h[n]w[n]$ is the (periodic) convolution of $H(F)$ and $W(F)$. Since $W(F)$ has the form of a Dirichlet kernel, this spectrum shows overshoot and ripples (the Gibbs effect). It is the abrupt truncation of the rectangular window that leads to overshoot and ripples in the magnitude spectrum. To reduce or eliminate the Gibbs effect, we use tapered windows.

## 10.3.1 Characteristics of Window Functions

The amplitude response of symmetric, finite-duration windows invariably shows a main lobe and decaying sidelobes that may be entirely positive or that may alternate in sign. The spectral measures for a typical window are illustrated in Figure 10.14.

Amplitude-based measures for a window include the peak sidelobe level (PSL), usually in decibels, and the decay rate $D_S$ in dB/dec. Frequency-based measures include the main lobe width $W_M$, the 3-dB and 6-dB widths ($W_3$ and $W_6$), and the width $W_S$ to reach the peak sidelobe level. The windows commonly used in FIR filter design and their spectral features are listed in Table 10.2 and illustrated in Figure 10.15. As the window length $N$ increases, the width parameters decrease, but the peak sidelobe level remains more or less constant. Ideally, the spectrum of a window should approximate an impulse and be confined to as narrow a main lobe as possible, with as little energy in the sidelobes as possible.

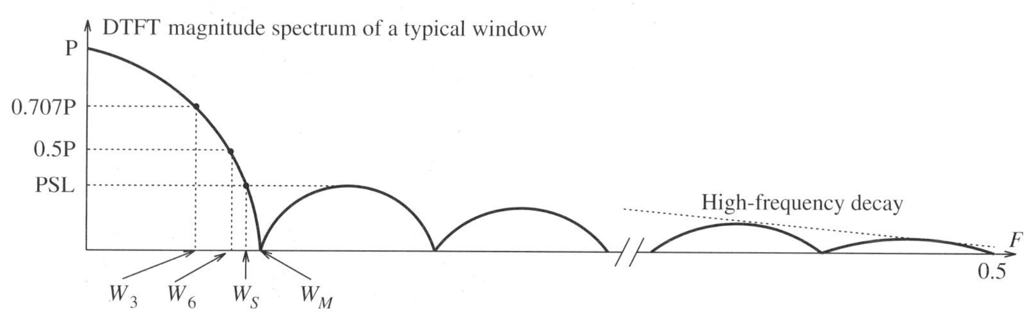


**FIGURE 10.14** The spectrum of a typical window. Measures include the peak gain P, the 3-dB and 6-dB gain 0.707P and 0.5P corresponding to the widths $W_3$ and $W_6$, the peak sidelobe level PSL corresponding to the width $W_S$, and the main lobe half-width $W_M$. The high-frequency decay $D_S$ is typically measured in dB/decade or dB/octave

**TABLE 10.2 ➤ Some Windows for FIR Filter Design**

| Window | Expression $w[n]$, $-0.5(N-1) \le n \le 0.5(N-1)$ |
|---|---|
| Boxcar | $1$ |
| Cosine | $\cos\left(\frac{n\pi}{N-1}\right)$ |
| Riemann | $\operatorname{sinc}^L\left(\frac{2n}{N-1}\right)$, $L > 0$ |
| Bartlett | $1 - \frac{2\lvert n\rvert}{N-1}$ |
| von Hann (Hanning) | $0.5 + 0.5\cos\left(\frac{2n\pi}{N-1}\right)$ |
| Hamming | $0.54 + 0.46\cos\left(\frac{2n\pi}{N-1}\right)$ |
| Blackman | $0.42 + 0.5\cos\left(\frac{2n\pi}{N-1}\right) + 0.08\cos\left(\frac{4n\pi}{N-1}\right)$ |
| Kaiser | $\frac{I_0(\pi\beta\sqrt{1-4[n/(N-1)]^2})}{I_0(\pi\beta)}$ |

*NOTE:* $I_0(x)$ is the modified Bessel function of order zero.

Spectral Characteristics of Window Functions

| Window | $G_P$ | $G_S/G_P$ | $A_{SL}$ (dB) | $W_M$ | $W_S$ | $W_6$ | $W_3$ | $D_S$ |
|---|---|---|---|---|---|---|---|---|
| Boxcar | 1 | 0.2172 | 13.3 | 1 | 0.81 | 0.6 | 0.44 | 20 |
| Cosine | 0.6366 | 0.0708 | 23 | 1.5 | 1.35 | 0.81 | 0.59 | 40 |
| Riemann | 0.5895 | 0.0478 | 26.4 | 1.64 | 1.5 | 0.86 | 0.62 | 40 |
| Bartlett | 0.5 | 0.0472 | 26.5 | 2 | 1.62 | 0.88 | 0.63 | 40 |
| von Hann (Hanning) | 0.5 | 0.0267 | 31.5 | 2 | 1.87 | 1.0 | 0.72 | 60 |
| Hamming | 0.54 | 0.0073 | 42.7 | 2 | 1.91 | 0.9 | 0.65 | 20 |
| Blackman | 0.42 | 0.0012 | 58.1 | 3 | 2.82 | 1.14 | 0.82 | 60 |
| Kaiser ($\beta = 2.6$) | 0.4314 | 0.0010 | 60 | 2.98 | 2.72 | 1.11 | 0.80 | 20 |

*NOTATION:*

$G_P$: Peak gain of main lobe
$G_S$: Peak sidelobe gain
$D_S$: High-frequency attenuation (dB/decade)
$A_{SL}$: Sidelobe attenuation ($\frac{G_P}{G_S}$) in dB
$W_6$: 6-dB half-width
$W_S$: Half-width of main lobe to reach $P_S$
$W_3$ 3-dB half-width
$W_M$: Half-width of main lobe

*NOTE:*

1. All widths ($W_M$, $W_S$, $W_6$, $W_3$) must be normalized (divided) by the window length $N$.
2. Values for the Kaiser window depend on the parameter $\beta$. Empirically determined relations are

$$G_P = \frac{|\operatorname{sinc}(j\beta)|}{I_0(\pi\beta)}, \quad \frac{G_S}{G_P} = \frac{0.22\pi\beta}{\sinh(\pi\beta)}, \quad W_M = (1+\beta^2)^{1/2}, \quad W_S = (0.661+\beta^2)^{1/2}$$

Most windows have been developed with some optimality criterion in mind. Ultimately, the trade-off is a compromise between the conflicting requirements of a narrow main lobe (or a small transition width), and small sidelobe levels. Some windows are based on combinations of simpler windows. For example, the von Hann (or Hanning) window is the sum of a rectangular and a cosine window, and the Bartlett window is the convolution of two rectangular windows. Other windows are designed to emphasize certain desirable features. The von Hann window improves the high-frequency decay (at the expense of a larger

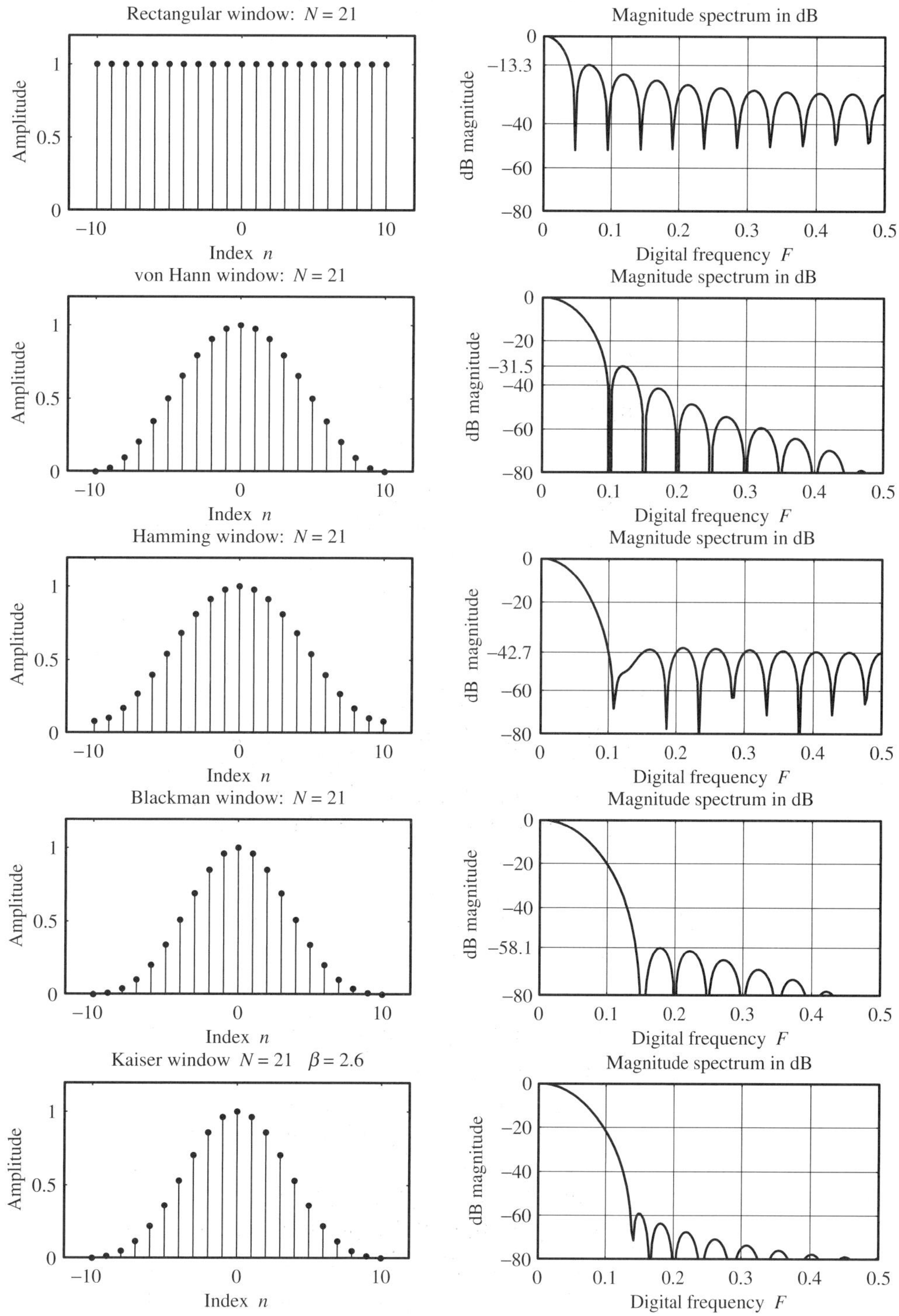


**FIGURE 10.15** Some DTFT windows for FIR filter design and their spectra. For purposes of comparison, all windows have 21 samples (some with zero-valued end samples). Note how some windows result in smaller sidelobe levels (but at the expense of a broader main lobe width). Typically, the main lobe width may be reduced by choosing a longer length, but the sidelobe levels are largely independent of the window length

peak sidelobe level). The Hamming window minimizes the sidelobe level (at the expense of a slower high-frequency decay). The Kaiser window has a variable parameter $\beta$ that controls the peak sidelobe level. Still other windows are based on simple mathematical forms or easy application. For example, the $\cos^a$ windows have easily recognizable transforms, and the von Hann window is easy to apply as a convolution in the frequency domain. An optimal time-limited window should maximize the energy in its spectrum over a given frequency band. In the continuous-time domain, this constraint leads to a window based on *prolate spheroidal wave functions* of the first order. The Kaiser window best approximates such an optimal window in the discrete domain.

### 10.3.2 Some Other Windows

The Dolph (also called Chebyshev) window corresponds to the inverse DTFT of a spectrum whose sidelobes remain constant (without decay) at a specified level and whose main lobe width is the smallest for a given length. The time-domain expression for this window is cumbersome. This window can be computed from

$$w[n] = \text{IDFT}\left\{T_{N-1}\left[\alpha\cos\left(\frac{n\pi}{N-1}\right)\right]\right\} \qquad \alpha = \cosh\left[\frac{\cosh^{-1}(10^{A/20})}{N-1}\right] \qquad (\text{odd } n \text{ only}) \tag{10.17}$$

Here, $T_n(x)$ is the Chebyshev polynomial of order $n$, and $A$ is the sidelobe attenuation in decibels.

**TABLE 10.3 ➤** Characteristics of Harris Windows

| Type | $b_0$ | $b_1$ | $b_2$ | $b_3$ | $A_{SL}$ (dB) | $W_S = K/N$ | $D_S$ |
|---|---|---|---|---|---|---|---|
| 0 | 0.375 | 0.5 | 0.125 | 0.0 | 46.74 | $K = 2.89$ | 30 |
| 1 | 10/32 | 15/32 | 6/32 | 1/32 | 60.95 | $K = 3.90$ | 42 |
| 2 | 0.40897 | 0.5 | 0.09103 | 0.0 | 64.19 | $K = 2.94$ | 18 |
| 3 | 0.42323 | 0.49755 | 0.07922 | 0.0 | 70.82 | $K = 2.95$ | 6 |
| 4 | 0.4243801 | 0.4973406 | 0.0782793 | 0.0 | 71.48 | $K = 2.95$ | 6 |
| 5 | 0.338946 | 0.481973 | 0.161054 | 0.018027 | 82.6 | $K = 3.95$ | 30 |
| 6 | 0.355768 | 0.487396 | 0.144232 | 0.012604 | 92.32 | $K = 3.96$ | 18 |
| 7 | 0.3635819 | 0.4891775 | 0.1365995 | 0.0106411 | 98.17 | $K = 3.96$ | 6 |

*NOTES:* $A_{SL}$ is the normalized peak sidelobe attenuation in dB (decibels).
$W_S$ is the half-width of the main lobe to reach $P_S$.
$D_S$ is the high-frequency decay rate in dB/octave (6 dB/oct = 20 dB/dec).

Harris windows offer a variety of peak sidelobe levels and high-frequency decay rates. Their characteristics are listed in Table 10.3. Harris windows have the general form

$$w[n] = b_0 - b_1\cos(m) + b_2\cos(2m) - b_3\cos(3m), \quad \text{where } m = \frac{2n\pi}{N-1}, \quad |n| \le \frac{N-1}{2} \tag{10.18}$$

### 10.3.3 What Windowing Means

Symmetric windowing of the impulse response of an ideal filter is accomplished by windows that are themselves symmetric, as illustrated in Figure 10.16.

An $N$-point FIR window uses $N-1$ intervals to generate $N$ samples (including both end samples). Once selected, the window sequence must be positioned symmetrically with

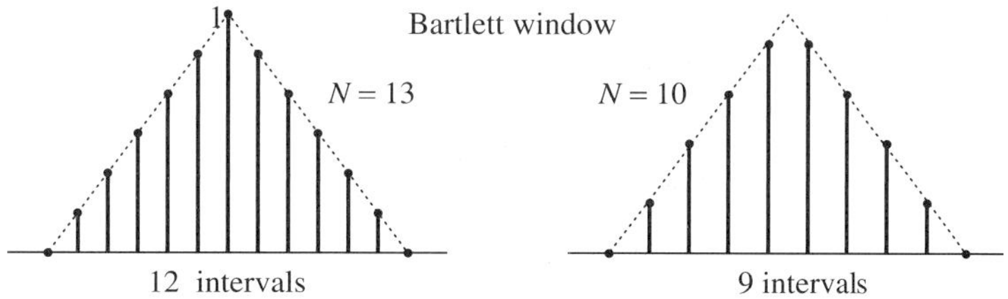


**FIGURE 10.16** The mechanics of windowing. If the window has odd length, its midpoint falls on a sample value. A window with even length has its midpoint between sample values

respect to the symmetric impulse-response sequence. Windowing is then simply a *pointwise multiplication* of the two sequences.

The symmetrically windowed impulse response of an ideal lowpass filter may be written as

$$h_W[n] = 2F_C\,\text{sinc}(2nF_C)w[n], \qquad -0.5(N-1) \le n \le 0.5(N-1) \tag{10.19}$$

For even $N$, the index $n$ is not an integer, even though it is incremented every unit, and we require a *non-integer* delay to produce a causal sequence. If the end samples of $w[n]$ equal zero, so will those of $h_W[n]$, and the true filter length will be less by 2.

**Example 10.2**

### Truncation and Windowing

Consider a lowpass filter with a cutoff frequency of 5 kHz and a sampling frequency of 20 kHz.

Then, $F_C = 0.25$, and $h[n] = 2F_C\,\text{sinc}(2nF_C) = 0.5\,\text{sinc}(0.5n)$.

**(a)** With $N = 9$ and a Bartlett window $w = 1 - \frac{2|n|}{N-1}$, $-4 \le n \le 4$, we find

$$h[n] = \{0,\ -0.1061,\ 0,\ 0.3183,\ \overset{\Downarrow}{0.5},\ 0.3183,\ 0,\ -0.1061,\ 0\}$$

$$w[n] = \{0,\ 0.25,\ 0.5,\ 0.75,\ \overset{\Downarrow}{1},\ 0.75,\ 0.5,\ 0.25,\ 0\}$$

Pointwise multiplication of $h_N$ and $w_N$ gives

$$h_W[n] = \{0,\ -0.0265,\ 0,\ 0.2387,\ \overset{\Downarrow}{0.5},\ 0.2387,\ 0,\ -0.0265,\ 0\}$$

If we include the end samples, the causal filter requires a delay of four samples to give

$$H_C(z) = 0 - 0.0265z^{-1} + 0.2387z^{-3} + 0.5z^{-4} + 0.2387z^{-5} - 0.0265z^{-7}$$

Since $S = 20$ kHz, we have $t_s = 1/S = 0.05$ ms, and a four-sample delay corresponds to an actual delay of $4(0.05) = 0.2$ ms. If we ignore the end samples, the causal filter requires a delay of only three (not four) samples to give

$$H_C(z) = -0.0265 + 0.2387z^{-2} + 0.5z^{-3} + 0.2387z^{-4} - 0.0265z^{-6}$$

This corresponds to a delay of 0.15 ms.

**(b)** With $N = 6$, we have $F_C = 0.25$ and $h_N[n] = 2F_C\,\text{sinc}(2nF_C) = 0.5\,\text{sinc}(0.5n)$, $-2.5 \le n \le 2.5$.

For a von Hann (Hanning) window, $w[n] = 0.5 + 0.5\cos(\frac{2n\pi}{N-1}) = 0.5 + 0.5\cos(0.4n\pi)$, $-2.5 \le n \le 2.5$.

We then compute values of $h[n]$, $w[n]$, and $h_W[n] = h[n]w[n]$ as

$$\begin{array}{rrrrrrr} h[n] = & -0.09 & 0.1501 & 0.4502 & 0.4502 & 0.1501 & -0.09 \\ w[n] = & 0.0 & 0.3455 & 0.9045 & 0.9045 & 0.3455 & 0.0 \\ h_W[n] = & 0.0 & 0.0518 & 0.4072 & 0.4072 & 0.0518 & 0.0 \end{array}$$

Since the first sample of $h_W[n]$ is zero, the *minimum delay* required for a causal filter is only 1.5 (and not 2.5) samples, and the transfer function of the causal sequence is

$$H_C(z) = 0.0518 + 0.4072z^{-1} + 0.4072z^{-2} + 0.0518z^{-3}$$

## 10.3.4 Some Design Issues

The gain of a typical lowpass filter using window-based design is illustrated in Figure 10.17. The gain specifications of an FIR filter are often based on the passband ripple $\delta_p$ (the maximum deviation from unit gain) and the stopband ripple $\delta_s$ (the maximum deviation from zero gain). Since the decibel gain is usually normalized with respect to the peak gain, the passband and stopband attenuation $A_p$ and $A_s$ (in decibels) are related to these ripple parameters by

$$A_p(\text{dB}) = -20\log\left(\frac{1-\delta_p}{1+\delta_p}\right) \qquad A_s(\text{dB}) = -20\log\left(\frac{\delta_s}{1+\delta_p}\right) \approx -20\log\delta_s, \quad \delta_p \ll 1 \tag{10.20}$$

To convert attenuation specifications (in decibels) to values for the ripple parameters, we use

$$\delta_p = \frac{10^{A_p/20}-1}{10^{A_p/20}+1} \qquad \delta_s = (1+\delta_p)10^{-A_s/20} \approx 10^{-A_s/20}, \quad \delta_p \ll 1 \tag{10.21}$$

Window-based design calls for normalization of the design frequencies by the sampling frequency, developing the lowpass filter by windowing the impulse response of an ideal filter, and converting to the required filter type using spectral transformations. The method may appear deceptively simple (and it is), but some issues can only be addressed qualitatively. For example, the choice of cutoff frequency is affected by the window length $N$. The smallest

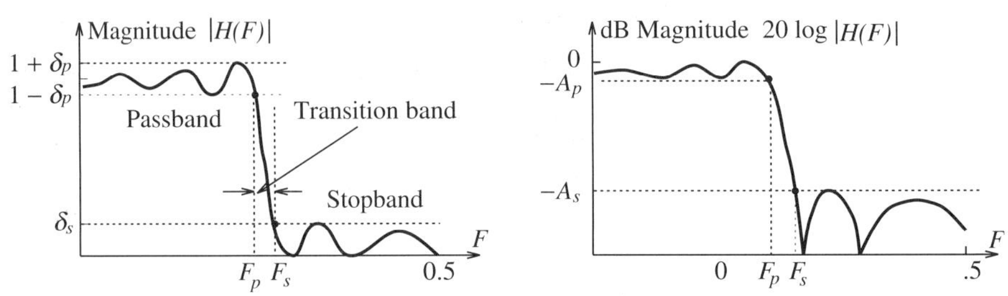


**FIGURE 10.17** The features of a typical lowpass filter. Up to the passband edge $F_p$, the gain is nearly unity. Past the stopband edge $F_s$, the gain is nearly zero. On a linear scale, measures for the gain include the passband ripple $d_p$ and stopband ripple $\delta_s$. On the decibel scale, unit gain corresponds to 0 dB. The *maximum* passband attenuation and *minimum* stopband attenuation are measured by $A_p$-dB and $A_s$-dB, respectively. These correspond to a *minimum* passband gain of $-A_p$-dB and a *maximum* stopband gain of $-A_s$-dB, respectively

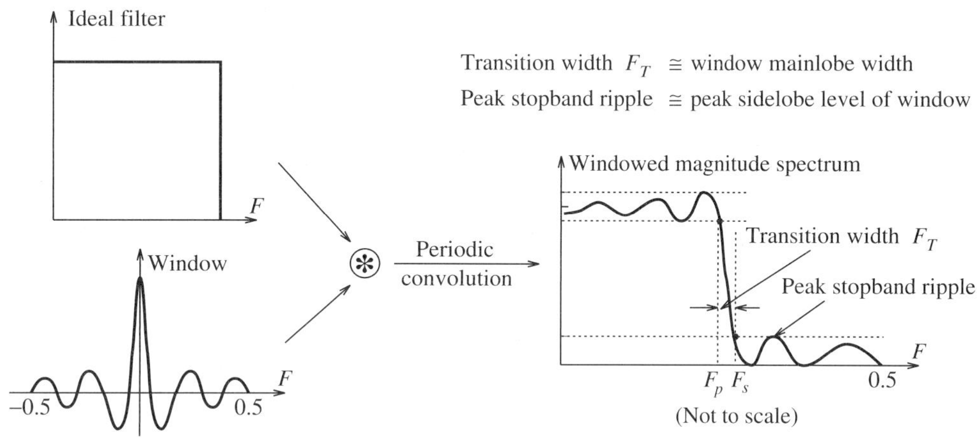


**FIGURE 10.18** The spectrum of a windowed ideal filter. Multiplying the impulse response of an ideal filter by a window function results in the convolution of their respective spectra, as shown. The spectrum of a typical window function shows a main lobe and sidelobes and leads to overshoot and oscillations in the spectrum of the windowed filter. To reduce the overshoot and oscillations, we choose windows whose spectra have small or positive sidelobes

length $N$ that meets specifications depends on the choice of window, and the choice of the window (in turn) depends on the (stopband) attenuation specifications.

## 10.3.5 Characteristics of the Windowed Spectrum

When we multiply the ideal impulse response $h[n] = 2F_C \operatorname{sinc}(2nF_C)$ by a window $w[n]$ of length $N$ in the time domain, the spectrum of the windowed impulse response $h_W[n] = h[n]w[n]$ is the (periodic) convolution of $H(F)$ and $W(F)$, as illustrated in Figure 10.18.

The ideal spectrum has a jump discontinuity at $F = F_C$. The windowed spectrum shows overshoot, ripples, and a finite transition width (but no abrupt jump). Its magnitude at $F = F_C$ equals 0.5 (corresponding to an attenuation of 6 dB).

Table 10.4 lists the characteristics of the *windowed spectrum* for various windows. It excludes windows (such as the Bartlett window) whose amplitude spectrum is entirely positive (because they result in a complete elimination of overshoot and the Gibbs effect). Since both the window function and the impulse response are symmetric sequences, the spectrum of the windowed filter is also endowed with symmetry. Here are some general observations about the windowed spectrum:

1. Even though the peak passband ripple equals the peak stopband ripple ($\delta_p = \delta_s$), the passband (or stopband) ripples are not of equal magnitude.
2. The peak stopband level of the windowed spectrum is typically slightly less than the peak sidelobe level of the window itself. In other words, the filter stopband attenuation (listed as $A_{WS}$ in Table 10.4) is typically greater (by a few decibels) than the peak sidelobe attenuation of the window (listed as $A_{SL}$ in Tables 10.2 and 10.3). The peak sidelobe level, the peak passband ripple, and the passband attenuation (listed as $A_{WP}$ in Table 10.4) remain more or less constant with $N$.

3. The peak-to-peak width across the transition band is roughly equal to the main lobe width of the window (listed as $W_M$ in Tables 10.2 and 10.3). The actual transition width (listed as $F_W$ in Table 10.4) of the windowed spectrum (when the response first reaches $1 - \delta_p$ and $\delta_s$) is less than this width. The transition width $F_{WS}$ is inversely related to the window length $N$ (with $F_{WS} \approx C/N$, where $C$ is more or less a constant for each window).

The numbers vary in the literature, and the values in Table 10.4 were found here by using an ideal impulse response $h[n] = 0.5\,\text{sinc}(0.5n)$, $F_C = 0.25$ and being windowed by a 51-point window. The magnitude specifications are normalized with respect to the peak magnitude. The passband and stopband attenuation are computed from the passband and stopband ripple (with $\delta_p = \delta_s$) using the relations already given.

**TABLE 10.4 ➤ Characteristics of the Windowed Spectrum**

| Window | Peak Ripple $\delta_p = \delta_s$ | Passband Attenuation $A_{WP}$ (dB) | Peak Sidelobe Attenuation $A_{WS}$ (dB) | Transition Width $F_{WS} \approx C/N$ |
|---|---|---|---|---|
| Boxcar | 0.0897 | 1.5618 | 21.7 | $C = 0.92$ |
| Cosine | 0.0207 | 0.36 | 33.8 | $C = 2.1$ |
| Riemann | 0.0120 | 0.2087 | 38.5 | $C = 2.5$ |
| von Hann (Hanning) | 0.0063 | 0.1103 | 44 | $C = 3.21$ |
| Hamming | 0.0022 | 0.0384 | 53 | $C = 3.47$ |
| Blackman | $(1.71)10^{-4}$ | $(2.97)10^{-3}$ | 75.3 | $C = 5.71$ |
| Dolph ($R = 40$ dB) | 0.0036 | 0.0620 | 49 | $C = 3.16$ |
| Dolph ($R = 50$ dB) | $(9.54)10^{-4}$ | 0.0166 | 60.4 | $C = 3.88$ |
| Dolph ($R = 60$ dB) | $(2.50)10^{-4}$ | 0.0043 | 72 | $C = 4.6$ |
| Harris (0) | $(8.55)10^{-4}$ | 0.0148 | 61.4 | $C = 5.36$ |
| Harris (1) | $(1.41)10^{-4}$ | $(2.44)10^{-3}$ | 77 | $C = 7.45$ |
| Harris (2) | $(1.18)10^{-4}$ | $(2.06)10^{-3}$ | 78.5 | $C = 5.6$ |
| Harris (3) | $(8.97)10^{-5}$ | $(1.56)10^{-3}$ | 81 | $C = 5.6$ |
| Harris (4) | $(9.24)10^{-5}$ | $(1.61)10^{-3}$ | 81 | $C = 5.6$ |
| Harris (5) | $(9.96)10^{-6}$ | $(1.73)10^{-4}$ | 100 | $C = 7.75$ |
| Harris (6) | $(1.94)10^{-6}$ | $(3.38)10^{-5}$ | 114 | $C = 7.96$ |
| Harris (7) | $(5.26)10^{-6}$ | $(9.15)10^{-5}$ | 106 | $C = 7.85$ |

### 10.3.6 Selection of Window and Design Parameters

The choice of a window is based primarily on the design stopband specification $A_s$. The peak sidelobe attenuation $A_{WS}$ of the windowed spectrum (listed in Table 10.4) should match (or exceed) the specified stopband attenuation $A_s$. Similarly, the peak passband attenuation $A_{WP}$ of the windowed spectrum should not exceed the specified passband attenuation $A_p$, a condition that is often ignored because it is usually satisfied for most practical specifications.

The windowed spectrum is the convolution of the spectra of the impulse response and the window function, and this spectrum changes as we change the filter length. An optimal (in the mean-square sense) impulse response and an optimal (in any sense) window may not together yield a windowed response with optimal features. Window selection is at best an art and at worst a matter of trial and error. Just so you know, the three most commonly used windows are the von Hann (Hanning), Hamming, and Kaiser windows.

### Choosing the Filter Length

The transition width of the windowed spectrum decreases with the length $N$. There is no accurate way to establish the minimum filter length $N$ that meets design specifications. However, empirical estimates are based on matching the given transition width specification $F_T$ to the transition width $F_{\text{WS}} = C/N$ of the windowed spectrum (as listed in Table 10.4)

$$F_T = F_s - F_p = F_{\text{WS}} \approx \frac{C}{N} \qquad N = \frac{C}{F_s - F_p} \tag{10.22}$$

Here, $F_p$ and $F_s$ are the *digital* passband and stopband frequencies. The window length depends on the choice of window (which dictates the choice of $C$). The closer the match between the stopband attenuation $A_s$ and the stopband attenuation $A_{\text{WS}}$ of the windowed spectrum, the smaller is the window length $N$. In any case, for a given window, this relation typically *overestimates* the smallest filter length, and we can often decrease this length and still meet design specifications.

### The Kaiser Window

Empirical relations have been developed to estimate the filter length $N$ of FIR filters based on the Kaiser window. We first compute the peak passband ripple $\delta_p$, the peak stopband ripple $\delta_s$, and choose the smallest of these as the *ripple parameter* $\delta$:

$$\delta_p = \frac{10^{A_p/20} - 1}{10^{A_p/20} + 1} \qquad \delta_s = 10^{-A_s/20} \qquad \delta = \min(\delta_p, \delta_s) \tag{10.23}$$

The ripple parameter $\delta$ is used to recompute the actual stopband attenuation $A_{s0}$ in decibels:

$$A_{s0} = -20 \log \delta \text{ dB} \tag{10.24}$$

Finally, the length $N$ is well approximated by

$$N \geq \begin{cases} \dfrac{A_{s0} - 7.95}{14.36(F_s - F_p)} + 1, & A_{s0} \geq 21 \text{ dB} \\[2ex] \dfrac{0.9222}{F_s - F_p} + 1, & A_{s0} < 21 \text{ dB} \end{cases} \tag{10.25}$$

The Kaiser window parameter $\beta$ is estimated from the actual stopband attenuation $A_{s0}$, as follows:

$$\beta = \begin{cases} 0.0351(A_{s0} - 8.7), & A_{s0} > 50 \text{ dB} \\ 0.186(A_{s0} - 21)^{0.4} + 0.0251(A_{s0} - 21), & 21 \text{ dB} \leq A_{s0} \leq 50 \text{ dB} \\ 0, & A_{s0} < 21 \text{ dB} \end{cases} \tag{10.26}$$

### Choosing the Cutoff Frequency

A common choice for the cutoff frequency (used in the expression for $h[n]$) is $F_C = 0.5(F_p + F_s)$. The actual frequency that meets specifications for the smallest length is often less than this value. The cutoff frequency is affected by the filter length $N$. A design that ensures the minimum length $N$ is based on starting with the above value of $F_C$, and then changing (typically reducing) the length and/or tweaking (typically decreasing) $F_C$ until we just meet specifications (typically at the passband edge).

### 10.3.7 Spectral Transformations

A useful approach to the design of FIR filters other than lowpass starts with a lowpass prototype developed from given specifications. This is followed by appropriate spectral transformations to convert the lowpass prototype to the required filter type. Finally, the impulse response may be windowed by appropriate windows. The spectral transformations are developed from the shifting and modulation properties of the DTFT. Unlike analog and IIR filters, these transformations do not change the filter order (or length). The starting point is an ideal lowpass filter, with *unit passband gain* and a cutoff frequency of $F_C = 0.5(F_p + F_s)$, whose noncausal impulse response $h_{\text{LP}}[n]$ is symmetrically truncated to length $N$ and given by

$$h_{\text{LP}}[n] = 2F_C \operatorname{sinc}(2nF_C), \qquad -0.5(N-1) \le n \le 0.5(N-1) \tag{10.27}$$

If $N$ is even, then $n$ takes on non-integer values. It is more useful to work with the causal version that has the form $h_{\text{LP}}[k],\ 0 \le k \le N-1$, where $k = n + 0.5(N-1)$ is always an integer. The sample values of the causal and noncausal versions are identical. Generating the causal version simply amounts to re-indexing the impulse response.

$$h_{\text{LP}}[k] = 2F_C \operatorname{sinc}\{2[k - 0.5(N-1)]F_C\}, \quad 0 \le k \le N-1$$

The lowpass-to-highpass (LP2HP) transformation may be achieved in two ways, as illustrated in Figure 10.19.

The first form of the LP2HP transformation of Figure 10.19 is based on the result

$$H_{\text{HP}}(F) = 1 - H_{\text{LP}}(F)$$

The noncausal impulse response has the form

$$h_{\text{HP}}[n] = \delta[n] - h_{\text{LP}}[n]$$

Note that this transformation is valid only if the filter length $N$ is odd. It also assumes unit passband gain. For a passband gain of $G$, it modifies to $h_{\text{HP}}[n] = G\delta[n] - h_{\text{LP}}[n]$. If $h_{\text{LP}}[n]$

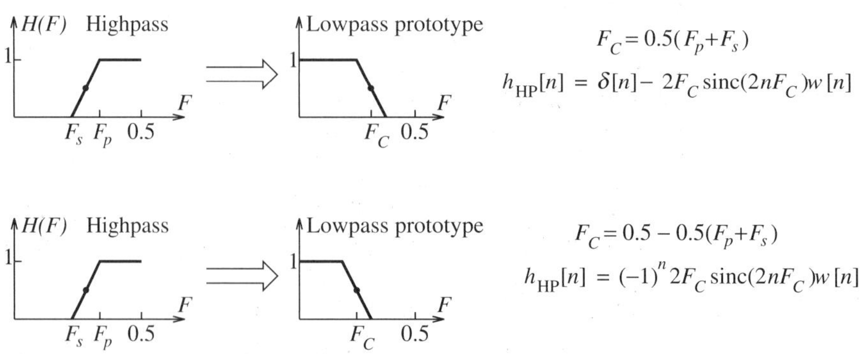


**FIGURE 10.19** Two ways of lowpass-to-highpass transformation. The first way is to shift the spectrum $H(F)$ of the lowpass filter by 0.5 to $H(F - 0.5)$. The second way is to subtract the lowpass spectrum $H(F)$ from a constant that equals the peak gain of $H(F)$ (assumed unity in the figure)

describes the (truncated) impulse response of an *ideal* lowpass filter with $F_C = 0.5(F_p + F_s)$, the cutoff frequency $F_H$ of the highpass filter also equals $F_C$. Upon delaying the sequence by $0.5(N-1)$ or re-indexing with $n = k - 0.5(N-1)$, we get the causal version of the highpass filter as

$$h_{\text{HP}}[k] = \delta[k - 0.5(N-1)] - 2F_C \operatorname{sinc}\{2[k - 0.5(N-1)]F_C\}, \quad 0 \le k \le N-1$$

As an example, if we wish to design a highpass filter of length $N = 15$ with a cutoff frequency of $F_H = 0.3$, we start with a lowpass prototype whose cutoff frequency is also $F_C = 0.3$ and whose (noncausal) impulse response is $h_{\text{LP}}[n] = 0.6\operatorname{sinc}(0.6n)$. This gives the noncausal impulse response of the highpass filter as

$$h_{\text{HP}}[n] = \delta[n] - h_{\text{LP}}[n] = \delta[n] - 0.6\operatorname{sinc}(0.6n), \quad -7 \le n \le 7$$

The causal impulse response has the form

$$h_{\text{HP}}[k] = \delta[k-7] - 0.6\operatorname{sinc}[0.6(k-7)], \quad 0 \le k \le 14$$

The second form of the LP2HP transformation uses the shifting property of the DTFT. Shifting the spectrum by $F = 0.5$ results in multiplication of the corresponding time signal by $(-1)^n$ (a change in the sign of every other sample value). Shifting the spectrum of a lowpass filter by $F = 0.5$ results in the highpass form

$$H_{\text{HP}}(F) = H_{\text{LP}}(F - 0.5)$$

Note that this transformation is valid for any lowpass filter (FIR or IIR) with any length $N$ (even or odd). If the lowpass filter has a cutoff frequency of $F_C$, a consequence of the frequency shift is that the cutoff frequency of the resulting highpass filter is $F_H = 0.5 - F_C$, regardless of the filter type. To design a highpass filter with a cutoff frequency of $F_H$, we must start with a lowpass prototype whose cutoff frequency is $F_C = 0.5 - F_H$. For an ideal causal filter, we start with the lowpass prototype

$$h_{\text{LP}}[k] = 2F_C \operatorname{sinc}\{2[k - 0.5(N-1)]F_C\}, \quad 0 \le k \le N-1 \quad \text{where } F_C = 0.5 - F_H$$

The causal impulse response of the highpass filter is then

$$h_{\text{HP}}[k] = (-1)^k h_{\text{LP}}[k], \qquad 0 \le k \le N-1 \tag{10.28}$$

As an example, if we wish to design a highpass filter of length $N = 12$ with a cutoff frequency of $F_H = 0.3$, we start with a causal lowpass prototype whose cutoff frequency is $F_C = 0.2$ and whose impulse response is

$$h_{\text{LP}}[k] = 0.4\operatorname{sinc}[0.4(k - 5.5)], \quad 0 \le k \le 11$$

The causal impulse response of the highpass filter is then given by

$$h_{\text{HP}}[k] = (-1)^k h_{\text{LP}}[k], \quad 0 \le k \le 11$$

The LP2BP and LP2BS transformations are based on arithmetic symmetry about the center frequency $F_0$. If the band edges are $[F_1,\ F_2,\ F_3,\ F_4]$ in increasing order, arithmetic symmetry means that $F_1 + F_4 = F_2 + F_3 = 2F_0$ and implies equal transition widths. If the transition widths are not equal, we must relocate a band edge to make both transition widths equal to the smaller transition width, as shown in Figure 10.20.

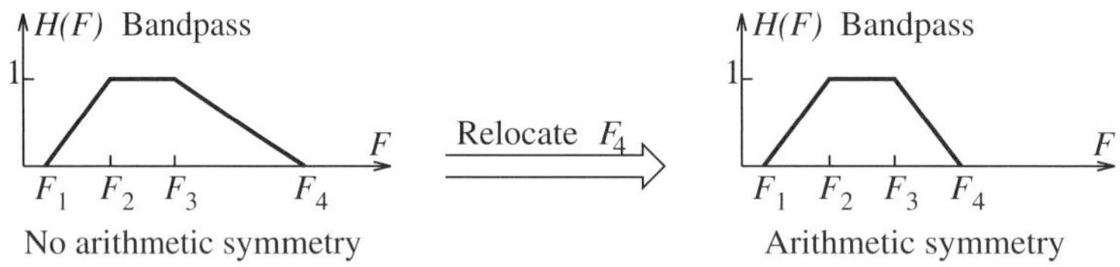


**FIGURE 10.20** How to ensure arithmetic symmetry of the band edges. The idea is to make the two transition widths equal to the smaller of the two by relocating a band edge ($F_4$ in the figure shown)

The LP2BP and LP2BS transformations are illustrated in Figure 10.21. In each transformation, the center frequency $F_0$ is given by

$$F_0 = 0.5(F_2 + F_3) = 0.5(F_1 + F_4) \tag{10.29}$$

The lowpass-to-bandpass (LP2BP) transformation results by shifting the spectrum of a lowpass filter by $\pm F_0$ to give

$$H_{\text{BP}}(F) = H_{\text{LP}}(F + F_0) + H_{\text{LP}}(F - F_0)$$

From Figure 10.21, the cutoff frequency $F_C$ of the lowpass prototype is

$$F_C = 0.5(F_3 + F_4) - F_0 \tag{10.30}$$

From the modulation property of the DTFT, shifting the spectrum $H_{\text{LP}}(F)$ of the lowpass filter by $\pm F_0$ results in multiplication of its impulse response $h_{\text{LP}}[n]$ by $2\cos(2\pi n F_0)$. We obtain the impulse response of the bandpass filter as

$$h_{\text{BP}}[n] = 2\cos(2n\pi F_0)h_{\text{LP}}[n] = 4F_C \operatorname{sinc}(2nF_C)\cos(2\pi n F_0), \quad -\frac{N-1}{2} \le n \le \frac{N-1}{2} \tag{10.31}$$

Upon re-indexing, its causal version assumes the form $h_{\text{BP}}[k]$, $0 \le k \le N-1$, where $k = n + 0.5(N-1)$.

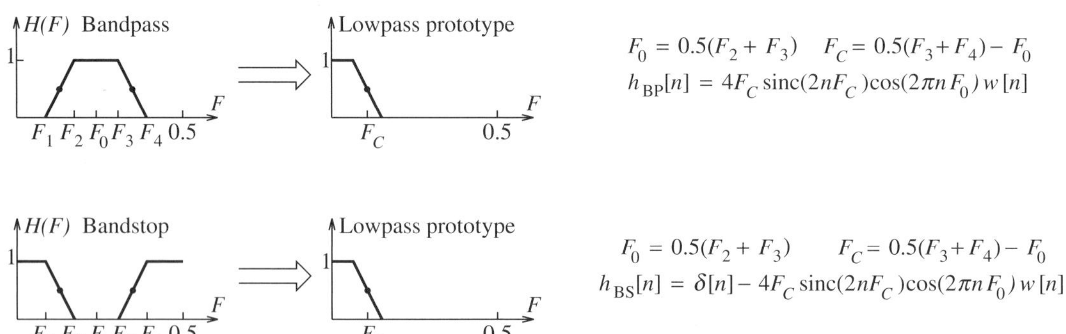


**FIGURE 10.21** Lowpass filter specifications from bandpass bandstop forms. The bandpass spectrum is shifted by $F_0$ to center it at the origin and get measures for the band edges of the lowpass filter. The bandstop spectrum is first subtracted from unity to create a bandpass form. The resulting lowpass-to-bandpass and lowpass-to-bandstop transformations are listed in the figure

A bandstop filter requires a type 4 sequence (with even symmetry and odd length $N$). The lowpass-to-bandstop (LP2BS) transformation of Figure 10.21 is described by $H_{BS}(F) = 1 - H_{BP}(F)$ and leads to the noncausal impulse response

$$h_{BS}[n] = \delta[n] - h_{BP}[n] = \delta[n] - 4F_C \operatorname{sinc}(2nF_C)\cos(2\pi nF_0), \quad -\frac{N-1}{2} \le n \le \frac{N-1}{2} \tag{10.32}$$

Upon re-indexing, its causal version has the form $h_{BS}[k]$, $0 \le k \le N-1$ where $k = n + 0.5(N-1)$.

A second form of the LP2BS transformation results by describing the bandstop filter as the sum of a lowpass filter with a cutoff frequency of $F_L = 0.5(F_1 + F_2)$ and a highpass filter with a cutoff frequency of $F_H = 0.5(F_3 + F_4)$. The noncausal impulse response of the bandstop filter is thus given by

$$h_{BS}[n] = 2F_L \operatorname{sinc}(2nF_L) + 2(-1)^n(0.5 - F_H)\operatorname{sinc}[2n(0.5 - F_H)], \quad -\frac{N-1}{2} \le n \le \frac{N-1}{2} \tag{10.33}$$

Upon re-indexing, its causal version has the form $h_{BS}[k]$, $0 \le k \le N-1$, where $k = n + 0.5(N-1)$.

---

### Recipe for Window-Based FIR Filter Design

- Normalize the analog design frequencies by the sampling frequency $S$.
- Obtain the band edges $F_p$ and $F_s$ of the lowpass prototype.
- Choose the lowpass prototype cutoff as $F_C = 0.5(F_p + F_s)$.
- Choose a window (from Table 10.4) that satisfies $A_{WS} \ge A_s$ and $A_{WP} \le A_p$.
- Compute the window length $N$ from $F_T = F_s - F_p = F_{WS} = \frac{C}{N}$ (with $C$ as in Table 10.4).
- Compute the prototype impulse response $h[n] = 2F_C \operatorname{sinc}[2nF_C]$, $|n| \le 0.5(N-1)$.
- Window $h[n]$ and apply spectral transformations (if needed) to convert to required filter type.

**Minimum-Length Design:** Adjust $N$ and/or $F_C$ until the design specifications are just met.

---

**Example 10.3**

### FIR Filter Design Using Windows

**(a)** Design an FIR filter to meet the following specifications:

- $f_p = 2$ kHz
- $f_s = 4$ kHz
- $A_p = 2$ dB
- $A_s = 40$ dB
- Sampling frequency: $S = 20$ kHz

This describes a lowpass filter. The digital frequencies are $F_p = f_p/S = 0.1$ and $F_s = f_s/S = 0.2$.

With $A_s = 40$ dB, possible choices for a window are (from Table 10.4) von Hann (with $A_{\mathrm{WS}} = 44$ dB) and Blackman (with $A_{\mathrm{WS}} = 75.3$ dB). Using $F_{\mathrm{WS}} \approx (F_s - F_p) = C/N$, the approximate filter lengths for these windows (using the values of $C$ from Table 10.4) are:

$$\text{von Hann:} N \approx \frac{3.21}{0.1} \approx 33 \qquad \text{Blackman:} N \approx \frac{5.71}{0.1} \approx 58$$

We choose the cutoff frequency as $F_C = 0.5(F_p + F_s) = 0.15$. The impulse then response equals

$$h_N[n] = 2F_C \operatorname{sinc}(2nF_C) = 0.3 \operatorname{sinc}(0.3n)$$

Windowing gives the impulse response of the required lowpass filter

$$h_{\mathrm{LP}}[n] = w[n]h_N[n] = 0.3w[n]\operatorname{sinc}(0.3n)$$

As Figure E10.3(1)(a) shows, the design specifications are indeed met by each filter (but the lengths are actually overestimated). The Blackman window requires a larger length because of the larger difference between $A_s$ and $A_{\mathrm{WS}}$. It also has the larger transition width.

**FIGURE E.10.3(1)** Lowpass FIR filters for Example 10.3(a and b)

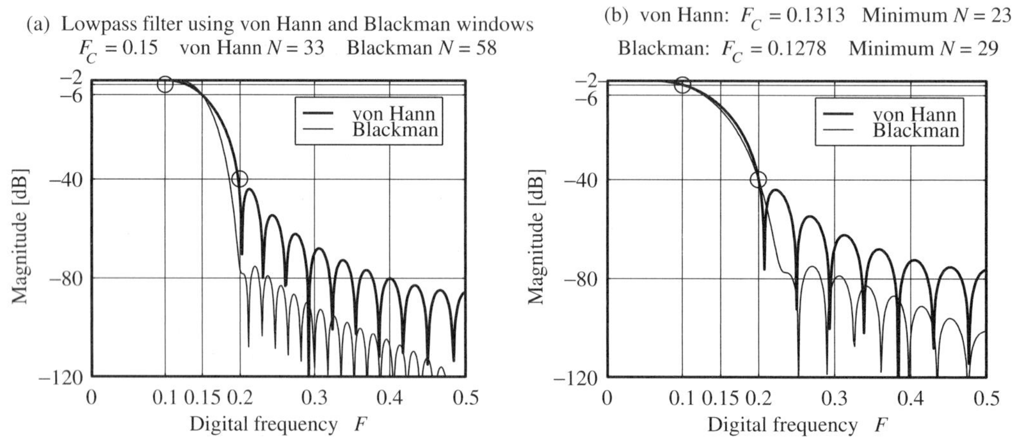


**(b) Minimum-Length Design**

By trial and error, the cutoff frequency and the smallest length that just meet specifications turn out to be $F_C = 0.1313$, $N = 23$ for the von Hann window, and $F_C = 0.1278$, $N = 29$ for the Blackman window. Figure E10.3(1)(b) shows the response of these minimum-length filters. The passband and stopband attenuation are [1.9, 40.5] dB for the von Hann window, and [1.98, 40.1] dB for the Blackman window. Even though the filter lengths are much smaller, each filter does meet the design specifications.

**(c)** Design an FIR filter to meet the following specifications:

- $f_p = 4$ kHz
- $f_s = 2$ kHz
- $A_p = 2$ dB

- $A_s = 40$ dB
- Sampling frequency: $S = 20$ kHz

The specifications describe a highpass filter. The digital frequencies are $F_p = f_p/S = 0.2$ and $F_s = f_s/S = 0.1$. The transition width is $F_T = 0.2 - 0.1 = 0.1$.

With $A_s = 40$ dB, possible choices for a window (see Table 10.4) are Hamming and Blackman. Using $F_T \approx F_{\text{WS}} = C/N$, the approximate filter lengths for these windows are

$$\text{Hamming:} N \approx \frac{3.47}{0.1} \approx 35 \qquad \text{Blackman:} N \approx \frac{5.71}{0.1} \approx 58$$

We can now design the highpass filter in one of two ways:

1. Choose the cutoff frequency of the lowpass filter as $F_C = 0.5(F_p + F_s) = 0.15$. The impulse response $h_N[n]$ then equals

$$h_N[n] = 2F_C \operatorname{sinc}(2nF_C) = 0.3 \operatorname{sinc}(0.3n)$$

   The windowed response is thus $h_W[n] = 0.3w[n]\operatorname{sinc}(0.3n)$.

   The impulse of the required highpass filter is then

$$h_{\text{HP}}[n] = \delta[n] - h_W[n] = \delta[n] - 0.3w[n]\operatorname{sinc}(0.3n)$$

2. Choose the cutoff frequency of the lowpass filter as $F_C = 0.5 - 0.5(F_p + F_s) = 0.35$.

   Then, the impulse response equals $h_N[n] = 2F_C \operatorname{sinc}(2nF_C) = 0.7 \operatorname{sinc}(0.7n)$.

   The windowed response is thus $h_W[n] = 0.7w[n]\operatorname{sinc}(0.7n)$.

   The impulse of the required highpass filter is then

$$h_{\text{HP}}[n] = (-1)^n h_W[n] = 0.7(-1)^n w[n]\operatorname{sinc}(0.7n)$$

The two methods yield identical results. As Figure E10.3(2)(a) shows, the design specifications are indeed met by each window, but the lengths are actually overestimated.

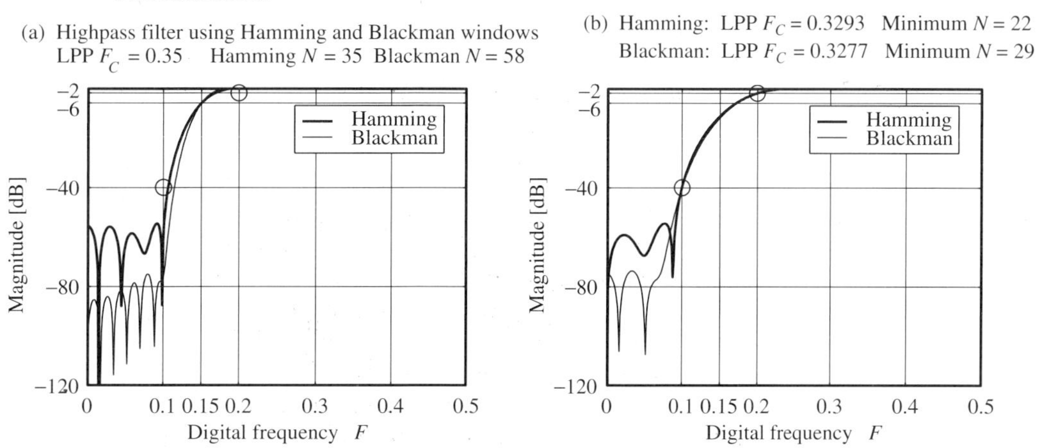


**FIGURE E.10.3(2)** Highpass FIR filters for Example 10.3(c and d)

**(d) Minimum-Length Design**

By trial and error, the cutoff frequency and the smallest length that just meet specifications, turn out to be $F_C = 0.3293,\ N = 22$ for the Hamming window, and $F_C = 0.3277,\ N = 29$ for the Blackman window. Figure E10.3(2)(b) shows the response of the minimum-length filters. The passband and stopband attenuation are [1.94, 40.01] dB for the Hamming window, and [1.99, 40.18] dB for the Blackman window. Each filter meets the design specifications, even though the filter lengths are much smaller than the values computed from the design relations.

**(e)** Design an FIR filter to meet the following specifications:

- $A_p = 3$ dB
- $A_s = 45$ dB
- Passband: [4, 8] kHz
- Stopband: [2, 12] kHz
- $S = 25$ kHz

The specifications describe a bandpass filter. If we assume a fixed passband, the center frequency lies at the center of the passband and is given by $f_0 = 0.5(4 + 8) = 6$ kHz.

The specifications do not show arithmetic symmetry. The smaller transition width is 2 kHz. For arithmetic symmetry, we therefore choose the band edges as [2, 4, 8, 10] kHz.

The digital frequencies are: passband [0.16, 0.32], stopband [0.08, 0.4], and $F_0 = 0.24$.

The lowpass band edges become $F_p = 0.5(F_{p2} - F_{p1}) = 0.08$ and $F_s = 0.5(F_{s2} - F_{s1}) = 0.16$.

With $A_s = 45$ dB, one of the windows we can use (from Table 10.4) is the Hamming window.

For this window, we estimate $F_{\mathrm{WS}} \approx (F_s - F_p) = C/N$ to obtain $N = 3.47/0.08 = 44$.

We choose the cutoff frequency as $F_C = 0.5(F_p + F_s) = 0.12$.

The lowpass impulse response is

$$h_N[n] = 2F_C \operatorname{sinc}(2nF_C) = 0.24 \operatorname{sinc}(0.24n), \quad -21.5 \le n \le 21.5$$

Windowing this gives $h_W[n] = h_N[n]w[n]$, and the LP2BP transformation gives

$$h_{\mathrm{BP}}[n] = 2\cos(2\pi nF_0)h_W[n] = 2\cos(0.48\pi n)h_W[n]$$

Its frequency response is shown in Figure E10.3(3)(a) and confirms that the specifications are met.

The cutoff frequency and smallest filter length that meets specifications turn out to be smaller. By decreasing $N$ and $F_C$, we find that the specifications are just met with $N = 27$ and $F_C = 0.0956$. For these values, the lowpass filter is

$$h_N[n] = 2F_C \operatorname{sinc}(2nF_C) = 0.1912 \operatorname{sinc}(0.1912n), \quad -13 \le n \le 13$$

Windowing and bandpass transformation yields the filter whose spectrum is shown in Figure E10.3(3)(b). The attenuation is 3.01 dB at 4 kHz and 8 kHz, 45.01 dB at 2 kHz, and 73.47 dB at 12 kHz.

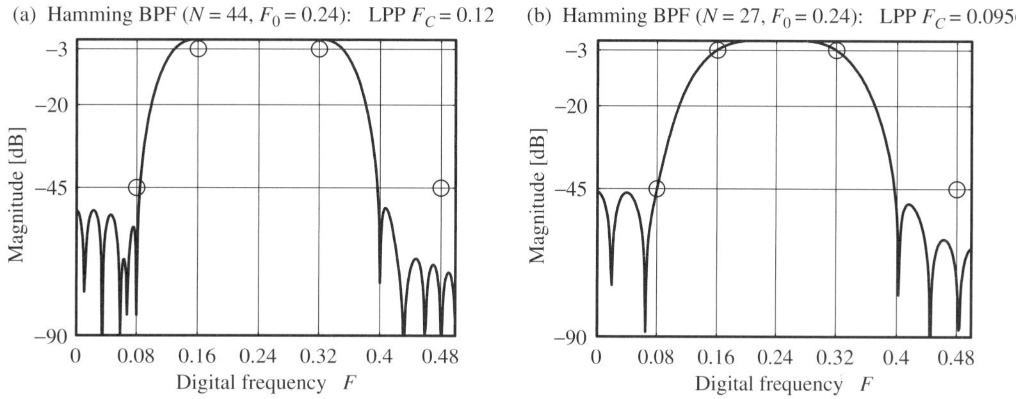


**FIGURE E.10.3(3)** Bandpass FIR filters for Example 10.3(e and f)

# 10.4 Half-Band FIR Filters

A **half-band** FIR filter has an odd-length impulse response $h[n]$ whose alternate samples are zero. The main advantage of half-band filters is that their realization requires only about half the number of multipliers. The impulse response of an ideal lowpass filter is $h[n] = 2F_C \operatorname{sinc}(2nF_C)$. If we choose $F_C = 0.25$, we obtain

$$h[n] = 2F_C \operatorname{sinc}(2nF_C) = 0.5 \operatorname{sinc}(0.5n), \quad |n| \le 0.5(N-1) \tag{10.34}$$

Thus, $h[n] = 0$ for even $n$, and the filter length $N$ is always odd. Being a type 1 sequence, its transfer function $H(F)$ displays even symmetry about $F = 0$. It is also antisymmetric about $F = 0.25$, with

$$H(F) = 1 - H(0.5 - F) \tag{10.35}$$

A highpass half-band filter also requires $F_C = 0.25$. If we choose $F_C = 0.5(F_p + F_s)$, the sampling frequency $S$ must equal $2(f_p + f_s)$ to ensure $F_C = 0.25$ and cannot be selected arbitrarily. Examples of lowpass and highpass half-band filters are shown in Figure 10.22. Note that the peak passband ripple and the peak stopband ripple are of equal magnitude, with $\delta_p = \delta_s = \delta$ (as they are for any symmetric window).

Since the impulse response of bandstop and bandpass filters contains the term $2\cos(2n\pi F_0)h_{\text{LP}}[n]$, a choice of $F_0 = 0.25$ (for the center frequency) ensures that the odd-indexed terms vanish. Once again, the sampling frequency $S$ cannot be arbitrarily chosen and must equal $4f_0$ to ensure $F_0 = 0.25$. Even though the choice of sampling rate may cause aliasing, the aliasing will be restricted primarily to the transition band between $f_p$ and $f_s$, where its effects are not critical.

Except for the restrictions in the choice of sampling rate $S$ (which dictates the choice of $F_C$ for lowpass and highpass filters or $F_0$ for bandpass and bandstop filters) and an odd length sequence, the design of half-band filters follows the same steps as window-based design.

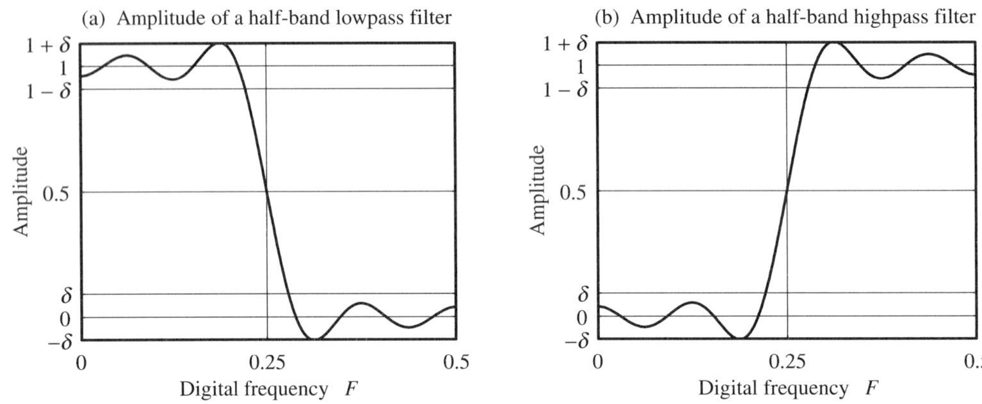


**FIGURE 10.22** Amplitude spectra of lowpass and highpass half-band filters. Note the odd symmetry of the spectra about $F = 0.25$ for each form

### Recipe for Design of Half-Band FIR Filters

Fix $S = 2(f_p + f_s)$ (for LPF and HPF) or $S = 4f_0$ (for BPF and BSF).
Find $F_p$ and $F_s$ for the lowpass prototype and pick $F_C = 0.5(F_p + F_s)$.
Choose an *odd* filter length $N$.
Compute the prototype impulse response $h[n] = 2F_C\,\text{sinc}[2nF_C],\ |n| \le 0.5(N-1)$.
Window the impulse response and apply spectral transformations to convert to required filter.
If the specifications are exceeded, decrease $N$ (in steps of 2) until specifications are just met.

**Example 10.4**

### Half-Band FIR Filter Design

**(a)** Design a lowpass half-band filter to meet the following specifications:

- Passband edge: 8 kHz
- Stopband edge: 16 kHz
- $A_p = 1$ dB
- $A_s = 50$ dB

We choose $S = 2(f_p + f_s) = 48$ kHz. The digital band edges are $F_p = \frac{1}{6}$, $F_s = \frac{1}{3}$, and $F_C = 0.25$.

The impulse response of the filter is $h[n] = 2F_C\,\text{sinc}(2nF_C) = 0.5\,\text{sinc}(0.5n)$.

**1.** If we use the Kaiser window, we compute the filter length $N$ and the Kaiser parameter $\beta$ as follows:

$$\delta_p = \frac{10^{A_p/20} - 1}{10^{A_p/20} + 1} = 0.0575 \qquad \delta_s = 10^{-A_s/20} = 0.00316$$

$$\delta = 0.00316 \qquad A_{s0} = -20\log\delta = 50$$

$$N = \frac{A_{s0} - 7.95}{14.36(F_s - F_p)} + 1 = 18.57 \approx 19 \qquad \beta = 0.0351(A_{s0} - 8.7) = 1.4431$$

The impulse response is therefore $h_N[n] = 0.5\,\mathrm{sinc}(0.5n)$, $-9 \le n \le 9$.

Windowing $h_N[n]$ gives the required impulse response $h_W[n]$.

Figure E10.4a(a) shows that this filter does meet specifications with an attenuation of 0.045 dB at 8 kHz and of 52.06 dB at 16 kHz.

**FIGURE E 10.4.a** Lowpass half-band filters for Example 10.4(a)

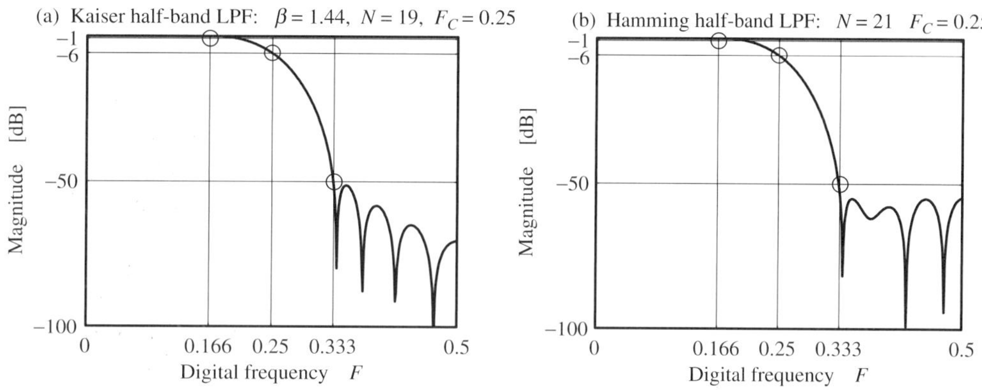


2. If we choose a Hamming window, we use Table 10.4 to approximate the *odd* filter length as

$$F_{\mathrm{WS}} = F_s - F_p = \frac{C}{N} \qquad N = \frac{C}{F_s - F_p} = \frac{3.47}{1/6} \approx 21$$

This value of $N$ meets specifications and also turns out to be the smallest length that does. Its response is plotted in Figure E10.4a(b). This filter shows an attenuation of 0.033 dB at 8 kHz and an attenuation of 53.9 dB at 16 kHz.

**(b)** Design a bandstop half-band filter to meet the following specifications:

- Stopband edges: [2, 3] kHz
- Passband edges: [1, 4] kHz
- $A_p = 1$ dB
- $A_s = 50$ dB

Since both the passband and the stopband are symmetric, we have $f_0 = 0.5(2+3) = 2.5$ kHz.

We then choose the sampling frequency as $S = 4f_0 = 10$ kHz. The digital frequencies are

$$\text{Stopband edges} = [0.2, 0.3], \quad \text{Passband edges} = [0.1, 0.4], \quad F_0 = 0.25$$

The specifications for the lowpass prototype are

$$F_p = 0.5(F_{s2} - F_{s1}) = 0.05, \quad F_s = 0.5(F_{p2} - F_{p1}) = 0.15, \quad F_C = 0.5(F_p + F_s) = 0.1$$

The impulse response of the prototype is $h[n] = 2F_C\,\mathrm{sinc}(2nF_C) = 0.2\,\mathrm{sinc}(0.2n)$.

1. If we use the Kaiser window, we must compute the filter length $N$ and the Kaiser parameter $\beta$ as follows:

$$\delta_p = \frac{10^{A_p/20} - 1}{10^{A_p/20} + 1} = 0.0575 \qquad \delta_s = 10^{-A_s/20} = 0.00316$$

$$\delta = 0.00316 \qquad A_{s0} = -20\log\delta = 50$$

$$N = \frac{A_{s0} - 7.95}{14.36(F_s - F_p)} + 1 = 30.28 \approx 31 \quad \beta = 0.0351(A_{s0} - 8.7) = 1.4431$$

The prototype impulse response is $h_N[n] = 0.2\,\text{sinc}(0.2n),\ -15 \le n \le 15$. Windowing $h_N[n]$ gives $h_W[n]$. We transform $h_W[n]$ to the bandstop form $h_{BS}[n]$ using $F_0 = 0.25$ to give

$$h_{BS}[n] = \delta[n] - 2\cos(2\pi n F_0)h_W[n] = \delta[n] - 2\cos(0.5\pi n)h_W[n]$$

Figure E10.4b(a) shows that this filter does meet specifications, with an attenuation of 0.046 dB at 2 kHz and 3 kHz and of 53.02 dB at 1 kHz and 4 kHz.

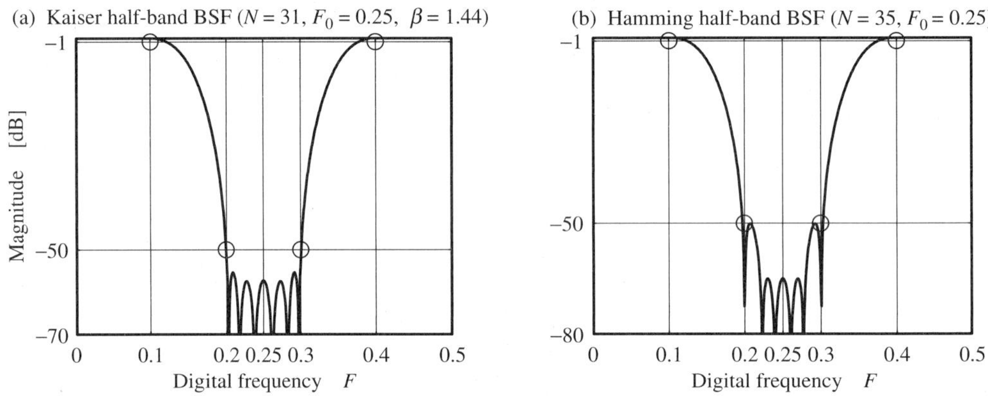


**FIGURE E 10.4.b** Bandstop half-band filters for Example 10.4(b)

2. For a Hamming window, we use Table 10.4 to approximate the *odd* filter length as

$$F_{WS} = F_s - F_p = \frac{C}{N} \qquad N = \frac{C}{F_s - F_p} = \frac{3.47}{0.1} \approx 35$$

This is also the smallest filter length that meets specifications. The magnitude response is shown in Figure E10.4b(b). We see an attenuation of 0.033 dB at [2, 3] kHz and of 69.22 dB at [1, 4] kHz.

## 10.5 FIR Filter Design by Frequency Sampling

In window-based design, we start with the impulse response $h[n]$ of an ideal filter and use truncation and windowing to obtain an acceptable frequency response $H(F)$. The design of FIR filters by frequency sampling starts with the required form for $H(F)$ and uses interpolation and the DFT to obtain $h[n]$. In this sense, it is more versatile, since arbitrary frequency-response forms can be handled with ease.

Recall that a continuous (but band limited) signal $h(t)$ can be perfectly reconstructed from its samples (taken above the Nyquist rate), using a sinc-interpolating function (that equals zero at the sampling instants). If $h(t)$ is not band limited, we get a perfect match only at the sampling instants.

By analogy, the continuous (but periodic) spectrum $H(F)$ can also be recovered from its frequency samples, using a periodic extension of the sinc interpolating function (that equals zero at the sampling points). The reconstructed spectrum $H_N(F)$ will show an exact match to a desired $H(F)$ at the sampling instants, even though $H_N(F)$ could vary wildly at other frequencies. This is the basis for FIR filter design by frequency sampling. Given the desired form for $H(F)$, we sample it at $N$ frequencies and find the IDFT of the $N$-point sequence $H[k]$, $k = 0, 1, \ldots, N-1$. The following design guidelines stem both from design aspects as well as computational aspects of the IDFT itself.

**1.** The $N$ samples of $H(F)$ must correspond to the digital frequency range $0 \le F < 1$, with

$$H[k] = H(F)|_{F=k/N}, \qquad k = 0,\ 1,\ 2,\ \ldots, N-1 \tag{10.36}$$

The reason is that most DFT and IDFT algorithms require samples in the range $0 \le k \le N-1$.

**2.** Since $h[n]$ must be real, its DFT $H[k]$ must possess conjugate symmetry about $k = 0.5N$ (this is a DFT *requirement*). Note that conjugate symmetry will always leave $H[0]$ unpaired. It can be set to any real value, in keeping with the required filter type (this is a design *requirement*). For example, we must choose $H[0] = 0$ for bandpass or highpass filters.

**3.** For even length $N$, the computed end samples of $h[n]$ may not turn out to be symmetric. To ensure symmetry, we must force $h[0]$ to equal $h[N]$ (setting both to $0.5h[0]$, for example).

**4.** For $h[n]$ to be causal, we must delay it (this is a design *requirement*). This translates to a linear-phase shift to produce the sequence $|H[k]|e^{j\phi[k]}$. In keeping with conjugate symmetry about the index $k = 0.5N$, the phase for the first $N/2$ samples of $H[k]$ will be given by

$$\phi[k] = \frac{-\pi k(N-1)}{N}, \qquad k = 0,\ 1,\ 2, \ldots, 0.5(N-1) \tag{10.37}$$

Note that for type 3 and type 4 (antisymmetric) sequences, we must also add a constant phase of $0.5\pi$ to $\phi[k]$ (up to $k = 0.5N$). The remaining samples of $H[k]$ are found by conjugate symmetry.

**5.** To minimize the Gibbs effect near discontinuities in $H(F)$, we may allow the sample values to vary gradually between jumps (this is a design *guideline*). This is equivalent to introducing a finite transition width. The choice of the sample values in the transition band can affect the response dramatically.

**Example 10.5** **FIR Filter Design by Frequency Sampling**

**(a)** Consider the design of a lowpass filter shown in Figure E10.5(a). Let us sample the ideal $H(F)$ (shown as a dark line) over $0 \le F < 1$ with $N = 10$ samples. The magnitude of the sampled sequence $H[k]$ is

$$|H[k]| = \{1,\ 1,\ 1,\ 0,\ 0,\ \underbrace{0,}_{k=5}\ 0,\ 0,\ 1,\ 1\}$$

The actual (phase-shifted) sequence is $H[k] = |H[k]|e^{j\phi[k]}$, where

$$\phi[k] = \frac{-\pi k(N-1)}{N} = -0.9\pi k, \quad k \le 5$$

Note that $H[k]$ must be conjugate symmetric about $k = 0.5N = 5$, with $H[k] = H^*[N-k]$.

Now, $H[k] = 0,\ k = 4, 5, 6, 7$, and the remaining samples are $H[0] = 1e^{j0} = 1$ and

$$H[1] = 1e^{j\phi[1]} = e^{-j0.9\pi} \qquad H[9] = H^*[1] = e^{j0.9\pi}$$
$$H[2] = 1e^{j\phi[3]} = e^{-j1.8\pi} \qquad H[8] = H^*[2] = e^{j1.8\pi}$$

The inverse DFT of $H[k]$ yields the symmetric, real impulse-response sequence $h_1[n]$ with

$$h_1[n] = \{0.0716,\ -0.0794,\ 0.1,\ 0.1558,\ 0.452,\ 0.452,\ 0.1558,\ 0.1,\ -0.0794,\ 0.0716\}$$

Its DTFT magnitude $H_1(F)$ (the light line shown in Figure E10.5(a)) reveals a perfect match at the sampling points but has a large overshoot near the cutoff frequency. To reduce the overshoot, let us pick

$$H[2] = 0.5e^{j\phi[2]} = 0.5e^{-1.8\pi} \qquad H[8] = H^*[2] = 0.5e^{1.8\pi}$$

The inverse DFT of this new set of samples yields the new impulse response sequence $h_2[n]$:

$$h_2[n] = \{-0.0093,\ -0.0485,\ 0,\ 0.1867,\ 0.3711,\ 0.3711,\ 0.1867,\ 0,\ -0.0485,\ -0.0093\}$$

Its frequency response $H_2(F)$, shown in Figure E10.5(a), not only shows a perfect match at the sampling points but also a reduced overshoot, which we obtain at the expense of a broader transition width.

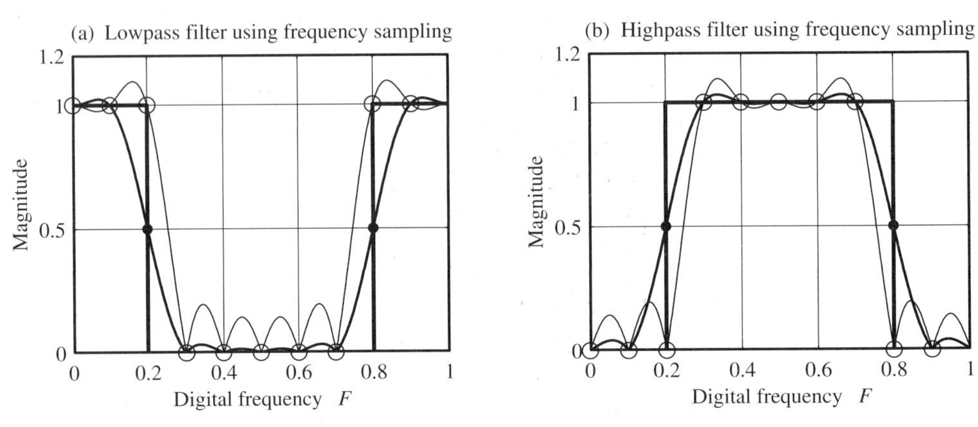


**FIGURE E.10.5** Lowpass and highpass filters for Example 10.5

**(b)** Consider the design of a highpass filter shown in Figure E10.5(b). Let us sample the ideal $H(F)$ (shown as a dark) over $0 \le F < 1$ with $N = 10$ samples. The magnitude of the sampled sequence $H[k]$ is

$$|H[k]| = \{0,\ 0,\ 0,\ 1,\ 1,\ \underbrace{1,}_{k=5}\ 1,\ 1,\ 0,\ 0\}$$

The actual (phase-shifted) sequence is $H[k] = |H[k]|e^{j\phi[k]}$. Since the impulse response $h[n]$ must be antisymmetric (for a highpass filter), $\phi[k]$ includes an additional phase of $0.5\pi$ and is given by

$$\phi[k] = \frac{-\pi k(N-1)}{N} + 0.5\pi = 0.5\pi - 0.9\pi k, \quad k \le 5$$

Note that $H[k]$ is conjugate symmetric about $k = 0.5N = 5$, with $H[k] = H^*[N-k]$. Now, $H[k] = 0,\ k = 0, 1, 2, 8, 9$, and $H[5] = 1e^{j\phi[5]} = 1$. The remaining samples are

$$H[3] = 1e^{j\phi[3]} = e^{-j2.2\pi} \qquad H[7] = H^*[3] = e^{j2.2\pi}$$
$$H[4] = 1e^{j\phi[4]} = e^{-j3.1\pi} \qquad H[6] = H^*[4] = e^{j3.1\pi}$$

The inverse DFT of $H[k]$ yields the antisymmetric real impulse response sequence $h_1[n]$, with

$$h_1[n] = \{0.0716,\ 0.0794,\ -0.1,\ -0.1558,\ 0.452,\ -0.452,\ 0.1558,\ 0.1,\ -0.0794,\ -0.0716\}$$

Its DTFT magnitude $H_1(F)$, shown as a light line in Figure E10.5(b), reveals a perfect match at the sampling points but a large overshoot near the cutoff frequency. To reduce the overshoot, let us choose

$$H[2] = 0.5e^{j\phi[2]} = 0.5e^{-1.3\pi} \qquad H[8] = 0.5e^{1.3\pi}$$

The inverse DFT of this new set of samples yields the new impulse response sequence $h_2[n]$:

$$h_2[n] = \{0.0128,\ -0.0157,\ -0.1,\ -0.0606,\ 0.5108,\ -0.5108,\ 0.0606,\ 0.1,\ 0.0157,\ -0.0128\}$$

Its frequency response $H_2(F)$, shown in Figure E10.5(b), not only shows a perfect match at the sampling points but also a reduced overshoot, which we obtain at the expense of a broader transition width.

### 10.5.1 Frequency Sampling and Windowing

The frequency-sampling method can be used to design filters with arbitrary frequency-response shapes. We can even combine this versatility with the advantages of window-based design. Given the response specification $H(F)$, we sample it at a *large* number of points $M$ and find the IDFT to obtain the $M$-point impulse response $h[n]$. The choice $M = 512$ is not unusual. Since $h[n]$ is unacceptably long, we truncate it to a smaller length $N$ by windowing $h[n]$. The choice of window is based on the same considerations that apply to window-based design. If the design does not meet specifications, we can change $N$ and/or adjust the sample values in the transition band and repeat the process. Naturally, this design is best carried out on a computer. The sample values around the transitions are adjusted to minimize the approximation error. This idea forms a special case of the more general optimization method called *linear programming*. But when it comes right down to choosing between the various computer-aided optimization methods, by far the most widely used is the *equiripple optimal approximation* method, which we describe in the next section.

### 10.5.2 Implementing Frequency-Sampling FIR Filters

We can readily implement frequency-sampling FIR filters by a nonrecursive structure once we know the impulse response $h[n]$. We can even implement a recursive realization without the need for finding $h[n]$, as follows. Suppose the frequency samples $H_N[k]$ of $H(F)$ are approximated by an $N$-point DFT of the filter impulse response $h_N[n]$. We may then write

$$H_N[k] = \sum_{k=0}^{N-1} h_N[n]e^{-j2\pi nk/N} \tag{10.38}$$

Its impulse response $h_N[n]$ may be found using the inverse DFT as

$$h_N[n] = \frac{1}{N}\sum_{k=0}^{N-1} H_N[k]e^{j2\pi nk/N} \tag{10.39}$$

The filter transfer function $H(z)$ is the $z$-transform of $h_N[n]$:

$$H(z) = \sum_{n=0}^{N-1} h_N[n]z^{-n} = \sum_{n=0}^{N-1} z^{-n}\left[\frac{1}{N}\sum_{k=0}^{N-1} H_N[k]e^{j2\pi nk/N}\right] \tag{10.40}$$

Interchanging summations, setting $z^{-n}e^{j2\pi nk/N} = [z^{-1}e^{j2\pi k/N}]^n$, and using the closed form for the finite geometric sum, we obtain

$$H(z) = \frac{1}{N}\sum_{k=0}^{N-1} H_N[k]\left[\frac{1-z^{-N}}{1-z^{-1}e^{j2\pi k/N}}\right] \tag{10.41}$$

The frequency response corresponding to $H(z)$ is

$$H(F) = \frac{1}{N}\sum_{k=0}^{N-1} H_N[k]\frac{1-e^{-j2\pi FN}}{1-e^{-j2\pi(F-k/N)}} \tag{10.42}$$

If we factor out $\exp(-j\pi FN)$ from the numerator, $\exp[-j\pi(F-k/N)]$ from the denominator, and use Euler's relation, we can simplify this result to

$$H(F) = \sum_{k=0}^{N-1} H_N[k]\frac{\text{sinc}[N(F-\frac{k}{N})]}{\text{sinc}[(F-\frac{k}{N})]}e^{-j\pi(N-1)(F-k/N)} = \sum_{k=0}^{N-1} H_N[k]W\left[F-\frac{k}{N}\right] \tag{10.43}$$

Here, $W[F-\frac{k}{N}]$ describes a sinc interpolating function, defined by

$$W\left[F-\frac{k}{N}\right] = \frac{\text{sinc}[N(F-\frac{k}{N})]}{\text{sinc}[(F-\frac{k}{N})]}e^{-j\pi(N-1)(F-k/N)} \tag{10.44}$$

and reconstructs $H(F)$ from its samples $H_N[k]$ taken at intervals $1/N$. It equals 1 when $F = k/N$, and zero otherwise. As a result, $H_N(F)$ equals the desired $H(F)$ at the sampling instants, even though $H_N(F)$ could vary wildly at other frequencies. This is the concept behind frequency sampling.

The transfer function $H(z)$ also may be written as the product of two transfer functions:

$$H(z) = H_1(z)H_2(z) = \left[\frac{1 - z^{-N}}{N}\right]\sum_{k=0}^{N-1}\left[\frac{H_N[k]}{1 - z^{-1}e^{j2\pi k/N}}\right] \tag{10.45}$$

This form of $H(z)$ suggests a method of recursive implementation of FIR filters. We cascade a comb filter (described by $H_1(z)$) with a parallel combination of $N$ first-order resonators (described by $H_2(z)$). Note that each resonator has a complex pole on the unit circle and the resonator poles actually lie at the same locations as the zeros of the comb filter. Each pair of terms corresponding to complex conjugate poles may be combined to form a second-order system with real coefficients for easier implementation.

Why implement FIR filters recursively? There are several reasons. In some cases, this may reduce the number of arithmetic operations. In other cases, it may reduce the number of delay elements required. Since the pole and zero locations depend only on $N$, such filters can be used for all FIR filters of length $L$ by changing only the multiplicative coefficients.

Even with these advantages, things can go wrong. In theory, the poles and zeros balance each other. In practice, quantization errors may move some poles outside the unit circle and lead to system instability. One remedy is to multiply the poles and zeros by a real number $\rho$ slightly smaller than unity to relocate the poles and zeros. The transfer function then becomes

$$H(z) = \frac{1 - (\rho z)^{-N}}{N}\sum_{k=0}^{N-1}\left[\frac{H_N[k]}{1 - (\rho z)^{-1}e^{j2\pi k/N}}\right] \tag{10.46}$$

With $\rho = 1 - \epsilon$, typically used values for $\epsilon$ range from $2^{-12}$ to $2^{-27}$ (roughly $10^{-4}$ to $10^{-9}$) and have been shown to improve stability with little change in the frequency response.

## 10.6 Design of Optimal Linear-Phase FIR Filters

Quite like analog filters, the design of optimal linear-phase FIR filters requires that we minimize the maximum error in the approximation. Optimal design of FIR filters is also based on a Chebyshev approximation. We should therefore expect such a design to yield the smallest filter length and a response that is equiripple in both the passband and the stopband. A typical spectrum is shown in Figure 10.23.

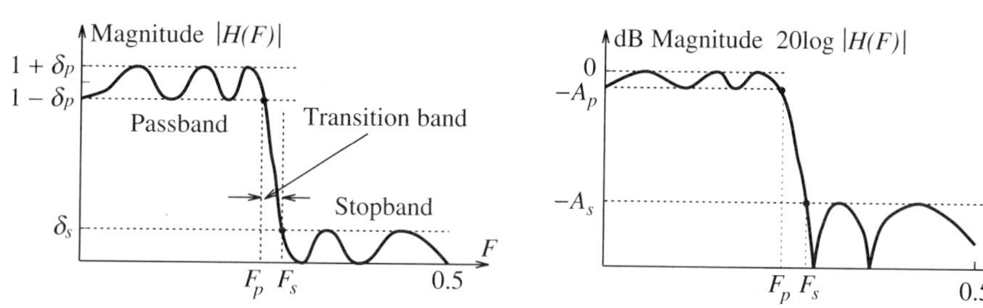


**FIGURE 10.23** An optimal filter has ripples of equal magnitude in the passband and ripples of equal magnitude in the stopband

There are three important concepts relevant to optimal design:

1. The error between the approximation $H(F)$ and the desired response $D(F)$ must be equiripple. The error curve must show equal maxima and minima with alternating zero crossings. The more the number of points where the error goes to zero (the zero crossings), the higher is the order of the approximating polynomial, and the longer is the filter length.
2. The frequency response $H(F)$ of a filter whose impulse response $h[n]$ is a symmetric sequence can always be put in the form

$$H(F) = Q(F)\sum_{n=0}^{M} \alpha_n \cos(2\pi nF) = Q(F)P(F) \tag{10.47}$$

Here, $Q(F)$ equals 1 (type 1), $\cos(\pi F)$ (type 2), $\sin(2\pi F)$ (type 3), or $\sin(\pi F)$ (type 4); $M$ is related to the filter length $N$ with $M = \text{int}(\frac{N-1}{2})$ (types 1, 2, 4) or $M = \text{int}(\frac{N-3}{2})$ (type 3); and the $\alpha_n$ are related to the impulse-response coefficients $h[n]$. The quantity $P(F)$ may also be expressed as a power series in $\cos(2\pi F)$ (or as a sum of Chebyshev polynomials). If we can select the $\alpha_n$ to best meet optimal constraints, we can design $H(F)$ as an optimal approximation to $D(F)$.
3. The *alternation theorem* offers the clue to selecting the $\alpha_n$.

### 10.6.1 The Alternation Theorem

We start by approximating $D(F)$ by the Chebyshev polynomial form for $H(F)$ and define the *weighted* approximation error $\epsilon(F)$ as

$$\epsilon(F) = W(F)[D(F) - H(F)] \tag{10.48}$$

Here, $W(F)$ represents a set of weight factors that can be used to select different error bounds in the passband and stopband. The nature of $D(F)$ and $W(F)$ depends on the type of the filter required. The idea is to select the $\alpha_k$ (in the expression for $H(F)$) so as to minimize the maximum absolute error $|\epsilon|_{\max}$. The alternation theorem points the way (though it does not tells us how). In essence, it says that we must be able to find at least $M+2$ frequencies $F_k$, $k = 1, 2, \ldots, M+2$, called the **extremal frequencies** or **alternations** where

1. The error alternates between two equal maxima and minima (extrema):

$$\epsilon(F_k) = -\epsilon(F_{k+1}), \qquad k = 1, 2, \ldots, M+1 \tag{10.49}$$

2. The error at the frequencies $F_k$ equals the *maximum absolute error:*

$$|\epsilon(F_k)| = |\epsilon(F)|_{\max}, \qquad k = 1, 2, \ldots, M+2 \tag{10.50}$$

In other words, we require $M+2$ extrema (including the band edges) where the error attains its maximum absolute value. These frequencies yield the smallest filter length (number of coefficients $\alpha_k$) for optimal design. In some instances, we may get $M+3$ extremal frequencies leading to *extra ripple* filters.

The design strategy to find the extremal frequencies invariably requires iterative methods. The most popular is the algorithm of Parks and McClellan, which in turn relies on the **Remez exchange algorithm**.

The Parks-McClellan algorithm requires the band edge frequencies $F_p$ and $F_s$, the ratio $K = \delta_p/\delta_s$ of the passband and stopband ripple, and the filter length $N$. It returns the coefficients $\alpha_k$ and the actual design values of $\delta_p$ and $\delta_s$ for the given filter length $N$. If these values of $\delta_p$ and $\delta_s$ are not acceptable (or do not meet requirements), we can increase $N$ (or change the ratio $K$) and repeat the design.

A good starting estimate for the filter length $N$ is given by a relation similar to the Kaiser relation for half-band filters and reads

$$N = 1 + \frac{-10\log(\delta_p\delta_s) - 13}{14.6F_T} \qquad \delta_p = \frac{10^{A_p/20} - 1}{10^{A_p/20} + 1} \qquad \delta_s = 10^{-A_s/20} \tag{10.51}$$

Here, $F_T$ is the digital transition width. More accurate (but more involved) design relations are also available.

To explain the algorithm, consider a lowpass filter design. To approximate an ideal lowpass filter, we choose $D(F)$ and $W(F)$ as

$$D(F) = \begin{cases} 1, & 0 \le F \le F_p \\ 0, & F_s \le F \le 0.5 \end{cases} \qquad W(F) = \begin{cases} 1, & 0 \le F \le F_p \\ K = \delta_p/\delta_s, & F_s \le F \le 0.5 \end{cases} \tag{10.52}$$

To find the $\alpha_k$ in $H(F)$, we use the **Remez exchange algorithm**. Here is how it works. We start with a trial set of $M+2$ frequencies $F_k$, $k = 1, 2, \ldots, M+2$. To force the alternation condition, we must satisfy $\epsilon(F_k) = -\epsilon(F_{k+1})$, $k = 1, 2, \ldots, M+1$. Since the maximum error is yet unknown, we let $\rho = \epsilon(F_k) = -\epsilon(F_{k+1})$. We now have $M+1$ unknown coefficients $\alpha_k$ and the unknown $\rho$, for a total of $M+2$. We solve for these by using the $M+2$ frequencies to generate the $M+2$ equations:

$$-(-1)^k\rho = W(F_k)[D(F_k) - H(F_k)], \quad k = 1, 2, \ldots, M+2 \tag{10.53}$$

Here, the quantity $(-1)^k$ brings out the alternating nature of the error.

Once the $\alpha_k$ are found, the right-hand side of this equation is known in its entirety and is used to compute the extremal frequencies. The problem is that these frequencies may no longer satisfy the alternation condition. So we must go back and evaluate a new set of $\alpha_k$ and $\rho$, using the computed frequencies. We continue this process until the computed frequencies also turn out to be the actual extremal frequencies (to within a given tolerance, of course).

Do you see why it is called the *exchange* algorithm? First, we exchange an old set of frequencies $F_k$ for a new one. Then we exchange an old set of $\alpha_k$ for a new one. Since the $\alpha_k$ and $F_k$ actually describe the impulse response and frequency response of the filter, we are in essence going back and forth between the two domains until the coefficients $\alpha_k$ yield a spectrum with the desired optimal characteristics.

Many time-saving steps have been suggested to speed the computation of the extremal frequencies, and the $\alpha_k$, at each iteration. The Parks-McClellan algorithm is arguably one of the most popular methods of filter design in the industry, and many of the better commercial software packages on signal processing include it in their stock list of programs. Having said that, we must also point out two disadvantages of this method. First, the filter length must still be estimated by empirical means. And second, we have no control over the actual ripple that the design yields. The only remedy, if this ripple is unacceptable, is to start afresh with a different set of weight functions or with a different filter length.

### 10.6.2 Optimal Half-Band Filters

Since about half the coefficients of a half-band filter are zero, the computational burden can be reduced by developing a filter that contains only the nonzero coefficients followed by zero interpolation. To design a half-band lowpass filter $h_{HB}[n]$ with band edges $F_p$ and $F_s$ and ripple $\delta$, an estimate of the filter length $N = 4k - 1$ is first found from the given design specifications. Next, we design a lowpass prototype $h_P[n]$ of *even length* $0.5(N+1)$ with band edges of $2F_p$ and 0.5 (in effect, with no stopband). This filter describes a type 2 sequence whose response is zero at $F = 0.5$ and whose ripple is twice the design ripple. Finally, we insert zeros between adjacent samples of $0.5h_P[n]$ and set the (zero-valued) center coefficient to 0.5 to obtain the required half-band filter $h_{HB}[n]$ with the required ripple.

**Example 10.6**

### Optimal FIR Filter Design

**(a)** We design an optimal bandstop filter with

- Stopband edges of [2, 3] kHz
- Passband edges of [1, 4] kHz
- $A_p = 1$ dB
- $A_s = 50$ dB
- Sampling frequency of $S = 10$ kHz

We find the digital passband edges as [0.2, 0.3] and the stopband edges as [0.1, 0.4]. The transition width is $F_T = 0.1$. We find the approximate filter length $N$, as follows:

$$\delta_p = \frac{10^{A_p/20} - 1}{10^{A_p/20} + 1} = 0.0575 \qquad \delta_s = 10^{-A_s/20} = 0.00316$$

$$N = 1 + \frac{-10\log(\delta_p\delta_s) - 13}{14.6F_T} \approx 17.7$$

Choosing the next odd length gives $N = 19$. The bandstop specifications are actually met by a filter with $N = 21$. The response of the designed filter is shown in Figure E10.6(a).

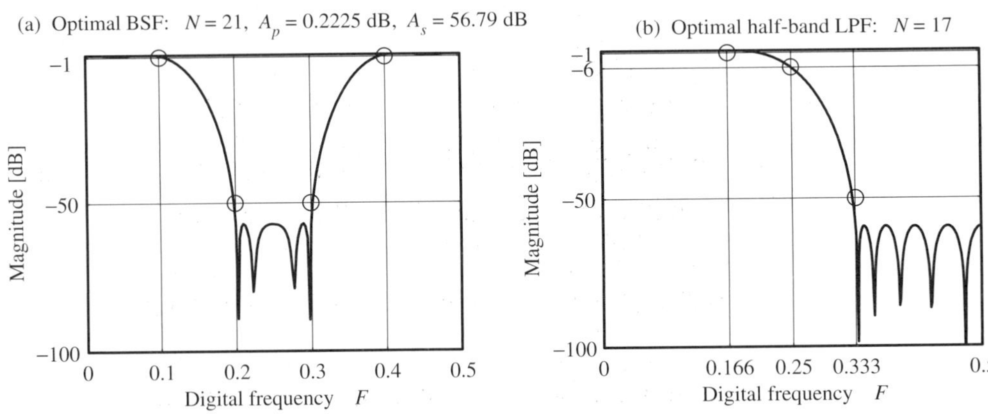


**FIGURE E 10.6.a** Optimal filters for Example 10.6

**(b) Optimal Half-Band Filter Design**

We design an optimal half-band filter to meet the following specifications:

- Passband edge: 8 kHz
- Stopband edge: 16 kHz

- $A_p = 1$ dB
- $A_s = 50$ dB

We choose $S = 2(f_p + f_s) = 48$ kHz. The digital band edges are $F_p = \frac{1}{6}$, $F_s = \frac{1}{3}$, and $F_C = 0.25$.

Next, we find the minimum ripple $\delta$ and the approximate filter length $N$ as

$$\delta_p = \frac{10^{A_p/20} - 1}{10^{A_p/20} + 1} = 0.0575 \qquad \delta_s = 10^{-A_s/20} = 0.00316 \qquad \delta = 0.00316$$

$$N = 1 + \frac{-10\log\delta^2 - 13}{14.6(F_s - F_p)} = 1 + \frac{-20\log(0.00316) - 13}{14.6(1/6)} \approx 16.2 \qquad \Rightarrow N = 19$$

The choice $N = 19$ is based on the form $N = 4k - 1$. Next, we design the lowpass prototype $h_P[n]$ of even length $M = 0.5(N + 1) = 10$ and band edges at $2F_p = 2(1/6) = 1/3$ and $F_s = 0.5$. The result is

$$h_P[n] = \{0.0074,\ -0.0267,\ 0.0708,\ -0.173,\ 0.6226,\ 0.6226,\ -0.173,\ 0.0708,\ -0.0267,\ 0.0074\}$$

Finally, we zero-interpolate $0.5h_P[n]$ and choose the center coefficient as 0.5 to obtain $H_{\text{HB}}[n]$:

$$\{0.0037, 0, -0.0133, 0, 0.0354,\ 0, -0.0865, 0,\ 0.3113,\ 0,\ 0.5,\ 0.3113, 0, -0.0865, 0, 0.0354, 0, -0.0133, 0, 0.0037\}$$

Its length is $N = 19$, as required. Figure E10.6(b) shows that this filter does meet specifications, with an attenuation of 0.02 dB at 8 kHz and 59.5 dB at 16 kHz.

# 10.7 Application: Multistage Interpolation and Decimation

Recall that a sampling rate increase, or interpolation, by $N$ involves up-sampling (inserting $N - 1$ zeros between samples) followed by lowpass filtering with a gain of $N$. If the interpolation factor $N$ is large, it is much more economical to carry out the interpolation in stages, because it results in a smaller overall filter length or order. For example, if $N$ can be factored as $N = I_1 I_2 I_3$, then interpolation by $N$ can be accomplished in three stages with individual interpolation factors of $I_1$, $I_2$, and $I_3$. At a typical $I$th stage, the output sampling rate is given by $S_{\text{out}} = IS_{\text{in}}$, as shown in Figure 10.24.

**FIGURE 10.24** One stage of a typical multistage interpolating filter

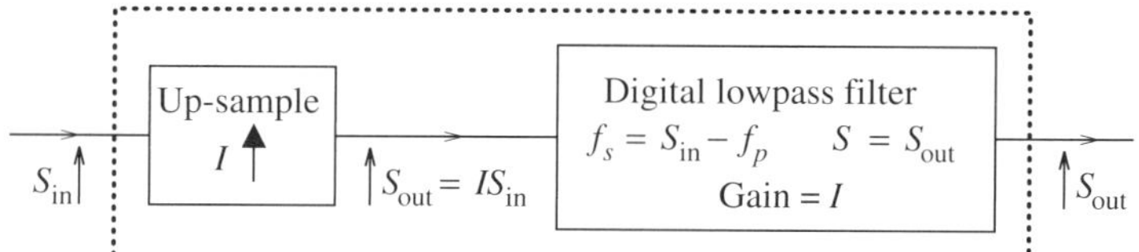

The filter serves to remove the spectral replications due to the zero interpolation by the up-sampler. These replicas occur at multiples of the input sampling rate. As a result, the filter stopband edge is computed from the input sampling rate $S_{\text{in}}$ as $f_s = S_{\text{in}} - f_p$, while the filter passband edge remains fixed (by the given specifications). The filter sampling rate corresponds to the output sampling rate $S_{\text{out}}$ and the filter gain equals the interpolation factor $I_k$. At each successive stage (except the first), the spectral images occur at higher and higher frequencies, and their removal requires filters whose transition bands get wider with each stage, leading to less complex filters with smaller filter lengths. Although it is not easy to establish the optimum values for the interpolating factors and their order for the smallest overall filter length, it turns out that interpolating factors in increasing order generally yield smaller overall lengths, and any multistage design results in a substantial reduction in the filter length as compared to a single-stage design.

**Example 10.7**

### The Concept of Multistage Interpolation

Consider a signal band-limited to 1.8 kHz and sampled at 4 kHz. It is required to raise the sampling rate to 48 kHz. This requires interpolation by 12. The value of the passband edge is $f_p = 1.8$ kHz for either a single-stage design or multistage design.

**(a)** For a single-stage interpolator, the output sampling rate is $S_{\text{out}} = 48$ kHz, and we thus require a filter with a stopband edge of $f_s = S_{\text{in}} - f_p = 4 - 1.8 = 2.2$ kHz, a sampling rate of $S = S_{\text{out}} = 48$ kHz, and a gain of 12. If we use a crude approximation for the filter length as $L \approx 4/F_T$, where $F_T = (f_s - f_p)/S$ is the digital transition width, we obtain $L = 4(48/0.4) = 480$.

**(b)** If we choose two-stage interpolation with $I_1 = 3$ and $I_2 = 4$, at each stage we compute the important parameters as follows:

| Stage | $S_{\text{in}}$ (kHz) | Interpolation Factor | $S_{\text{out}} = S$ (kHz) | $f_p$ (kHz) | $f_s = S_{\text{in}} - f_p$ (kHz) | Filter Length $L \approx 4S/(f_s - f_p)$ |
|---|---|---|---|---|---|---|
| 1 | 4 | $I_1 = 3$ | 12 | 1.8 | 2.2 | $48/0.4 = 120$ |
| 2 | 12 | $I_2 = 4$ | 48 | 1.8 | 10.2 | $192/8.4 = 23$ |

The total filter length is thus 143.

**(c)** If we choose three-stage interpolation with $I_1 = 2,\ I_2 = 3$, and $I_3 = 2$, at each stage we compute the important parameters, as follows:

| Stage | $S_{\text{in}}$ (kHz) | Interpolation Factor | $S_{\text{out}} = S$ (kHz) | $f_p$ (kHz) | $f_s = S_{\text{in}} - f_p$ (kHz) | Filter Length $L \approx 4S/(f_s - f_p)$ |
|---|---|---|---|---|---|---|
| 1 | 4 | $I_1 = 2$ | 8 | 1.8 | 2.2 | $32/0.4 = 80$ |
| 2 | 8 | $I_2 = 3$ | 24 | 1.8 | 6.2 | $96/4.4 = 22$ |
| 3 | 24 | $I_2 = 2$ | 48 | 1.8 | 22.2 | $192/20.4 = 10$ |

The total filter length is thus 112.

**(d)** If we choose three-stage interpolation but with the different order $I_1 = 3,\ I_2 = 2$, and $I_3 = 2$, at each stage we compute the important parameters, as follows:

| Stage | $S_{\text{in}}$ (kHz) | Interpolation Factor | $S_{\text{out}} = S$ (kHz) | $f_p$ (kHz) | $f_s = S_{\text{in}} - f_p$ (kHz) | Filter Length $L \approx 4S/(f_s - f_p)$ |
|---|---|---|---|---|---|---|
| 1 | 4 | $I_1 = 3$ | 12 | 1.8 | 2.2 | $48/0.4 = 120$ |
| 2 | 12 | $I_2 = 2$ | 24 | 1.8 | 10.2 | $96/8.4 = 12$ |
| 3 | 24 | $I_2 = 2$ | 48 | 1.8 | 22.2 | $192/20.4 = 10$ |

The total filter length is thus 142.

Any multistage design results in a substantial reduction in the filter length as compared to a single-stage design, and a smaller interpolation factor in the first stage of a multistage design does seem to yield smaller overall lengths. Also, remember that the filter lengths are only a crude approximation to illustrate the relative merits of each design, and the actual filter lengths will depend on the attenuation specifications.

---

For multistage operations, the actual filter lengths depend not only on the order of the interpolating factors, but also on the given attenuation specifications $A_p$ and $A_s$. Since the attenuation in decibels is additive in a cascaded system, the passband attenuation $A_p$ is usually distributed among the various stages to ensure an overall attenuation that matches specifications. The stopband specification needs no such adjustment. Since the signal is attenuated even further at each stage, the overall stopband attenuation always exceeds specifications. For interpolation, we require a filter whose gain is scaled (multiplied) by the interpolation factor of the stage.

**Example 10.8**

### Design of Interpolating Filters

Consider a signal band limited to 1.8 kHz and sampled at 4 kHz. It is required to raise the sampling rate to 48 kHz. The passband attenuation should not exceed 0.6 dB, and the minimum stopband attenuation should be 50 dB. Design an interpolation filter using a single-stage interpolator.

**(a)** The single stage requires interpolation by 12. The value of the passband edge is $f_p = 1.8$ kHz. The output sampling rate is $S_{\text{out}} = 48$ kHz, and we thus require a filter with a stopband edge of $f_s = S_{\text{in}} - f_p = 4 - 1.8 = 2.2$ kHz and a sampling rate of $S = S_{\text{out}} = 48$ kHz. We ignore the filter gain of 12 in the following computations. To compute the filter length, we first find the ripple parameters

$$\delta_p = \frac{10^{A_p/20} - 1}{10^{A_p/20} + 1} = \frac{10^{0.6/20} - 1}{10^{0.6/20} + 1} = 0.0345 \qquad \delta_s = 10^{-A_s/20} = 10^{-50/20} = 0.00316$$

and then approximate the length $N$ by

$$N \approx \frac{-10\log(\delta_p\delta_s) - 13}{14.6(F_s - F_p)} + 1 = \frac{S[-10\log(\delta_p\delta_s) - 13]}{14.6(f_s - f_p)} + 1 = 230$$

The actual filter length of the optimal filter turns out to be $N = 233$, and the filter shows a passband attenuation of 0.597 dB and a stopband attenuation of 50.05 dB.

**(b)** Repeat the design using a three-stage interpolator with $I_1 = 2,\ I_2 = 3$, and $I_3 = 2$. How do the results compare with those of the single-stage design?

We distribute the passband attenuation (in decibels) equally among the three stages. Thus, $A_p = 0.2$ dB for each stage, and the ripple parameters for each stage are

$$\delta_p = \frac{10^{0.2/20} - 1}{10^{0.2/20} + 1} = 0.0115 \qquad \delta_s = 10^{-50/20} = 0.00316$$

For each stage, the important parameters and the filter length are listed in the following table:

| Stage | $S_{in}$ (kHz) | Interpolation Factor | $S_{out} = S$ (kHz) | $f_p$ (kHz) | $f_s = S_{in} - f_p$ (kHz) | Filter Length $L$ |
|---|---|---|---|---|---|---|
| 1 | 4 | $I_1 = 2$ | 8 | 1.8 | 2.2 | 44 |
| 2 | 8 | $I_2 = 3$ | 24 | 1.8 | 6.2 | 13 |
| 3 | 24 | $I_2 = 2$ | 48 | 1.8 | 22.2 | 7 |

In this table, for example, we compute the filter length for the first stage as

$$L \approx \frac{-10\log(\delta_p\delta_s) - 13}{14.6(F_s - F_p)} + 1 = \frac{S[-10\log(\delta_p\delta_s) - 13]}{14.6(f_s - f_p)} + 1 = 44$$

The actual filter lengths of the optimal filters turn out to be 47 (with design attenuations of 0.19 dB and 50.31 dB), 13 (with design attenuations of 0.18 dB and 51.09 dB), and 4 (with design attenuations of 0.18 dB and 50.91 dB). The overall filter length is only 64. This is about four times less than the filter length for single-stage design.

### 10.7.1 Multistage Decimation

Decimation by $M$ involves lowpass filtering (with a gain of unity), followed by down-sampling (discarding $M - 1$ samples and retaining every $M$th sample). The process is essentially the inverse of interpolation. If the decimation factor $M$ is large, decimation in stages results in a smaller overall filter length or order. If $M$ can be factored as $M = D_1 D_2 D_3$, then decimation by $M$ can be accomplished in three stages with individual decimation factors of $D_1$, $D_2$, and $D_3$. At a typical stage, the output sampling rate is given by $S_{out} = S_{in}/D$, where $D$ is the decimation factor for that stage. This is illustrated in Figure 10.25.

The decimation filter has a gain of unity and operates at the input sampling rate $S_{in}$, and its stopband edge is computed from the output sampling rate as $f_s = S_{out} - f_p$. At each successive stage (except the first), the transition bands get narrower with each stage.

The overall filter length does depend on the order in which the decimating factors are used. Although it is not easy to establish the optimum values for the decimation factors and their order for the smallest overall filter length, it turns out that decimation factors in decreasing order generally yield smaller overall lengths, and any multistage design results in a substantial reduction in the filter length as compared to a single-stage design.

The actual filter lengths also depend on the given attenuation specifications. Since attenuations in decibels add in a cascaded system, the passband attenuation $A_p$ is usually distributed among the various stages to ensure an overall value that matches specifications.

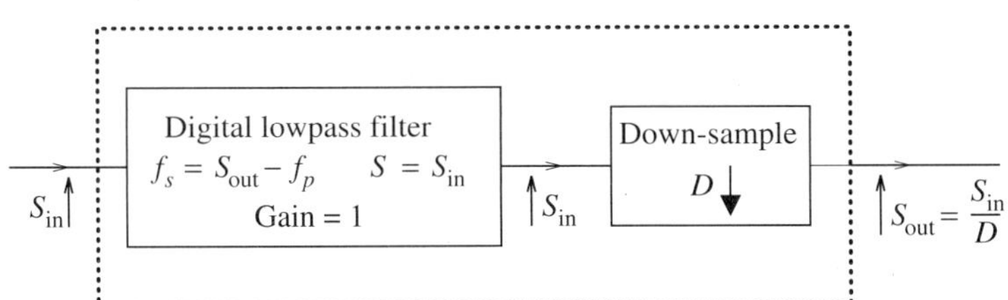


**FIGURE 10.25** One stage of a typical multistage decimating filter

**Example 10.9**

### The Concept of Multistage Decimation

Consider a signal band-limited to 1.8 kHz and sampled at 48 kHz. It is required to reduce the sampling rate to 4 kHz. This requires decimation by 12. The passband edge is $f_p = 1.8$ kHz and remains unchanged for a single stage-design or multistage design.

**(a)** For a single-stage decimator, the output sampling rate is $S_{\text{out}} = 4$ kHz, and we thus require a filter with a stopband edge of $f_s = S_{\text{out}} - f_p = 4 - 1.8 = 2.2$ kHz, a sampling rate of $S = S_{\text{in}} = 48$ kHz, and a gain of unity. If we use a crude approximation for the filter length as $L \approx 4/F_T$, where $F_T = (f_s - f_p)/S$ is the digital transition width, we obtain $L = 48/0.4 = 120$.

**(b)** If we choose two-stage decimation with $D_1 = 4$ and $D_2 = 3$, at each stage we compute the important parameters, as follows:

| Stage | $S_{\text{in}} = S$ (kHz) | Decimation Factor | $S_{\text{out}}$ (kHz) | $f_p$ (kHz) | $f_s = S_{\text{out}} - f_p$ (kHz) | Filter Length $L \approx 4S/(f_s - f_p)$ |
|---|---|---|---|---|---|---|
| 1 | 48 | $D_1 = 4$ | 12 | 1.8 | 10.2 | $192/8.4 = 23$ |
| 2 | 12 | $D_2 = 3$ | 4 | 1.8 | 2.2 | $48/0.4 = 120$ |

The total filter length is thus 143.

**(c)** If we choose three-stage decimation with $D_1 = 2,\ D_2 = 3$, and $D_3 = 2$, at each stage we compute the important parameters, as follows:

| Stage | $S_{\text{in}} = S$ (kHz) | Decimation Factor | $S_{\text{out}}$ (kHz) | $f_p$ (kHz) | $f_s = S_{\text{out}} - f_p$ (kHz) | Filter Length $L \approx 4S/(f_s - f_p)$ |
|---|---|---|---|---|---|---|
| 1 | 48 | $D_1 = 2$ | 24 | 1.8 | 22.2 | $192/20.4 = 10$ |
| 2 | 24 | $D_2 = 3$ | 8 | 1.8 | 6.2 | $96/4.4 = 22$ |
| 3 | 8 | $D_3 = 2$ | 4 | 1.8 | 2.2 | $32/0.4 = 80$ |

The total filter length is thus 112.

**(d)** If we choose three-stage decimation but with the different order $D_1 = 2,\ D_2 = 2$, and $D_3 = 3$, at each stage we compute the important parameters, as follows:

| Stage | $S_{\text{in}} = S$ (kHz) | Decimation Factor | $S_{\text{out}}$ (kHz) | $f_p$ (kHz) | $f_s = S_{\text{out}} - f_p$ (kHz) | Filter Length $L \approx 4S/(f_s - f_p)$ |
|---|---|---|---|---|---|---|
| 1 | 48 | $D_1 = 2$ | 24 | 1.8 | 22.2 | $192/20.4 = 10$ |
| 2 | 24 | $D_2 = 2$ | 12 | 1.8 | 10.2 | $96/8.4 = 12$ |
| 3 | 12 | $D_3 = 3$ | 4 | 1.8 | 2.2 | $48/0.4 = 120$ |

The total filter length is thus 142.

Note how decimation uses the same filter frequency specifications as does interpolation for a given split, except in reversed order. Any multistage design results in a substantial reduction in the filter length as compared to a single-stage design. Also, remember that the filter lengths chosen here are only a crude approximation (in order to illustrate the relative merits of each design), and the actual filter lengths will depend on the attenuation specifications.

## 10.8 Maximally Flat FIR Filters

Linear-phase FIR filters can also be designed with **maximally flat** frequency response. Such filters are usually used in situations where accurate filtering is needed at low frequencies (near dc). The design of lowpass maximally flat filters uses a closed form for the transfer function $H(F)$ given by

$$H(F) = \cos^{2K}(\pi F)\sum_{n=0}^{L-1} d_n \sin^{2n}(\pi F) \qquad d_n = \frac{(K+n-1)!}{(K-1)!\,n!} = C_n^{K+n-1} \tag{10.54}$$

Here, the $d_n$ have the form of binomial coefficients as indicated. Note that $2L-1$ derivatives of $|H(F)|^2$ are zero at $F=0$, and $2K-1$ derivatives are zero at $F=0.5$. This is the basis for the maximally flat response of the filter. The filter length equals $N = 2(K+L)-1$ and is thus odd. The integers $K$ and $L$ are determined from the passband and stopband frequencies $F_p$ and $F_s$ that correspond to gains of 0.95 and 0.05 (or attenuations of about 0.5 dB and 26 dB), respectively. Here is an empirical design method:

1. Define the cutoff frequency as $F_C = 0.5(F_p + F_s)$ and let $F_T = F_s - F_p$.
2. Obtain a first estimate of the odd filter length as $N_0 = 1 + 0.5/F_T^2$.
3. Define the parameter $\alpha$ as $\alpha = \cos^2(\pi F_C)$.
4. Find the best rational approximation $\alpha \approx K/M_{\min}$, with $0.5(N_0 - 1) \le M_{\min} \le (N_0 - 1)$.
5. Evaluate $L$ and the true filter length $N$ from $L = M_{\min} - K$ and $N = 2M_{\min} - 1$.
6. Find $h[n]$ as the $N$-point inverse DFT of $H(F)$, $F = 0, 1/N, \ldots, (N-1)/N$.

**Example 10.10**

### Maximally Flat FIR Filter Design

Consider the design of a maximally flat lowpass FIR filter with normalized frequencies $F_p = 0.2$ and $F_s = 0.4$.

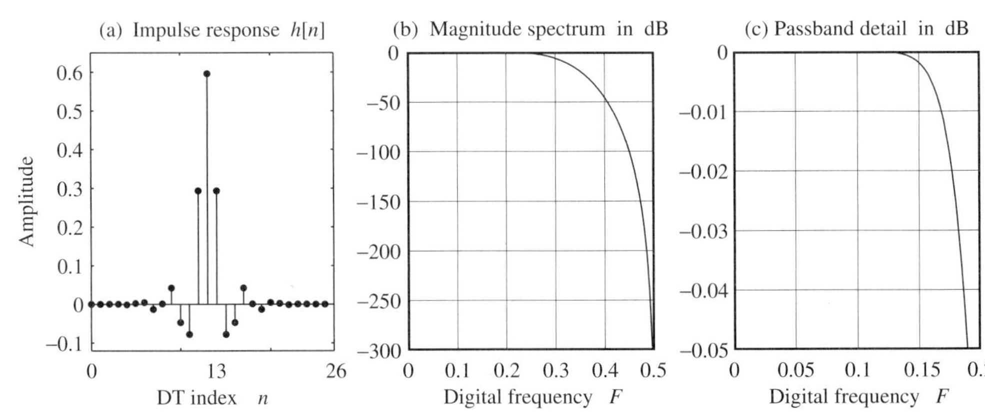


**FIGURE E.10.10** Features of the maximally flat lowpass filter for Example 10.10

We have $F_C = 0.3$ and $F_T = 0.2$. We compute $N_0 = 1 + 0.5/F_T^2 = 13.5 \approx 15$, and $\alpha = \cos^2(\pi F_C) = 0.3455$. The best rational approximation to $\alpha$ works out to be $\alpha \approx 5/14 = K/M_{\min}$.

With $K = 5$ and $M_{\min} = 14$, we get $L = M_{\min} - K = 9$, and the filter length $N = 2M_{\min} - 1 = 27$.

Figure E10.10 shows the impulse response and $|H(F)|$ of the designed filter. The response is maximally flat, with no ripples in the passband.

## 10.9 FIR Differentiators and Hilbert Transformers

An ideal digital differentiator is described by $H(F) = j2\pi F$, $|F| \le 0.5$. In practical situations, we seldom require filters that differentiate for the full frequency range up to $|F| = 0.5$. If we require differentiation only up to a cutoff frequency of $F_C$, then

$$H(F) = j2\pi F, \qquad |F| \le F_C \tag{10.55}$$

The magnitude and phase spectrum of such a differentiator are shown in Figure 10.26. Since $H(F)$ is odd, $h[0] = 0$. To find $h[n]$, $n \ne 0$, we use the inverse DTFT to obtain

$$h[n] = j\int_{-F_C}^{F_C} 2\pi F e^{j2n\pi F}\, dF \tag{10.56}$$

Invoking Euler's relation and symmetry, this simplifies to

$$h[n] = -4\pi \int_0^{F_C} F\sin(2n\pi F)\, dF = \frac{2n\pi F_C\cos(2n\pi F_C) - \sin(2n\pi F_C)}{\pi n^2} \tag{10.57}$$

For $F_C = 0.5$, this yields $h[0] = 0$ and $h[n] = \dfrac{\cos(n\pi)}{n}$, $n \ne 0$.

### 10.9.1 Hilbert Transformers

A **Hilbert transformer** is used to shift the phase of a signal by $-\frac{\pi}{2}$ (or $-90^\circ$) over the frequency range $|F| \le 0.5$. In practice, we seldom require filters that shift the phase for the full frequency range up to $|F| = 0.5$. If we require phase shifting only up to a cutoff frequency of $F_C$, the Hilbert transformer may be described by

$$H(F) = -j\,\mathrm{sgn}(F), \qquad |F| \le F_C \tag{10.58}$$

Its magnitude and phase spectrum are shown in Figure 10.27.

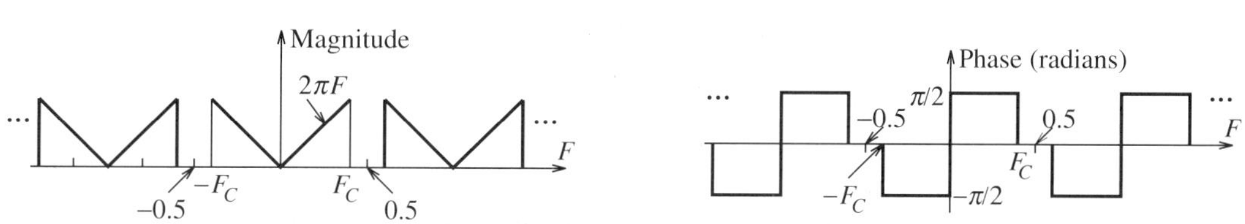


**FIGURE 10.26** Magnitude and phase spectrum of an ideal differentiator

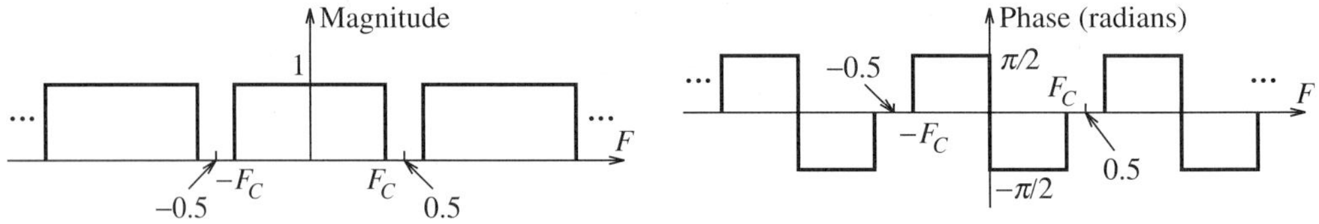


**FIGURE 10.27** Magnitude and phase spectrum of an ideal Hilbert transformer

With $h[0] = 0$, we use the inverse DTFT to find its impulse response $h[n]$, $n \neq 0$, as

$$h[n] = \int_{-F_C}^{F_C} -j\,\mathrm{sgn}(F)e^{j2n\pi F}\,dF = 2\int_0^{F_C} \sin(2n\pi F)\,dF = \frac{1-\cos(2n\pi F_C)}{n\pi} \quad (10.59)$$

For $F_C = 0.5$, this reduces to $h[n] = \dfrac{1-\cos(n\pi)}{n\pi}$, $n \neq 0$.

### 10.9.2 Design of FIR Differentiators and Hilbert Transformers

To design an FIR differentiator, we must truncate $h[n]$ to $h_N[n]$ and choose a type 3 or type 4 sequence, since $H(F)$ is purely imaginary. To ensure odd symmetry, the filter coefficients may be computed only for $n > 0$, and the same values (in reversed order) are used for $n < 0$. If $N$ is odd, we must also include the sample $h[0]$. We may window $h_N[n]$ to minimize the overshoot and ripple in the spectrum $H_N(F)$. And, finally, to ensure causality, we must introduce a delay of $(N-1)/2$ samples. Figure 10.28(a) shows the magnitude response of Hamming-windowed FIR differentiators for both even and odd lengths $N$. Note that $H(0)$ is always zero, but $H(0.5) = 0$ only for type 3 (odd-length) sequences.

The design of Hilbert transformers closely parallels the design of FIR differentiators. The sequence $h[n]$ is truncated to $h_N[n]$. The chosen filter must correspond to a type 3 or type 4 sequence, since $H(F)$ is imaginary. To ensure odd symmetry, the filter coefficients may be computed only for $n > 0$ with the same values (in reversed order) used for $n < 0$. If $N$ is odd, we must also include the sample $h[0]$. We may window $h_N[n]$ to minimize the ripples (due to the Gibbs effect) in the spectrum $H_N(F)$. The Hamming window is a common choice, but others may also be used. To make the filter causal, we introduce a

**FIGURE 10.28** Magnitude spectra of ideal differentiators and Hilbert transformers using a Hamming window with different lengths

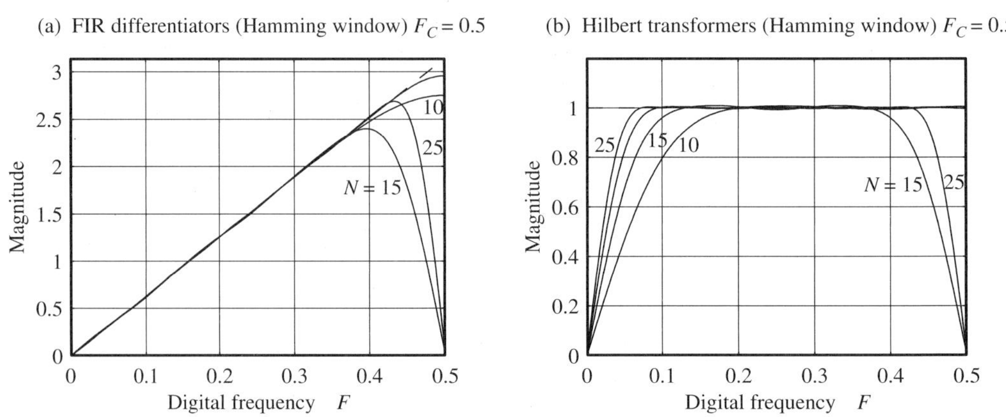

delay of $(N-1)/2$ samples. The magnitude spectrum of a Hilbert transformer becomes flatter with increasing filter length $N$. Figure 10.28(b) shows the magnitude response of Hilbert transformers for both even and odd lengths $N$. Note that $H(0)$ is always zero, but $H(0.5)=0$ only for type 3 (odd-length) sequences.

# 10.10 Least Squares and Adaptive Signal Processing

The concept of least squares can be formulated as the solution to a set of algebraic equations expressed in matrix form as

$$\mathbf{Xb} = \mathbf{Y} \tag{10.60}$$

If the number of unknowns equals the number of equations, and the matrix $\mathbf{X}$ is non-singular such that its inverse $\mathbf{X}^{-1}$ exists (the unique case), the solution is simply $\mathbf{b}=\mathbf{X}^{-1}\mathbf{Y}$. However, if the number of equations exceeds the number of unknowns (the over-determined case), no solution is possible, and only an approximate result may be obtained using, for example, least-squares minimization. If the number of unknowns exceeds the number of equations (the underdetermined case), many possible solutions exist, including the one where the mean-square error is minimized.

In many practical situations, the set of equations is overdetermined and thus amenable to a least-squares solution. To solve for $\mathbf{b}$, we simply premultiply both sides by $\mathbf{X}^T$ to obtain

$$\mathbf{X}^T\mathbf{Xb} = \mathbf{X}^T\mathbf{Y} \tag{10.61}$$

If the inverse of $\mathbf{X}^T\mathbf{X}$ exists, the solution is given by

$$\mathbf{b} = (\mathbf{X}^T\mathbf{X})^{-1}\mathbf{X}^T\mathbf{Y} \tag{10.62}$$

The matrix $\mathbf{X}^T\mathbf{X}$ is called the **covariance matrix**.

## 10.10.1 Adaptive Filtering

The idea of least squares finds widespread application in **adaptive signal processing**—a field that includes adaptive filtering, deconvolution, and system identification and encompasses many disciplines. Typically, the goal is to devise a digital filter whose coefficients $b[k]$ can be adjusted to optimize its performance (in the face of changing signal characteristics or to combat the effects of noise, for example). A representative system for adaptive filtering is shown in Figure 10.29. The measured signal $x[n]$ (which may be contaminated by noise) is fed to an FIR filter whose output $\hat{y}[n]$ is compared with the desired signal $y[n]$ to generate the error signal $e[n]$. The error signal is used to update the filter coefficients $b[k]$ through an adaptation algorithm in a way that minimizes the error $e[n]$ and thus provides an optimal estimate of the desired signal $y[n]$.

In convolution form, the output $\hat{\mathbf{Y}}$ of the $(M+1)$-point adaptive FIR filter is described by the model

$$\hat{y}[n] = \sum_{k=0}^{M} b[k]x[n-k], \qquad n = 0, 1, \ldots, M+N+1 \tag{10.63}$$

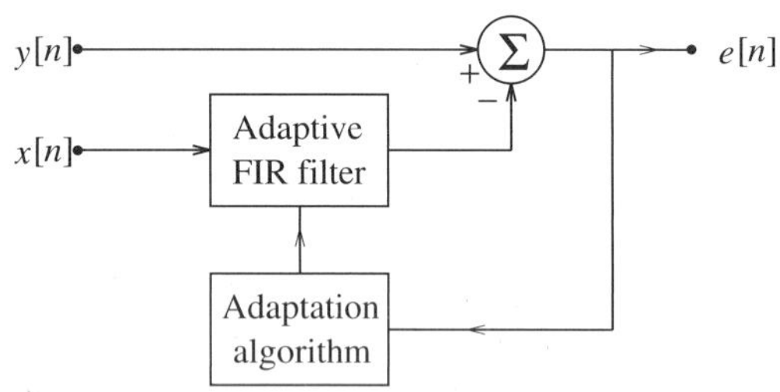


**FIGURE 10.29** The structure of a typical adaptive filtering system

This set of $M + N + 1$ equations may be cast in matrix form as

$$\begin{bmatrix} \hat{y}[0] \\ \vdots \\ \hat{y}[N] \end{bmatrix} = \begin{bmatrix} x[0] & 0 & \cdots & \cdots & \cdots \\ x[1] & x[0] & 0 & \cdots & \cdots \\ x[2] & x[1] & x[0] & 0 & \cdots \\ \vdots & \vdots & \vdots & \ddots & \vdots \\ x[N] & x[N-1] & x[N-2] & \cdots & x[N-M] \\ 0 & \cdots & \cdots & \cdots & \cdots \\ 0 & 0 & \cdots & \cdots & \cdots \\ 0 & 0 & 0 & \cdots & \cdots \end{bmatrix} \begin{bmatrix} b[0] \\ \vdots \\ b[M] \end{bmatrix} \tag{10.64}$$

In vector notation, it has the form $\hat{\mathbf{Y}} = \mathbf{Xb}$, where $\mathbf{b}$ is an $(M + 1) \times 1$ column matrix of the filter coefficients $b[0]$ through $b[M]$, $\hat{\mathbf{Y}}$ is an $(M + N + 1) \times 1$ matrix of the output samples, and $\mathbf{X}$ is an $(M + N + 1) \times (M + 1)$ **Toeplitz** (constant diagonal) matrix whose columns are successively shifted replicas of the input sequence. The $(M + 1)$ filter coefficients in $\mathbf{b}$ are chosen to ensure that the output $\hat{\mathbf{Y}}$ of the model provides an optimal estimate of the desired output $\mathbf{Y}$. Clearly, this problem is amenable to a least squares solution.

Since the data is being acquired continuously in many practical situations, the least-squares solution is implemented on-line or in real time using iterative or recursive numerical algorithms. Two common approaches are the **recursive least squares** (RLS) algorithm and the **least mean squares** (LMS) algorithm. Both start with an assumed set of filter coefficients and update this set as each new input sample arrives.

**The RLS Algorithm:** In the RLS algorithm, the filter coefficients are updated by weighting the input samples (typically in an exponential manner) so as to emphasize more recent inputs.

**The LMS Algorithm:** The LMS algorithm, though not directly related to least squares, uses the *method of steepest descent* to generate updates that converge about the least-squares solution. Although it may not converge as rapidly as the RLS algorithm, it is far more popular due to its ease of implementation. In fact, the updating equation (for the $n$th update) has the simple form

$$b_n[k] = b_{n-1}[k] + 2\mu(y[n] - \hat{y}[n])x[n-k] = b_{n-1}[k] + 2\mu e[n]x[n-k], \quad 0 \le k \le M \tag{10.65}$$

The parameter $\mu$ governs both the rate of convergence and the stability of the algorithm. Larger values result in faster convergence, but the filter coefficients tend to oscillate about the optimum values. Typically, $\mu$ is restricted to the range $0 < \mu < \frac{1}{\sigma_x}$, where $\sigma_x$ (the variance of $x[n]$) provides a measure of the power in the input signal $x[n]$.

### 10.10.2 Applications of Adaptive Filtering

Adaptive filtering forms the basis for many signal-processing applications including system identification, noise cancellation, and channel equalization. We conclude with a brief introduction.

#### System Identification

In system identification, the goal is to identify the transfer function (or impulse response) of an unknown system. Both the adaptive filter and the unknown system are excited by the same input, and the signal $y[n]$ represents the output of the unknown system. Minimizing $e[n]$ implies that the output of the unknown system and the adaptive filter are very close, and the adaptive filter coefficients describe an FIR approximation to the unknown system.

#### Noise Cancellation

In adaptive noise-cancellation systems, the goal is to improve the quality of a desired signal $y[n]$ that may be contaminated by noise. The signal $x[n]$ is a noise signal, and the adaptive filter minimizes the power in $e[n]$. Since the noise power and signal power add (if they are uncorrelated), the signal $e[n]$ (with its power minimized) also represents a cleaner estimate of the desired signal $y[n]$.

#### Channel Equalization

In adaptive channel equalization, the goal is to allow a modem to adapt to different telephone lines (so as to prevent distortion and inter-symbol interference). A known *training* signal $y[n]$ is transmitted at the start of each call, and $x[n]$ is the output of the telephone channel. The error signal $e[n]$ is used to generate an FIR filter (an inverse system) that cancels out the effects of the telephone channel. Once found, the filter coefficients are fixed, and the modem operates with the fixed filter.

## 10.11 Problems

10.1. **(Ideal Filter Concepts)** Let $H_1(F)$ be an ideal lowpass filter with a cutoff frequency of $F_1 = 0.2$ and let $H_2(F)$ be an ideal lowpass filter with a cutoff frequency of $F_2 = 0.4$. Make a block diagram of how you could use these filters to implement the following and express their transfer function in terms of $H_1(F)$ and/or $H_2(F)$.

**(a)** an ideal highpass filter with a cutoff frequency of $F_C = 0.2$
**(b)** an ideal highpass filter with a cutoff frequency of $F_C = 0.4$
**(c)** an ideal bandpass filter with a passband covering $0.2 \le F \le 0.4$
**(d)** an ideal bandstop filter with a stopband covering $0.2 \le F \le 0.4$

[**Hints and Suggestions:** In (a), use $H(F) = 1 - H_1(F)$ to set up the block diagram.]

10.2. **(Symmetric Sequences)** Find $H(z)$ and $H(F)$ for each sequence and establish the type of FIR filter it describes by checking values of $H(F)$ at $F = 0$ and $F = 0.5$.

(a) $h[n] = \{\overset{\Downarrow}{1}, 0, 1\}$ (b) $h[n] = \{\overset{\Downarrow}{1}, 2, 2, 1\}$

(c) $h[n] = \{\overset{\Downarrow}{1}, 0, -1\}$ (d) $h[n] = \{\overset{\Downarrow}{-1}, 2, -2, 1\}$

**10.3.** **(Symmetric Sequences)** What types of sequences can we use to design the following filters?

**(a)** lowpass **(b)** highpass **(c)** bandpass **(d)** bandstop

**10.4.** **(Linear-Phase Sequences)** The first few values of the impulse response sequence of a linear-phase filter are $h[n] = \{2,\ -3,\ 4,\ 1,\ \ldots\}$. Determine the complete sequence (assuming the smallest length for $h[n]$) if the sequence is to be:

**(a)** type 1 **(b)** type 2 **(c)** type 3 **(d)** type 4

[**Hints and Suggestions:** For part (a), the even symmetry will be about the fourth sample. Part (c) requires odd length, odd symmetry and a zero-valued sample at the midpoint.]

**10.5.** **(Linear Phase and Symmetry)** Assume a finite length impulse-response sequence $h[n]$ with real coefficients and argue for or against the following statements.

**(a)** If all the zeros lie on the unit circle, $h[n]$ must be linear phase.
**(b)** If $h[n]$ is linear phase, its zeros must *always* lie on the unit circle.
**(c)** If $h[n]$ is odd symmetric, there must be an odd number of zeros at $z = 1$.

[**Hints and Suggestions:** Use the following facts. Each pair of reciprocal zeros, such as $(z - \alpha)$ and $(z - 1/\alpha)$ yields an even, symmetric impulse response of the form $\{1, \beta, 1\}$. Multiplication in the $z$-domain means convolution in the time domain. The convolution of symmetric sequences is also symmetric.]

**10.6.** **(Linear Phase and Symmetry)** Assume a linear-phase sequence $h[n]$ with real coefficients and refute the following statements by providing simple examples using zero locations only at $z = \pm 1$.

**(a)** If $h[n]$ has zeros at $z = -1$, it must be a type 2 sequence.
**(b)** If $h[n]$ has zeros at $z = -1$ and $z = 1$, it must be a type 3 sequence.
**(c)** If $h[n]$ has zeros at $z = 1$, it must be a type 4 sequence.

**10.7.** **(Linear Phase and Symmetry)** The locations of the zeros at $z = \pm 1$ and their number provides useful clues about the type of a linear-phase sequence. What is the sequence type for the following zero locations at $z = \pm 1$? Other zero locations are in keeping with linear phase and real coefficients.

**(a)** no zeros at $z = \pm 1$
**(b)** one zero at $z = -1$, none at $z = 1$
**(c)** two zeros at $z = -1$, one zero at $z = 1$
**(d)** one zero at $z = 1$, none at $z = -1$
**(e)** two zeros at $z = 1$, none at $z = -1$
**(f)** one zero at $z = 1$, one zero at $z = -1$
**(g)** two zeros at $z = 1$, one zero at $z = -1$

[**Hints and Suggestions:** For (a), the length is odd and the symmetry is even (so, type 1). For the rest, each zero increases the length by one, and each zero at $z = 1$ toggles the symmetry.]

**10.8.** **(Linear-Phase Sequences)** What is the smallest length linear-phase sequence with real coefficients that meets the requirements listed? Identify all of the zero locations and the type of linear-phase sequence.

**(a)** zero location: $z = e^{j0.25\pi}$; even symmetry; odd length
**(b)** zero location: $z = 0.5e^{j0.25\pi}$; even symmetry; even length
**(c)** zero location: $z = e^{j0.25\pi}$; odd symmetry; even length
**(d)** zero location: $z = 0.5e^{j0.25\pi}$; odd symmetry; odd length

[**Hints and Suggestions:** For (a) and (c), the given zero will be paired with its conjugate. For (b) and (d), the given zero will be part of a conjugate reciprocal quadruple. For all parts, no additional zeros give an odd length and even symmetry. Each additional zero at $z = 1$ or $z = -1$ will increase the length by one. Each zero at $z = 1$ will toggle the symmetry.]

**10.9.** **(Linear-Phase Sequences)** Partial details of various filters are listed. Zero locations are in keeping with linear-phase and real coefficients. Assuming the smallest length, identify the sequence type and find the transfer function of each filter.

**(a)** zero location: $z = 0.5e^{j0.25\pi}$
**(b)** zero location: $z = e^{j0.25\pi}$
**(c)** zero locations: $z = 1,\ z = e^{j0.25\pi}$
**(d)** zero locations: $z = 0.5,\ z = -1$; odd symmetry
**(e)** zero locations: $z = 0.5,\ z = 1,\ z = -1$; even symmetry

[**Hints and Suggestions:** Linear phase requires conjugate reciprocal zeros. No zeros at $z = 1$ yield even symmetry. Each additional zero at $z = 1$ toggles the symmetry.]

**10.10.** **(Truncation and Windowing)** Consider a windowed lowpass FIR filter with cutoff frequency 5 kHz and sampling frequency $S = 20$ kHz. Find the truncated, windowed sequence, the minimum delay (in samples and in seconds) to make the filter causal, and the transfer function $H(z)$ of the causal filter if

**(a)** $N = 7$, and we use a Bartlett window.
**(b)** $N = 8$, and we use a von Hann (Hanning) window.
**(c)** $N = 9$, and we use a Hamming window.

[**Hints and Suggestions:** For (a) through (b), find the results after discarding any zero-valued end-samples of the windowed sequence.]

**10.11.** **(Spectral Transformations)** Assuming a sampling frequency of 40 kHz and a fixed passband, find the specifications for a digital lowpass FIR prototype and the subsequent spectral transformation to convert to the required filter type for the following filters.

**(a)** Highpass: passband edge at 10 kHz, stopband edge at 4 kHz.
**(b)** Bandpass: passband edges at 6 kHz and 10 kHz, stopband edges at 2 kHz and 12 kHz.
**(c)** Bandstop: passband edges 8 kHz and 16 kHz, stopband edges 12 kHz and 14 kHz.

[**Hints and Suggestions:** For (b) through (c), ensure arithmetic symmetry by relocating some band edges (assuming a fixed passband, for example) to maintain the smallest transition width.]

**10.12.** **(Window-Based FIR Filter Design)** We wish to design a window-based linear-phase FIR filter. What is the *approximate* filter length $N$ required if the filter to be designed is

**(a)** Lowpass: $f_p = 1$ kHz, $f_s = 2$ kHz, $S = 10$ kHz, using a von Hann (Hanning) window?

**(b)** Highpass: $f_p = 2$ kHz, $f_s = 1$ kHz, $S = 8$ kHz, using a Blackman window?

**(c)** Bandpass: $f_p = [4, 8]$ kHz, $f_s = [2, 12]$ kHz, $S = 25$ kHz, using a Hamming window?

**(d)** Bandstop: $f_p = [2, 12]$ kHz, $f_s = [4, 8]$ kHz, $S = 25$ kHz, using a Hamming window?

[**Hints and Suggestions:** This requires table look-up and the digital transition width. Round up the length to the next highest integer (an odd integer for bandstop filters). ]

10.13. **(Half-Band FIR Filter Design)** A lowpass half-band FIR filter is to be designed using a von Hann window. Assume a filter length $N = 11$ and find its windowed, causal impulse-response sequence and the transfer function $H(z)$ of the causal filter.

10.14. **(Half-Band FIR Filter Design)** Design the following half-band FIR filters, using a Kaiser window.

**(a)** Lowpass filter: 3-dB frequency 4 kHz, stopband edge 8 kHz, and $A_s = 40$ dB.

**(b)** Highpass filter: 3-dB frequency 6 kHz, stopband edge 3 kHz, and $A_s = 50$ dB.

**(c)** Bandpass filter: passband edges at [2, 3] kHz, stopband edges at [1, 4] kHz, $A_p = 1$ dB, and $A_s = 35$ dB.

**(d)** Bandstop filter: stopband edges at [2, 3] kHz, passband edges at [1, 4] kHz, $A_p = 1$ dB, and $A_s = 35$ dB.

[**Hints and Suggestions:** For (a) through (b), pick $S = 2(f_p + f_s)$ and $F_C = 0.25$. For (b), design a Kaiser lowpass prototype $h[n]$ to get $h_{\text{HP}}[n] = (-1)^n h[n]$. For (c) through (d), design a lowpass prototype $h[n]$ with $S = 4f_0$ and $F_C = 0.5(F_p + F_s)$. Then, $h_{\text{BP}}[n] = 2\cos(0.5n\pi)h[n]$ and $h_{\text{BS}}[n] = \delta[n] - 2\cos(0.5n\pi)h[n]$. You will need to compute the filter length $N$ (an odd integer), the Kaiser parameter $\beta$, and values of the $N$-sample Kaiser window for each part.]

10.15. **(Frequency-Sampling FIR Filter Design)** Consider the frequency-sampling design of a lowpass FIR filter with $F_C = 0.25$.

**(a)** Sketch the gain of an ideal filter with $F_C = 0.25$. Pick eight samples over the range $0 \le F < 1$ and set up the sampled frequency response $H[k]$ of the filter assuming a real, causal $h[n]$.

**(b)** Compute the impulse response $h[n]$ for the filter.

**(c)** To reduce the overshoot, modify $H[3]$ and recompute $h[n]$.

[**Hints and Suggestions:** For (a), set $|H[k]| = \{1, 1, 1, 0, 0, 0, 1, 1\}$ and find $H[k] = |H[k]|e^{j\phi[k]}$ where $k = 0, 1, 2, \ldots, 7$, $H[k] = H^*[N-k]$ and $\phi[k] = -k\pi(N-1)/N$ for a real and causal $h[n]$. For (b), find $h[n]$ as the IDFT. For (c), pick $|H[3]| = 0.5$ and $H[3] = 0.5e^{j\phi[3]}$ and $H[5] = H^*[3]$.]

10.16. **(Maximally Flat FIR Filter Design)** Design a maximally flat lowpass FIR filter with normalized frequencies $F_p = 0.1$ and $F_s = 0.4$, and find its frequency response $H(F)$.

10.17. **(FIR Differentiators)** Find the impulse response of a digital FIR differentiator with

**(a)** $N = 6$, cutoff frequency $F_C = 0.4$, and no window.

**(b)** $N = 6$, cutoff frequency $F_C = 0.4$, and a Hamming window.

**(c)** $N = 5$, cutoff frequency $F_C = 0.5$, and a Hamming window.

[**Hints and Suggestions:** For (a) through (b), $N = 6$ and so $n = -2.5, -1.5, -0.5, 0.5, 1.5, 2.5$.]

**10.18. (FIR Hilbert Transformers)** Find the impulse response of an FIR Hilbert transformer with

**(a)** $N = 6$, cutoff frequency $F_C = 0.4$, and no window.
**(b)** $N = 6$, cutoff frequency $F_C = 0.4$, and a von Hann window.
**(c)** $N = 7$, cutoff frequency $F_C = 0.5$, and a von Hann window.

[**Hints and Suggestions:** For (a) through (b), $N = 6$ and so $n = -2.5, -1.5, -0.5, 0.5, 1.5, 2.5$.]

**10.19. (IIR Filters and Linear Phase)** Even though IIR filters cannot be designed with linear phase, it is possible to implement systems containing IIR filters to eliminate phase distortion. A signal $x[n]$ is flipped and passed through a filter $H(F)$, and the filter output is then flipped to obtain the signal $y_1[n]$. The signal $x[n]$ is also passed directly through the filter $H(F)$ to get the signal $y_2[n]$. The signals $y_1[n]$ and $y_2[n]$ are summed to obtain the overall output $y[n]$.

**(a)** Sketch a block diagram of this system.
**(b)** How is $Y(F)$ related to $X(F)$?
**(c)** Does the system provide freedom from phase distortion?

[**Hints and Suggestions:** If $h[n] \Leftrightarrow H(F)$, then $h[-n] \Leftrightarrow H(-F)$. Also, $H(F)+H(-F)$ is purely real.]

**10.20. (Filter Specifications)** A hi-fi audio signal band limited to 20 kHz is contaminated by high-frequency noise between 70 kHz and 110 kHz. We wish to design a digital filter that reduces the noise by a factor of 100, while causing no appreciable signal loss. One way is to design the filter at a sampling rate that exceeds the Nyquist rate. However, we can also make do with a smaller sampling rate that avoids aliasing of the noise spectrum into the signal spectrum. Pick such a sampling rate and develop the frequency and attenuation specifications for the digital filter.
[**Hints and Suggestions:** The image of the aliased noise spectrum does not overlap the signal spectrum as long as it lies between 20 kHz and 70 kHz.]

**10.21. (FIR Filter Specifications)** Figure P10.21 shows the magnitude and phase characteristics of a causal FIR filter designed at a sampling frequency of 10 kHz.

**FIGURE P.10.21**
Filter characteristics for Problem 10.21

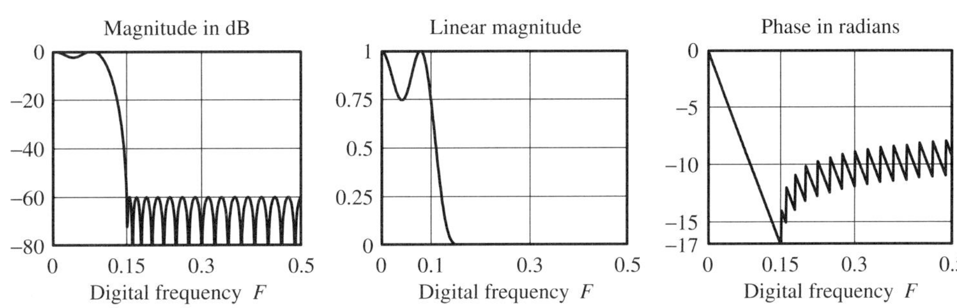


**(a)** What are the values of the passband ripple $\delta_p$ and stopband ripple $\delta_s$?
**(b)** What are the values of the attenuation $A_s$ and $A_p$ in decibels?

**(c)** What are the frequencies (in Hz) of the passband edge and stopband edge?
**(d)** Does this filter show linear phase? What is the group delay?
**(e)** What is the filter length $N$?
**(f)** Could this filter have been designed using the window method? Explain.
**(g)** Could this filter have been designed using the optimal method? Explain.

[**Hints and Suggestions:** The dB magnitude gives $\delta_s$ and $F_s$. The linear magnitude gives $\delta_p$ and $F_p$. The slope of the phase plot gives the delay as $D = -\frac{\Delta\theta}{2\pi\,\Delta F}$ (an integer). For part (e), $N = 2D + 1$. The size of the stopband and passband ripples provides a clue to parts (f) through (g).]

10.22. **(Multistage Interpolation)** It is required to design a three-stage interpolator. The interpolating filters are to be designed with identical passband and stopband attenuation and are required to provide an overall attenuation of no more than 3 dB in the passband and at least 50 dB in the stopband. Specify the passband and stopband attenuation of each filter.

## Computation and Design

10.23. **(FIR Filter Design)** It is desired to reduce the frequency content of a hi-fi audio signal band limited to 20 kHz and sampled at 44.1 kHz for purposes of AM transmission. Only frequencies up to 10 kHz are of interest. Frequencies past 15 kHz are to be attenuated by at least 55 dB, and the passband loss is to be less than 10%. Design a digital filter using the Kaiser window that meets these specifications.

10.24. **(FIR Filter Design)** It is desired to eliminate 60-Hz interference from an ECG signal whose significant frequencies extend to 35 Hz.

**(a)** What is the minimum sampling frequency we can use to avoid in-band aliasing?
**(b)** If the 60-Hz interference is to be suppressed by a factor of at least 100 (with no appreciable signal loss) what should be the filter specifications?
**(c)** Design the filter using a Hamming window and plot its frequency response.
**(d)** Test your filter on the signal $x(t) = \cos(40\pi t) + \cos(70\pi t) + \cos(120\pi t)$. Plot and compare the frequency response of the sampled test signal and the filtered signal to confirm that your design objectives are met.

10.25. **(Digital Filter Design)** We wish to design a lowpass filter for processing speech signals. The specifications call for a passband of 4 kHz and a stopband of 5 kHz. The passband attenuation is to be less than 1 dB, and the stopband gain is to be less than 0.01. The sampling frequency is 40 kHz.

**(a)** Design FIR filters using the window method (with Hamming and Kaiser windows) and using optimal design. Which of these filters has the minimum length?
**(b)** Design IIR Butterworth and elliptic filters using the bilinear transformation to meet the same set of specifications. Which of these filters has the minimum order? Which has the best delay characteristics?
**(c)** How does the complexity of the IIR filters compare with that of the FIR filters designed with the same specifications? What are the trade-offs in using an IIR filter over an FIR filter?

10.26. **(The Effect of Group Delay)** The nonlinear phase of IIR filters is responsible for signal distortion. Consider a lowpass filter with a 1-dB passband edge at $f_p = 1$ kHz, a 50-dB stopband edge at $f_p = 2$ kHz, and a sampling frequency of $S = 10$ kHz.

**(a)** Design a Butterworth filter $H_B(z)$, an elliptic filter $H_E(z)$, and an optimal FIR filter $H_O(z)$ to meet these specifications. Using the MATLAB routine **grpdelay** (or otherwise), compute and plot the group delay of each filter. Which filter has the best (most nearly constant) group delay in the passband? Which filter would cause the least phase distortion in the passband? What are the group delays $N_B,\ N_E$, and $N_O$ (expressed as the number of samples) of the three filters?

**(b)** Generate the signal $x[n] = 3\sin(0.03n\pi) + \sin(0.09n\pi) + 0.6\sin(0.15n\pi)$ over $0 \le n \le 100$. Use the MATLAB routine **filter** to compute the response $y_B[n],\ y_E[n]$, and $y_O[n]$ of each filter. Plot the filter outputs $y_B[n],\ y_E[n]$, and $y_O[n]$ (delayed by $N_B,\ N_E$, and $N_O$, respectively) and the input $x[n]$ on the same plot to compare results. Which filter results in the smallest signal distortion?

**(c)** Are all the frequency components of the input signal in the filter passband? If so, how can you justify that what you observe as distortion is actually the result of the non-constant group delay and not the filter attenuation in the passband?

10.27. **(Raised Cosine Filters)** The impulse response of a raised cosine filter has the form

$$h_R[n] = h[n]\frac{\cos(2n\pi RF_C)}{1-(4nRF_C)^2}$$

where the *roll-off factor* $R$ satisfies $0 < R < 1$ and $h[n] = 2F_C\,\text{sinc}(2nF_C)$ is the impulse response of an ideal lowpass filter.

**(a)** Let $F_C = 0.2$. Generate the impulse response of an ideal lowpass filter with length 21 and the impulse response of the corresponding raised cosine filter with $R = 0.2,\ 0.5,\ 0.9$. Plot the magnitude spectra of each filter over $0 \le F \le 1$ on the same plot, using linear and decibel scales. How does the response in the passband and stopband of the raised-cosine filter differ from that of the ideal filter? How does the transition width and peak sidelobe attenuation of the raised-cosine filter compare with that of the ideal filter for different values of $R$? What is the effect of increasing $R$ on the frequency response?

**(b)** Compare the frequency response of $h_R[n]$ with that of an ideal lowpass filter with $F_C = 0.25$. Is the raised-cosine filter related to this ideal filter?

10.28. **(Interpolation)** The signal $x[n] = \cos(2\pi F_0 n)$ is to be interpolated by 5, using up-sampling followed by lowpass filtering. Let $F_0 = 0.4$.

**(a)** Generate and plot 20 samples of $x[n]$ and up-sample by 5.

**(b)** What must be the cutoff frequency $F_C$ and gain $A$ of a lowpass filter that follows the up-sampler to produce the interpolated output?

**(c)** Design an FIR filter (using the window method or optimal design) to meet these specifications.

**(d)** Filter the up-sampled signal through this filter and plot the result.

**(e)** Is the filter output an interpolated version of the input signal? Do the peak amplitude of the interpolated signal and original signal match? Should they? Explain.

10.29. **(Multistage Interpolation)** To relax the design requirements for the analog reconstruction filter, many compact disc systems employ oversampling during the DSP stages. Assume that audio signals are band limited to 20 kHz and sampled at 44.1 kHz. Assume a maximum passband attenuation of 1 dB and a minimum stopband attenuation of 50 dB.

**(a)** Design a single-stage optimal interpolating filter that increases the sampling rate to 176.4 kHz.

**(b)** Design multistage optimal interpolating filters that increase the sampling rate to 176.4 kHz.

**(c)** Which of the two designs would you recommend?

**(d)** For each design, explain how you might incorporate compensating filters during the DSP stage to offset the effects of the sinc distortion caused by the zero-order-hold reconstruction device.

10.30. **(Multistage Interpolation)** The sampling rate of a speech signal band limited to 3.4 kHz and sampled at 8 kHz is to be increased to 48 kHz. Design three different schemes that will achieve this rate increase and compare their performance. Use optimal FIR filter design where required and assume a maximum passband attenuation of 1 dB and a minimum stopband attenuation of 45 dB.

10.31. **(Multistage Decimation)** The sampling rate of a speech signal band limited to 3.4 kHz and sampled at 48 kHz is to be decreased to 8 kHz. Design three different schemes that will achieve this rate decrease and compare their performance. Use optimal FIR filter design where required and assume a maximum passband attenuation of 1 dB and a minimum stopband attenuation of 45 dB. How do these filters compare with the filters designed for multistage interpolation in the previous problem?

10.32. **(Filtering Concepts)** This problem deals with time-frequency plots of a combination of sinusoids and their filtered versions.

**(a)** Generate 600 samples of the signal $x[n] = \cos(0.1n\pi) + \cos(0.4n\pi) + \cos(0.7n\pi)$ comprising the sum of three pure cosines at $F = 0.05,\ 0.2,\ 0.35$. Use the MATLAB command **fft** to plot its DFT magnitude. Use the routine **timefreq** (from the author's website) to display its time-frequency plot. What do the plots reveal? Now design an optimal lowpass filter with a 1-dB passband edge at $F = 0.1$ and a 50-dB stopband edge at $F = 0.15$ and filter $x[n]$ through this filter to obtain the filtered signal $x_f[n]$. Plot its DFT magnitude and display its time-frequency plot. What do the plots reveal? Does the filter perform its function? Plot $x_f[n]$ over a length that enables you to identify its period. Does the period of $x_f[n]$ match your expectations?

**(b)** Generate 200 samples each of the three signals $y_1[n] = \cos(0.1n\pi)$, $y_2[n] = \cos(0.4n\pi)$, and $y_3[n] = \cos(0.7n\pi)$. Concatenate them to form the 600-sample signal $y[n] = \{y_1[n],\ y_2[n],\ y_3[n]\}$. Plot its DFT magnitude and display its time-frequency plot. What do the plots reveal? In what way does the DFT magnitude plot differ from part (a)? In what way does the time-frequency plot differ from part (a)? Use the optimal lowpass filter designed in part (a) to filter $y[n]$, obtain the filtered signal $y_f[n]$, plot its DFT magnitude, and display its time-frequency plot. What do the plots reveal? In what way does the DFT magnitude plot differ from part (a)? In what way does the time-frequency plot differ from part (a)? Does the filter perform its function? Plot $y_f[n]$ over a length that enables you to identify its period. Does the period of $y_f[n]$ match your expectations?

10.33. **(Decoding a Mystery Message)** During transmission, a message signal gets contaminated by a low-frequency signal and high-frequency noise. The message can be decoded only by displaying it in the time domain. The contaminated signal $x[n]$ is provided on the author's website as **mystery1**. Load this signal into MATLAB (use the command **load mystery1**). In an effort to decode the message, try the following methods and determine what the decoded message says.

**(a)** Display the contaminated signal. Can you "read" the message? Display the DFT of the signal to identify the range of the message spectrum.

**(b)** Design an optimal FIR bandpass filter capable of extracting the message spectrum. Filter the contaminated signal and display the filtered signal to decode the message. Use both the **filter** (linear-phase filtering) and **filtfilt** (zero-phase filtering) commands.

**(c)** As an alternative method, first zero out the DFT component corresponding to the low-frequency contamination and obtain its IDFT $y[n]$. Next, design an optimal lowpass FIR filter to reject the high-frequency noise. Filter the signal $y[n]$ and display the filtered signal to decode the message. Use both the **filter** and **filtfilt** commands.

**(d)** Which of the two methods allows better visual detection of the message? Which of the two filtering routines (in each method) allows better visual detection of the message?

10.34. **(Filtering of a Chirp Signal)** This problem deals with time-frequency plots of a chirp signal and its filtered versions.

**(a)** Generate 500 samples of a chirp signal $x[n]$ whose frequency varies from $F = 0$ to $F = 0.12$. Then, compute its DFT and plot the the DFT magnitude. Use the routine **timefreq** (from the author's website) to display its time-frequency plot. What do the plots reveal? Plot $x[n]$ and confirm that its frequency is increasing with time.

**(b)** Design an optimal lowpass filter with a 1-dB passband edge at $F = 0.04$ and a 40-dB stopband edge at $F = 0.1$ and use the MATLAB command **filtfilt** to obtain the zero-phase filtered signal $y_1[n]$. Plot its DFT magnitude and display its time-frequency plot. What do the plots reveal? Plot $y_1[n]$ and $x[n]$ on the same plot and compare. Does the filter perform its function?

**(c)** Design an optimal highpass filter with a 1-dB passband edge at $F = 0.06$ and a 40-dB stopband edge at $F = 0.01$ and use the MATLAB command **filtfilt** to obtain the zero-phase filtered signal $y_2[n]$. Plot its DFT magnitude and display its time-frequency plot. What do the plots reveal? Plot $y_2[n]$ and $x[n]$ on the same plot and compare. Does the filter perform its function?

10.35. **(Filtering of a Chirp Signal)** This problem deals with time-frequency plots of a chirp signal and a sinusoid and its filtered versions.

**(a)** Generate 500 samples of a signal $x[n]$ that consists of the sum of $\cos(0.6n\pi)$ and a chirp whose frequency varies from $F = 0$ to $F = 0.05$. Then, compute its DFT and plot the DFT magnitude. Use the routine **psdwelch** (from the author's website) to display its power spectral-density plot. Use the routine **timefreq** (from the author's website) to display its time-frequency plot. What do the plots reveal?

**(b)** Design an optimal lowpass filter with a 1-dB passband edge at $F = 0.08$ and a 40-dB stopband edge at $F = 0.25$ and use the MATLAB command **filtfilt** to obtain the zero-phase filtered signal $y_1[n]$. Plot its DFT magnitude and display its PSD and time-frequency plot. What do the plots reveal? Plot $y_1[n]$. Does it look like a signal whose frequency is increasing with time? Do the results confirm that the filter performs its function?

**(c)** Design an optimal highpass filter with a 1-dB passband edge at $F = 0.25$ and a 40-dB stopband edge at $F = 0.08$ and use the MATLAB command **filtfilt** to obtain the zero-phase filtered signal $y_2[n]$. Plot its DFT magnitude and display its PSD and time-frequency plot. What do the plots reveal? Plot $y_2[n]$. Does it look like a sinusoid? Can you identify its period from the plot? Do the results confirm that the filter performs its function?

10.36. **(A Multiband Filter)** A requirement exists for a multiband digital FIR filter operating at 140 Hz with the following specifications:

**Passband 1:** from dc to 5 Hz

Maximum passband attenuation = 2 dB (from peak)
Minimum attenuation at 10 Hz = 40 dB (from peak)

**Passband 2:** from 30 Hz to 40 Hz

Maximum passband attenuation = 2 dB (from peak)
Minimum attenuation at 20 Hz and 50 Hz = 40 dB (from peak)

**(a)** Design the first stage as an odd-length optimal filter, using the routine **firpm** (from the author's website).
**(b)** Design the second stage as an odd-length half-band filter, using the routine **firhb** (from the author's website) with a Kaiser window.
**(c)** Combine the two stages to obtain the impulse response of the overall filter.
**(d)** Plot the overall response of the filter. Verify that the attenuation specifications are met at each design frequency.

10.37. **(Audio Equalizers)** Many hi-fi systems are equipped with graphic equalizers to tailor the frequency response. Consider the design of a four-band graphic equalizer. The first section is to be a lowpass filter with a passband edge of 300 Hz. The next two sections are bandpass filters with passband edges of [300, 1000] Hz and [1000, 3000] Hz, respectively. The fourth section is a highpass filter with a passband edge at 3 kHz. The sampling rate is to be 20 kHz. Implement this equalizer, using FIR filters based on window design. Repeat the design using an optimal FIR filter. Repeat the design, using an IIR filter based on the bilinear transformation. For each design, superimpose plots of the frequency response of each section and their parallel combination. What are the differences between IIR and FIR design? Which design would you recommend?

CHAPTER 11

# MATLAB Examples

## 11.0 Introduction

In this last chapter, we provide examples of MATLAB code illustrating the concepts covered in the book. The code is easy to customize for use in other similar examples. You may wish to consult the extensive help facility within MATLAB to learn more about the commands used.

### 11.0.1 MATLAB Tips and Pointers

We assume you have a passing familiarity with MATLAB. However, the following tips and pointers may prove useful if you are new to MATLAB and want to get started.

1. MATLAB works in double precision. It is case sensitive. All built-in commands are in lowercase.
2. Use **format long** to see results in double precision, and **format** to get back to the default display.
3. In MATLAB, the indexing starts at 1 (not 0).
4. A semicolon (;) after a statement or command suppresses on-screen display of results.
5. Comments (which follow the % sign) are not executed.
6. To abort an executing MATLAB command or on-screen data display, use Ctrl-C.
7. Use the up/down "arrow keys" to scroll through commands and re-edit/execute them.
8. To exit MATLAB, type **quit** at the MATLAB prompt.

### 11.0.2 Array Operations

Most operations you will perform will be element-wise operations on arrays. Be aware of the following:

```
>> x=0:0.01:2;     % Generate an array x at intervals of 0.01
>> y=x .^ 2;       % Generate array of squared values; uses " .^ "
>> z=x.';          % Transpose x from a row to column or vice versa; uses " .' "
>> x.*y;           % Pointwise product of equal size arrays x and y; uses " .* "
>> x./y;           % Pointwise division of equal size arrays x and y; uses " ./ "
>> l=length(x)     % Returns the length of an array or vector x
>> p=x(3)          % Third element of x (MATLAB indexing starts with 1)
```

### 11.0.3 A List of Useful Commands

Some of the MATLAB commands for mathematical operations are **exp, sin, cos, tan, acos, asin, atan, pi** (for $\pi$), **eps** (for default tolerance = $(2.22)10^{-16}$), **i** or **j** (for $\sqrt{-1}$), **real, imag, sqrt** (for square root), **abs** (for absolute value or magnitude), **angle** (for the angle or phase in *radians*), **log2** and **log10** (for logarithms to the base 2 or 10), **length** (of a vector or array), and **size** (of a matrix). Other useful commands are **min, max, int, fix, find, conj, sum, cumsum, diff, prod, eval, fliplr**, and **sign**. Use the MATLAB help facility to learn more about them.

#### String Functions and Logical Operations

A string function (symbolic expression) is entered in single quotes, for example **x = '4*t.*exp(-2*t)'**. MATLAB displays it on screen without quotes. To use its results, or plot the expression, it must first be evaluated using **eval**. For example:

```
>> x = '4*t.*exp(-2*t)'   % String or symbolic expression
>> t=0:0.05:4;            % Pick a time range t
>> xe = eval(x);          % Evaluate the expression x
>> plot(t,xe)             % Plot
```

Logical operations are useful in many situations. Here are some examples:

```
>> i=find(xe>=0.7);       % Index of values in xe that equal or exceed 0.7
>> k=find(xe==max(xe));   % Index of maximum in xe. Note the ==
>> tm=t(k);               % Time of maximum
```

#### Frequently Asked Questions

**Q.** How many data points do I need to obtain a smooth curve when plotting analog signals?
**A.** A good rule of thumb is at least 200 points.

**Q.** Why do I get an error while using **plot(t,x)**, with **x** and **t** previously defined?
**A.** Both **t** and **x** must have the same dimensions (size). To check, use **size(t)** and **size(x)**.

**Q.** Why do I get an error while generating $x(t) = 2^{-t}\sin(2t)$, with the array **t** previously defined?
**A.** You probably did not use array operations (such as **.*** or **.^**). Try **>> x=(2 .^(-t)).*sin(2*t)**.

**Q.** How can I plot several graphs on the same plot?
**A.** Example: use **>>plot(t,x),hold on,plot(t1,x1),hold off** OR **>>plot(t,x,t1,x1)**.

**Q.** How do I use **subplot** to generate six plots in the same window (two rows of three subplots)?
**A.** Example: use **>>subplot(2,3,n);plot(t,x)** for the *n*th (count by rows) plot (**n** cycles from 1 to 6). A subplot command *precedes* each plot command and has its own xlabel, ylabel, axis, title, etc.

## 11.1 Examples of MATLAB Code

This extended section provides examples of MATLAB *script files*. In all examples of MATLAB usage presented here, note that

1. The MATLAB statements are *case sensitive* and entered at the prompt.
2. Comments following the % sign are not executed and need not be typed.
3. Sequences will be assumed to start at $n = 0$ unless specified otherwise.

To save space, we have entered several MATLAB statements on the same line (separated by commas to display results or by semicolons to suppress the display). MATLAB statements for titles, axis labels, axis limits, and other embellishments for plotting have also been omitted for the most part. The plots have not been reproduced.

**Example 11.1**

### Plotting Discrete Signals

Plot $\delta(n-3)$, $u[n-3]$, $r[n-3]$, $\text{sinc}(n/4)$ and $4(0.8)^n \cos(0.2n\pi)u[n]$ over the range $-10 \le n \le 10$.

Here are the MATLAB statements with remarks:

```
>> n=-10:10;                                   % discrete integer index (-10 to 10)
>> xa=(n==3); subplot(3,2,1),stem(n,xa)        % δ[n-3].
>> xb=(n>=3); subplot(3,2,2), stem(n,xb)       % Step u[n-3].
>> xc=(n-3).*(n>=3); subplot(3,2,3),stem(n,xc) % Ramp r[n-3]
>> xd=sinc(n/4); subplot(3,2,4),stem(n,xd)     % The sinc.
>> xe=4*(0.8 .^ n).*cos(0.2*n*pi).*(n>=0);     % Damped cosine.
>> subplot(3,2,5), stem(n,xe)                  %
```

**Example 11.2**

### Signal Measures

Let $x[n] = r[n] - r[n-5] - 5u[n-10]$.

**(a)** Sketch $x[n]$, $x[n+2]$, $x[-n]$, $x_e[n]$, and $x_o[n]$.
**(b)** Find the signal energy in $x[n]$
**(c)** Is $x[n]$ absolutely summable? square summable?
**(d)** Sketch the periodic extension of $x[n]$ with period $N = 7$ and find its signal power.

Here are the MATLAB statements:

```
>> n=-15:15;                                    % Symmetric discrete index
>> x=n.*(n>=0)-(n-5).*(n>=5)-5*(n>=10);         % Generate x[n]
>> subplot(3,2,1),stem(n,x)                     % Plot.
>> xd= (n+2).*(n>=-2)-(n-3).*(n>=3)-5*(n>=8);   % n ⇒ n+2
>> subplot(3,2,2),stem(n,xd)                    % Plot.
>> xf=fliplr(x); subplot(3,2,3),stem(n,xf)      % x[-n].
>> xe=0.5*(x+xf);subplot(3,2,4), stem(n,xe)     % Even part of x[n].
>> xo=0.5*(x-xf);subplot(3,2,5), stem(n,xo)     % Odd part of x[n].
>> sa=sum(abs(x)); e=sum(x.*x);                 % Absolute sum and energy.
>> nx=length(x);r=rem(nx,7);m=fix(nx/7);        % m one-period sections
>> xpe=x(1:7);for j=1:m-1                       % Add one-period sections
>> xpe=xpe+x(7*j+1:7*j+7);                      % to get periodic extension
>> end                                          %
```

```
>> if r>0,xpe(1:r)=xpe(1:r)+x(nx-r+1:nx);end    % Add any leftovers
>> subplot(3,2,6),stem(0:20,[xpe xpe xpe]);      % 3 periods.
>> pwr=sum(xpe.*xpe)/7                           % Display power in xpe[n]
```

**Example 11.3**

### Random Distributions

Recall that random signals whose precise values cannot be predicted in advance are characterized by their probability distribution and density functions. In a uniform distribution, every value is equally likely, and the density function is just a rectangular pulse. The Gaussian probability density function is bell shaped. The **mean** and **variance** of a random signal are typically estimated directly from the observations $x_k$, $k = 0, 1, 2, \ldots, N-1$ as

$$m_x = \frac{1}{N}\sum_{k=0}^{N-1} x_k \qquad \sigma_x^2 = \frac{1}{N-1}\sum_{k=0}^{N-1}(x_k - m_x)^2$$

The mean is a measure of where the distribution is *centered*. The **variance** is a measure of the spread of the distribution about its mean. It is also a measure of the ac signal power. In practice, a probability distribution is often estimated as a series of values $f_k$ by constructing a **histogram** from a large number of observations. A histogram is a bar graph of the number of observations falling within specified amplitude levels (or *bins*).

Generate about 500 points each of a uniform and Gaussian random signal.

**(a)** Plot their first 100 values.
**(b)** Plot their histograms using 20 bins.
**(c)** Compute their mean and variance.

Here are the MATLAB statements:

```
>> n=0:499;p=length(n);                       % discrete index and length
>> xu=rand(size(n));                          % Uniform (mean=0.5, var=0)
>> xr=randn(size(n));                         % Gaussian (mean=0, var=1)
>> subplot(2,2,1),plot(n(1:100),xu(1:100)),   % (a) Plot uniform
>> subplot(2,2,2),plot(n(1:100),xr(1:100)),   % (b) Plot Gaussian
>> subplot(2,2,3),hist(xu,20);                % (c) Histogram of uniform
>> subplot(2,2,4),hist(xr,20);                % (d) Histogram of Gaussian
>> mu=sum(xu)/p,mr=sum(xr)/p                  % Display means
>> vu=sum((xu-mu).*(xu-mu))/(p-1)             % And variance of uniform
>> vr=sum((xr-mr).*(xr-mr))/(p-1)             % And variance of Gaussian
```

**Example 11.4**

### The Central Limit Theorem

The probability distribution of combinations of statistically independent, random phenomena often tends to a Gaussian distribution. This is the **central limit theorem**. Demonstrate the central limit theorem by generating five realizations of a uniformly distributed random signal and plotting the histogram of the individual signals and their sum.

Here are the MATLAB statements:

```
>> n=0:499;p=length(n);                       % discrete index and length
>> x1=rand(size(n));x2=rand(size(n));         % Uniform signals x1 and x2
>> x3=rand(size(n));x4=rand(size(n));         % And x3 and x4
>> x5=rand(size(n));                          % And x5. All have mean=0.5
>> x=x1+x2+x3+x4+x5;                          % Their sum x
>> subplot(2,2,1),plot(n(1:100),x1(1:100)),   % Plot x1
>> subplot(2,2,2),plot(n(1:100),x(1:100)),    % (b) Plot sum x
>> subplot(2,2,3),hist(x1,20);                % (c) Histogram of x1
>> subplot(2,2,4),hist(x,20);                 % (d) Histogram of sum x
>> m1=sum(x1)/p, m=sum(x)/p                   % Display means of x1 and x
>> v1=sum((x1-m1).*(x1-m1))/(p-1)             % And variance of x1
>> v=sum((x-m).*(x-m))/(p-1)                  % And variance of sum x
```

*Comment:* The mean of the sum equals the sum of the individual means.

**Example 11.5**

### Signal-to-Noise Ratio

For a noisy signal $x(t) = s(t) + An(t)$ with a signal component $s(t)$ and a noise component $An(t)$, the **signal-to-noise ratio** (SNR) is the ratio of the signal power $\sigma_s^2$ and noise power $A^2\sigma_n^2$ and defined in **decibels** (dB) as SNR=$10\log\frac{\sigma_s^2}{A^2\sigma_n^2}$ dB. We can adjust the SNR by varying the noise amplitude $A$. Use this result to generate a noisy sinusoid with a signal-to-noise ratio of 18 dB.

From the SNR result, $A = \dfrac{\sigma_s/\sigma_n}{10^{\mathrm{SNR}/20}}$.

We use the MATLAB routine **std** to compute $\sigma$ and the following commands:

```
>> t=0:0.01:2;                       % Time array
>> s=sin(2*pi*t);                    % Signal
>> n=randn(size(s));                 % Gaussian noise
>> snr=18;                           % Desired SNR
>> a=std(s)/std(n)/(10^(snr/20));    % Compute a
>> x=s+a*n;                          % Generate noisy signal
```

**Example 11.6**

### Signal Averaging

Extraction of signals from noise is an important signal-processing application. **Coherent signal averaging** assumes that (*i*) the experiment can be repeated and (*ii*) the noise corrupting the signal is random (and uncorrelated). Averaging the results of many runs tends to average out the noise to zero and the signal quality (or signal-to-noise ratio) improves.

**(a)** Sample $x = \sin(40\pi t)$ at 1000 Hz for 0.2 s to obtain the discrete signal $x[n]$.
**(b)** Generate 16 runs (*realizations*) of a noisy signal by adding uniformly distributed random noise (with zero mean) to $x[n]$ and average the results.

**(c)** Repeat Part(b) for 64 runs and compare results.
**(d)** Does averaging improve the quality of the noisy signal?

We use the following MATLAB statements:

```
>> ts=0.001;t=0:ts:0.2;           % Time array
>> x=sin(40*pi*t);                % Pure sinusoid
>> xn=x+rand(size(x))-0.5;        % Signal+uniform noise (mean=0)
>> z=0; for n=1:64;               % Initialize z and start loop
>> z=z+x+rand(size(x))-0.5;       % Keep summing each run
>> if n==16, a16=z/16; end        % Save average of 16 runs
>> end                            % End of main loop
>> a64=z/64;                      % Average of 64 runs
>> subplot(2,2,1),plot(t,x)       % (a) Pure signal
>> subplot(2,2,2),plot(t,xn)      % (b) Typical noisy signal
>> subplot(2,2,3),plot(t,a16)     % (c) Averaged signal 16 runs
>> subplot(2,2,4),plot(t,a64)     % (d) Averaged signal 64 runs
```

*Comments:*

1. We subtract 0.5 from *rand* because it returns results with a mean of 0.5.
2. Signal averaging is not very attractive because it requires *time coherence* (time alignment of the pure signal for each run) and because the noise distribution can affect the results of averaging.
3. On systems with sound capability, one could create a longer signal (2 s or so) at the natural sampling rate and listen to the pure, noisy, and averaged signal for audible differences using the MATLAB command *sound*.

**Example 11.7**

### Discrete System Response

Consider the second-order system $y[n]-0.64y[n-2]=x[n]+2x[n-1]$ with zero initial conditions and $x[n]=20(0.8)^n u[n]$.

**(a)** Find its response using **dlsim** and **filter** and compare the results.
**(b)** Is this system BIBO stable?

Here are the MATLAB statements to generate and compare the response:

```
>> n=0:60;                           % Create a discrete index
>> xin=20*(0.8 .^n); xin=xin(:);     % Create input as a vector
>> a=[1 0 -0.64]; b=[1 2 0];         % Create coefficient arrays
>> y1 = filter(b, a, xin);           % System Response to n=60 using filter
>> y2=dlsim(b,a,xin);                % OR using dlsim
>> subplot(2,1,1),stem(n,y1);        % Plot Response.
>> subplot(2,1,2),plot(n,y1-y2);     % Plot the difference.
>> r=abs(roots(a));                  % Abs value of Ch Eq roots
>> if any(r>=1),disp('Unstable')     % Display if unstable
>> else disp('Stable'),end           % or stable
```

*Comments:* The routine **dlsim** will not accept initial conditions in the difference equation format. With **filter**, the initial condition array **ic** must be *modified* and used as follows:

```
>> nic = filtic(b,a,ic); y = filter(b, a, xin, nic);        %
>> nic = fliplr(filter(-a(2:length(a)), 1, ic))/a(1);       % Method 2
```

**Example 11.8**

## Smoothing Effects of a Moving Average Filter

Consider the 20-point *moving average* filter $y[n] = \frac{1}{20}\{x[n] + x[n-1] + \cdots + x[n-19]\}$. It is also called a *smoothing filter* because it tends to smooth out the rapid variations in a signal. To confirm this, try the following:

**(a)** Generate 200 samples of a 1-Hz sine wave sampled at 40 Hz.
**(b)** Add some noise to generate a noisy signal.
**(c)** Filter the noisy signal through the 20-point MA filter.
**(d)** Plot each signal to display the effects of noise and smoothing.

Here are the MATLAB statements:

```
>> n=0:199;                                  % Create a discrete index
>> x=sin(2*pi*n*0.025); x=x(:);              % 1 Hz sine sampled at 40 Hz
>> xn=x+0.5*rand(size(x));                   % Add (uniform) random noise
>> a=[1 zeros(1,19)]; b=0.05*ones(1,20);     % Coefficient arrays of MA filter
>> y=filter(b, a, xn);                       % Response using filter
>> subplot(2,2,1),plot(n,x);                 % Plot input.
>> subplot(2,2,2),plot(n,xn);                % Plot noisy input.
>> subplot(2,2,3),plot(n,y);                 % Plot smoothed output.
```

**Example 11.9**

## Convolution and Convolution Indices

An input $x[n] = \{2, \overset{\Downarrow}{-1}, 3\}$ is applied to an FIR filter whose impulse response is given by $h[n] = \{1, 2, \overset{\Downarrow}{2}, 3\}$. The markers correspond to the index $n = 0$ (the origin). Find the response $y[n]$ and sketch all three signals using the same axis limits.

The starting indices are $n = -1$ for $x[n]$, $n = -2$ for $h[n]$, and thus, $n = -3$ for $y[n]$. Here are the MATLAB statements to evaluate and plot the results:

```
>> x=[2 -1 3];h=[1 2 2 3];                   % Signals x[n] and h[n]
>> y=conv(x,h)                               % Display convolution y[n]
>> nx=-1:1;nh=-2:1;                          % Indices of x[n] and y[n]
>> ns=nx(1)+nh(1);                           % Starting index of y[n]
>> ne=nx(length(nx))+nh(length(nh));         % Ending index of y[n]
>> ny=ns:ne;                                 % Indices of y[n]
>> amax=max([x h y]);amin=min([x h y]);      % Max and min values
>> ax=[ns ne amin amax];                     % Generate axis limits
>> subplot(3,1,1),stem(nx,x),axis(ax);       % (a) Signal x
```

```
>> subplot(3,1,2),stem(nh,h),axis(ax);    % (b) and h (Same limits)
>> subplot(3,1,3),stem(ny,y),axis(ax);    % (c) and y (Same limits)
>> s=sum(y)-sum(x)*sum(h)                 % Consistency check: s=0
```

**Example 11.10**

### Approximating Analytical Convolution

The impulse response of a digital filter is described by $h[n] = (0.4)^n u[n]$. Evaluate and plot the response $y[n]$ of this filter to the input $x[n] = (0.8)^n u[n]$ over the range $0 \le n \le 20$.

If $x[n]$ and $h[n]$ are defined over $0 \le n \le 20$, their convolution will extend over $0 \le n \le 40$ but match the analytical results only over $0 \le n \le 20$. So, we use the following MATLAB statements:

```
>> n=0:20;                              % discrete index
>> x=(0.8 .^ n);h=(0.4 .^ n);           % Signals x[n] and h[n]
>> y=conv(x,h);                         % Convolution y[n]
>> y=y(1:length(n));                    % Truncate y to length of x and h
>> subplot(2,2,1),stem(n,x)             % (a) Signal x
>> subplot(2,2,2),stem(n,h)             % (b) and h
>> subplot(2,2,3),stem(n,y)             % (c) and y (conv)
>> ya=(0.4 .^ n).*((2 .^ (n+1))-1);     % Exact result
>> subplot(2,2,4),plot(n,ya-y)          % Plot error.
>> maxerr=max(abs(ya-y))                % Display maximum error
```

**Example 11.11**

### System Response to Sinusoidal Inputs

We claim that the response of LTI systems to a sinusoidal input is a sinusoid at the input frequency. Justify this statement using an input $x[n] = \cos(0.2\pi n)$ to a digital filter whose impulse response is described by $h[n] = \{\overset{\Downarrow}{1}, 2, 3, 4, 5, 6, 7, 8\}$.

The period of $x[n] = \cos(0.2\pi n)$ is $N = 10$. To find and plot the convolution result, we generate $x[n]$ for, say, $0 \le n \le 50$, and use the following MATLAB commands:

```
>> n=0:50;                      % Create a discrete index
>> x=cos(0.2*pi*n);h=1:8;       % Signals x and h
>> y=conv(x,h);                 % Response
>> y=y(1:length(n));            % Truncate to length of x
>> subplot(2,1,1),stem(n,x)     % Plot input.
>> subplot(2,1,2),stem(n,y)     % Plot output.
```

*Comments:* From the plots, you should observe the periodic nature of the response *except for start-up transients*. You should also be able to discern the period of both the input and output as $N = 10$.

**Example 11.12**

### Convolution and Filtering

The convolution response is equivalent to the *zero-state* response obtained from the difference equation that describes the impulse response $h[n]$ and may be found using the MATLAB routine **filter**. Note that **filter** returns a result that equals the input length while the results of **conv** are longer. Over the input length, however, the results of **conv** and **filter** are identical.

The difference equation describing the digital filter of the previous example may be written as

$$y[n] = x[n] + 2x[n-1] + \ldots + 8x[n-7]$$

Use this to find the response to $x[n] = \cos(0.2\pi n)$ and compare with the previous example.

**Solution:** We use the routine **filter** and invoke the following MATLAB commands:

```
>> n=0:50;x=cos(0.2*pi*n);h=1:8;          % Signals x and h
>> a=[1 zeros(1,7)];                        % Filter coefficients
>> yf=filter(h,a,x);                        % Response using filter
>> lf=length(yf);                           % same length as x and h
>> yc=conv(x,h);                            % yc is longer than x and h
>> lc=length(yc);                           % Length of yc
>> yt=yc(1:length(n));err=yf-yt;            % So, truncate and find error
>> subplot(3,1,1),stem(0:lc-1,yc)           % (a) Plot yc.
>> ax=axis;                                 % Axis limits of longer result
>> subplot(3,1,2),stem(n,yf),axis(ax);      % Plot yf. Same limits as (a).
>> subplot(3,1,3),plot(n,err),axis(ax)      % Plot error.
```

**Example 11.13**

### Deconvolution or System Identification

If $y[n] = x[n] * h[n]$ and the response $y[n]$ and input $x[n]$ are known, we can identify the system impulse response $h[n]$ using **deconvolution**. Just as convolution is equivalent to polynomial multiplication, deconvolution is equivalent to polynomial division. For perfect results, the remainder (after division) should be zero.

Given $y[n] = \{\overset{\Downarrow}{3}, 9, 17, 21, 19, 13, 6, 2\}$ and $x[n] = \{\overset{\Downarrow}{3}, 3, 2, 2\}$, identify $h[n]$.

We use the routine **deconv** and invoke the following MATLAB commands:

```
>> y=[3 9 17 21 19 13 6 2];   % Output y
>> x=[3 3 2 2];                % Input x
>> [h,r]=deconv(y,x)           % Display impulse response and remainder
```

*Comment:* Deconvolution is quite sensitive to noise and much easier to implement in the frequency domain.

**Example 11.14**

## Periodic or Circular Convolution

Consider two periodic signals described over one period by

$$x_p[n] = \{\overset{\Downarrow}{1},\ 2,\ -1,\ 0,\ 2,\ 3\} \qquad h_p[n] = \{\overset{\Downarrow}{2},\ 1,\ 0,\ -1,\ -2,\ -3\}$$

Find their periodic convolution.

Here are the MATLAB statements:

```
>> x=[1 2 -1 0 2 3]; h=2:-1:-3;   % One period of x and h
>> N=length(x);                   % Length of x or h
>> y=[conv(x,h) 0];               % Regular conv to 2N samples
>> yw=y(1:N)+y(N+1:2*N)           % Display wraparound result
```

**Example 11.15**

## Normalization and Periodic Convolution

Let $x_p[n] = \{\overset{\Downarrow}{1},\ 2,\ -1,\ 0,\ 2,\ 3\}$ $\quad h_p[n] = \{\overset{\Downarrow}{2},\ 1,\ 0,\ -1,\ -2,\ -3\}$.

**(a)** Find the periodic convolution $y_1[n]$ using one period of $x$ and $h$.
**(b)** Find the periodic convolution $y_5[n]$ using 5 periods of $x$ and $h$.
**(c)** How is the period of $y_5[n]$ related to that of $y_1[n]$?
**(d)** How are the convolution values of $y_5[n]$ and $y_1[n]$ related?

Here are the statements using the wraparound method:

```
>> x=[1 2 -1 0 2 3];h=2:-1:-3;            % 1 period of x and h
>> y1=conv(x,h);N=length(x);              % Regular conv
>> y1=y1(1:N)+[y1(N+1:2*N-1) 0];          % Periodic conv y1[n]
>> x5=[x x x x x];h5=[h h h h h];         % 5 periods of x and h
>> y5=conv(x5,h5);M=length(x5);           % Regular conv
>> y5=y5(1:M)+[y5(M+1:2*M-1) 0];          % Periodic conv y5[n]
>> subplot(2,1,1),stem(0:M-1,y5),ax=axis; % (a) Plot y5.
>> subplot(2,1,2),stem(0:N-1,y1),axis(ax); % Plot y1. Same limits as (a)
>> scale=y5(1:N)./y1                      % Ratio over 1 period
```

*Comment:* We note that $y_5[n]$ is periodic with period $N$ and also 5 times longer than $y_1[n]$. Also, over one period $y_5[n] = 5y_1[n]$. Normalizing (dividing) $y_1[n]$ by $N$ or $y_5[n]$ by $5N$ produces *identical* results over one period.

**Example 11.16**

## Regular Convolution from Periodic Convolution

Let $x[n] = \{\overset{\Downarrow}{1},\ 2,\ -1,\ 0,\ 2\}$ and $h[n] = \{\overset{\Downarrow}{2},\ 1,\ 0,\ -1,\ -2,\ -3\}$. Find their regular convolution using zero padding and periodic convolution.

Here are the MATLAB statements using the wraparound method:

```
>> x=[1 2 -1 0 2]; h=2:-1:-3;      % Signals x and h
>> Nx=length(x);Nh=length(h);      % Signal lengths
>> Ny=Nx+Nh-1;                     % Convolution length
>> xz=zeros(1,Ny);hz=xz;           % Initialize zero-padded signals
>> xz(1:Nx)=x;hz(1:Nh)=h;          % Zero-padded x and h
>> N=length(xz);                   % Length of xz or hz
>> y=[conv(x,h) 0];                % Regular conv to 2N samples
>> yr=y(1:N)+y(N+1:2*N)            % Wraparound gives regular conv
```

**Example 11.17**

### Autocorrelation and Cross-correlation

Consider the sequences $x[n] = n,\ 0 \le n \le 8$ and $h[n] = n,\ 0 \le n \le 3$.

**(a)** Evaluate and plot $r_{xx}[n]$ and $r_{hh}[n]$ and find where they attain their maximum.
**(b)** Evaluate and plot $r_{xh}[n]$ and $r_{hx}[n]$.
**(c)** Evaluate and plot the correlation of $h[n]$ and $h[n-4]$ and find where it attains a maximum.

**Solution:** We use MATLAB as follows:

```
>> x=0:8;h=0:3;                            % Signals x and h
>> nx=0:8;nh=0:3;                          % discrete index for x and h
>> rxx=conv(x,fliplr(x));                  % Autocorrelation rxx
>> rhh=conv(h,fliplr(h));                  % Autocorrelation rhh
>> nhh=-3:3;nxx=-8:8;                      % discrete index for rhh and rxx
>> [mx, i1] =find(rxx==max(rxx))           % Max rxx and MATLAB index
>> [mh, i2] =find(rhh==max(rhh))           % Max rhh and MATLAB index
>> nxxm=nxx(i1),nhhm=nhh(i2)               % discrete index of max rxx and rhh
>> rxh=conv(x,fliplr(h));                  % Cross-correlation rxh
>> rhx=conv(h,fliplr(x));                  % Cross-correlation rhx
>> nxh=-3:8;nhx=-8:3;                      % discrete index for rxh and rhx
>> ax=[-8 8 0 inf];                        % Axis limits for all plots
>> subplot(3,2,1),stem(nxx,rxx),axis(ax);  % Plot rxx.
>> subplot(3,2,2),stem(nhh,rhh),axis(ax);  % Plot rhh.
>> subplot(3,2,3),stem(nxh,rxh),axis(ax);  % Plot rxh.
>> subplot(3,2,4),stem(nhx,rhx),axis(ax);  % Plot rhx.
>> h1=[0 0 0 0 h];n=0:7;                   % Shifted h and its discrete index
>> rhd=conv(h,fliplr(h1));                 % Correlation of h and h1
>> nhd=-7:3;                               % discrete index for rhh1
>> rhd2=conv(h1,fliplr(h));                % Correlation of h1 and h
>> nhd2=-3:7;                              % discrete index for rhh2
>> subplot(3,2,5),stem(nhd,rhd),axis(ax);  % Plot rhd.
>> subplot(3,2,6),stem(nhd2,rhd2),axis(ax); % Plot rhd2.
```

*Comments:* Note that the correlation $r_{hd}[n] = h[n] ** h[n-4]$ equals $r_{hh}[n-4]$. The delay is indicated by the location of the peak in $r_{hd}[n]$. This result is the basis for target ranging (in radar or ultrasound applications, for example). We send a sampled signal $x[n]$, which is reflected by the target (an airplane say). Ideally, the received signal is just a delayed version $x[n-n_0]$. The delay $n_0 t_s = 2Rc$ where $R$ is the target distance and $c$ is the propagation velocity. We correlate $x[n]$ and $x[n-n_0]$ and locate the peak. This location gives $n_0$. From $n_0$, we estimate the target range $R$.

---

**Example 11.18**

### Signals Buried in Noise

Correlation is an effective method of detecting signals buried in noise. Noise is essentially uncorrelated. This means that, if we correlate a noisy signal with itself, the correlation will be due only to the signal (if present). This will exhibit a sharp peak at $n = 0$.

The spectrum of the correlated noisy signal can be used to estimate the frequency of the signal.

Generate two noisy signals by adding noise to a 20-Hz sinusoid sampled at $t_s = 0.01$ s for 2 s.

**(a)** Verify the presence of the signal by correlating the two noisy signals.
**(b)** Estimate the frequency of the signal from the FFT spectrum of the correlation.

We use the following MATLAB statements:

```
>> ts=0.01;t=0:ts:2;N1=length(t);              % Set up time array
>> x=sin(2*20*pi*t);                           % Signal x
>> z1=2*randn(size(x));z2=2*randn(size(x));;   % Random noise
>> x1=x+z1;x2=x+z2;                            % Noisy signals
>> cc=conv(x1,fliplr(x2));                     % cross-correlation of x1 and x2
>> cn=conv(z1,fliplr(z2));                     % Correlation of noise
>> N=length(cn);                               % Length of corr
>> tc=-(N1-1):N1-1;f=(0:N-1)/N/ts;             % corr and FFT axis values
>> ccspec=abs(fft(cc))/N;                      % FFT of cross-correlation
>> cnspec=abs(fft(cn))/N;                      % FFT of noise
>> subplot(2,2,1),plot(tc,cc);ax=axis;         % Cross-correlation cc.
>> subplot(2,2,2),plot(tc,cn);axis(ax);        % Noise correlation cn.
>> subplot(2,2,3),plot(f,ccspec);ay=axis;      % spectrum of cc.
>> subplot(2,2,4),plot(f,cnspec);axis(ay);     % spectrum of cn.
```

---

**Example 11.19**

### Bandstop Filtering of a Chirp Signal

Frequency-domain filtering constitutes a very versatile application of signal processing. Digital filters can be designed either in hardware or software, and much of what DSP is all about can, in fact, be brought down to the level of filtering in one form or another!

A linear chirp signal is a signal whose frequency *varies* linearly with time. Such a *swept frequency* signal can give us a nice visual indication of filtering effects.

**(a)** Generate and plot a chirp whose frequency varies from $F = 0$ to $F = 0.5$ over 400 samples.
**(b)** Find and plot the response of a bandstop FIR filter whose impulse response is described by $h[n] = \delta[n] - 0.2\cos(0.5n\pi)\text{sinc}(0.1n)$ to this chirp signal.
**(c)** Do the plots reveal the bandstop nature of the filter?

We generate use the following MATLAB statements:

```
>> nx=0:399;                                      % Create discrete index for x
>> x=cos(pi*nx.*nx/800);                          % Chirp x from F=0 to F=0.5
>> n=-50:50;                                      % Index for bandstop filter
>> hbs=(n==0)-0.2*cos(0.5*n*pi).*sinc(0.1*n);     % BS filter coeffs
>> y=filter(hbs,1,x);                             % Filter response
>> subplot(2,1,1),plot(nx,x);ax=axis;             % Plot chirp signal.
>> subplot(2,1,2),plot(nx,y);axis(ax);            % Plot filter Response.
```

*Comments:*

1. Except for end effects, the response does reveal the bandstop nature of the filter.
2. On machines with sound capability, one could use the natural sampling rate to generate and listen to the chirp signal and the filter response for audible differences using **sound**.

**Example 11.20**

### The Gibbs Effect and Periodic Convolution

An ideal lowpass filter with cutoff frequency $F_C$ has the impulse response $h[n] = 2F_C\text{sinc}(2F_Cn)$. Symmetric truncation (about $n = 0$) leads to the **Gibbs effect** (overshoot and oscillations in the frequency response). Truncation of $h[n]$ to $|n| \leq N$ is equivalent to multiplying $h[n]$ by a rectangular **window** $w_D[n] = \text{rect}(n/2N)$. In the frequency domain, it is equivalent to the *periodic* convolution of $H(F)$ and $W_D(F)$ (the Dirichlet kernel). It is the *slowly* decaying sidelobes of $W_D(F)$, which lead to the overshoot and oscillations. To eliminate overshoot and reduce the oscillations, we choose a window whose DTFT has sidelobes that decay much faster. The triangular window (whose DTFT describes the Fejer kernel) is one familiar example.

Consider an ideal filter with $F_C = 0.2$ whose impulse response $h[n]$ is truncated over $-20 \leq n \leq 20$.

**(a)** Show that its periodic convolution with the Dirichlet kernel results in the Gibbs effect.
**(b)** Show that the Gibbs effect is absent in its periodic convolution with the Fejer kernel.

We choose $F_C = 0.2$, a truncated $h[n]$ over $-20 \leq n \leq 20$, and use the following commands:

```
>> n=-20:20;                            % Create a discrete index
>> Fc=0.2; h=2*Fc*sinc(2*Fc*n);         % Create impulse response
>> a=[zeros(1,20) 1 zeros(1,20)];       % Denominator array of DTFT
>> f=(-200:199)/400;w=2*pi*f;           % Freq array (F=-0.5 to F=-0.5)
>> HI=(abs(f)<=0.2);                    % Ideal Filter Spectrum, Fc = 0.2
>> HT=abs(freqz(h,a,w));                % DTFT of truncated ideal filter
>> f1=f+0.5;                            % Freq array (F=0 to F=1)
>> x=sinc(41*f1)./sinc(f1);Wd=41*x;     % Dirichlet kernel
>> x=sinc(20*f1)./sinc(f1);Wf=20*x.*x;  % Fejer kernel
>> y1=[conv(HI,Wd) 0];                  % 799 point conv and one zero
>> HD=(y1(1:400)+y1(401:800))/400;      % Periodic conv normalized by 400
>> y2=[conv(HI,Wf) 0];                  % 799 point conv and one zero
>> HF=(y2(1:400)+y2(401:800))/400;      % Periodic conv normalized by 400
>> subplot(2,2,1),plot(f,HI)            % (a) Spectrum of ideal filter
>> subplot(2,2,2),plot(f,HD)            % (b) and rect windowed h[n]
>> subplot(2,2,3),plot(f,HF)            % (c) and tri windowed h[n]
>> subplot(2,2,4),plot(f,HT)            % (d) DTFT of truncated h[n]
```

*Comment:* Since $h[n]$ is generated over $-0.5 \le F < 0.5$, one period of the periodic convolution will cover the same range only if the Dirichlet and Fejer kernels are generated over $0 \le F < 1$.

**Example 11.21**

### Signal Reconstruction and Resampling

Consider the signal $x(t) = \sin(2\pi ft)$ sampled at 0.25 s for one period to obtain the sampled signal $x[n] = \{\overset{\Downarrow}{0}, \ 1, \ 0, \ -1\}$. Sketch the signals that result from:

**(a)** A 16-fold interpolation of $x[n]$ using sinc, constant, and linear interpolation.
**(b)** Resampling $x[n]$ at 3.2 times the original sampling rate.

Since $3.2 = \frac{16}{5}$, the resampled signal of part(b) can be generated by decimating the 16-fold interpolated signal of part(a) by 5. We use the following MATLAB commands:

```
>> T=2;tc=0:0.01:T;xc=sin(2*pi*tc);     % Analog Signal
>> ts=0.25;L=round(T/ts);td=ts*(0:L-1); % Discrete array
>> xd=sin(2*pi*td);                     % Sampled signal
>> N=16;ti=ts/N;                        % N times finer interval
>> m=L*N;tn=(0:m-1)*ti;                 % Finer time axis
>> x1=0*tn;x1(1:N:m)=xd;                % Signal for finer axis
>> t=-ts*L:ti:ts*L-ti;h=sinc(t/ts);d=m;  % Sinc interpolator
 %NOTE: The above line generates code for the sinc interpolator
 %Replace it by one of the next two lines for constant or tri
>> %t=0:ti:ts-ti;h=0*t+1;d=0;           % Constant interpolator
>> %t=-ts:ti:ts-ti;h=tri(t/ts);d=N;      % Tri interpolator
>> y=conv(h,x1);                        % Interpolated signal of part(a)
```

```
>> xn=y(d+1:d+m);                    % Over limits of finer time axis
>> subplot(2,1,1),plot(tc,xc);       % Plot analog signal and sampled signal
>> hold on;                          %
>> stem(tn,xn);                      % Overplot interpolated signal xn
>> plot(td,xd,'*m');                 % And discrete samples.
>> hold off;                         %
>> M=5;xr=xn(1:M:m);                 % Decimate xn by 5 for (b)
>> tr=tn(1:M:m);                     % discrete array for resampled xr
>> subplot(2,1,2),stem(tr,xr);       % Plot xr.
```

*Comments:*

1. In practice, the interpolated signal would have been lowpass filtered before decimation.
2. In theory, perfect recovery is possible for a bandlimited signal using sinc interpolation. Does sinc interpolation confirm this for the sine wave? (No!) If not, why? (Because signal values outside the one period range are assumed to be zero during the interpolation).

**Example 11.22**

## Signal Quantization and SNR

The signal $x(t) = \sin(2\pi f t)$ is sampled at 0.01 s and 200 samples are obtained, starting at $t = 0$. The sampled signal $x[n]$ is quantized by *rounding* using a 3-bit quantizer to obtain the quantized signal $x_Q[n]$.

**(a)** Plot the CT signal $x(t)$ and the quantized signal $x_Q[n]$.
**(b)** Plot the histogram of the error signal $x[n] - x_Q[n]$.
**(c)** Find the quantization SNR.
**(d)** Find the statistical estimate of the quantization SNR.
**(e)** Find the SNR using the expression valid for sinusoids: SNR=$6B + 1.76$.

We use the following MATLAB commands. You should observe that the error signal (**er** in the MATLAB code) is more or less uniformly distributed by examining its histogram.

```
>> N=200;t=(0:N-1)*0.01;                  % Time array
>> x=sin(2*pi*t);                         % discrete sampled signal
>> d=max(x)-min(x);                       % Dynamic range
>> b=3;L=2^b;                             % Number of levels
>> s=d/L;                                 % Step size
>> y=s*round(x/s);                        % Quantized signal by rounding
>> % y=s*floor(x/s);                      % Use this for Truncation
>> % y=s*sign(x).*floor(abs(x/s));        % Or sign magnitude truncation
>> subplot(211),plot(t,x);                % Overplot the CT signal
>> hold on,stem(t,y);hold off             % And quantized signal.
>> er=x-y;                                % Error signal
>> subplot(212),hist(er,L);               % And its histogram.
>> snra=10*log10(sum(x.*x)/sum(er.*er));  % SNR estimate of (c)
```

```
>> sigpwr=10*log10(sum(x.*x)/length(x));      % Signal power
>> snr2=sigpwr+20*log10(L)-20*log10(d)+10.8   % SNR estimate of (d)
>> snre=6*b+1.76                              % SNR estimate of (e)
```

**Example 11.23**

### Convolution by the FFT

The FFT can be used to perform periodic and regular convolution. For periodic convolution, we multiply the FFT of $x_p[n]$ and $h_p[n]$ pointwise and take the inverse FFT of the product. For regular convolution, we first zero pad $x[n]$ and $h[n]$ to a length that equals the length of regular convolution before taking the FFT. Use the FFT to find:

**(a)** The periodic convolution of $x_p[n] = \{\overset{\Downarrow}{1}, 2, -1, 0, 2, 3\}$ and $h_p[n] = \{\overset{\Downarrow}{2}, 1, 0, -1, -2, -3\}$.

**(b)** The regular convolution of $x[n] = \{\overset{\Downarrow}{1}, 2, -1, 0, 2\}$ and $h[n] = \{\overset{\Downarrow}{2}, 1, 0, -1, -2, -3\}$.

Here are the MATLAB statements using the MATLAB routines **fft** and **ifft**:

```
>> xp=[1 2 -1 0 2 3]; hp=2:-1:-3;    % One period of xp and hp
>> N=length(xp);                     % Length of xp or hp
>> YP=fft(hp).*fft(xp);              % Product of FFT of x and h
>> yp=ifft(YP);yp=real(yp)           % Real result of Periodic Convolution
>> x=[1 2 -1 0 2]; h=2:-1:-3;        % Signals x and h
>> Nx=length(x);Nh=length(h);        % Signal lengths
>> Ny=Nx+Nh-1;                       % Convolution length
>> xz=zeros(1,Ny);hz=xz;             % Initialize zero-padded signals
>> xz(1:Nx)=x;hz(1:Nh)=h;            % Zero-padded x and h
>> Y=fft(xz).*fft(hz);               %
>> y=real(ifft(Y))                   % Display regular conv
```

*Comment:* For real sequences, **ifft** may show an imaginary part which is zero (to within machine roundoff).

**Example 11.24**

### IIR Filter Design by the Bilinear Transformation

Design a Butterworth lowpass filter using the bilinear transformation to meet the following specifications.

- Passband edge = 1 kHz
- Stopband edge = 1.5 kHz
- $S = 10$ kHz
- $A_p = 0.5$ dB
- $A_s = 25$ dB.

We use MATLAB routines for the direct design of IIR filters as follows:

```
>> fp=1;fs=1.5;S=10;SN=0.5S;            % Specifications
>> Fp=fp/SN;Fs=fs/SN;                   % MATLAB requires normalization by SN=0.5S
>> [N,Wn]=buttord(Fp, Fs, 0.5, 25)      % Digital lowpass prototype
>> [n,d]=butter(N,Wn)                   % Num and den of bilinear IIR lowpass filter
```

*Comments:* MATLAB routines for other classical filters include **cheb1ord, cheb2ord, ellipord** and **cheby1, cheby2, ellip** whose syntax is similar.

**Example 11.25**

### IIR Filter Design by Impulse Invariance

Design a Chebyshev bandpass filter using impulse invariance to meet the following specifications.

- Passband = [2 3] kHz
- Stopband edges = [1 6] kHz
- $S = 18$ kHz
- $A_p = 0.5$ dB
- $A_s = 25$ dB.

We use MATLAB routines for analog filter design followed by the impulse invariant transformation:

```
>> fs=[1000 6000];fp=[2000 3000];       % Specifications
>> S=18000;Fp=fp/S;Fs=fs/S;             % Normailze frequencies
>> fplpp=1;fslpp=diff(Fs)/diff(Fp);     % Specs for analog lowpass prototype
>> [N,Wn]=cheb1ord(1,fslpp,0.5,25,'s')  % Find order
>> [n,d]=cheby1(N,0.5,Wn,'s')           % Analog lowpass prototype
>> WC=2*pi*sqrt(prod(Fp));              % Normalized Center freq
>> BW=2*pi*abs(diff(Fp));               % and normalized Bandwidth
>> [nb,db]=lp2bp(n,d,WC,BW)             % Convert lowpass prototype to analog
                                          bandpass
>> [ni,di]=impinvar(nb,db,1)            % Convert to IIR with ts=1
```

*Comment:* MATLAB routines for other transformations include **lp2lp, lp2hp, lp2bs**. For other mappings, try the routines **c2d, c2dm**.

# Appendix A Useful Concepts from Analog Theory

## A.0 Scope and Objectives

This appendix collects useful concepts and results from the area of analog signals and systems that are relevant to the study of digital signal processing. It includes short descriptions of signals and systems, convolution, Fourier series, Fourier transforms, Laplace transforms, and analog filters. The material presented in this appendix should lead to a better understanding of some of the techniques of digital signal processing described in the text.

## A.1 Signals

An analog signal $x(t)$ is a continuous function of the time variable $t$. The signal energy is defined as

$$E = \int_{-\infty}^{\infty} p_i(t)\,dt = \int_{-\infty}^{\infty} |x(t)|^2\,dt \tag{A.1}$$

The absolute value $|x(t)|$ is required only for complex-valued signals. Signals of finite duration and amplitude have finite energy.

A periodic signal $x_p(t)$ is characterized by several measures. Its **duty ratio** equals the ratio of its pulse width and period. Its **average value** $x_{\text{av}}$ equals the average area per period. Its signal power $P$ equals the average energy per period. Its **rms value** $x_{\text{rms}}$ equals $\sqrt{P}$.

$$x_{\text{av}} = \frac{1}{T}\int_T x(t)\,dt \qquad P = \frac{1}{T}\int_T |x(t)|^2\,dt \qquad x_{\text{rms}} = \sqrt{P} \tag{A.2}$$

### Signal Operations and Symmetry

A **time shift** displaces a signal $x(t)$ in time without changing its shape. The signal $y(t) = x(t-\alpha)$ is a delayed (shifted right by $\alpha$) replica of $x(t)$. A **time scaling** results in signal compression or stretching. The signal $f(t) = x(2t)$ describes a two-fold compression, $g(t) = x(t/2)$ describes a two-fold stretching, and $p(t) = x(-t)$ describes a reflection about the vertical axis. Shifting or folding a signal $x(t)$ will not change its area or energy, but time scaling $x(t)$ to $x(\alpha t)$ will reduce both its area and energy by $|\alpha|$.

A signal possesses **even symmetry** if $x(t) = x(-t)$, and possesses **odd symmetry** if $x(t) = -x(-t)$. The area of an odd symmetric signal is always zero.

## Sinusoids and Complex Harmonics

An analog sinusoid or harmonic signal is *always periodic and unique* for any choice of period or frequency. For the sinusoid $x_p(t) = A\cos(\omega_0 t + \theta) = A\cos[\omega_0(t - t_p)]$, the quantity $t_p = -\theta/\omega_0$ is called the **phase delay** and describes the time delay in the signal caused by a phase shift of $\theta$. The various time and frequency measures are related by

$$f_0 = \frac{1}{T} \qquad \omega_0 = \frac{2\pi}{T} = 2\pi f_0 \qquad \theta = \omega_0 t_p = 2\pi f_0 t_p = 2\pi \frac{t_p}{T} \tag{A.3}$$

If $x(t) = A\cos(2\pi f_0 t + \theta)$, then $P = 0.5A^2$ and $x_{\text{rms}} = A\sqrt{2} = 0.707A$. If $x(t) = Ae^{\pm j(2\pi f_0 t + \theta)}$, then $P = A^2$. The **common period** or **time period** $T$ of a combination of sinusoids is given by the least common multiple (LCM) of the individual periods. The **fundamental frequency** $f_0$ is the reciprocal of $T$ and also equals the greatest common divisor (GCD) of the individual frequencies. For a combination of *sinusoids at different frequencies*, say $y(t) = x_1(t) + x_2(t) + \cdots$, the signal power $P_y$ equals the sum of the individual powers, and the rms value equals $\sqrt{P_y}$.

## Useful Signals

The **unit step** $u(t)$, **unit ramp** $r(t)$, and **signum function** $\text{sgn}(t)$ are shown in Figure A.1. All are piecewise linear. The unit step is discontinuous at $t = 0$, where its value is undefined (sometimes chosen as 0.5). The value of the signum function is also undefined at $t = 0$ and chosen as zero. The unit ramp may also be written as $r(t) = tu(t)$.

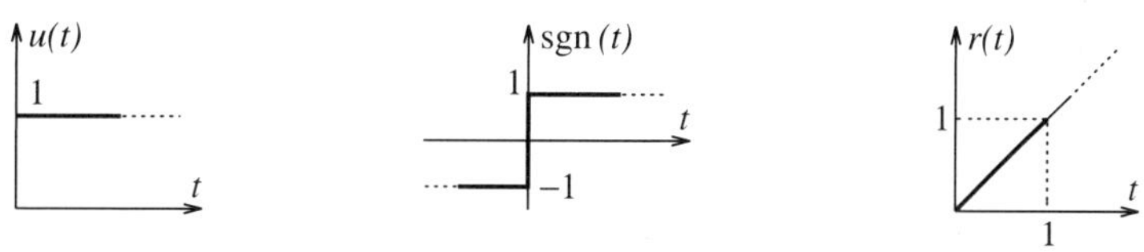


**FIGURE A.1** The unit step, unit ramp, and signum functions are piecewise linear

The rectangular pulse **rect**$(t)$ and triangular pulse **tri**$(t)$ are even symmetric and possess unit area and unit height, as shown in Figure A.2. The signal $f(t) = \text{rect}(\frac{t-\beta}{\alpha})$ is a rectangular pulse of width $\alpha$, which is centered at $t = \beta$. The signal $g(t) = \text{tri}(\frac{t-\beta}{\alpha})$ is a triangular pulse of width $2\alpha$ centered at $t = \beta$.

**FIGURE A.2** The signals rect($t$) and tri($t$) have unit area and unit height

**Example A.1**

### Signal Representation

Consider the following signals

- $f(t) = u(1-t)$
- $g(\lambda) = u(t-\lambda)$ (versus $\lambda$)
- $x(t) = \text{rect}[0.5(t-1)]$
- $y(t) = \text{tri}[\frac{1}{3}(t-1)]$

These signals are sketched in Figure EA.1 as follows:

- The signal $f(t) = u(1-t)$ is a folded step, delayed by 1 unit.
- The signal $g(\lambda) = u(t-\lambda)$ versus $\lambda$ is a folded step, shifted $t$ units to the right. Pay particular attention to this form, which will rear its head again when we study convolution.
- The signal $x(t) = \text{rect}[0.5(t-1)]$ is a stretched rectangular pulse (of width 2 units) delayed by (or centered at) 1 unit. It may also be expressed as $x(t) = u(t) - u(t-2)$.
- The signal $y(t) = \text{tri}[\frac{1}{3}(t-1)]$ is a stretched triangular pulse (of width 6 units) delayed by (or centered at) 1 unit. We may also write $y(t) = \frac{1}{3}[r(t+2) - r(t-1) + r(t-4)]$

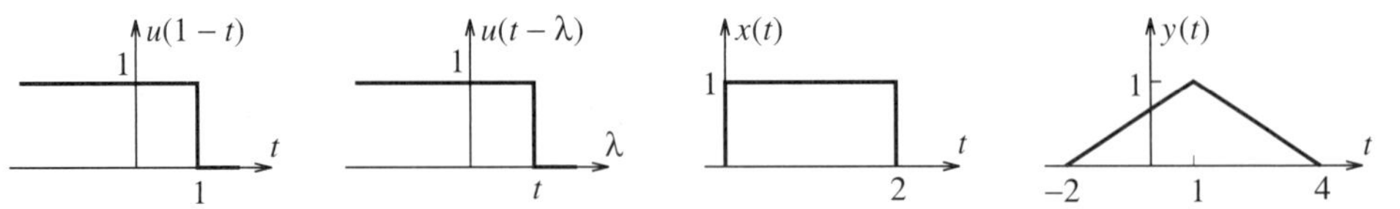


**FIGURE E.A.1** The signals for Example A.1

## The Sinc Function

The sinc function sinc($t$), shown in Figure A.3 is defined as

$$\text{sinc}(t) = \frac{\sin(\pi t)}{\pi t} \tag{A.4}$$

It is an even symmetric function with unit area. It has unit height at the origin, zero crossings (which occur at $t = \pm 1, \pm 2, \pm 3, \ldots$), a central **main lobe** between the first zero on either

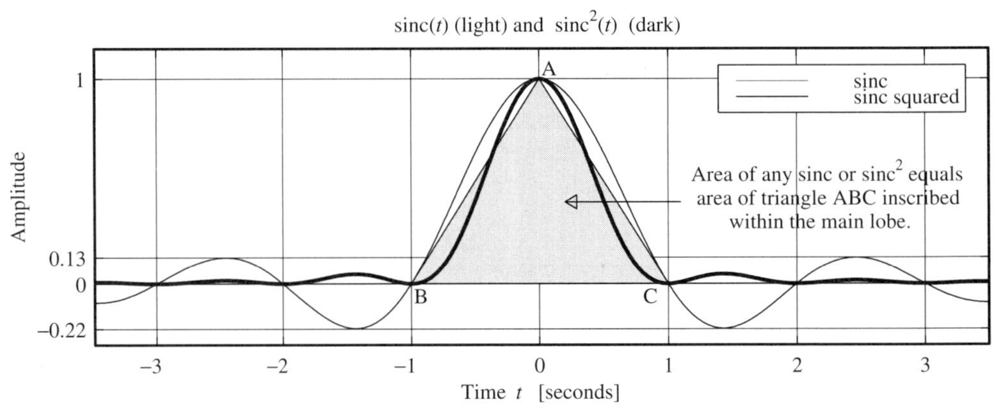


**FIGURE A.3** The sinc function and the sinc-squared function

side of the origin, and progressively decaying negative and positive **sidelobes**. The minimum peak equals $-0.2178$, and the next positive peak equals 0.1284. The signal $\text{sinc}^2(t)$ also has unit area with zeros at unit intervals but is entirely positive. The area of *any sinc or sinc-squared* function equals the area of the triangle inscribed within its central lobe as shown.

### The Impulse Function

The unit **impulse** is a tall, narrow spike with unit area defined by

$$\int_{-\infty}^{\infty} \delta(\tau)\,d\tau = 1 \qquad\qquad \delta(t) = \begin{cases} 0, & t \neq 0 \\ \infty, & t = 0 \end{cases}$$

The function $A\delta(t)$ is shown as an arrow with its area (or strength) $A$ labeled next to the tip. For visual appeal, we make its height proportional to $A$. Signals such as the rect pulse $\frac{1}{\tau}\,\text{rect}(t/\tau)$, tri pulse $\frac{1}{\tau}\,\text{tri}(t/\tau)$, the exponential $\frac{1}{\tau}e^{-t/\tau}u(t)$, and the sinc $\frac{1}{\tau}\,\text{sinc}(t/\tau)$ all possess unit area and give rise to the unit impulse $\delta(t)$ as $\tau \to 0$. The unit impulse $\delta(t)$ also may be regarded as the derivative of $u(t)$.

$$\delta(t) = \frac{du(t)}{dt} \qquad\qquad u(t) = \int_{-\infty}^{t} \delta(t)\,dt$$

Three useful properties of the impulse function relate to scaling, products, and sifting.

| **Scaling** | **Product** | **Sifting** |
|---|---|---|
| $\delta(\alpha t) = \dfrac{1}{\lvert\alpha\rvert}\delta(t)$ | $x(t)\delta(t-\alpha) = x(\alpha)\delta(t-\alpha)$ | $\displaystyle\int_{-\infty}^{\infty} x(t)\delta(t-\alpha)\,dt = x(\alpha)$ |

Time scaling implies that, since $\delta(t)$ has unit area, its compressed version $\delta(\alpha t)$ has an area of $\frac{1}{|\alpha|}$. The product property says that an arbitrary signal $x(t)$ multiplied by the impulse $\delta(t-\alpha)$ is still an impulse (whose strength equals $x(\alpha)$). The sifting property says that the area of the product $x(t)\delta(t-\alpha)$ is just $x(\alpha)$.

**Example A.2**

#### Properties of the Impulse Function

**(a)** Consider the signal $x(t) = 2r(t) - 2r(t-2) - 4u(t-3)$. Sketch $x(t)$, $f(t) = x(t)\delta(t-1)$, and $g(t) = x'(t)$. Also evaluate $I = \int_{-\infty}^{\infty} x(t)\delta(t-2)\,dt$.
Refer to Figure EA.2a for the sketches.

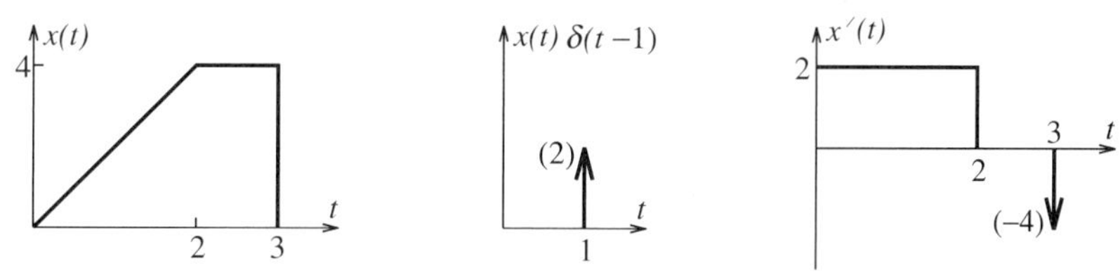


**FIGURE E A.2.a** The signals for Example A.2(a)

From the product property, $f(t) = x(t)\delta(t-1) = x(1)\delta(t-1)$. This is an impulse function with strength $x(1) = 2$.

The derivative $g(t) = x'(t)$ includes the ordinary derivative (slopes) of $x(t)$ and an impulse function of strength $-4$ at $t = 3$.

By the sifting property,

$$I = \int_{-\infty}^{\infty} x(t)\delta(t-2)\,dt = x(2) = 4$$

**(b)** Evaluate $I = \int_0^{\infty} 4t^2\delta(t-3)\,dt$.
Using the sifting property, we get $I = 4(3)^2 = 36$.

## Signal Approximation by Impulses

A signal $x(t)$ multiplied by a periodic-unit impulse train with period $t_s$ yields the **ideally sampled signal**, $x_I(t)$, as shown in Figure A.4. The ideally sampled signal is an impulse train described by

$$x_I(t) = x(t)\sum_{k=-\infty}^{\infty}\delta(t-kt_s) = \sum_{k=-\infty}^{\infty} x(kt_s)\delta(t-kt_s) \tag{A.5}$$

Note that $x_I(t)$ is nonperiodic, and the strength of each impulse equals the signal value $x(kt_s)$. This form actually provides a link between analog and digital signals.

## Moments

Moments are general measures of signal "size" based on area and are defined as shown.

| ***n*th moment** | **Mean** | **Central moment** |
|---|---|---|
| $m_n = \int_{-\infty}^{\infty} t^n x(t)\,dt$ | $m_x(\text{mean}) = \dfrac{m_1}{m_0}$ | $\mu_n = \int_{-\infty}^{\infty}(t-m_x)^n x(t)\,dt$ |

The zeroth moment $m_0 = \int x(t)\,dt$ is the area of $x(t)$. The normalized first moment $m_x = m_1/m_0$ is the mean. Moments about the mean are called **central moments**. The second central moment $\mu_2$ is called the **variance**. It is denoted $\sigma^2$, and its square root $\sigma$ is called the *standard deviation*.

$$\mu_2 = \sigma^2 = \frac{m_2}{m_0} - m_x^2 \tag{A.6}$$

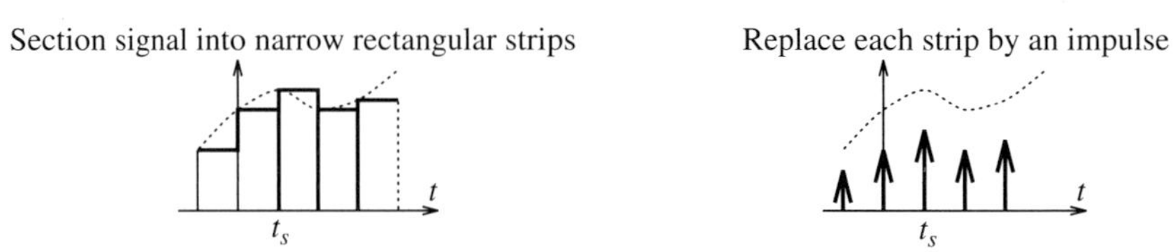


**FIGURE A.4** Signal approximation by impulses

# A.2 System Analysis

Analog linear, time-invariant (LTI) systems may be described by differential equations with constant coefficients. An $n$th-order differential has the general form

$$y^{(n)}(t) + a_1 y^{(n-1)}(t) + \cdots + a_{n-1}y^{(1)}(t) + a_n y(t) = b_0 x^{(m)}(t) + b_1 x^{(m-1)}(t) + \cdots + b_{m-1}x^{(1)}(t) + b_m x(t) \qquad \text{(A.7)}$$

To solve for the output $y(t)$ for $t > 0$, we require the $n$ initial conditions $y(0),\ y^{(1)}(0),\ \ldots,\ y^{(n-1)}(0)$ (the response and its $n-1$ successive derivatives at $t=0$). A convenient technique for solving a linear constant-coefficient differential equation (LCCDE) is the **method of undetermined coefficients**, which yields the total response as the sum of the **forced response** $y_F(t)$ and the **natural response** $y_N(t)$.

The **forced response** is determined by the input terms (left-hand side of the differential equation) and has the same form as the input as summarized in Table A.1. The constants are found by satisfying the given differential equation.

The form of the natural response depends only on the system details and is independent of the nature of the input. It is a sum of exponentials whose exponents are the roots (real or complex) of the *characteristic equation* or *characteristic polynomial* defined by

$$a_0 s^n + a_1 s^{n-1} + a_2 s^{n-2} + \cdots + a_{n-2}s^2 + a_{n-1}s + a_n = 0 \qquad \text{(A.8)}$$

Its $n$ roots $s_1, s_2, \ldots, s_n$ define the form of the natural response as summarized in Table A.2. The constants are evaluated (*after setting up the total response*) using the specified initial conditions.

**TABLE A.1 ➤** Form of the Forced Response for Analog LTI Systems

| Entry | Forcing Function (RHS) | Form of Forced Response |
|---|---|---|
| 1 | $C_0$ (constant) | $C_1$ (another constant) |
| 2 | $e^{\alpha t}$ (see note above) | $Ce^{\alpha t}$ |
| 3 | $\cos(\omega t + \beta)$ | $C_1\cos(\omega t) + C_2\sin(\omega t)$ or $C\cos(\omega t + \theta)$ |
| 4 | $e^{\alpha t}\cos(\omega t + \beta)$ (see note above) | $e^{\alpha t}[C_1\cos(\omega t) + C_2\sin(\omega t)]$ |
| 5 | $t$ | $C_0 + C_1 t$ |
| 6 | $te^{\alpha t}$ (see note above) | $e^{\alpha t}(C_0 + C_1 t)$ |

*NOTE:* If the right-hand side (RHS) is $e^{\alpha t}$, where $\alpha$ is also a root of the characteristic equation repeated $r$ times, the forced response form must be multiplied by $t^r$.

**TABLE A.2 ➤** Form of the Natural Response for Analog LTI Systems

| Entry | Root of Characteristic Equation | Form of Natural Response |
|---|---|---|
| 1 | Real and distinct: $r$ | $Ke^{rt}$ |
| 2 | Complex conjugate: $\beta \pm j\omega$ | $e^{\beta t}[K_1\cos(\omega t) + K_2\sin(\omega t)]$ |
| 3 | Real, repeated: $r^{p+1}$ | $e^{rt}(K_0 + K_1 t + K_2 t^2 + \cdots + K_p t^p)$ |
| 4 | Complex, repeated: $(\beta \pm j\omega)^{p+1}$ | $e^{\beta t}\cos(\omega t)(A_0 + A_1 t + A_2 t^2 + \cdots + A_p t^p)$ $+ e^{\beta t}\sin(\omega t)(B_0 + B_1 t + B_2 t^2 + \cdots + B_p t^p)$ |

**Example A.3**

**Natural and Forced Response**

Consider the first-order system $y'(t) + 2y(t) = x(t)$. Find its response if
**(a)** $x(t) = 6,\ y(0) = 8$ **(b)** $x(t) = \cos(2t),\ y(0) = 2$ **(c)** $x(t) = e^{-2t},\ y(0) = 3$

The characteristic equation $s + 2 = 0$ has the root $s = -2$.

The natural response is thus $y_N(t) = Ke^{-2t}$.

**(a)** Since $x(t) = 6$ is a constant, we choose $y_F(t) = C$.

Then $y'_F(t) = 0$ and $y'_F(t) + 2y_F(t) = 2C = 6$, and thus $y_F(t) = C = 3$.

The total response is

$$y(t) = y_N(t) + y_F(t) = Ke^{-2t} + 3$$

With $y(0) = 8$, we find

$$8 = K + 3 \text{ (or } K = 5\text{) and } y(t) = 5e^{-2t} + 3,\ t \ge 0 \text{ or } y(t) = (5e^{-2t} + 3)u(t)$$

**(b)** Since $x(t) = \cos(2t)$, we choose $y_F(t) = A\cos(2t) + B\sin(2t)$.

Then $y'_F(t) = -2A\sin(2t) + 2B\cos(2t)$, and $y'_F(t) + 2y_F(t) = (2A + 2B)\cos(2t) + (2B - 2A)\sin(2t) = \cos(2t)$.

Comparing the coefficients of the sine and cosine terms on either side, we obtain $2A + 2B = 1,\ \ 2B - 2A = 0$ or $A = 0.25,\ \ B = 0.25$. This gives $y_F(t) = 0.25\cos(2t) + 0.25\sin(2t)$.

The total response is

$$y(t) = y_N(t) + y_F(t) = Ke^{-2t} + 0.25\cos(2t) + 0.25\sin(2t)$$

With $y(0) = 2$, we find

$$2 = K + 0.25 \text{ (or } K = 1.75\text{) and}$$

$$y(t) = [1.75e^{-2t} + 0.25\cos(2t) + 0.25\sin(2t)]u(t)$$

The steady-state response is $y_{ss}(t) = 0.25\cos(2t) + 0.25\sin(2t)$, a sinusoid at the input frequency.

**(c)** Since $x(t) = e^{-2t}$ has the same form as $y_N(t)$, we must choose $y_F(t) = Cte^{-2t}$.

Then $y'_F(t) = Ce^{-2t} - 2Cte^{-2t}$, and $y'_F(t) + 2y_F(t) = Ce^{-2t} - 2Cte^{-2t} + 2Cte^{-2t} = e^{-2t}$.

This gives $C = 1$, and thus $y_F(t) = te^{-2t}$ and

$$y(t) = y_N(t) + y_F(t) = Ke^{-2t} + te^{-2t}$$

With $y(0) = 3$, we find

$$3 = K + 0 \text{ and } y(t) = (3e^{-2t} + te^{-2t})u(t)$$

### A.2.1 The Zero-State Response and Zero-Input Response

It is often more convenient to describe the response $y(t)$ of an LTI system as the sum of its zero-state response (ZSR) $y_{zs}(t)$ (assuming zero initial conditions) and zero-input response (ZIR) $y_{zi}(t)$ (assuming zero input). Each component is found using the method of undetermined coefficients. Note that the natural and forced components $y_N(t)$ and $y_F(t)$ do not, in general, correspond to the zero-input and zero-state response, respectively, even though each pair adds up to the total response.

**Example A.4**

**Zero-Input and Zero-State Response for the Single-Input Case**

Let $y''(t) + 3y'(t) + 2y(t) = x(t)$ with $x(t) = 4e^{-3t}$ and initial conditions $y(0) = 3$ and $y'(0) = 4$.

Find its zero-input response and zero-state response.

The characteristic equation is $s^2 + 3s + 2 = 0$ with roots $s_1 = -1$ and $s_2 = -2$.

Its natural response is $y_N(t) = K_1e^{s_1t} + K_2e^{s_2t} = K_1e^{-t} + K_2e^{-2t}$.

1. The zero-input response is found from $y_N(t)$ and the prescribed initial conditions:

$$y_{zi}(t) = K_1e^{-t} + K_2e^{-2t} \qquad y_{zi}(0) = K_1 + K_2 = 3 \qquad y'_{zi}(0) = -K_1 - 2K_2 = 4$$

   This yields $K_2 = -7$, $K_1 = 10$, and $y_{zi}(t) = 10e^{-t} - 7e^{-2t}$.

2. Similarly, $y_{zs}(t)$ is found from the general form of $y(t)$ but with zero initial conditions. Since $x(t) = 4e^{-3t}$, we select the forced response as $y_F(t) = Ce^{-3t}$.

   Then, $y'_F(t) = -3Ce^{-3t}$, $y''_F(t) = 9Ce^{-3t}$, and $y''_F(t) + 3y'_F(t) + 2y_F(t) = (9C - 9C + 2C)e^{-3t} = 4e^{-3t}$.

   Thus, $C = 2$, $y_F(t) = 2e^{-3t}$, and $y_{zs}(t) = K_1e^{-t} + K_2e^{-2t} + 2e^{-3t}$.

   With zero initial conditions, we obtain

$$y_{zs}(0) = K_1 + K_2 + 2 = 0 \qquad y'_{zs}(0) = -K_1 - 2K_2 - 6 = 0$$

   This yields $K_2 = -4$, $K_1 = 2$, and $y_{zs}(t) = 2e^{-t} - 4e^{-2t} + 2e^{-3t}$.

3. The total response is the sum of $y_{zs}(t)$ and $y_{zi}(t)$:

$$y(t) = y_{zi}(t) + y_{zs}(t) = 12e^{-t} - 11e^{-2t} + 2e^{-3t}, \qquad t \geq 0$$

## A.2.2 Step Response and Impulse Response

The **step response**, $s(t)$ is the response of a *relaxed* LTI system to a unit step $u(t)$. The **impulse response** $h(t)$ is the response to a unit impulse $\delta(t)$ and also equals $s'(t)$. The differential equation for the RC lowpass filter sketched in the following panel is

$$y'(t) + \frac{1}{\tau}y(t) = \frac{1}{\tau}x(t) \tag{A.9}$$

where $\tau = RC$ defines the **time constant**. The characteristic equation $s + \frac{1}{\tau} = 0$ yields the natural response $y_N(t) = Ke^{-t/\tau}$. To find the step response, we let $x(t) = u(t) = 1,\ t \geq 0$. So the forced response, $y_F(t) = B$, is a constant. Then, with $y'_F(t) = 0$, we obtain

$$y'_F(t) + \frac{1}{\tau}y_F(t) = \frac{1}{\tau} = 0 + B \qquad \text{or} \qquad B = 1$$

Thus, $y(t) = y_F(t) + y_N(t) = 1 + Ke^{-t/\tau}$. With $y(0) = 0$, we get $0 = 1 + K$ and

$$s(t) = y(t) = (1 - e^{-t/\tau})u(t) \qquad \text{(step response)} \tag{A.10}$$

The impulse response $h(t)$ equals the derivative of the step response. Thus,

$$h(t) = s'(t) = \frac{1}{\tau}e^{-t/\tau}u(t) \qquad \text{(impulse response)} \tag{A.11}$$

**Unit Step Response and Unit Impulse Response of an RC Lowpass Filter**

The output is the capacitor voltage. The time constant is $\tau = RC$.

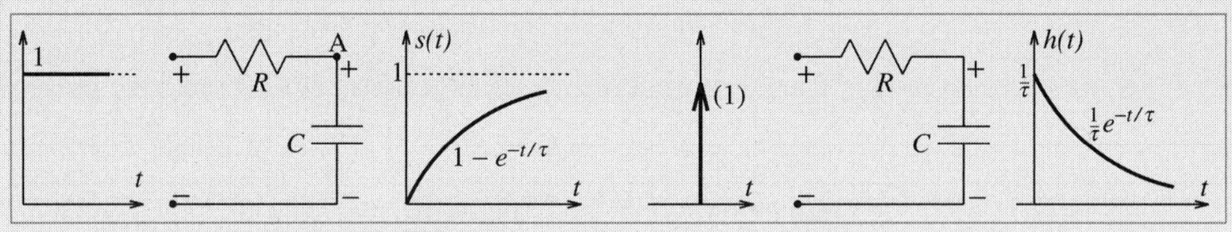


Step response: $s(t) = (1 - e^{-t/\tau})u(t)$ Impulse response: $h(t) = s'(t) = \frac{1}{\tau}e^{-t/\tau}u(t)$

# A.3 Convolution

Convolution finds the *zero-state* response $y(t)$ of an LTI system to an input $x(t)$ and is defined by

$$y(t) = x(t) * h(t) = \int_{-\infty}^{\infty} x(\lambda)h(t-\lambda)\,d\lambda \tag{A.12}$$

The shorthand notation $x(t) * h(t)$ describes the convolution of the signals $x(t)$ and $h(t)$.

## Useful Convolution Properties

The starting time of the convolution equals the sum of the starting times of $x(t)$ and $h(t)$. The ending time of the convolution equals the sum of the ending times of $x(t)$ and $h(t)$. The convolution duration equals the sum of the durations of $x(t)$ and $h(t)$. The area of the convolution equals the product of the areas of $x(t)$ and $h(t)$. The convolution of an odd-symmetric and an even-symmetric signal is odd symmetric, whereas the convolution of two even symmetric (or two odd symmetric) signals is even symmetric. Interestingly, the convolution of $x(t)$ with its folded version $x(-t)$ is also even symmetric, with a maximum at $t = 0$. The convolution $x(t) * x(-t)$ is called the **autocorrelation** of $x(t)$. The convolution of a large number of functions approaches a Gaussian form. This is the **central limit theorem**.

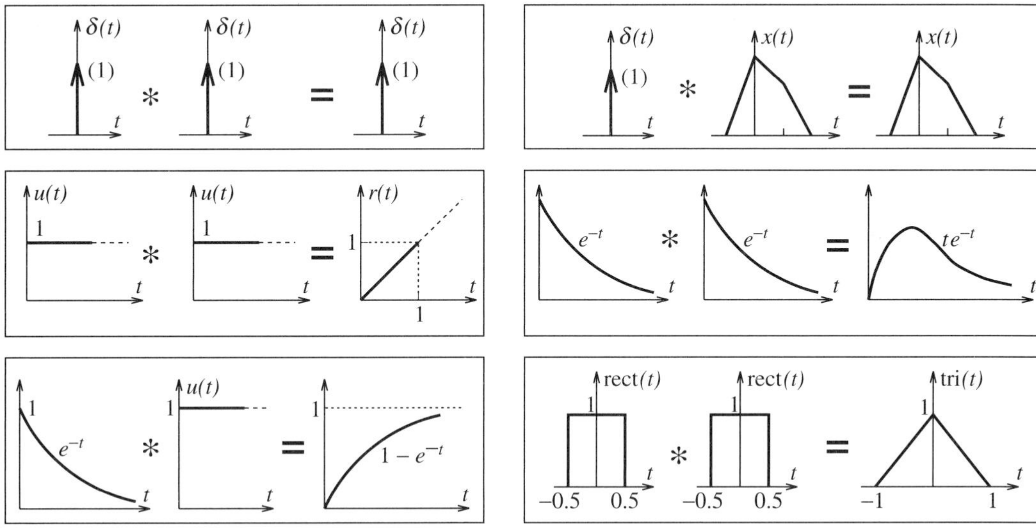


**FIGURE A.5** The convolution of some useful signals

### A.3.1 Useful Convolution Results

The convolution of any signal $x(t)$ with the impulse $\delta(t)$ reproduces the signal $x(t)$. If the impulse is shifted, so is convolution.

$$x(t) * \delta(t) = x(t) \qquad x(t) * \delta(t-\alpha) = x(t-\alpha)$$

Other useful convolution results are illustrated in Figure A.5. By way of an example,

$$e^{-\alpha t}u(t) * e^{-\alpha t}u(t) = \int_{-\infty}^{\infty} e^{-\alpha\lambda}e^{-\alpha(t-\lambda)}u(\lambda)u(t-\lambda)\,d\lambda = e^{-\alpha t}\int_{0}^{t} d\lambda = te^{-\alpha t}u(t)$$

# A.4 The Laplace Transform

The **Laplace transform** $X(s)$ of a causal signal $x(t)$ is defined as

$$X(s) = \int_{0-}^{\infty} x(t)e^{-(\sigma+j\omega)t}\,dt = \int_{0-}^{\infty} x(t)e^{-st}\,dt \qquad \text{(A.13)}$$

The complex quantity $s = \sigma + j\omega$ generalizes the concept of frequency to the complex domain. Some useful transform pairs and properties are listed in Table A.3 and Table A.4.

### A.4.1 The Inverse Laplace Transform

The inverse transform of $H(s)$ may be found by resorting to partial-fraction expansion and a table look-up. If $H(s) = P(s)/Q(s)$, the form of the expansion depends on the nature of the factors in $Q(s)$ and summarized below. Once the partial-fraction expansion is established, the inverse transform for each term can be found with the help of Table A.5.

**TABLE A.3 ➤** A Short Table of Laplace Transforms

| Entry | $x(t)$ | $X(s)$ | Entry | $x(t)$ | $X(s)$ |
|---|---|---|---|---|---|
| 1 | $\delta(t)$ | 1 | 2 | $u(t)$ | $\frac{1}{s}$ |
| 3 | $e^{-\alpha t}u(t)$ | $\frac{1}{s+\alpha}$ | 4 | $e^{-\alpha t}\sin(\beta t)u(t)$ | $\frac{\beta}{(s+\alpha)^2+\beta^2}$ |
| 5 | $te^{-\alpha t}u(t)$ | $\frac{1}{(s+\alpha)^2}$ | 6 | $e^{-\alpha t}\cos(\beta t)u(t)$ | $\frac{s+\alpha}{(s+\alpha)^2+\beta^2}$ |
| 7 | $\cos(\beta t)u(t)$ | $\frac{s}{s^2+\beta^2}$ | 8 | $\sin(\beta t)u(t)$ | $\frac{\beta}{s^2+\beta^2}$ |

**TABLE A.4 ➤** Operational Properties of the Laplace Transform

| Entry | Property | $x(t)$ | $X(s)$ |
|---|---|---|---|
| 1 | Superposition | $\alpha x_1(t)+\beta x_2(t)$ | $\alpha X_1(s)+\beta X_2(s)$ |
| 2 | Times-exp | $e^{-\alpha t}x(t)$ | $X(s+\alpha)$ |
| 3 | Time scaling | $x(\alpha t),\ \alpha>0$ | $\frac{1}{\alpha}X(\frac{s}{\alpha})$ |
| 4 | Time shift | $x(t-\alpha)u(t-\alpha),\ \alpha>0$ | $e^{-\alpha s}X(s)$ |
| 5 | Times-$t$ | $tx(t)$ | $-\frac{dX(s)}{ds}$ |
| 6 | Derivative | $x'(t)$ | $sX(s)-x(0-)$ |
| 7 | $n$th derivative | $x^{(n)}(t)$ | $s^nX(s)-s^{n-1}x(0-)-\cdots-x^{n-1}(0-)$ |
| 8 | Convolution | $x(t)*h(t)$ | $X(s)H(s)$ |

*NOTE:* $x(t)$ is to be regarded as the *causal* signal $x(t)u(t)$.

**TABLE A.5 ➤** Inverse Laplace Transforms of Partial Fraction Expansion Terms

| Entry | Partial Fraction Expansion Term | Inverse Transform |
|---|---|---|
| 1 | $\frac{K}{s+\alpha}$ | $Ke^{-\alpha t}u(t)$ |
| 2 | $\frac{K}{(s+\alpha)^n}$ | $\frac{K}{(n-1)!}t^{n-1}e^{-\alpha t}u(t)$ |
| 3 | $\frac{Cs+D}{(s+\alpha)^2+\beta^2}$ | $e^{-\alpha t}[C\cos(\beta t)+\frac{D-\alpha C}{\beta}\sin(\beta t)]u(t)$ |
| 4 | $\frac{M\angle\theta}{s+\alpha+j\beta}+\frac{M\angle-\theta}{s+\alpha-j\beta}$ | $2Me^{-\alpha t}\cos(\beta t-\theta)u(t)$ |

### Inverse Transformation of $X(s)$ Relies on Partial Fractions and Table Look-Up

**Distinct Roots:** $X(s) \;=\; \prod_{m=1}^{N}\frac{K_m}{s+p_m} \;=\; \sum_{m=1}^{N}\frac{P(s)}{s+p_m}$, where

$$K_m = (s+p_m)X(s)|_{s=-p_m}$$

**Repeated:** $\frac{1}{(s+r)^k}\prod_{m=1}^{N}\frac{P(s)}{s+p_m} = \sum_{m=1}^{N}\frac{K_m}{s+p_m}+\sum_{n=0}^{k-1}\frac{A_n}{(s+r)^{k-n}}$

$$A_n = \frac{1}{n!}\frac{d^n}{ds^n}[(s+r)^kX(s)]|_{s=-r}$$

**Example A.5**

### Partial Fraction Expansion

**(a) Non-Repeated Poles**

Let $X(s) = \dfrac{2s^3 + 8s^2 + 4s + 8}{s(s+1)(s^2+4s+8)}$. This can be factored as

$$X(s) = \frac{K_1}{s} + \frac{K_2}{s+1} + \frac{A}{s+2+j2} + \frac{A^*}{s+2-j2}$$

We successively evaluate

$$K_1 = sX(s)|_{s=0} = \left.\frac{2s^3 + 8s^2 + 4s + 8}{(s+1)(s^2+4s+8)}\right|_{s=0} = \frac{8}{8} = 1$$

$$K_2 = (s+1)X(s)|_{s=-1} = \left.\frac{2s^3 + 8s^2 + 4s + 8}{s(s^2+4s+8)}\right|_{s=-1} = \frac{10}{-5} = -2$$

$$A = (s+2+j2)X(s)|_{s=-2-j2} = \left.\frac{2s^3 + 8s^2 + 4s + 8}{s(s+1)(s+2-j2)}\right|_{s=-2-j2} = 1.5 + j0.5$$

The partial-fraction expansion thus becomes

$$X(s) = \frac{1}{s} - \frac{2}{s+1} + \frac{1.5+j0.5}{s+2+j2} + \frac{1.5-j0.5}{s+2-j2}$$

With $1.5 + j0.5 = 1.581\angle 18.4^\circ = 1.581\angle 0.1024\pi = M\angle\theta$, we find $x(t)$ as

$$x(t) = u(t) - 2e^{-t}u(t) + 3.162e^{-2t}\cos(2t - 0.1024\pi)u(t)$$

**(b) Repeated Poles**

Let $X(s) = \dfrac{4}{(s+1)(s+2)^3}$. Its partial-fraction expansion is

$$X(s) = \frac{K_1}{s+1} + \frac{A_0}{(s+2)^3} + \frac{A_1}{(s+2)^2} + \frac{A_2}{(s+2)}$$

We compute $K_1 = (s+1)X(s)|_{s=-1} = \dfrac{4}{(s+2)^3}|_{s=-1} = 4.$

Since $(s+2)^3X(s) = \dfrac{4}{s+1}$, we also successively compute

$$A_0 = \left.\frac{4}{s+1}\right|_{s=-2} = -4$$

$$A_1 = \frac{d}{ds}\left[\frac{4}{s+1}\right]\Bigg|_{s=-2} = -\left.\frac{4}{(s+1)^2}\right|_{s=-2} = -4$$

$$A_2 = \frac{1}{2}\frac{d^2}{ds^2}\left[\frac{4}{s+1}\right]\Bigg|_{s=-2} = \left.\frac{4}{(s+1)^3}\right|_{s=-2} = -4$$

This gives the result

$$X(s) = \frac{4}{s+1} - \frac{4}{(s+2)^3} - \frac{4}{(s+2)^2} - \frac{4}{s+2}$$

We then find $x(t) = 4e^{-t}u(t) - 2t^2e^{-2t}u(t) - 4te^{-2t}u(t) - 4e^{-2t}u(t)$.

### A.4.2 Interconnected Systems

The impulse response $h(t)$ of cascaded LTI systems is the convolution of the individual impulse responses. The impulse response of systems in parallel equals the sum of the individual impulse responses.

$$h_C(t) = h_1(t) * h_2(t) * \cdots * h_N(t) \quad \text{(cascade)}$$

$$h_P(t) = h_1(t) + h_2(t) + \cdots + h_N(t) \quad \text{(parallel)}$$

The overall transfer function of cascaded systems is the product of the individual transfer functions (assuming ideal cascading and no loading effects). The overall transfer function of systems in parallel is the algebraic sum of the individual transfer functions.

$$H_C(s) = H_1(s)H_2(s)\cdots H_N(s) \quad \text{(cascade)}$$

$$H_P(s) = H_1(s) + H_2(s) + \cdots + H_N(s) \quad \text{(parallel)}$$

### A.4.3 Stability

In the time domain, bounded-input, bounded-output (BIBO) stability of an LTI system requires a differential equation in which the highest derivative of the input never exceeds the highest derivative of the output and a characteristic equation whose roots have negative real parts. Equivalently, we require the impulse response $h(t)$ to be absolutely integrable. In the $s$-domain, we require a *proper* transfer function $H(s)$ (with common factors canceled) whose poles lie in the left half of the $s$-plane (excluding the $j\omega$-axis).

#### Minimum-Phase Filters

A minimum-phase system has all of its poles and zeros in the left half-plane of the $s$-plane. It has the smallest group delay and smallest deviation from zero phase, at every frequency and among all transfer functions with the same magnitude spectrum $|H(\omega)|$.

### A.4.4 The Laplace Transform and System Analysis

The Laplace transform is a useful tool for the analysis of LTI systems. To find the zero-state response of an electric circuit, we transform a circuit to the $s$-domain by replacing the elements $R$, $L$, and $C$ by their impedances $Z_R$, $Z_L$, and $Z_C$ and replacing sources by their Laplace transforms. We may also transform the system differential equation. For a relaxed LTI system, the zero-state response $Y(s)$ to an input $X(s)$ is $H(s)X(s)$. If the system is not relaxed, the effect of initial conditions is easy to include.

**Example A.6** **Solving Differential Equations**

Let $y''(t) + 3y'(t) + 2y(t) = 4e^{-2t}$, with $y(0) = 3$ and $y'(0) = 4$. Solve for $y(t)$.

Transformation to the $s$-domain, using the derivative property, yields

$$s^2Y(s) - sy(0) - y'(0) + 3[sY(s) - y(0)] + 2Y(s) = \frac{4}{s+2}$$

**(a) Total Response**

Substitute for the initial conditions and rearrange:

$$(s^2+3s+2)Y(s) = 3s+4+9+\frac{4}{s+2} = \frac{3s^2+19s+30}{s+2}$$

Upon simplification, we obtain $Y(s)$ and its partial fraction form as

$$Y(s) = \frac{3s^2+19s+30}{(s+1)(s+2)^2} = \frac{K_1}{s+1}+\frac{A_0}{(s+2)^2}+\frac{A_1}{s+2}$$

Solving for the constants, we obtain

$$K_1 = \left.\frac{3s^2+19s+30}{(s+2)^2}\right|_{s=-1} = 14$$

$$A_0 = \left.\frac{3s^2+19s+30}{s+1}\right|_{s=-2} = -4$$

$$A_1 = \left.\frac{d}{ds}\left[\frac{3s^2+19s+30}{s+1}\right]\right|_{s=-2} = -11$$

Upon inverse transformation, $y(t) = (14e^{-t} - 4te^{-2t} - 11e^{-2t})u(t)$.

As a check, we confirm that $y(0) = 3$ and $y'(0) = -14+22-4 = 4$.

**(b) Zero-State Response**

For the zero-state response, we assume zero initial conditions to obtain

$$(s^2+3s+2)Y_{zs}(s) = \frac{4}{s+2}$$

This gives

$$Y_{zs}(s) = \frac{4}{(s+2)(s^2+3s+2)} = \frac{4}{s+1} - \frac{4}{(s+2)^2} - \frac{4}{s+2}$$

Inverse transformation gives $y_{zs}(t) = (4e^{-t} - 4te^{-2t} - 4e^{-2t})u(t)$.

**(c) Zero-Input Response**

For the zero-input response, we assume zero input to obtain

$$(s^2+3s+2)Y_{zi}(s) = 3s+13 \qquad Y_{zi}(s) = \frac{3s+13}{s^2+3s+2} = \frac{10}{s+1} - \frac{7}{s+2}$$

Upon inverse transformation, $y_{zi}(t) = (10e^{-t} - 7e^{-2t})u(t)$. The total response equals

$$y(t) = y_{zs}(t) + y_{zi}(t) = (14e^{-t} - 4te^{-2t} - 11e^{-2t})u(t)$$

This matches the result found from the direct solution.

## A.4.5 The Steady-State Response to Harmonic Inputs

The steady-state response of an LTI system to a sinusoid or harmonic input is also a harmonic at the input frequency. To find the steady-state response $y_{ss}(t)$ to the sinusoidal input $x(t) =$

$A\cos(\omega_0 t + \theta)$, we first evaluate the transfer function at the input frequency $\omega_0$ to obtain $H(\omega_0) = K\angle\phi$. The output is then given by $y_{ss}(t) = KA\cos(\omega_0 t + \theta + \phi)$.

**Example A.7**

**Steady-State Response to Harmonic Inputs**

Let $H(s) = \dfrac{s-2}{s^2+4s+4}$. Find the steady-state response $y_{ss}(t)$ if $x(t) = 8\cos(2t) + 4$.

Let $x_1(t) = 8\cos(2t)$. Then, $\omega = 2$, and $H(2) = \dfrac{2j-2}{-4+j8+4} = 0.3536\angle 45^\circ$.

So, its steady-state output is

$$y_1(t) = 8(0.3536)\cos(2t + 45^\circ) = 2.8284\cos(2t + 45^\circ)$$

Let $x_2(t) = 4$. Then, $\omega = 0$ and $H(0) = -0.5$. So, its steady-state output is $y_2(t) = (4)(-0.5) = -2$.

By superposition

$$y_{ss}(t) = y_1(t) + y_2(t) = 2.8284\cos(2t + 60^\circ) - 2$$

# A.5 The Fourier Transform

The Fourier transform provides a frequency-domain representation of a signal $x(t)$ and is defined by

$$X(f) = \int_{-\infty}^{\infty} x(t)e^{-j2\pi ft}\,dt \quad \text{(the } f\text{-form)} \qquad X(\omega) = \int_{-\infty}^{\infty} x(t)e^{-j\omega t}\,dt \quad \text{(the } \omega\text{-form)}$$

The **inverse Fourier transform** allows us to obtain $x(t)$ from its spectrum and is defined by

$$x(t) = \int_{-\infty}^{\infty} X(f)e^{j2\pi ft}\,df \quad \text{(the } f\text{-form)} \qquad x(t) = \frac{1}{2\pi}\int_{-\infty}^{\infty} X(\omega)e^{j\omega t}\,d\omega \quad \text{(the } \omega\text{-form)}$$

The Fourier transform is, in general, complex. For real signals, $X(f)$ is *conjugate symmetric* with $X(-f) = X^*(f)$. This means that the magnitude $|X(f)|$ or $\mathrm{Re}\{X(f)\}$ displays even symmetry and the phase $\phi(f)$ or $\mathrm{Im}\{X(f)\}$ displays odd symmetry. It is customary to plot the magnitude and phase of $X(f)$ as two-sided functions. The effect of signal symmetry on the Fourier transform are summarized.

**Effect of Signal Symmetry on the Fourier Transform of Real-Valued Signals**

**Even Symmetry in $x(t)$:** The Fourier transform $X(f)$ is real and even symmetric.
**Odd Symmetry in $x(t)$:** The Fourier transform $X(f)$ is imaginary and odd symmetric.
**No Symmetry in $x(t)$:** $\mathrm{Re}\{X(f)\}$ is even symmetric, and $\mathrm{Im}\{X(f)\}$ is odd symmetric.

**TABLE A.6 ➤** Some Useful Fourier Transform Pairs

| Entry | $x(t)$ | $X(f)$ | $X(\omega)$ |
|---|---|---|---|
| 1 | $\delta(t)$ | $1$ | $1$ |
| 2 | $\text{rect}(t)$ | $\text{sinc}(f)$ | $\text{sinc}(\frac{\omega}{2\pi})$ |
| 3 | $\text{tri}(t)$ | $\text{sinc}^2(f)$ | $\text{sinc}^2(\frac{\omega}{2\pi})$ |
| 4 | $\text{sinc}(t)$ | $\text{rect}(f)$ | $\text{rect}(\frac{\omega}{2\pi})$ |
| 5 | $\cos(2\pi\alpha t)$ | $0.5[\delta(f+\alpha)+\delta(f-\alpha)]$ | $\pi[\delta(\omega+2\pi\alpha)+\delta(\omega-2\pi\alpha)]$ |
| 6 | $\sin(2\pi\alpha t)$ | $j0.5[\delta(f+\alpha)-\delta(f-\alpha)]$ | $j\pi[\delta(\omega+2\pi\alpha)-\delta(\omega-2\pi\alpha)]$ |
| 7 | $e^{-\alpha t}u(t)$ | $\dfrac{1}{\alpha+j2\pi f}$ | $\dfrac{1}{\alpha+j\omega}$ |
| 8 | $te^{-\alpha t}u(t)$ | $\dfrac{1}{(\alpha+j2\pi f)^2}$ | $\dfrac{1}{(\alpha+j\omega)^2}$ |
| 9 | $e^{-\alpha\|t\|}$ | $\dfrac{2\alpha}{\alpha^2+4\pi^2 f^2}$ | $\dfrac{2\alpha}{\alpha^2+\omega^2}$ |
| 10 | $e^{-\pi t^2}$ | $e^{-\pi f^2}$ | $e^{-\omega^2/4\pi}$ |
| 11 | $\text{sgn}(t)$ | $\dfrac{1}{j\pi f}$ | $\dfrac{2}{j\omega}$ |
| 12 | $u(t)$ | $0.5\delta(f)+\dfrac{1}{j2\pi f}$ | $\pi\delta(\omega)+\dfrac{1}{j\omega}$ |
| 13 | $e^{-\alpha t}\cos(2\pi\beta t)u(t)$ | $\dfrac{\alpha+j2\pi f}{(\alpha+j2\pi f)^2+(2\pi\beta)^2}$ | $\dfrac{\alpha+j\omega}{(\alpha+j\omega)^2+(2\pi\beta)^2}$ |
| 14 | $e^{-\alpha t}\sin(2\pi\beta t)u(t)$ | $\dfrac{2\pi\beta}{(\alpha+j2\pi f)^2+(2\pi\beta)^2}$ | $\dfrac{2\pi\beta}{(\alpha+j\omega)^2+(2\pi\beta)^2}$ |
| 15 | $\sum_{n=-\infty}^{\infty}\delta(t-nT)$ | $\frac{1}{T}\sum_{k=-\infty}^{\infty}\delta\left(f-\frac{k}{T}\right)$ | $\frac{2\pi}{T}\sum_{k=-\infty}^{\infty}\delta\left(\omega-\frac{2\pi k}{T}\right)$ |
| 16 | $x_p(t)=\sum_{k=-\infty}^{\infty}X[k]e^{j2\pi kf_0t}$ | $\sum_{k=-\infty}^{\infty}X[k]\delta(f-kf_0)$ | $\sum_{k=-\infty}^{\infty}2\pi X[k]\delta(\omega-k\omega_0)$ |

The three most useful transform pairs are listed in the following panel.

**Three Basic Fourier Transform Pairs**

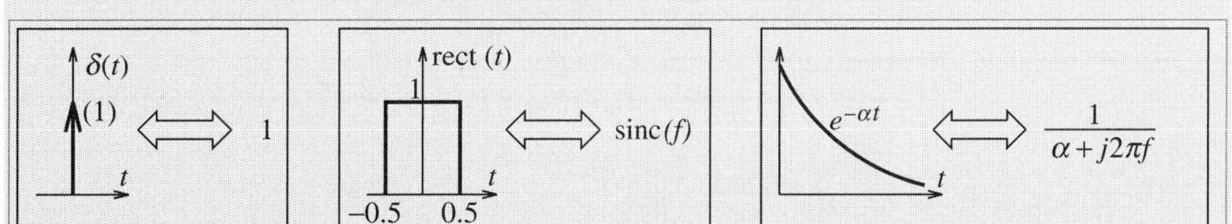


Table A.6 lists the Fourier transform of various signals while Table A.7 lists various properties and theorems useful in problem solving.

## A.5.1 Connections between Laplace and Fourier Transforms

The following panel shows the connection between the Fourier and Laplace transform. We can always find the Laplace transform of causal signals from their Fourier transform, but not the other way around.

**TABLE A.7** ➤ Operational Properties of the Fourier Transform

| Property | $x(t)$ | $X(f)$ | $X(\omega)$ |
|---|---|---|---|
| Similarity | $X(t)$ | $x(-f)$ | $2\pi x(-\omega)$ |
| Time scaling | $x(\alpha t)$ | $\frac{1}{\lvert\alpha\rvert}X\left(\frac{f}{\alpha}\right)$ | $\frac{1}{\lvert\alpha\rvert}X\left(\frac{\omega}{\alpha}\right)$ |
| Folding | $x(-t)$ | $X(-f)$ | $X(-\omega)$ |
| Time shift | $x(t-\alpha)$ | $e^{-j2\pi f\alpha}X(f)$ | $e^{-j\omega\alpha}X(\omega)$ |
| Frequency shift | $e^{j2\pi\alpha t}x(t)$ | $X(f-\alpha)$ | $X(\omega-2\pi\alpha)$ |
| Convolution | $x(t)*h(t)$ | $X(f)H(f)$ | $X(\omega)H(\omega)$ |
| Multiplication | $x(t)h(t)$ | $X(f)*H(f)$ | $\frac{1}{2\pi}X(\omega)*H(\omega)$ |
| Modulation | $x(t)\cos(2\pi\alpha t)$ | $0.5[X(f+\alpha)+X(f-\alpha)]$ | $0.5[X(\omega+2\pi\alpha)+X(\omega-2\pi\alpha)]$ |
| Times-$t$ | $-j2\pi t x(t)$ | $X'(f)$ | $2\pi X'(\omega)$ |
| Conjugation | $x^*(t)$ | $X^*(-f)$ | $X^*(-\omega)$ |
| Correlation | $x(t)**y(t)$ | $X(f)Y^*(f)$ | $X(\omega)Y^*(\omega)$ |
| Autocorrelation | $x(t)**x(t)$ | $X(f)X^*(f)=\lvert X(f)\rvert^2$ | $X(\omega)X^*(\omega)=\lvert X(\omega)\rvert^2$ |
| | | Fourier Transform Theorems | |
| Central ordinates | $x(0)=\int_{-\infty}^{\infty}X(f)\,df=\frac{1}{2\pi}\int_{-\infty}^{\infty}X(\omega)\,d\omega$ | | $X(0)=\int_{-\infty}^{\infty}x(t)\,dt$ |
| Parseval's theorem | $E=\int_{-\infty}^{\infty}x^2(t)\,dt=\int_{-\infty}^{\infty}\lvert X(f)\rvert^2\,df=\frac{1}{2\pi}\int_{-\infty}^{\infty}\lvert X(\omega)\rvert^2\,d\omega$ | | |
| Plancherel'stheorem | $\int_{-\infty}^{\infty}x(t)y^*(t)\,dt=\int_{-\infty}^{\infty}X(f)Y^*(f)\,df=\frac{1}{2\pi}\int_{-\infty}^{\infty}X(\omega)Y^*(\omega)\,d\omega$ | | |

### Relating the Laplace Transform and the Fourier Transform

**From $X(s)$ to $X(f)$:** If $x(t)$ is *causal* and *absolutely integrable*, simply replace $s$ by $j2\pi f$.

**From $X(f)$ to $X(s)$:** If $x(t)$ is *causal*, delete impulsive terms in $X(f)$ and replace $j2\pi f$ by $s$.

## Some Useful Properties

**Scaling** of $x(t)$ to $x(\alpha t)$ leads to a stretching of $X(f)$ by $\alpha$ and an amplitude reduction by $|\alpha|$. If a signal $x(t)$ is modulated (multiplied) by the high-frequency sinusoid $\cos(2\pi f_0 t)$, its spectrum $X(f)$ gets halved and centered at $f=\pm f_0$. This **modulation property** shifts the spectrum of $x(t)$ to higher frequencies.

$$x(t)\cos(2\pi f_0 t) \Leftarrow\boxed{\text{FT}}\Rightarrow 0.5[X(f+f_0)+X(f-f_0)] \qquad \text{(A.14)}$$

**Parseval's Theorem** says that the Fourier transform is an *energy-conserving* relation, and the energy may be found from the time signal $x(t)$ or its magnitude spectrum $|X(f)|$.

$$E=\int_{-\infty}^{\infty}x^2(t)\,dt=\int_{-\infty}^{\infty}|X(f)|^2\,df \qquad \text{(A.15)}$$

**Example A.8**

## Some Transform Pairs using Properties

**(a)** For $x(t) = \text{tri}(t) = \text{rect}(t) * \text{rect}(t)$, use $\text{rect}(t) \Leftarrow\boxed{\text{FT}}\Rightarrow \text{sinc}(f)$ and convolution (see Figure EA.8(1)):

$$\text{tri}(t) = \text{rect}(t) * \text{rect}(t) \Leftarrow\boxed{\text{FT}}\Rightarrow \text{sinc}(f)\text{sinc}(f) = \text{sinc}^2(f)$$

**FIGURE E.A.8(1)** The signals for Example A.8(a and b)

**(b)** For $x(t) = te^{-\alpha t}u(t)$, start with $e^{-\alpha t} \Leftarrow\boxed{\text{FT}}\Rightarrow \frac{1}{\alpha + j2\pi f}$ and use convolution (see Figure EA.8(1)):

$$te^{-\alpha t}u(t) = e^{-\alpha t}u(t) * e^{-\alpha t}u(t) \Leftarrow\boxed{\text{FT}}\Rightarrow \frac{1}{(\alpha + j2\pi f)^2}$$

**(c)** For $x(t) = e^{-\alpha|t|} = e^{\alpha t}u(-t) + e^{-\alpha t}u(t)$, start with $e^{-\alpha t}u(t) \Leftarrow\boxed{\text{FT}}\Rightarrow \frac{1}{\alpha + j2\pi f}$ and use the folding property and superposition:

$$e^{-\alpha|t|} \Leftarrow\boxed{\text{FT}}\Rightarrow \frac{1}{\alpha + j2\pi f} + \frac{1}{\alpha - j2\pi f} = \frac{2\alpha}{\alpha^2 + 4\pi^2 f^2}$$

**(d)** For $x(t) = \text{sgn}(t)$, use the limiting form for $y(t) = e^{-\alpha t}u(t) - e^{\alpha t}u(-t)$ as $\alpha \to 0$ to give

$$e^{-\alpha t}u(t) - e^{\alpha t}u(-t) \Leftarrow\boxed{\text{FT}}\Rightarrow \frac{1}{\alpha + j2\pi f} - \frac{1}{\alpha - j2\pi f} = \frac{-4j\pi f}{\alpha^2 + 4\pi^2 f^2}$$

$$\text{sgn}(t) \Leftarrow\boxed{\text{FT}}\Rightarrow \frac{1}{j\pi f}$$

**(e)** For $x(t) = \cos(2\pi\alpha t) = 0.5e^{j2\pi\alpha t} + 0.5e^{-j2\pi\alpha t}$, start with $1 \Leftarrow\boxed{\text{FT}}\Rightarrow \delta(f)$ and use the dual of the time-shift property:

$$\cos(2\pi\alpha t) = 0.5e^{j2\pi\alpha t} + 0.5e^{-j2\pi\alpha t} \Leftarrow\boxed{\text{FT}}\Rightarrow 0.5\delta(f - \alpha) + 0.5\delta(f + \alpha)$$

Its magnitude spectrum (see Figure EA.8(2)(a)) is an impulse pair at $f = \pm\alpha$ with strengths of 0.5.

**(f)** For $x(t) = \cos(2\pi\alpha t + \theta)$, start with $\cos(2\pi\alpha t) \Leftarrow\boxed{\text{FT}}\Rightarrow 0.5\delta(f - \alpha) + 0.5\delta(f + \alpha)$ and use the shifting property with $t \to t + \theta/2\pi\alpha$ (and the product property of impulses) to get

$$\cos(2\pi\alpha t + \theta) \Leftarrow\boxed{\text{FT}}\Rightarrow 0.5e^{j\theta f/\alpha}[\delta(f - \alpha) + \delta(f + \alpha)]$$
$$= 0.5e^{j\theta}\delta(f - \alpha) + 0.5e^{-j\theta}\delta(f - \alpha)$$

Its magnitude spectrum is an impulse pair at $f = \pm\alpha$ with strengths of 0.5. Its phase spectrum shows a phase of $\theta$ at $f = \alpha$ and $-\theta$ at $f = -\alpha$. The spectra are shown in Figure EA.8(2)(b).

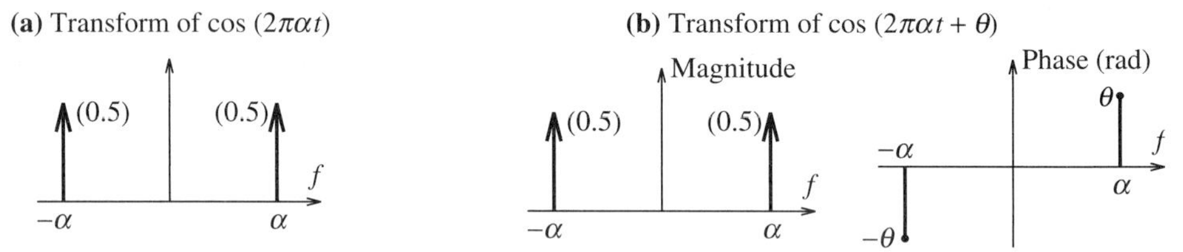


**FIGURE E.A.8(2)** The Fourier transforms of $\cos(2\pi\alpha t)$ and $\cos(2\pi\alpha t + \theta)$ for Example A.8(e and f)

## A.5.2 Amplitude Modulation

Communication of information often requires transmission of signals (or *messages*) through space by converting electrical signals to electromagnetic waves using antennas. Efficient radiation requires antenna dimensions comparable to the wavelength of the signal. The wavelength $\lambda$ of a radiated signal equals $c/f$, where $c$ is the speed of light (at which electromagnetic waves travel) and $f$ is the signal frequency. Transmission of low-frequency messages, such as speech and music, would require huge antenna sizes measured in kilometers. A practical approach is to shift low-frequency messages to a much higher frequency using *modulation* and to transmit the modulated signal instead. In turn, this requires a means of recovering the original message at the receiving end—a process called *demodulation*. Amplitude modulation schemes typically modulate the amplitude of a high-frequency carrier $x_C(t) = \cos(2\pi f_C t)$ by a message signal $x_S(t)$ band limited to a frequency $B \ll f_C$. In double-sideband suppressed-carrier (DSBSC) AM, the carrier $x_C(t)$ is multiplied by the message $x_S(t)$. The modulated signal $x_M(t)$ and its spectrum $X_M(f)$ are given by

$$x_M(t) = x_S(t)x_C(t) = x_S(t)\cos(2\pi f_C t) \qquad X_M(f) = 0.5[X(f + f_C) + X(f - f_C)] \tag{A.16}$$

The message spectrum is centered about $\pm f_C$, as illustrated in Figure A.6. It exhibits symmetry about $\pm f_C$. The portion for $f > |f_C|$ is called the **upper sideband** (USB) and that for $f < |f_C|$, the **lower sideband** (LSB). Since $x_S(t)$ is band limited to $B$, the spectrum occupies a band of $2B$ (centered about $\pm f_C$).

### Detection

At the receiver, the message $x_S(t)$ may be recovered using **demodulation** or **detection**. In **synchronous** or **coherent** detection, the AM signal is multiplied by $\cos(2\pi f_C t)$ and then lowpass filtered, as shown in Figure A.7.

**FIGURE A.6** DSBSC amplitude modulation

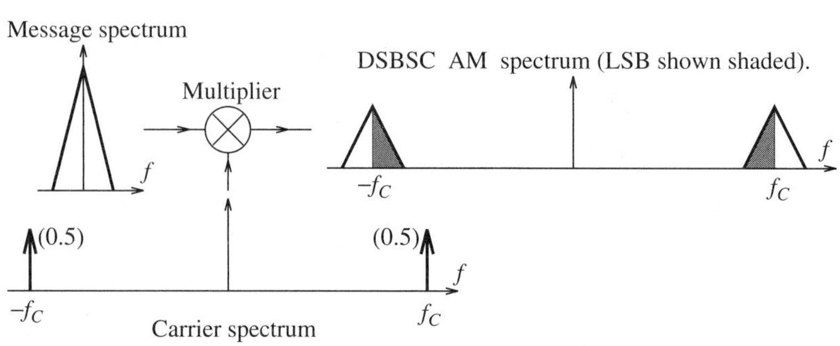

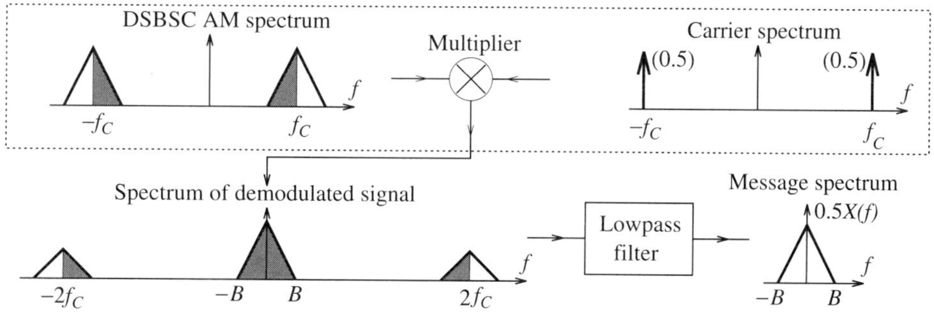


**FIGURE A.7** Synchronous demodulation of DSBSC AM

With $x_M(t) = x_S(t)\cos(2\pi f_C t)$, the demodulated signal equals

$$x_D(t) = x_M(t)\cos(2\pi f_C t) = x_S(t)\cos(2\pi f_C t)\cos(2\pi f_C t) \tag{A.17}$$

Using the identity $\cos^2(\alpha) = 0.5(1 + \cos 2\alpha)$, we obtain

$$x_D(t) = 0.5x_S(t) + 0.5x_S(t)\cos(4\pi f_C t) \tag{A.18}$$

If $x_D(t)$ is passed through a lowpass filter with a gain of 2 and a bandwidth of $B$, the high-frequency component (centered at $\pm 2f_C$) is blocked, and we recover the message $x_S(t)$.

This method is called synchronous or coherent because it requires a signal whose frequency and phase (but not amplitude) matches the carrier. For DSBSC AM, this coherent signal may be obtained by transmitting a small fraction of the carrier (a *pilot signal*) along with $x_M(t)$.

## A.5.3 Fourier Series

The Fourier series is the best least squares fit to a periodic signal $x_p(t)$. It describes $x_p(t)$ as a sum of sinusoids at its fundamental frequency $f_0 = 1/T$ and multiples $kf_0$ whose weights (magnitude and phase) are selected to minimize the *mean square error*. The trigonometric, polar, and exponential forms of the Fourier series are listed in the following panel.

---

**Three Forms of the Fourier Series for a Periodic Signal $x_p(t)$**

**Trigonometric**

$$a_0 + \sum_{k=1}^{\infty} a_k \cos(2\pi k f_0 t) + b_k \sin(2\pi k f_0 t)$$

**Polar**

$$c_0 + \sum_{k=1}^{\infty} c_k \cos(2\pi k f_0 t + \theta_k)$$

**Exponential**

$$\sum_{k=-\infty}^{\infty} X[k] e^{j2\pi k f_0 t}$$

---

The Exponential Fourier series coefficients are given by

$$X[k] = \frac{1}{T}\int_T x(t)e^{-j2\pi k f_0 t}\,dt \tag{A.19}$$

These coefficients display conjugate symmetry with $X[-k] = X^*[k]$. The connection between the three forms of the coefficients is given by

$$X[0] = a_0 = c_0 \qquad \text{and} \qquad X[k] = 0.5(a_k - jb_k) = 0.5c_k\angle\theta_k, \quad k \geq 1 \qquad \text{(A.20)}$$

The **magnitude spectrum** and **phase spectrum** describe plots of the magnitude and phase of each harmonic. They are plotted as discrete signals and are sometimes called **line spectra**. For real periodic signals, the $X[k]$ display conjugate symmetry.

### A.5.4 Fourier Series Coefficients from the Fourier Transform

The Fourier series coefficients $X[k]$ of a periodic signal $x_p(t)$ may be found by evaluating (or sampling) the Fourier transform $X_1(f)$ of its one period $x_1(t)$ at integer multiples of the fundamental frequency $f_0$ (i.e., at $f = kf_0$) and dividing the result by the period $T$.

$$X[k] = \frac{1}{T}X_1(f)\big|_{f=kf_0} \quad \text{(where } X_1(f) \text{ is the Fourier transform of one period)} \qquad \text{(A.21)}$$

**Example A.9**

**Some Fourier Series Results**

**(a) An Impulse Train**

From Figure EA.9A, and by the sifting property of impulses, we get

$$X[k] = \frac{1}{T}\int_{-T/2}^{T/2} \delta(t)e^{-j2k\pi f_0 t}\,dt = \frac{1}{T}$$

All the coefficients of an impulse train are constant!

FIGURE E A.9.a Impulse train and its Fourier series coefficients for Example A.9(a)

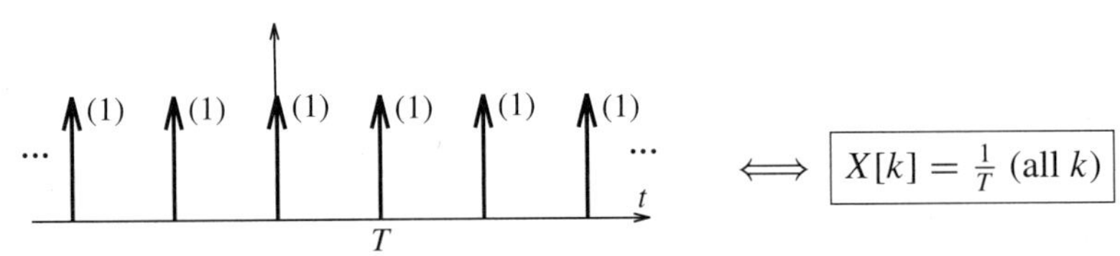


**(b) A Rectangular Pulse Train**

The coefficients $X[k]$ of a train of the rectangular pulse train shown in Figure EA.9B are

$$X[k] = \frac{At_0}{T}\,\text{sinc}(kf_0t_0)$$

FIGURE E A.9.b Rectangular pulse train and its one period for Example A.9(b)

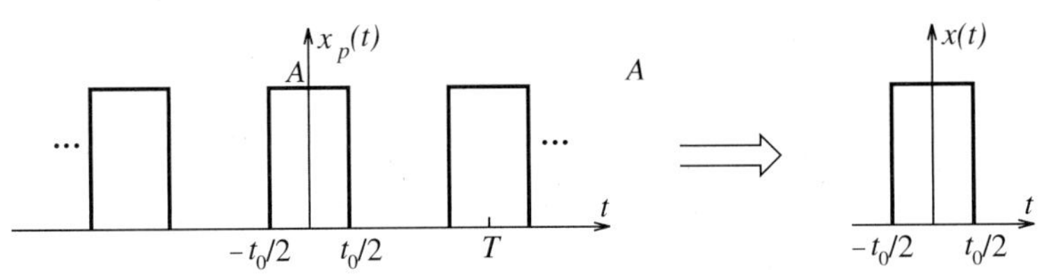

**(c) A Pure Sinusoid**

If $x(t) = \cos(2\pi f_0 t)$, its exponential Fourier series coefficients are $X[1] = 0.5$, $X[-1] = 0.5$. Its two-sided magnitude spectrum shows sample values of 0.5 at $f = \pm f_0$.

If $x(t) = \cos(2\pi f_0 t + \theta)$, its exponential coefficients are

$$X[1] = 0.5e^{j\theta}, \; X[-1] = 0.5e^{-j\theta}$$

So, Its two-sided magnitude spectrum has sample values of 0.5 at $f = \pm f_0$, and its phase spectrum shows a phase of $\theta$ at $f_0$ and of $-\theta$ at $-f_0$.

### A.5.5 Some Useful Results

A time shift changes $x_p(t)$ to $y_p(t) = x_p(t \pm \alpha)$ and changes the Fourier series coefficients from $X[k]$ to $Y[k] = X[k]e^{-j2\pi k f_0 \alpha}$. So, the magnitude spectrum shows no change.

#### Zero Interpolation of the Spectrum

A *zero interpolation* of the spectral coefficients $X[k]$ to $X[k/N]$ leads to an $N$-fold signal compression (with $N$ compressed copies) within one time period. Figure A.8 illustrates this result for compression by 2.

The signal power may be found in the time domain from $x_p(t)$ or in the frequency domain from $|X[k]|$ (its magnitude spectrum). This is Parseval's theorem for periodic signals.

$$P = \frac{1}{T}\int_T x_p^2(t)\,dt = \sum_{k=-\infty}^{\infty} |X[k]|^2 \tag{A.22}$$

#### The Gibbs Effect

The Fourier series coefficients of periodic signals with jumps decay as $1/k$. The Gibbs effect arises because perfect reconstruction from harmonics is impossible for such signals (since we are, in effect, trying to reconstruct a signal with sudden jumps from sinusoids that

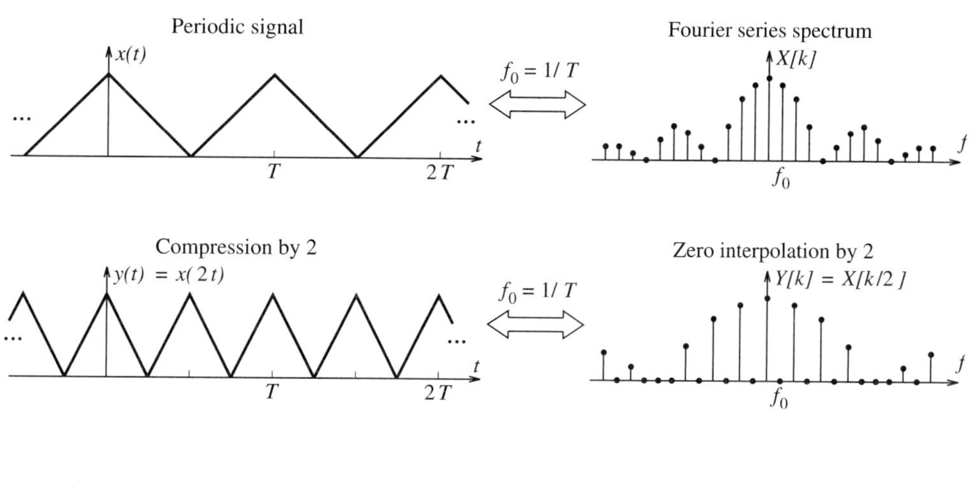


**FIGURE A.8** Signal compression leads to zero interpolation of the spectrum

are smooth functions with no sudden jumps). The reconstructed signal shows imperfections in the form of overshoot and undershoot (about 9% of jump) near each jump location. The reconstructed value equals the midpoint value of the jump at each jump location.

### A.5.6 Fourier Transform of Periodic Signals

The Fourier transform of a periodic signal is an impulse train whose spectral envelope corresponds to the Fourier transform of its one period

$$x_p(t) = \sum_{k=-\infty}^{\infty} X[k]e^{j2\pi k f_0 t} \Leftarrow\boxed{\text{FT}}\Rightarrow \sum_{k=-\infty}^{\infty} X[k]\delta(f - kf_0) \qquad \text{(A.23)}$$

The impulses are located at the harmonic frequencies $kf_0$, and their strengths are given by the Fourier series coefficients $X[k]$. The impulse train is periodic only if $X[k] = C$ (i.e., has the same value for every $k$).

**Example A.10**

**Fourier Transform of an Impulse Train**

For a unit impulse train with period $T$, as shown in Figure EA.10, we get

$$X[k] = \frac{1}{T}\int_{-T/2}^{T/2} \delta(t)e^{-j2k\pi f_0 t}\,dt = \frac{1}{T}$$

**FIGURE E A.10.3** Impulse train and its Fourier series coefficients for Example A.10

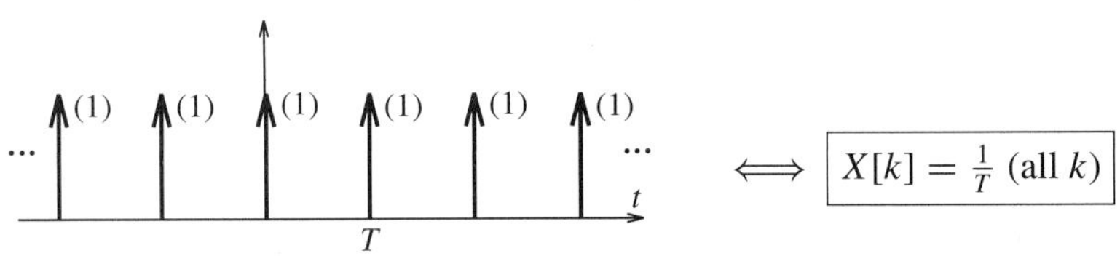


The Fourier transform of this function is

$$X(f) = \sum_{k=-\infty}^{\infty} X[k]\delta(f - kf_0) = \frac{1}{T}\sum_{k=-\infty}^{\infty} \delta(f - kf_0), \qquad f_0 = \frac{1}{T}$$

The Fourier transform of a periodic impulse train is also a periodic impulse train. For other periodic signals, the Fourier transform will is not periodic (even though it is an impulse train).

### A.5.7 Spectral Density

The spectral density is the Fourier transform of the autocorrelation function.

$$\text{(autocorrelation) } r_{xx}(t) \Leftarrow\boxed{\text{FT}}\Rightarrow R_{xx}(f) \text{ (spectral density)} \qquad \text{(A.24)}$$

This is the celebrated **Wiener-Khintchine theorem**. For power signals (and periodic signals), we must use averaged measures consistent with power (and not energy). This leads to the concept of **power spectral density** (PSD). The PSD of a periodic signal is a train of impulses at $f = kf_0$ with strengths $|X[k]|^2$ whose sum equals the total signal power.

$$R_{xx}(f) = \sum_{k=-\infty}^{\infty} |X[k]|^2 \delta(f - kf_0) \qquad \text{(A.25)}$$

### White Noise and Colored Noise

A signal with zero mean whose PSD is constant (with frequency) is called **white noise**. The autocorrelation function of such a signal is an impulse. A signal whose PSD is constant only over a finite frequency range is called **band-limited white noise**.

## A.5.8 Ideal Filters

The transfer function $H_{\text{LP}}(f)$ and impulse response $h_{\text{LP}}(t)$ of an ideal lowpass filter (LPF) with unit gain, zero phase, and cutoff frequency $f_C$ may be written as

$$H_{\text{LP}}(f) = \text{rect}(f/2f_C) \qquad h_{\text{LP}}(t) = 2f_C \,\text{sinc}(2f_C t) \quad \text{(ideal LPF)} \qquad \text{(A.26)}$$

Ideal filters are noncausal and unrealizable, they form the yardstick by which the design of practical filters is measured.

## A.5.9 Measures for Real Filters

The **phase delay** and **group delay** of a system whose transfer function is $H(\omega) = |H(\omega)| \angle\theta(\omega)$ are defined as

$$t_p = -\frac{\theta(\omega)}{\omega} \quad \text{(phase delay)} \qquad t_g = -\frac{d\,\theta(\omega)}{d\omega} \quad \text{(group delay)} \qquad \text{(A.27)}$$

If $\theta(\omega)$ varies linearly with frequency, $t_p$ and $t_g$ are not only constant but also equal. For LTI systems (with rational transfer functions), the phase $\theta(\omega)$ is a transcendental function but *the group delay is always a rational function* of $\omega^2$ and is much easier to work with in many filter applications.

The **time-limited/band-limited theorem** asserts that *no signal can be both time-limited and band limited simultaneously*. In other words, the spectrum of a finite-duration signal is always of infinite extent. The narrower a time signal, the more spread out its spectrum. Measures of duration in the time domain are inversely related to measures of bandwidth in the frequency domain, and their product is a constant. A sharper frequency response $|H(\omega)|$ can be achieved only at the expense of a slower time response. Typical time-domain and frequency-domain measures are listed in Table A.8.

## A.5.10 A First-Order Lowpass Filter

Consider the RC circuit shown in Figure A.9. If we assume that the output is the capacitor voltage, its transfer function may be written as

$$H(f) = \frac{Y(f)}{X(f)} = \frac{1/j2\pi fC}{R + (1/j2\pi fC)} = \frac{1}{1 + j2\pi fRC} = \frac{1}{1 + j2\pi f\tau} \qquad \text{(A.28)}$$

The quantity $\tau = RC$ is the circuit time constant. The magnitude $|H(f)|$ and phase $\theta(f)$ of the transfer function are sketched in Figure A.10 and given by

$$|H(f)| = \frac{1}{\sqrt{1 + 4\pi^2 f^2 \tau^2}} \qquad \theta(f) = -\tan^{-1}(2\pi f\tau) \qquad \text{(A.29)}$$

**TABLE A.8 ➤** Typical Measures for Real Filters

| Measure | Explanation |
|---|---|
| Time delay | Time between application of input and appearance of response<br>Typical measure: Time to reach 50% of final value |
| Rise time | Measure of the steepness of initial slope of response<br>Typical measure: Time to rise from 10% to 90% of final value |
| Overshoot | Deviation (if any) beyond the final value<br>Typical measure: Peak overshoot |
| Settling time | Time for oscillations to settle to within a specified value<br>Typical measure: Time to settle to within 5% or 2% of final value |
| Speed of response | Depends on the largest time constant $\tau_{\max}$ in $h(t)$<br>Typical measure: Steady state is reached in about $5\tau_{\max}$ |
| Damping | Rate of change toward final value<br>Typical measure: Damping factor $\zeta$ or quality factor $Q$ |
| Bandwidth | Frequency range over which gain exceeds a given value<br>Typical measure: $B_{3\text{dB}}$ for which $\lvert H(f)\rvert \geq 0.707\lvert H(f)\rvert_{\max}$ |

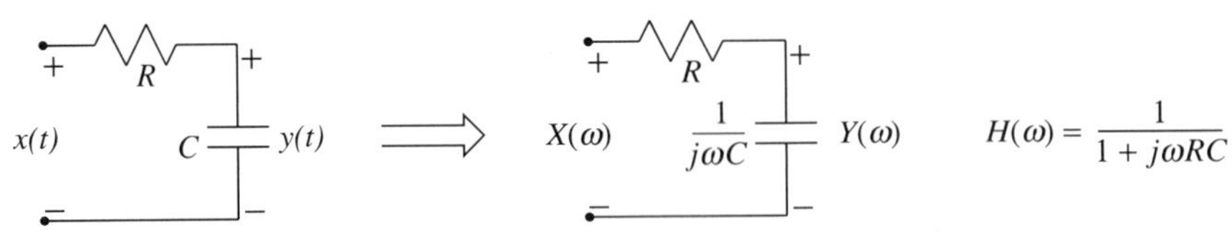


**FIGURE A.9** An RC lowpass filter

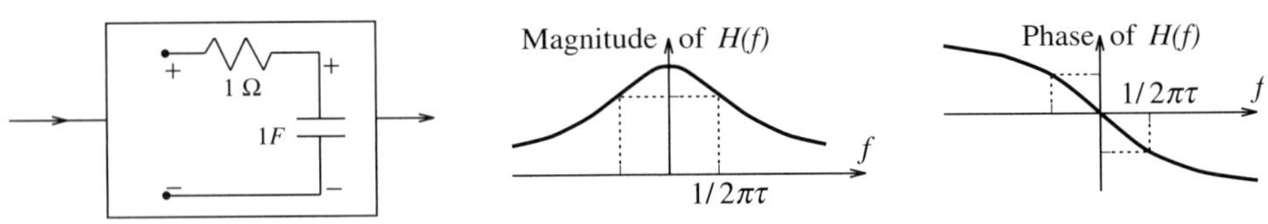


**FIGURE A.10** Frequency response of the RC lowpass filter

The system is called a *lowpass* filter because $|H(f)|$ decays monotonically with positive $f$, leading to a reduced-output amplitude at higher frequencies. At $f = \frac{1}{2\pi\tau}$, the magnitude equals $1/\sqrt{2}$ (or 0.707) and the phase equals $-45^\circ$. The frequency $f = \frac{1}{2\pi\tau}$ is called the **half-power frequency** because the output power of a sinusoid at this frequency is only half the input power. The frequency range $0 \leq f \leq \frac{1}{2\pi\tau}$ defines the **half-power bandwidth** over which the gain exceeds 0.707 times the peak gain.

The time-domain performance of this system is measured either by its impulse response $h(t) = e^{-t/\tau}u(t)$ or by its step response $s(t)$, as plotted in Figure A.11 and described by

$$h(t) = e^{-t/\tau}u(t) \qquad\qquad s(t) = (1 - e^{-t/\tau})u(t) \tag{A.30}$$

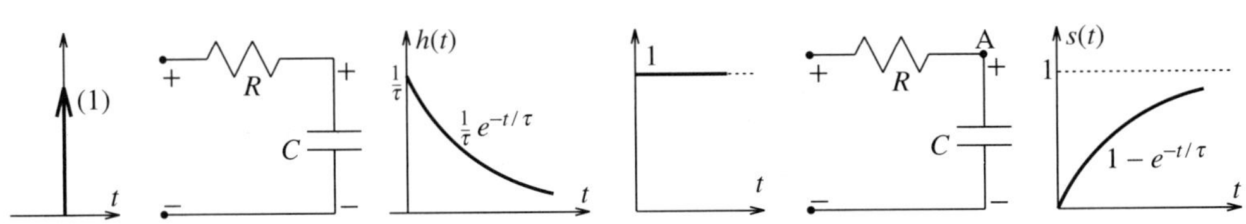


**FIGURE A.11** Impulse response $h(t)$ and step response $s(t)$ of the RC lowpass filter

The step response rises smoothly to unity and is within 1% of the final value in about $5\tau$. A smaller $\tau_{max}$ implies a faster response—a shorter time to reach steady state. Other performance measures include the rise time $t_r$ (the time taken to rise from 10 to 90% of the final value).The 10 to 90% rise time $T_r$ is computed by finding

$$s(t_{10}) = 1 - e^{-t_{10}/\tau} = 0.1 \qquad s(t_{90}) = 1 - e^{-t_{90}/\tau} = 0.9$$

We obtain $t_{10} = \tau \ln(10/9)$, $t_{90} = \tau \ln 10$, and $T_r = t_{90} - t_{10} = \tau \ln 9$.

With $B_{3\text{dB}} = \frac{1}{2\pi\tau}$, the time-bandwidth product gives

$$T_r B_{3\text{dB}} = \frac{\tau \ln 9}{2\pi\tau} \approx 0.35$$

This result is often used as an approximation for higher-order systems also.

The delay time is $t_d$ (the time taken to rise to 50% of the final value), and the settling time is $t_{P\%}$ (the time taken to settle to within $P\%$ of its final value). Commonly used measures are the 5% settling time and the 2% settling time. Exact expressions for these measures are found to be

$$t_d = \tau \ln 2 \qquad t_{5\%} = \tau \ln 20 \qquad t_{2\%} = \tau \ln 50 \tag{A.31}$$

A smaller time constant $\tau$ implies a faster rise time and a larger bandwidth. The phase delay and group delay of the RC lowpass filter are given by

$$t_p = -\frac{\theta(f)}{2\pi f} = \frac{1}{2\pi f}\tan^{-1}(2\pi f\tau) \qquad t_g = -\frac{1}{2\pi}\frac{d\theta(f)}{df} = \frac{\tau}{1 + 4\pi^2 f^2 \tau^2} \tag{A.32}$$

**Frequency Response of an RC Lowpass Filter with Time Constant $\tau = RC$**

$f_{3\text{dB}} = \dfrac{1}{2\pi\tau}$ Hz **10%–90% Rise-Time:** $\tau \ln 9$ **Time-Bandwidth Product:** $\approx 0.35$

## A.5.11 A Second-Order Lowpass Filter

A second-order lowpass filter with unit dc gain may be described by

$$H(s) = \frac{\omega_p^2}{s^2 + \frac{\omega_p}{Q}s + \omega_p^2} \tag{A.33}$$

Here, $\omega_p$ is called the **undamped natural frequency** (or **pole frequency**) and $Q$ represents the **quality factor** and is a measure of the losses in the circuit. The quantity $\zeta = 1/2Q$ is called the **damping factor**.

### Frequency Domain Performance

If $Q > 0.5$, the poles are complex conjugates and lie on a circle of radius $\omega_p$ in the $s$-plane. The higher the $Q$, the closer the poles are to the $j\omega$-axis. For $Q < 1/\sqrt{2}$, the magnitude $|H(\omega)|$ is monotonic, but for $Q > 1/\sqrt{2}$, it shows overshoot and a peak near $\omega_p$, and the peaking increases with $Q$ as shown in Figure A.12. The phase delay and group delay of

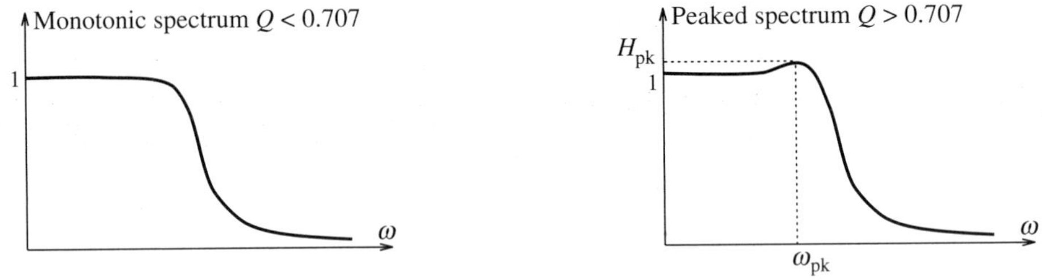


**FIGURE A.12** Frequency response of a second-order system

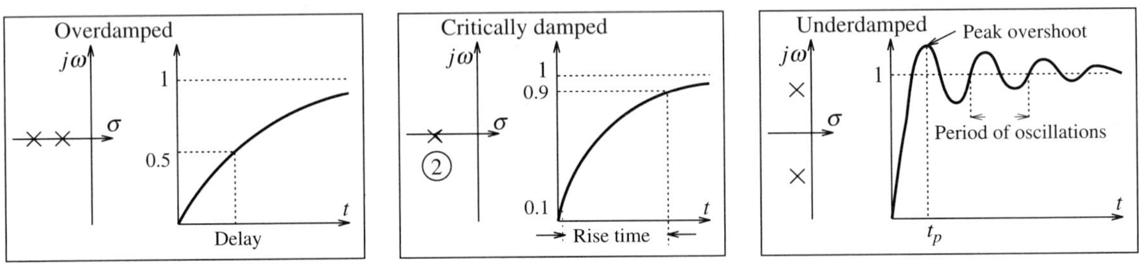


**FIGURE A.13** Step response of a second-order system

second-order filters are given by

$$\theta(\omega) = -\tan^{-1}\left[\frac{(\omega/\omega_p)/Q}{1-(\omega/\omega_p)^2}\right] \qquad t_g(\omega) = \frac{2(\omega_p/Q)(\omega_p^2+\omega^2)}{(\omega_p^2-\omega^2)^2+(\omega_p^2/Q^2)\omega^2} \tag{A.34}$$

## Time-Domain Performance

The step response of second-order lowpass filters also depends on $Q$, as shown in Figure A.13. If $Q < 0.5$, the poles are real and distinct, and the step response shows a smooth, monotonic rise to the final value (**overdamped**) with a large time constant. If $Q > 0.5$, the poles are complex conjugates, and the step response is **underdamped** with overshoot and decaying oscillations (ringing) about the final value. This results in a small rise time but a large settling time. The frequency of oscillations increases with $Q$. For $Q = 0.5$, the poles are real and equal, and the response is **critically damped** and yields the fastest monotonic approach to the steady-state value with no overshoot.

**Example A.11**

### Results for Some Analog Filters

The step response and impulse response of the filters shown in Figure EA.11 are summarized below.

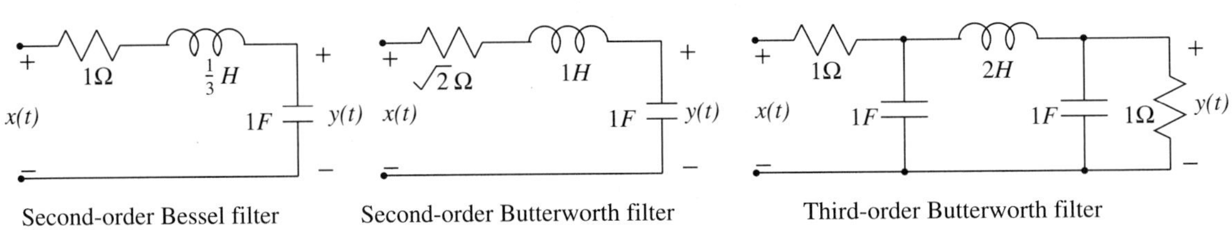


**FIGURE E A.11.3** Filters for Example

**(a) A Second-Order Bessel Filter**

$$H(s) = \frac{3}{s^2 + 3s + 3}$$

Differential equation: $y''(t) + 3y'(t) + 3y(t) = 3x(t)$

Step response: $s(t) = u(t) - e^{-3t/2}[\cos(\sqrt{3}t/2) + \sqrt{3}\sin(\sqrt{3}t/2)]u(t)$

Impulse response: $h(t) = 2\sqrt{3}e^{-3t/2}\sin(\sqrt{3}t/2)u(t)$

**(b) A Second-Order Butterworth Filter**

$$H(s) = \frac{1}{s^2 + \sqrt{2}s + 1}$$

Differential equation: $y''(t) + \sqrt{2}y'(t) + y(t) = x(t)$

Step response: $s(t) = u(t) - e^{-t/\sqrt{2}}[\cos(t/\sqrt{2}) + \sin(t/\sqrt{2})]u(t)$

Impulse response: $h(t) = y'(t) = \sqrt{2}e^{-t/\sqrt{2}}\sin(t/\sqrt{2})u(t)$

**(c) A Third-Order Butterworth Filter**

$$H(s) = \frac{0.5}{s^3 + 2s^2 + 2s + 1}$$

Differential equation: $y'''(t) + 2y''(t) + 2y'(t) + y(t) = \frac{1}{2}x(t)$

Step response: $s(t) = \frac{1}{2}u(t) - \frac{1}{\sqrt{3}}e^{-t/2}\sin(\sqrt{3}t/2)u(t) - \frac{1}{2}e^{-t}u(t)$

Impulse response: $h(t) = e^{-t/2}[\frac{1}{2\sqrt{3}}\sin(\sqrt{3}t/2) - \frac{1}{2}\cos(\sqrt{3}t/2)]u(t) - \frac{1}{2}e^{-t}u(t)$

# A.6 Bode Plots

Bode plots allow us to plot the frequency response over a wide frequency and amplitude range by using logarithmic *scale compression*. The magnitude or gain is plotted in decibels (with $H_{\text{dB}} = 20\log|H(f)|$) against $\log(f)$. For LTI systems whose transfer function is a ratio of polynomials in $j\omega$, a rough sketch can be quickly generated using linear approximations called **asymptotes** over different frequency ranges to obtain **asymptotic Bode plots**. The numerator and denominator are factored into linear and quadratic factors in $j\omega$ with real coefficients. A **standard form** is obtained by setting the real part of each factored term to unity. A summary appears in the following panel.

**Some Asymptotic Bode Plots**

*Note:* 20 dB/dec = 6 dB/oct

**Term** = $j\omega$: Straight line with slope = 20 dB/dec for *all* $\omega$ with $H_{\text{dB}} = 0$ at $\omega = 1$ rad/s.

**Term** = $1 + j\frac{\omega}{\alpha}$: For $\omega \le \alpha$, $H_{\text{dB}} = 0$. For $\omega \ge \alpha$, straight line with slope = 20 dB/dec.

**If Repeated** $k$ **Times:** Multiply slopes by $k$.

**If in Denominator:** Slopes are negative.

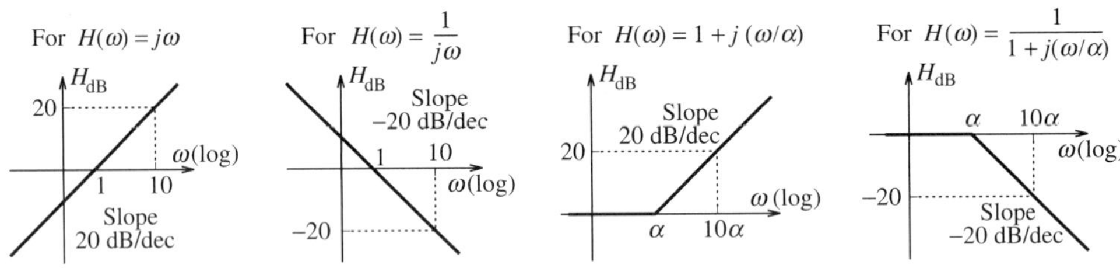


**FIGURE A.14** Asymptotic Bode magnitude plots for for some standard forms

The frequency $\alpha$ where the slope changes is called **break frequency**. For a first-order filter, the break frequency is also called the 3-dB frequency (or the half-power frequency), because the true value differs from the asymptotic value by 3 dB at this frequency. Figure A.14 shows asymptotic magnitude plots of various first-order terms. The decibel value of a constant $H(\omega) = K$ is $H_{\text{dB}} = 20 \log |K|$ dB—a constant for all $\omega$. The Bode magnitude plot for a transfer function with several terms is the sum of similar plots for each of its individual terms. The composite plot can be sketched directly using the guidelines of the following panel.

### Guidelines for Sketching a Composite Asymptotic Magnitude Plot for $H(\omega)$

**Initial Slope:** 20 dB/dec (with $H_{\text{dB}} = 0$ at $\omega = 1$ rad/s) if a term $j\omega$ is present; 0 dB/dec if absent.

**At a Break Frequency:** The slope *increases* by $20k$ dB/dec due to a (repeated by $k$) numerator term and *decreases* by $20k$ dB/dec due to a denominator term.

**Example A.12**

### Bode Magnitude Plots

**(a) Linear Factors**

Let $H(\omega) = \dfrac{40(0.25 + j\omega)(10 + j\omega)}{j\omega(20 + j\omega)}$. Sketch its bode magnitude.

We write this in the standard form

$$H(\omega) = \frac{5(1 + j\omega/0.25)(1 + j\omega/10)}{j\omega(1 + j\omega/20)}$$

The break frequencies, in ascending order, are:

$$\omega_1 = 0.25 \text{ rad/s (numerator)}$$
$$\omega_2 = 10 \text{ rad/s (numerator)}$$
$$\omega_3 = 20 \text{ rad/s (denominator)}$$

The term $1/j\omega$ provides a starting asymptote $-20$ dB/dec whose value is 0 dB at $\omega = 1$ rad/s. We can now sketch a composite plot by including the other terms:

- At $\omega_1 = 0.25$ rad/s (numerator), the slope increases by (by $+20$ dB/dec) to 0 dB/dec.
- At $\omega_2 = 10$ rad/s (numerator), the slope increases to 20 dB/dec.
- At $\omega_3 = 20$ rad/s (denominator), the slope decreases to 0 dB/dec.

Finally, the constant 5 shifts the plot by $20 \log 5 = 14$ dB. Its Bode plot is shown in Figure EA.12(a).

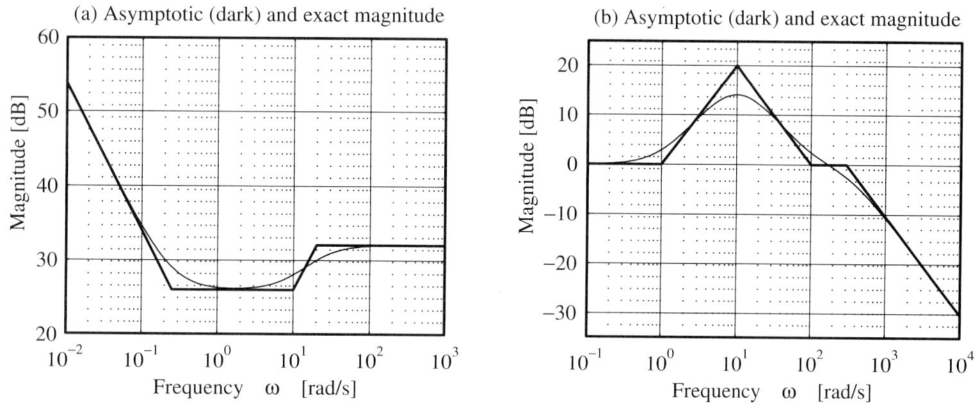


**FIGURE E A.12.a** Bode magnitude plots for Example A.12

**(b) Repeated Factors**

Let $H(\omega) = \dfrac{(1 + j\omega)(1 + j\omega/100)}{(1 + j\omega/10)^2(1 + j\omega/300)}$. Sketch its Bode magnitude.

Its Bode plot is sketched in Figure EA.12(b). We make the following remarks:

- The starting slope is 0 dB/dec since a term of the form $(j\omega)^k$ is absent.
- The slope changes by −40 dB/dec at $\omega_B = 10$ rad/s due to the repeated factor.

# A.7 Classical Analog Filter Design

Classical analog filters include Butterworth (maximally flat passband), Chebyshev I (rippled passband), Chebyshev II (rippled stopband), and elliptic (rippled passband and stopband), as shown in Figure A.15. The design of these classical analog filters typically relies on frequency specifications (passband and stopband edge(s)) and magnitude specifications (maximum

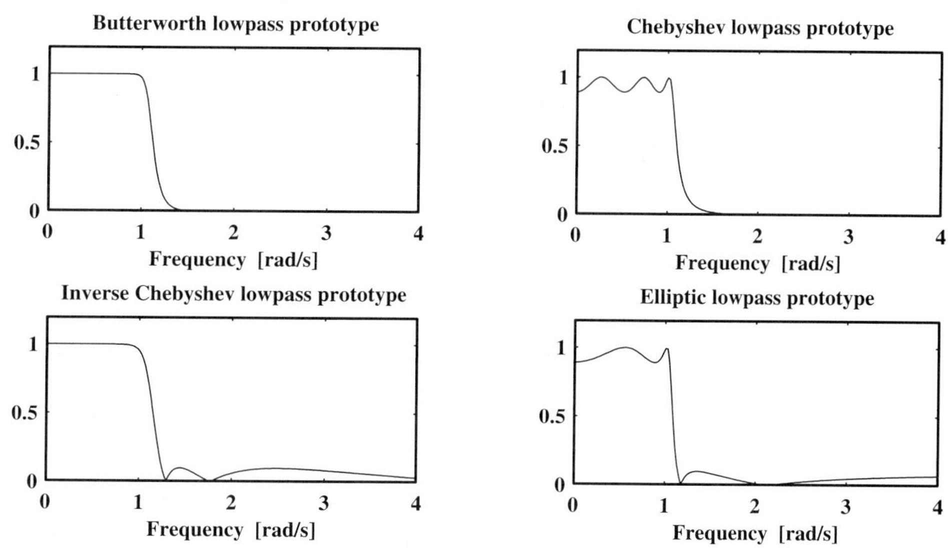


**FIGURE A.15** Classical analog lowpass prototypes

**TABLE A.9** ➤ Analog Filter Design

| | Butterworth | Chebyshev |
|---|---|---|
| Ripple | $\epsilon^2 = 10^{0.1A_p} - 1$ | $\epsilon^2 = 10^{0.1A_p} - 1$ |
| Order | $n = \dfrac{\log[(10^{0.1A_s} - 1)/\epsilon^2]^{1/2}}{\log(\nu_s)}$ | $n = \dfrac{\cosh^{-1}[(10^{0.1A_s} - 1)/\epsilon^2]^{1/2}}{\cosh^{-1}(\nu_s)}$ |
| 3-dB Frequency | $\nu_3 = (1/\epsilon)^{1/n} = R$ | $\nu_3 = \cosh\left[\frac{1}{n}\cosh^{-1}(1/\epsilon)\right]$ |
| Poles of $H(s)$ | $p_k = -R\sin\theta_k + jR\cos\theta_k$<br>$\theta_k = (2k-1)\pi/2n$<br>$k = 1, 2, \ldots, n$ | $p_k = -\sin\theta_k \sinh(\alpha) + j\cos\theta_k \cosh(\alpha)$<br>$\theta_k = (2k-1)\pi/2n \quad \alpha = \frac{1}{n}\sinh^{-1}(\frac{1}{\epsilon})$<br>$k = 1, 2, \ldots, n$ |
| Denominator $Q_P(s)$ | $(s-p_1)(s-p_2)\ldots(s-p_n)$ | $(s-p_1)(s-p_2)\ldots(s-p_n)$ |
| $K$ for unit peak gain | $K = Q_P(0) = \prod_{k=1}^{n} p_k$ | $K = \begin{cases} Q_P(0)/\sqrt{1+\epsilon^2} & n \text{ even} \\ Q_P(0) & n \text{ odd} \end{cases}$ |

*NOTES:*

For LP filters: $\nu_s = \dfrac{\text{Stopband edge}}{\text{Passband edge}}$  For HP filters: $\nu_s = \dfrac{\text{Passband edge}}{\text{Stopband edge}}$

For BP and BS filters, with band edges $[f_1, f_2, f_3, f_4]$ in increasing order: $\nu_s = \dfrac{f_4 - f_1}{f_3 - f_2}$.

If $f_0$ is the center frequency, we require geometric symmetry with $f_1 f_4 = f_2 f_3 = f_0^2$.

For **transformation** of the LPP transfer function $H(s)$ to other forms, use:

**To LPF** $s \Rightarrow s/\omega_p$  **To HPF** $s \Rightarrow \omega_p/s$  **To BPF** $s \Rightarrow \dfrac{s^2+\omega_0^2}{sB_{BP}}$  **To BSF** $s \Rightarrow \dfrac{sB_{BS}}{s^2+\omega_0^2}$

where $\omega_0 = 2\pi f_0$, $B_{BP} = 2\pi(f_3 - f_2)$ for bandpass, and $B_{BS} = 2\pi(f_4 - f_1)$ for bandstop

For numerical computation:

$$\cosh^{-1}(x) = \ln[x + (x^2-1)^{1/2}],\ x \ge 1$$

passband attenuation and minimum stopband attenuation) to generate a minimum-phase filter transfer function with the smallest order that meets or exceeds specifications.

Most design strategies are based on converting the given frequency specifications to those applicable to a *lowpass prototype* with *unit radian* cutoff frequency, designing the lowpass prototype and converting to the required filter type using frequency transformations. We only concentrate on the design of the lowpass prototype. Table A.9 describes how to obtain the prototype specifications from given filter specifications, the design recipe for Butterworth and Chebyshev filters, and frequency transformations to convert the lowpass prototype back to the required form.

For bandpass and bandstop filters, if the given specifications $[f_1, f_2, f_3, f_4]$ are not geometrically symmetric, one of the stopband edges must be relocated (increased or decreased) in a way that the new transition widths do not exceed the original. The quadratic transformations to bandpass and bandstop filters yield transfer functions with *twice* the order of the lowpass prototype.

The poles of a Butterworth lowpass prototype lie equispaced on a circle of radius $R = (1/\epsilon)^{1/n}$ in the $s$-plane, while the poles of a Chebyshev prototype lie on an ellipse. The high-frequency attenuation of an $n$th-order Butterworth or Chebyshev lowpass filter is $20n$ dB/dec.

**Example A.13**

## Analog Filter Design

**(a) Butterworth Lowpass Filter**

Design a Butterworth filter to meet the following specifications.

- $A_p \le 1$ dB for $f \le 4$ kHz
- $A_s \ge 20$ dB for $f \ge 8$ kHz

From the design equations, $\nu_s = f_s/f_p = 2$ and $\epsilon^2 = 10^{0.1A_p} - 1 = 0.2589$

$$n = \frac{\log[(10^{0.1A_s} - 1)/\epsilon^2]^{1/2}}{\log(\nu_s)} = \frac{\log[(10^2 - 1)/\epsilon^2]^{1/2}}{\log(2)} = 4.289 \quad \Rightarrow \quad n = 5$$

$\nu_3 = (1/\epsilon)^{1/n} = 1.1447 = R$

We have $\theta_k = (2k-1)\pi/2n = 0.1(2k-1)\pi \qquad k = 1, 2, \ldots, 5.$

This gives $\theta_k = 0.1\pi,\ 0.3\pi,\ 0.5\pi,\ 0.7\pi,\ 0.9\pi$ rad.

The pole locations $p_k = -R\sin(\theta_k) + jR\cos(\theta_k)$ are then

$$p_k = -1.1447, \quad -0.3537 \pm j1.0887, \quad -0.9261 \pm j0.6728$$

The denominator $Q(s)$ may be written as

$Q(s) = s^5 + 3.7042s^4 + 6.8607s^3 + 7.8533s^2 + 5.5558s + 1.9652$

The numerator is given by $K = Q(0) = 1.9652$. The transfer function of the analog lowpass prototype is then

$$H(s) = \frac{K}{Q(s)} = \frac{1.9652}{s^5 + 3.7042s^4 + 6.8607s^3 + 7.8533s^2 + 5.5558s + 1.9652}$$

**(b) Butterworth Bandstop Filter**

Design a Butterworth bandstop filter with 2-dB passband edges of 30 Hz and 100 Hz, and 40-dB stopband edges of 50 Hz and 70 Hz.

The band edges are $[f_1,\ f_2,\ f_3,\ f_4] = [30,\ 50,\ 70,\ 100]$ Hz. Since $f_1 f_4 = 3000$ and $f_2 f_3 = 3500$, the specifications are not geometrically symmetric. Assuming a fixed passband, we relocate the upper stopband edge $f_3$ to ensure geometric symmetry $f_2 f_3 = f_1 f_4$. This gives $f_3 = (30)(100)/50 = 60$ Hz.

The lowpass prototype band edges are $\nu_p = 1$ rad/s and $\nu_s = \frac{f_4 - f_1}{f_3 - f_2} = 7$ rad/s.

We compute $\epsilon^2 = 10^{0.1A_p} - 1 = 0.5849$ and the lowpass prototype order as

$$n = \frac{\log[(10^{0.1A_s} - 1)/\epsilon^2]^{1/2}}{\log \nu_s} \Rightarrow n = 3$$

The pole radius is $R = (1/\epsilon)^{1/n} = 1.0935$. The pole angles are $\theta_k = [\frac{\pi}{6},\ \frac{\pi}{2},\ \frac{5\pi}{6}]$ rad.

Thus, $p_k = -R\sin\theta_k + jR\cos\theta_k = -1.0935,\ 0.5468 \pm j0.9470$, and the lowpass prototype becomes

$$H_P(s) = \frac{1.3076}{s^3 + 2.1870s^2 + 2.3915s + 1.3076}$$

With $\omega_0^2 = \omega_1\omega_4 = 4\pi^2(3000)$ and $B = 2\pi(f_4 - f_1) = 2\pi(70)$ rad/s, the LP2BS transformation $s \to sB/(s^2 + \omega_0^2)$ gives

$$H(s) = \frac{s^6 + 3.55(10)^5 s^4 + 4.21(10)^{10} s^2 + 1.66(10)^{15}}{s^6 + 8.04(10)^2 s^5 + 6.79(10)^5 s^4 + 2.56(10)^8 s^3 + 8.04(10)^{10} s^2 + 1.13(10)^{13} s + 1.66(10)^{15}}$$

The linear and decibel magnitude of this filter is sketched in Figure EA.13b.

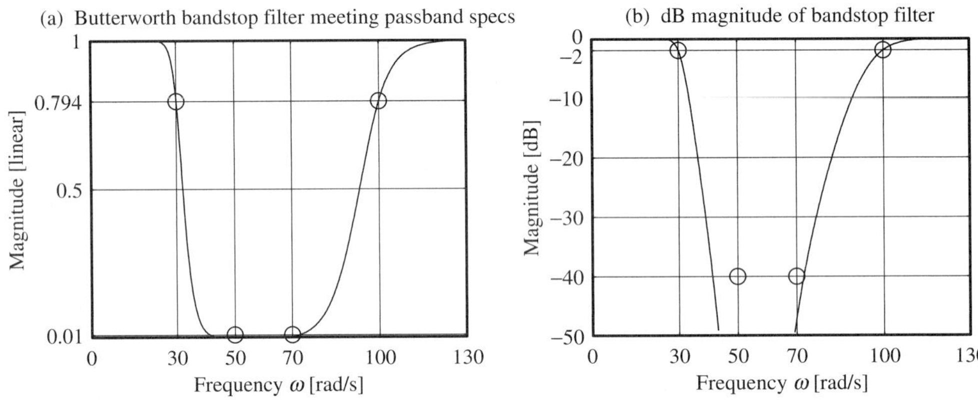


**FIGURE E A.13.b** Butterworth bandstop filter of Example A.13(b)

**(c) Chebyshev Lowpass Filter**

Design a Chebyshev filter to meet the following specifications.

- $A_p \le 1$ dB for $f \le 4$ kHz
- $A_s \ge 20$ dB for $f \ge 8$ kHz

From the design equations:

$$\epsilon^2 = 10^{0.1A_p} - 1 = 0.2589$$

$$n = \frac{\cosh^{-1}[(10^{0.1A_s} - 1)/\epsilon^2]^{1/2}}{\cosh^{-1}(f_s/f_p)} = \frac{\cosh^{-1}[(10^2 - 1)/\epsilon^2]^{1/2}}{\cosh^{-1}(2)} = 2.783 \quad \rightarrow \quad n = 3$$

$$\nu_3 = \cosh\left[\frac{1}{n}\cosh^{-1}(1/\epsilon)\right] = 1.0949$$

To find the poles, we first compute

$$\alpha = (1/n)\sinh^{-1}(1/\epsilon) = 0.4760$$

$$\theta_k = (2k-1)\pi/2n = (2k-1)\pi/6 \quad k = 1, 2, 3$$

The poles $p_k = -\sinh(\alpha)\sin(\theta_k) + j\cosh(\alpha)\cos(\theta_k)$ then yield

$$p_k = -0.4942, \quad -0.2471 \pm j0.966$$

The denominator $Q(s)$ equals

$$Q(s) = s^3 + 0.9883s^2 + 1.2384s + 0.4913$$

Since $n$ is odd, the numerator is $K = Q(0) = 0.4913$ for unity peak gain. The transfer function of the analog lowpass prototype is then

$$H(s) = \frac{K}{Q(s)} = \frac{0.4913}{s^3 + 0.9883s^2 + 1.2384s + 0.4913}$$

**(d) Chebyshev Bandpass Filter**

Design a Chebyshev bandpass filter for which we are given:

- Passband edges: $[\omega_1, \omega_2, \omega_3, \omega_4] = [0.89, 1.019, 2.221, 6.155]$ rad/s
- Maximum passband attenuation: $A_p = 2$ dB
- Minimum stopband attenuation: $A_s = 20$ dB

The frequencies are not geometrically symmetric. So, we assume fixed passband edges and compute $\omega_0^2 = \omega_2\omega_3 = 1.5045$. Since $\omega_1\omega_4 > \omega_0^2$, we decrease $\omega_4$ to $\omega_4 = \omega_0^2/\omega_1 = 2.54$ rad/s.

Then, $B = \omega_3 - \omega_2 = 1.202$ rad/s, $\nu_p = 1$ rad/s, and $\nu_s = \dfrac{\omega_4 - \omega_1}{B} = \dfrac{2.54 - 0.89}{1.202} = 1.3738$ rad/s.

The value of $\epsilon^2$ and the order $n$ is given by

$$\epsilon^2 = 10^{0.1A_p} - 1 = 0.5849 \qquad n = \frac{\cosh^{-1}[(10^{0.1A_s} - 1)/\epsilon^2]^{1/2}}{\cosh^{-1}\nu_s} = 3.879 \Rightarrow n = 4$$

The half-power frequency is $\nu_3 = \cosh[\frac{1}{3}\cosh^{-1}(1/\epsilon)] = 1.018$. To find the LHP poles of the prototype filter, we need

$$\alpha = (1/n)\sinh^{-1}(1/\epsilon) = 0.2708$$

$$\theta_k(\text{rad}) = \frac{(2k-1)\pi}{2n} = \frac{(2k-1)\pi}{8}, \qquad k = 1, 2, 3, 4$$

From the LHP poles $p_k = -\sinh\alpha\sin\theta_k + j\cosh\alpha\cos\theta_k$, we compute $p_1,\ p_3 = -0.1049 \pm j0.958$ and $p_2,\ p_4 = -0.2532 \pm j0.3968$. The denominator $Q_P(s)$ of the prototype $H_P(s) = K/Q_P(s)$ is thus

$$Q_P(s) = (s-p_1)(s-p_2)(s-p_3)(s-p_4) = s^4 + 0.7162s^3 + 1.2565s^2 + 0.5168s + 0.2058$$

Since $n$ is even, we choose $K = Q_P(0)/\sqrt{1+\epsilon^2} = 0.1634$ for *peak unit gain*, and thus

$$H_P(s) = \frac{0.1634}{s^4 + 0.7162s^3 + 1.2565s^2 + 0.5168s + 0.2058}$$

We transform this using the LP2BP transformation $s \longrightarrow (s^2 + \omega_0^2)/sB$ to give the eighth-order analog bandpass filter $H(s)$ as

$$H(s) = \frac{0.34s^4}{s^8 + 0.86s^7 + 10.87s^6 + 6.75s^5 + 39.39s^4 + 15.27s^3 + 55.69s^2 + 9.99s + 26.25}$$

The linear and decibel magnitude of this filter is shown in Figure EA.13D.

**FIGURE E A.13.d** Chebyshev I bandpass filter for Example A.13(d)

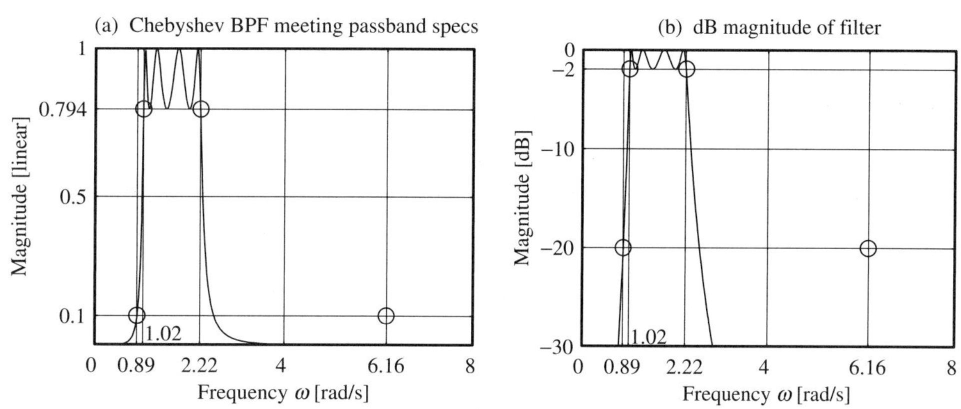

# References

## Books on Digital Signal Processing

A good source of current references is the Web site of the MathWorks, Inc. (makers of MATLAB): **http://www.mathworks.com**

Many books cover more or less the same topics as this text and at about the same level. The following list (arranged alphabetically by author) is representative.


AMBARDAR, A., AND BORGHESANI, C., *Mastering DSP Concepts Using MATLAB* (Prentice Hall, 1998).

ANTONIOU, A., *Digital Filters* (McGraw-Hill, 1993).

BURRUS, C. S., *et al.*, *Computer-Based Exercises for Digital Signal Processing* (Prentice Hall, 1994).

CADZOW, J. A., *Foundations of Digital Signal Processing and Data Analysis* (Macmillan, 1987).

CUNNINGHAM, E. P., *Digital Filtering* (Houghton Mifflin, 1992).

DEFATTA, D. J., LUCAS, J. G., and Hodgkiss, W. S., *Digital Signal Processing* (Wiley, 1988).

IFEACHOR, E. C., AND JERVIS, B. W., *Digital Signal Processing* (Addison-Wesley, 1993).

INGLE, V. K., AND PROAKIS, J. G., *Digital Signal Processing Using MATLAB* (PWS, 1996).

JACKSON, L. B., *Digital Filters and Signal Processing* (Kluwer, 1995).

JOHNSON, J. R., *Introduction to Digital Signal Processing* (Prentice Hall, 1989).

KUC, R., *Introduction to Digital Signal Processing* (McGraw-Hill, 1988).

LUDEMAN, L. C., *Fundamentals of Digital Signal Processing* (Wiley, 1986).

MITRA, S. K., *Digital Signal Processing* (Wiley, 1998).

OPPENHEIM, A. V., AND SCHAFER, R. W., *Digital Signal Processing* (Prentice Hall, 1975).

OPPENHEIM, A. V., AND SCHAFER, R. W., *Discrete-Time Signal Processing* (Prentice Hall, 1989).

ORFANIDIS, S. J., *Introduction to Signal Processing* (Prentice Hall, 1996).

PORAT, B., *A Course in Digital Signal Processing* (Wiley, 1997).

PROAKIS, J. G., AND MANOLAKIS, D. G., *Digital Signal Processing* (Prentice Hall, 1996).

RABINER, L. R., AND GOLD, B., *Theory and Application of Digital Signal Processing* (Prentice Hall, 1974).

STEIGLITZ, K., *A Digital Signal Processing Primer* (Addison-Wesley, 1996).

STRUM, R. D., AND KIRK, D. E., *First Principles of Discrete Systems and Digital Signal Processing* (Addison-Wesley, 1988).

# Other References

The following list contains more exhaustive discussions of specific topics.


## Impulse Functions, Convolution, and Fourier Transforms

BRACEWELL, R. N., *The Fourier Transform and Its Applications* (McGraw-Hill, 1978).
GASKILL, J. D., *Linear Systems, Fourier Transforms and Optics* (Wiley, 1978).

## Window Functions

GECKINLI, N. C., AND YAVUZ, D., *Discrete Fourier Transformation and Its Applications to Power Spectra Estimation* (Elsevier, 1983).
HARRIS, F. J., "On the Use of Windows for Harmonic Analysis with the Discrete Fourier Transform," *Proceedings of the IEEE*, vol. 66, no. 1, pp. 51–83, Jan. 1978.
NUTTALL, A. H., "Some Windows with Very Good Sidelobe Behavior," *IEEE Transactions on Acoustics, Speech and Signal Processing*, vol. 29, no. 1, pp. 84–91, Feb. 1981 (containing corrections to the paper by Harris).

## Analog Filters

DANIELS, R. W., *Approximation Methods for Electronic Filter Design* (McGraw-Hill, 1974).
WEINBERG, L., *Network Analysis and Synthesis* (McGraw-Hill, 1962).

## Digital Filters and Mappings

KUNT, M., *Digital Signal Processing* (Artech House, 1986).
BOSE, N. K., *Digital Filters* (Elsevier, 1985).
HAMMING, R. W., *Digital Filters* (Prentice Hall,1983).
SCHEIDER, A. M., *et al.*, "Higher Order $s$-to-$z$ Mapping Functions and Their Application in Digitizing Continuous-Time Filters," *IEEE Proceedings*, vol. 79, no. 11, pp. 1661–1674, Nov. 1991.

## Applications of DSP

ELLIOTT, D. F. (ED.), *Handbook of Digital Signal Processing* (Academic Press, 1987).
A good resource for audio applications on the Internet is **http://www.harmonycentral.com/Effects/**

## Mathematical Functions

ABRAMOWITZ, M., AND STEGUN I. A., (EDS.), *Handbook of Mathematical Functions* (Dover, 1964).

## Numerical Methods

PRESS, W. H., *et al.*, *Numerical Recipes* (Cambridge University Press, 1986).

# Index

## G



## H

## K



## L

## N



## O

## P

## Q



## R

## T

## U



## V



## W

## Z